RADIATION DETECTION
AND MEASUREMENT

RADIATION DETECTION AND MEASUREMENT

SECOND EDITION

GLENN F. KNOLL

Professor of Nuclear Engineering
The University of Michigan
Ann Arbor, Michigan

WILEY

JOHN WILEY & SONS

New York • Chichester • Brisbane • Toronto • Singapore

Copyright © 1979, 1989, by John Wiley & Sons, Inc.

All rights reserved. Published simultaneously in Canada.

Reproduction or translation of any part of
this work beyond that permitted by Sections
107 and 108 of the 1976 United States Copyright
Act without the permission of the copyright
owner is unlawful. Requests for permission
or further information should be addressed to
the Permissions Department, John Wiley & Sons.

Library of Congress Cataloging in Publication Data:

Knoll, Glenn F.
　Radiation detection and measurement.

　Includes bibliographical references and index.
　1. Nuclear counters.　2. Radiation–Measurement.
I. Title.

QC787.C6K56　1989　　　　539.7'7　　　　88-26142
ISBN 0-471-81504-7

Printed in the United States of America

20　19　18　17　16　15　14　13　12

Dedicated in memory of my parents,
Oswald Herman Knoll and Clara Bernthal Knoll

About the Author

GLENN FREDERICK KNOLL is currently Professor and Chairman of the Department of Nuclear Engineering at The University of Michigan. Following his undergraduate education at Case Institute of Technology, he earned a master's degree from Stanford University and a doctorate in nuclear engineering from The University of Michigan. During his graduate work, he held national fellowships from the Atomic Energy Commission and the National Science Foundation.

He joined the Michigan faculty in 1962, and was named Department Chairman in 1979. He has held appointments as Visiting Scientist at the Nuclear Research Center in Karlsruhe, Germany, and as Senior Fellow in the Department of Physics at the University of Surrey, U.K. His research interests have centered on radiation measurements, nuclear instrumentation, and radiation imaging. He is author or coauthor of more than 80 technical publications, 4 patents, and 2 textbooks.

He is a Fellow of both the American Nuclear Society and the Institute of Electrical and Electronics Engineers. In 1979, he was selected for the Glenn Murphy Award, which is given by the American Society for Engineering Education to an outstanding educator in Nuclear Engineering. He is a member of the Editorial Board for *Nuclear Instruments and Methods in Physics Research* and *IEEE Transactions on Medical Imaging*. He has served as consultant to 16 industrial organizations in technical areas related to radiation measurements, and is a Registered Professional Engineer in the State of Michigan.

Preface

In this second edition, I have maintained the dual objective of the original: to serve as a textbook for those new to the field and to provide sufficient substance so that the book may also be helpful to those actively involved in radiation measurements. In the decade since the first edition was published, there has been significant development in most areas of the subject matter. As a result, those familiar with the original will notice modifications, additions, and updates throughout this version. Inevitably, there have been more additions than deletions, reflecting the growing breadth of topics important in radiation measurements. Therefore, it is even more important that the instructor using this book as a text provide guidance to the students on which sections are most essential for the purposes of the course. The illustrative problems at the ends of the chapters have been doubled, and the companion solutions manual should be helpful in the tutorial use of the book.

The organization by chapters has remained unchanged, with one exception. The sequencing of the chapters on statistics and detector general properties (Chapters 3 and 4) is reversed from the first edition. This change was made to facilitate the use of statistical concepts in the discussions throughout Chapter 4.

The most extensive changes will be found in Chapter 12, reflecting the replacement of Ge(Li) detectors by the newer HPGe type. Developments in scintillation materials and in the use of photodiodes to convert the scintillation light have led to significant revisions in Chapters 8 and 9. Elsewhere in this edition, pulse-type ion chambers, long in dormancy, are now described more fully because of their enhanced importance in heavy ion measurements. A section has been added on the self-quenched streamer mode of operation in gas detectors. The description of silicon detectors now emphasizes the fully depleted configurations, and the discussions of CdTe and HgI_2 detectors have been expanded. Activation counters used in pulsed fast neutron measurements are now described, and a section has been added on developmental cryogenic and superconducting detectors.

In the areas that are related to pulse processing, full derivations are now provided for the pulse shape from coaxial germanium detectors and from proportional counters. Reflecting a growing interest in high counting rate applications, discussions of reset preamplifiers, gated integrators, and fast pulse-processing methods are now included. A section on the pileup contribution to recorded spectra also is new. Because of the increasing interest in position-sensitive detectors in many applications, more attention has been given to techniques for determining the location of the radiation interaction in the major detector types. Finally, many small updates and refinements were made throughout the book.

References to the literature have been updated where needed to keep up with current practice. The citations are intended to lead to more detailed descriptions than are possible in this text and to provide starting points for literature searches. No attempt has been made to compile a comprehensive bibliography.

I acknowledge a number of individuals who provided significant assistance in the preparation of the second edition. Don Miller of Ohio State University contributed substantially to an update of the discussion of reactor instrumentation in Chapter 14, and also to Chapter 4. Dennis Persyk of Siemens Gammasonics and Mario Martini of EG & G ORTEC provided critical reviews of early drafts of Chapters 9 and 12, respectively, and made many suggestions that have now been incorporated. The Japanese translators of the first edition, Itsuru Kimura of Kyoto University and the late Eiji Sakai of JAERI, corrected a number of errors and provided valuable points of clarification throughout the text. Detailed reviews of the manuscript by Stephen Binney of Oregon State University, Gary Catchen of Penn State, and Bradley Micklich of Illinois caught many errors and made helpful suggestions for improvements. Faculty colleague David Wehe, departmental students Tim DeVol, Alison Stolle, Yuji Fujii and Richard Kruger, and my son Tom Knoll were also extremely helpful during stages of the manuscript preparation. Finally, many authors have been kind enough to provide original art for figures that have been taken from previously published articles. To all, I express sincere thanks.

The loving support of my wife Gladys throughout this endeavor has been essential and greatly appreciated.

Glenn F. Knoll

Ann Arbor, Michigan
July 1988

Preface *To The First Edition*

This book serves two purposes. First, as a textbook, it is appropriate for use in a first course in nuclear instrumentation or radiation measurements. Such courses are taught at levels ranging from the junior undergraduate year through the first year of graduate programs and are an important part of most curricula in nuclear engineering or radiation physics. Students in health physics, radiation biology, and nuclear chemistry often will also include a similar course in their program. Substantially more material is included, however, than can possibly be covered in the usual one-term course. I have intentionally done so in order that the book remain useful to the student after completion of the course, and so that it may also serve its second purpose as a general review of radiation detection techniques for scientists and engineers actively involved in radiation measurements. The instructor using this book as a text will therefore need to select only those portions deemed most relevant to the purposes of the course. I feel that this inconvenience is offset by the larger scope and more lasting value of the book. Problems intended as student exercises are provided at the end of most chapters.

The level of the discussions assumes an elementary background in radioactivity, radiation properties, and basic electrical circuits. Some topics from these categories are reviewed in the first two chapters, but only in the limited context of laboratory radiation sources and the more important interaction mechanisms. Readers who would like information beyond the scope of the text are referred to the current scientific literature that is cited at the end of each chapter. These references have largely been limited to fairly recent publications, and I apologize in advance to my colleagues whose important but older work may not be referenced.

The important detection techniques for ionizing radiations with energies below about 20 MeV are covered in various chapters of the book. These are the radiations of primary interest in fission and fusion energy systems, as well as in medical, environmental, and industrial applications of radioisotopes. I have concentrated on the basic detector configurations most frequently encountered by the typical user and excluded more complex or specialized detection systems that may be found in many research laboratories. Also not included are the instruments such as bubble chambers, spark chambers, and calorimeters of principal use in high-energy particle physics research. The sections on electronic components and pulse processing aspects of detector signals are based on functional descriptions rather than detailed circuit analyses. This approach reflects the usual interests of the user rather than those of the circuit designer.

Although illustrative applications are included, the discussions emphasize the principles of operation and basic characteristics of the various detector systems. Other publications are available to the reader who seeks more detailed and complete description of specific applications of these instruments. A good example is NCRP Report No. 58, "A Handbook of Radioactivity Measurement Procedures," published by the National Council on Radiation Protection and Measurements, Washington, DC, in 1978 [a second edition of this report published in 1985 is now available].

The SI system of units is used throughout the text. Many traditional radiation units familiar to experienced users are destined to be phased out, so that those not familiar with the newer SI units of activity, gamma-ray exposure, absorbed dose, and dose equivalent should review the definitions as introduced in Chapters 1 and 2.

I have been primarily responsible for teaching nuclear instrumentation courses at The University of Michigan since 1962. Parts of the manuscript have evolved from lecture notes developed over this time, and a preliminary version of the text has been in use for several semesters. This book therefore reflects considerable student feedback that has been essential in improving the clarity of presentation in many areas. I particularly thank present or former graduate students George Baldwin, Hadi Bozorgmanesh, John Engdahl, Dan Grady, Bill Halsey, Bill Martin, Warren Snapp, and Jay Williams, who provided valuable input in many forms. I also thank those faculty colleagues who reviewed portions of the manuscript and offered many helpful comments and suggestions: David Bach, Chihiro Kikuchi, John Lee, Craig Robertson, and Dieter Vincent, and also Lou Costrell and Ron Fleming of the National Bureau of Standards. Pam Hale carried out the formidable task of typing the manuscript with great skill and patience.

I owe a special debt to Jim Duderstadt, who provided the initial encouragement that transformed good intentions into a definite commitment toward this project. The steadfast support, understanding, and help of my wife Gladys and sons Tom, John, and Peter throughout the several years required for its completion have been an essential contribution for which I will always be grateful.

Glenn F. Knoll

Ann Arbor, Michigan
September 1978

Credits

Many figures and tables in this text have been reproduced from previously published and copyrighted sources. The cooperation of the publishers in granting permission for the use of this material has been a major contribution.

Except for those sources directly acknowledged in captions and footnotes, most are identified by citing a reference in a scientific journal where the table or figure first appeared. A compilation of these citations, arranged by publication, is given below.

Contents

CHAPTER · 1

Radiation Sources

The radiations of primary concern in this text originate in atomic or nuclear processes. They are conveniently categorized into four general types as follows:

Charged particulate radiation $\begin{cases} \text{Fast electrons} \\ \text{Heavy charged particles} \end{cases}$

Uncharged radiation $\begin{cases} \text{Electromagnetic radiation} \\ \text{Neutrons} \end{cases}$

Fast electrons include beta particles (positive or negative) emitted in nuclear decay, as well as energetic electrons produced by any other process. *Heavy charged particles* denote a category that encompasses all energetic ions with mass of one atomic mass unit or greater, such as alpha particles, protons, fission products, or the products of many nuclear reactions. The *electromagnetic radiation* of interest includes X-rays emitted in the rearrangement of electron shells of atoms, and gamma rays which originate from transitions within the nucleus itself. *Neutrons* generated in various nuclear processes comprise the final major category, which is often further divided into *slow neutron* and *fast neutron* subcategories (see Chapter 14).

The energy range of interest spans over six decades, ranging from about 10 eV to 20 MeV. (Slow neutrons are technically an exception but are included because of their technological importance.) The lower energy bound is set by the minimum energy required to produce ionization in typical materials by the radiation or the secondary products of its interaction. Radiations with energy greater than this minimum are classified as *ionizing radiations*. The upper bound is chosen to limit the topics in this coverage to those of primary concern in nuclear science and technology.

The main emphasis in this chapter will be the laboratory-scale sources of these radiations, which are likely to be of interest either in the calibration and testing of radiation detectors described in the following chapters, or as objects of the measurements themselves. Natural background radiation is an important additional source and is discussed separately in Chapter 20.

The radiations of interest differ in their "hardness" or ability to penetrate thicknesses of material. Although this property is discussed in greater detail in Chapter 2, it is also of considerable concern in determining the physical form of radiation sources. Soft radiations, such as alpha particles or low-energy X-rays, penetrate only small thicknesses of material. Radioisotope sources must therefore be deposited in very thin layers if a large fraction of these radiations is to escape from the source itself. Sources that are physically thicker are subject to "self-absorption," which is likely to affect both the number and the energy spectrum of the radiations that emerge from its surface. Typical thicknesses for such sources are therefore measured in micrometers. Beta particles are generally more penetrating, and sources up to a few tenths of a millimeter in thickness can usually be tolerated. Harder radiations, such as gamma rays or neutrons, are much less affected by self-absorption and sources can be millimeters or centimeters in dimension without seriously affecting the radiation properties.

I. UNITS AND DEFINITIONS

A. Radioactivity

The *activity* of a radioisotope source is defined as its rate of decay and is given by the fundamental law of radioactive decay

$$\left.\frac{dN}{dt}\right|_{\text{decay}} = -\lambda N \tag{1-1}$$

where N is the number of radioactive nuclei and λ is defined as the *decay constant*.[†] The historical unit of activity has been the *curie* (Ci), defined as exactly 3.7×10^{10} disintegrations/second, which owes its definition to its origin as the best available estimate of the activity of 1 gram of pure ^{226}Ra. Its submultiples, the millicurie (mCi) or microcurie (μCi), generally are more suitable units for laboratory-scale radioisotope sources.

   Although still widely used in the literature, the curie is destined to be gradually replaced by its SI equivalent, the becquerel (Bq). At its 1975 meeting, the General Conference of Weights and Measures (GCPM) adopted a resolution declaring that the becquerel, defined as one disintegration per second, has become the standard unit of activity. Thus

$$1 \, \text{Bq} = 2.703 \times 10^{-11} \, \text{Ci}$$

Radioactive sources of convenient size in the laboratory are more reasonably measured in kilobecquerels (kBq) or megabecquerels (MBq).

   It should be emphasized that activity measures the source disintegration rate, which is not synonymous with the emission rate of radiation produced in its decay. Frequently, a given radiation will be emitted in only a fraction of all the decays, so a knowledge of the decay scheme of the particular isotope is necessary to infer a radiation emission rate from its activity. Also, the decay of a given radioisotope may lead to a daughter product whose activity also contributes to the radiation yield from the source. A complete listing of radioisotope decay schemes is tabulated in Ref. 1.

   The *specific activity* of a radioactive source is defined as the activity per unit mass of the radioisotope sample. If a pure or "carrier-free" sample is obtained that is unmixed

---

[†]One should be aware that Eq. (1-1) represents the decay rate only, and the *net* value of dN/dt may be altered by other production or disappearance mechanisms. As one example, the radioisotope may be produced as the daughter product of the decay of a parent species also present in the sample. Then a production term is present for the daughter that is given by the decay rate of the parent multiplied by the fraction of such decays that lead to the daughter species. If the half-life of the parent is very long, the number of daughter nuclei increases until the daughter activity reaches an equilibrium value (after many daughter half-lives have passed) when the production and decay rates are equal, and $dN/dt = 0$ for the number of daughter nuclei.

with any other nuclear species, its specific activity can be calculated from

$$\text{specific activity} \equiv \frac{\text{activity}}{\text{mass}} = \frac{\lambda N}{NM/A_v} = \frac{\lambda A_v}{M} \qquad (1\text{-}2)$$

where
$\qquad M$ = molecular weight of sample

$\qquad A_v$ = Avogadro's number $\left(= 6.02 \times 10^{23} \text{ nuclei/mole}\right)$

$\qquad \lambda$ = radioisotope decay constant $\left(= \ln 2/\text{half-life}\right)$

Radioisotopes are seldom obtained in carrier-free form, however, and are usually diluted in a much larger concentration of stable nuclei of the same element. Also, if not prepared in pure elemental form, additional stable nuclei may be included from other elements that are chemically combined with those of the source. For sources in which self-absorption is a problem, there is a premium on obtaining a sample with high specific activity to maximize the number of radioactive nuclei within a given thickness. From Eq. (1-2), high specific activity is most readily obtained using radionuclides with large λ (or small half-life).

B. Energy

The traditional unit for the measurement of radiation energy is the *electron volt* or eV, defined as the kinetic energy gained by an electron by its acceleration through a potential difference of 1 volt. The multiples of kiloelectron volt (keV) and megaelectron volt (MeV) are more common in the measurement of energies for ionizing radiation. The electron volt is a convenient unit when dealing with particulate radiation because the energy gained from an electric field can easily be obtained by multiplying the potential difference by the number of electronic charges carried by the particle. For example, an alpha particle that carries an electronic charge of $+2$ will gain an energy of 2 keV when accelerated by a potential difference of 1000 volts.

The SI unit of energy is the *joule* (J). When dealing with radiation energies, the submultiple femtojoule (fJ) is more convenient and is related to the electron volt by the conversion

$$1 \text{ fJ} \left(= 10^{-15} \text{ J}\right) = 6.241 \times 10^3 \text{ eV}$$

It is not clear to what extent the electron volt will be phased out in future usage because its physical basis and universal use in the literature are strong arguments for its continued application to radiation measurements.

The energy of an X- or gamma-ray photon is related to the radiation frequency by

$$E = h\nu \qquad (1\text{-}3)$$

where
$\qquad h$ = Planck's constant $\left(6.626 \times 10^{-34} \text{ J} \cdot \text{s, or } 4.135 \times 10^{-15} \text{ eV} \cdot \text{s}\right)$

$\qquad \nu$ = frequency

The wavelength λ is related to the photon energy by

$$\lambda = \frac{1.240 \times 10^{-6}}{E}$$

where λ is in meters and E in eV.

II. FAST ELECTRON SOURCES

A. Beta Decay

The most common source of fast electrons in radiation measurements is a radioisotope that decays by beta-minus emission. The process is written schematically

$$_Z^A X \rightarrow _{Z+1}^A Y + \beta^- + \bar{\nu} \qquad (1\text{-}4)$$

where X and Y are the initial and final nuclear species, and $\bar{\nu}$ is the antineutrino. Because neutrinos and antineutrinos have an extremely small interaction probability with matter, they are undetectable for all practical purposes. The recoil nucleus Y appears with a very small recoil energy, which is ordinarily below the ionization threshold, and therefore it cannot be detected by conventional means. Thus, the only significant ionizing radiation produced by beta decay is the fast electron or beta particle itself.

Because most radionuclides produced by neutron bombardment of stable materials are beta-active, a large assortment of beta emitters are readily available through production in a reactor flux. Species with many different half-lives can be obtained, ranging from thousands of years down to as short a half-life as is practical in the application. Most beta decays populate an excited state of the product nucleus, so that the subsequent deexcitation gamma rays are emitted together with the beta particles in many common beta sources. Some examples of nuclides that decay directly to the ground state of the product and are therefore "pure beta emitters" are shown in Table 1-1.

Each specific beta decay transition is characterized by a fixed decay energy or Q-value. Because the energy of the recoil nucleus is virtually zero, this energy is shared between the beta particle and the "invisible" neutrino. The beta particle thus appears with an energy that varies from decay to decay and can range from zero to the "beta endpoint energy," which is numerically equal to the Q-value. A representative beta energy spectrum is illustrated in Fig. 1-1. The Q-value for a given decay is normally quoted assuming that the transition takes place between the ground states of both the parent and daughter nuclei. If the transition involves an excited state of either the parent or daughter, the endpoint energy of the corresponding beta spectrum will be changed by the difference in excitation energies. Since several excited states can be populated in some decay

TABLE 1-1 Some "Pure" Beta-Minus Sources

Nuclide	Half-Life	Endpoint Energy (MeV)
^{3}H	12.26 y	0.0186
^{14}C	5730 y	0.156
^{32}P	14.28 d	1.710
^{33}P	24.4 d	0.248
^{35}S	87.9 d	0.167
^{36}Cl	3.08×10^5 y	0.714
^{45}Ca	165 d	0.252
^{63}Ni	92 y	0.067
^{90}Sr/^{90}Y	27.7 y/64 h	0.546/2.27
^{99}Tc	2.12×10^5 y	0.292
^{147}Pm	2.62 y	0.224
^{204}Tl	3.81 y	0.766

Data from Lederer and Shirley.[1]

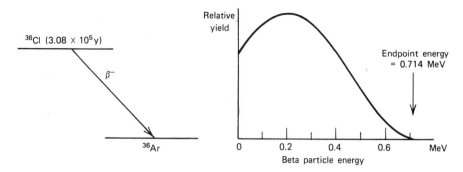

Figure 1-1 The decay scheme of ^{36}Cl and the resulting beta particle energy distribution.

schemes, the measured beta particle spectrum may then consist of several components with different endpoint energies.

B. Internal Conversion

The continuum of energies produced by any beta source is inappropriate for some applications. For example, if an energy calibration is to be carried out for an electron detector, it is much more convenient to use a source of monoenergetic electrons. The nuclear process of *internal conversion* can be the source of *conversion electrons*, which are, under some circumstances, nearly monoenergetic.

The internal conversion process begins with an excited nuclear state, which may be formed by a preceding process—often beta decay of a parent species. The common method of deexcitation is through emission of a gamma-ray photon. For some excited states, gamma emission may be somewhat inhibited, and the alternative of internal conversion can become significant. Here the nuclear excitation energy E_{ex} is transferred directly to one of the orbital electrons of the atom. This electron then appears with an

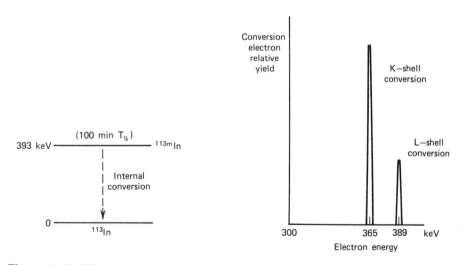

Figure 1-2 The conversion electron spectrum expected from internal conversion of the isomeric level at 393 keV in ^{113m}In.

TABLE 1-2   Some Common Conversion Electron Sources

Parent Nuclide	Parent Half-Life	Decay Mode	Decay Product	Transition Energy of Decay Product (keV)	Conversion Electron Energy (keV)
^{109}Cd	453 d	EC	^{109m}Ag	88	62
					84
^{113}Sn	115 d	EC	^{113m}In	393	365
					389
^{137}Cs	30.2 y	β^-	^{137m}Ba	662	624
					656
^{139}Ce	137 d	EC	^{139m}La	166	126
					159
^{207}Bi	38 y	EC	^{207m}Pb	$\left\{ \begin{array}{l} 570 \\ \\ 1064 \end{array} \right.$	482
					554
					976
					1048

Data from Lederer and Shirley.[1]

energy given by

$$E_{e^-} = E_{ex} - E_b \tag{1-5}$$

where E_b is its binding energy in the original electron shell.

An example of a conversion electron spectrum is shown in Fig. 1-2. Because the conversion electron can originate from any one of a number of different electron shells within the atom, a single nuclear excitation level generally leads to several groups of electrons with different energies. The spectrum may be further complicated in those cases in which more than one excited state within the nucleus is converted. Furthermore, the electron energy spectrum may also be superimposed on a continuum consisting of the beta spectrum of the parent nucleus which leads to the excited state. Despite these shortcomings, conversion electrons are the only practical laboratory-scale source of monoenergetic electron groups in the high keV to MeV energy range. Several useful radioisotope sources of conversion electrons are compiled in Table 1-2.

C. Auger Electrons

Auger electrons are roughly the analogue of internal conversion electrons when the excitation energy originates in the atom rather than in the nucleus. A preceding process (such as electron capture) may leave the atom with a vacancy in a normally complete electron shell. This vacancy is often filled by electrons from the outer shells of the atom with the emission of a characteristic X-ray photon. Alternatively, the excitation energy of the atom may be transferred directly to one of the outer electrons, causing it to be ejected from the atom. This electron is called an Auger electron and appears with an energy given by the difference between the original atomic excitation energy and the binding energy of the shell from which the electron was ejected. Auger electrons therefore produce a discrete energy spectrum, with different groups corresponding to different initial and final states. In all cases, their energy is relatively low compared with beta particles or conversion electrons, particularly because Auger electron emission is favored only in low-Z elements for which electron binding energies are small. Typical Auger electrons with a few keV initial energy are subject to pronounced self-absorption within the source and are easily stopped by very thin source covers or detector entrance windows.

III. HEAVY CHARGED PARTICLE SOURCES

A. Alpha Decay

Heavy nuclei are energetically unstable against the spontaneous emission of an alpha particle (or ^{4}He nucleus). The probability of decay is governed by the *barrier penetration* mechanism described in most texts on nuclear physics, and the half-life of useful sources varies from days to many thousands of years. The decay process is written schematically as

$$^A_Z X \rightarrow ^{A-4}_{Z-2} Y + ^4_2 \alpha$$

where X and Y are the initial and final nuclear species. A representative alpha decay scheme is shown in Fig. 1-3, together with the expected energy spectrum of the corresponding alpha particles emitted in the decay.

The alpha particles appear in one or more energy groups that are, for all practical purposes, monoenergetic. For each distinct transition between initial and final nucleus (e.g., between ground state and ground state), a fixed energy difference or Q-value characterizes the decay. This energy is shared between the alpha particle and the recoil nucleus in a unique way, so that each alpha particle appears with the same energy given by $Q(A - 4)/A$. There are many practical instances in which only one such transition is involved and for which the alpha particles are therefore emitted with a unique single energy. Other examples, such as that shown in Fig. 1-3, may involve more than one transition energy so that the alpha particles appear in groups with differing relative intensities.

Table 1-3 lists some properties of the more common radioisotope sources of alpha particles. It is no accident that most alpha particle energies are limited to between about 4 and 6 MeV. There is a very strong correlation between alpha particle energy and half-life of the parent isotope, and those with the highest energies are those with shortest half-life. Beyond about 6.5 MeV, the half-life can be expected to be less than a few days, and therefore the source is of very limited utility. On the other hand, if the energy drops below 4 MeV, the barrier penetration probability becomes very small and the half-life of the isotope is very large. If the half-life is exceedingly long, the specific activity attainable in a practical sample of the material becomes very small and the source is of no interest because its intensity is too low. Probably the most common calibration source for alpha particles is ^{241}Am, and an example of its application to the calibration of silicon solid-state detectors is shown in Fig. 11-12.

Because alpha particles lose energy rapidly in materials, alpha particle sources that are to be nearly monoenergetic must be prepared in very thin layers. In order to contain the radioactive material, typical sources are covered with a metallic foil or other material which must also be kept very thin if the original energy and monoenergetic nature of the alpha emission is to be preserved.

B. Spontaneous Fission

The fission process is the only spontaneous source of energetic heavy charged particles with mass greater than that of the alpha particle. Fission fragments are therefore widely used in the calibration and testing of detectors intended for general application to heavy ion measurements.

All heavy nuclei are, in principle, unstable against spontaneous fission into two lighter fragments. For all but the extremely heavy nuclei, however, the process is inhibited

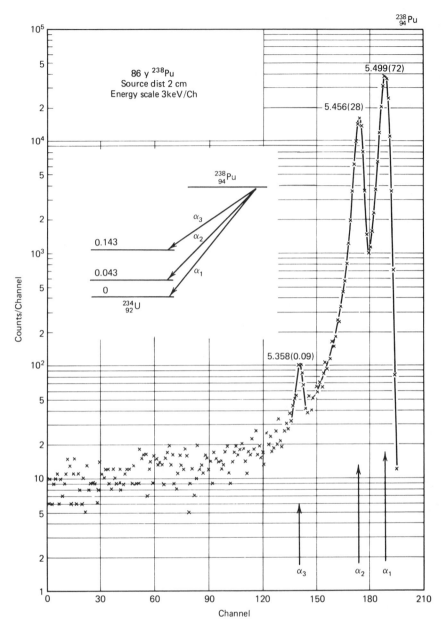

Figure 1-3 Alpha particle groups produced in the decay of ^{238}Pu. The pulse height spectrum shows the three groups as measured by a silicon surface barrier detector. Each peak is identified by its energy in MeV and percent abundance (in parentheses). The insert shows the decay scheme, with energy levels in the product nucleus labeled in MeV. (Spectrum from Chanda and Deal.[2])

TABLE 1-3 Common Alpha-Emitting Radioisotope Sources

Source	Half-Life	Alpha Particle Kinetic Energy (with Uncertainty) in MeV		Percent Branching
^{148}Gd	93 y	3.182787	± 0.000024	100
^{232}Th	1.4×10^{10} y	4.012	± 0.005	77
		3.953	± 0.008	23
^{238}U	4.5×10^9 y	4.196	± 0.004	77
		4.149	± 0.005	23
^{235}U	7.1×10^8 y	4.598	± 0.002	4.6
		4.401	± 0.002	56
		4.374	± 0.002	6
		4.365	± 0.002	12
		4.219	± 0.002	6
^{236}U	2.4×10^7 y	4.494	± 0.003	74
		4.445	± 0.005	26
^{230}Th	7.7×10^4 y	4.6875	± 0.0015	76.3
		4.6210	± 0.0015	23.4
^{234}U	2.5×10^5 y	4.7739	± 0.0009	72
		4.7220	± 0.0009	28
^{231}Pa	3.2×10^4 y	5.0590	± 0.0008	11
		5.0297	± 0.0008	20
		5.0141	± 0.0008	25.4
		4.9517	± 0.0008	22.8
^{239}Pu	2.4×10^4 y	5.1554	± 0.0007	73.3
		5.1429	± 0.0008	15.1
		5.1046	± 0.0008	11.5
^{240}Pu	6.5×10^3 y	5.16830	± 0.00015	76
		5.12382	± 0.00023	24
^{243}Am	7.4×10^3 y	5.2754	± 0.0010	87.4
		5.2335	± 0.0010	11
^{210}Po	138 d	5.30451	± 0.00007	99 +
^{241}Am	433 y	5.48574	± 0.00012	85.2
		5.44298	± 0.00013	12.8
^{238}Pu	88 y	5.49921	± 0.00020	71.1
		5.4565	± 0.0004	28.7
^{244}Cm	18 y	5.80496	± 0.00005	76.4
		5.762835	± 0.000030	23.6
^{243}Cm	30 y	6.067	± 0.003	1.5
		5.992	± 0.002	5.7
		5.7847	± 0.0009	73.2
		5.7415	± 0.0009	11.5
^{242}Cm	163 d	6.11292	± 0.00008	74
		6.06963	± 0.00012	26
^{254m}Es	276 d	6.4288	± 0.0015	93
^{253}Es	20.5 d	6.63273	± 0.00005	90
		6.5916	± 0.0002	6.6

Data from Rytz.[3]

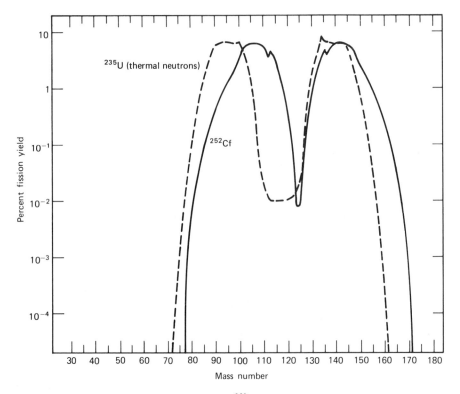

Figure 1-4 (a) The mass distribution of ^{252}Cf spontaneous fission fragments. Also shown is the corresponding distribution from fission of ^{235}U induced by thermal neutrons. (From Nervik.[4])

by the large potential barrier that must be overcome in the distortion of the nucleus from its original near-spherical shape. Spontaneous fission is therefore not a significant process except for some transuranic isotopes of very large mass number. The most widely used example is ^{252}Cf, which undergoes spontaneous fission with a half-life (if it were the only decay process) of 85 years. However, most transuranic elements also undergo alpha decay, and in ^{252}Cf the probability for alpha emission is considerably higher than that for spontaneous fission. Therefore, the actual half-life for this isotope is 2.65 years, and a sample of 1 microgram of ^{252}Cf will emit 1.92×10^7 alpha particles and undergo 6.14×10^5 spontaneous fissions per second.

Each fission gives rise to two fission fragments, which, by the conservation of momentum, are emitted in opposite directions. Because the normal physical form for a spontaneous fission source is a thin deposit on a flat backing, only one fragment per fission can escape from the surface, whereas the other is lost by absorption within the backing. As described later in this chapter, each spontaneous fission in ^{252}Cf also liberates a number of fast neutrons.

The fission fragments are medium-weight positive ions with a mass distribution illustrated in Fig. 1-4a. The fission is predominantly asymmetric so that the fragments are clustered into a "light group" and "heavy group," with average mass numbers of 108 and 143. The fragments appear initially as positive ions for which the net charge approaches the atomic number of the fragment. As the fragment slows down by interacting with the

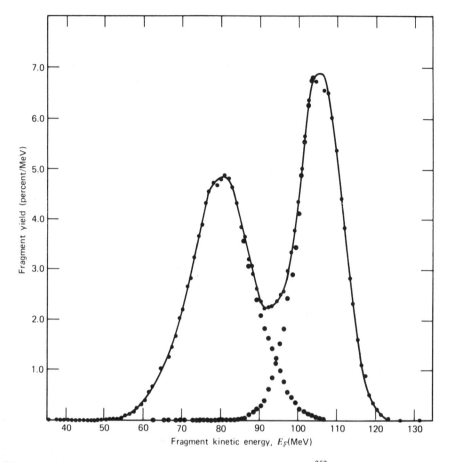

Figure 1-4 (b) The distribution in kinetic energy of the ^{252}Cf spontaneous fission fragments. The peak on the left corresponds to the heavy fragments, and that on the right to the light fragments. (From Whetstone.[5])

matter through which it passes, additional electrons are picked up by the ion, reducing its effective charge.

The energy shared by the two fragments averages about 185 MeV. The distribution of this energy is also asymmetric with the light fragment receiving the greater fraction. A plot of their initial energy distribution is also shown in Fig. 1-4b. Because they also lose energy readily in solid materials, self-absorption and energy loss of the fragments are important considerations unless the source is prepared in a very thin layer. The type of degradation of a fission fragment energy spectrum observed from thicker sources (for the case of neutron-induced ^{235}U fission) is illustrated in Fig. 14-6.

IV. SOURCES OF ELECTROMAGNETIC RADIATION

A. Gamma Rays Following Beta Decay

Gamma radiation is emitted by excited nuclei in their transition to lower-lying nuclear levels. In most practical laboratory sources, the excited nuclear states are created in the decay of a parent radionuclide. Four common examples widely used as gamma-ray

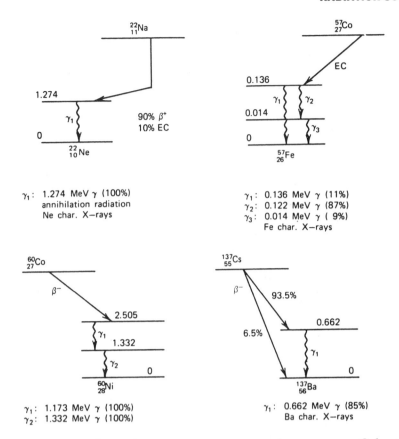

Figure 1-5 Decay schemes for some common gamma reference sources. Only major transitions are shown. The energies and yields per disintegration of X- and gamma rays emitted in each decay are listed below the diagram. (Data from Lederer and Shirley.[1])

calibration sources are illustrated in the decay schemes in Fig. 1-5. In each case, a form of beta decay leads to the population of the excited state in the daughter nucleus. For the examples shown, the beta decay is a relatively slow process characterized by a half-life of hundreds of days or greater, whereas the excited states in the daughter nucleus have a much shorter average lifetime (typically of the order of picoseconds or less). Deexcitation takes place through the emission of a gamma-ray photon whose energy is essentially equal to the difference in energy between the initial and final nuclear states. The gamma rays therefore appear with a half-life characteristic of the parent beta decay, but with an energy that reflects the energy level structure of the daughter nucleus. For example, although "^{60}Co gamma rays" decrease in intensity with the 5.26 year half-life characteristic of ^{60}Co, they actually arise from transitions in the ^{60}Ni nucleus. Decay schemes of the type shown in Fig. 1-5 are compiled for all radioactive nuclei in Ref. 1. From the probabilities of various deexcitation transitions (or "branching ratios") given in these decay schemes, the number of gamma-ray photons per disintegration of the parent nucleus can be deduced. Some specific radionuclide gamma-ray sources useful in the precise energy calibration and efficiency calibration of gamma-ray detectors are listed in Tables 12-1 and 12-2.

Because nuclear states have very well-defined energies, the energies of gamma rays emitted in state-to-state transitions are also very specific. The gamma rays from any one transition are nearly monoenergetic, and the inherent line width of the photon energy distribution is always small compared with the energy resolution of any of the detectors described later in this text. A measurement of the detector response is therefore indicative of its own limiting resolution rather than any variation in the incident gamma-ray energy.[†]

The common gamma-ray sources based on beta decay are generally limited to energies below about 2.8 MeV. One nuclide, ^{56}Co, has gained recent attention[6] as a potential source for gamma rays of higher energy. The decay scheme of this isotope, which proceeds both by electron capture and beta-plus decay, gives rise to a complex spectrum of gamma rays whose energies extend to as high as 3.55 MeV. However, its short half-life of 77 days largely limits its use to facilities with access to accelerators necessary to carry out its production through the ^{56}Fe(p, n) reaction.

Gamma-ray reference sources are an essential accessory in any radiation measurements laboratory in which gamma-ray measurements are carried out. They normally consist of samples of radioisotopes of a few microcuries (around 10^5 Bq) encased in plastic disks or rods. The thickness of the encapsulation is generally large enough to stop any particulate radiation from the decay of the parent nucleus, and the only primary radiation emerging from the surface is the gamma radiation produced in the daughter decay. However, secondary radiations such as annihilation photons or bremsstrahlung can be significant at times (see below). Although the radiation hazard of such sources is minimal, the gamma-ray emission rate is sufficiently high to permit ready energy calibration of most types of gamma-ray detectors.

If the sources are to be used to carry out accurate efficiency calibration as well, their absolute activity must also be known. In these cases, the radioisotope deposits are generally prepared on much thinner backings with a minimum of overlying cover to reduce gamma-ray attenuation and scattering within the source structure. External absorbers must then be used to eliminate any particulate emission if its presence will interfere with the application.

B. Annihilation Radiation

When the parent nucleus undergoes beta-plus decay, additional electromagnetic radiation is generated. The origin lies in the fate of the positrons emitted in the primary decay process. Because they generally travel only a few millimeters before losing their kinetic energy (see Chapter 2), the inherent encapsulation around the source is often sufficiently thick to fully stop the positrons. When their energy is very low, near the end of their range, they combine with normal negative electrons in the absorbing materials in the process of *annihilation*.[‡] The original positron and electron disappear and are replaced by two oppositely directed 0.511 MeV electromagnetic photons known as *annihilation radiation*. This radiation is then superimposed on whatever gamma radiation may be emitted in the subsequent decay of the daughter product. For example, in the decay of ^{22}Na, shown in Fig. 1-5, photons of both 0.511 and 1.274 MeV energy are emitted from encapsulated sources.

[†]A rare exception may occur if the emitting nuclei have large velocities. The Doppler effect can then introduce an energy spread that may be significant in detectors with excellent energy resolution. An example is given later in this chapter in which gamma rays are emitted from nuclei that are still moving after being formed in a nuclear reaction.

[‡]This step can take place either directly or through an intermediate stage in which the positron and electron form a quasistable combination, known as positronium, which may exist for a few nanoseconds.

C. Gamma Rays Following Nuclear Reactions

If gamma rays with energies higher than those available from beta-active isotopes are needed, some other process must lead to the population of higher-lying nuclear states. One possibility is the nuclear reaction

$$\frac{4}{2}\alpha + \frac{9}{4}\text{Be} \rightarrow \frac{12}{6}\text{C*} + \frac{1}{0}\text{n}$$

where the product nucleus ^{12}C is left in an excited state. Its decay gives rise to a gamma-ray photon of 4.44 MeV energy. Unfortunately, the average lifetime of this state is so short that the recoil carbon atom does not have time to come to rest before the gamma ray is emitted. The resulting photon energies are therefore broadened by Doppler effects, depending on the relative orientation of the recoil atom velocity and the photon direction, and there is an inherent spread of about 1% in the gamma-ray energies. This line width is adequate for many calibration purposes, but it is too large for detectors with very good energy resolution (such as the germanium detectors of Chapter 12).

Another possibility is the reaction

$$\frac{4}{2}\alpha + \frac{13}{6}\text{C} \rightarrow \frac{16}{8}\text{O*} + \frac{1}{0}\text{n}$$

Here the product nucleus ^{16}O can be formed in an excited state at 6.130 MeV above the ground state and with a lifetime of about 2×10^{-11} s. This lifetime is sufficiently long to eliminate almost all Doppler effects, and the resulting 6.130 MeV gamma-ray photons are essentially monoenergetic.

Both the above reactions can be exploited by combining a radioisotope that decays by alpha emission with the appropriate target material (either ^{9}Be or ^{13}C). Because sources of this type are more commonly used to produce neutrons, further discussion of the choice of alpha emitter and other aspects of the source design will be postponed until the following section on neutron sources. Because most alpha particles do not lead to a reaction before losing their energy in the target material, large activities of the alpha emitter must be used to produce a gamma-ray source of practical intensity. For example, a typical source[7] fabricated from 6×10^9 Bq of ^{238}PuO$_2$ and 200 mg of isotopically separated ^{13}C produces a 6.130 MeV gamma-ray yield of 770 photons/s.

D. Bremsstrahlung

When fast electrons interact in matter, part of their energy is converted into electromagnetic radiation in the form of *bremsstrahlung*. (This process is discussed in somewhat more detail in Chapter 2.) The fraction of the electron energy converted into bremsstrahlung increases with increasing electron energy and is largest for absorbing materials of high atomic number. The process is important in the production of X-rays from conventional X-ray tubes.

For monoenergetic electrons that slow down and stop in a given material, the bremsstrahlung energy spectrum is a continuum with photon energies that extend as high as the electron energy itself. The shape of a typical spectrum produced by monoenergetic electrons is shown in Fig. 1-6. The emission of low-energy photons predominates, and the average photon energy is a small fraction of the incident electron energy. Because these spectra are continua, they cannot be applied directly to the energy calibration of radiation detectors.

The shape of the energy spectrum from an X-ray tube can be beneficially altered by *filtration* or passage through appropriate absorber materials. Through the use of absorbers

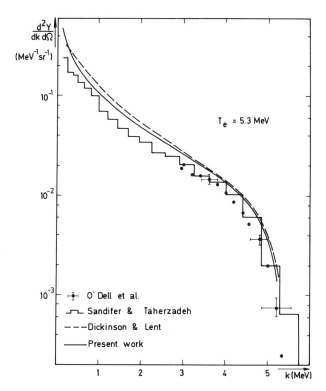

Figure 1-6 The bremsstrahlung energy spectrum emitted in the forward direction by 5.3 MeV electrons incident on a Au-W target. A 7.72 g/cm^2 aluminum filter also was present. (From Ferdinande et al.[8])

that preferentially remove the lower-energy photons, a peaked spectrum can be produced which, although far from monoenergetic, can be useful in the energy calibration of detectors whose response changes only gradually with energy. Some examples of filtered spectra from X-ray tubes are plotted in Fig. 1-7. At lower energies, the abrupt change in filter transmission at its K-absorption edge (see Chapter 2) can produce a prominent peak at the corresponding energy in the filtered spectrum.[10]

Bremsstrahlung is also produced by other sources of fast electrons, including beta particles. Therefore, some bremsstrahlung photons are generated by any beta-active isotope encapsulated to stop the beta particles. Some examples of bremsstrahlung spectra from specific isotopes are plotted in Fig. 10-5.

In addition to bremsstrahlung, characteristic X-rays (see following section) are also produced when fast electrons pass through an absorber. Therefore, the spectra from X-ray tubes or other bremsstrahlung sources also show characteristic X-ray emission lines superimposed on the continuous bremsstrahlung spectrum.

E. Characteristic X-Rays

If the orbital electrons in an atom are disrupted from their normal configuration by some excitation process, the atom may exist in an excited state for a short period of time. There is a natural tendency for the electrons to rearrange themselves to return the atom to its lowest energy or ground state within a time that is characteristically a nanosecond or less in a solid material. The energy liberated in the transition from the excited to the ground

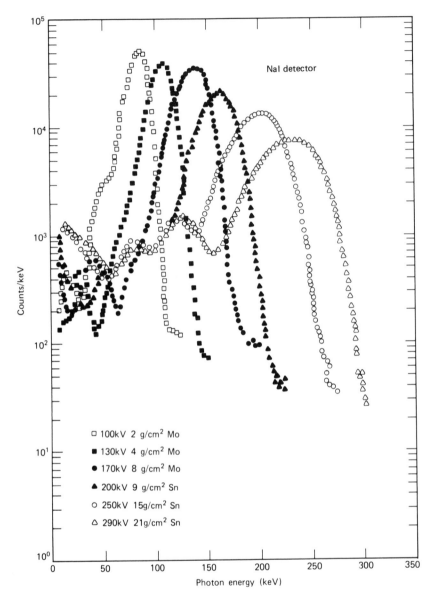

Figure 1-7 Examples of measured pulse height spectra [using a NaI(Tl) scintillator] after filtration of an X-ray tube output using the indicated absorbers and tube voltages. (From Storm et al.[9])

state takes the form of a *characteristic X-ray* photon whose energy is given by the energy difference between the initial and final states. For example, if a vacancy is temporarily created in the K shell of an atom, then a characteristic K *X-ray* is liberated when that vacancy is subsequently filled. If that electron comes from the L shell, then a K_α photon is produced whose energy is equal to the difference in binding energies between the K and L shells. If the filling electron originated in the M shell instead, then a K_β photon is produced with slightly larger energy, and so on. The maximum K-series photon is produced when the vacancy is filled by a free or unbound electron, and the corresponding photon energy is then simply given by the K shell binding energy. Vacancies created in outer shells by the filling of a K shell vacancy are subsequently filled with the emission of L-, M-,... series characteristic X-rays.

Because their energy is greatest, the K-series X-rays are generally of most practical significance. Their energy increases regularly with atomic number of the element and is, for example, about 1 keV for sodium with $Z = 11$, 10 keV for gallium with $Z = 31$, and 100 keV for radium with $Z = 88$. The L series X-rays do not reach 1 keV until $Z = 28$ and 10 keV at $Z = 74$. Extensive tables of precise characteristic X-rays can be found in Ref. 1. Because the energy of the characteristic X-ray is unique to each individual element, they are often used in the elemental analysis of unknown samples (see Fig. 13-17).

For an atom in an excited state, the ejection of an Auger electron is a competitive process to the emission of characteristic X-rays. The *fluorescent yield* is defined as the fraction of all cases in which the excited atom emits a characteristic X-ray photon in its deexcitation. Values for the fluorescent yield are often tabulated as part of spectroscopic data.

A large number of different physical processes can lead to the population of excited atomic states from which characteristic X-rays originate. In general, the relative yields of the K, L, and subsequent series will depend on the excitation method, but the energy of the characteristic photons is fixed by the basic atomic binding energies. We list below those excitation mechanisms that are of most practical importance for compact laboratory sources of characteristic X-rays.

1. EXCITATION BY RADIOACTIVE DECAY

In the nuclear decay process of electron capture, the nuclear charge is decreased by one unit by the capture of an orbital electron, most often a K-electron. The resulting atom still has the right number of orbital electrons, but the capture process has created a vacancy in one of the inner shells. When this vacancy is subsequently filled, X-rays are generated which are characteristic of the product element. The decay may populate either the ground state or an excited state in the product nucleus, so that the characteristic X-rays may also be accompanied by gamma rays from subsequent nuclear deexcitation.[†]

Internal conversion is another nuclear process that can lead to characteristic X-rays. As defined earlier in this chapter, internal conversion results in the ejection of an orbital electron from the atom leaving behind a vacancy. Again, it is the K-electrons that are most readily converted, and therefore the K-series characteristic X-rays are the most prominent. Because gamma ray deexcitation of the nuclear state is always a competing process to internal conversion, radioisotope sources of this type usually emit gamma rays in addition to the characteristic X-rays. The conversion electrons may also lead to a measurable bremsstrahlung continuum, particularly when their energy is high.

In Table 13-2, some examples are given of radioisotopes that involve either electron capture or internal conversion and are possible sources of characteristic X-rays. In all

---

[†]Electron capture can lead to another form of continuous electromagnetic radiation known as *inner bremsstrahlung*. In the decay process, an orbital electron is captured by the nucleus and therefore must undergo some acceleration. From classical theory, an accelerated charge must emit electromagnetic radiation. Because the acceleration may vary over a wide range depending on the specifics of the capture process, the resulting emission spectrum is a continuum ranging up to a maximum photon energy given by the Q-value of the electron capture decay (the maximum energy available in the nuclear transition). Inner bremsstrahlung therefore adds a continuous electromagnetic spectrum to the other radiations normally expected as a product of the electron capture decay, although in many cases the intensity of this spectrum may be negligibly small.

TABLE 1-4   Some Radioisotope Sources of Low-Energy X-Rays

Nuclide	Half-Life	Weighted K_α X-Ray Energy	Fluorescent Yield	Other Radiations
^{37}Ar	35.1 d	2.957 keV	0.086	Some IB[a]
^{41}Ca	8×10^4 y	3.690	0.129	Pure
^{44}Ti	48 y	4.508	0.174	γ Rays at 68 and 78 keV
^{49}V	330 d	4.949	0.200	IB
^{55}Fe	2.60 y	5.895	0.282	Weak IB

[a] IB represents inner bremsstrahlung.
Data from Amlauer and Tuohy.[11]

these examples, the yield of high-energy gamma rays from a nuclear transition is large compared with that of the characteristic X-rays. If a pure X-ray source free of gamma-ray contamination is required, a radioisotope that decays by electron capture leading directly to the ground nuclear state of the daughter must be chosen. Table 1-4 gives some examples for the lower energy range. Of these, ^{55}Fe is most widely used because of its convenient half-life and high available specific activity. It is very nearly a pure source of manganese K-series X-rays at about 5.9 keV, and the inner bremsstrahlung associated with the electron capture process is very weak.

Self-absorption is a significant technical problem in the preparation of radioisotope X-ray sources. As the thickness of the radioisotope deposit is increased, the X-ray flux emerging from its surface approaches a limiting value because only those atoms near the surface can contribute photons that escape. The number of emitting atoms within a given distance of the deposit surface is maximized by increasing the specific activity of the radioisotope, and carrier-free samples will exhibit the maximum attainable source intensity per unit area.

2. EXCITATION BY EXTERNAL RADIATION

The general scheme sketched in Fig. 1-8 may also be used to generate characteristic X-rays. Here an external source of radiation (X-rays, electrons, alpha particles, etc.) is caused to strike a target, creating excited or ionized atoms in the target through the processes detailed in Chapter 2. Because many excited atoms or ions in the target subsequently deexcite to the ground state through the emission of characteristic X-rays, the target can serve as a localized source of these X-rays.

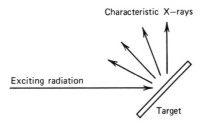

Figure 1-8   General method for the generation of characteristic X-rays from a specific target element. The exciting radiation can be X-rays, electrons, alpha particles, or any other ionizing radiation.

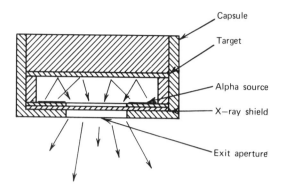

Figure 1-9   Cross-sectional view of a compact source of characteristic X-rays which utilizes alpha particle excitation of a target. (From Amlauer and Tuohy.[11])

The energy of the X-rays emitted depends on the choice of target material. Targets with low atomic number result in soft characteristic X-rays, and high-Z targets produce harder or higher energy X-rays. The incident radiation must have an energy larger than the maximum photon energy expected from the target, because the excited states leading to the corresponding atomic transition must be populated by the incident radiation.

As one example, the incident radiation may consist of X-rays generated in a conventional X-ray tube. These X-rays may then interact in the target through photoelectric absorption, and the subsequent deexcitation of the target ions creates their characteristic X-ray spectrum. In this case, the process is called *X-ray fluorescence*. Although the characteristic X-ray spectrum can be contaminated by scattered photons from the incident X-ray beam, this component can be kept below about 10 or 20% of the total photon yield with proper choice of target and geometry.[9]

As an alternative to bulky X-ray tubes, radioisotopes that emit low-energy photons may also be used as the source of the excitation. An example is ^{241}Am, in which gamma rays of 60 keV energy are emitted in 36% of the decays. Characteristic X-ray sources using this isotope for excitation of targets in the geometry shown in Fig. 1-9 are currently available using up to 10 mCi of ^{241}Am activity.

Another method of exciting the target is through the use of an external electron beam. For targets of low atomic number, accelerating potentials of only a few thousand volts are required so that relatively compact electron sources can be devised. In the case of electron excitation, the characteristic X-ray spectrum from the target will be contaminated by the continuous bremsstrahlung spectrum also generated by interactions of the incident electrons in the target. For thin targets, however, the bremsstrahlung photons are preferentially emitted in the forward direction, whereas the characteristic X-rays are emitted isotropically.[12] Placing the exit window at a large angle (120–180°) with respect to the incident electron direction will therefore minimize the bremsstrahlung contamination.

The incident radiation in Fig. 1-8 can also consist of heavy charged particles. Again, the interactions of these particles in the target will give rise to the excited atoms necessary for the emission of characteristic X-rays. For compact and portable sources, alpha particles emitted by radioisotope sources are the most convenient source of the incident particles. Of the various potential alpha emitters, the most useful are ^{210}Po and ^{244}Cm because of their convenient half-lives and relative freedom from contaminant electromagnetic radiation (see Table 1-5). Alpha particle excitation avoids the complication of

TABLE 1-5 Alpha Particle Sources Useful for Excitation of Characteristic X-Rays

	^{210}Po	^{244}Cm
Half-life	138 d	17.6 y
Alpha emissions	5.305 MeV (100%)	5.81, 5.77 MeV
Gamma rays	803 keV (0.0011%)	43 keV (0.02%) 100 keV (0.0015%) 150 keV (0.0013%) 262 keV (1.4×10^{-4}%) 590 keV (2.5×10^{-4}%) 820 keV (7×10^{-5}%)
X-rays	Pb characteristic L and M (trace)	Pu characteristic L and M

Data from Amlauer and Tuohy.[11]

bremsstrahlung associated with electron excitation and is therefore capable of generating a relatively "clean" characteristic X-ray spectrum. A cross-sectional diagram of a typical source of this type is shown in Fig. 1-9. The X-ray yields into one steradian solid angle per mCi (37 MBq) of ^{244}Cm ranges from 1.7×10^5 photons/second for a beryllium target to about 10^4 photons/second for targets of higher Z (Ref. 11).

V. NEUTRON SOURCES

Although nuclei created with excitation energy greater than the neutron binding energy can decay by neutron emission, these highly excited states are not produced as a result of any convenient radioactive decay process. Consequently, practical isotope sources of neutrons do not exist in the same sense that gamma-ray sources are available from many different nuclei populated by beta decay.[†] The possible choices for radioisotope neutron sources are much more limited and are based on either spontaneous fission or on nuclear reactions for which the incident particle is the product of a conventional decay process.

A. Spontaneous Fission

Many of the transuranic heavy nuclides have an appreciable spontaneous fission decay probability. Several fast neutrons are promptly emitted in each fission event, so a sample of such a radionuclide can be a simple and convenient isotopic neutron source. Other products of the fission process are the heavy fission products described earlier, prompt fission gamma rays, and the beta and gamma activity of the fission products accumulated within the sample. When used as a neutron source, the isotope is generally encapsulated in a sufficiently thick container so that only the fast neutrons and gamma rays emerge from the source.

The most common spontaneous fission source is ^{252}Cf. Its half-life of 2.65 years is long enough to be reasonably convenient, and the isotope is one of the most widely

---

[†] The specific beta decay with the longest half-life which leads to an excited state that does deexcite by neutron emission is

$$^{87}\text{Br} \rightarrow {}^{87}\text{Kr}^* + \beta^-$$
$$\quad\quad\quad\quad\quad \longmapsto {}^{86}\text{Kr} + \text{n}$$

Because the half-life of this beta decay is only 55 s, ^{87}Br is not practical as a laboratory neutron source.

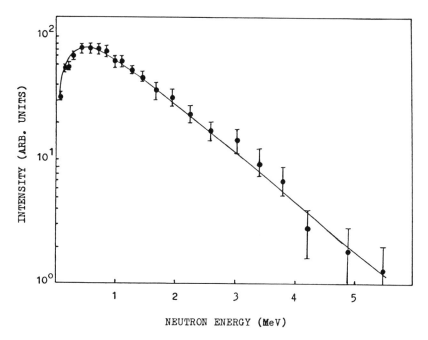

Figure 1-10 Measured neutron energy spectrum from the spontaneous fission of ^{252}Cf. (From Batenkov et al.[12])

produced of all the transuranics. The dominant decay mechanism is alpha decay, and the alpha emission rate is about 32 times that for spontaneous fission. The neutron yield is 0.116 n/s per Bq, where the activity is the combined alpha and spontaneous fission decay rate. On a unit mass basis, 2.30×10^6 n/s are produced per microgram of the sample. Compared with the other isotopic neutron sources described below, ^{252}Cf sources involve very small amounts of active material (normally of the order of micrograms) and can therefore be made in very small sizes dictated only by the encapsulation requirements.

The energy spectrum of the neutrons is plotted in Fig. 1-10. The spectrum is peaked between 0.5 and 1 MeV, although a significant yield of neutrons extends to as high as 8 or 10 MeV. The shape of a typical fission spectrum is approximated by the expression

$$\frac{dN}{dE} = E^{1/2} e^{-E/T} \qquad (1\text{-}6)$$

For the spontaneous fission of ^{252}Cf, the constant T in Eq. (1-6) has a value of 1.3 MeV (Ref. 13).

B. Radioisotope (α, n) Sources

Because energetic alpha particles are available from the direct decay of a number of convenient radionuclides, it is possible to fabricate a small self-contained neutron source by mixing an alpha-emitting isotope with a suitable target material. Several different target materials can lead to (α, n) reactions for the alpha particle energies which are readily available in radioactive decay. The maximum neutron yield is obtained when beryllium is chosen as the target, and neutrons are produced through the reaction

$$^{4}_{2}\alpha + ^{9}_{4}\text{Be} \rightarrow ^{12}_{6}\text{C} + ^{1}_{0}\text{n}$$

which has a Q-value of $+5.71$ MeV.

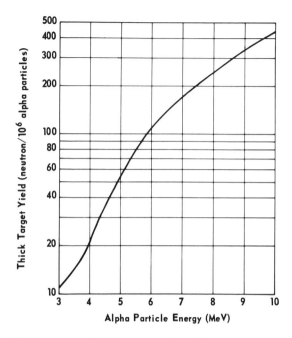

Figure 1-11 Thick target yield of neutrons for alpha particles on beryllium. (From Anderson and Hertz.[14])

The neutron yield from this reaction when a beam of alpha particles strikes a target that is thick compared with their range is plotted in Fig. 1-11. Most of the alpha particles simply are stopped in the target, and only 1 in about 10^4 reacts with a beryllium nucleus. Virtually the same yield can be obtained from an intimate mixture of the alpha particle emitter and beryllium, provided the alpha emitter is homogeneously distributed throughout the beryllium in a small relative concentration. All the alpha emitters of practical interest are actinide elements, and investigations have shown that a stable alloy can be formed between the actinides and beryllium of the form MBe_{13}, where M represents the actinide metal. Most of the sources described below therefore are metallurgically prepared in the form of this alloy, and each alpha particle has an opportunity to interact with beryllium nuclei without any intermediate energy loss.

Some of the common choices for alpha emitters and properties of the resulting neutron sources are listed in Table 1-6. Several of these isotopes, notably ^{226}Ra and ^{227}Ac, lead to long chains of daughter products that, although adding to the alpha particle yield, also contribute a large gamma-ray background. These sources are therefore inappropriate for some applications in which the intense gamma-ray background interferes with the measurement. Also, these Ra-Be and Ac-Be sources require more elaborate handling procedures because of the added biological hazard of the gamma radiation.

The remaining radioisotopes in Table 1-6 involve simpler alpha decays and the gamma-ray background is much lower. The choice between these alternatives is made primarily on the basis of availability, cost, and half-life. Because the physical size of the sources is no longer negligible, one would like the half-life to be as short as possible, consistent with the application, so that the specific activity of the emitter is high.

The ^{239}Pu/Be source is probably the most widely used of the (α, n) isotopic neutron sources. However, because about 16 g of the material is required for 1 Ci (3.7×10^{10} Bq)

TABLE 1-6   Characteristics of Be(α, n) Neutron Sources

Source	Half-Life	E_α (MeV)	Neutron Yield per 10^6 Primary Alpha Particles		Percent Yield with $E_n < 1.5$ MeV	
			Calculated	Experimental	Calculated	Experimental
^{239}Pu/Be	24000 y	5.14	65	57	11	9–33
^{210}Po/Be	138 d	5.30	73	69	13	12
^{238}Pu/Be	87.4 y	5.48	79^a	—	—	—
^{241}Am/Be	433 y	5.48	82	70	14	15–23
^{244}Cm/Be	18 y	5.79	100^b	—	18	29
^{242}Cm/Be	162 d	6.10	118	106	22	26
^{226}Ra/Be + daughters	1602 y	Multiple	502	—	26	33–38
^{227}Ac/Be + daughters	21.6 y	Multiple	702	—	28	38

^{a}From Anderson and Hertz.[14] All other data as calculated or cited in Geiger and Van der Zwan.[15].
^{b}Does not include a 4% contribution from spontaneous fission of ^{244}Cm.

of activity, sources of this type of a few centimeters in dimension are limited to about 10^7 n/s. In order to increase the neutron yield without increasing the physical source size, alpha emitters with higher specific activities must be substituted. Therefore, sources incorporating ^{241}Am (half-life of 433 years) and ^{238}Pu (half-life of 87.4 years) are also widely used if high neutron yields are needed. Although limited experience has been gained to date, sources utilizing ^{244}Cm (half-life of 18 years) might well represent the near ideal compromise between specific activity and source lifetime.

The neutron energy spectra from all such alpha/Be sources are similar, and any differences reflect only the small variations in the primary alpha energies. A plot of the spectrum from a ^{239}Pu/Be source is shown in Fig. 1-12. The various peaks and valleys in this energy distribution can be analyzed in terms of the excitation state in which the ^{12}C product nucleus is left.[15,16] The alpha particles lose a variable amount of energy before reacting with a beryllium nucleus, however, and their continuous energy distribution

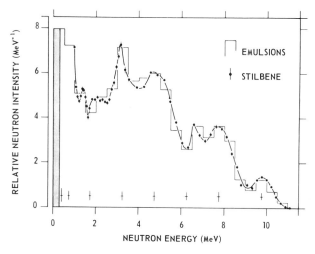

Figure 1-12   Measured energy spectra for neutrons from a ^{239}Pu/Be source containing 80g of the isotope. (From Anderson and Neff.[17])

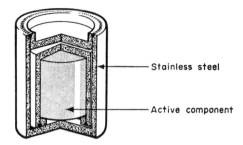

Figure 1-13   Typical double-walled construction for Be(α, n) sources. (From Lorch.[13])

washes out much of the structure that would be observed if the alpha particles were monoenergetic. For sources that contain only a few grams of material, the spectrum of neutrons which emerges from the source surface is essentially the same as that created in the (α, n) reactions. For larger sources, the secondary processes of neutron scattering within the source, (n, 2n) reactions in beryllium, and (n, fission) events within the plutonium or other actinide can introduce some dependence of the energy spectrum on the source size.[17]

Because large activities of the actinide isotope are involved in these neutron sources, special precautions must be taken in their fabrication to ensure that the material remains safely encapsulated. The actinide–beryllium alloy is usually sealed within two individually welded stainless steel cylinders in the arrangement shown in Fig. 1-13. Some expansion space must be allowed within the inner cylinder to accommodate the slow evolution of helium gas formed when the alpha particles are stopped and neutralized.

When applied to the efficiency calibration of detectors, some caution must be used in assuming that the neutron yield for these sources decays exactly as the half-life of the principal actinide alpha emitter. Small amounts of contaminant alpha activity, present in either the original radioisotope sample or produced through the decay of a precursor, can influence the overall neutron yield. For example, many ^{239}Pu/Be sources have been prepared from plutonium containing small amounts of other plutonium isotopes. The isotope ^{241}Pu is particularly significant, because it beta decays with a half-life of 13.2 years to form ^{241}Am, an alpha emitter. The neutron yield of these sources can therefore gradually increase with time as the ^{241}Am accumulates in the source. An original ^{241}Pu isotopic fraction of 0.7% will result in an initial growth rate of the neutron yield of 2% per year.[18]

TABLE 1-7   Alternative (α, n) Isotopic Neutron Sources

Target	Reaction	Q-Value	Neutron Yield per 10^6 Alpha Particles
Natural B	^{10}B(α, n)	+ 1.07 MeV	13 for ^{241}Am alpha particles
	^{11}B(α, n)	+ 0.158 MeV	
F	^{19}F(α, n)	− 1.93 MeV	4.1 for ^{241}Am alpha particles
Isotopically separated ^{13}C	^{13}C(α, n)	+ 2.2 MeV	11 for ^{238}Pu alpha particles
Natural Li	^{7}Li(α, n)	− 2.79 MeV	
Be (for comparison)	^{9}Be(α, n)	+ 5.71 MeV	70 for ^{241}Am alpha particles

Data from Lorch[13] and Geiger and Van der Zwan.[19]

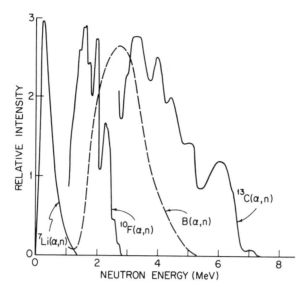

Figure 1-14 Neutron energy spectra from alternative (α, n) sources. (^{7}Li data from Geiger and Van der Zwan,[19] remainder from Lorch.[13])

A number of other alpha-particle-induced reactions have occasionally been employed as neutron sources, but all have a substantially lower neutron yield per unit alpha activity compared with the beryllium reaction. Some of the potential useful reactions are listed in Table 1-7. Because all the Q-values of these reactions are less than that of the beryllium reaction, the resulting neutron spectra shown in Fig. 1-14 have a somewhat lower average energy. In particular, the ^{7}Li(α, n) reaction with its highly negative Q-value leads to a neutron spectrum with a low 0.5 MeV average energy that is especially useful in some applications. Details of this spectrum shape are given in Ref. 20.

C. Photoneutron Sources

Some radioisotope gamma-ray emitters can also be used to produce neutrons when combined with an appropriate target material. The resulting *photoneutron sources* are based on supplying sufficient excitation energy to a target nucleus by absorption of a gamma-ray photon to allow the emission of a free neutron. Only two target nuclei, ^{9}Be and ^{2}H, are of any practical significance for radioisotope photoneutron sources. The corresponding reactions can be written

$$Q$$
$$^9_4\text{Be} + h\nu \rightarrow {}^8_4\text{Be} + {}^1_0\text{n} \qquad -1.666 \text{ MeV}$$
$$^2_1\text{H} + h\nu \rightarrow {}^1_1\text{H} + {}^1_0\text{n} \qquad -2.226 \text{ MeV}$$

A gamma-ray photon with an energy of at least the negative of the Q-value is required to make the reactions energetically possible, so that only relatively high-energy gamma rays can be applied. For gamma-ray energies that exceed this minimum, the corresponding neutron energy can be calculated from

$$E_n(\theta) \cong \frac{M(E_\gamma + Q)}{m + M} + \frac{E_\gamma \left[(2mM)(m + M)(E_\gamma + Q)\right]^{1/2}}{(m + M)^2} \cos\theta \qquad (1\text{-}7)$$

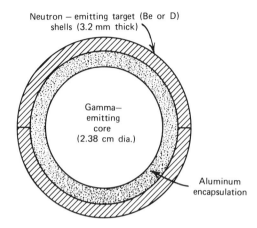

Figure 1-15   Construction of a simple spherical photoneutron source.

where θ = angle between gamma photon and neutron direction

   E_γ = gamma energy (assumed $\ll 931$ MeV)

   M = mass of recoil nucleus $\times c^2$

   m = mass of neutron $\times c^2$

One decided advantage of photoneutron sources is that if the gamma rays are monoen-ergetic, the neutrons are also nearly monoenergetic. The relatively small kinematic spread obtained from Eq. (1-7) by letting the angle θ vary between 0 and π broadens the neutron energy spectrum by only a few percent. For large sources, the spectrum is also somewhat degraded by the scattering of some neutrons within the source before their escape.

The main disadvantage of photoneutron sources is the fact that very large gamma-ray activities must be used in order to produce neutron sources of attractive intensity. For the type of source sketched in Fig. 1-15, only about 1 gamma ray in 10^5 or 10^6 interacts to produce a neutron, and therefore the neutrons appear in a much more intense gamma-ray background. Some of the more common gamma-ray emitters are ^{226}Ra, ^{124}Sb, ^{72}Ga, ^{140}La, and ^{24}Na. Some properties and energy spectra of corresponding photoneutron sources are shown in Table 1-8 and Fig. 1-16. For many of these sources, the half-lives of the gamma emitters are short enough to require their reactivation in a nuclear reactor between uses.

D. Reactions from Accelerated Charged Particles

Because alpha particles are the only heavy charged particles with low Z conveniently available from radioisotopes, reactions involving incident protons, deuterons, and so on, must rely on artificially accelerated particles. Two of the most common reactions of this type used to produce neutrons are

$$Q$$

The D-D reaction	$^2_1\text{H} + ^2_1\text{H} \rightarrow ^3_2\text{He} + ^1_0\text{n}$	$+3.26$ MeV
The D-T reaction	$^2_1\text{H} + ^3_1\text{H} \rightarrow ^4_2\text{He} + ^1_0\text{n}$	$+17.6$ MeV

TABLE 1-8 Photoneutron Source Characteristics

Gamma-Ray Emitter	Half-Life[a]	Gamma Energy[a] (MeV)	Target	Neutron Energy[b] (keV)	Neutron Yield (n/s) for 10^{10} Bq Activity[c]
^{24}Na	15.0 h	2.7541	Be	967	340,000
		2.7541	D	263	330,000
^{28}Al	2.24 min	1.7787	Be	101	32,600
^{38}Cl	37.3 min	2.1676	Be	446	43,100
^{56}Mn	2.58 h	1.8107	Be	129⎫	
		2.1131		398⎬	91,500
		2.9598		1,149⎭	
		2.9598	D	365	162
^{72}Ga	14.1 h	1.8611	Be	174⎫	
		2.2016		476⎬	64,900
		2.5077		748⎭	
		2.5077	D	140	25,100
^{76}As	26.3 h	1.7877	Be	109⎫	3,050
		2.0963		383⎭	
^{88}Y	107 d	1.8361	Be	152⎫	229,000
		2.7340		949⎭	
		2.7340	D	253	160
^{116m}In	54.1 min	2.1121	Be	397	15,600
^{124}Sb	60.2 d	1.6910	Be	23	210,000
^{140}La	40.3 h	2.5217	Be	760	10,200
		2.5217	D	147	6,600
^{144}Pr	17.3 min	2.1856	Be	462	690

[a] Decay data from Ref. 1.

[b] Calculated for $\theta = \pi/2$, approximate midpoint of primary spectrum.

[c] Monte Carlo calculations for the source dimensions given in Fig. 1-15. Outer target shells are either metallic Be or deuterated polyethylene. Core materials assumed to be NaF, Al, CCl_4, MnO_2, Ga_2O_3, As_2O_3, Y_2O_3, In, Sb, La_2O_3, and Pr_2O_3.

Source: G. F. Knoll, "Radioisotope Neutron Sources," Chap. II in *Neutron Sources for Basic Physics and Applications*, Pergamon Press, New York, 1983.

Because the coulomb barrier between the incident deuteron and the light target nucleus is relatively small, the deuterons need not be accelerated to a very high energy in order to create a significant neutron yield. These reactions are widely exploited in "neutron generators" in which deuterium ions are accelerated by a potential of about 100–300 kV. Because the incident particle energy is then small compared with the Q-value of either reaction, all the neutrons produced are of about the same energy (near 3 MeV for the D-D reaction and 14 MeV for the D-T reaction). A 1 mA beam of deuterons will produce about 10^9 n/s from a thick deuterium target, and about 10^{11} n/s from a tritium target. Somewhat smaller yields are produced in compact neutron generators consisting of a sealed tube containing the ion source and target, together with a portable high-voltage generator.

A number of other charged-particle-induced reactions that involve either a negative Q-value or a target with higher atomic number are also applied to neutron generation. Some common examples are ^{9}Be(d, n), ^{7}Li(p, n), and ^{3}H(p, n). In these cases, a higher incident particle energy is required, and large accelerator facilities such as cyclotrons or Van de Graaff accelerators are needed to produce the incident particle beam.

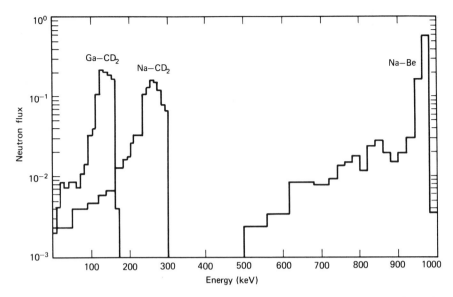

Figure 1-16 Neutron spectra calculated for the photoneutron source dimensions shown in Fig. 1-15. The gamma emitters are either ^{72}Ga or ^{24}Na. The outer shells are either deuterated polyethylene (CD_2) or beryllium (Be).

PROBLEMS

1-1. Radiation energy spectra can be categorized into two main groups: those that consist of one or more discrete energies (line spectra) and those that consist of a broad distribution of energies (continuous spectra). For each of the radiation sources listed below, indicate whether "line" or "continuous" is a better description:

(a) Alpha particles.
(b) Beta particles.
(c) Gamma rays.
(d) Characteristic X-rays.
(e) Conversion electrons.
(f) Auger electrons.
(g) Fission fragments.
(h) Bremsstrahlung.
(i) Annihilation radiation.

1-2. Which has the higher energy: a conversion electron from the L shell or from the M shell, if both arise from the same nuclear excitation energy?

1-3. By simultaneously conserving energy and momentum, find the alpha particle energy emitted in the decay of a nucleus with mass number 210 if the Q-value of the decay is 5.50 MeV.

1-4. What is the lowest wavelength limit of the X-rays emitted by a tube operating at a potential of 195 kV?

1-5. From a table of atomic mass values, find the approximate energy released by the spontaneous fission of ^{235}U into two equal-mass fragments.

1-6. Calculate the specific activity of pure tritium (^{3}H) with a half-life of 12.26 years.

1-7. What is the highest energy to which doubly ionized helium atoms (alpha particles) can be accelerated in a dc accelerator with 3 MV maximum voltage?

1-8. What is the minimum gamma-ray energy required to produce photoneutrons in water from the trace heavy water content?

1-9. By simultaneously conserving energy and momentum, calculate the energy of the neutrons emitted in the forward direction by a beam of 150 keV deuterons undergoing the D-T reaction in a tritium target.

REFERENCES

1. C. M. Lederer and V. S. Shirley, *Table of Isotopes*, 7th ed., Wiley-Interscience, New York, 1978.

2. R. N. Chanda and R. A. Deal, *Catalog of Semiconductor Alpha-Particle Spectra*, IN-1261, (1970).

3. A. Rytz, *Atomic Data and Nuclear Data Tables* **12**, 479 (1973).

4. W. E. Nervik, *Phys. Rev.* **119**, 1685 (1960).

5. S. L. Whetstone, *Phys. Rev.* **131**, 1232 (1963).

6. R. J. Gehrke, J. E. Cline, and R. L. Heath, *Nucl. Instrum. Meth.* **91**, 349 (1971).

7. J. P. Mason, *Nucl. Instrum. Meth. Phys. Res.* **A241**, 207 (1985).

8. H. Ferdinande, G. Knuyt, R. Van De Vijver, and R. Jacobs, *Nucl. Instrum. Meth.* **91**, 135 (1971).

9. E. Storm, D. W. Lier, and H. I. Israel, *Health Phys.* **26**, 179 (1974).

10. J. L.-H. Chan and A. Macovski, *IEEE Trans. Nucl. Sci.* **NS-24**, No. 4, 1968 (1977).

11. K. Amlauer and I. Tuohy, *Proceedings, ERDA X- and Gamma Ray Symposium*, Ann Arbor, CONF-760539, p. 19 (1976).

12. O. I. Batenkov et al., INDC (NDS)-146, (1983).

13. E. A. Lorch, *Int. J. Appl. Radiat. Isotopes* **24**, 585 (1973).

14. M. E. Anderson and M. R. Hertz, *Nucl. Sci. Eng.* **44**, 437 (1971).

15. K. W. Geiger and L. Van der Zwan, *Nucl. Instrum. Meth.* **131**, 315 (1975).

16. A. Kumar and P. S. Nagarajan, *Nucl. Instrum. Meth.* **140**, 175 (1977).

17. M. E. Anderson and R. A. Neff, *Nucl. Instrum. Meth.* **99**, 231 (1972).

18. M. E. Anderson, *Nucl. Appl.* **4**, 142 (1968).

19. K. W. Geiger and L. Van der Zwan, *Health Phys.* **21**, 120 (1971).

20. D. R. Weaver, J. G. Owen, and J. Walker, *Nucl. Instrum. Meth.* **198**, 599 (1982).

CHAPTER · 2

Radiation Interactions

The operation of any radiation detector basically depends on the manner in which the radiation to be detected interacts with the material of the detector itself. An understanding of the response of a specific type of detector must therefore be based on a familiarity with the fundamental mechanisms by which radiations interact and lose their energy in matter. Many general reference works are available concerning this broad topic; the classic text by Evans,[1] to mention only one, has served as a standard reference over several decades.

To organize the discussions that follow, it is convenient to arrange the four major categories of radiations introduced in Chapter 1 into the following matrix:

Charged Particulate Radiations		*Uncharged Radiations*
Heavy charged particles (characteristic distance $\cong 10^{-5}$ m)	$\Leftarrow$	Neutrons (characteristic length $\cong 10^{-1}$ m)
Fast electrons (characteristic distance $\cong 10^{-3}$ m)	$\Leftarrow$	X-rays and gamma rays (characteristic length $\cong 10^{-1}$ m)

The entries in the left column represent the charged particulate radiations that, because of the electric charge carried by the particle, continuously interact through the coulomb force with the electrons present in any medium through which they pass. The radiations in the right column are uncharged and therefore are not subject to the coulomb force. Instead, these radiations must first undergo a "catastrophic" interaction (often involving the nucleus of constituent atoms) that radically alters the properties of the incident radiation in a single encounter. In all cases of practical interest, the interaction results in the full or partial transfer of energy of the incident radiation to electrons or nuclei of the constituent atoms, or to charged particle products of nuclear reactions. If the interaction does not occur within the detector, these uncharged radiations (e.g., neutrons or gamma rays) can pass completely through the detector volume without revealing the slightest hint that they were ever there.

The horizontal arrows shown in the diagram illustrate the results of such catastrophic interactions. An X- or gamma ray, through the processes described in this chapter, can transfer all or part of its energy to electrons within the medium. The resulting *secondary electrons* bear a close similarity to the fast electron radiations (such as the beta particle) discussed in Chapter 1. Devices designed to detect gamma rays are tailored to promote such interactions and to fully stop the resulting secondary electrons so that their entire energy may contribute to the output signal. In contrast, neutrons may interact in such a way as to produce secondary heavy charged particles, which then serve as the basis of the detector signal.

Also listed in the diagram are order-of-magnitude numbers for the characteristic distance of penetration or average path length (range or mean free path) in solids for typical energy radiations in each category.

I. INTERACTION OF HEAVY CHARGED PARTICLES

A. Nature of the Interaction

Heavy charged particles, such as the alpha particle, interact with matter primarily through coulomb forces between their positive charge and the negative charge of the orbital electrons within the absorber atoms. Although interactions of the particle with nuclei (as in Rutherford scattering or alpha-particle-induced reactions) also are possible, such encounters occur only rarely and they are not normally significant in the response of radiation detectors. Instead, charged particle detectors must rely on the results of interactions with electrons for their response.

Upon entering any absorbing medium, the charged particle immediately interacts simultaneously with many electrons. In any one such encounter, the electron feels an impulse from the attractive coulomb force as the particle passes its vicinity. Depending on the proximity of the encounter, this impulse may be sufficient either to raise the electron to a higher-lying shell within the absorber atom (*excitation*) or to remove completely the electron from the atom (*ionization*). The energy that is transferred to the electron must come at the expense of the charged particle, and its velocity is therefore decreased as a result of the encounter. The maximum energy that can be transferred from a charged particle of mass m with kinetic energy E to an electron of mass m_0 in a single collision is $4Em_0/m$, or about 1/500 of the particle energy per nucleon. Because this is a small fraction of the total energy, the primary particle must lose its energy in many such interactions during its passage through an absorber. At any given time, the particle is interacting with many electrons, so the net effect is to decrease its velocity continuously until the particle is stopped.

Representative paths taken by heavy charged particles in their slowing down process are schematically represented in Fig. 2-1. Except at their very end, the tracks tend to be quite straight because the particle is not greatly deflected by any one encounter, and interactions occur in all directions simultaneously. Charged particles are therefore characterized by a definite *range* in a given absorber material. The range, to be defined more precisely below, represents a distance beyond which no particles will penetrate.

The products of these encounters in the absorber are either excited atoms or *ion pairs*. Each ion pair is made up of a free electron and the corresponding positive ion of an absorber atom from which an electron has been totally removed. The ion pairs have a natural tendency to recombine to form neutral atoms, but in some types of detectors, this

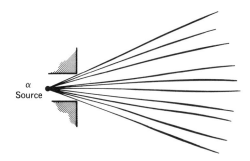

Figure 2-1 Representation of the tracks of alpha particles from a monoenergetic source.

recombination is suppressed so that the ion pairs may be used as the basis of the detector response.

In particularly close encounters, an electron may undergo a large enough impulse so that after having left its parent atom, it still may have sufficient kinetic energy to create further ions. These energetic electrons are sometimes called *delta rays* and represent an indirect means by which the charged particle energy is transferred to the absorbing medium. Under typical conditions, the majority of the energy loss of the charged particle occurs via these delta rays. The range of the delta rays is always very small compared with the range of the incident charged particle, and in most practical situations it is immaterial whether the energy is deposited directly by the primary particle itself or by the secondary delta rays.

B. Stopping Power

The *linear stopping power S* for charged particles in a given absorber is simply defined as the differential energy loss for that particle within the material divided by the corresponding differential path length:

$$S = -\frac{dE}{dx} \qquad (2\text{-}1)$$

The value of $-dE/dx$ along a particle track is also called its *specific energy loss* or, more casually, its "rate" of energy loss.

For particles with a given charge state, S increases as the particle velocity is decreased. The classical expression that describes the specific energy loss is known as the *Bethe formula* and is written

$$-\frac{dE}{dx} = \frac{4\pi e^4 z^2}{m_0 v^2} NB \times \frac{1}{4\pi\varepsilon_0} \qquad (2\text{-}2)$$

where

$$B \equiv Z\left[\ln\frac{2m_0 v^2}{I} - \ln\left(1 - \frac{v^2}{c^2}\right) - \frac{v^2}{c^2}\right]$$

In these expressions, v and ze are the velocity and charge of the primary particle, N and Z are the number density and atomic number of the absorber atoms, m_0 is the electron rest mass, and e is the electronic charge. The parameter I represents the average excitation and ionization potential of the absorber and is normally treated as an experimentally determined parameter for each element. For nonrelativistic charged particles ($v \ll c$), only the first term in B is significant. Equation (2-2) is generally valid for different types of charged particles provided their velocity remains large compared with the velocities of the orbital electrons in the absorbing atoms.

The expression for B in Eq. (2-2) varies slowly with particle energy. Thus, the general behavior of dE/dx can be inferred from the behavior of the multiplicative factor. For a given nonrelativistic particle, dE/dx therefore varies as $1/v^2$, or inversely with particle energy. This behavior can be heuristically explained by noting that because the charged particle spends a greater time in the vicinity of any given electron when its velocity is low, the impulse felt by the electron, and hence the energy transfer, is largest. When comparing different charged particles of the same velocity, the only factor that may change outside the logarithmic term in Eq. (2-2) is z^2, which occurs in the numerator of the expression. Therefore, particles with the greatest charge will have the largest specific energy loss. Alpha particles, for example, will lose energy at a rate that is greater than protons of the same velocity, but less than that of more highly charged ions. In comparing different materials as absorbers, dE/dx depends primarily on the product NZ, which is outside the logarithmic term. High atomic number, high-density materials will consequently result in the greatest linear stopping power.

The Bethe formula begins to fail at low particle energies where charge exchange between the particle and absorber becomes important. The positively charged particle will then tend to pick up electrons from the absorber, which effectively reduce its charge and consequent linear energy loss. At the end of its track, the particle has accumulated z electrons and becomes a neutral atom.

C. Energy Loss Characteristics

1. THE BRAGG CURVE

A plot of the specific energy loss along the track of a charged particle such as that shown in Fig. 2-2 is known as a Bragg curve. The example is shown for an alpha particle of several MeV initial energy. For most of the track, the charge on the alpha particle is two electronic charges, and the specific energy loss increases roughly as $1/E$ as predicted by Eq. (2-2). Near the end of the track, the charge is reduced through electron pickup and the curve falls off. Plots are shown both for a single alpha particle track and for the average behavior of a parallel beam of alpha particles of the same initial energy. The two curves differ somewhat due to the effects of straggling, to be discussed below.

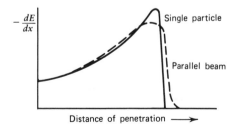

Figure 2-2 The specific energy loss along an alpha track.

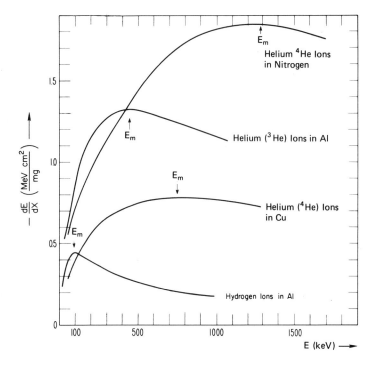

Figure 2-3 Specific energy loss as a function of energy for hydrogen and helium ions. E_m indicates the energy at which dE/dx is maximized. (From Wilken and Fritz.[2])

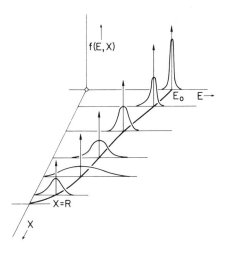

Figure 2-4 Plots of the energy distribution of a beam of initially monoenergetic charged particles at various penetration distances. E is the particle energy and X is the distance along the track. (From Wilken and Fritz.[2])

Related plots showing $-dE/dx$ versus particle energy for a number of different heavy charged particles are given in Fig. 2-3. These examples illustrate the energy at which charge pickup by the ion becomes significant. Charged particles with the greatest number of nuclear charges begin to pick up electrons early in their slowing down process. Note that in an aluminum absorber, singly charged hydrogen ions (protons) show strong effects of charge pickup below about 100 keV, but doubly charged ^{3}He ions show equivalent effects at about 400 keV.

2. ENERGY STRAGGLING

Because the details of the microscopic interactions undergone by any specific particle vary somewhat randomly, its energy loss is a statistical or stochastic process. Therefore, a spread in energies always results after a beam of monoenergetic charged particles has passed through a given thickness of absorber. The width of this energy distribution is a measure of *energy straggling*, which varies with the distance along the particle track.

Figure 2-4 shows a schematic presentation of the energy distribution of a beam of initially monoenergetic particles at various points along its range. Over the first portion, the distribution becomes wider (and more skewed) with penetration distance, showing the increasing importance of energy straggling. Near the end of the range, the distribution narrows again because the mean particle energy has greatly been reduced.

D. Particle Range

1. DEFINITIONS OF RANGE

In order to quantify the definition of particle range, we refer to the conceptual experiment sketched in Fig. 2-5. Here a collimated source of monoenergetic alpha particles is counted by a detector after passing through an absorber of variable thickness. (We later contrast the behavior of other types of radiation when observed under similar conditions.) For alpha particles, the results are also plotted in Fig. 2-5. For small values of the absorber thickness, the only effect is to cause an energy loss of the alpha particles in the absorber as they pass through. Because the tracks through the absorber are quite straight, the total number that reach the detector remains the same. No attenuation in the number of alpha particles takes place until the absorber thickness approaches the length of the shortest track in the absorbing material. Increasing the thickness then stops more and more of the alpha particles, and the intensity of the detected beam drops rapidly to zero.

The range of the alpha particles in the absorber material can be determined from this curve in several ways. The *mean range* is defined as the absorber thickness that reduces the alpha particle count to exactly one-half of its value in the absence of the absorber.

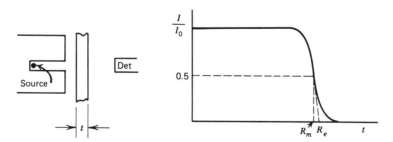

Figure 2-5 An alpha particle transmission experiment. I is the detected number of alpha particles through an absorber thickness, t, whereas I_0 is the number detected without the absorber. The mean range R_m and extrapolated range R_e are indicated.

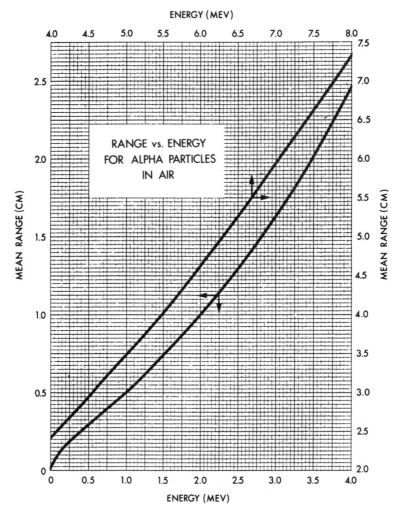

Figure 2-6 Range–energy plot for alpha particles in air at 15°C and 760 mm Hg pressure. (From *Radiological Health Handbook*, U.S. Department of Health, Education and Welfare, Washington, DC, 1970.)

This definition is most commonly used in tables of numerical range values. Another version that often appears in the literature is the *extrapolated range*, which is obtained by extrapolating the linear portion of the end of the transmission curve to zero.

The range of charged particles of a given energy is thus a fairly unique quantity in a specific absorber material. In the early days of radiation measurement, experiments of the type sketched in Fig. 2-5 were widely used to measure the energy of alpha particles indirectly by determining the absorber thickness equivalent to their mean range. With the availability of detectors that provide an output signal directly related to the alpha particle energy, such indirect measurements are no longer necessary.

Some graphs of the range of various charged particles in materials of interest in detectors are given in Figs. 2-6 through 2-8. As one obvious application of these curves, any detector that is to measure the full incident energy of a charged particle must have an active thickness that is greater than the range of that particle in the detector material.

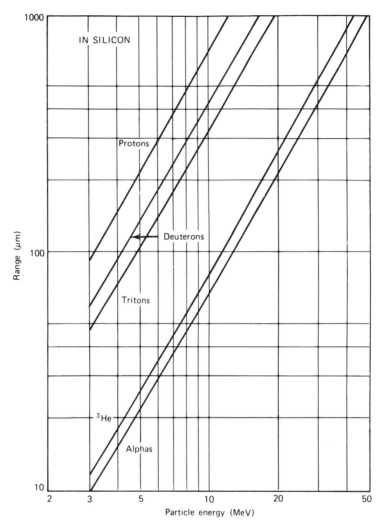

Figure 2-7 Range–energy curves calculated for different charged particles in silicon. The near-linear behavior of the log–log plot over the energy range shown suggests an empirical relation of the form $R = aE^b$, where the slope-related parameter b is not greatly different for the various particles. (From Skyrme.[3])

2. RANGE STRAGGLING

Charged particles are also subject to *range straggling*, defined as the fluctuation in path length for individual particles of the same initial energy. The same stochastic factors that lead to energy straggling at a given penetration distance also result in slightly different total path lengths for each particle. For heavy charged particles such as protons or alphas, the straggling amounts to a few percent of the mean range. The degree of straggling is evidenced by the sharpness of the cutoff at the end of the average transmission curve plotted in Fig. 2-2. Differentiating this curve leads to a peak whose width is often taken as a quantitative measure of the importance of range straggling for the particles and absorber used in the measurement.

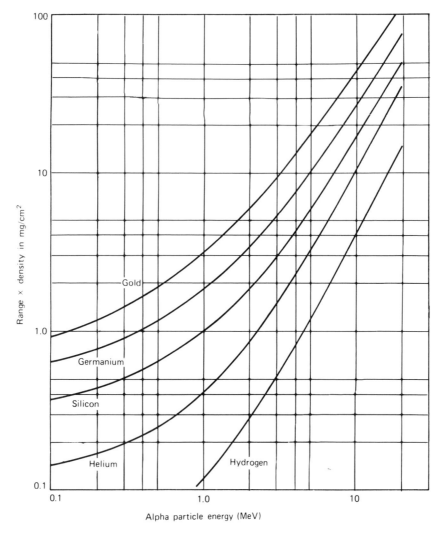

Figure 2-8 Range–energy curves calculated for alpha particles in different materials. Units of the range are given in mass thickness (see Section III.B.2) to minimize the differences in these curves. (Data from Williamson et al.[4])

3. STOPPING TIME

The time required to stop a charged particle in an absorber can be deduced from its range and average velocity. For nonrelativistic particles of mass m and kinetic energy E, the velocity is

$$v = \sqrt{\frac{2E}{m}} = c\sqrt{\frac{2E}{mc^2}} = \left(3.00 \times 10^8 \frac{\text{m}}{\text{s}}\right)\sqrt{\frac{2E}{(931 \text{ MeV/amu})m_A}}$$

where m_A is the particle mass in atomic mass units. If we assume that the average particle velocity as it slows down is $\langle v \rangle = Kv$, where v is evaluated at the initial energy, then the stopping time T can be calculated from the range R as

$$T = \frac{R}{\langle v \rangle} = \frac{R}{Kc} \sqrt{\frac{mc^2}{2E}} = \frac{R}{K\left(3.00 \times 10^8 \frac{m}{s}\right)} \sqrt{\frac{931 \text{ MeV/amu}}{2}} \sqrt{\frac{m_A}{E}}$$

If the particle were uniformly decelerated, then $\langle v \rangle$ would be given by $v/2$ and K would be $\frac{1}{2}$. However, charged particles generally lose energy at a greater rate near the end of their range, and K should be a somewhat higher fraction. By assuming $K = 0.60$, the stopping time can be estimated as

$$T \cong 1.2 \times 10^{-7} R \sqrt{\frac{m_A}{E}} \tag{2-3}$$

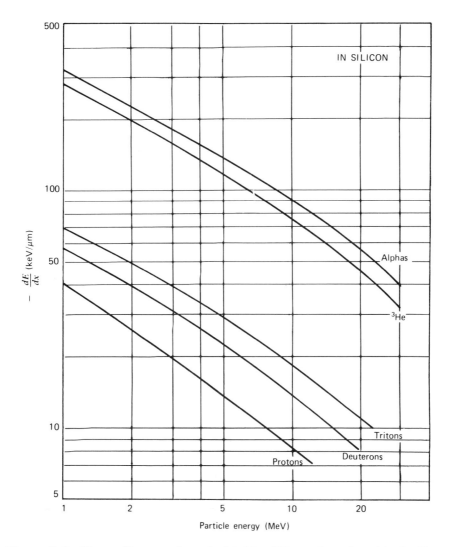

Figure 2-9 The specific energy loss calculated for different charged particles in silicon. (From Skyrme.[3])

where T is in seconds, R in meters, m_A in amu, and E in MeV. This approximation is expected to be reasonably accurate for light charged particles (protons, alpha particles, etc.) over much of the energy range of interest here. It is not, however, to be used for relativistic particles such as fast electrons.

Using typical range values, stopping times calculated from Eq. (2-3) for charged particles are a few picoseconds in solids or liquids and a few nanoseconds in gases.

E. Energy Loss in Thin Absorbers

For thin absorbers (or detectors) that are penetrated by a given charged particle, the energy deposited within the absorber can be calculated from

$$\Delta E = \left(-\frac{dE}{dx} \right)_{\text{avg}} t \qquad (2\text{-}4)$$

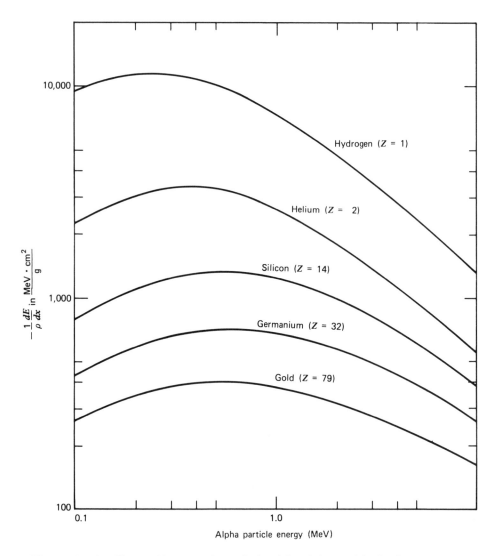

Figure 2-10   The specific energy loss calculated for alpha particles in different materials. Values are normalized by the density of the absorber material. (Data from Williamson et al.[4])

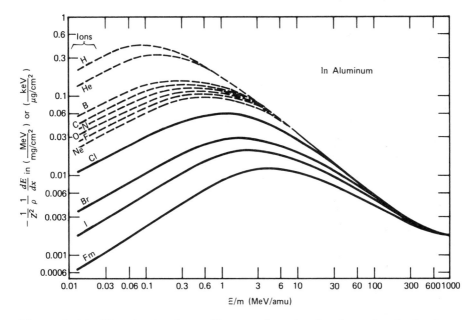

Figure 2-11 Plots showing the specific energy loss of various heavy ions in aluminum. The abscissa is the ion energy divided by its mass, and the ordinate is $-dE/dx$ divided by the density of aluminum and the square of the ion atomic number. Typical fission fragments (e.g., iodine) show a continuously decreasing $-dE/dx$ while slowing from their initial energy (~ 1 MeV/amu). (From Northcliffe and Schilling.[7])

where t is the absorber thickness and $(-dE/dx)_{avg}$ is the linear stopping power averaged over the energy of the particle while in the absorber. If the energy loss is small, the stopping power does not change much and it can be approximated by its value at the incident particle energy. Tabular values for dE/dx for a number of different charged particles in a variety of absorbing media are given in Refs. 4–8. Some graphs for materials of interest in specific detectors are shown in Figs. 2-9 through 2-11.

For absorber thicknesses through which the energy loss is not small, it is not easy to obtain a properly weighted $(-dE/dx)_{avg}$ value directly from such data. In these cases, it is easier to obtain the deposited energy in a way that makes use of range–energy data of the type plotted in Figs. 2-6 through 2-8. The basis of the method is as follows: Let R_1 represent the full range of the incident particle with energy E_0 in the absorber material. By subtracting the physical thickness of the absorber t from R_1, a value R_2 is obtained which represents the range of those alpha particles that emerge from the opposite surface of the absorber. By finding the energy corresponding to R_2, the energy of the transmitted charged particles E_t is obtained. The deposited energy ΔE is then given simply by $E_0 - E_t$. These steps are illustrated below:

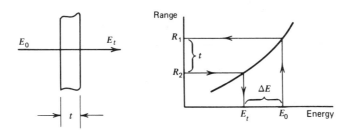

The procedure is based on the assumption that the charged particle tracks are perfectly linear in the absorber, and the method does not apply in situations where the particle can be significantly deflected (such as for fast electrons).

The combined effects of particle range and the decrease in dE/dx with increasing energy are illustrated in Fig. 2-12. Here the energy loss of protons in a thin detector is plotted versus the incident proton energy. For low energies, the proton range is less than the detector thickness. Therefore, as the energy is increased, the energy deposited in the detector (which is just equal to the incident energy) increases linearly. At a proton energy of 425 keV, the range is exactly equal to the detector thickness. For higher energies, only a portion of the incident energy is deposited, and the transmitted proton carries off the remainder. Under these conditions, the energy deposited in the detector is given by Eq. (2-4), or simply the product of the detector thickness and the average linear stopping power. Because the stopping power continuously decreases with increasing energy in this region (see Fig. 2-3), the deposited energy therefore *decreases* with further increases in the incident proton energy. The second curve in Fig. 2-12 plots the transmitted energy (E_t on the diagram above) as recorded by a second thick detector.

F. Scaling Laws

Sometimes data are not available on the range or energy loss characteristics of precisely the same particle–absorber combination needed in a given experiment. Recourse must then be made to various approximations, most of which are derived based on the Bethe formula [Eq. (2-2)] and on the assumption that the stopping power per atom of compounds or mixtures is additive. This latter assumption, known as the *Bragg–Kleeman*

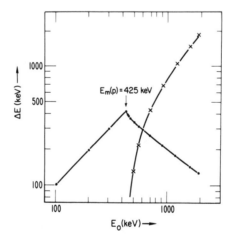

Figure 2-12   Energy loss of protons of initial energy E_0 in a silicon detector of 4.6 μm thickness (shown as dots). The transmitted energy for penetrating protons is also shown (as crosses). (From Wilken and Fritz.[2])

rule, may be written

$$\frac{1}{N_c}\left(\frac{dE}{dx}\right)_c = \sum_i W_i \frac{1}{N_i}\left(\frac{dE}{dx}\right)_i \qquad (2\text{-}5)$$

In this expression, N is the atomic density, dE/dx is the linear stopping power, and W_i represents the atom fraction of the ith component in the compound (subscript c). As an example of the application of Eq. (2-5), the linear stopping power of alpha particles in a metallic oxide could be obtained from separate data on the stopping power in both the pure metal and in oxygen. Some caution should be used in applying such results, however, since some measurements[10-12] for compounds have indicated a stopping power differing by as much as 10–20% from that calculated from Eq. (2-5).

It can be shown[9] that the range of a charged particle in a compound material can also be estimated provided its range is known in all the constituent elements. In this derivation, it is necessary to assume that the shape of the dE/dx curve is independent of the stopping medium. Under these conditions, the range in the compound is given by

$$R_c = \frac{M_c}{\sum_i n_i(A_i/R_i)} \qquad (2\text{-}6)$$

where R_i is the range in element i, n_i is the number of atoms of element i in the molecule, A_i is the atomic weight of element i, and M_c is the molecular weight of the compound.

If range data are not available for all the constituent elements, estimates can be made based on a semiempirical formula (commonly called the Bragg–Kleeman rule as well)

$$\frac{R_1}{R_0} \cong \frac{\rho_0\sqrt{A_1}}{\rho_1\sqrt{A_0}} \qquad (2\text{-}7)$$

where ρ and A represent density and atomic weight, and subscripts 0 and 1 refer to different absorbing materials. The accuracy of this estimate diminishes when the two materials are of widely different atomic weights, so it is always best to use range data from a material that is as close as possible in A to the absorber of interest.

Range data can also be generalized to different charged particles within a given absorber material. By integration of Eq. (2-2), it can be shown that the range of a particle of mass m and charge z can be represented by

$$R(v) = \frac{m}{z^2}F(v) \qquad (2\text{-}8)$$

where $F(v)$ represents a unique function of the particle initial velocity v. For particles of the same initial velocity, this factor will be identical and therefore we can write

$$R_a(v) = \frac{m_a z_b^2}{m_b z_a^2}R_b(v) \qquad (2\text{-}9)$$

where the subscripts a and b refer to different charged particles. Thus, the range of a particle for which data are not available can be estimated by calculating its initial velocity, finding the range of any other particle of the same initial velocity in the same material, and applying Eq. (2-9). It should be emphasized that these estimates are only approximate, because no account is taken of the change in charge state of the particle as it nears the end of its path. Correction factors necessary to compensate for this effect and predict the range more accurately are presented by Evans.[1]

G. Behavior of Fission Fragments

The heavy fragments produced as a result of neutron-induced or spontaneous fission of heavy nuclei are energetic charged particles with properties somewhat different from those discussed up to this point. Because the fragments start out stripped of many electrons, their very large effective charge results in a specific energy loss greater than that encountered with any other radiations discussed in this text. Because the initial energy is also very high (see Fig. 1-4), however, the range of a typical fission fragment is approximately half that of a 5 MeV alpha particle.

   An important feature of a fission fragment track is the fact that the specific energy loss ($-dE/dx$) decreases as the particle loses energy in the absorber. This behavior is in marked contrast to the lighter particles, such as alpha particles or protons, and is a result of the continuous decrease in the effective charge carried by the fragment as its velocity is reduced. The pickup of electrons begins immediately at the start of the track, and therefore the factor z in the numerator of Eq. (2-2) continuously drops. The resulting decrease in $-dE/dx$ is large enough to overcome the increase that normally accompanies a reduction in velocity. For particles with much lower initial charge state, such as the alpha particle, electron pickup does not become significant until near the end of the range.

II. INTERACTION OF FAST ELECTRONS

When compared with heavy charged particles, fast electrons lose energy at a lower rate and follow a much more tortuous path through absorbing materials. A series of tracks from a source of monoenergetic electrons might appear as in the sketch below:

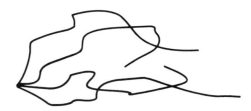

Large deviations in the electron path are now possible because its mass is equal to that of the orbital electrons with which it is interacting, and a much larger fraction of its energy can be lost in a single encounter. In addition, electron–nuclear interactions, which can abruptly change the electron direction, sometimes occur.

A. Specific Energy Loss

An expression similar to that of Eq. (2-2) has also been derived by Bethe to describe the specific energy loss due to ionization and excitation (the "collisional losses") for fast electrons:

$$-\left(\frac{dE}{dx}\right)_c = \frac{2\pi e^4 NZ}{m_0 v^2}\left(\ln\frac{m_0 v^2 E}{2I^2(1-\beta^2)} - (\ln 2)\left(2\sqrt{1-\beta^2} - 1 + \beta^2\right)\right.$$

$$\left. + (1-\beta^2) + \frac{1}{8}\left(1 - \sqrt{1-\beta^2}\right)^2\right) \quad (2\text{-}10)$$

where the symbols have the same meaning as in Eq. (2-2), and $\beta \equiv v/c$.

Electrons also differ from heavy charged particles in that energy may be lost by radiative processes as well as by coulomb interactions. These radiative losses take the form of *bremsstrahlung* or electromagnetic radiation, which can emanate from any position along the electron track. From classical theory, any charge must radiate energy when accelerated, and the deflections of the electron in its interactions with the absorber correspond to such acceleration. The linear specific energy loss through this radiative process is

$$-\left(\frac{dE}{dx}\right)_r = \frac{NEZ(Z+1)e^4}{137 m_0^2 c^4}\left(4\ln\frac{2E}{m_0 c^2} - \frac{4}{3}\right) \quad (2\text{-}11)$$

The factors of E and Z^2 in the numerator of Eq. (2-11) show that radiative losses are most important for high electron energies and for absorber materials of large atomic number. For typical electron energies, the average bremsstrahlung photon energy is quite low (see Fig. 1-6) and is therefore normally reabsorbed fairly close to its point of origin. In some cases, however, the escape of bremsstrahlung can influence the response of small detectors.

The total linear stopping power for electrons is the sum of the collisional and radiative losses:

$$\frac{dE}{dx} = \left(\frac{dE}{dx}\right)_c + \left(\frac{dE}{dx}\right)_r \quad (2\text{-}12)$$

The ratio of the specific energy losses is given approximately by

$$\frac{(dE/dx)_r}{(dE/dx)_c} \cong \frac{EZ}{700} \quad (2\text{-}13)$$

where E is in units of MeV. For the electrons of interest here (such as beta particles or secondary electrons from gamma-ray interactions), typical energies are less than a few MeV. Therefore, radiative losses are always a small fraction of the energy losses due to ionization and excitation and are significant only in absorber materials of high atomic number.

B. Electron Range and Transmission Curves

1. ABSORPTION OF MONOENERGETIC ELECTRONS

An attenuation experiment of the type discussed earlier for alpha particles is sketched in Fig. 2-13 for a source of monoenergetic fast electrons. Even small values of the absorber thickness lead to the loss of some electrons from the detected beam because scattering of

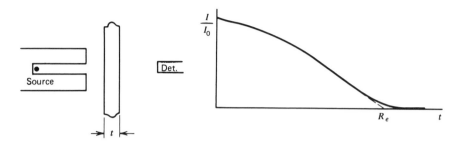

Figure 2-13 Transmission curve for monoenergetic electrons. R_e is the extrapolated range.

the electron effectively removes it from the flux striking the detector. A plot of the detected number of electrons versus absorber thickness therefore begins to drop immediately and gradually approaches zero for large absorber thicknesses. Those electrons that penetrate the greatest absorber thickness will be the ones whose initial direction has changed least in their path through the absorber.

The concept of range is less definite for fast electrons than for heavy charged particles, because the electron total path length is considerably greater than the distance of penetration along the initial velocity vector. Normally, the electron range is taken from a plot, such as that given in Fig. 2-13, by extrapolation of the linear portion of the curve to zero and represents the absorber thickness required to assure that almost no electrons can penetrate the entire thickness.

For equivalent energy, the specific energy loss of electrons is much lower than that of heavy charged particles, so their path length in typical absorbers is hundreds of times greater. As a very crude estimate, electron ranges tend to be about 2 mm per MeV in low-density materials, or about 1 mm per MeV in materials of moderate density.

Tabular data are given in Refs. 13 and 14 for the stopping power and range of electrons and positrons in elements and compounds, covering a large region of energy. To a fair degree of approximation, the product of the range times the density of the absorber is a constant for different materials for electrons of equal initial energy. Plots of the range of electrons in several common detector materials are given in Fig. 2-14.

2. ABSORPTION OF BETA PARTICLES

The transmission curve for beta particles emitted by a radioisotope source, because of the continuous distribution in their energy, differs significantly from that sketched in Fig. 2-13 for monoenergetic electrons. The "soft" or low-energy beta particles are rapidly absorbed even in small thicknesses of the absorber, so that the initial slope on the attenuation curve is much greater. For the majority of beta spectra, the curve happens to have a near-exponential shape and is therefore nearly linear on the semilog plot of the type shown in Fig. 2-15. This exponential behavior is only an empirical approximation and does not have a fundamental basis as does the exponential attenuation of gamma rays [see Eq. (2-20)]. An *absorption coefficient n* is sometimes defined by

$$\frac{I}{I_0} = e^{-nt} \tag{2-14}$$

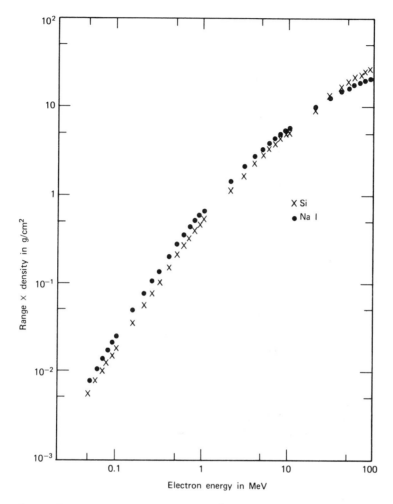

Figure 2-14 Range–energy plots for electrons in silicon and sodium iodide. If units of mass thickness (distance × density) are used for the range as shown, values at the same electron energy are similar even for materials with widely different physical properties or atomic number. (Data from Mukoyama.[15])

where $\qquad\qquad\qquad$ I_0 = counting rate without absorber

$\qquad\qquad\qquad\qquad$ I = counting rate with absorber

$\qquad\qquad\qquad\qquad$ t = absorber thickness in g/cm^2

The coefficient n correlates well with the endpoint energy of the beta emitter for a specific absorbing material. This dependence is shown in Fig. 2-16 for aluminum. Through the use of such data, attenuation measurements can be used to identify indirectly endpoint energies of unknown beta emitters, although direct energy measurements are more common.

3. BACKSCATTERING

The fact that electrons often undergo large-angle deflections along their track leads to the phenomenon of *backscattering*. An electron entering one surface of an absorber may

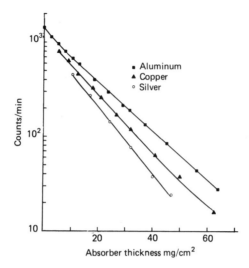

Figure 2-15 Transmission curves for beta particles from ^{185}W (endpoint energy of 0.43 MeV). (From Baltakmens.[16])

undergo sufficient deflection so that it reemerges from the surface through which it entered. These backscattered electrons do not deposit all their energy in the absorbing medium and therefore can have a significant effect on the response of detectors designed to measure the energy of externally incident electrons. Electrons that backscatter in the detector "entrance window" or dead layer will escape detection entirely.

Backscattering is most pronounced for electrons with low incident energy and absorbers with high atomic number. Figure 2-17 shows the fraction of monoenergetic

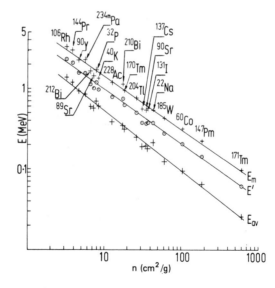

Figure 2-16 Beta particle absorption coefficient n in aluminum as a function of the endpoint energy E_m, average energy E_{av}, and $E' \equiv 0.5(E_m + E_{av})$ of different beta emitters. (From Baltakmens.[17])

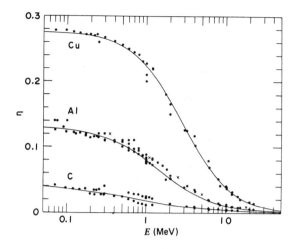

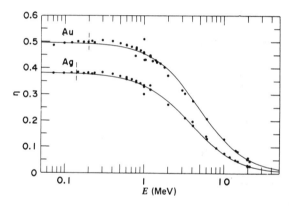

Figure 2-17 Fraction η of normally incident electrons that are backscattered from thick slabs of various materials, as a function of incident energy E. (From Tabata et al.[18])

electrons that are backscattered when normally incident on the surface of various absorbers. Additional data for materials commonly used as electron detectors are given in Table 10-1.

Backscattering can also influence the apparent yield from radioisotope sources of beta particles or conversion electrons. If the source is deposited on a thick backing, electrons that are emitted initially into this backing may backscatter and reemerge from the surface of the source.

C. Positron Interactions

The coulomb forces that constitute the major mechanism of energy loss for both electrons and heavy charged particles are present for either positive or negative charge on the particle. Whether the interaction involves a repulsive or attractive force between the incident particle and orbital electron, the impulse and energy transfer for particles of equal mass are about the same. Therefore, the tracks of positrons in an absorber are similar to those of normal negative electrons, and their specific energy loss and range are about the same for equal initial energies.

Positrons differ significantly, however, in that the annihilation radiation described in Chapter 1 is generated at the end of the positron track. Because these 0.511 MeV photons are very penetrating compared with the range of the positron, they can lead to the deposition of energy far from the original positron track.

III. INTERACTION OF GAMMA RAYS

Although a large number of possible interaction mechanisms are known for gamma rays in matter, only three major types play an important role in radiation measurements: *photoelectric absorption*, *Compton scattering*, and *pair production*. All these processes lead to the partial or complete transfer of the gamma-ray photon energy to electron energy. They result in sudden and abrupt changes in the gamma-ray photon history, in that the photon either disappears entirely or is scattered through a significant angle. This behavior is in marked contrast to the charged particles discussed earlier in this chapter, which slow down gradually through continuous, simultaneous interactions with many absorber atoms. The fundamentals of the gamma-ray interaction mechanisms are introduced here but are again reviewed at the beginning of Chapter 10 in the context of their influence on the response of gamma-ray detectors.

A. Interaction Mechanisms

1. PHOTOELECTRIC ABSORPTION

In the photoelectric absorption process, a photon undergoes an interaction with an absorber atom in which the photon completely disappears. In its place, an energetic *photoelectron* is ejected by the atom from one of its bound shells. The interaction is with the atom as a whole and cannot take place with free electrons. For gamma rays of sufficient energy, the most probable origin of the photoelectron is the most tightly bound or K shell of the atom. The photoelectron appears with an energy given by

$$E_{e^-} = h\nu - E_b \tag{2-15}$$

where E_b represents the binding energy of the photoelectron in its original shell. For gamma-ray energies of more than a few hundred keV, the photoelectron carries off the majority of the original photon energy.

In addition to the photoelectron, the interaction also creates an ionized absorber atom with a vacancy in one of its bound shells. This vacancy is quickly filled through capture of a free electron from the medium and/or rearrangement of electrons from other shells of the atom. Therefore, one or more characteristic X-ray photons may also be generated. Although in most cases these X-rays are reabsorbed close to the original site through photoelectric absorption involving less tightly bound shells, their migration and possible escape from radiation detectors can influence their response (see Chapter 10). In

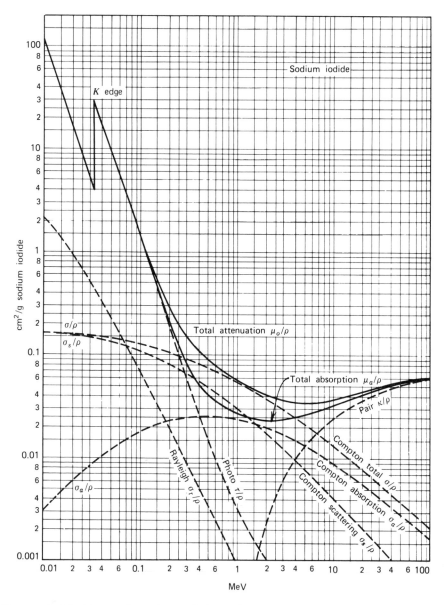

Figure 2-18 Energy dependence of the various gamma-ray interaction processes in sodium iodide. (From *The Atomic Nucleus* by R. D. Evans. Copyright 1955 by the McGraw-Hill Book Company. Used with permission.)

some fraction of the cases, the emission of an Auger electron may substitute for the characteristic X-ray in carrying away the atomic excitation energy.

The photoelectric process is the predominant mode of interaction for gamma rays (or X-rays) of relatively low energy. The process is also enhanced for absorber materials of high atomic number Z. No single analytic expression is valid for the probability of photoelectric absorption per atom over all ranges of E_γ and Z, but a rough approximation is

$$\tau \cong \text{constant} \times \frac{Z^n}{E_\gamma^{3.5}} \tag{2-16}$$

where the exponent n varies between 4 and 5 over the gamma-ray energy region of interest.[1] This severe dependence of the photoelectric absorption probability on the atomic number of the absorber is a primary reason for the preponderance of high-Z materials (such as lead) in gamma-ray shields. As further detailed in Chapter 10, many detectors used for gamma-ray spectroscopy are chosen from high-Z constituents for the same reason.

A plot of the photoelectric absorption cross section for a popular gamma-ray detection material, sodium iodide, is shown in Fig. 2-18. In the low-energy region, discontinuities in the curve or "absorption edges" appear at gamma-ray energies that correspond to the binding energies of electrons in the various shells of the absorber atom. The edge lying highest in energy therefore corresponds to the binding energy of the K-shell electron. For gamma-ray energies slightly above the edge, the photon energy is just sufficient to undergo a photoelectric interaction in which a K-electron is ejected from the atom. For gamma-ray energies slightly below the edge, this process is no longer energetically possible and therefore the interaction probability drops abruptly. Similar absorption edges occur at lower energies for the $L, M, \ldots$ electron shells of the atom.

2. COMPTON SCATTERING

The interaction process of Compton scattering takes place between the incident gamma-ray photon and an electron in the absorbing material. It is most often the predominant interaction mechanism for gamma-ray energies typical of radioisotope sources.

In Compton scattering, the incoming gamma-ray photon is deflected through an angle θ with respect to its original direction. The photon transfers a portion of its energy to the electron (assumed to be initially at rest), which is then known as a *recoil electron*. Because all angles of scattering are possible, the energy transferred to the electron can vary from zero to a large fraction of the gamma-ray energy.

The expression that relates the energy transfer and the scattering angle for any given interaction can simply be derived by writing simultaneous equations for the conservation of energy and momentum. Using the symbols defined in the sketch below

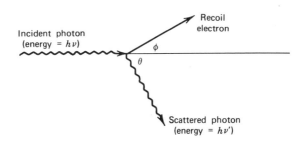

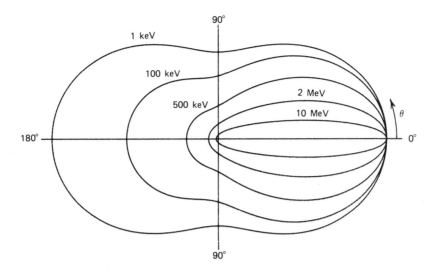

Figure 2-19 A polar plot of the number of photons (incident from the left) Compton scattered into a unit solid angle at the scattering angle θ. The curves are shown for the indicated initial energies.

we can show[1] that

$$hv' = \frac{hv}{1 + \dfrac{hv}{m_0c^2}(1 - \cos \theta)}$$ (2-17)

where m_0c^2 is the rest-mass energy of the electron (0.511 MeV). For small scattering angles θ, very little energy is transferred. Some of the original energy is always retained by the incident photon, even in the extreme of $\theta = \pi$. Equations (10-2) through (10-6) describe some properties of the energy transfer for limiting cases. A plot of the scattered photon energy predicted from Eq. (2-17) is also shown in Fig. 10-7.

The probability of Compton scattering per atom of the absorber depends on the number of electrons available as scattering targets and therefore increases linearly with Z. The dependence on gamma-ray energy is illustrated in Fig. 2-18 for the case of sodium iodide and generally falls off gradually with increasing energy.

The angular distribution of scattered gamma rays is predicted by the *Klein–Nishina formula* for the differential scattering cross section $d\sigma/d\Omega$:

$$\frac{d\sigma}{d\Omega} = Zr_0^2 \left(\frac{1}{1 + \alpha(1 - \cos \theta)}\right)^2 \left(\frac{1 + \cos^2 \theta}{2}\right)\left(1 + \frac{\alpha^2(1 - \cos \theta)^2}{(1 + \cos^2 \theta)[1 + \alpha(1 - \cos \theta)]}\right)$$ (2-18)

where $\alpha \equiv hv/m_0c^2$ and r_0 is the classical electron radius. The distribution is shown graphically in Fig. 2-19 and illustrates the strong tendency for forward scattering at high values of the gamma-ray energy.

3. PAIR PRODUCTION
If the gamma-ray energy exceeds twice the rest-mass energy of an electron (1.02 MeV), the process of pair production is energetically possible. As a practical matter, the probability of this interaction remains very low until the gamma-ray energy approaches

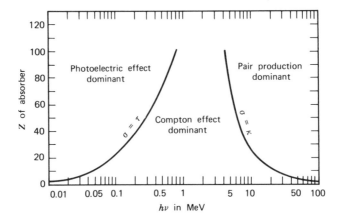

Figure 2-20 The relative importance of the three major types of gamma-ray interaction. The lines show the values of Z and $h\nu$ for which the two neighboring effects are just equal. (From *The Atomic Nucleus* by R. D. Evans. Copyright 1955 by the McGraw-Hill Book Company. Used with permission.)

several MeV and therefore pair production is predominantly confined to high-energy gamma rays. In the interaction (which must take place in the coulomb field of a nucleus), the gamma-ray photon disappears and is replaced by an electron–positron pair. All the excess energy carried in by the photon above the 1.02 MeV required to create the pair goes into kinetic energy shared by the positron and electron. Because the positron will subsequently annihilate after slowing down in the absorbing medium, two annihilation photons are normally produced as secondary products of the interaction. The subsequent fate of this annihilation radiation has an important effect on the response of gamma-ray detectors, as described in Chapter 10.

No simple expression exists for the probability of pair production per nucleus, but its magnitude varies approximately as the square of the absorber atomic number.[1] The importance of pair production rises sharply with energy, as indicated in Fig. 2-18.

The relative importance of the three processes described above for different absorber materials and gamma-ray energies is conveniently illustrated in Fig. 2-20. The line at the left represents the energy at which photoelectric absorption and Compton scattering are equally probable as a function of the absorber atomic number. The line at the right represents the energy at which Compton scattering and pair production are equally probable. Three areas are thus defined on the plot within which photoelectric absorption, Compton scattering, and pair production each predominate.

B. Gamma-Ray Attenuation

1. ATTENUATION COEFFICIENTS

If we again picture a transmission experiment as in Fig. 2-21, where monoenergetic gamma rays are collimated into a narrow beam and allowed to strike a detector after passing through an absorber of variable thickness, the result should be simple exponential attenuation of the gamma rays as also shown in Fig. 2-21. Each of the interaction processes removes the gamma-ray photon from the beam either by absorption or by scattering away from the detector direction and can be characterized by a fixed probability of occurrence per unit path length in the absorber. The sum of these probabilities is

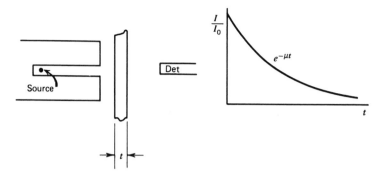

Figure 2-21 The exponential transmission curve for gamma rays measured under "good geometry" conditions.

simply the probability per unit path length that the gamma-ray photon is removed from the beam:

$$\mu = \tau(\text{photoelectric}) + \sigma(\text{Compton}) + \kappa(\text{pair}) \qquad (2\text{-}19)$$

and is called the *linear attenuation coefficient*. The number of transmitted photons I is then given in terms of the number without an absorber I_0 as

$$\frac{I}{I_0} = e^{-\mu t} \qquad (2\text{-}20)$$

The gamma-ray photons can also be characterized by their *mean free path* λ, defined as the average distance traveled in the absorber before an interaction takes place. Its value can be obtained from

$$\lambda = \frac{\int_0^\infty x e^{-\mu x}\, dx}{\int_0^\infty e^{-\mu x}\, dx} = \frac{1}{\mu} \qquad (2\text{-}21)$$

and is simply the reciprocal of the linear attenuation coefficient. Typical values of λ range from a few mm to tens of cm in solids for common gamma-ray energies.

Use of the linear attenuation coefficient is limited by the fact that it varies with the density of the absorber, even though the absorber material is the same. Therefore, the *mass attenuation coefficient* is much more widely used and is defined as

$$\text{mass attenuation coefficient} = \frac{\mu}{\rho} \qquad (2\text{-}22)$$

where ρ represents the density of the medium. For a given gamma-ray energy, the mass attenuation coefficient does not change with the physical state of a given absorber. For example, it is the same for water whether present in liquid or vapor form. The mass attenuation coefficient of a compound or mixture of elements can be calculated from

$$\left(\frac{\mu}{\rho}\right)_c = \sum_i w_i \left(\frac{\mu}{\rho}\right)_i \qquad (2\text{-}23)$$

where the w_i factors represent the weight fraction of element i in the compound or mixture.

2. ABSORBER MASS THICKNESS

In terms of the mass attenuation coefficient, the attenuation law for gamma rays now takes the form

$$\frac{I}{I_0} = e^{-(\mu/\rho)\rho t} \qquad (2\text{-}24)$$

The product ρt, known as the *mass thickness* of the absorber, is now the significant parameter that determines its degree of attenuation. Units of mass thickness have historically been mg/cm^2, and this convention is retained in this text. The thickness of absorbers used in radiation measurements is therefore often measured in mass thickness rather than physical thickness, because it is a more fundamental physical quantity.

The mass thickness is also a useful concept when discussing the energy loss of charged particles and fast electrons. For absorber materials with similar neutron/proton ratios, a particle will encounter about the same number of electrons passing through absorbers of equal mass thickness. Therefore, the stopping power and range, when expressed in units of ρt, are roughly the same for materials that do not differ greatly in Z.

3. BUILDUP

The gamma-ray attenuation experiment of Fig. 2-21, in which the gamma rays are collimated to a narrow beam before striking the absorber, is sometimes characterized as a "narrow beam" or "good geometry" measurement. The essential characteristic is that only gamma rays from the source which escape interaction in the absorber can be counted by the detector. Real measurements are often carried out under different circumstances (as sketched below) in which the severe collimation of the gamma rays is absent.

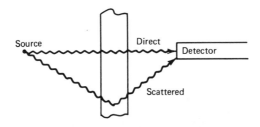

Now the detector can respond either to gamma rays directly from the source, to gamma rays that reach the detector after having scattered in the absorber, or to other types of secondary photon radiation. Many types of detectors will be unable to distinguish between these possibilities, so that the measured detector signal will be larger than that recorded under equivalent "good geometry" conditions. The conditions that lead to the simple exponential attenuation of Eq. (2-20) are therefore violated in this "broad beam" or "bad geometry" measurement because of the additional contribution of the secondary gamma rays. This situation is usually handled by replacing Eq. (2-20) by the following:

$$\frac{I}{I_0} = B(t, E_\gamma) e^{-\mu t} \qquad (2\text{-}25)$$

where the factor $B(t, E_\gamma)$ is called the *buildup factor*. The exponential term is retained to describe the major variation of the gamma-ray counting rate with absorber thickness, and the buildup factor is introduced as a simple multiplicative correction. The magnitude of the buildup factor depends on the type of gamma-ray detector used, because this will affect the relative weight given to the direct and secondary gamma rays. (With a detector that responds only to the direct gamma rays, the buildup factor is unity.) The buildup also depends on the specific geometry of the experiment. As a rough rule of thumb, the buildup factor for thick slab absorbers tends to be about equal to the thickness of the absorber measured in units of mean free path of the incident gamma rays, provided the detector responds to a broad range of gamma-ray energies.

IV. INTERACTION OF NEUTRONS

A. General Properties

In common with gamma rays, neutrons carry no charge and therefore cannot interact in matter by means of the coulomb force, which dominates the energy loss mechanisms for charged particles and electrons. Neutrons can also travel through many centimeters of matter without any type of interaction and thus can be totally invisible to a detector of common size. When a neutron does undergo interaction, it is with a nucleus of the absorbing material. As a result of the interaction, the neutron may either totally disappear and be replaced by one or more secondary radiations, or else the energy or direction of the neutron is changed significantly.

In contrast to gamma rays, the secondary radiations resulting from neutron interactions are almost always heavy charged particles. These particles may be produced either as a result of neutron-induced nuclear reactions, or they may be the nuclei of the absorbing material itself, which have gained energy as a result of neutron collisions. Most neutron detectors utilize some type of conversion of the incident neutron into secondary charged particles, which can then be detected directly. Specific examples of the most useful conversion processes are detailed in Chapters 14 and 15.

The relative probabilities of the various types of neutron interaction change dramatically with neutron energy. In somewhat of an oversimplification, we will divide neutrons into two categories on the basis of their energy, either "fast neutrons" or "slow neutrons," and discuss their interaction properties separately. The dividing line will be at about 0.5 eV, or about the energy of the abrupt drop in absorption cross section in cadmium (the *cadmium cutoff* energy).

B. Slow Neutron Interactions

For slow neutrons, the significant interactions include elastic scattering with absorber nuclei and a large set of neutron-induced nuclear reactions. Because of the small kinetic energy of slow neutrons, very little energy can be transferred to the nucleus in elastic scattering. Consequently, this is not an interaction on which detectors of slow neutrons can be based. Elastic collisions tend to be very probable, however, and often serve to bring the slow neutron into thermal equilibrium with the absorber medium before a different type of interaction takes place. Much of the population in the slow neutron energy range will therefore be found among these *thermal neutrons*, which, at room temperature, have an average energy of about 0.025 eV.

The slow neutron interactions of real importance are neutron-induced reactions that can create secondary radiations of sufficient energy to be detected directly. Because the incoming neutron energy is so low, all such reactions must have a positive Q-value to be energetically possible. In most materials, the radiative capture reaction [or (n, γ) reaction] is the most probable and plays an important part in the attenuation or shielding of neutrons. Radiative capture reactions can be useful in the indirect detection of neutrons using activation foils as described in Chapter 19, but they are not widely applied in active neutron detectors because the secondary radiation takes the form of gamma rays, which are also difficult to detect. Instead, reactions such as (n, α), (n, p), and (n, fission) are much more attractive because the secondary radiations are charged particles. A number of specific reactions of this type are detailed in Chapter 14.

C. Fast Neutron Interactions

The probability of most neutron-induced reactions potentially useful in detectors drops off rapidly with increasing neutron energy. The importance of scattering becomes greater, however, because the neutron can transfer an appreciable amount of energy in one collision. The secondary radiations in this case are *recoil nuclei*, which have picked up a detectable amount of energy from neutron collisions. At each scattering site, the neutron loses energy and is thereby *moderated* or slowed to lower energy. The most efficient moderator is hydrogen because the neutron can lose up to all its energy in a single collision with a hydrogen nucleus. For heavier nuclei, only a partial energy transfer is possible [see Eq. (15-4) and the associated discussion].

If the energy of the fast neutron is sufficiently high, *inelastic scattering* with nuclei can take place in which the recoil nucleus is elevated to one of its excited states during the collision. The nucleus quickly deexcites, emitting a gamma ray, and the neutron loses a greater fraction of its energy than it would in an equivalent elastic collision. Inelastic scattering and the subsequent secondary gamma rays play an important role in the shielding of high-energy neutrons but are an unwanted complication in the response of most fast neutron detectors based on elastic scattering.

D. Neutron Cross Sections

For neutrons of a fixed energy, the probability per unit path length is a constant for any one of the interaction mechanisms. It is conventional to express this probability in terms of the *cross section* σ per nucleus for each type of interaction. The cross section has units of area and has traditionally been measured in units of the *barn* (10^{-28} m^2). For example, each nuclear species will have an elastic scattering cross section, a radiative capture cross section, and so on, each of which will be a function of the neutron energy. Some examples of cross section plots are given in Figs. 14-1 and 14-2.

When multiplied by the number of nuclei N per unit volume, the cross section σ is converted into the macroscopic cross section Σ

$$\Sigma = N\sigma \tag{2-26}$$

which now has dimensions of inverse length. Σ has the physical interpretation of the probability per unit path length for the specific process described by the "microscopic" cross section σ. When all processes are combined by adding together the cross sections for each individual interaction

$$\Sigma_{\text{tot}} = \Sigma_{\text{scatter}} + \Sigma_{\text{rad. capture}} + \cdots \tag{2-27}$$

the resulting Σ_{tot} is the probability per unit path length that any type of interaction will occur. This quantity has the same significance for neutrons as the linear absorption coefficient for gamma rays defined earlier. If a narrow beam attenuation experiment is carried out for neutrons, as sketched earlier for gamma rays, the result will be the same: The number of detected neutrons will fall off exponentially with absorber thickness. In this case the attenuation relation is written

$$\frac{I}{I_0} = e^{-\Sigma_{tot}t} \tag{2-28}$$

The neutron mean free path λ is, by analogy with the gamma-ray case, given by $1/\Sigma_{tot}$. In solid materials, λ for slow neutrons may be of the order of a centimeter or less, whereas for fast neutrons, it is normally tens of centimeters.

Under most circumstances, neutrons are not narrowly collimated so that typical shielding situations involve broad beam or "bad geometry" conditions. Just as in the case of gamma rays, the exponential attenuation of Eq. (2-28) is no longer an adequate description because of the added importance of scattered neutrons reaching the detector. A more complex neutron transport computation is then required to predict the number of transmitted neutrons and their distribution in energy.

When discussing the rate of reactions induced by neutrons, it is convenient to introduce the concept of neutron flux. If we first consider neutrons with a single energy or fixed velocity v, the product $v\Sigma$ gives the interaction frequency for the process for which Σ is the macroscopic cross section. The reaction rate density (reactions per unit time and volume) is then given by $n(r)v\Sigma$, where $n(r)$ is the neutron number density at the vector position r, and $n(r)v$ is defined as the *neutron flux* $\varphi(r)$ with dimensions of length^{-2} time^{-1}. Thus, the reaction rate density is given by the product of the neutron flux and the macroscopic cross section for the reaction of interest:

$$\text{reaction rate density} = \varphi(r)\Sigma \tag{2-29}$$

This relation can be generalized to include an energy-dependent neutron flux $\varphi(r, E)$ and cross section $\Sigma(E)$:

$$\text{reaction rate density} = \int_0^\infty \varphi(r, E)\Sigma(E)\, dE \tag{2-30}$$

V. RADIATION EXPOSURE AND DOSE

Because of their importance in radiation protection and the safe handling of radiation sources, the concepts of radiation exposure and dose play prominent roles in radiation measurements. In the following sections, we introduce the basic quantities and units of importance in this area. For further definitions and more extensive discussions on these topics, the reader is referred to publications of the International Commission on Radiation Units and Measurements. In particular, the ICRU Report #19 [*Radiation Quantities and Units* (1971)] contains a complete set of definitions and notes that are the basis of radiation dose measurements.

A. Gamma-Ray Exposure

The concept of gamma-ray exposure was introduced early in the history of radioisotope research and is a quantity that is roughly analogous to the strength of an electric field created by a point charge. Defined only for sources of X- or gamma rays, a fixed exposure rate exists at every point in space surrounding a source of fixed intensity. The exposure is linear, in that doubling the source intensity also doubles the exposure rate everywhere around the source.

The basic unit of gamma-ray exposure is defined in terms of the charge dQ due to ionization created by the secondary electrons (negative electrons and positrons) formed within a volume element of air and mass dm, when these secondary electrons are completely stopped in air. The exposure value X is then given by dQ/dm. The SI unit of gamma-ray exposure is thus the coulomb per kilogram (C/kg), which has not been given a special name. The historical unit has been the *roentgen* (R), defined as the exposure that results in the generation of one electrostatic unit of charge per 0.001293 g (1 cm³ at STP) of air. The two units are related by

$$1 \text{ R} = 2.58 \times 10^{-4} \text{ C/kg}$$

The exposure is therefore defined in terms of the effect of a given flux of gamma rays on a test volume of air and is a function only of the intensity of the source, the geometry between the source and test volume, and any attenuation of the gamma rays which may take place between the two. Its measurement fundamentally requires the determination of the charge due to ionization produced in air under specific conditions. Inherent in the above definition is the obligation to track each secondary electron created by primary gamma-ray interactions in the test volume under consideration, and to add up the ionization charges formed by that secondary electron until it reaches the end of its path. This requirement is often difficult or impossible to achieve in actual practice, and therefore instruments that are designed to measure gamma-ray exposure usually employ approximations that involve the principle of compensation introduced in Chapter 5.

The gamma-ray exposure, although not directly tied to physical phenomena, is often of interest in gamma-ray dosimetry. It is therefore often convenient to be able to calculate the exposure rate at a known distance from a point radioisotope source. If we assume that the yield per disintegration of X- and gamma rays is accurately known for the radioisotope of interest, the exposure per unit activity of the source at a known distance can simply be expressed under the following conditions:

1. The source is sufficiently small so that spherical geometry holds (i.e., the photon flux diminishes as $1/d^2$, where d is the distance to the source).

2. No attenuation of the X- or gamma rays takes place in the air or other material between the source and measuring point.

3. Only photons passing directly from the source to the measuring point contribute to the exposure, and any gamma rays scattered in surrounding materials may be neglected.

The exposure rate $\dot{X}$ is then

$$\dot{X} = \Gamma_\delta \frac{\alpha}{d^2} \tag{2-31}$$

where α is the activity of the source, and Γ_δ is defined as the *exposure rate constant* for the specific radioisotope of interest. The subscript δ implies that the assumption has been made that all X- and gamma rays emitted by the source above an energy δ contribute to the dose, whereas those below this energy are not sufficiently penetrating to be of practical interest. The value of Γ_δ for a particular radioisotope can be calculated from its

TABLE 2-1 Exposure Rate Constant for Some
Common Radioisotope Gamma-Ray Sources

Nuclide	Γ^a
Antimony-124	9.8
Cesium-137	3.3
Cobalt-57	0.9
Cobalt-60	13.2
Iodine-125	~ 0.7
Iodine-131	2.2
Manganese-54	4.7
Radium-226	8.25
Sodium-22	12.0
Sodium-24	18.4
Technetium-99m	1.2
Zinc-65	2.7

^{a}The exposure rate constant Γ is in units of $R \cdot cm^2/hr \cdot mCi$.

*Source: The Health Physics and Radiological Health
Handbook*, Nucleon Lectern Associates, Olney, MD, 1984.

gamma-ray yield and the energy-dependent absorption properties of air. Some particular values for $\delta = 0$ are listed in Table 2-1.

B. Absorbed Dose

Two different materials, if subjected to the same gamma-ray exposure, will in general absorb different amounts of energy. Because many important phenomena, including changes in physical properties or induced chemical reactions, would be expected to scale as the energy absorbed per unit mass of the material, a unit that measures this quantity is of fundamental interest. The energy absorbed from any type of radiation per unit mass of the absorber is defined as the *absorbed dose*. The historical unit of absorbed dose has been the *rad*, defined as 100 ergs/gram. As with other historical radiation units, the rad is being gradually replaced by its SI equivalent, the *gray* (Gy) defined as 1 joule/kilogram. The two units are therefore simply related by:

$$1 \text{ Gy} = 100 \text{ rad}$$

The absorbed dose should be a reasonable measure of the chemical or physical effects created by a given radiation exposure in an absorbing material. Careful measurements have shown that the absorbed dose in air corresponding to a gamma-ray exposure of 1 coulomb/kilogram amounts to 33.8 joules/kilogram, or 33.8 Gy. If water is substituted for the air, its absorption properties per unit mass do not differ greatly because the average atomic number of water is much the same as that of air. For absorbing materials of greatly dissimilar atomic numbers, however, the interaction mechanisms have different relative importance, and therefore the absorbed dose per unit exposure would show greater differences.

In order to measure absorbed dose in a fundamental manner, some type of energy measurement must be carried out. One possibility is a calorimetric measurement in which the rate of rise of the temperature in a sample of absorber is used to calculate the rate of

TABLE 2-2   Quality Factors for Different
Radiations (from ICRU Report #19)

L in Water (keV/μm)	Q
3.5 or less	1
7.0	2
23	5
53	10
175	20

energy deposition per unit mass. Because of their difficulty, these measurements are not commonplace for routine application since the thermal effects created even by large doses of radiation are very small. Instead, indirect measurements of absorbed dose are much more common, in which its magnitude is inferred from ionization measurements carried out under proper conditions (see Chapter 5).

C. Dose Equivalent

When the effects of radiation on living organisms are evaluated, the absorption of equal amounts of energy per unit mass under different irradiation conditions does not guarantee the same biological effect. In fact, the extent of the effects can differ by as much as an order of magnitude depending on whether the energy is deposited in the form of heavy charged particles or electrons.

The biological damage created by ionizing radiation is traceable to the chemical alteration of the biological molecules that are influenced by the ionization or excitation caused by the radiation. The severity and permanence of these changes are directly related to the local rate of energy deposition along the particle track, known as the *linear energy transfer*[†] L. Those radiations with large values of L (such as heavy charged particles) tend to result in greater biological damage than those with lower L (such as electrons), even though the total energy deposited per unit mass is the same.

The concept of *dose equivalent* has therefore been introduced to more adequately quantify the probable biological effect of a given radiation exposure. A unit of dose equivalent is defined as that amount of any type of radiation which, when absorbed in a biological system, results in the same biological effect as one unit of absorbed dose delivered in the form of low-L radiation. The dose equivalent H is the product of the absorbed dose D and the *quality factor* Q that characterizes the specific radiation:

$$H = DQ \qquad (2\text{-}32)$$

The quality factor increases with linear energy transfer L as shown in Table 2-2.

For the fast electron radiations of interest in this text, L is sufficiently low so that Q is essentially unity in all applications. Therefore, the dose equivalent is numerically equal to the absorbed dose for beta particles or other fast electrons. The same is true for X-rays and gamma rays because their energy is also delivered in the form of fast secondary electrons. Charged particles have a much higher linear energy transfer and the dose

---

[†]The linear energy transfer is nearly identical to the specific energy loss ($-dE/dx$) defined earlier. The only differences arise when a substantial portion of the radiation energy is liberated in the form of bremsstrahlung, which may travel a substantial distance from the particle track before depositing its energy. The specific energy loss includes the bremsstrahlung as part of the energy loss of the particle, but the linear energy transfer L counts only the energy that is deposited along the track and therefore excludes the bremsstrahlung.

equivalent is larger than the absorbed dose. For example, Q is approximately 20 for alpha particles of typical energies. Because neutrons deliver most of their energy in the form of heavy charged particles, their effective quality factor is also considerably greater than unity and varies significantly with the neutron energy.

The units used for dose equivalent H depend on the corresponding units of absorbed dose D in Eq. (2-32). If D is expressed in the historical unit of the rad, H is defined to be in units of the *rem*. Until very recently, the rem (or millirem) was universally used to quantify dose equivalent. Under the SI convention, D is instead expressed in grays, and a corresponding unit of dose equivalent called the *sievert* (Sv) has been introduced. For example, an absorbed dose of 2 Gy delivered by radiation with Q of 5 will result in a dose equivalent of 10 Sv. Because 1 Gy = 100 rad, the units are interrelated by

$$1 \text{ Sv} = 100 \text{ rem}$$

Guidelines for radiation exposure limits to personnel are quoted in units of dose equivalent in order to place exposures to different types and energies of radiation on a common basis.

PROBLEMS

2-1. Estimate the time required for a 5 MeV alpha particle to slow down and stop in silicon. Repeat for the same particle in hydrogen gas.

2-2. With the aid of Fig. 2-7, estimate the energy remaining in a beam of 5 MeV protons after passing through 100 μm of silicon.

2-3. Using Fig. 2-10, find the approximate energy loss of 1 MeV alpha particles in a thickness of 5 μm of gold.

2-4. Estimate the range of 1 MeV electrons in aluminum with the aid of Fig. 2-14.

2-5. Calculate the energy of a 1 MeV gamma-ray photon after Compton scattering through 90°.

2-6. Give a rough estimate of the ratio of the probability per atom for photoelectric absorption in silicon to that in germanium.

2-7. Indicate which of the three major interaction processes (photoelectric absorption, Compton scattering, pair production) is dominant in the following situations:

(a) 1 MeV gamma rays in aluminum.
(b) 100 keV gamma rays in hydrogen.
(c) 100 keV gamma rays in iron.
(d) 10 MeV gamma rays in carbon.
(e) 10 MeV gamma rays in lead.

2-8. (a) From Fig. 2-18, calculate the mean free path of 1 MeV gamma rays in sodium iodide (specific gravity = 3.67).

(b) What is the probability that a 600 keV gamma ray undergoes photoelectric absorption in 1 cm of sodium iodide?

2-9. Define the following terms:

(a) Absorber mass thickness.
(b) Buildup.
(c) Neutron inelastic scattering.
(d) Macroscopic cross section.
(e) Neutron flux

2-10. For 140 keV gamma rays, the mass attenuation coefficients for hydrogen and oxygen are 0.26 and 0.14 cm^2/g, respectively. What is the mean free path in water at this energy?

2-11. How many 5 MeV alpha particles are required to deposit a total energy of 1 J?

2-12. A beam of 1 MeV electrons strikes a thick target. For a beam current of 100 microamperes (μA), find the power dissipated in the target.

2-13. Using the data in Table 2-1, estimate the exposure rate 5 m from a 1 Ci source of ^{60}Co.

2-14. Calculate the rate of temperature rise in a sample of liquid water adiabatically exposed to radiation that results in an absorbed dose rate of 10 mrad/h.

REFERENCES

1. R. D. Evans, *The Atomic Nucleus*, Krieger, New York, 1982.
2. B. Wilken and T. A. Fritz, *Nucl. Instrum. Meth.* **138**, 331 (1976).
3. D. J. Skyrme, *Nucl. Instrum. Meth.* **57**, 61 (1967).
4. C. F. Williamson, J. P. Boujot, and J. Picard, CEA-R3042 (1966).
4. C. Hanke and J. Laursen, *Nucl. Instrum. Meth.* **151**, 253 (1978).
6. W. H. Barkas and M. J. Berger, National Academy of Sciences, National Research Council, Publication 1133, 103 (1964).
7. L. C. Northcliffe and R. F. Schilling, *Nuclear Data Tables* **A7**, 233 (1970).
8. J. F. Ziegler, *The Stopping and Ranges of Ions in Matter*, Pergamon Press, New York, 1977.
9. G. Cesini, G. Lucarini, and F. Rustichelli, *Nucl. Instrum. Meth.* **127**, 579 (1975).
10. P. D. Bourland and D. Powers, *Phys. Rev. B* **3**, 3635 (1971).
11. J. S.-Y. Feng, W. K. Chu, and M-A. Nicolet, *Phys. Rev. B* **10**, 3781 (1974).
12. S. Matteson, E. K. L. Chau, and D. Powers, *Phys. Rev. A* **14**, 169 (1976).
13. M. J. Berger and S. M. Seltzer, National Bureau of Standards Publication NBSIR 82-2550-A (1982).
14. L. Pages et al., *Atomic Data* **4**, 1 (1972).
15. T. Mukoyama, *Nucl. Instrum. Meth.* **134**, 125 (1976).
16. T. Baltakmens, *Nucl. Instrum. Meth.* **82**, 264 (1970).
17. T. Baltakmens, *Nucl. Instrum. Meth.* **142**, 535 (1977).
18. T. Tabata, R. Ito, and S. Okabe, *Nucl. Instrum. Meth.* **94**, 509 (1971).

CHAPTER · 3

Counting Statistics
and Error Prediction

Radioactive decay is a random process. Consequently, any measurement based on observing the radiation emitted in nuclear decay is subject to some degree of statistical fluctuation. These inherent fluctuations represent an unavoidable source of uncertainty in all nuclear measurements and often can be the predominant source of imprecision or error. The term *counting statistics* includes the framework of statistical analysis required to process the results of nuclear counting experiments and to make predictions about the expected precision of quantities derived from these measurements.

The value of counting statistics falls into two general categories. The first is to serve as a check on the normal functioning of a piece of nuclear counting equipment. Here a set of measurements is recorded under conditions in which all aspects of the experiment are held as constant as possible. Because of the influence of statistical fluctuations, these measurements will not all be the same but will show some degree of internal variation. The amount of this fluctuation can be quantified and compared with predictions of statistical models. If the amount of observed fluctuation is not consistent with predictions, one can conclude that some abnormality exists in the counting system. The second application is generally more valuable and deals with the situation in which we have only one measurement. We can then use counting statistics to predict its inherent statistical uncertainty and thus estimate a precision that should be associated with that single measurement.

The distinctions made in the organization of this chapter are a critical part of the topic. The confusion that often arises when the student is first introduced to counting statistics originates more from a failure to keep separate the concepts presented in Sections I and II below than from any other single cause. In Section I we are careful to limit the discussion to methods used in the characterization or organization of experimental data. We are not particularly concerned where these data come from but rather are interested only in presenting the formal methods by which we can describe the amount of fluctuation displayed by the data. In Section II, we discuss the separate topic of probabilistic mathematical models, which can sometimes represent real measurement systems. For purposes of the discussion in Section II, however, we are concerned only with the structure and predictions of these models as mathematical entities. We reserve, until Section III, the demonstration of how the statistical models can be matched to experimental data, resulting in the two common applications of counting statistics mentioned above. Finally, in Section IV, we examine how the predicted statistical uncertainties contribute to the overall uncertainty in a numerical result which is calculated from nuclear counting data.

I. CHARACTERIZATION OF DATA

We begin by assuming that we have a collection of N independent measurements of the same physical quantity:

$$x_1, x_2, x_3, \ldots, x_i, \ldots, x_N$$

We further assume that a single typical value x_i from this set can only assume integer values so that the data might represent, for example, a number of successive readings from a radiation counter for repeated time intervals of equal length. Two elementary properties of this data set are

Sum
$$\Sigma \equiv \sum_{i=1}^{N} x_i \tag{3-1}$$

Experimental mean
$$\bar{x}_e \equiv \Sigma/N \tag{3-2}$$

The experimental mean is written with the subscript to distinguish it from the mean of a particular statistical model which will be introduced later.

It is often convenient to represent the data set by a corresponding *frequency distribution function* $F(x)$. The value of $F(x)$ is the relative frequency with which the number appears in the collection of data. By definition

$$F(x) \equiv \frac{\text{number of occurrences of the value } x}{\text{number of measurements } (= N)} \tag{3-3}$$

The distribution is automatically normalized, that is,

$$\sum_{x=0}^{\infty} F(x) = 1 \tag{3-4}$$

TABLE 3-1   **Example of Data Distribution Function**

Data		Frequency Distribution Function	
8	14	$F(3) = 1/20$	= 0.05
5	8	$F(4)$	= 0
12	8	$F(5)$	= 0.05
10	3	$F(6)$	= 0.10
13	9	$F(7)$	= 0.10
7	12	$F(8)$	= 0.20
9	6	$F(9)$	= 0.10
10	10	$F(10)$	= 0.15
6	8	$F(11)$	= 0.05
11	7	$F(12)$	= 0.10
		$F(13)$	= 0.05
		$F(14)$	= 0.05
		$\sum_{x=0}^{\infty} F(x)$	= 1.00

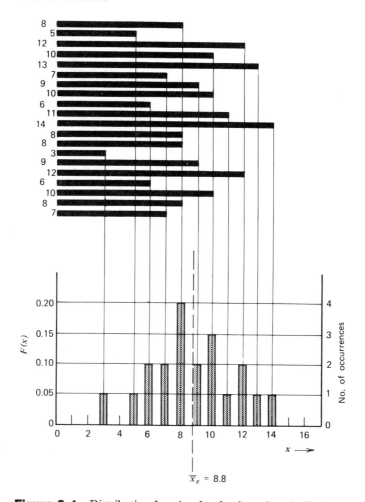

Figure 3-1 Distribution function for the data given in Table 3-1.

As long as we do not care about the specific sequence of the numbers, the complete data distribution function $F(x)$ represents all the information contained in the original data set.

For purposes of illustration, Table 3-1 gives a hypothetical set of data consisting of 20 entries. Because these entries range from a minimum of 3 to a maximum of 14, the data distribution function will have nonzero values only between these extreme values of the argument x. The corresponding values of $F(x)$ are also shown in Table 3-1.

A plot of the data distribution function for the example is given in Fig. 3-1. Also shown directly above the plot is a horizontal bar graph of the original 20 numbers from which the distribution was derived. These data show an experimental mean of 8.8, and the distribution function is in some sense centered about that value. Furthermore, the relative shape of the distribution function indicates qualitatively the amount of internal fluctuation in the data set. For example, Fig. 3-2 shows the shape of the distribution functions corresponding to two extreme sets of data: one with large amounts of scatter about the mean and one with little. An obvious conclusion is that the width of the distribution function is a relative measure of the amount of fluctuation or scattering about the mean inherent in a given set of data.

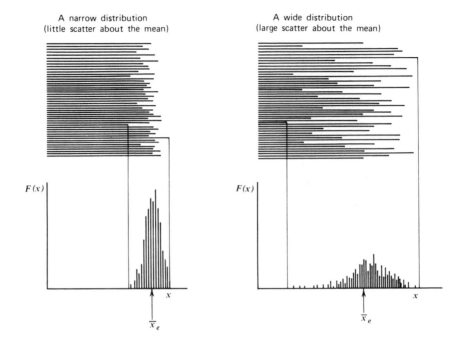

Figure 3-2 Distribution functions for two sets of data with differing amounts of internal fluctuation.

It is possible to calculate the experimental mean by using the data distribution function, because the mean of any distribution is simply its first moment

$$\bar{x}_e = \sum_{x=0}^{\infty} x F(x) \tag{3-5}$$

It is also possible to derive another parameter, known as the *sample variance*, which will serve to quantify the amount of internal fluctuation in the data set. The first step is to define the *deviation* of any data point as the amount by which it differs from the mean value

$$\epsilon_i \equiv x_i - \bar{x}_e \tag{3-6}$$

To illustrate, the example of the 20 numbers given in Table 3-1 is shown as the bar graph of Fig. 3-3a. The deviation of each of these values from the mean has been separately plotted on part (b) of the figure. There must be an equal contribution of positive and negative deviations, so that

$$\sum_{1}^{N} \epsilon_i = 0 \tag{3-7}$$

If we take the square of each deviation, however, a positive number will always result. These are plotted for the example in Fig. 3-3c. We can now introduce the sample variance s^2 as

$$s^2 = \frac{1}{N-1} \sum_{1}^{N} \epsilon_i^2 \tag{3-8}$$

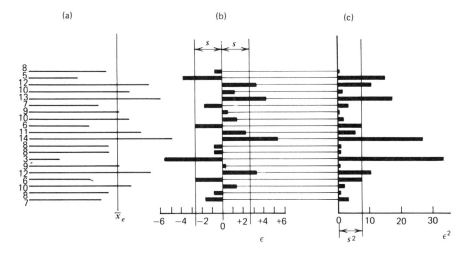

Figure 3-3 (a) Plot of the data given in Table 3-1. Corresponding values for the deviation ϵ and ϵ^2 are shown in parts (b) and (c).

which will now serve as a single index of the degree of fluctuation inherent in the original data. As long as the number of data entries N is reasonably large, the sample variance is essentially the average value of the squared deviation of each data point. To be precise, the sample variance is more fundamentally defined as the average value of the squared deviation of each data point from the *true mean value* $\bar{x}$ that would be derived if an infinite number of data points were accumulated.

$$s^2 \equiv \frac{1}{N} \sum_{i=1}^{N} (x_i - \bar{x})^2 \tag{3-9}$$

Because we cannot know $\bar{x}$ from a finite set of measurements, we use instead the value $\bar{x}_e$ derived from the data set itself to calculate values for the deviations. The true mean value is independent of the set of data, while the experimental mean is not. Use of the experimental rather than the true theoretical mean value will tend to reduce the average deviation and therefore result in a smaller than normal variance. In statistical parlance, the number of degrees of freedom of the system has been reduced by one. In Eq. (3-8), the experimental mean is used to calculate the ϵ_i values. Therefore, the sum is divided by $N - 1$ (rather than by N) to make the calculated value of s^2 slightly larger, accounting for the lost degree of statistical freedom.

The sample variance s^2 for the example of 20 numbers is shown graphically in Fig. 3-3c. Because it is essentially a measure of the average value of the squared deviations of each point, s^2 is an effective measure of the amount of fluctuation in the original data. A data set with a narrow distribution will have a small typical deviation from the mean, and therefore the value for the sample variance will be small. On the other hand, data with a large amount of fluctuation will have a wide distribution and a large value for typical deviations, and the corresponding sample variance will also be large. It is important to note that the sample variance is an absolute measure of the amount of internal scatter in the data and does not, to first approximation, depend on the number of values in the data set. For example, if the data shown in Fig. 3-3 were extended by simply collecting an additional 20 values by the same process, we would not expect the sample variance calculated for the extended collection of 40 numbers to be substantially different from that shown in Fig. 3-3.

We can also calculate the sample variance directly from the data distribution function $F(x)$. Because Eq. (3-9) indicates that s^2 is simply the average value of $(x - \bar{x})^2$, we can write that same average as

$$s^2 = \sum_{x=0}^{\infty} (x - \bar{x})^2 F(x) \qquad (3\text{-}10)$$

Equation (3-10) is not introduced so much for its usefulness in computation as for the parallel it provides to a similar expression, Eq. (3-17), which will be introduced in a later discussion of statistical models. An expansion of Eq. (3-10) will yield the well-known result

$$s^2 = \overline{x^2} - (\bar{x})^2 \qquad (3\text{-}11)$$

We now end our discussion of the organization of experimental data with two important conclusions:

1. Any set of data can be completely described by its frequency distribution function $F(x)$.

2. Two properties of this frequency distribution function are of particular interest: the experimental mean and the sample variance.

The experimental mean is given by Eq. (3-5) and is the value about which the distribution is centered. The sample variance is given by Eq. (3-10) and is a measure of the width of the distribution, or the amount of internal fluctuation in the data.

II. STATISTICAL MODELS

Under certain conditions, we can predict the distribution function that will describe the results of many repetitions of a given measurement. We define a measurement as counting the number of *successes* resulting from a given number of *trials*. Each trial is assumed to be a *binary* process in that only two results are possible: The trial is either a success or it is not a success. For everything that follows, we also assume that the probability of success is a constant for all trials.

To show how these conditions apply to real situations, Table 3-2 gives three separate examples. The third example indicates the basis for applying the theoretical framework that follows to the case of counting nuclear radiation events. In this case a trial consists of observing a given radioactive nucleus for a period of time t, the number of trials is equivalent to the number of nuclei in the sample under observation, and the measurement

TABLE 3-2 Examples of Binary Processes

Trial	Definition of Success	Probability of Success $\equiv p$
Tossing a coin	Heads	1/2
Rolling a die	A six	1/6
Observing a given radioactive nucleus for a time t	The nucleus decays during the observation	$1 - e^{-\lambda t}$

consists of counting those nuclei that undergo decay. We identify the probability of success of any one trial as p. In the case of radioactive decay, that probability is equal to $(1 - e^{-\lambda t})$, where λ is the decay constant of the radioactive sample.

Three specific statistical models are introduced:

1. *The Binomial Distribution.* This is the most general model and is widely applicable to all constant-p processes. It is, unfortunately, computationally cumbersome in radioactive decay, where the number of nuclei is always very large, and is used only rarely in nuclear applications. One example in which the binomial distribution must be used is in the examination of data acquired by counting a very short-lived radioisotope with high counting efficiency. In this case the criteria for applications of the models that follow are not met.

2. *The Poisson Distribution.* This model is a direct mathematical simplification of the binomial distribution under conditions that the success probability p is small. In practical terms, the condition implies that we have chosen an observation time that is small compared with the half-life of the source, or that the detection efficiency is small. Then if we single out any given radioactive nucleus, the probability that it results in a recorded count in the observation time will be a very small number and the Poisson distribution will apply.

3. *The Gaussian or Normal Distribution.* The third important distribution is the Gaussian, which is a further simplification if the average number of successes is relatively large (say greater than 20 or 30). This condition will apply for any situation in which we accumulate more than a few counts during the course of the measurement. This is most often the case so that the Gaussian model is widely applicable to many problems in counting statistics.

It should be emphasized that all the above models become identical for processes with a small individual success probability p but with a large enough number of trials so that the expected mean number of successes is large.

A. The Binomial Distribution

The binomial distribution is the most general of the statistical models discussed. If n is the number of trials for which each trial has a success probability p, then the predicted probability of counting exactly x successes can be shown[1] to be

$$P(x) = \frac{n!}{(n - x)!\, x!} p^x (1 - p)^{n - x}$$
(3-12)

$P(x)$ is the predicted probability distribution function, as given by the binomial distribution, and is defined only for integer values of n and x.

We show one example of an application of the binomial distribution. Imagine that we have an honest die so that the numbers 1 through 6 are all equally probable. Let us define a successful roll as one in which any of the numbers 3, 4, 5, or 6 appear. Because these are four of the six possible results, the individual probability of success p is equal to $\frac{4}{6}$ or 0.667. We now roll the die a total of 10 times and record the number of rolls that result in success as defined above. The binomial distribution now allows us to calculate the probability that exactly x out of the 10 trials will be successful, where x can vary between

TABLE 3-3 Values of the Binomial
Distribution for the Parameters
$p = \frac{4}{6}$ or $\frac{2}{3}$, $n = 10$

x	$P(x)$
0	0.00002
1	0.00034
2	0.00305
3	0.01626
4	0.05690
5	0.13656
6	0.22761
7	0.26012
8	0.19509
9	0.08671
10	0.01734

$$\sum_{x=0}^{10} P(x) = 1.00000$$

0 and 10. Table 3-3 gives the values of the predicted probability distribution from Eq. (3-12) for the parameters $p = \frac{2}{3}$ and $n = 10$. The results are also plotted in Fig. 3-4. We see that 7 is the most probable number of successes from the 10 rolls of the die, with a probability of occurrence slightly greater than 1 out of 4. From the value of $P(0)$ we see that only twice out of 100,000 tests would we expect to see no successes from 10 rolls of the die.

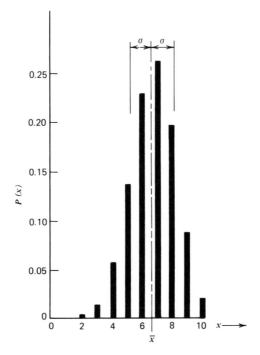

Figure 3-4 A plot of the binomial distribution for $p = \frac{2}{3}$ and $n = 10$.

Some properties of the binomial distribution are important. First, the distribution is normalized:

$$\sum_{x=0}^{n} P(x) = 1 \qquad (3\text{-}13)$$

Also, we know that the average or mean value of the distribution is given by

$$\bar{x} = \sum_{x=0}^{n} xP(x) \qquad (3\text{-}14)$$

If we now substitute Eq. (3-12) for $P(x)$ and carry out the summation, a remarkably simple result is derived:

$$\bar{x} = pn \qquad (3\text{-}15)$$

Thus, we can calculate the expected average number of successes by multiplying the number of trials n by the probability p that any one trial will result in a success. In the example just discussed, we calculate an average number of successes as

$$\bar{x} = pn = \left(\tfrac{2}{3}\right)(10) = 6.67 \qquad (3\text{-}16)$$

The mean value is obviously a very fundamental and important property of any predicted distribution.

It is also important to derive a single parameter that can describe the amount of fluctuation predicted by a given distribution. We have already defined such a parameter, called the sample variance, for a set of experimental data as defined in Eq. (3-10). By analogy we now define a *predicted variance* σ^2, which is a measure of the scatter about the mean predicted by a specific statistical model $P(x)$:

$$\sigma^2 \equiv \sum_{x=0}^{n} (x - \bar{x})^2 P(x) \qquad (3\text{-}17)$$

Conventionally, σ^2 is called the variance, and we emphasize the fact that it is associated with a predicted probability distribution function by calling it a *predicted variance*. It is also conventional to define the *standard deviation* as the square root of σ^2. Recall that the variance is in some sense a typical value of the squared deviation from the mean. Therefore, σ represents a typical value for the deviation itself, hence the name "standard deviation."

Now if we carry out the summation indicated in Eq. (3-17) for the specific case of $P(x)$ given by the binomial distribution, the following result is obtained:

$$\sigma^2 = np(1 - p) \qquad (3\text{-}18)$$

Because $\bar{x} = np$, we can also write

$$\sigma^2 = \bar{x}(1 - p) \qquad (3\text{-}19)$$

$$\sigma = \sqrt{\bar{x}(1 - p)} \qquad (3\text{-}20)$$

We now have an expression that gives an immediate prediction of the amount of fluctuation inherent in a given binomial distribution in terms of the basic parameters of the distribution, n and p, where $\bar{x} = np$.

To return to the example of rolling a die given earlier, we defined success in such a way that $p = \frac{2}{3}$. We also assumed 10 rolls of the die for each measurement so that $n = 10$. For this example, the predicted mean number of successes is 6.67 and we can proceed to calculate the predicted variance

$$\sigma^2 = np(1 - p) = (10)(0.667)(0.333) = 2.22 \tag{3-21}$$

By taking the square root we get the predicted standard deviation:

$$\sigma = \sqrt{\sigma^2} = \sqrt{2.22} = 1.49 \tag{3-22}$$

The significance of the standard deviation is illustrated in Fig. 3-4. The mean value of the distribution is shown as the dashed line, and one value of the standard deviation is shown on either side of this mean. Because σ is a typical value for the difference between a given measurement and the true value of the mean, wide distributions will have large values for σ and narrow distributions will correspond to small values. The plot illustrates that the association of σ with the width of the distribution is not inconsistent with the example shown in Fig. 3-4.

B. The Poisson Distribution

Many categories of binary processes can be characterized by a low probability of success for each individual trial. Included are most nuclear counting experiments in which large numbers of radioactive nuclei make up the sample or number of trials, whereas a relatively small fraction of these give rise to recorded counts. Similarly, in a nuclear beam experiment, many nuclear particles from an accelerator might strike a target for every one recorded reaction product. Under these conditions the approximation that $p \ll 1$ will hold and some mathematical simplifications can be applied to the binomial distribution. It can be shown[2] that in this limit the binomial distribution reduces to the form

$$P(x) = \frac{(pn)^x e^{-pn}}{x!} \tag{3-23}$$

Because $pn = \bar{x}$ holds for this distribution as well as for the parent binomial distribution,

$$\boxed{P(x) = \frac{(\bar{x})^x e^{-\bar{x}}}{x!}} \tag{3-24}$$

which is now the familiar form of the Poisson distribution.

Recall that the binomial distribution requires values for two parameters: the number of trials n and the individual success probability p. We note from Eq. (3.24) that a significant simplification has occurred in deriving the Poisson distribution—only one parameter is required, which is the product of n and p. This is a very useful simplification because now we need only know the mean value of the distribution in order to reconstruct its amplitude at all other values of the argument. That is a great help for processes in which we can in some way measure or estimate the mean value, but for which we have no idea of either the individual probability or the size of the sample. Such is usually the case in nuclear measurements.

Some properties of the Poisson distribution follow directly. First, it is also a normalized distribution, or

$$\sum_{x=0}^{n} P(x) = 1 \tag{3-25}$$

We can also calculate the first moment or mean value of the distribution:

$$\bar{x} = \sum_{x=0}^{n} xP(x) = pn \tag{3-26}$$

which is the intuitively obvious result also obtained for the binomial distribution. The predicted variance of the distribution, however, differs from that of the binomial and can be evaluated from our prior definition

$$\sigma^2 \equiv \sum_{x=0}^{n} (x - \bar{x})^2 P(x) = pn \tag{3-27}$$

or noting the result from Eq. (3-26)

$$\sigma^2 = \bar{x} \tag{3-28}$$

The predicted standard deviation is just the square root of the predicted variance, or

$$\sigma = \sqrt{\bar{x}} \tag{3-29}$$

Thus, we see that the predicted standard deviation of any Poisson distribution is just the square root of the mean value that characterizes that same distribution. Note that the corresponding result obtained earlier for the binomial distribution [Eq. (3-20)] reduces to the above result in the limit of $p \ll 1$ already incorporated into the Poisson assumptions.

We again illustrate with an example. Suppose we randomly select a group of 1000 people and define our measurement as counting the number of current birthdays found among all members of that group. The measurement then consists of 1000 trials, each of which is a success only if a particular individual has his or her birthday today. If we assume a random distribution of birthdays, then the probability of success p is equal to $1/365$. Because p is much less than one in this example, we can immediately turn to the Poisson distribution to evaluate the probability distribution function that will describe the expected results from many such samplings of 1000 people. Thus, for our example,

$$p = \frac{1}{365} = 0.00274 \qquad \bar{x} = pn = 2.74$$

$$n = 1000 \qquad\qquad \sigma = \sqrt{\bar{x}} = 1.66$$

$$P(x) = \frac{\bar{x}^x e^{-\bar{x}}}{x!} = \frac{(2.74)^x e^{-2.74}}{x!}$$

x	$P(x)$
0	0.064
1	0.177
2	0.242
3	0.221
4	0.152
5	0.083
6	0.038
7	0.014
⋮	⋮

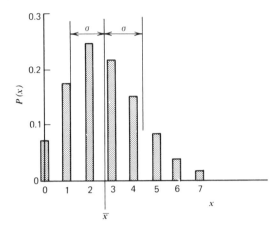

Figure 3-5   The Poisson distribution for a mean value $\bar{x} = 2.74$.

Recall that $P(x)$ gives the predicted probability that exactly x birthdays will be observed from a random sampling of 1000 people. The numerical values are plotted in Fig. 3-5 and show that $x = 2$ is the most probable result. The mean value of 2.74 is also shown in the figure, together with one value of the standard deviation of 1.66 on either side of the mean. The distribution is roughly centered about the mean value, although considerable asymmetry is evident for this low value of the mean. Again the size of the standard deviation gives some indication of the width of the distribution or the amount of scatter predicted by the distribution.

C. The Gaussian or Normal Distribution

The Poisson distribution holds as a mathematical simplification to the binomial distribution in the limit $p \ll 1$. If, in addition, the mean value of the distribution is large (say greater than 20), additional simplifications can generally be carried out[2] which lead to the *Gaussian* distribution:

$$P(x) = \frac{1}{\sqrt{2\pi\bar{x}}} \exp\left(-\frac{(x - \bar{x})^2}{2\bar{x}} \right)$$

(3-30)

This is again a pointwise distribution function defined only for integer values of x. It shares the following properties with the Poisson distribution:

1. It is normalized: $\sum\limits_{x=0}^{\infty} P(x) = 1$.

2. The distribution is characterized by a single parameter $\bar{x}$, which is given by the product np.

3. The predicted variance σ^2 as defined in Eq. (3-17) is again equal to the mean value $\bar{x}$.

We can again illustrate an example of a physical situation in which the Gaussian distribution is applicable. Suppose we return to the previous example of counting birthdays out of a group of randomly selected individuals, but now consider a much larger group of 10,000 people. For this example, $p = \frac{1}{365}$ and $n = 10,000$, so the predicted

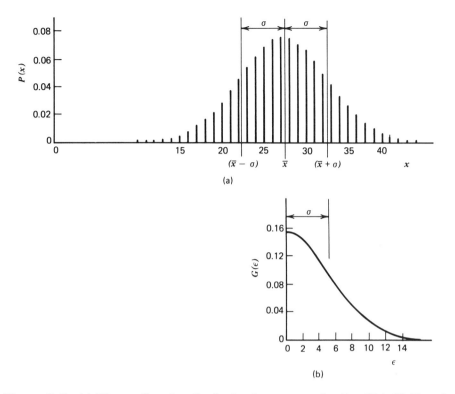

Figure 3-6 (a) Discrete Gaussian distribution for a mean value $\bar{x} = 27.4$. (b) Plot of the corresponding continuous form of the Gaussian.

mean value of the distribution $\bar{x} = np = 27.4$. Because the predicted mean is larger than 20, we can turn to the Gaussian distribution for the predicted distribution of the results of many measurements, each of which consists of counting the number of birthdays found in a different group of 10,000 people. The predicted probability of observing a specific count x is then given by

$$P(x) = \frac{1}{\sqrt{2\pi \cdot 27.4}} \exp\left(-\frac{(x - 27.4)^2}{2 \cdot 27.4}\right) \tag{3-31}$$

and the predicted standard deviation for the example is

$$\sigma = \sqrt{\bar{x}} = \sqrt{27.4} = 5.23 \tag{3-32}$$

These results are shown graphically in Fig. 3-6a.

Two important observations can be made at this point about the Gaussian distribution:

1. The distribution is symmetric about the mean value $\bar{x}$. Therefore, $P(x)$ depends only on the absolute value of the deviation of any value x from the mean, defined as $\epsilon \equiv |x - \bar{x}|$.

2. Because the mean value $\bar{x}$ is large, values of $P(x)$ for adjacent values of x are not greatly different from each other. In other words, the distribution is slowly varying.

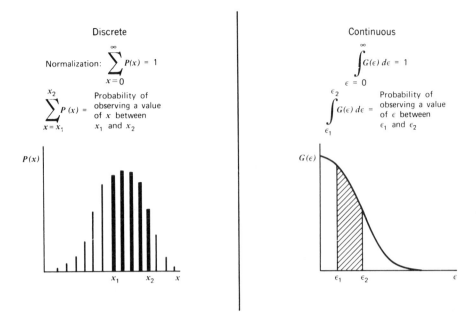

Figure 3-7 Comparison of the discrete and continuous forms of the Gaussian distribution.

These two observations suggest a recasting of the distribution as an explicit function of the deviation ϵ (rather than of x) and as a continuous function (rather than a pointwise discrete function). These changes are accomplished by rewriting the Gaussian distribution as

$$G(\epsilon) = \sqrt{\frac{2}{\pi \bar{x}}}\, e^{-\epsilon^2/2\bar{x}} \qquad (3\text{-}33)$$

where $G(\epsilon)\, d\epsilon$ is now defined as the differential probability of observing a deviation in $d\epsilon$ about ϵ. Comparing Eq. (3-33) with Eq. (3-30), we note a factor of 2 that has entered in $G(\epsilon)$ because there are two values of x for every value of the deviation ϵ.

Figure 3-6b shows the continuous form of the Gaussian distribution for the same example chosen to illustrate the discrete case. Comparing Fig. 3-6a and 3-6b, the scale factors for each abscissa are the same but the origin for Fig. 3-6b has been shifted to illustrate that a value of zero for the deviation ϵ corresponds to the position of the mean value $\bar{x}$ on Fig. 3-6a. If a factor of 2 difference in the relative ordinate scales is included as shown, then the continuous distribution $G(\epsilon)$ represents the smooth curve that connects the pointwise values plotted in Fig. 3-6a.

Because we are now dealing with a continuous function, we must redefine some properties of the distribution as shown in Fig. 3-7. It should be particularly noted that quantities of physical interest now involve integrals of the distribution between set limits, or *areas* under the curve, rather than sums of discrete values.

Equation (3-33) can be rewritten in a more general form by incorporating several observations. We have already seen that the standard deviation σ of a Gaussian distribution is given by $\sigma = \sqrt{\bar{x}}$, or $\bar{x} = \sigma^2$. With this substitution in Eq. (3-33), the value of the

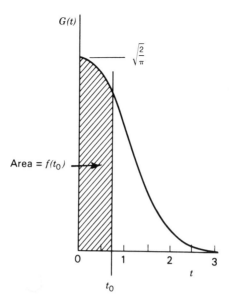

Figure 3-8 A plot of the general Gaussian curve.

exponential factor now depends only on the ratio of ϵ to σ. Formally defining this ratio as

$$t \equiv \frac{\epsilon}{\sigma}$$

the Gaussian distribution can be rewritten in terms of this new variable t:

$$G(t) = G(\epsilon)\frac{d\epsilon}{dt} = G(\epsilon)\sigma$$

$$G(t) = \sqrt{\frac{2}{\pi}}\, e^{-t^2/2} \tag{3-34}$$

where $0 \le t \le \infty$. We now have a universal form of the Gaussian distribution, shown in Fig. 3-8, that is valid for all values of the mean $\bar{x}$. Recall that t is just the observed deviation $\epsilon \equiv |x - \bar{x}|$ normalized in units of the standard deviation σ.

From the definitions illustrated in Fig. 3-7, the probability that a typical normalized deviation t predicted by a Gaussian distribution will be less than a specific value t_0 is given by the integral

$$\int_0^{t_0} G(t)\, dt \equiv f(t_0)$$

Tabulated values of this function can be found in most collections of statistical tables. Some selected values are given in Table 3-4. The value of $f(t_0)$ gives the probability that a random sample from a Gaussian distribution will show a normalized deviation t ($\equiv \epsilon/\sigma$) that is less than the assumed value t_0. For example, we can conclude that about 68% of all samples will deviate from the true mean by less than one value of the standard deviation.

TABLE 3-4 Probability of Occurrence of Given
Deviations Predicted by the Gaussian Distribution

ϵ_0	t_0	$f(t_0)$
0	0	0
0.674σ	0.674	0.500
σ	1.00	0.683
$1.64\ \sigma$	1.64	0.900
$1.96\ \sigma$	1.96	0.950
$2.58\ \sigma$	2.58	0.990
$3.00\ \sigma$	3.00	0.997

III. APPLICATIONS OF STATISTICAL MODELS

The first two sections of this chapter dealt with independent topics: the organization of experimental data in Section I, and the structure of certain statistical models in Section II. The practical uses of statistical analysis are now illustrated by bringing together these two separate topics.

There are two major applications of counting statistics in nuclear measurements. The first of these, "Application A," involves the use of statistical analysis to determine whether a set of multiple measurements of the same physical quantity shows an amount of internal fluctuation that is consistent with statistical predictions. The usual motivation here is to determine whether a particular counting system is functioning normally. Although this is a useful application, a far more valuable contribution of counting statistics arises in situations in which we have only a single experimental measurement. In "Application B" we examine the methods available to make a prediction about the uncertainty one should associate with that single measurement to account for the unavoidable effects of statistical fluctuations.

Application A: Checkout of the Counting System to See if Observed Fluctuations Are Consistent with Expected Statistical Fluctuation

A common quality control procedure in many counting laboratories is to periodically (perhaps once a month) record a series of 20–50 successive counts from the detector system while keeping all experimental conditions as constant as possible. By applying the analytical procedures to be described here, it can be determined whether the internal fluctuation shown by these multiple measurements is consistent with the amount of fluctuation expected if statistical fluctuations were the only origin. In this way abnormal amounts of fluctuation can be detected which could indicate malfunctioning of some portion of the counting system.

Figure 3-9 shows the chain of events that characterizes this application of counting statistics. Properties of the experimental data are confined to the left half of the figure, whereas on the right-hand side are listed properties of an appropriate statistical model. We start in the upper left corner with the collection of N independent measurements of the same physical quantity. These might be, for example, successive 1 min counts from a detector. Using the methods outlined in Section I, we can characterize the data in several ways. The data distribution function $F(x)$ as defined in Eq. (3-3) can be compiled. From this distribution, the mean value $\bar{x}_e$ and the sample variance s^2 can be computed by the formulas given in Eqs. (3-5) and (3-10). Recall that the mean value $\bar{x}_e$ gives the value

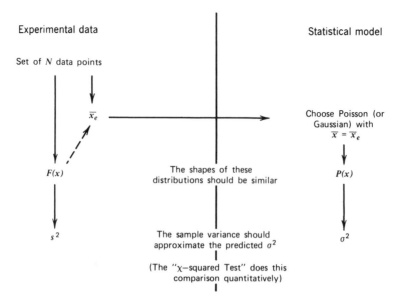

Figure 3-9 An illustration of Application A of counting statistics—the inspection of a set of data for consistency with a statistical model.

about which the distribution is centered, whereas the sample variance s^2 is a quantitative measure of the amount of fluctuation present in the collection of data.

We now are faced with the task of matching these experimental data with an appropriate statistical model. Almost universally we will want to match to either a Poisson or Gaussian distribution (depending on how large the mean value is), either of which is fully specified by its own mean value $\bar{x}$. What should we choose for $\bar{x}$? We would be rather foolish if we chose any value other than $\bar{x}_e$, which is our only estimate of the mean value for the distribution from which the data have been drawn. Setting $\bar{x} = \bar{x}_e$ then provides the bridge from left to right in the figure, so that we now have a fully specified statistical model. If we let $P(x)$ represent the Poisson or Gaussian distribution with $\bar{x} = \bar{x}_e$, then the measured data distribution function $F(x)$ should be an approximation to $P(x)$ provided the statistical model accurately describes the distribution from which the data have arisen. One method of carrying out a comparison at this level is simply to make a superimposed plot of $F(x)$ and $P(x)$ and then to compare the shape and amplitude of the two distributions.

But such a comparison of two functions is, as yet, only qualitative. It is desirable to extract a single parameter from each distribution so that they can be compared quantitatively. The most fundamental parameter is the mean value, but these have already been matched and are the same by definition. A second parameter of each distribution is the variance, and we can carry out the desired quantitative comparison by noting the predicted variance σ^2 of the statistical model and comparing with the measured sample variance s^2 of the collection of data. If the data are actually characterized by the statistical model and show a degree of internal fluctuation that is consistent with statistical prediction, these two variance values should be about the same.

To illustrate the direct comparison of the data distribution function with the predicted probability distribution function, we return to the example of data given in Table 3-1. In Fig. 3-10 the data distribution function has been replotted as the solid

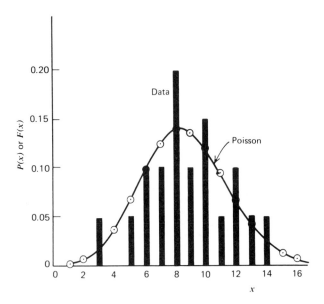

Figure 3-10   A direct comparison of experimental data (from Table 3-1) with predictions of a statistical model (the Poisson distribution for $\bar{x} = 8.8$).

vertical bars. The mean value for these data was calculated to be $\bar{x}_e = 8.8$, so the transition to the appropriate statistical model is made by assuming its mean value to be $\bar{x} = 8.8$. Because the mean value is not large, we are prohibited from using the Gaussian distribution and we therefore use the Poisson as the assumed statistical model. The points on Fig. 3-10 are the values of the predicted distribution function of the Poisson distribution for a mean value of 8.8. Because the Poisson is defined only for discrete values of x, the continuous curve is drawn only to connect the points for visual reference.

At this point a comparison of the two distributions is difficult. Because relatively little experimental data was gathered (20 measurements) the value of $F(x)$ at each point is subject to rather large fluctuations. One would expect that, if more data were gathered, the fluctuations would diminish and the data distribution function $F(x)$ would adhere more and more closely to the predicted probability distribution function $P(x)$, provided the data are indeed a true sample from the predicted statistical model. From Fig. 3-10 we can only say that the experimental data are not grossly at variance with the prediction.

To take the comparison one step further, we would now like to compare the value of the sample variance and the predicted variance from the statistical model. The sample variance calculated from Eq. (3-8) for the same set of data is found to be

$$s^2 = 7.36$$

Because the assumed statistical model is the Poisson distribution, the predicted variance is given by

$$\sigma^2 = \bar{x} = 8.80$$

These two results show that there is less fluctuation in the data than would be predicted if the data were a perfect sample from a Poisson distribution of the same mean. With a limited sample size, however, one would not expect these two parameters to be precisely

TABLE 3-5 Portion of a Chi-Squared Distribution Table

Statistical Degrees of Freedom	Number of Measurements N	$p = 0.8$	0.7	0.6	0.5
18	19	12.85	14.44	15.89	17.33
19	20	13.72	15.35	16.85	18.33
20	21	14.58	16.26	17.80	19.34

the same and a more quantitative test is required to determine whether the observed difference is really significant. This function is provided by the "chi-squared test."

Chi-squared is simply another parameter of the experimental data distribution and is defined as

$$\chi^2 \equiv \frac{1}{\bar{x}_e} \sum_{i=1}^{N} (x_i - \bar{x}_e)^2 \tag{3-35}$$

where the summation is taken over each individual data point x_i. Chi-squared is closely related to the sample variance and the two are related by

$$\chi^2 = \frac{(N-1)s^2}{\bar{x}_e} \tag{3-36}$$

Now if the amount of fluctuation present in the data is closely modeled by the Poisson distribution, then $s^2 \cong \sigma^2$. But we know that for the Poisson distribution, $\sigma^2 = \bar{x}$. Furthermore, we have chosen $\bar{x}$ to be equal to $\bar{x}_e$. Therefore, the degree to which the ratio $s^2/\bar{x}_e$ deviates from unity is a direct measure of the extent to which the observed sample variance differs from the predicted variance. Now referring to Eq. (3-36), the degree to which χ^2 differs from $N - 1$ is a corresponding measure of the departure of the data from predictions of the Poisson distribution. Chi-squared distribution tables may be found (e.g., Ref. 3) which are generally cast in the form shown in Table 3-5. The column on the left indicates the number of statistical degrees of freedom in the system. (This is one less than the number of independent measurements used to derive the value of χ^2 because $\bar{x}_e$ has been calculated from the same set of data.) Each column in the table is headed by a specific value of p, defined as the probability that a random sample from a true Poisson distribution would have a larger value of χ^2 than the specific value shown in the table. Very low probabilities (say less than 0.02) indicate abnormally large fluctuations in the data, whereas very high probabilities (greater than 0.98) indicate abnormally small fluctuations. A perfect fit to the Poisson distribution for large samples would yield a probability of 0.50, whereas the somewhat arbitrary limits listed above indicate situations in which the counting system may be displaying either abnormally large fluctuations (which is the usual type of malfunction) or data that are too regular and show abnormally small fluctuations. Figure 3-11 gives a plot of the χ^2 distribution for a wider range of the parameters involved.

For the illustrative example given above, we calculate a χ^2 value of 15.89. From Table 3-5 for $N = 20$ we find (by interpolation) a value of $p = 0.66$. Because that probability is neither very large nor very small, we would conclude that the equipment used to generate the set of numbers originally shown does not give rise to abnormal fluctuations.

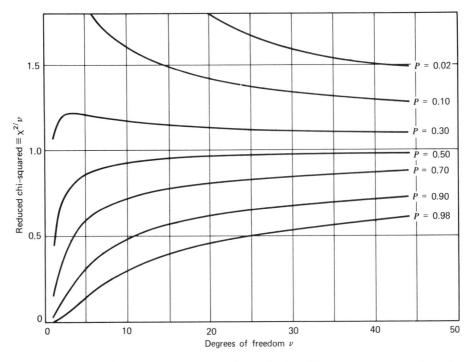

Figure 3-11 A plot of the chi-squared distribution. For each curve, p gives the probability that a random sample of N numbers from a true Poisson distribution would have a larger value of χ^2/ν than that of the ordinate. For data for which the experimental mean is used to calculate χ^2, the number of degrees of freedom $\nu = N - 1$.

Application B: Estimation of the Precision of a Single Measurement

A more valuable application of counting statistics applies to the case in which we have only a single measurement of a particular quantity and wish to associate a given degree of uncertainty with that measurement. To state the objective in another way, we would like to have some estimate of the sample variance to be expected if we were to repeat the measurement many times. The square root of the sample variance should be a measure of the typical deviation of any one measurement from the true mean value and thus will serve as a single index of the degree of precision one should associate with a typical measurement from that set. Because we have only a single measurement, however, the sample variance cannot be calculated directly but must be estimated by analogy with an appropriate statistical model.

The process is illustrated in Fig. 3-12. Again, the left half of the figure deals only with experimental data, whereas the right half deals only with the statistical model. We start in the upper-left corner with a single measurement, x. If we make the assumption that the measurement has been drawn from a population whose theoretical distribution function is predicted by either a Poisson or Gaussian distribution, then we must match an appropriate theoretical distribution to the available data. For either model we must start with a value for the mean $\bar{x}$ of the distribution. Because the value of our single measurement x is the only information we have about the theoretical distribution from which it has been drawn, we have no real choice other than to assume that the mean of the distribution is equal to the single measurement, of $\bar{x} = x$. Having now obtained an

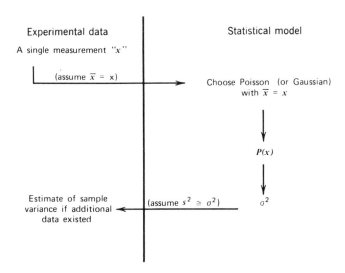

Figure 3-12 An illustration of Application B of counting statistics—prediction of the precision to be associated with a single measurement.

assumed value for $\bar{x}$, the entire predicted probability distribution function $P(x)$ is defined for all values of x. We can also immediately find a value for the predicted variance σ^2 of that distribution. We can then use the association that, if the data are drawn from the same distribution, an estimate of the sample variance s^2 of a collection of such data should be given by σ^2. Through this process we have therefore obtained an estimate for the sample variance of a repeated set of measurements that do not exist but that represent the expected results if the single measurement were to be repeated many times.

The conclusion we reach can then be stated as follows:

The expected sample variance	$s^2 \cong \sigma^2$	of the statistical model from which we think the measurement x is drawn
	$= \bar{x}$	provided the model is either Poisson or Gaussian
	$\cong x$	because x is our only measurement on which to base an estimate of $\bar{x}$

We therefore conclude that

$$\sqrt{s^2} \cong \sigma = \sqrt{x}$$

is our best estimate of the deviation from the true mean which should typify our single measurement x.

This conclusion can be stated somewhat more quantitatively provided the assumed probability distribution function is a Gaussian (x is large). Then the range of values

TABLE 3-6   Examples of Error Intervals for a Single
Measurement $x = 100$

	Interval	Probability that the True Mean $\bar{x}$ Is Included
$x \pm 0.67\sigma$	93.3–106.7	50%
$x \pm \sigma$	90 –110	68%
$x \pm 1.64\sigma$	83.6–116.4	90%
$x \pm 2.58\sigma$	74.2–125.8	99%

$x \pm \sigma$ or $x \pm \sqrt{x}$ will contain the true mean $\bar{x}$ with 68% probability. This conclusion follows directly from earlier statements about the shape of the Gaussian curve.[†] It is conventional to quote the uncertainty or "error" of a single measurement as simply one value of the standard deviation σ. If we quote a larger uncertainty, then the probability of including the true mean within the quoted interval is increased, and vice versa.

To illustrate, assume we have a single measurement $x = 100$. Then

$$\sigma = \sqrt{x} = \sqrt{100} = 10$$

Because our best estimate of the mean value of the distribution from which this measurement was drawn (the measurement itself) is large, we can assume that the parent distribution is a Gaussian. From the shape of the Gaussian curve (see Table 3-4) we can then construct Table 3-6 for the specific example. The table gives various options available in quoting the uncertainty to be associated with our single measurement. The conventional choice is to quote the measurement plus or minus one value of the standard deviation σ, or 100 ± 10. This interval is expected to contain the true mean value $\bar{x}$ with a probability of 68%. If we wish to increase the probability that the true mean is included, we can do so only by expanding the interval or error associated with the measurement. For example, to achieve a 99% probability that the true mean is included, the interval must be expanded to 2.58σ, or the range 100 ± 25.8 for our example. Unless otherwise stated, the errors quoted with a particular nuclear measurement normally represent one standard deviation.

The *fractional standard deviation*, defined as σ/x, of a simple counting measurement is given by $\sqrt{x}/x$, or $1/\sqrt{x}$. Thus, the total number of recorded counts x completely determines the fractional error to be associated with that measurement. If 100 counts are recorded, the fractional standard deviation is 10%, whereas it can be reduced to 1% only by increasing the total counts recorded to 10,000. For events occurring at a constant rate, this relation implies that the time required to achieve a given fractional error will increase as the inverse square of the desired statistical precision.

When a set of measurements is presented graphically, the estimated errors associated with each measurement are often also displayed on the same graph. Figure 3-13 gives a hypothetical set of measurements of a quantity x as a function of some other variable or parameter z. The measured data are presented as points, whereas the error associated with each point is indicated by the length of the "error bar" drawn around each point. It is conventional to show the length of the error bar equal to one value of σ on either side of

[†] Technically, we can only say that the interval $\bar{x} \pm \sqrt{x}$ has a 68% probability of containing our single measurement x. However, since the mean value has already been assumed to be large, we do not expect a large difference between $\bar{x}$ and any typical measurement x. Thus, we can interchange $\bar{x}$ and x in the statement above without seriously affecting the general conclusions.

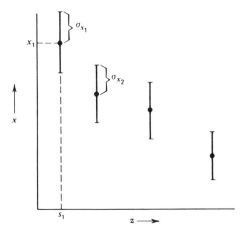

Figure 3-13 A graphical display of error bars associated with experimental data.

the point, or the total length of the error bar equal to 2σ. Under these conditions, if one were to attempt a fit of an assumed functional behavior $x = f(z)$, the fitted function should at best pass through 68% (or roughly two-thirds) of all the error bars associated with the data.

Caution

All the conclusions we have drawn apply *only* to a measurement of a number of successes (number of heads in coin tossing, number of birthdays, etc.). In radioactive decay or nuclear counting, we may directly apply $\sigma = \sqrt{x}$ *only* if x represents a counted number of successes, that is, a number of events over a given observation time recorded from a detector. The vast majority of mistakes made in the use of counting statistics results from the misapplication of the above relation.

One *cannot* associate the standard deviation σ with the square root of any quantity that is not a directly measured number of counts. For example, the association does *not* apply to:

1. counting *rates*,
2. *sums* or *differences* of counts,
3. *averages* of independent counts,
4. any *derived* quantity.

In all these cases the quantity is *calculated* as a function of the number of counts recorded in a given experiment. The error to be associated with that quantity must then be calculated according to the methods outlined in the next section.

IV. ERROR PROPAGATION

In a typical nuclear measurement, one is seldom interested in the unprocessed data consisting of the number of counts over a particular interval. More often the data are processed through multiplication, addition, or other functional manipulation to arrive at a derived number of more direct interest. We must then be concerned with the manner in which the error associated with the original number of counts propagates through these calculations and is reflected as a corresponding uncertainty in the derived quantity. It can be shown[3] that if the errors are individually small and symmetric about zero, a general result can be obtained for the expected error to be associated with any quantity that is calculated as a function of any number of independent variables. If $x, y, z, \ldots$ are directly measured counts or related variables for which we know $\sigma_x, \sigma_y, \sigma_z, \ldots$, then the standard deviation for any quantity u derived from these counts can be calculated from

$$\sigma_u^2 = \left(\frac{\partial u}{\partial x}\right)^2 \sigma_x^2 + \left(\frac{\partial u}{\partial y}\right)^2 \sigma_y^2 + \left(\frac{\partial u}{\partial z}\right)^2 \sigma_z^2 + \cdots \qquad (3\text{-}37)$$

where $u = u(x, y, z, \ldots)$ represents the derived quantity. Equation (3-37) is generally known as the *error propagation formula* and is applicable to almost all situations in nuclear measurements. The variables $x, y, z, \ldots$, however, must be chosen so that they are truly independent in order to avoid the effects of correlation. The same specific count should not contribute to the value of more than one such variable. The use of Eq. (3-37) can be illustrated by application to some simple cases.

Case 1. Sums or Differences of Counts

If we define

$$u = x + y \quad \text{or} \quad u = x - y$$

then

$$\frac{\partial u}{\partial x} = 1 \quad \text{and} \quad \frac{\partial u}{\partial y} = \pm 1$$

Application of Eq. (3-37) yields

$$\sigma_u^2 = (1)^2 \sigma_x^2 + (\pm 1)^2 \sigma_y^2$$

or

$$\boxed{\sigma_u = \sqrt{\sigma_x^2 + \sigma_y^2}} \qquad (3\text{-}38)$$

A common application of this case arises when counts due to a radioactive source must be corrected by subtracting an appropriate background count. If we assume equal counting times, then

net counts = total counts − background counts

or

$$u = x - y$$

Because both x and y are directly measured numbers of counts (or successes), the expected standard deviation of each is known to be its own square root. The object is to deduce the expected standard deviation of the net counts, a derived number. Because a simple difference is involved, the answer will be given by Eq. (3-38).

To illustrate by example, suppose we have recorded the following data for equal counting times

$$\text{total counts} \quad = x = 1071$$
$$\text{background counts} = y = \underline{521}$$

then

$$\text{net counts} \quad\quad = u = 550$$

We know a priori

$$\sigma_x = \sqrt{x} = \sqrt{1071}$$
$$\sigma_y = \sqrt{y} = \sqrt{521}$$

Thus

$$\sigma_u = \sqrt{\sigma_x^2 + \sigma_y^2} = \sqrt{x + y} = \sqrt{1592} = 39.9$$

We would then quote the result plus or minus one standard deviation as

$$\text{net counts} = 550 \pm 39.9$$

Case 2. Multiplication or Division by a Constant

If we define

$$u = Ax$$

where A is a constant (no associated uncertainty), then

$$\frac{\partial u}{\partial x} = A$$

and application of Eq. (3-37) gives

$$\boxed{\sigma_u = A\sigma_x} \tag{3-39}$$

Similarly, if

$$v = \frac{x}{B}$$

where B is also a constant, then

$$\boxed{\sigma_v = \frac{\sigma_x}{B}} \tag{3-40}$$

Note that, in either case, the final "fractional error" (σ_u/u or σ_v/v) is the same as the original fractional error (σ_x/x). As we would expect intuitively, multiplying or dividing a value by a constant does not change its relative error.

A familiar example of the above case is the calculation of a counting rate. If x counts are recorded over a time t, then

$$\text{counting rate} \equiv r = \frac{x}{t}.$$

The usual assumption is that the time is measured with very small uncertainty, so that t can be considered a constant. Then Eq. (3-40) can be used to calculate the expected standard deviation in r corresponding to the known standard deviation in the number of counts x.

As an example, suppose

$$x = 1120 \text{ counts and } t = 5 \text{ s}$$

Then

$$r = \frac{1120}{5 \text{ s}} = 224 \text{ s}^{-1}$$

The associated standard deviation is

$$\sigma_r = \frac{\sigma_x}{t} = \frac{\sqrt{1120}}{5 \text{ s}} = 6.7 \text{ s}^{-1}$$

Therefore, the counting rate is

$$r = 224 \pm 6.7 \text{ counts/s}$$

Case 3. Multiplication or Division of Counts

For the case

$$u = xy, \qquad \frac{\partial u}{\partial x} = y \qquad \frac{\partial u}{\partial y} = x$$

$$\sigma_u^2 = y^2 \sigma_x^2 + x^2 \sigma_y^2$$

Dividing both sides by $u^2 = x^2 y^2$

$$\boxed{\left(\frac{\sigma_u}{u}\right)^2 = \left(\frac{\sigma_x}{x}\right)^2 + \left(\frac{\sigma_y}{y}\right)^2} \tag{3-41}$$

Similarly, if

$$u = \frac{x}{y}, \qquad \frac{\partial u}{\partial x} = \frac{1}{y} \qquad \frac{\partial u}{\partial y} = -\frac{x}{y^2}$$

$$\sigma_u^2 = \left(\frac{1}{y}\right)^2 \sigma_x^2 + \left(-\frac{x}{y^2}\right)^2 \sigma_y^2$$

Again, dividing both sides by $u^2 = x^2/y^2$

$$\boxed{\left(\frac{\sigma_u}{u}\right)^2 = \left(\frac{\sigma_x}{x}\right)^2 + \left(\frac{\sigma_y}{y}\right)^2}$$ (3-41')

Thus, for either $u = xy$ or $u = x/y$, the *fractional errors* in x and y (σ_x/x and σ_y/y) combine in quadrature sum to give the *fractional error* in u.

As an example, suppose we wish to calculate the ratio of two source activities from independent counts taken for equal counting times (background is neglected). Assume

$$\text{counts from source } 1 \equiv N_1 = 16{,}265$$

$$\text{counts from source } 2 \equiv N_2 = 8192$$

$$\text{activity ratio:} \quad R \equiv \frac{N_1}{N_2} = \frac{16{,}265}{8192} = 1.985$$

From Eq. (3-41')

$$\left(\frac{\sigma_R}{R}\right)^2 = \left(\frac{\sigma_{N_1}}{N_1}\right)^2 + \left(\frac{\sigma_{N_2}}{N_2}\right)^2 = \frac{N_1}{N_1^2} + \frac{N_2}{N_2^2}$$

$$= 1.835 \times 10^{-4}$$

$$\frac{\sigma_R}{R} = 0.0135$$

and multiplying by the value of R

$$\sigma_R = 0.027$$

Therefore, the reported result would be

$$R = 1.985 \pm 0.027$$

Case 4. Mean Value of Multiple Independent Counts

Suppose we have recorded N repeated counts from the same source for equal counting times. Let the results of these multiple counts be designated $x_1, x_2, \ldots, x_N$ and their sum be Σ. Then

$$\Sigma = x_1 + x_2 + \cdots + x_N$$

If we formally apply the error propagation formula [Eq. (3-37)] to find the expected error in Σ, we find $\partial\Sigma/\partial x_i = 1$ for all independent counts x_i, and therefore

$$\sigma_\Sigma^2 = \sigma_{x_1}^2 + \sigma_{x_2}^2 + \cdots + \sigma_{x_N}^2$$

But because $\sigma_{x_i} = \sqrt{x_i}$ for each independent count,

$$\sigma_\Sigma^2 = x_1 + x_2 + \cdots + x_N = \Sigma$$

$$\sigma_\Sigma = \sqrt{\Sigma}$$ (3-42)

This result shows that the standard deviation expected for the sum of all the counts is the same as if the measurement had been carried out by performing a single count, extending over the entire period represented by all the independent counts.

Now if we proceed to calculate a mean value from these N independent measurements,

$$\bar{x} = \frac{\Sigma}{N} \tag{3-43}$$

Equation (3-43) is an example of dividing an error-associated quantity (Σ) by a constant (N). Therefore, Eq. (3-40) applies and the expected standard deviation of this mean value is given by

$$\sigma_{\bar{x}} = \frac{\sigma_{\Sigma}}{N} = \frac{\sqrt{\Sigma}}{N} = \frac{\sqrt{N\bar{x}}}{N}$$

$$\boxed{\sigma_{\bar{x}} = \sqrt{\frac{\bar{x}}{N}}} \tag{3-44}$$

Note that the expected standard deviation of any single measurement x_i is

$$\sigma_{x_i} = \sqrt{x_i}$$

Because any typical count will not differ greatly from the mean, $x_i \cong \bar{x}$, and we therefore conclude that the mean value based on N independent counts will have an expected error that is smaller by a factor $\sqrt{N}$ compared with any single measurement on which the mean is based. A general conclusion is that, if we wish to improve the statistical precision of a given measurement by a factor of 2, we must invest four times the initial counting time.

Case 5. Combination of Independent Measurements with Unequal Errors

If N independent measurements of the same quantity have been carried out and they do not all have nearly the same associated precision, then a simple average (as discussed in Case 4) no longer is the optimal way to calculate a single "best value." We instead want to give more weight to those measurements with small values for σ_{x_i} (the standard deviation associated with x_i) and less weight to measurements for which this estimated error is large.

Let each individual measurement x_i be given a weighting factor a_i and the best value $\langle x \rangle$ computed from the linear combination

$$\langle x \rangle = \frac{\sum\limits_{i=1}^{N} a_i x_i}{\sum\limits_{i=1}^{N} a_i} \tag{3-45}$$

We now seek a criterion by which the weighting factors a_i should be chosen in order to minimize the expected error in $\langle x \rangle$.

For brevity, we write

$$\alpha \equiv \sum_{i=1}^{N} a_i$$

so that

$$\langle x \rangle = \frac{1}{\alpha} \sum_{i=1}^{N} a_i x_i$$

Now apply the error propagation formula [Eq. (3-37)] to this case:

$$\sigma_{\langle x \rangle}^2 = \sum_{i=1}^{N} \left(\frac{\partial \langle x \rangle}{\partial x_i} \right)^2 \sigma_{x_i}^2$$

$$= \sum_{i=1}^{N} \left(\frac{a_i}{\alpha} \right)^2 \sigma_{x_i}^2$$

$$= \frac{1}{\alpha^2} \sum_{i=1}^{N} a_i^2 \sigma_{x_i}^2$$

$$\sigma_{\langle x \rangle}^2 = \frac{\beta}{\alpha^2} \qquad (3\text{-}46)$$

where

$$\beta \equiv \sum_{i=1}^{N} a_i^2 \sigma_{x_i}^2$$

In order to minimize $\sigma_{\langle x \rangle}$, we must minimize $\sigma_{\langle x \rangle}^2$ from Eq. (3-46) with respect to a typical weighting factor a_j:

$$0 = \frac{\partial \sigma_{\langle x \rangle}^2}{\partial a_j} = \frac{\alpha^2 \dfrac{\partial \beta}{\partial a_j} - 2\alpha\beta \dfrac{\partial \alpha}{\partial a_j}}{\alpha^4} \qquad (3\text{-}47)$$

Note that

$$\frac{\partial \alpha}{\partial a_j} = 1 \qquad \frac{\partial \beta}{\partial a_j} = 2 a_j \sigma_{x_j}^2$$

Putting these results into Eq. (3-47), we obtain

$$\frac{1}{\alpha^4} \left(2\alpha^2 a_j \sigma_{x_j}^2 - 2\alpha\beta \right) = 0$$

and solving for a_j, we find

$$a_j = \frac{\beta}{\alpha} \cdot \frac{1}{\sigma_{x_j}^2} \qquad (3\text{-}48)$$

If we choose to normalize the weighting coefficients,

$$\sum_{i=1}^{N} a_i \equiv \alpha = 1$$

$$a_j = \frac{\beta}{\sigma_{x_j}^2}$$

Putting this into the definition of β, we obtain

$$\beta = \sum_{i=1}^{N} a_i^2 \sigma_{x_i}^2 = \sum_{i=1}^{N} \left(\frac{\beta}{\sigma_{x_i}^2} \right)^2 \sigma_{x_i}^2$$

or

$$\beta = \left(\sum_{i=1}^{N} \frac{1}{\sigma_{x_i}^2} \right)^{-1} \tag{3-49}$$

Therefore, the proper choice for the normalized weighting coefficient for x_j is

$$\boxed{a_j = \frac{1}{\sigma_{x_j}^2} \left(\sum_{i=1}^{N} \frac{1}{\sigma_{x_i}^2} \right)^{-1}} \tag{3-50}$$

We therefore see that *each data point should be weighted inversely as the square of its own error.*

Assuming that this optimal weighting is followed, what will be the resultant (minimum) error in $\langle x \rangle$? Because we have chosen $\alpha = 1$ for normalization, Eq. (3-46) becomes

$$\sigma_{\langle x \rangle}^2 = \beta$$

In the case of optimal weighting, β is given by Eq. (3-49). Therefore,

$$\boxed{\frac{1}{\sigma_{\langle x \rangle}^2} = \sum_{i=1}^{N} \frac{1}{\sigma_{x_i}^2}} \tag{3-51}$$

From Eq. (3-51), the expected standard deviation $\sigma_{\langle x \rangle}$ can be calculated from the standard deviations σ_{x_i} associated with each individual measurement.

V. OPTIMIZATION OF COUNTING EXPERIMENTS

The principle of error propagation can be applied in the design of counting experiments to minimize the associated statistical uncertainty. To illustrate, consider the simple case of measurement of the net counting rate from a long-lived radioactive source in the presence of a steady-state background. Define the following:

$S \equiv$ counting rate due to the source alone without background

$B \equiv$ counting rate due to background

The measurement of S is normally carried out by counting the source plus background (at an average rate of $S + B$) for a time T_{S+B} and then counting background alone for a time T_B. The net rate due to the source alone is then

$$S = \frac{N_1}{T_{S+B}} - \frac{N_2}{T_B} \tag{3-52}$$

where N_1 and N_2 are the total counts in each measurement.

Applying the results of error propagation analysis to Eq. (3-52), we obtain

$$\sigma_S = \left[\left(\frac{\sigma_{N_1}}{T_{S+B}} \right)^2 + \left(\frac{\sigma_{N_2}}{T_B} \right)^2 \right]^{1/2}$$

$$\sigma_S = \left(\frac{N_1}{T_{S+B}^2} + \frac{N_2}{T_B^2} \right)^{1/2}$$

$$\sigma_S = \left(\frac{S+B}{T_{S+B}} + \frac{B}{T_B} \right)^{1/2} \tag{3-53}$$

If we now assume that a fixed total time $T = T_{S+B} + T_B$ is available to carry out both measurements, the above uncertainty can be minimized by optimally choosing the fraction of T allocated to T_{S+B} (or T_B). We square Eq. (3-53) and differentiate

$$2\sigma_S \, d\sigma_S = - \frac{S+B}{T_{S+B}^2} dT_{S+B} - \frac{B}{T_B^2} dT_B$$

and set $d\sigma_S = 0$ to find the optimum condition. Also, because T is a constant, $dT_{S+B} + dT_B = 0$. The optimum division of time is then obtained by meeting the condition

$$\left. \frac{T_{S+B}}{T_B} \right|_{\text{opt}} = \sqrt{\frac{S+B}{B}} \tag{3-54}$$

A figure of merit that can be used to characterize this type of counting experiment is the inverse of the total time, or $1/T$, required to determine S to within a given statistical accuracy. If certain parameters of the experiment (such as detector size and pulse acceptance criteria) can be varied, the optimal choice should correspond to maximizing this figure of merit.

In the following analysis, we assume that the optimal division of counting times given by Eq. (3-54) is chosen. Then we can combine Eqs. (3-53) and (3-54) to obtain an expression for the figure of merit in terms of the fractional standard deviation of the source rate, defined as $\epsilon \equiv \sigma_S/S$

$$\frac{1}{T} = \epsilon^2 \frac{S^2}{\left(\sqrt{S+B} + \sqrt{B} \right)^2} \tag{3-55}$$

Equation (3-55) is a useful result that can be applied to analyze the large category of radiation measurements in which a signal rate S is to be measured in the presence of a steady-state background rate B. For example, it predicts the attainable statistical accuracy (in terms of the fractional standard deviation ϵ) when a total time T is available to measure the signal plus background and the background alone. The assumption has been made that this time is subdivided optimally between the two counts. Note that, in common with simple counting measurements, the time required varies as the inverse square of the fractional standard deviation desired for the net signal rate precision. Cutting the predicted statistical error of a measurement in half requires increasing the available time by a factor of 4.

It is instructive to examine two extreme cases in the application of Eq. (3-55). If the source-induced rate is much greater than the background, $S \gg B$ and Eq. (3-55) reduces to

$$\frac{1}{T} \cong \epsilon^2 S \qquad (3\text{-}56)$$

In this limit, the statistical influence of background is negligible. The figure of merit $1/T$ is maximized simply by choosing all experiment parameters to maximize S, or the rate due to the source alone.

The opposite extreme of a small source rate in a much larger background ($S \ll B$) is typical of low-level radioactivity measurements. In this case, Eq. (3-55) reduces to

$$\frac{1}{T} \cong \epsilon^2 \frac{S^2}{4B} \qquad (3\text{-}57)$$

For such applications, the figure of merit is maximized by choosing experimental conditions so that the ratio S^2/B is maximized. As an example of the application of Eq. (3-57), assume that changing the detector configuration in a low-level counting experiment increases the rate due to the source alone by a factor of 1.5, but also increases the background by a factor of 2.0. The ratio S^2/B is then $(1.5)^2/2.0 = 1.125$ times its former value. Because this ratio exceeds unity, the change will slightly improve the overall statistical accuracy of the net source rate determination if the total measurement time is held constant.

VI. DISTRIBUTION OF TIME INTERVALS

The time intervals separating random events are often of practical interest in nuclear measurements. We present some results that apply to any random process characterized by a constant probability of occurrence per unit time. In most cases, these results will adequately describe the behavior of a radiation detector undergoing irradiation by a steady-state or long-lived radiation source.

In the following discussion, r represents the average rate at which events are occurring. It follows that $r\,dt$ is the differential probability that an event will take place in the differential time increment dt. For a radiation detector with unity efficiency counting a single radioisotope,

$$r = \left| \frac{dN}{dt} \right| = \lambda N$$

where N is the number of radioactive nuclei and λ is their decay constant.

A. Intervals Between Successive Events

In order to derive a distribution function to describe the time intervals between adjacent random events, first assume that an event has occurred at time $t = 0$. What is the differential probability that the next event will take place within a differential time dt after a time interval of length t? Two independent processes must take place: No events may occur within the time interval from 0 to t, but an event must take place in the next differential time increment dt. The overall probability will then be given by the product of the probabilities characterizing the two processes, or

$$
\begin{matrix}
\text{probability of next} & & \text{probability of} & & \text{probability} \\
\text{event taking place in} & = & no \text{ events during} & \times & \text{of an event} \\
dt \text{ after delay of } t & & \text{time from 0 to } t & & \text{during } dt
\end{matrix}
$$

$$
I_1(t)\,dt \quad = \quad P(0) \quad \times \quad r\,dt \tag{3-58}
$$

The first factor on the right-hand side follows directly from the earlier discussion of the Poisson distribution. We seek the probability that no events will be recorded over an interval of length t for which the average number of recorded events should be rt. From Eq. (3-24)

$$
P(0) = \frac{(rt)^0 e^{-rt}}{0!}
$$

$$
P(0) = e^{-rt} \tag{3-59}
$$

Substituting Eq. (3-59) into Eq. (3-58) leads to

$$
\boxed{I_1(t)\,dt = re^{-rt}\,dt} \tag{3-60}
$$

$I_1(t)$ is now the distribution function for intervals between adjacent random events. The plot below shows the simple exponential shape of this distribution.

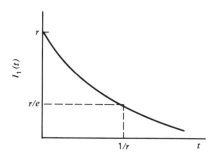

Note that the *most probable* interval is zero. The *average* interval length is calculated by

$$
\bar{t} = \frac{\displaystyle\int_0^\infty t I_1(t)\,dt}{\displaystyle\int_0^\infty I_1(t)\,dt} = \frac{\displaystyle\int_0^\infty t e^{-rt}\,dt}{\displaystyle\int_0^\infty e^{-rt}\,dt} = \frac{1}{r} \tag{3-61}
$$

which is the expected result.

In some radiation applications, the counting rate is low enough so that each individual count can be visually observed as the data are being collected. Experienced observers soon learn what a true exponential interval distribution looks like and occasionally can spot a malfunctioning detector by noting a deviation from an expected random input signal.

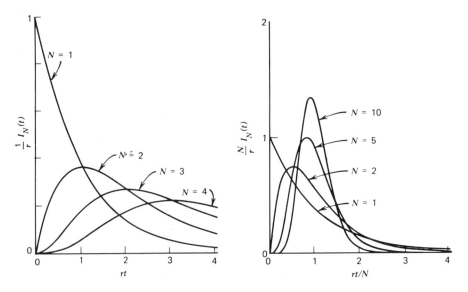

Figure 3-14 Graphical representations of the scaled interval distribution $I_N(t)$. (a) Four distributions for scaling factors of 1, 2, 3, and 4. (b) Interval distributions for $N = 1$ through $N = 10$ normalized to the same average interval N/r.

B. Intervals Between Scaled Events

There are some occasions in which a digital "scaler" may be employed to reduce the rate at which data are recorded from a detector system. A scaler functions as a data buffer by producing an output pulse only when N input pulses have been accumulated.

A general form for the distribution which describes scaled intervals can be derived using arguments parallel to those given earlier for unscaled intervals. Again, two independent processes must occur: A time interval of length t must be observed over which exactly $N - 1$ events are presented to the scaler, and an additional event must occur in the increment dt following this time interval. Under these conditions, a scaled interval of length within dt about t will take place. The parallel expression to Eq. (3-58) then becomes

$$I_N(t)\, dt = P(N - 1) r\, dt \tag{3-62}$$

Again using the Poisson form for $P(N - 1)$, Eq. (3-62) becomes

$$\boxed{I_N(t)\, dt = \frac{(rt)^{N-1} e^{-rt}}{(N - 1)!}\, r\, dt} \tag{3-63}$$

$I_N(t)$ is the interval distribution for N-scaled intervals. A plot is given in Fig. 3-14 for various scaling factors and shows the more uniform intervals that accompany larger values of N. The average interval is

$$\bar{t} = \frac{\displaystyle\int_0^\infty t I_N(t)\, dt}{\displaystyle\int_0^\infty I_N(t)\, dt} = \frac{N}{r} \tag{3-64}$$

whereas the most probable interval is evaluated by setting

$$\frac{dI_n(t)}{dt} = 0$$

and leads to

$$t\Big|_{\substack{\text{most} \\ \text{probable}}} = \frac{N-1}{r} \tag{3-65}$$

PROBLEMS

3-1. A series of 100 measurements of a physical quantity have been made that show a random fluctuation characterized by a sample variance of 2% of the mean value. If the series is lengthened to 1000 measurements made under the same conditions, estimate the sample variance of the larger set of data.

3-2. Find the probability that exactly 8 heads will occur in 12 tosses of a coin.

3-3. A given large population consists of 75% males. Random samples of 15 people are taken from this population, and the number of males tallied for each sample. Find the predicted mean, variance, and standard deviation of the expected results.

3-4. Find the probability that no sixes turn up in 10 rolls of a die.

3-5. A computer programmer averages one error per 60 program statements.

    (a) Find the mean and standard deviation of the number of errors expected in a 250 statement program.

    (b) Find the probability that a 100 statement program is free of errors.

3-6. From the following list, single out those measurements for which the square root of a typical measurement is a proper estimate of the standard deviation of the distribution from which the measurement is drawn:

    (a) A 1 min count from a detector.

    (b) A 5 min count from a detector.

    (c) The net counts from a detector over a 1 min period after background subtraction.

    (d) The counting rate expressed as counts per second based on a 100 s measurement.

    (e) The average of five sequential 1 min counts.

    (f) The sum of five sequential 1 min counts.

3-7. A source is counted for 1 min and gives 561 counts. The source is removed and a 1 min background count gives 410 counts. What is the net count due to the source alone and its associated standard deviation?

3-8. A 10 min count of a source + background gives a total of 846 counts. Background alone counted for 10 min gives a total of 73 counts. What is the net counting rate due to source alone, and what is its associated standard deviation?

3-9. The measurement described in Problem 3-8 is to be repeated, but in this case the available 20 min is to be subdivided optimally between the two separate counts. Find the optimal allocation of time that minimizes the expected standard deviation in the net source counting rate. By what factor has the expected statistical error been reduced from the situation of Problem 3-8?

3-10. In a given application, a 10 min measurement resulted in a statistical uncertainty of 2.8%. How much additional time must be allocated to reduce the statistical uncertainty to 1.0%?

3-11. A designer has the choice in a specific application of either doubling the signal from the source or reducing the background by a factor of 2. From the standpoint of counting statistics, which should be chosen under the following conditions:

(a) The signal is large compared with background.
(b) The signal is small compared with background.

3-12. A flow counter shows an average background rate of 2.87 counts/min. What is the probability that a given 2 min count will contain (a) exactly five counts and, (b) at least one count? What length of counting time is required to ensure with $> 99\%$ probability that at least one count is recorded?

3-13. The following data are obtained from sources A and B of the same isotope:

		Timing Period
Source A + background	251 counts	5 min
Source B + background	717 counts	2 min
Background	51 counts	10 min

What is the ratio of the activity of source B to source A, and what is the percent standard deviation in this ratio?

3-14. The background count from a detector was measured to be 845 over a 30 min period. A source to be measured increases the total counting rate to about 80 counts/min. Estimate the time the source should be counted to determine the counting rate due to the source alone to within a fractional standard deviation of 3%.

3-15. Thirty different students have measured the background counting rate with the same apparatus. Each used the same procedure, consisting of recording the number of counts in five 1 min intervals and taking their average. A set of numbers from a typical student is shown below:

$$25 = \text{count in first minute}$$
$$35 = \text{count in second minute}$$
$$30 = \text{count in third minute}$$
$$23 = \text{count in fourth minute}$$
$$27 = \text{count in fifth minute}$$
$$\text{total} = \overline{140}$$

$$\text{mean} = \frac{140}{5} = 28.0 \frac{\text{counts}}{\text{min}}$$

(a) Do these data seem reasonable assuming all the fluctuations are statistical? Substantiate your conclusion quantitatively.
(b) Based on the above data, what is the expected standard deviation of the mean?
(c) Estimate the sample variance of the 30 numbers representing a similar calculation of the mean background rate by each of the 30 students.

(d) Again assuming only statistical variations, estimate the standard deviation of the final answer for the mean obtained by averaging all 30 independent values.

3-16. The following set of 25 counts was recorded under identical detector conditions and counting times. Apply the chi-squared test to determine whether the observed fluctuations are consistent with expectations from Poisson statistics.

3626	3711	3677	3678	3465
3731	3617	3630	3624	3574
3572	3572	3615	3652	3601
3689	3578	3605	3595	3540
3625	3569	3591	3636	3629

3-17. An average of five sequential 2 min counts of a constant source by Lab Group A gave a resulting value of 2162.4 counts/min. Lab Group B then used the same source and detector in identical conditions and arrived at a value of 2081.5 counts/min based on four sequential 5 min counts. Is the difference between these two results statistically significant?

3-18. You are asked to calibrate the activity of a Cs-137 gamma-ray source by comparison with a standard Cs-137 reference source of approximately the same activity. The standard source has a quoted activity of 3.50 ± 0.05 μCi ($\pm$one standard deviation) and either source alone gives rise to a counting rate of about 1000 per second in the available counter. Background rates are negligible. Assuming that each source is counted separately for equal counting times, how much *total* time will be required to determine the unknown activity to within a 2% expected standard deviation?

3-19. A particular counting system has a stable average background rate (measured over a long time) of 50 counts/min. A decaying radioisotope source was introduced and a 10 min count showed a total of 1683 counts. After a delay of 24 h, the 10 min count was repeated, this time giving a total of 914 counts.

(a) What is the half-life of the source?

(b) What is the expected standard deviation of the half-life value due to counting statistics?

3-20. An engine wear test is to be carried out in which the weight of radioactive piston ring particles in an oil sample is to be determined. A sample of the used oil gives 13,834 counts over a 3 min period. A standard has been prepared using exactly 100 μg of the same activity material which gives 91,396 counts over a 10 min period. Background for the detector has been determined to be 281 counts/min, measured over a very long counting period ($\sim$ 24 h). Find the weight of particles in the sample and its expected fractional standard deviation.

3-21. The decay constant λ of a radioisotope sample is to be determined by counting in a detector system with negligible background. An approximate value λ' is already known. The procedure will be to count for a short time τ at $t = 0$, wait a time Δt, and then count again for the same time τ. Assuming that $\tau \ll 1/\lambda'$, what value of the waiting time Δt will minimize the expected statistical error in the value of λ derived from these measurements?

3-22. The thickness of nominal 1 cm sheet aluminum is to be monitored by noting the attenuation of a gamma-ray parallel beam passing perpendicularly through the sheet. The source and detector are well shielded, so background and scattering into the detector are negligible. Any given sample will spend 1 s in the beam. The detector counting rate with no sheet in place has a mean value (measured over a long time) of 10,000 per second.

(a) Find the optimum value of the linear attenuation coefficient μ which will minimize the uncertainty in the derived sheet thickness value due to statistical fluctuations. (What is the corresponding gamma-ray energy?)

(b) What is the lowest attainable fractional standard deviation under these conditions?

3-23. At an average rate of 100 per second, what fraction of the intervals between randomly occurring events are shorter than 10 ms?

REFERENCES

1. R. D. Evans, *The Atomic Nucleus*, Krieger, New York, 1982, Chaps. 26–28.

2. W. Feller, *An Introduction to Probability Theory and Its Applications*, 2nd ed., Wiley, New York, 1957.

3. P. R. Bevington, *Data Reduction and Error Analysis for the Physical Sciences*, McGraw-Hill, New York, 1969.

CHAPTER · 4

General Properties
of Radiation Detectors

Before discussing the different types of radiation detectors individually, we first outline some general properties that apply to all types. Included will be some basic definitions of detector properties, such as efficiency and energy resolution, together with some general modes of operation and methods of recording data which will be helpful in categorizing detector applications.

I. SIMPLIFIED DETECTOR MODEL

We begin with a hypothetical detector that is subject to some type of irradiation. Attention is first focused on the interaction of a *single* particle or quantum of radiation in the detector, which might, for example, be a single alpha particle or an individual gamma-ray photon. In order for the detector to respond at all, the radiation must undergo interaction through one of the mechanisms discussed in Chapter 2. As indicated by Eq. (2-3), the interaction or stopping time is very small (typically a few nanoseconds in gases or a few picoseconds in solids). In most practical situations, these times are so short that the deposition of the radiation energy can be considered instantaneous.

The net result of the radiation interaction in a wide category of detectors is the appearance of a given amount of electric charge within the detector active volume.[†] Our simplified detector model thus assumes that a charge Q appears within the detector at time $t = 0$ due to the interaction of a single particle or quantum of radiation. Next, this charge must be collected to form the basic electrical signal. Typically, collection of the charge is accomplished through the imposition of an electric field within the detector which causes the positive and negative charges created by the radiation to flow in opposite directions. The time required to fully collect the charge varies greatly from one detector to another. For example, in ion chambers the collection time can be as long as a few milliseconds, whereas in semiconductor diode detectors the time is a few nanoseconds. These times reflect both the mobility of the charge carriers within the detector active volume and the average distance that must be traveled before arrival at the collection electrodes.

[†]Strictly true only for detectors such as ion chambers, proportional tubes, G-M tubes, or semiconductor diode detectors. The discussion is also useful for detector types in which the charge is formed indirectly, as from a photomultiplier tube used with a scintillation crystal.

We therefore begin with a model of a prototypical detector whose response to a single particle or quantum of radiation will be a current that flows for a time equal to the charge collection time. The sketch below illustrates one example for the time dependence the detector current might assume, where t_c represents the charge collection time.

The time integral over the duration of the current must simply be equal to Q, the total amount of charge generated in that specific interaction.

In any real situation, many quanta of radiation will interact over a period of time. If the irradiation rate is high, situations can arise in which current is flowing in the detector from more than one interaction at a given time. For purposes of the present discussion, we assume that the rate is low enough so that each individual interaction gives rise to a current that is distinguishable from all others. The magnitude and duration of each current pulse may vary depending on the type of interaction, and a sketch of the instantaneous current flowing in the detector might then appear as shown in the sketch below.

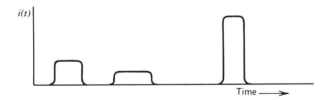

It is important to recall that, since the arrival of radiation quanta is a random phenomenon governed by Poisson statistics, the time intervals between successive current pulses are also randomly distributed.

II. MODES OF DETECTOR OPERATION

We can now introduce a fundamental distinction between three general modes of operation of radiation detectors. The three modes are called *pulse mode*, *current mode*, and *mean square voltage mode* (abbreviated MSV mode, or sometimes called *Campbelling mode*). Pulse mode is easily the most commonly applied of these, but current mode also finds many applications. MSV mode is limited to some specialized applications that make use of its unique characteristics. Although the three modes are operationally distinct, they are interrelated through their common dependence on the sequence of current pulses that are the output of our simplified detector model.

In pulse mode operation, the measurement instrumentation is designed to record each individual quantum of radiation that interacts in the detector. In most common applications, the time integral of each burst of current, or the total charge Q, is recorded since the energy deposited in the detector is directly related to Q. All detectors used to measure the energy of individual radiation quanta must be operated in pulse mode. Such applications are generally categorized as *radiation spectroscopy* and are the subject of much of the remainder of this text.

At very high event rates, pulse mode operation becomes impractical or even impossible. The time between adjacent events may become too short to carry out an adequate analysis, or the current pulses from successive events may overlap in time. In such cases, one can revert to alternative measurement techniques that respond to the time average taken over many individual events. This approach leads to the remaining two modes of operation: current mode and MSV mode.

A. Current Mode

In the sketch below, we show a current-measuring device (an ammeter or, more practically, a picoammeter) connected across the output terminals of a radiation detector.

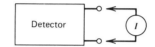

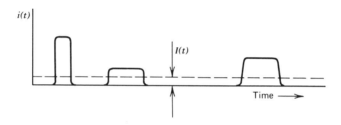

If we assume that the measuring device has a fixed response time T, then the recorded signal from a sequence of events will be a time-dependent current given by

$$I(t) = \frac{1}{T} \int_{t-T}^{t} i(t') \, dt' \tag{4-1}$$

Because the response time T is typically a fraction of a second or greater, the effect is to average out many of the fluctuations in the intervals between individual radiation interactions and to record an average current that depends on the product of the interaction rate and the charge per interaction. In current mode, this time average of the individual current bursts serves as the basic signal that is recorded.

At any instant of time, however, there is a statistical uncertainty in this signal due to the random fluctuations in the arrival time of the event. In many ways, the integration time T is analogous to the measurement time discussed in the statistical analysis of the previous chapter. Thus, the choice of large T will minimize statistical fluctuations in the signal but will also slow the response to rapid changes in the rate or nature of the radiation interactions.

The *average* current is given by the product of the average event rate and the charge produced per event,

$$I_0 = rQ = r\frac{E}{w}q \tag{4-2}$$

where r = event rate

$Q = Eq/w$ = charge produced for each event

E = average energy deposited per event

$w = \dfrac{\text{average energy required to produce a unit}}{\text{charge pair (e.g., electron–ion pair)}}$

$q = 1.6 \times 10^{-19}$ C.

For steady-state irradiation of the detector, this average current can also be rewritten as the sum of a constant current I_0 plus a time-dependent fluctuating component $\sigma_i(t)$, as sketched below.

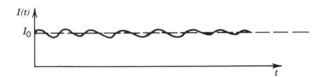

Here $\sigma_i(t)$ is a random time-dependent variable that occurs as a consequence of the random nature of the radiation events interacting within the detector.

A statistical measure of this random component is the variance or mean square value, defined as the time average of the square of the difference between the fluctuating current $I(t)$ and the average current I_0. This mean square value is given by

$$\overline{\sigma_I^2(t)} = \frac{1}{T}\int_{t-T}^{t}\left[I(t') - I_0\right]^2 dt' = \frac{1}{T}\int_{t-T}^{t}\sigma_i^2(t')\, dt' \qquad (4\text{-}3)$$

and the standard deviation follows as

$$\overline{\sigma_I(t)} = \sqrt{\overline{\sigma_I^2(t)}} \qquad (4\text{-}4)$$

Recall from Poisson statistics that the standard deviation in the number of recorded events n over a given observation period is expected to be

$$\sigma_n = \sqrt{n} \qquad (4\text{-}5)$$

Therefore, the standard deviation in the number of events occurring at a rate r in an effective measurement time T is simply

$$\sigma_n = \sqrt{rT} \qquad (4\text{-}6)$$

Thus, the *fractional* standard deviation in the measured signal due to random fluctuations in pulse arrival time is given by

$$\frac{\overline{\sigma_I(t)}}{I_0} = \frac{\sigma_n}{n} = \frac{1}{\sqrt{rT}} \qquad (4\text{-}7)$$

Here $\overline{\sigma_I(t)}$ is the time average of the standard deviation in the measured current, where T is the response time of the picoammeter and I_0 is the average current read on the meter.

This result is useful in estimating the uncertainty associated with a given current mode measurement.

B. Mean Square Voltage Mode

An extension of this discussion of the statistical properties of the signal in current mode leads us to the next general mode of operation: the mean square voltage (MSV) mode. Suppose that we send the current signal through a circuit element that blocks the average current I_0 and only passes the fluctuating component $I(t)$. By providing additional signal-processing elements, we now compute the time average of the squared amplitude of $I(t)$. (The details of these circuits are not important here; for further discussion see Ref. 1.) The processing steps are illustrated below:

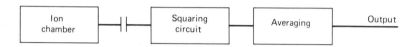

The result corresponds to the quantity $\overline{\sigma_I^2(t)}$ defined previously in Eq. (4-3). Combining Eqs. (4-2) and (4-7), we predict the magnitude of the signal derived in this way to be

$$\overline{\sigma_I^2(t)} = \frac{rQ^2}{T} \tag{4-8}$$

We see that this mean square signal is directly proportional to the event rate r and, more significantly, proportional to the *square of the charge Q produced in each event*. An analysis of this mode of operation was first given by Campbell,[2] and the term Campbelling mode is therefore substituted for MSV mode in some usage.

It should be noted that, in the derivation of Eq. (4-8), the charge produced in each event (Q) was assumed to be constant. Therefore, the result accounts for only the random fluctuations in pulse arrival time, but not for fluctuations in pulse amplitude. In many applications, however, this second source of variance in the signal is small in comparison with the first, and the general character of the results given above remains applicable.

The MSV mode of operation is most useful when making measurements in mixed radiation environments when the charge produced by one type of radiation is much different from that from a second type. If simple current mode operation is chosen, the measured current will linearly reflect the charges contributed by each type. In MSV mode, however, the derived signal is proportional to the *square* of the charge per event. This operational mode will therefore further weight the detector response in favor of the type of radiation giving the larger average charge per event. As one example of the useful application of the MSV mode, in Chapter 14 we describe its use with neutron detectors in reactor instrumentation to enhance the neutron signal compared with the response due to smaller-amplitude gamma-ray events.

C. Pulse Mode

In reviewing various applications of radiation detectors, we find that current mode operation is used with many detectors when event rates are very high. Detectors that are applied to radiation dosimetry are also normally operated in current mode for reasons that will be discussed in Chapter 5. MSV mode is useful in enhancing the relative response to large-amplitude events and finds widespread application in reactor instrumentation. Most applications, however, are better served by preserving information on the amplitude and timing of individual events that only pulse mode can provide. Consequently, the remainder of this chapter deals with various aspects of pulse mode operation.

The nature of the signal pulse produced from a single event depends on the input characteristics of the circuit to which the detector is connected (usually a preamplifier). The equivalent circuit can often be represented as shown below.

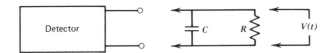

Here R represents the input resistance of the circuit, and C represents the equivalent capacitance of both the detector itself and the measuring circuit. If, for example, a preamplifier is attached to the detector, then R is its input resistance and C is the summed capacitance of the detector, the cable used to connect the detector to the preamplifier, and the input capacitance of the preamplifier itself. In most cases the time-dependent voltage $V(t)$ across the load resistance is the fundamental signal voltage on which pulse mode operation is based. Two separate extremes of operation can be identified which depend on the relative value of the time constant of the measuring circuit. From simple circuit analysis, this time constant is given by the product of R and C, or $\tau = RC$.

Case 1. Small *RC* ($\tau \ll t_c$)

In this extreme the time constant of the external circuit is kept small compared with the charge collection time, so that the current flowing through the load resistance R is essentially equal to the instantaneous value of the current flowing in the detector. The signal voltage $V(t)$ produced under these conditions has a shape nearly identical to the time dependence of the current produced within the detector as illustrated in Fig. 4-1b. Radiation detectors are sometimes operated under these conditions when high event rates or timing information is more important than accurate energy information.

Case 2. Large *RC* ($\tau \gg t_c$)

It is far more common to operate detectors in the opposite extreme in which the time constant of the external circuit is much larger than the detector charge collection time. In this case, very little current will flow in the load resistance during the charge collection time and the detector current is momentarily integrated on the capacitance. If we assume that the time between pulses is sufficiently large, the capacitance will then discharge through the resistance, returning the voltage across the load resistance to zero. The corresponding signal voltage $V(t)$ is illustrated in Fig. 4-1c.

Because the latter case is by far the most common means of pulse-type operation of detectors, it is important to draw some general conclusions. First, the time required for the signal pulse to reach its maximum value is determined by the charge collection time within the detector itself. No properties of the external or load circuit influence the rise time of the pulses. On the other hand, the decay time of the pulses, or the time required to

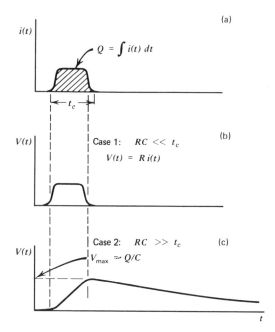

Figure 4-1 (a) The assumed current output from a hypothetical detector. (b) The signal voltage $V(t)$ for the case of a small time constant load circuit. (c) The signal voltage $V(t)$ for the case of a large time constant load circuit.

restore the signal voltage to zero, is determined only by the time constant of the load circuit. The conclusion that the leading edge is detector dependent and the trailing edge circuit dependent is a generality that will hold for a wide variety of radiation detectors operated under conditions in which $RC \gg t_c$. Second, the amplitude of the signal pulse shown as V_{max} in Fig. 4-1c is determined simply by the ratio of the total charge Q created within the detector during one radiation interaction divided by the capacitance C of the equivalent load circuit. Because this capacitance is normally fixed, *the amplitude of the signal pulse is directly proportional to the corresponding charge generated within the detector* and is given by the simple expression

$$V_{max} = \frac{Q}{C} \tag{4-9}$$

Thus, the output of a detector operated in pulse mode normally consists of a string of individual signal pulses, each representing the results of the interaction of a single quantum of radiation within the detector. A measurement of the rate at which such pulses occur will give the corresponding rate of radiation interactions within the detector. Furthermore, the amplitude of each individual pulse reflects the amount of charge generated due to each individual interaction. We shall see that a very common analytical method is to record the distribution of these amplitudes from which some information can often be inferred about the incident radiation. An example is that set of conditions in which the charge Q is directly proportional to the energy of the incident quantum of radiation. Then, a recorded distribution of pulse amplitudes will reflect the corresponding distribution in energy of the incident radiation.

As shown by Eq. (4-9), the proportionality between V_{max} and Q holds only if the capacitance C remains constant. In most detectors, the inherent capacitance is set by its size and shape, and the assumption of constancy is fully warranted. In other types (notably the semiconductor diode detector), the capacitance may change with variations in normal operating parameters. In such cases, voltage pulses of different amplitude may result from events with the same Q. In order to preserve the basic information carried by the magnitude of Q, a type of preamplifier circuit known as a *charge-sensitive* configuration has come into widespread use. As described in Chapter 17, this type of circuit uses feedback to largely eliminate the dependence of the output amplitude on the value of C, and restores proportionality to the charge Q even in cases in which C may change.

Pulse mode operation is the more common choice for most radiation detector applications because of several inherent advantages over current mode. First, the sensitivity that is achievable is often many factors greater than when using current or MSV mode because each individual quantum of radiation can be detected as a distinct pulse. Lower limits of detectability are then normally set by background radiation levels. In current mode, the minimum detectable current may represent an average interaction rate in the detector which is many times greater. The second and more important advantage is that each pulse amplitude carries some information that is often a useful or even necessary part of a particular application. In both current and MSV mode operations, this information on individual pulse amplitudes is lost and all interactions, regardless of amplitude, contribute to the average measured current. Because of these inherent advantages of pulse mode, the emphasis in nuclear instrumentation is largely in pulse circuits and pulse-processing techniques.

III. PULSE HEIGHT SPECTRA

When operating a radiation detector in pulse mode, each individual pulse amplitude carries important information regarding the charge generated by that particular radiation interaction in the detector. If we examine a large number of such pulses, their amplitudes will not all be the same. Variations may be due either to differences in the radiation energy or to fluctuations in the inherent response of a detector to monoenergetic radiation. The pulse amplitude distribution is a fundamental property of the detector output which is routinely used to deduce information about the incident radiation or the operation of the detector itself.

The most common way of displaying pulse amplitude information is through the *differential pulse height distribution*. Figure 4-2a gives a hypothetical distribution for purposes of example. The abscissa is a linear pulse amplitude scale that runs from zero to a value larger than the amplitude of any pulse observed from the source. The ordinate is the differential number dN of pulses observed with an amplitude within the differential amplitude increment dH, divided by that increment, or dN/dH. The horizontal scale then has units of pulse amplitude (volts), whereas the vertical scale has units of inverse amplitude (volts^{-1}). The number of pulses whose amplitude lies between two specific values, H_1 and H_2, can be obtained by integrating the area under the distribution between those two limits, as shown by the cross-hatched area in Fig. 4-2a:

$$\text{number of pulses with amplitude between } H_1 \text{ and } H_2 = \int_{H_1}^{H_2} \frac{dN}{dH} \, dH \qquad (4\text{-}10)$$

The total number of pulses N_0 represented by the distribution can be obtained by

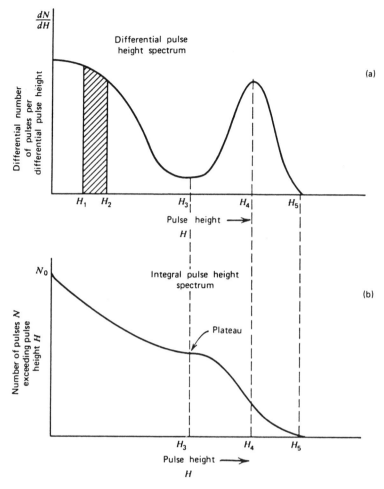

Figure 4-2 Examples of differential and integral pulse height spectra for an assumed source of pulses.

integrating the area under the entire spectrum:

$$N_0 = \int_0^\infty \frac{dN}{dH} \, dH \tag{4-11}$$

Most users of radiation instrumentation are accustomed to looking at the shape of the differential pulse height distribution to display significant features about the source of the pulses. The maximum pulse height observed (H_5) is simply the point along the abscissa at which the distribution goes to zero. Peaks in the distribution, such as at H_4, indicate pulse amplitudes about which a large number of pulses may be found. On the other hand, valleys or low points in the spectrum, such as at pulse height H_3, indicate values of the pulse amplitude around which relatively few pulses occur. The physical interpretation of differential pulse height spectra always involves *areas* under the spectrum between two given limits of pulse height. The value of the ordinate itself (dN/dH) has no physical significance until multiplied by an increment of the abscissa H.

A less common way of displaying the same information about the distribution of pulse amplitudes is through the *integral pulse height distribution*. Figure 4-2b shows the integral distribution for the same pulse source displayed as a differential spectrum in Fig. 4-2a. The abscissa in the integral case is the same pulse height scale shown for the differential distribution. The ordinate now represents the number of pulses whose amplitude exceeds that of a given value of the abscissa H. The ordinate N must always be a monotonically decreasing function of H because fewer and fewer pulses will lie above an amplitude H that is allowed to increase from zero. Because all pulses have some finite amplitude, the value of the integral spectrum at $H = 0$ must be the total number of pulses observed (N_0). The value of the integral distribution must decrease to zero at the maximum observed pulse height (H_5).

The differential and integral distributions convey exactly the same information and one can be derived from the other. The amplitude of the differential distribution at any pulse height H is given by the absolute value of the slope of the integral distribution at the same value. Where peaks appear in the differential distribution, such as H_4, local maxima will occur in the magnitude of the slope of the integral distribution. On the other hand, where minima appear in the differential spectrum, such as H_3, regions of minimum magnitude of the slope are observed in the integral distribution. Because it is easier to display subtle differences by using the differential distribution, it has become the predominant means of displaying pulse height distribution information.

IV. COUNTING CURVES AND PLATEAUS

When radiation detectors are operated in pulse mode, a common situation often arises in which the pulses from the detector are fed to a counting device with a fixed discrimination level. Signal pulses must exceed a given level H_d in order to be registered by the counting circuit. Sometimes it is possible to vary the level H_d during the course of the measurement to provide information about the amplitude distribution of the pulses. Assuming that H_d can be varied between 0 and H_5 in Fig. 4-2, a series of measurements can be carried out in which the number of pulses N per unit time is measured as H_d is changed through a sequence of values between 0 and H_5. This series of measurements is just an experimental determination of the integral pulse height distribution, and the measured counts should lie directly on the curve shown in Fig. 4-2b.

In setting up a nuclear counting measurement, it is often desirable to establish an operating point that will provide maximum stability over long periods of time. For example, small drifts in the value of H_d could be expected in any real application, and one would like to establish conditions under which these drifts would have minimal influence on the measured counts. One such stable operating point can be achieved at a discrimination point set at the level H_3 in Fig. 4-2. Because the slope of the integral distribution is a minimum at that point, small changes in the discrimination level will have minimum impact on the total number of pulses recorded. In general, regions of minimum slope on the integral distribution are called *counting plateaus* and represent areas of operation in which minimum sensitivity to drifts in discrimination level are achieved. It should be noted that plateaus in the integral spectrum correspond to valleys in the differential distribution.

Plateaus in counting data can also be observed with a different procedure. For a particular radiation detector it is often possible to vary the gain or amplification provided for the charge produced in radiation interactions. This variation could be accomplished by varying the amplification factor of a linear amplifier between the detector and counting circuit, or in many cases more directly by changing the applied voltage to the

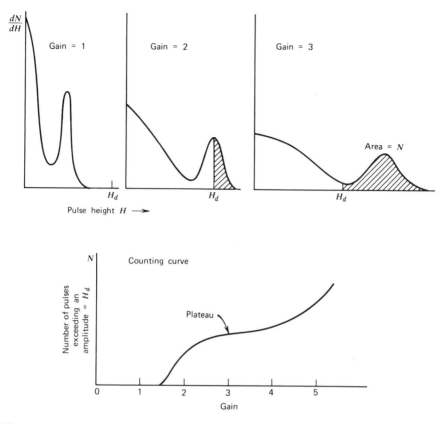

Figure 4-3 Example of a counting curve generated by varying gain under constant source conditions. The three plots at the top give the corresponding differential pulse height spectra.

detector itself. Figure 4-3 shows the differential pulse height distribution corresponding to three different values of voltage gain applied to the same source of pulses. Here the value of gain can be defined as the ratio of the voltage amplitude for a given event in the detector to the same amplitude before some parameter (such as amplification or detector voltage) was changed. The highest voltage gain will result in the largest maximum pulse height, but in all cases the area under the differential distribution will be a constant. In the example shown in Fig. 4-3, no counts will be recorded for a gain $G = 1$ because under those conditions all pulses will be smaller than H_d. Pulses will begin to be recorded somewhere between a gain $G = 1$ and $G = 2$. An experiment can be carried out in which the number of pulses recorded is measured as a function of the gain applied, sometimes called the *counting curve*. Such a plot is shown also in Fig. 4-3 and in many ways resembles an integral pulse height distribution. We now have a mirror image of the integral distribution, however, because small values of gain will record no pulses, whereas large values will result in counting nearly all the pulses. Again, plateaus can be anticipated in this counting curve for values of the gain in which the effective discrimination pulse height H_d passes through minima in the differential pulse height distribution. In the example shown in Fig. 4-3, the minimum slope in the counting curve should correspond to a gain of about 3, in which case the discrimination point is near the minimum of the valley in the differential pulse height distribution.

In some types of radiation detector, such as Geiger–Mueller tubes or scintillation counters, the gain can conveniently be varied by changing the applied voltage to the detector. Although the gain may not change linearly with voltage, the qualitative features of the counting curve can be traced by a simple measurement of the detector counting rate as a function of voltage. In order to select an operating point of maximum stability, plateaus are again sought in the counting curve that results, and the voltage is often selected to lie at a point of minimum slope on this counting curve. We shall discuss these plateau measurements more specifically in Chapters 6 and 7 in connection with proportional counters and Geiger–Mueller detectors.

V. ENERGY RESOLUTION

In many applications of radiation detectors, the object is to measure the energy distribution of the incident radiation. These efforts are classified under the general term *radiation spectroscopy*, and later chapters give examples of the use of specific detectors for spectroscopy involving alpha particles, gamma rays, and other types of nuclear radiation. At this point we discuss some general properties of detectors when applied to radiation spectroscopy and introduce some definitions that will be useful in these discussions.

One important property of a detector in radiation spectroscopy can be examined by noting its response to a monoenergetic source of that radiation. Figure 4-4 illustrates the differential pulse height distribution that might be produced by a detector under these conditions. This distribution is called the *response function* of the detector for the energy used in the determination. The curve labeled "good resolution" illustrates one possible distribution around an average pulse height H_0. The second curve, labeled "poor resolution," illustrates the response of a detector with inferior performance. Provided the same number of pulses are recorded in both cases, the areas under each peak are equal. Although both distributions are centered at the same average value H_0, the width of the distribution in the poor resolution case is much greater. This width reflects the fact that a large amount of fluctuation was recorded from pulse to pulse even though the same energy was deposited in the detector for each event. If the amount of these fluctuations is made smaller, the width of the corresponding distribution will also become smaller and the peak will approach a sharp spike or a mathematical delta function. The ability of a given measurement to resolve fine detail in the incident energy of the radiation is obviously improved as the width of the response function (illustrated in Fig. 4-4) becomes smaller and smaller.

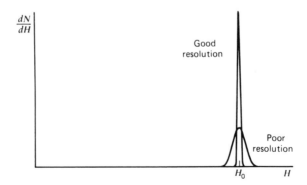

Figure 4-4 Examples of response functions for detectors with relatively good resolution and relatively poor resolution.

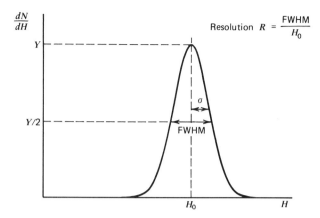

Figure 4-5 Definition of detector resolution. For peaks whose shape is Gaussian with standard deviation σ, the FWHM is given by 2.35σ.

A formal definition of detector *energy resolution* is shown in Fig. 4-5. The differential pulse height distribution for a hypothetical detector is shown under the same assumption that only radiation of a single energy is being recorded. The full width at half maximum (FWHM) is illustrated in the figure and is defined as the width of the distribution at a level that is just half the maximum ordinate of the peak. This definition assumes that any background or continuum on which the peak may be superimposed is negligible or has been subtracted away. The energy resolution of the detector is conventionally defined as the FWHM divided by the location of the peak centroid H_0. The energy resolution R is thus a dimensionless fraction conventionally expressed as a percentage. Semiconductor diode detectors used in alpha spectroscopy can have an energy resolution less than 1%, whereas scintillation detectors used in gamma-ray spectroscopy normally show an energy resolution in the range of 5–10%. It should be clear that the smaller the figure for the energy resolution, the better the detector will be able to distingish between two radiations whose energies lie near each other. An approximate rule of thumb is that one should be able to resolve two energies that are separated by more than one value of the detector FWHM.

There are a number of potential sources of fluctuation in the response of a given detector which result in imperfect energy resolution. These include any drift of the operating characteristics of the detector during the course of the measurements, sources of random noise within the detector and instrumentation system, and statistical noise arising from the discrete nature of the measured signal itself. The third source is in some sense the most important because it represents an irreducible minimum amount of fluctuation that will always be present in the detector signal no matter how perfect the remainder of the system is made. In a wide category of detector applications, the statistical noise represents the dominant source of fluctuation in the signal and thus sets an important limit on detector performance.

The statistical noise arises from the fact that the charge Q generated within the detector by a quantum of radiation is not a continuous variable but instead represents a discrete number of charge carriers. For example, in an ion chamber the charge carriers are the ion pairs produced by the passage of the charged particle through the chamber, whereas in a scintillation counter they are the number of electrons collected from the photocathode of the photomultiplier tube. In all cases the number of carriers is discrete and subject to random fluctuation from event to event even though exactly the same amount of energy is deposited in the detector.

An estimate can be made of the amount of inherent fluctuation by assuming that the formation of each charge carrier is a Poisson process. Under this assumption, if a total number N of charge carriers is generated on the average, one would expect a standard deviation of $\sqrt{N}$ to characterize the inherent statistical fluctuations in that number [see Eq. (3-29)]. If this were the only source of fluctuation in the signal, the response function, as shown in Fig. 4-5, should have a Gaussian shape, because N is typically a large number. In this case, the Gaussian function introduced in the previous chapter is most conveniently written

$$G(H) = \frac{A}{\sigma\sqrt{2\pi}} \exp\left(-\frac{(H - H_0)^2}{2\sigma^2}\right) \qquad (4\text{-}12)$$

The width parameter σ determines the FWHM of any Gaussian through the relation FWHM $= 2.35\sigma$. (The remaining two parameters, H_0 and A, represent the centroid and area, respectively.)

The response of many detectors is approximately linear, so that the average pulse amplitude $H_0 = KN$, where K is a proportionality constant. The standard deviation σ of the peak in the pulse height spectrum is then $\sigma = K\sqrt{N}$ and its FWHM is $2.35K\sqrt{N}$. We then would calculate a limiting resolution R due only to statistical fluctuations in the number of charge carriers as

$$R\bigg|_{\substack{\text{Poisson} \\ \text{limit}}} \equiv \frac{\text{FWHM}}{H_0} = \frac{2.35K\sqrt{N}}{KN} = \frac{2.35}{\sqrt{N}} \qquad (4\text{-}13)$$

Note that this limiting value of R depends only on the number of charge carriers N, and the resolution improves (R will decrease) as N is increased. From Eq. (4-13) we see that in order to achieve an energy resolution better than 1%, one must have N greater than 55,000. An ideal detector would have as many charge carriers generated per event as possible, so that this limiting resolution would be as small a percentage as possible. The great popularity of semiconductor diode detectors stems from the fact that a very large number of charge carriers are generated in these devices per unit energy lost by the incident radiation.

Careful measurements of the energy resolution of some types of radiation detectors have shown that the achievable values for R can be lower by a factor as large as 3 or 4 than the minimum predicted by the statistical arguments given above. These results would indicate that the processes that give rise to the formation of each individual charge carrier are not independent, and therefore the total number of charge carriers cannot be described by simple Poisson statistics. The *Fano factor* has been introduced in an attempt to quantify the departure of the observed statistical fluctuations in the number of charge carriers from pure Poisson statistics and is defined as

$$F \equiv \frac{\text{observed variance in } N}{\text{Poisson predicted variance}(= N)} \qquad (4\text{-}14)$$

Because the variance is given by σ^2, the equivalent expression to Eq. (4-13) is now

$$R \bigg|_{\substack{\text{Statistical} \\ \text{limit}}} = \frac{2.35 K\sqrt{N}\sqrt{F}}{KN} = 2.35\sqrt{\frac{F}{N}} \qquad (4\text{-}15)$$

Although the Fano factor is substantially less than unity for semiconductor diode detectors and proportional counters, other types such as scintillation detectors appear to show a limiting resolution consistent with Poisson statistics and the Fano factor would, in these cases, be unity.

Any other source of fluctuations in the signal chain will combine with the inherent statistical fluctuations from the detector to give the overall energy resolution of the measuring system. It is sometimes possible to measure the contribution to the overall FWHM due to a single component alone. For example, if the detector is replaced by a stable pulse generator, the measured response of the remainder of the system will show a fluctuation due primarily to electronic noise. If there are several sources of fluctuation present and each is symmetric and independent, statistical theory predicts that the overall response function will always tend toward a Gaussian shape, even if the individual sources are characterized by distributions of different shape. As a result, the Gaussian function given in Eq. (4-12) is widely used to represent the response function of detector systems in which many different factors may contribute to the overall energy resolution. Then the total FWHM will be the quadrature sum of the FWHM values for each individual source of fluctuation:

$$(\text{FWHM})^2_{\text{overall}} = (\text{FWHM})^2_{\text{statistical}} + (\text{FWHM})^2_{\text{noise}} + (\text{FWHM})^2_{\text{drift}} + \cdots$$

Each term on the right is the square of the FWHM that would be observed if all other sources of fluctuation were zero.

VI. DETECTION EFFICIENCY

All radiation detectors will, in principle, give rise to an output pulse for each quantum of radiation that interacts within its active volume. For primary charged radiation such as alpha or beta particles, interaction in the form of ionization or excitation will take place immediately upon entry of the particle into the active volume. After traveling a small fraction of its range, a typical particle will form enough ion pairs along its path to ensure that the resulting pulse is large enough to be recorded. Thus, it is often easy to arrange a situation in which a detector will see every alpha or beta particle that enters its active volume. Under these conditions, the detector is said to have a counting efficiency of 100%.

On the other hand, uncharged radiations such as gamma rays or neutrons must first undergo a significant interaction in the detector before detection is possible. Because these radiations can travel large distances between interactions, detectors are often less than 100% efficient. It then becomes necessary to have a precise figure for the detector efficiency in order to relate the number of pulses counted to the number of neutrons or photons incident on the detector.

It is convenient to subdivide counting efficiencies into two classes: *absolute* and *intrinsic*. Absolute efficiencies are defined as

$$\epsilon_{\text{abs}} = \frac{\text{number of pulses recorded}}{\text{number of radiation quanta emitted by source}} \qquad (4\text{-}16)$$

and are dependent not only on detector properties but also on the details of the counting geometry (primarily the distance from the source to the detector). The intrinsic efficiency is defined as

$$\epsilon_{int} = \frac{\text{number of pulses recorded}}{\text{number of quanta incident on detector}} \qquad (4\text{-}17)$$

and no longer includes the solid angle subtended by the detector as an implicit factor. The two efficiencies are simply related for isotropic sources by $\epsilon_{int} = \epsilon_{abs}(4\pi/\Omega)$, where Ω is the solid angle of the detector seen from the actual source position. It is much more convenient to tabulate values of intrinsic rather than absolute efficiencies because the geometric dependence is much milder for the former. The intrinsic efficiency of a detector usually depends primarily on the detector material, the radiation energy, and the physical thickness of the detector in the direction of the incident radiation. A slight dependence on distance between the source and the detector does remain, however, because the average path length of the radiation through the detector will change somewhat with this spacing.

Counting efficiencies are also categorized by the nature of the event recorded. If we accept all pulses from the detector, then it is appropriate to use *total* efficiencies. In this case all interactions, no matter how low in energy, are assumed to be counted. In terms of a hypothetical differential pulse height distribution shown in Fig. 4-6, the entire area under the spectrum is a measure of the number of all pulses that are recorded, regardless of amplitude, and would be counted in defining the total efficiency. In practice, any measurement system always imposes a requirement that pulses be larger than some finite threshold level set to discriminate against very small pulses from electronic noise sources. Thus, one can only approach the theoretical total efficiency by setting this threshold level as low as possible. The *peak* efficiency, however, assumes that only those interactions that deposit the full energy of the incident radiation are counted. In a differential pulse height distribution, these full energy events are normally evidenced by a peak that appears at the highest end of the spectrum. Events that deposit only part of the incident radiation energy then will appear farther to the left in the spectrum. The number of full energy events can be obtained by simply integrating the total area under the peak, which is shown as the cross-hatched area in Fig. 4-6. The total and peak efficiencies are related by the *peak-to-total* ratio r

$$r = \frac{\epsilon_{peak}}{\epsilon_{total}} \qquad (4\text{-}18)$$

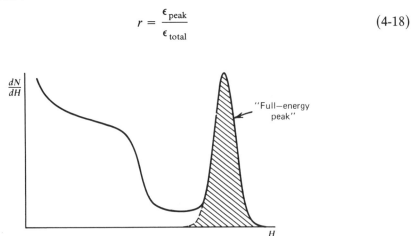

Figure 4-6   Example of the full-energy peak in a differential pulse height spectrum.

which is sometimes tabulated separately. It is often preferable from an experimental standpoint to use only peak efficiencies, because the number of full energy events is not sensitive to some perturbing effects such as scattering from surrounding objects or spurious noise. Therefore, values for the peak efficiency can be compiled and universally applied to a wide variety of laboratory conditions, whereas total efficiency values may be influenced by variable conditions.

To be complete, a detector efficiency should be specified according to both of the above criteria. For example, the most common type of efficiency tabulated for gamma-ray detectors is the *intrinsic peak efficiency*.

A detector with known efficiency can be used to measure the absolute activity of a radioactive source. In the following discussion, we assume that a detector with an intrinsic peak efficiency ϵ_{ip} has been used to record N events under the full energy peak in the detector spectrum. For simplicity, we also assume that the source emits radiation isotropically, and that no attenuation takes place between the source and detector. From the definition of intrinsic peak efficiency, the number of radiation quanta S emitted by the source over the measurement period is then given by

$$S = N \frac{4\pi}{\epsilon_{ip}\Omega} \qquad (4\text{-}19)$$

where Ω represent the solid angle (in steradians) subtended by the detector at the source position. The solid angle is defined by an integral over the detector surface that faces the source, of the form

$$\Omega = \int_A \frac{\cos \alpha}{r^2} dA \qquad (4\text{-}20)$$

where r represents the distance between the source and a surface element dA, and α is the angle between the normal to the surface element and the source direction. If the volume of the source is not negligible, then a second integration must be carried out over all volume elements of the source. For the common case of a point source located along the axis of a right circular cylindrical detector, Ω is given by

$$\Omega = 2\pi \left(1 - \frac{d}{\sqrt{d^2 + a^2}} \right) \qquad (4\text{-}21)$$

where the source–detector distance d and detector radius a are shown in the sketch below:

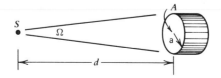

For $d \gg a$, the solid angle reduces to the ratio of the detector plane frontal area A visible at the source to the square of the distance:

$$\Omega \cong \frac{A}{d^2} = \frac{\pi a^2}{d^2} \qquad (4\text{-}22)$$

Published values for Ω can sometimes be found for more complicated geometric arrangements involving off-axis or volumetric sources, or detectors with more complex shapes. Some specific examples of data or descriptions of algorithms useful in solid angle computations are given in Refs. 3–14.

VII. DEAD TIME

In nearly all detector systems, there will be a minimum amount of time that must separate two events in order that they be recorded as two separate pulses. In some cases the limiting time may be set by processes in the detector itself, and in other cases the limit may arise in the associated electronics. This minimum time separation is usually called the *dead time* of the counting system. Because of the random nature of radioactive decay, there is always some probability that a true event will be lost because it occurs too quickly following a preceding event. These "dead time losses" can become rather severe when high counting rates are encountered, and any accurate counting measurements made under these conditions must include some correction for these losses. In this section we discuss some simple models of dead time behavior of counting systems, together with two experimental methods of determining system dead time.

A. Models for Dead Time Behavior

Two models of dead time behavior of counting systems have come into common usage: *paralyzable* and *nonparalyzable* response. These models represent idealized behavior, one or the other of which often adequately resembles the response of a real counting system. The fundamental assumptions of the two models are illustrated in Fig. 4-7. At the center of the figure, a time scale is shown on which six randomly spaced events in the detector are indicated. At the bottom of the figure is the corresponding dead time behavior of a detector assumed to be nonparalyzable. A fixed time τ is assumed to follow each true event that occurs during the "live period" of the detector. True events that occur during

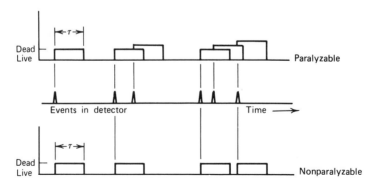

Figure 4-7 Illustration of two assumed models of dead time behavior for radiation detectors.

the dead period are lost and assumed to have no effect whatsoever on the behavior of the detector. In the example shown, the nonparalyzable detector would record four counts from the six true interactions. In contrast, the behavior of a paralyzable detector is shown along the top line of Fig. 4-7. The same dead time τ is assumed to follow each true interaction that occurs during the live period of the detector. True events that occur during the dead period, however, although still not recorded as counts, are assumed to extend the dead time by another period τ following the lost event. In the example shown, only three counts are recorded for the six true events.

The two models predict the same first-order losses and differ only when true event rates are high. They are in some sense two extremes of idealized system behavior, and real counting systems will often display a behavior that is intermediate between these extremes. The detailed behavior of a specific counting system may depend on the physical processes taking place in the detector itself or on delays introduced by the pulse processing and recording electronics.

In the discussion that follows, we examine the response of a detector system to a steady-state source of radiation, and we adopt the following definitions:

$$n = \text{true interaction rate}$$
$$m = \text{recorded count rate}$$
$$\tau = \text{system dead time}$$

We assume that the counting time is long so that both n and m may be regarded as average rates. In general, we would like to obtain an expression for the true interaction rate n as a function of the measured rate m and the system dead time τ, so that appropriate corrections can be made to measured data to account for the dead time losses.

In the nonparalyzable case, the fraction of all time that the detector is dead is given simply by the product $m\tau$. Therefore, the rate at which true events are lost is simply $nm\tau$. But because $n - m$ is another expression for the rate of losses,

$$n - m = nm\tau \tag{4-23}$$

Solving for n, we obtain

$$\boxed{n = \frac{m}{1 - m\tau}} \quad \begin{array}{c} \text{Nonparalyzable} \\ \text{model} \end{array} \tag{4-24}$$

In the paralyzable case, dead periods are not always of fixed length, so we cannot apply the same argument. Instead, we note that rate m is identical to the rate of occurrences of time intervals between true events which exceed τ. The distribution of intervals between random events occurring at an average rate n was previously shown [Eq. (3-60)] to be

$$P_1(T)\, dT = ne^{-nT}\, dT \tag{4-25}$$

where $P_1(T)\, dT$ is the probability of observing an interval whose length lies within dT about T. The probability of intervals larger than τ can be obtained by integrating this distribution between τ and ∞

$$P_2(\tau) = \int_\tau^\infty P_1(T)\, dT = e^{-n\tau} \tag{4-26}$$

The rate of occurrence of such intervals is then obtained by simply multiplying the above expression by the true rate n

$$\boxed{m = ne^{-n\tau}} \qquad \begin{array}{l}\text{Paralyzable}\\ \text{model}\end{array} \qquad (4\text{-}27)$$

The paralyzable model leads to a more cumbersome result because we cannot solve explicitly for the true rate n. Instead, Eq. (4-27) must be solved iteratively if n is to be calculated from measurements of m and knowledge of τ.

A plot of the observed rate m versus the true rate n is given in Fig. 4-8 for both models. When the rates are low the two models give virtually the same result, but the behavior at high rates is markedly different. A nonparalyzable system will approach an asymptotic value for the observed rate of $1/\tau$, which represents the situation in which the counter barely has time to finish one dead period before starting another. For paralyzable behavior, the observed rate is seen to go through a maximum. Very high true interaction rates result in a multiple extension of the dead period following an initial recorded count, and very few true events can be recorded. One must always be careful when using a counting system that may be paralyzable to ensure that ostensibly low observed rates actually correspond to low interaction rates rather than very high rates on the opposite side of the maximum. Mistakes in the interpretation of nuclear counting data from paralyzable systems have occurred in the past by overlooking the fact that there are always two possible true interaction rates corresponding to a given observed rate. As shown in Fig. 4-8, the observed rate m_1 can correspond to either true rates n_1 or n_2. The ambiguity can be resolved only by changing the true rate in a known direction while observing whether the observed rate increases or decreases.

For low rates $(n \ll 1/\tau)$ the following approximations can be written:

Nonparalyzable $\qquad\qquad m = \dfrac{n}{1 + n\tau} \cong n(1 - n\tau) \qquad (4\text{-}28)$

Paralyzable $\qquad\qquad m = ne^{-n\tau} \cong n(1 - n\tau) \qquad (4\text{-}29)$

Thus, the two models lead to identical results in the limit of small dead time losses.

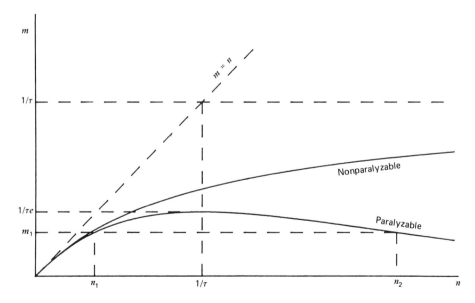

Figure 4-8 Variation of the observed rate m as a function of the true rate n for two models of dead time losses.

If possible, one should avoid measurement conditions under which dead time losses are high because of the errors that inevitably occur in making corrections for the losses. The value of τ may be uncertain or subject to variation, and the system behavior may not follow exactly either of the models described above. When losses are greater than 30 or 40%, the calculated true rate becomes very sensitive to small changes in the measured rate and the assumed system behavior. Instead, the user should seek to reduce the losses by changing the conditions of the measurement or by choosing a counting system with smaller dead time.

B. Methods of Dead Time Measurement

In order to make dead time corrections using either model, prior knowledge of the dead time τ is required. Sometimes this dead time can be associated with a known limiting property of the counting system (e.g., a fixed resolving time of an electronic circuit). More often, the dead time will not be known or may vary with operating conditions and must therefore be measured directly. Common measurement techniques are based on the fact that the observed rate varies nonlinearly with the true rate. Therefore, by assuming that one of the specific models is applicable, and by measuring the observed rate for at least two different true rates that differ by a known ratio, the dead time can be calculated.

The common example is the *two-source method*. The method is based on observing the counting rate from two sources individually and in combination. Because the counting losses are nonlinear, the observed rate due to the combined sources will be less than the sum of the rates due to the two sources counted individually, and the dead time can be calculated from the discrepancy.

To illustrate the method, let n_1, n_2, and n_{12} be the true counting rates (sample plus background) with source 1, source 2, and the combined sources, respectively, in place. Let m_1, m_2, and m_{12} represent the corresponding observed rates. Also, let n_b and m_b be the true and measured background rates with both sources removed. Then

$$n_{12} - n_b = (n_1 - n_b) + (n_2 - n_b)$$
$$n_{12} + n_b = n_1 + n_2 \tag{4-30}$$

Now assuming the nonparalyzable model [Eq. (4-24)] and substituting, we obtain

$$\frac{m_{12}}{1 - m_{12}\tau} + \frac{m_b}{1 - m_b\tau} = \frac{m_1}{1 - m_1\tau} + \frac{m_2}{1 - m_2\tau} \tag{4-31}$$

Solving this equation explicitly for τ gives the following result:

$$\tau = \frac{X(1 - \sqrt{1 - Z})}{Y} \tag{4-32}$$

where

$$X \equiv m_1 m_2 - m_b m_{12}$$
$$Y \equiv m_1 m_2 (m_{12} + m_b) - m_b m_{12}(m_1 + m_2)$$
$$Z \equiv \frac{Y(m_1 + m_2 - m_{12} - m_b)}{X^2}$$

A number of approximations to this general solution are often recommended in textbooks. For example, in the case of zero background ($m_b = 0$)

$$\tau = \frac{m_1 m_2 - [m_1 m_2 (m_{12} - m_1)(m_{12} - m_2)]^{1/2}}{m_1 m_2 m_{12}} \tag{4-33}$$

Other simplifications of Eq. (4-32) have appeared which are based on various mathematical approximations. However, the use of any type of approximation should be discouraged because significant errors can be introduced under typical experimental conditions.[15] Because the two-source method involves a substantial amount of experimental time and effort, it is difficult to justify the use of any expression other than Eq. (4-32) in analyzing the results.

Because the method is essentially based on observing the difference between two nearly equal large numbers, careful measurements are required in order to get reliable values for the dead time. The measurement is usually carried out by counting source 1, placing source 2 nearby and measuring the combined rate, and then removing source 1 to measure the rate due to source 2 alone. During this operation, care must be exercised not to move the source already in place, and consideration must be given to the possibility that the presence of a second source will scatter radiation into the detector which would not ordinarily be counted from the first source alone. In order to keep the scattering unchanged, a dummy second source without activity is normally put in place when the sources are counted individually. Best results are obtained by using sources active enough to result in a fractional dead time $m_{12}\tau$ of at least 20%.

A second method can be carried out if a short-lived radioisotope source is available.[†] In this case the departure of the observed counting rate from the known exponential decay of the source can be used to calculate the dead time. The technique, known as the *decaying source method*, is based on the known behavior of the true rate n:

$$n = n_0 e^{-\lambda t} + n_b \tag{4-34}$$

where n_0 is the true rate at the beginning of the measurement and λ is the decay constant of the particular isotope used for the measurement.

In the limit of negligible background, a simple graphical procedure can be applied to analyze the resulting data. Then Eq. (4-34) becomes

$$n \cong n_0 e^{-\lambda t} \tag{4-35}$$

By inserting Eq. (4-35) into Eq. (4-24) and carrying out some algebra, we get the following relation for the nonparalyzable model:

$$me^{\lambda t} = -n_0\tau m + n_0 \tag{4-36}$$

If we identify, as in Fig. 4-9a, the abscissa as m and the ordinate as the product $me^{\lambda t}$, then Eq. (4-36) is that of a straight line. The experimental procedure consists of measuring the observed rate m as a function of time t, and thus defining points that should lie on this line starting from the right and moving left as the source decays. By fitting the best straight line to the data, the intercept will give n_0, the true rate at the beginning of the measurement, and the negative slope will give the product of $n_0\tau$. The dead time τ then follows directly from the ratio of the slope to the intercept.

For the paralyzable model, inserting Eq. (4-35) into Eq. (4-27) gives the following result:

$$\lambda t + \ln m = -n_0\tau e^{-\lambda t} + \ln n_0 \tag{4-37}$$

[†]For laboratories with access to neutron irradiation facilities, a convenient isotope is [116m]In (half-life of 54.0 min), which is readily produced by neutron absorption in an indium foil.

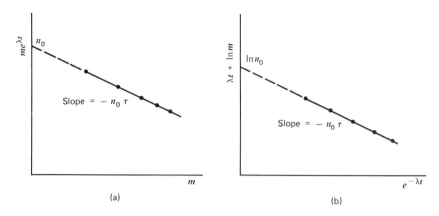

Figure 4-9 Application of the decaying source method to determine dead time.

Again by choosing the abscissa and ordinate as shown in Fig. 4-9b, the equation of a straight line results. In this case the intercept gives the value $\ln n_0$, whereas the slope again gives the negative of the product $n_0\tau$. The dead time can simply be derived from these two values. The decaying source method offers the advantage of not only being able to measure the value of the dead time but also being able to test the validity of the assumed models. If the counting system is best described by a nonparalyzable model, the data will most closely fit a straight line for the format shown in Fig. 4-9a. On the other hand, if a paralyzable model is more appropriate, the format shown in Fig. 4-9b will result in a more nearly linear plot of the data. In order to be effective, the measurements should be carried out for a time period at least equal to the half-life of the decaying radioisotope, and the initial loss fraction $m\tau$ should be at least 20%.

If the background is more than a few percent of the smallest measured rate, the graphical procedure can lead to significant errors. Although some improvement will result from subtracting the observed background rate from all the measured m values, this correction is not rigorous and an exact analysis can only be made by reverting back to Eq. (4-34). It then becomes necessary to use numerical computation techniques to derive values for n_0 and τ which, when inserted in an assumed model of the dead time behavior, result in a best fit to the measured data.

C. Statistics of Dead Time Losses

When measuring radiation from steady-state sources, we normally assume that the true events occurring in the detector follow Poisson statistics in which the probability of an event occurring per unit time is a constant. The effect of system dead time is to remove selectively some of the events before they are recorded as counts. Specifically, events occurring after short time intervals following preceding events are preferentially discarded, and the interval distribution [Eq. (3-60)] is modified from its normal exponential shape. Dead time losses therefore distort the statistics of the recorded counts away from a true Poisson behavior. If the losses are small ($n\tau$ less than 10 or 20%), however, this distortion has little practical effect on the validity of the statistical formulas for the prediction of statistical counting uncertainties developed in Chapter 3.

If the dead time losses are not small, the deviations from Poisson statistics become more significant. The discarding of events that occur after short time intervals causes the sequence of recorded counts to become somewhat more regular, and the variance expected in repeated measurements is reduced. Detailed analyses of the statistics of counts distorted by dead time are beyond the present scope, but can be found in Refs. 16–19.

With either paralyzable or nonparalyzable behavior, there is some chance that more than one true event is lost per dead period. A recorded count therefore can correspond to the occurrence of any number of true events, from one to many. The relative probability that multiple true events are contained in a typical dead period will increase as the true event rate becomes higher. Because the true events still obey Poisson statistics, relatively simple analyses can be made to predict the probability that a typical recorded count results from the combination of exactly x true events. These analyses are given later in this text, beginning on p. 614, in connection with the closely related topic of *pulse pileup*.

D. Dead Time Losses from Pulsed Sources

The analysis of counting losses due to detector dead time given in the previous sections assumed that the detector was irradiated by a steady-state source of constant intensity. There are many applications in which the source of radiation is not continuous but instead consists of short pulses repeated at a constant frequency. For example, electron linear accelerators used to generate high-energy X-rays can be operated to produce pulses of a few microsecond width with a repetition frequency of several kilohertz. In such cases, the results given previously may not be applicable to correct properly for the effects of dead time losses. Substitute analytical methods must now be applied that make use of some of the statistical principles introduced in Chapter 3.

We confine our analysis to a radiation source that can be represented by the time-dependent intensity sketched below:

It is assumed that the radiation intensity is constant throughout the duration T of each pulse, and that the pulses occur at a constant frequency f. Depending on the relative value of the detector dead time τ, several conditions may apply:

1. If τ is much smaller than T, the fact that the source is pulsed has little effect, and the results given earlier in this section for steady-state sources may be applied with reasonable accuracy.

2. If τ is less than T but not by a large factor, only a small number of counts may be registered by the detector during a single pulse. This is the most complicated circumstance and is beyond the scope of the present discussion. Detailed analysis of this case may be found in Refs. 20 and 21.

3. If τ is larger than T but less than the "off" time between pulses (given by $1/f - T$), the following analysis applies. Note that under these conditions, we can have a maximum of only one detector count per source pulse. Also, the detector will be fully recovered at the start of each pulse.

We carry through the following definitions:

$$\tau = \text{dead time of the detector system}$$
$$m = \text{observed counting rate}$$
$$n = \text{true counting rate (if } \tau \text{ were 0)}$$
$$T = \text{source pulse length}$$
$$f = \text{source pulse frequency}$$

Since there can be at most a single count per pulse, the probability of an observed count per source pulse is given by m/f.

The average number of true events per source pulse is by definition equal to n/f. (Note that this average can be greater than 1.) We can apply the Poisson distribution [Eq. (3-24)] to predict the probability that *at least one* true event occurs per source pulse:

$$P(>0) = 1 - P(0)$$
$$= 1 - e^{-\bar{x}}$$
$$= 1 - e^{-n/f} \tag{4-38}$$

Since the detector is "live" at the start of each pulse, a count will be recorded if at least one true event occurs during the pulse. Only one such count can be recorded, so the above expression is also the probability of recording a count per source pulse. Equating the two expressions for this probability, we obtain

$$\frac{m}{f} = 1 - e^{-n/f}$$

or

$$m = f(1 - e^{-n/f}) \tag{4-39}$$

A plot of this behavior is shown below:

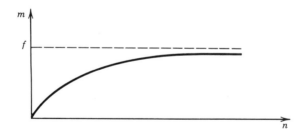

Under these conditions, the maximum observable counting rate is just the pulse repetition frequency, since no more than a single count can be recorded per pulse. Also, neither the specific length of the dead time nor the detailed dead time behavior of the system (whether it is paralyzable or nonparalyzable) has any influence on the losses.

Under normal circumstances, we are more interested in a correction formula to predict the true rate from the measured rate and the system dead time. Solving Eq. (4-39) for n, we derive

$$n = f \ln\left(\frac{f}{f - m}\right) \tag{4-40}$$

Recall that this correction is valid only under the conditions $T < \tau < (1/f - T)$.

In this case, the dead time losses are small under the condition $m \ll f$. Expanding the logarithmic term above for this limit, we find that a first-order correction is then given by

$$n \cong \frac{m}{1 - m/2f} \qquad (4\text{-}41)$$

This result, because of its similarity to Eq. (4-24), can be viewed as predicting an *effective dead time* value of $1/2f$ in this low-loss limit. Since this value is now one-half the source pulsing period, it can be many times larger than the actual physical dead time of the detector system.

PROBLEMS

4-1. Calculate the amplitude of the voltage pulse produced by collecting a charge equal to that carried by 10^6 electrons on a capacitance of 100 pF. ($e = 1.602 \times 10^{-19}$ C.)

4-2. Compare the characteristics of pulse, MSV, and current mode operations as they are applied in radiation measurement systems. Include a table that lists the advantages and disadvantages of each.

4-3. Derive Eq. (4-8).

4-4. A detector with charge collection time of 150 ns is used with a preamplifier whose input circuit can be represented by the parallel combination of 300 pF and 10,000 ohms. Does this situation fall in the category of small or large collection circuit time constant?

4-5. A scintillation counter operated at a given voltage produces a differential pulse height spectrum as sketched below:

(a) Draw the corresponding integral pulse height spectrum.
(b) Sketch the expected counting curve obtained by varying the voltage to the detector while counting above a fixed threshold.

4-6. Sketch both the differential and integral pulse height spectra (using the same horizontal scale) for the following cases:

(a) Pulses with a single amplitude of 1 V.
(b) Pulses uniformly distributed in amplitude between 0 and 1 V.
(c) Pulses distributed around an average amplitude of 1.5 V with a pulse height resolution of 8%.

4-7. A gamma-ray spectrometer records peaks corresponding to two different gamma-ray energies of 435 and 490 keV. What must be the energy resolution of the system (expressed as a percentage) in order just to distinguish these two peaks?

4-8. In a detector with a Fano factor of 0.1, what should be the minimum number of charge carriers per pulse to achieve a statistical energy resolution limit of 0.5%?

4-9. A pulse-processing system operated over a long period of time shows a typical drift that broadens single-amplitude pulses into a distribution with pulse height resolution of 2%. If this system is used with a detector with an intrinsic pulse height resolution of 4%, what will be the expected overall pulse height resolution?

4-10. Find the solid angle subtended by the circular end surface of a cylindrical detector (diameter of 10 cm) for a point source located 20 cm from the surface along the cylindrical axis.

4-11. The diameter of the moon as seen from the earth subtends an angle of about 0.5°. Find the probability that a laser beam aimed in a random direction from the earth's surface will strike the moon.

4-12. The detector of Problem 4-10 has an intrinsic peak efficiency at 1 MeV of 12%. The point source emits a 1 MeV gamma ray in 80% of its decays and has an activity of 20 kBq. Neglecting attenuation between the source and detector, calculate the number of counts that will appear under the 1 MeV full-energy peak in the pulse height spectrum from the detector over a 100 s count.

4-13. A source of ^{116m}In (half-life = 54.0 min) is counted using a G-M tube. Successive 1 min observations gave 131,340 counts at 12:00 noon and 93,384 counts at 12:40. Neglecting background and using a reasonable model for dead time losses, calculate the true interaction rate in the G-M tube at 12:00.

4-14. Counters A and B are nonparalyzable with dead time of 30 and 100 μs, respectively. At what *true* event rate will the dead time losses in counter B be twice as great as those for counter A?

4-15. A counter with negligible background gives exactly 10,000 counts in a 1 s period when a standard source is in place. An identical source is placed beside the first, and the counter now records 19,000 counts in 1 s. What is the counter dead time?

4-16. A paralyzable detector system has a dead time of 1.5 μs. If a counting rate of 10^5 per second is recorded, find the two possible values for the true interaction rate.

4-17. As a source is brought progressively closer to a paralyzable detector, the measured counting rate rises to a maximum and then decreases. If a maximum counting rate of 50,000 per second is reached, find the dead time of the detector.

REFERENCES

1. J. M. Harrer and J. G. Beckerley, *Nuclear Power Reactor Instrumentation Systems Handbook*, Vol. 1, Chap. 5, TID-25952-P1 (1973).

2. N. R. Campbell and V. J. Francis, *IEEE* **93**, Part III (1946).

3. R. P. Gardner and K. Verghese, *Nucl. Instrum. Meth.* **93**, 163 (1971).

4. H. Gotoh and H. Yagi, *Nucl. Instrum. Meth.* **96**, 485 (1971).

5. P. Oblozinsky and I. Ribansky, *Nucl. Instrum. Meth.* **94**, 187 (1971).

6. K. Verghese, R. P. Gardner, and R. M. Felder, *Nucl. Instrum. Meth.* **101**, 391 (1972).

7. M. Belluscio, R. De Leo, A. Pantaleo, and A. Vox, *Nucl. Instrum. Meth.* **114**, 145 (1974).

8. M. V. Green, R. L. Aamodt, and G. S. Johnston, *Nucl. Instrum. Meth.* **117**, 409 (1974).

9. R. Carchon, E. Van Camp, G. Knuyt, R. Van de Vyver, J. Devos, and H. Ferdinande, *Nucl. Instrum. Meth.* **128**, 195 (1975).

10. L. Wielopolski, *Nucl. Instrum. Meth.* **143**, 577 (1977).

11. R. P. Gardner, K. Verghese, and H.-M. Lee, *Nucl. Instrum. Meth.* **176**, 615 (1980).

12. J. Cook, *Nucl. Instrum. Meth.* **178**, 561 (1980).

13. L. Wielopolski, *Nucl. Instrum. Meth. Phys. Res.* **226**, 436 (1984).

14. R. A. Rizk, A. M. Hathout, and A.-R. Z. Hussein, *Nucl. Instrum. Meth. Phys. Res.* **A245**, 162 (1986).

15. W. S. Diethorn, *Int. J. Appl. Radial. Isotopes* **25**, 55 (1974).

16. J. W. Mueller, *Nucl. Instrum. Meth.* **112**, 47 (1973).

17. J. W. Mueller, *Nucl. Instrum. Meth.* **117**, 401 (1974).

18. J. Libert, *Nucl. Instrum. Meth.* **151**, 555 (1978).

19. G. Faraci and A. R. Pennisi, *Nucl. Instrum. Meth.* **212**, 307 (1983).

20. A. M. Cormack, *Nucl. Instrum. Meth.* **15**, 268 (1962).

21. C. H. Westcott, *Proc. R. Soc. London Ser. A* **194**, 508 (1948).

CHAPTER · 5

Ionization Chambers

Several of the oldest and most widely used types of radiation detectors are based on the effects produced when a charged particle passes through a gas. The primary modes of interaction involve ionization and excitation of gas molecules along the particle track. Although the excited molecules can at times be used to derive an appropriate signal (as in the gas scintillators discussed in Chapter 8), the majority of gas-filled detectors are based on sensing the direct ionization created by the passage of the radiation. The detectors that are the topic of the following three chapters (ion chambers, proportional counters, Geiger tubes) all derive, in somewhat different ways, an electronic output signal that originates with the ion pairs formed within the gas filling the detector.

Ion chambers in principle are the simplest of all gas-filled detectors. Their normal operation is based on collection of all the charges created by direct ionization within the gas through the application of an electric field. As with other detectors, ion chambers can be operated in current or pulse mode. In most common applications, ion chambers are used in current mode as dc devices, although some examples of pulse mode applications will be given at the end of this chapter. In contrast, proportional counters or Geiger tubes are almost always used in pulse mode.

The term *ionization chamber* has conventionally come to be used exclusively for the type of detector in which ion pairs are collected from gases. The corresponding process in solids is the collection of electron–hole pairs in the semiconductor detectors described in Chapters 11–13. Direct ionization is only rarely exploited in liquids, although some recent developments of this type are described in Chapter 19.

Many details that are omitted in the following discussions can be found in the classic books on ionization chambers by Rossi and Staub[1] and by Wilkinson.[2] More specific descriptions of chamber design and construction are included in other books, of which Refs. 3–5 are representative examples.

I. THE IONIZATION PROCESS IN GASES

As a fast charged particle passes through a gas, the types of interaction detailed in Chapter 2 create both excited molecules and ionized molecules along its path. After a neutral molecule is ionized, the resulting positive ion and free electron are called an *ion pair*, and it serves as the basic constituent of the electrical signal developed by the ion chamber. Ions can be formed either by direct interaction with the incident particle, or

TABLE 5-1   Values of the Energy Dissipation per Ion Pair
(the W-Value) for Different Gases[a]

Gas	W-Value (eV/ion pair)	
	Fast Electrons	Alpha Particles
Ar	26.4	26.3
He	41.3	42.7
H_2	36.5	36.4
N_2	34.8	36.4
Air	33.8	35.1
O_2	30.8	32.2
CH_4	27.3	29.1

[a]Data from ICRU Report 31, "Average Energy Required to Produce an Ion Pair," International Commission on Radiation Units and Measurements, Washington, DC, 1979.

through a secondary process in which some of the particle energy is first transferred to an energetic electron or "delta ray" (see Chapter 2). Regardless of the detailed mechanisms involved, the practical quantity of interest is the total number of ion pairs created along the track of the radiation.

A. Number of Ion Pairs Formed

At a minimum, the particle must transfer an amount of energy equal to the ionization energy of the gas molecule to permit the ionization process to occur. In most gases of interest for radiation detectors, the ionization energy for the least tightly bound electron shells is between 10 and 20 eV. However, there are other mechanisms by which the incident particle may lose energy within the gas that do not create ions. Examples are excitation processes in which an electron may be elevated to a higher bound state in the molecule without being completely removed. Therefore, the average energy lost by the incident particle per ion pair formed (defined as the *W-value*) is always substantially greater than the ionization energy. The W-value is in principle a function of the species of gas involved, the type of radiation, and its energy. Empirical observations, however, show that it is not a strong function of any of these variables and is a remarkably constant parameter for many gases and different types of radiation. Some specific data are shown in Table 5-1, and a typical value is 30–35 eV/ion pair. Therefore, an incident 1 MeV particle, if it is fully stopped within the gas, will create about 30,000 ion pairs. Assuming that W is constant for a given type of radiation, the deposited energy will be proportional to the number of ion pairs formed and can be determined if a corresponding measurement of the number of ion pairs is carried out.

B. The Fano Factor

In addition to the mean number of ion pairs formed by each incident particle, the fluctuation in their number for incident particles of identical energy is also of interest. These fluctuations will set a fundamental limit on the energy resolution that can be achieved in any detector based on collection of the ions. In the simplest model, the formation of each ion pair would be considered a Poisson process. The total number of ion pairs formed should therefore be subject to fluctuations characterized by a standard deviation equal to the square root of the average number formed. As discussed in Chapter 4, many radiation detectors show an inherent fluctuation that is less than predicted by this simplified model. The *Fano factor* is introduced as an empirical constant by which the predicted variance must be multiplied to give the experimentally observed variance [see Eq. (4-14)].

The Fano factor reflects to some degree the fraction of all the incident particle energy that is converted into information carriers within the detector. If the entire energy of the incident radiation were always converted into ion pairs, the number of pairs produced would always be precisely the same and there would be no statistical fluctuation. Under these conditions the Fano factor would be zero. If only a very small fraction of the incident radiation is converted, however, then the ion pairs would be formed far apart and with a relatively low probability, and there would be good reason to expect that the distribution in their number should follow a Poisson distribution. In gases, the Fano factor is empirically observed to be less than 1 so that the fluctuations are smaller than would be predicted based on Poisson statistics alone.

The Fano factor has significance only when the detector is operated in pulse mode. We therefore postpone further discussions of its magnitude in gases to the following chapter on proportional counters, where pulse mode operation and good energy resolution are more important considerations.

C. Diffusion, Charge Transfer, and Recombination

The neutral atoms or molecules of the gas are in constant thermal motion, characterized by a mean free path for typical gases under standard conditions of about 10^{-6}–10^{-8} m. Positive ions or free electrons created within the gas also take part in the random thermal motion and therefore have some tendency to diffuse away from regions of high density. Of the many types of collision that normally take place, several are of importance in gas-filled radiation detectors. *Charge transfer collisions* can occur when a positive ion encounters another neutral gas molecule. In such a collision, an electron is transferred from the neutral molecule to the ion, thereby reversing the roles of each. This charge transfer is particularly significant in gas mixtures containing several different molecular species. There will then be a tendency to transfer the net positive charge to the gas with the lowest ionization energy because energy is liberated in collisions which leave that species as the positive ion.

The free electron member of the original ion pair also undergoes many collisions in its normal diffusion. In some species of gas, there may be a tendency to form *negative ions* by the attachment of the free electron to a neutral gas molecule. This negative ion then shares many properties with the original positive ion formed in the ionization process, but with opposite electric charge. Oxygen is an example of a gas that readily attaches electrons, so that free electrons diffusing in air are rapidly converted to negative ions. In contrast, nitrogen, hydrogen, hydrocarbon gases, and noble gases all are characterized by relatively low electron attachment coefficients, and therefore the electron continues to migrate in these gases as a free electron under normal conditions.

Collisions between positive ions and free electrons may result in *recombination* in which the electron is captured by the positive ion and returns it to a state of charge neutrality. Alternatively, the positive ion may undergo a collision with a negative ion in which the extra electron is transferred to the positive ion and both ions are neutralized. In either case, the charge represented by the original ion pair is lost and cannot contribute further to the signal in detectors based on collection of the ionization charge.

Because the collision frequency is proportional to the product of the concentrations of the two species involved, the recombination rate can be written

$$\frac{dn^+}{dt} = \frac{dn^-}{dt} = -\alpha n^+ n^- \tag{5-1}$$

where
$n^+ =$ number density of positive species
$n^- =$ number density of negative species
$\alpha =$ recombination coefficient.

The recombination coefficient is normally orders of magnitude larger between positive ions and negative ions compared with that between positive ions and free electrons. In gases that readily form negative ions through electron attachment, virtually all the recombination takes place between positive and negative ions.

There are two general types of recombination loss: *columnar* recombination and *volume* recombination. The first type (sometimes also called *initial* recombination) arises from the fact that ion pairs are first formed in a column along the track of the ionizing particle. The local density of ion pairs is therefore high along the track until the ion pairs are caused to drift or diffuse away from their point of formation. Columnar recombination is most severe for densely ionizing particles such as alpha particles or fission fragments compared with fast electrons that deposit their energy over a longer track. This loss mechanism is dependent only on the local conditions along individual tracks and does not depend on the rate at which such tracks are formed within the detector volume. In contrast, volume recombination is due to encounters between ions and/or electrons after they have left the immediate location of the track. Since many tracks are typically formed over the time it takes for ions to drift all the way to the collecting electrodes, it is possible for ions and/or electrons from independent tracks to collide and recombine. Volume recombination therefore increases in importance with irradiation rate.

II. CHARGE MIGRATION AND COLLECTION

A. Charge Mobility

If an external electric field is applied to the region in which ions or electrons exist in the gas, electrostatic forces will tend to move the charges away from their point of origin. The net motion consists of a superposition of a random thermal velocity together with a net drift velocity in a given direction. The drift velocity for positive ions is in the direction of the conventional electric field, whereas free electrons and negative ions drift in the opposite direction.

For ions in a gas, the drift velocity can be fairly accurately predicted from the relation

$$v = \mu \frac{\mathscr{E}}{p} \tag{5-2}$$

where
$v =$ drift velocity
$\mathscr{E} =$ electric field strength
$p =$ gas pressure

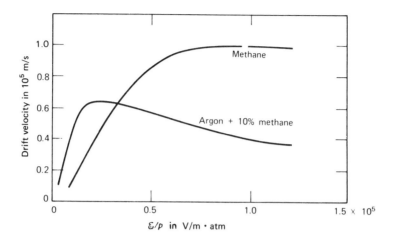

Figure 5-1 Electron drift velocity as a function of electric field $\mathscr{E}$ divided by gas pressure p. (Data from Bortner et al.[6])

The *mobility* μ tends to remain fairly constant over wide ranges of electric field and gas pressure and does not differ greatly for either positive or negative ions in the same gas. Tabulated values for the mobility of ions in various gases can be found in Ref. 2. Typical values are between 1 and 1.5×10^{-4} $m^2 \cdot atm/V \cdot s$ for detector gases of medium atomic number. Therefore, at 1 atm pressure, a typical electric field of 10^4 V/m will result in a drift velocity of the order of 1 m/s. Ion transit times over typical detector dimensions of a centimeter will therefore be approximately 10 ms. By most standards, this is a very long time.

Free electrons behave quite differently. Their much lower mass allows a greater acceleration between encounters with neutral gas molecules, and the value of the mobility in Eq. (5-2) is typically 1000 times greater than that for ions. Typical collection times for electrons are therefore of the order of microseconds, rather than milliseconds. In some gases, for example, hydrocarbons and argon–hydrocarbon mixtures, there is a saturation effect in the electron drift velocity (see Fig. 5-1). Its value approaches a maximum for high values of the electric field and may even decrease slightly if the field is further increased. In many other gases (see Fig. 6-13), the electron drift velocity continues to increase for the largest $\mathscr{E}/p$ values likely to be used in gas-filled counters.

B. The Ionization Current

In the presence of an electric field, the drift of the positive and negative charges represented by the ions and electrons constitutes an electric current. If a given volume of gas is undergoing steady-state irradiation, the rate of formation of ion pairs is constant.

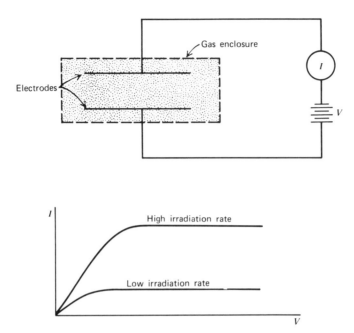

Figure 5-2 The basic components of an ion chamber and the corresponding current–voltage characteristics.

For any small test volume of the gas, this rate of formation will be exactly balanced by the rate at which ion pairs are lost from the volume, either through recombination or by diffusion or migration from the volume. Under the conditions that recombination is negligible and all the charges are efficiently collected, the steady-state current produced is an accurate measure of the rate at which ion pairs are formed within the volume. Measurement of this *ionization current* is the basic principle of the dc ion chamber.

Figure 5-2 illustrates the basic elements of a rudimentary ion chamber. A volume of gas is enclosed within a region in which an electric field can be created by the application of an external voltage. At equilibrium, the current flowing in the external circuit will be equal to the ionization current collected at the electrodes, and a sensitive ammeter placed in the external circuit can therefore measure the ionization current.

The current–voltage characteristics of such a chamber are also sketched in Fig. 5-2. Neglecting some subtle effects related to differences in diffusion characteristics between ions and electrons, no net current should flow in the absence of an applied voltage because no electric field will then exist within the gas. Ions and electrons that are created ultimately disappear either by recombination or by diffusion from the active volume. As the voltage is increased, the resulting electric field begins to separate the ion pairs more rapidly, and columnar recombination is diminished. The positive and negative charges are also swept toward the respective electrodes with increasing drift velocity, reducing the equilibrium concentration of ions within the gas and therefore further suppressing volume recombination between the point of origin and the collecting electrodes. The measured current thus increases with applied voltage as these effects reduce the amount of the original charge that is lost. At a sufficiently high applied voltage, the electric field is large enough to effectively suppress recombination to a negligible level, and all the original charges created through the ionization process contribute to the ion current. Increasing

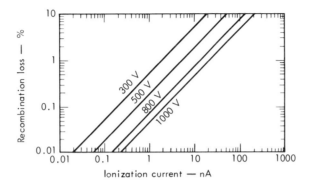

Figure 5-3 Losses due to recombination in an ion chamber filled with argon at 1 atm. These losses are minimized at high values of the applied voltage as shown. (From Colmenares.[7])

the voltage further cannot increase the current because all charges are already collected and their rate of formation is constant. This is the region of *ion saturation* in which ion chambers are conventionally operated. Under these conditions, the current measured in the external circuit is a true indication of the rate of formation of all charges due to ionization within the active volume of the chamber.

C. Factors Affecting Saturation

Several factors can detract from saturation in an ion chamber. The most important of these is recombination, which is minimized by ensuring that a large value of the electric field exists everywhere within the ion chamber volume. Columnar recombination along the track of heavy charged particles (such as alpha particles or fission fragments) is particularly significant, so that larger values of the applied voltage are required in these cases to achieve saturation, compared with electron or gamma-ray irradiation. Also, the effects of volume recombination are more important for higher intensity of the irradiation (and thus higher ion currents), as illustrated in Fig. 5-3. Here the percentage by which the measured current falls short of true saturation is seen to increase as a function of the measured ion current. At the lower irradiation levels, the density of ions and electrons (or negative ions) is correspondingly low, and from Eq. (5-1), the recombination rate is less significant than at high rates. Thus higher voltages are required at high irradiation rates to approach the true saturated ion current. This dependence is evident in the current–voltage characteristics shown in Fig. 5-2 and also is apparent in the example of an ionization chamber used for neutron monitoring shown in Fig. 14-13.

If the production of ion pairs is uniform throughout the volume of the ion chamber, the drift of positive charges toward the cathode and negative charges toward the anode will create some imbalance in the steady-state concentrations of the two charge carriers. The concentration of positive ions will be greatest near the cathode; the opposite is true for the negative charges. Wherever a gradient exists for a species that is free to migrate, some net diffusion must take place in the direction of decreasing concentration. The direction of diffusion is therefore opposite that of the charge carrier flow caused by the electric field, and the effect can be to reduce the measured ion current. From an analysis given by Rossi and Staub[1] the perturbation in the measured ion current is given by

$$-\frac{\Delta I}{I} = \frac{\epsilon kT}{eV} \tag{5-3}$$

where parallel planar electrodes have been assumed. In this expression,

ϵ = ratio of the average energy of charge carrier with electric
   field present to that without the electric field
k = Boltzmann constant
T = absolute temperature
e = electronic charge
V = applied voltage between electrodes

The size of this loss in the saturation current depends primarily on the applied voltage and the size of the quantity ϵ defined above. At room temperature, the factor kT/e is approximately 2.5×10^{-2} V. For ions, ϵ is not much larger than unity, and Eq. (5-3) therefore predicts that the loss due to diffusion is negligible even for low values of the applied voltage. When the negative charge is carried by free electrons, however, ϵ can be of the order of several hundred and a significant loss in saturation current due to electron diffusion is possible. Because ϵ will tend toward a limiting saturated value as the voltage V is raised to high values, the loss is minimized by operating the chamber at high values of the applied voltage.

In many cases, losses due to diffusion and volume recombination can be reduced to insignificant levels at reasonable values of the applied voltage. Columnar recombination is more difficult to eliminate completely, and small losses may persist even at the highest available voltages. It is then helpful to make a number of measurements of the ionization current as a function of voltage in order to determine the true saturated current. It has been shown[8] that under these conditions a plot of $1/I$ versus $1/V$ can be used to extrapolate the measurements to zero on the $1/V$ axis (infinite electric field) to make an accurate prediction of the saturated value of the current.

III. DESIGN AND OPERATION OF DC ION CHAMBERS

A. General Considerations

When an ion chamber is used in direct current mode, it is possible to collect the negative charges either as free electrons or as negative ions. Therefore, virtually any fill gas can be used for the chamber, including those that have high electron attachment coefficients. Although recombination is more significant when negative ions are formed, diffusion losses are less important. Conditions of saturation can normally be achieved over dimensions of a few centimeters using applied voltages that are no greater than tens or hundreds of volts. Air is the most common fill gas and is one in which negative ions are readily formed. Air is required in those chambers designed for the measurement of gamma-ray exposure (see p. 142). Denser gases such as argon are sometimes chosen in other applications to increase the ionization density within a given volume. The fill gas pressure is often 1 atm, although higher pressures are sometimes used to increase the sensitivity.

The geometry chosen for an ion chamber can be varied greatly to suit the application, provided the electric field throughout the active volume can be maintained high enough to lead to saturation of the ion current. Parallel plate or planar geometry leads to a uniform electric field between the plates. Also common is a cylindrical geometry in which the outer shell of the cylinder is operated at ground potential and a central conducting rod carries the applied voltage. In this case, a field that varies inversely with radius is created. Analytical methods that can be used to predict the current–voltage characteristics of chambers of various geometries are described in Ref. 9.

B. Insulators and Guard Rings

With any design, some sort of supporting insulator must be provided between the two electrodes. Because typical ionization currents in most applications are extremely small (of the order of 10^{-12} A or less), the leakage current through these insulators must be kept very small. In the simple scheme of Fig. 5-2, any leakage through the insulator will add to the measured ionization current and cause an unwanted component of the signal. In order to keep this component below 1% of an ionization current of 10^{-12} A for an applied voltage of 100 V, the resistance of the insulator would need to be greater than 10^{16} ohms. Although it may be possible to find materials with bulk resistivity sufficiently high to meet this criterion, leakage paths across the surface of the insulator due to absorbed moisture or other contaminants almost always represent a lower resistance. A different design is therefore often used in low-current applications of ion chambers in which a *guard ring* is employed to reduce the effects of insulator leakage.

A diagram illustrating the use of a guard ring is shown in Fig. 5-4. The insulator is now segmented into two parts, one part separating the conducting guard ring from the negative electrode and the other part separating it from the positive electrode. Most of the voltage drop occurs across the outer segment in which the resulting leakage current does not pass through the measuring instrument. The voltage drop across the inner segment is only the voltage difference across the ammeter terminals and can be very small. Therefore, the component of leakage current which is added to the signal is greatly reduced compared with the case without the guard ring.

Insulators for applications that do not involve high radiation fields are normally manufactured from one of the high-resistivity synthetic plastics. Care is taken to keep the surface smooth and as free of defects as possible in order to minimize the amount of moisture absorption. Radiation damage in these materials in high radiation applications (such as in reactor instruments) can lead to rapid deterioration of the insulation properties. In such cases, inorganic materials such as ceramics are preferred because of their higher resistance to radiation damage.

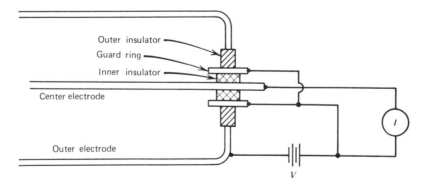

Figure 5-4 Cross-sectional view of one end of a cylindrical ion chamber that utilizes guard ring construction. Most of the applied voltage V appears across the outer insulator, for which the resulting leakage current does not contribute to the measured current I.

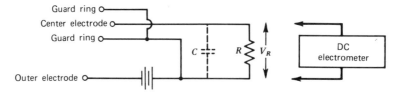

Figure 5-5 Measurement of small ion currents through the use of a series resistance R and an electrometer to record the resulting dc voltage V_R. The chamber capacitance plus any parallel stray capacitance is represented by C. Provided the ion current does not change for several values of the measurement time constant RC, its steady-state value is given by $I = V_R/R$.

C. Measurement of the Ion Current

The magnitude of the ionization current under typical conditions is much too small to be measured using standard galvanometer techniques. Instead, some active amplification of the current must be carried out to allow its indirect measurement. An *electrometer* indirectly measures the current by sensing the voltage drop across a series resistance placed in the measuring circuit (see Fig. 5-5). The voltage developed across the resistor (typically with a value of 10^9–10^{12} ohms) can be applied directly to the grid of a vacuum tube selected to have low noise characteristics and small quiescent grid current. The amplified voltage then serves as the basis for the measured signal. One weakness of standard electrometer circuits is that they must be dc coupled throughout. Any small drift or gradual change in component values therefore results in a corresponding change in the measured output current. Thus, circuits of this type must frequently be balanced by shorting the input and resetting the scale to zero.

An alternative approach is to convert the signal from dc to ac at an early stage, which then allows a more stable amplification of the ac signal in subsequent stages. This conversion is accomplished in the *dynamic-capacitor* or *vibrating reed* electrometer by collecting the ion current across an RC circuit with long time constant, as shown in Fig. 5-6. At equilibrium, a constant voltage is developed across this circuit, which is given by

$$V = IR \tag{5-4}$$

where I is the steady-state ionization current. A charge Q is stored on the capacitance, which is given by

$$Q = CV \tag{5-5}$$

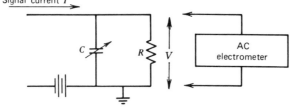

Figure 5-6 Principle of the vibrating reed electrometer. Oscillations of the capacitance induce an ac voltage that is proportional to the steady-state signal current I [see Eq. (5-6)].

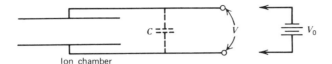

Ion chamber

Figure 5-7 Measurement of the total charge due to ionization ΔQ over a finite period of time by integration of the current. The voltage across the chamber is initially set to V_0, and the voltage source is disconnected. By measuring V at a later time, the ionization charge is given by $\Delta Q = C \Delta V$, where $\Delta V \equiv V_0 - V$, and C is the capacitance of the chamber plus any parallel capacitance.

If the capacitance is now caused to change rapidly compared with the time constant of the circuit, a corresponding change will be induced in the voltage across C given by

$$\Delta V = \frac{Q}{C^2} \Delta C$$

$$\Delta V = I \frac{R}{C} \Delta C \tag{5-6}$$

If the value of the capacitance is varied sinusoidally about an average value C, the amplitude of the ac voltage that is induced is therefore proportional to the ionization current.

The average ionization current can also be measured over finite periods of time by integration methods. If the value of R in Fig. 5-5 is made infinite, any ionization current from the chamber is simply integrated across the capacitance C. By noting the change in the voltage across the capacitance over the measurement period, the total integrated ionization current or ionization charge can be deduced. If the amount of natural leakage across the capacitance can be kept small, this integration technique has the potential of being able to measure much smaller ionization currents than through direct dc current measurement.

To illustrate, suppose that the ion chamber of Fig. 5-7 is originally charged to a voltage V_0. If the leakage across the chamber insulators and external capacitor is negligible, this voltage would be maintained at its original value indefinitely in the absence of ionizing radiation. If radiation is present, the ions will act to partially discharge the capacitance and reduce the voltage from its original value. If a charge ΔQ (defined either as the positive charge of the positive ions or the negative charge of the electrons) is created by the radiation, then the total charge stored on the capacitance will be reduced by ΔQ. The voltage will therefore drop from its original value of V_0 by amount ΔV given by

$$\Delta V = \frac{\Delta Q}{C} \tag{5-7}$$

A measurement of ΔV thus gives the total ionization charge or the integrated ionization current over the period of the measurement.

D. The Electret

The previous section described the measurement of ionization current integrated over a period of time by noting the drop in voltage measured across the chamber capacitance after an initial charging to a known voltage. A similar measurement can be carried out by noting the drop in surface voltage on an element known as an *electret* that is placed inside an ion chamber. An electret is a sample of dielectric material (commonly Teflon) that carries a quasi-permanent electrical charge. A disk of the material of approximately 1 cm diameter and 1 mm thickness is fabricated by heating in the presence of an electric field and then cooling to "freeze" electric dipoles in place. A voltage of up to 1000 V may then appear between the opposite surfaces of the electret. With proper encapsulation, this stored charge may be stable over periods of a year or more, even in the presence of high humidity.

An electret placed with one of its surfaces in contact with the conducting walls of an ion chamber will create an electric field throughout the chamber volume because of the voltage difference on its opposite surface. Any ion pairs created by radiation within the chamber will be separated and collected by this field, serving to partially neutralize the charge carried by the electret. Measurements of the electret voltage before and after the exposure can then be calibrated in terms of the total ionization charge produced in the chamber.

IV. RADIATION DOSE MEASUREMENT WITH ION CHAMBERS

A. Gamma-Ray Exposure

One of the most important applications of ion chambers is in the measurement of gamma-ray exposure. An air-filled ion chamber is particularly well-suited for this application because exposure is defined in terms of the amount of ionization charge created in air.[†] Under the proper conditions a determination of the ionization charge in an air-filled ionization chamber can give an accurate measure of the exposure, and a measurement of the ionization current will indicate the exposure rate.

The task of measuring exposure is somewhat more complicated than it might first appear, because it is defined in terms of the ionization created by all the secondary electrons generated at the point at which the dose is to be measured. Strictly speaking, one would need to follow each of these secondary electrons over its entire range and measure all the ionization created along its track. Because the range in air of secondary electrons created by typical gamma-ray energies can be several meters, it is impractical to design an instrument that would carry out such a measurement directly. Instead, the principle of compensation is used.

If the test volume of air is surrounded by an infinite sea of equivalent air that is also subject to the same exposure over the course of the measurement, an exact compensation will occur. That is, all the ionization charge created outside the test volume from secondary electrons that were formed within the volume is exactly balanced by charge created within the test volume from secondary electrons formed in the surrounding air. This situation is illustrated in Fig. 5-8.

One popular design based on this compensation is diagrammed in Fig. 5-9 and is called the "free-air" ionization chamber. Each end of the chamber is rendered insensitive by grounding the guard electrodes shown in the diagram. The parallel plate geometry creates electric field lines that are perpendicular to the plates in the space between them, but only the volume defined by the central electrode collects ionization current that is

---

[†] The fundamental SI unit of exposure corresponds to that amount of X-ray or gamma-ray radiation whose associated secondary electrons create an ionization charge of 1 coulomb per kilogram of dry air at STP (see Chapter 2).

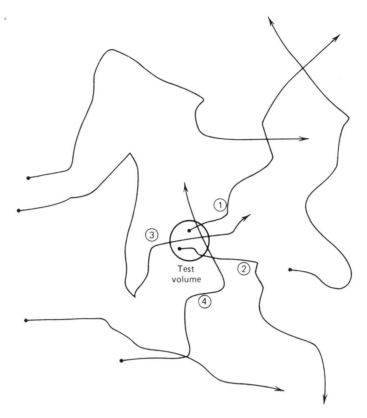

Figure 5-8 The principle of compensation in the measurement of gamma-ray exposure. If the density of gamma-ray interactions is uniform, the test volume will record an amount of ionization that is just equal to that produced along the extended tracks of all the secondary electrons formed within the test volume (such as tracks ① and ② above). The ionization produced by these electrons outside the volume is compensated by ionization within the volume which is produced by tracks originating elsewhere (such as tracks ③ and ④).

registered by the external circuit. The incident gamma-ray beam is collimated so that it is confined to a region that is far from the chamber electrodes, and secondary electrons created in the sensitive volume cannot reach either electrode. Compensation is therefore not required in the vertical dimension but will take place in the horizontal dimension, provided the intensity of the incident radiation beam is not appreciably reduced in its passage through the chamber. Free-air ionization chambers are widely used for accurate exposure measurements for gamma-ray energies below about 100 keV.

At higher energies, the larger range of the secondary electrons creates some difficulties. In order to prevent ionization loss from secondary electrons reaching the electrodes, the dimensions of the chamber must become impractically large. Therefore, gamma-ray exposure measurements at higher energies are conventionally carried out in *cavity chambers* in which a small volume of air is surrounded by a solid material (chosen because its properties are as similar as possible to air).

To see how compensation can take place in such a situation, first consider the hypothetical arrangement sketched in Fig. 5-10 in which a small volume of air is defined by ideal electrodes that are transparent to both gamma rays and electrons. If this test

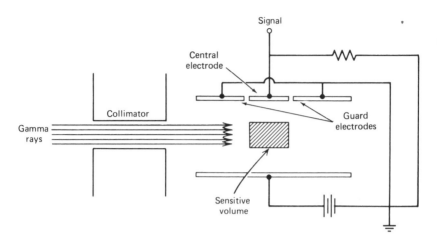

Figure 5-9 The free-air ionization chamber. Because secondary electrons created in the sensitive volume cannot reach the electrodes before stopping, compensation is required only in the dimension parallel to the incident radiation.

volume is at the center of a large volume of air that is subject to the same exposure, the compensation previously described will take place. For gamma-ray energies greater than a few hundred keV, however, the extended range of the secondary electrons makes the required surrounding volume of air much too large. The test volume would have to be at the center of a room-sized volume of air throughout which the exposure is the same as at the test volume itself. In practical situations uniform exposure conditions seldom exist over such large volumes.

The situation could be improved by compressing the air surrounding the test volume into a shell, which could then be no more than a centimeter or two in thickness. Because the same number of air molecules would still be present, none of the compensation properties would change and the test volume would still register an accurate measure of gamma-ray exposure. Now the demands of uniformity could be greatly relaxed to requiring only that the exposure be uniform over the much smaller volume defined by the shell of compressed air. Under these conditions, all ionization lost from the test volume will be compensated by ionization from secondary electrons created within the compressed air shell.

It is now only one step further to replace the hypothetical compressed air shell with a more practical wall of solid material (see Fig. 5-10). The wall is said to be *air equivalent* if its compensation properties are similar to those of an air layer. This condition will be met provided the secondary electron yield and rate of electron energy loss per unit mass are similar to those of air. Because both these phenomena depend largely on atomic number, it turns out that virtually any material with atomic number close to that of air (such as aluminum or plastics) is reasonably air equivalent.

If the walls are sufficiently thick compared with secondary electron ranges, a condition of *electronic equilibrium* is established in which the flux of secondary electrons leaving the inner surface of the wall becomes independent of the wall thickness. Neglecting attenuation of the incident gamma-ray beam in the wall, the measured ion current from a chamber of constant air volume will then be the same, independent of the wall thickness. Table 5-2 lists the minimum air-equivalent wall thickness required to establish electronic equilibrium. For ordinary gamma-ray energies, this thickness is 1 cm or less.

For an air-equivalent ion chamber, the exposure rate R in C/kg · s is simply given by the ratio of the saturated ion current I_s (in amperes) to the mass M (in kg) contained in

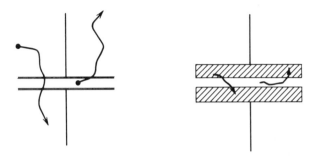

Figure 5-10 Compensation in an ideal ion chamber on the left with fully transparent electrodes could be accomplished by secondary electrons created outside the chamber. However, uniform irradiation conditions seldom exist over the required large surrounding volume of air. On the right, the chamber is provided with air-equivalent solid walls that are thicker than the maximum secondary electron range. Now compensation can take place from secondary electrons formed within the much smaller volume of the walls.

the active volume:

$$R = \frac{I_s}{M} \tag{5-8}$$

The air mass M is normally calculated from a measurement of the chamber volume and the density at STP,

$$M = 1.293 \frac{\text{kg}}{\text{m}^3} \cdot V \cdot \frac{P}{P_0} \cdot \frac{T_0}{T} \tag{5-9}$$

where
V = chamber volume $\left(\text{in m}^3\right)$

P = air pressure within the chamber

P_0 = standard pressure $\left(760 \text{ mm Hg, or } 1.013 \times 10^5 \text{ Pa}\right)$

T = air temperature within the chamber

T_0 = standard temperature $\left(273.15 \text{ K}\right)$

TABLE 5-2 Thicknesses of Ionization Chamber
Walls Required for Establishment of Electronic Equilibrium[a]

Photon Energy (MeV)	Thickness[b] (g/cm²)
0.02	0.0008
0.05	0.0042
0.1	0.014
0.2	0.044
0.5	0.17
1	0.43
2	0.96
5	2.5
10	4.9

[a]From International Commission on Radiation Units and Measurements Report ICRU #20 (1971).
[b]The thicknesses quoted are based on the range of electrons in water. The values will be substantially correct for tissue-equivalent ionization chamber walls and also for air. Half of the above thickness will give an ionization current within a few percent of its equilibrium value.

In routine monitoring, exposure rates of the order of 10^{-3} roentgens/hour (7.167×10^{-11} C/kg · s) are of typical interest. For an ion chamber of 1000 cm^3 volume, the saturated ion current at standard temperature and pressure calculated from Eqs. (5-8) and (5-9) is 9.27×10^{-14} A. Because this signal current is very low, sensitive electrometers and careful chamber design are required to minimize leakage currents.

B. Absorbed Dose

Gas-filled ionization chambers can also be applied indirectly to the measurement of absorbed dose (the energy absorbed per unit mass) in arbitrary materials. The technique is based on application of the *Bragg–Gray principle*, which states that the absorbed dose D_m in a given material can be deduced from the ionization produced in a small gas-filled cavity within that material as follows:

$$D_m = WS_mP \tag{5-10}$$

where W = average energy loss per ion pair formed in the gas

           S_m = relative mass stopping power (energy loss per unit density) of the material to that of the gas

           P = number of ion pairs per unit mass formed in the gas.

In order for the dose D_m to be measured in grays (J/kg), W must be expressed in J/ion pair and P in ion pairs/kg. Equation (5-10) holds to good approximation for different types of radiation provided several geometric conditions are met. The cavity should be small compared with the range of the primary or secondary charged particles associated with the radiation so that its presence does not greatly affect the particle flux. In the case of gamma rays, the solid medium should also be large compared with the range of the secondary electrons so that electronic equilibrium is established at the inner walls of the cavity.

For an ion chamber, the solid medium is the wall material, and the cavity is its internal gas-filled volume. If the gas is air and the wall is air-equivalent material, the factor S_m in Eq. (5-10) is unity. These are the conditions necessary for measurement of absorbed dose in air, which is equivalent to measurement of the gamma-ray exposure discussed in the previous section. If other wall materials or gases are used, the absorbed dose in the wall can be calculated by inserting appropriate values for W and S_m in Eq. (5-10). The dose in biological tissue is of particular interest in radiation protection, so that "tissue-equivalent" ion chambers are widely applied in which the wall is made from a material with a chemical composition similar to that of tissue.

In some applications, there is a problem in keeping the cavity small enough so that the Bragg–Gray principle applies. For example, if the spectrum of gamma rays under measurement extends to very low photon energies, the corresponding secondary electron ranges will be short. An ion chamber with small internal dimensions compared with these ranges would result in very small ion currents, which may cause measurement difficulties. In these applications, use can be made of the "extrapolation chamber," which consists of a pair of electrodes made from the material of interest mounted with variable spacing between them. A series of measurements can be carried out with different electrode spacings and the measured ion current per unit volume extrapolated to the case of zero spacing. With a vanishingly small cavity, the Bragg–Gray principle is once again applicable and the absorbed dose rate can be calculated from this extrapolated ion current.

V. APPLICATIONS OF DC ION CHAMBERS

A. Radiation Survey Instruments

Portable ion chambers of various designs are commonly used as survey instruments for radiation monitoring purposes. These typically consist of a closed air volume of several hundred cm^3 from which the saturated ion current is measured using a battery-powered electrometer circuit. Walls are approximately air equivalent and may be fabricated from plastic or aluminum. These instruments give relatively accurate measurements of the exposure for gamma-ray energies high enough to avoid significant attenuation in the walls or entrance window of the chamber, but low enough so that electronic equilibrium is established in the walls. Figure 5-11 shows a calibration curve for two such instruments and illustrates the drop-off in sensitivity for gamma-ray energies that are less than 50–100 keV due to window attenuation.

Other types of portable ion chamber used for dose measurements are based on the charge integration principle. As described previously, the chamber is initially charged, placed in the radiation field for a period of time, and the resulting drop in chamber voltage is used as a measure of the total integrated ionization charge. The "Condenser R-meter" consists of a charger–reader mechanism and several interchangeable ion chambers of different sizes and wall thicknesses. Each is designed to cover a different range of gamma-ray energy and maximum dose. When used in the proper range, these devices can give results that are accurate to within a few percent. Another related instrument is a small-size ion chamber known as a "pocket chamber." These are often provided with an integral quartz fiber electroscope that can be read on an internal scale by holding the pocket chamber up to the light. An initial charging of the chamber zeros the scale of the electroscope. The pocket chamber is then worn during the period of monitoring. The total integrated dose can be checked periodically simply by noting the degree of discharge of the chamber as indicated on the electroscope. The ultimate sensitivity of such devices is normally limited by leakage current, which inevitably occurs across the insulator surface over long periods of time.

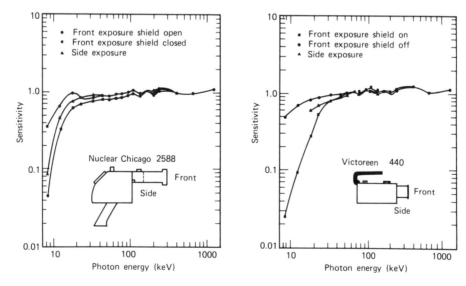

Figure 5-11 Energy calibration of two different gamma-ray survey meters. (From Storm et al.[10])

B. Radiation Source Calibrators

Many of the common applications of ion chambers take advantage of their excellent long-term stability. When operated in the saturated region, the ion current depends only on the geometry of the source and detector and can remain stable over very long periods of time. For example, the standardization of gamma-ray emitters is often carried out by comparing the ion current from an unknown source with that generated by a standard source under identical geometry. Typical operating characteristics can remain stable to within $\pm 0.1\%$ over several years,[11] eliminating the need for frequent recalibration.

One specific design of such a chamber is shown in Fig. 5-12. Typical chamber volumes are several thousand cm^3 and the walls are usually made of brass or steel. The inner collecting electrode is made of a thin foil of aluminum or copper to avoid unnecessary radiation attenuation. The geometry is chosen to avoid low-field regions that could lead to changes in the effective active volume with applied voltage. In such a chamber with a 10^4 cm^3 active volume, the saturation current produced by 1 μCi (3.7×10^4 Bq) of ^{60}Co is on the order of 10^{-13} A, about five times the background current.[13] If higher sensitivity is needed, the gas within the ion chamber may be pressurized. Raising the pressure to 20 atm will increase the ion current by a factor of 20, but the background will increase by a lesser factor. For large chambers, the component of background arising from alpha activity of the chamber walls should be independent of pressure, provided the alpha particles are already fully stopped within the gas. Pressurizing the chamber, however, requires the use of relatively thick entrance windows, which, although not usually a problem for gamma rays, can interfere with the extension of the applications to include pure beta-emitting isotopes.

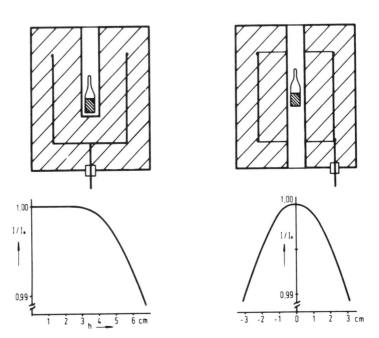

Figure 5-12 Ionization chambers used for calibration of gamma-ray sources. The curve below each configuration shows the behavior of the ionization current for small displacements of the source position. (From Weiss.[12])

C. Measurement of Radioactive Gases

Radioactive gases can conveniently be measured by incorporating them as a constituent of the fill gas of an ionization chamber. Entrance and exit ports are provided to the chamber volume to allow introduction of the gas to be sampled on a continuous flow basis. Examples of specific ion chamber designs of this type are given in Refs. 7 and 13.

The ionization current expected from a given quantity of radioactive gas within the chamber is given by

$$I = \frac{\bar{E}\alpha e}{W} \tag{5-11}$$

where I = ionization current (in A)

$\bar{E}$ = average energy deposited in the gas per disintegration (in eV)

α = total activity (in Bq)

e = electronic charge (in C)

W = average energy deposited per ion pair in the gas (in eV).

The quantity $\bar{E}$ is simple to predict only in the case of very small radiation energies that are nearly always fully absorbed within the chamber gas. For example, the soft beta particles emitted in the decay of tritium have an average energy of 5.65 keV, and $\bar{E}$ can be taken as equal to this value for chambers that are large compared with the corresponding beta particle range in the gas. Equation (5-11) then predicts a sensitivity of about 1 pA/μCi, or 2.7×10^{-17} A/Bq. Once the beta particle energy is large enough so that typical ranges are comparable to chamber dimensions, more complex procedures are required to estimate the average energy deposited (e.g., see Ref. 14).

A common application is one in which samples of air that contain trace quantities of a radioactive gas are to be continuously monitored. When the gas that flows through the chamber is subject to atmospheric changes, a number of difficulties can arise. These perturbing influences can include effects due to moisture, aerosols, ions, and smoke,[†] which may be found in the sampled air. Prior treatment of the incoming air by filtration or electrostatic precipitation can help to control many of these influences.[15]

The chamber will also be sensitive to the ambient gamma-ray background radiation. If constant, this background can be eliminated by simple subtraction of the signal recorded when the chamber is filled with pure air. In other situations in which the gamma-ray background may change during the course of the measurement, a twin chamber filled with pure air can be used to generate a compensating signal that is subtracted from that produced by the chamber through which the sample gas is circulated.[13]

VI. PULSE MODE OPERATION

A. General Considerations

Most applications of ionization chambers involve their use in current mode in which the average rate of ion formation within the chamber is measured. Like many other radiation detectors, ion chambers can also be operated in pulse mode in which each separate

---

[†]The decrease in ionization current from an internal alpha particle source due to smoke particles is the operational basis of the familiar ionization-type home smoke detector.

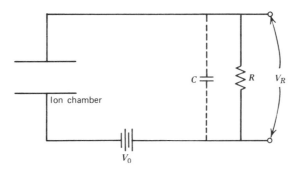

Figure 5-13 Equivalent circuit of an ion chamber operated in pulse mode. Here C represents the capacitance of the chamber plus any parallel capacitance. V_R is the output pulse waveform.

radiation quantum gives rise to a distinguishable signal pulse. As outlined in Chapter 4, pulse mode operation can offer significant advantages in sensitivity or ability to measure the energy of the incident radiation. Pulse mode ion chambers are used to some extent in radiation spectroscopy, although they have largely been replaced by semiconductor diode detectors (see Chapter 11) for many such applications. However, pulse ion chambers remain important instruments in certain specialized applications such as in neutron detectors of the type described in Chapter 14.

The equivalent circuit of an ion chamber operated in pulse mode is shown in Fig. 5-13. The voltage across the load resistance R is the basic electrical signal. In the absence of any ionization charge within the ion chamber, this signal voltage is zero, and all the applied voltage V_0 appears across the ionization chamber itself. When an ionizing particle passes through the chamber, the ion pairs that are created begin to drift under the influence of the electric field. As will be shown in the analysis below, these drifting charges give rise to *induced charges* on the electrodes of the ion chamber which reduce the ion chamber voltage from its equilibrium value V_0. A voltage then appears across the load resistance, which is equal to the amount by which the chamber voltage has dropped. When all the charges within the chamber have been collected at the opposite electrodes, this voltage reaches its maximum value. There then follows a slow return to equilibrium conditions on a time scale determined by the time constant RC of the external circuit. During this period, the voltage across the load resistance gradually drops to zero and the chamber voltage returns to its original value V_0. If the time constant of the external circuit is long compared with the time required to collect the charges within the chamber, a signal pulse is produced whose amplitude indicates the magnitude of the original charge generated within the ion chamber.

As indicated in the earlier discussion of electron and ion mobilities in gases, typical times required to collect free electrons over several centimeters are a few microseconds. On the other hand, ions (either positive or negative) drift much more slowly and typically require collection times of the order of milliseconds. Therefore, if a signal that accurately reflects the original charge of both the ions and electrons is to be generated, the collection circuit time constant and subsequent pulse-shaping time constants must be long compared with a millisecond. Under these conditions, the ion chamber must be restricted to very low pulse rates to avoid excessive pulse pile up. Furthermore, the sensitivity of the output of the shaping circuits to low frequencies makes the system susceptible to interference from microphonic signals generated by mechanical vibrations within the ion chamber.

For these reasons, pulse-type ion chambers are more often operated in the electron-sensitive mode. Here a time constant is chosen that is intermediate between the electron collection time and the ion collection time. The amplitude of the pulse that is produced then reflects only the drift of the electrons and will have much faster rise and fall times. Shorter shaping time constants and much higher rates can therefore be tolerated. A significant sacrifice has been made, however, in that the amplitude of the output pulse now becomes sensitive to the position of the original radiation interaction within the chamber and no longer reflects only the total number of ion pairs formed. The use of more complex gridded chambers, described in a later section, can overcome this disadvantage to a large degree. The fill gas of any electron-sensitive ion chamber must of course be chosen from those gases in which the electrons remain as free electrons and do not form negative ions.

B. Derivation of the Pulse Shape

In the derivation that follows, we assume that a sufficient electric field is applied so that electron–ion recombination is insignificant, and also that the negative charges remain as free electrons. We first derive an expression for the pulse shape for the case in which the collecting circuit time constant is much longer than both the ion and electron collection times.

The pulse shape depends on the configuration of the electric field and the position at which the ion pairs are formed with respect to the equipotential surfaces that characterize the field geometry. To simplify the following analysis, we assume that the chamber electrodes are parallel plates, for which the equipotential surfaces are uniformly spaced planes parallel to the electrode surfaces, and the constant electric field intensity is given by

$$\mathscr{E} = \frac{V}{d} \tag{5-12}$$

Here V is the voltage across the chamber electrodes and d is their spacing. As a further simplification, we assume that all ion pairs are formed at an equal distance x from the positive electrode where the electric potential is equal to $\mathscr{E}x$. This situation is sketched in Fig. 5-14.

The pulse shape is most easily derived based on arguments involving the conservation of energy. Because the time constant of the external circuit is assumed to be large, no appreciable current can flow during the relatively short time required to collect the charges within the ion chamber. Therefore, the energy required to move the charges from their place of origin must come from the energy originally stored across the capacitance C, represented by the ion chamber and associated stray capacitance. This energy is $\frac{1}{2}CV_0^2$, where V_0 is the applied voltage.

After a time t, the ions will have drifted a distance v^+t toward the cathode, where v^+ is the ion drift velocity. Similarly, the electrons will have moved a distance v^-t toward the anode. Both of these motions represent the movement of charge to a region of lower electric potential, and the difference in potential energy is absorbed in the gas through the multiple collisions the charge carriers undergo with gas molecules during their motion. This energy is equal to $Q\,\Delta\varphi$ for both the ions and electrons, where Q is the total charge and $\Delta\varphi$ is the change in electric potential. The charge $Q = n_0e$, where n_0 is the number of original ion pairs and e is the electronic charge. The potential difference $\Delta\varphi$ is the product of the electric field $\mathscr{E}$ and the distance traveled toward the electrode. Conserva-

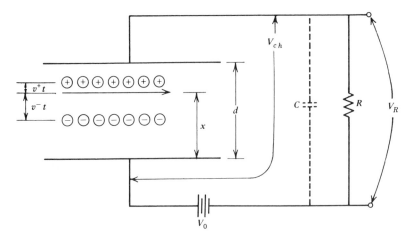

Figure 5-14 Diagram for the derivation of the pulse shape $V_R(t)$ for the signal from an ion chamber. See text for identification.

tion of energy can therefore be written

$$
\begin{array}{cccc}
\text{original} & \text{energy} & \text{energy} & \text{remaining} \\
\text{stored} = & \text{absorbed} + & \text{absorbed} + & \text{stored} \\
\text{energy} & \text{by ions} & \text{by electrons} & \text{energy}
\end{array}
$$

$$\tfrac{1}{2}CV_0^2 = n_0 e \mathscr{E} v^+ t + n_0 e \mathscr{E} v^- t + \tfrac{1}{2}CV_{\text{ch}}^2$$

$$\tfrac{1}{2}C(V_0^2 - V_{\text{ch}}^2) = n_0 e \mathscr{E}(v^+ + v^-)t$$

$$\tfrac{1}{2}C(V_0 + V_{\text{ch}})(V_0 - V_{\text{ch}}) = n_0 e \left(\frac{V_{\text{ch}}}{d}\right)(v^+ + v^-)t \tag{5-13}$$

The signal voltage is measured across R in Fig. 5-14 and is denoted as V_R. Its magnitude is almost always small compared with V_0 and is given by $V_0 - V_{\text{ch}}$. We can therefore make the approximations

$$V_0 + V_{\text{ch}} \cong 2V_0 \quad \text{and} \quad \frac{V_{\text{ch}}}{d} \cong \frac{V_0}{d}$$

Putting these substitutions in Eq. (5-13), we obtain

$$\tfrac{1}{2}C(2V_0)V_R = n_0 e \left(\frac{V_0}{d}\right)(v^+ + v^-)t$$

$$V_R = \frac{n_0 e}{dC}(v^+ + v^-)t \tag{5-14}$$

This result describes the initial portion of the signal pulse and predicts a linear rise with time. It is valid only for the period that both the ions and electrons are drifting within the chamber.

The concept of induced charge is sometimes used to describe the changes caused by the drifting charge carriers. By drifting a distance $v^+ t$, the ions cause the chamber voltage to drop by an amount equal to $n_0 e v^+ t / dC$. The same effect would be caused by the reduction of the charge stored across the capacitance C by an amount $n_0 e v^+ t / d$. Therefore, the ion motion can be thought of as inducing a charge of this magnitude. A similar induced charge is created by the electron motion. It should be emphasized that the induced charge results only from the motion of the charge carriers within the chamber volume and does not require their collection at either electrode.

After a time $t^- \equiv x / v^-$, the electrons reach the anode. Their drift has then contributed the maximum possible to the signal voltage, and the second term in Eq. (5-14) becomes a constant equal to its value at t^-. This constant value is $n_0 e v^- t^- / dC$ or $n_0 e x / dC$. For the next period of time, only the ions are still drifting, and Eq. (5-14) takes the form

$$V_R = \frac{n_0 e}{dC} (v^+ t + x)$$ (5-15)

The ions reach the cathode after a time $t^+ \equiv (d - x) / v^+$. At this point, the signal voltage no longer increases, and Eq. (5-14) becomes

$$V_R = \frac{n_0 e}{dC} [(d - x) + x]$$

or

$$V_R = \frac{n_0 e}{C}$$ (5-16)

The shape of the signal pulse predicted by Eqs. (5-14), (5-15), and (5-16) is shown in Fig. 5-15. When the collection circuit time constant is very large, or $RC \gg t^+$, the maximum amplitude of the signal pulse is given by

$$V_{max} = \frac{n_0 e}{C}$$ (5-17)

and is independent of the position at which the ion pairs were formed within the chamber. Under these conditions, a measurement of the pulse amplitude V_{max} gives a direct indication of the original number of ion pairs n_0 that contributed to the pulse.

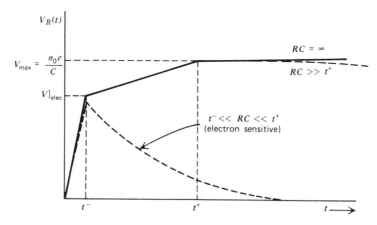

Figure 5-15 Output pulse shape $V_R(t)$ for various time constants RC in the schematic diagram of Fig. 5-14.

In electron-sensitive operation, however, the portion of the pulse derived above which corresponds to drift of the ions is almost entirely lost by choosing a collection time constant that is much shorter than the ion collection time. The pulse that remains then reflects only the drift of the electrons and will have an amplitude given by Eq. (5-15) (neglecting ion drift)

$$V|_{elec} = \frac{n_0 e}{C} \cdot \frac{x}{d} \qquad (5\text{-}18)$$

The shape of this pulse is also sketched in Fig. 5-15. Only the fast rising portion of the pulse is preserved, and the amplitude is now dependent on the position x at which the electrons were originally formed within the chamber.

In any real situation, the incident radiation creates ion pairs over a range of positions within the chamber. The sharp discontinuities shown in Fig. 5-15 are then somewhat "washed out" in the resulting pulse shape. Electron-sensitive operation will also lead to a situation in which a range of pulse amplitudes will be produced for monoenergetic incident radiation, often a decided disadvantage.

C. The Gridded Ion Chamber

The dependence of the pulse amplitude on position of interaction in electron-sensitive ion chambers can be removed through the use of an arrangement sketched in Fig. 5-16. Here the volume of the ion chamber is divided into two parts by a *Frisch grid*, named after the originator of the design.[16] Through the use of external collimation or preferential location of the radiation source, all the radiation interactions are confined to the volume between the grid and the cathode of the chamber. Positive ions simply drift from this volume to the cathode. The grid is maintained at an intermediate potential between the two electrodes and is made to be as transparent as possible to electrons. Electrons are therefore drawn initially from the interaction volume toward the grid. Because of the location of the load resistor in the circuit, neither the downward drift of the ions nor the upward drift of the electrons as far as the grid produces any measured signal voltage. However, once the electrons pass through the grid on their way to the anode, the grid–anode voltage begins to drop and a signal voltage begins to develop across the resistor. The same type of argument that led to Eq. (5-14) predicts that, for a circuit time constant large compared with the electron collection time, the time-dependent signal voltage across the resistor is

$$V_R = \frac{n_0 e}{dC} v^- t \qquad (5\text{-}19)$$

where d is now the grid–anode spacing. This linear rise continues until the electrons reach the anode (see Fig. 5-16b). The maximum signal voltage is therefore

$$V_{max} = \frac{n_0 e}{C} \qquad (5\text{-}20)$$

which is identical to Eq. (5-17). However, now the signal is derived only from the drift of electrons rather than from the motion of both electrons and positive ions. The slow rise corresponding to the drift of ions is eliminated, and the circuit time constant can therefore be set at a much shorter value typical of the electron-sensitive mode of operation described in the previous section. Since each electron passes through the same potential difference and contributes equally to the signal pulse, the pulse amplitude is now

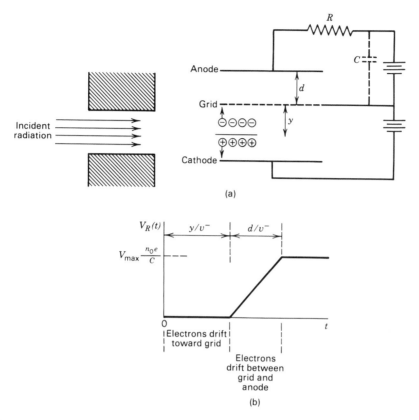

Figure 5-16 (a) Operational principle of the gridded ion chamber. All ion pairs must be formed in the lower region of the chamber between the cathode and grid. (b) The pulse shape that results from the formation of n_0 ion pairs at a distance y from the grid, where d is the grid–anode spacing [see part (a)]. The rise of the pulse results from the drift of the electrons across the grid–anode region. The pulse will decay back to zero with a time constant given by RC.

independent of the position of formation of the original ion pairs and is simply proportional to the total number of ion pairs formed along the track of the incident particle.

D. Pulse Amplitude

The typical amplitude of an ion chamber pulse is relatively small. From the analysis given in the previous sections, the maximum pulse amplitude to be expected from the creation of n_0 ion pairs in either a standard or gridded ion chamber is given by

$$V_{max} = \frac{n_0 e}{C} \tag{5-21}$$

If a 1 MeV charged particle loses all its energy within the chamber, n_0 can be estimated as

$$n_0 = \frac{E_0}{W} \cong \frac{10^6 \text{ ev}}{35 \text{ eV/ion pair}} = 2.86 \times 10^4$$

For typical ion chambers and associated wiring, the capacitance C will be of the order of 10^{-10} farads. We then calculate a pulse amplitude of

$$V_{max} = \frac{(2.86 \times 10^4)(1.60 \times 10^{-19} \text{ C})}{(10^{-10} \text{ F})} = 4.58 \times 10^{-5} \text{ V}$$

Signals of this size can be processed successfully but require the use of relatively sophisticated preamplifiers and pulse-processing electronics to avoid loss of their inherent resolution.

Pulses from electron or gamma-ray interactions typically correspond to fewer original ion pairs, and n_0 can be as much as a factor of 100 smaller than in the previous example. Now it becomes very difficult, if not impossible, to amplify these pulses successfully without severe deterioration due to various sources of noise in the signal chain. For these reasons, the detectors described in the following two chapters, which take advantage of internal gas multiplication of the charge, are often preferred because of their relative ease of use.

E. Charged Particle Spectroscopy

There are applications in which pulse-type ion chambers can be used to good advantage for the measurement of charged particle energies. Although it is much more common to use semiconductor detectors for this purpose (see Chapter 11), ion chambers offer features that may be attractive in some circumstances. They can be constructed with almost arbitrary size and geometry, and the gas pressure can be chosen to tailor the stopping power or effective thickness of the active volume. Ion chambers are also far less subject to performance degradation due to radiation damage than are semiconductor detectors. Their design is generally quite simple, and they can be fabricated by many users with standard shop facilities.

After a period of some neglect, pulse-type ion chambers have received renewed interest in recent years.[17] They have proved useful in low-level alpha particle measurements[18] where parallel-plate-type gridded ion chambers have been constructed with cross-sectional areas up to 500 cm^2. Even larger chambers can conveniently be constructed by using cylindrical geometry. In order to achieve good energy resolution, a Frisch grid is normally incorporated into the chamber design. Because of the small amplitude of typical pulses, special care must be taken in choosing a low noise preamplifier and in minimizing mechanical vibrations, which, by modulating the detector capacitance, can give rise to interfering microphonic noise.[19] Using standard techniques, one can achieve energy resolution for 5 MeV alpha particles of 35–45 keV for such chambers.[18] By using a cooled preamplifier to reduce further the electronic noise, Bertolini[20] demonstrated an energy resolution of 11.5 keV for alpha particles, approaching the ultimate limit set by charge carrier statistics. This energy resolution is comparable to the best that can currently be achieved using silicon semiconductor detectors.

In the conventional configuration, an entrance window for the chamber is provided so that the charged particles travel perpendicular to the electric field lines as shown in Fig. 5-16a. In that case, all the electrons formed along a given track are approximately equidistant from the grid, and each will have approximately the same drift time. The voltage pulse therefore has the general characteristics shown in Fig. 5-16b, and all particle tracks give signal pulses of similar shape. It has recently been demonstrated[21] that there can be some advantage in designing the chamber so that the particle tracks are parallel rather than perpendicular to the field lines. Then the drift time of electrons to the grid will be different for those formed at the beginning of the track compared with those near the end. The detailed shape of the rise of the output pulse will therefore reflect the spatial distribution of ion pairs as they were formed along the track of the incident particle. Because this spatial distribution is conventionally called a *Bragg curve* (see p. 33), an analysis of the pulse shape to yield information on this spatial distribution is conventionally called *Bragg curve spectroscopy*. This approach has proved to be very useful in extending the information obtained from gridded ion chambers beyond a simple measurement of the energy of the particle. Appropriate analysis of the pulse shape[22, 23] can distinguish between particles of different type (atomic number and/or charge state) through differences in the shape of their Bragg curve.

PROBLEMS

5-1. Calculate the charge represented by the positive ions (or free electrons) created when a 5.5 MeV alpha particle is stopped in helium. Find the corresponding saturated current if 300 alpha particles per second enter a helium-filled ion chamber.

5-2. The following data were taken from an ion chamber under constant irradiation conditions:

Voltage (V)	Current (pA)
10	18.72
20	19.41
50	19.93
100	20.12

Using the extrapolation procedure outlined on p. 138, find the saturated ionization current. What minimum voltage must be applied in order to reach a current within 0.5% of saturation?

5-3. An air-equivalent pocket chamber having a capacitance of 75 pF is initially charged to a voltage of 25 V. If the active volume contains 50 cm^3 of air at STP, what value of gamma-ray exposure will reduce the chamber voltage to 20 V?

5-4. An ion chamber is constructed using parallel plate electrodes with a spacing of 5.0 cm. It is filled with pure methane gas at a pressure of 1 atm and operated at an applied voltage of 1000 V. From the data given in Fig. 5-1, calculate the maximum electron collection time.

5-5. An ion chamber of the type shown in Fig. 5-7 has an associated capacitance of 250 pF. It is charged initially to a voltage of 1000 V, exposed to a gamma-ray flux for a period of 30 min, at which time a second voltage measurement indicates a value of 850 V. Calculate the average current that would have been measured over the exposure period under these conditions.

5-6. With the aid of Fig. 2-14, estimate the maximum distance of penetration or "range" of an electron with 0.5 MeV initial energy in atmospheric air.

5-7. A free-air ionization chamber of the type sketched in Fig. 5-9 has a sensitive volume of very small dimensions. Estimate the minimum electrode spacing if true compensation is to be maintained up to a maximum gamma-ray energy of 5 MeV.

5-8. The average beta particle energy emitted by ^{14}C is 49 keV. Calculate the saturated ion current if 150 kBq of the isotope in the form of CO_2 gas is introduced into a large-volume ion chamber filled with pressurized argon.

5-9. Calculate the minimum current that must be measured if a 1 liter ion chamber is to be used as a gamma-ray survey meter down to dose rates of 0.5 mR/h (35.8 pC/kg · s).

5-10. A parallel plate ion chamber with 150 pF capacitance is operated in electron-sensitive mode. Calculate the pulse amplitude expected from 1000 ion pairs formed 2 cm from the anode, if the total spacing between the plates is 5 cm.

5-11. A small air-filled ionization chamber of fixed interior dimensions is located 10 m from a 1 MeV gamma-ray source. Imagine that the walls of the chamber are made of aluminum and are of variable thickness. Neglecting background radiation, plot the relative ionization current from the chamber as a function of wall thickness for the two conditions listed below:

  (a) The experiment is performed in a normal air-filled laboratory.
  (b) The experiment is performed in an earth satellite and the surrounding space is essentially a vacuum.

5-12. An air-equivalent ion chamber is constructed using aluminum walls. What is the minimum thickness of these walls if compensation is to be maintained up to a maximum gamma-ray energy of 10 MeV?

5-13. An air-filled ion chamber is operated at a pressure of 3 atm and a temperature of 100°C. If its active volume is 2500 cm^3, find the saturated ion current corresponding to a gamma-ray exposure rate of 100 pC/kg · s.

5-14. A pocket chamber is constructed with an internal volume of 10 cm^3 and a capacitance of 20 pF. If the smallest voltage change that can be sensed is 50 mV, calculate the minimum gamma-ray exposure to which the pocket chamber will be sensitive.

5-15. Using estimates for the mobility of ions and electrons in air, estimate the ratio of the slopes of the two segments of the rise of the pulse shown in Fig. 5-15.

REFERENCES

1. B. B. Rossi and H. H. Staub, *Ionization Chambers and Counters*, McGraw-Hill, New York, 1949.

2. D. H. Wilkinson, *Ionization Chambers and Counters*, Cambridge University Press, Cambridge, 1950.

3. J. Sharpe, *Nuclear Radiation Detectors*, 2nd ed., Methuen and Co., London, 1964.

4. W. J. Price, *Nuclear Radiation Detection*, 2nd ed., McGraw-Hill, New York, 1964.

5. J. W. Boag, "Ionization Chambers," in *Radiation Dosimetry*, Vol. II (F. H. Attix and W. C. Roesch, eds.), Academic Press, New York, 1966.

6. T. E. Bortner, G. S. Hurst, and W. G. Stone, *Rev. Sci. Instrum.* **28**, 103 (1957).

7. C. A. Colmenares, *Nucl. Instrum. Meth.* **114**, 269 (1974).

8. S. M. Mustafa and K. Mahesh, *Nucl. Instrum. Meth.* **150**, 549 (1978).

9. F. Hajnal and J. Pane, *IEEE Trans. Nucl. Sci.* **NS-25**(1), 550 (1978).

10. E. Storm, D. W. Lier, and H. I. Israel, *Health Phys.* **26**, 179 (1974).

11. G. C. Lowenthal, *Int. J. Appl. Radiat. Isotopes* **20**, 559 (1969).

12. H. M. Weiss, *Nucl. Instrum. Meth.* **112**, 291 (1973).

13. R. A. Jalbert and R. D. Hiebert, *Nucl. Instrum. Meth.* **96**, 61 (1971).

14. P. Leboleux, Report CEA-2215, (1962).

15. J. R. Waters, *Nucl. Instrum. Meth.* **117**, 39 (1974).

16. O. Frisch, *British Atomic Energy Report* BR-49 (1944).

17. H. W. Fulbright, *Nucl. Instrum. Meth.* **162**, 21 (1979).

18. H. Hoetzl and R. Winkler, *Nucl. Instrum. Meth.* **223**, 290 (1984).

19. G. F. Nowack, *Nucl. Instrum. Meth.* **A255**, 217 (1987).

20. G. Bertolini, *Nucl. Instrum. Meth.* **223**, 285 (1984).

21. C. R. Gruhn et al., *Nucl. Instrum. Meth.* **196**, 33 (1982).

22. N. J. Shenhav and H. Stelzer, *Nucl. Instrum. Meth.* **228**, 359 (1985).

23. R. Kotte, H.-J. Keller, H.-G. Ortlepp, and F. Stary, *Nucl. Instrum. Meth.* **A257**, 244 (1987).

CHAPTER · 6

Proportional Counters

The proportional counter is a type of gas-filled detector that was introduced in the late 1940s. In common with the Geiger–Mueller tubes described in Chapter 7, proportional tubes are almost always operated in pulse mode and rely on the phenomenon of *gas multiplication* to amplify the charge represented by the original ion pairs created within the gas. Pulses are therefore considerably larger than those from ion chambers used under the same conditions, and proportional counters can be applied to situations in which the number of ion pairs generated by the radiation is too small to permit satisfactory operation in pulse-type ion chambers. One important application of proportional counters has therefore been in the detection and spectroscopy of low-energy X-radiation (where their principal competitor is the lithium-drifted silicon detector described in Chapter 13). Proportional counters are also widely applied in the detection of neutrons, and specific examples are given in Chapters 14 and 15.

The general reference texts listed at the beginning of Chapter 5 also contain detailed discussions of many topics dealing with the design and operation of proportional counters. References 1 and 2 at the end of this chapter are journal articles that contain extensive reviews of the literature. A number of basic papers on various aspects of proportional counters have also been published as part of the Proceedings of the First International Summer School on Radionuclide Metrology [*Nuclear Instruments and Methods*, Vol. 112 (1973)].

I. GAS MULTIPLICATION

A. Avalanche Formation

Gas multiplication is a consequence of increasing the electric field within the gas to a sufficiently high value. At low values of the field, the electrons and ions created by the incident radiation simply drift to their respective collecting electrodes. During the migration of these charges, many collisions normally occur with neutral gas molecules. Because of their low mobility, positive or negative ions achieve very little average energy between collisions. Free electrons, on the other hand, are easily accelerated by the applied field and may have significant kinetic energy when undergoing such a collision. If this energy is greater than the ionization energy of the neutral gas molecule, it is possible for an additional ion pair to be created in the collision. Because the average energy of the

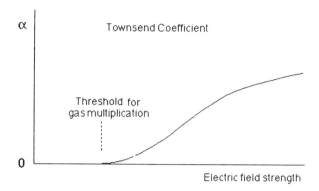

Figure 6-1   A plot of the first Townsend coefficient as a function of electric field for a typical gas.

electron between collisions increases with increasing electric field, there is a threshold value of the field above which this secondary ionization will occur. In typical gases, at atmospheric pressure, the threshold field is of the order of 10^6 V/m.

The electron liberated by this secondary ionization process will also be accelerated by the electric field. During its subsequent drift, it undergoes collisions with other neutral gas molecules and thus can create additional ionization. The gas multiplication process therefore takes the form of a cascade, known as a *Townsend avalanche*, in which each free electron created in such a collision can potentially create more free electrons by the same process. The fractional increase in the number of electrons per unit path length is governed by the Townsend equation:

$$\frac{dn}{n} = \alpha \, dx \tag{6-1}$$

Here α is called the first Townsend coefficient for the gas. Its value is zero for electric field values below the threshold and generally increases with increasing field strength above this minimum (see Fig. 6-1). For a spatially constant field (as in parallel plate geometry), α is a constant in the Townsend equation. Its solution then predicts that the density of electrons grows exponentially with distance as the avalanche progresses:

$$n(x) = n(0)e^{\alpha x} \tag{6-2}$$

For the cylindrical geometry used in most proportional counters, the electric field increases in the direction that the avalanche progresses, and the growth with distance is even steeper. In the proportional counter, the avalanche terminates when all free electrons have been collected at the anode. Under proper conditions, the number of secondary ionization events can be kept proportional to the number of primary ion pairs formed, but the total number of ions can be multiplied by a factor of many thousands. This charge amplification within the detector itself reduces the demands on external amplifiers and can result in significantly improved signal-to-noise characteristics compared with pulse-type ion chambers. The formation of an avalanche involves many energetic electron–atom collisions in which a variety of excited atomic or molecular states may be formed. The performance of proportional counters is therefore much more sensitive to the composition of trace impurities in the fill gas than is the case for ion chambers.

B. Regions of Detector Operation

The differences between various types of gas counter operated in pulse mode are illustrated in Fig. 6-2. The amplitude of the observed pulse from the detector is plotted versus the applied voltage or electric field within the detector. At very low values of the voltage, the field is insufficient to prevent recombination of the original ion pairs, and the collected charge is less than that represented by the original ion pairs. As the voltage is raised, recombination is suppressed and the region of ion saturation discussed in Chapter 5 is achieved. This is the normal mode of operation for ionization chambers. As the voltage is increased still further, the threshold field at which gas multiplication begins is reached. The collected charge then begins to multiply, and the observed pulse amplitude will increase. Over some region of the electric field, the gas multiplication will be linear, and the collected charge will be proportional to the number of original ion pairs created by the incident radiation. This is the region of *true proportionality* and represents the mode of operation of conventional proportional counters, which are the topic of this chapter. Under constant operating conditions, the observed pulse amplitude still indicates the number of ion pairs created within the counter, although their charge has been greatly amplified.

Increasing the applied voltage or electric field still further can introduce nonlinear effects. The most important of these is related to the positive ions, which are also created in each secondary ionization process. Although the free electrons are quickly collected, the positive ions move much more slowly and, during the time it takes to collect the electrons, they barely move at all. Therefore, each pulse within the counter creates a cloud of positive ions, which is slow to disperse as it drifts toward the cathode. If the concentration of these ions is sufficiently high, they represent a space charge that can significantly alter the shape of the electric field within the detector. Because further gas

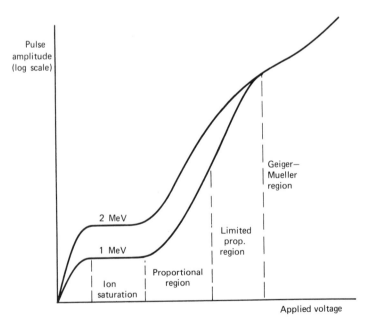

Figure 6-2 The different regions of operation of gas-filled detectors. The observed pulse amplitude is plotted for events depositing two different amounts of energy within the gas.

multiplication is dependent on the magnitude of the electric field, some nonlinearities will begin to be observed. These effects mark the onset of the region of *limited proportionality* in which the pulse amplitude still increases with increasing number of initial ion pairs, but not in a linear fashion.

If the applied voltage is made sufficiently high, the space charge created by the positive ions can become completely dominant in determining the subsequent history of the pulse. Under these conditions, the avalanche proceeds until a sufficient number of positive ions have been created to reduce the electric field below the point at which additional gas multiplication can take place. The process is then self-limiting and will terminate when the same total number of positive ions have been formed regardless of the number of initial ion pairs created by the incident radiation. Then each output pulse from the detector is of the same amplitude and no longer reflects any properties of the incident radiation. This is the *Geiger–Mueller* region of operation, and detectors operated under these conditions are discussed in Chapter 7. A related mode of operation, known as *self-quenched streamer* mode, is of interest in position-sensitive detectors and is examined in Chapter 19.

C. Choice of Geometry

Typical proportional counters are constructed with the cylindrical geometry illustrated in Fig. 6-3. The anode consists of a fine wire that is positioned along the axis of a large hollow tube that serves as the cathode. The polarity of the applied voltage in this configuration is important, because the electrons must be attracted toward the center axial wire. This polarity is necessary from two standpoints:

1. Gas multiplication requires large values of the electric field. In cylindrical geometry, the electric field at a radius r is given by

$$\mathscr{E}(r) = \frac{V}{r \ln(b/a)} \tag{6-3}$$

where V = voltage applied between anode and cathode

a = anode wire radius

b = cathode inner radius.

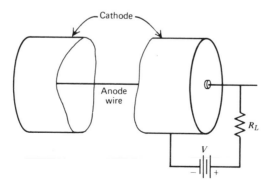

Figure 6-3 Basic elements of a proportional counter. The outer cathode must also provide a vacuum-tight enclosure for the fill gas. The output pulse is developed across the load resistance R_L.

Large values of the electric field therefore occur in the immediate vicinity of the anode wire where r is small. Because electrons are attracted to the anode, they will be drawn toward the high-field region.

To illustrate, suppose a voltage V of 2000 V is applied to a cylindrical counter with $a = 0.008$ cm and $b = 1.0$ cm. The electric field at the anode surface is then 5.18×10^6 V/m. In parallel plate geometry, the field would be uniform between the planar electrodes. With a spacing of 1.0 cm, an applied voltage of 51,800 V would be required to achieve the same electric field. Such a high voltage is practically unworkable, and therefore small-diameter anode wires are used in proportional counters to generate the high field required.

2. If uniform multiplication is to be achieved for all ion pairs formed by the original radiation interaction, the region of gas multiplication must be confined to a very small volume compared with the total volume of the gas. Under these conditions, almost all primary ion pairs are formed outside the multiplying region, and the primary electron simply drifts to that region before multiplication takes place. Therefore, each electron undergoes the same multiplication process regardless of its original position of formation, and the multiplication factor will be the same for all original ion pairs.

In the previous example, suppose that the threshold electric field for multiplication in the fill gas is 10^6 V/m. From Eq. (6-3), the field exceeds this value only for radii less than 0.041 cm, or about five times the anode radius. The volume contained within this radius is only about 0.17% of the total counter volume.

To help visualize the avalanche formation in the neighborhood of the wire surface, Fig. 6-4 shows results obtained from a Monte Carlo modeling of the electron multiplication and diffusion processes that are important in a typical avalanche. It was assumed

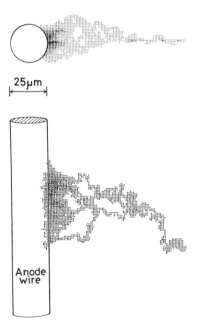

Figure 6-4 Orthogonal views of an avalanche triggered by a single electron as simulated by a Monte Carlo calculation. The density of the shading indicates the concentration of electrons formed in the avalanche. (From Matoba et al.[3])

that a single free electron drifted into the vicinity of the wire. The resulting avalanche is confined to a small distance along the length of the wire equivalent to only several times its diameter. As a consequence, methods for sensing its position along the wire (discussed later in this chapter) can accurately measure the axial position of the incident electron. The avalanche also covers only a limited range of the wire circumference, oriented generally toward the incident electron direction.

II. DESIGN FEATURES OF PROPORTIONAL COUNTERS

A. Sealed Tubes

A sketch of a proportional counter incorporating many common design features is shown in Fig. 6-5. The thin axial wire anode is supported at either end by insulators that provide a vacuum-tight electrical feedthrough for connection to the high voltage. The outer cathode is conventionally grounded, so that positive high voltage must be applied to assure that electrons are attracted toward the high-field region in the vicinity of the anode wire. For applications involving neutrons or high-energy gamma rays, the cathode wall can be several millimeters thick to provide adequate structural rigidity. For low-energy gamma rays, X-rays, or particulate radiation, a thin entrance "window" can be provided either in one end of the tube or at some point along the cathode wall.

Good energy resolution in a proportional counter is critically dependent on assuring that each electron formed in an original ionization event is multiplied by the same factor in the gas multiplication process. The most important mechanical effect that can upset this proportionality is a distortion of the axially uniform electric field predicted by Eq. (6-3). One potential source is any variation in the diameter of the anode wire along its length. Then the value of a in Eq. (6-3) will no longer be constant everywhere along the length of the tube and the degree of gas multiplication will vary from point to point. Because it is easier to make a large diameter wire uniform to within a given fractional tolerance, this effect is minimized by avoiding the use of extremely fine anode wires. However, the applied voltage required to generate a given minimum field increases with wire diameter, and therefore some compromise must be struck.

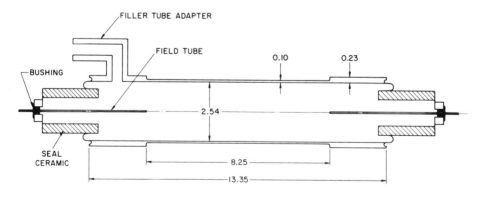

Figure 6-5 Cross-sectional view of a specific proportional tube design used in fast neutron detection. The anode is a 0.025 mm diameter stainless steel wire. The field tubes consist of 0.25 mm diameter hypodermic needles fitted around the anode at either end of the tube. (From Bennett and Yule.[4])

Other common causes of field distortion are effects that occur near the ends of the tube. The electric field can become extremely distorted near the point at which the anode wire enters the insulator, due to the presence of the end wall of the tube or other conducting structures in the vicinity. Unless precautions are taken to avoid these difficulties, ion pairs created in these end regions will undergo a degree of gas multiplication different from those created more generally throughout the volume of the tube.

The most common solution to the end effect problem is to design the counter so that events occurring near the ends do not undergo any gas multiplication and so that an abrupt transition takes place between these "dead" regions and the remainder of the active volume of the proportional tube. One method of achieving this condition is through the use of *field tubes*. As illustrated in Fig. 6-5, field tubes consist of short lengths of conducting tubing (with a diameter many times greater than that of the anode), which are positioned around the anode wire at each end of the counter. The voltage applied to the field tube can be maintained at the anode voltage, but because its diameter is large, no gas multiplication will take place in the region between the field tube and the cathode. The end of the field tube then marks the beginning of the active volume of the proportional counter, which can be some distance away from either end of the overall counter. Alternatively, the field tube may be operated at an intermediate potential between that of the anode and cathode. By setting this potential to that which would exist in its absence at the same radial position, the field within the active volume of the counter is less disturbed by the presence of the field tube, and Eq. (6-3) will describe the radial dependence of the electric field to a very good approximation everywhere within the active volume. Detailed analyses of the electric field in the vicinity of the field tubes can be found in Refs. 4 and 5. Other solutions to the end effect problem involve correction of the field through the use of a semiconducting end plate[6] or by conducting rings on an insulating end plate which are maintained at the proper potential to avoid distortion of the field near the tube end.[7]

B. Windowless Flow Counters

Another common configuration for the proportional counter is sketched in Fig. 6-6. Here it is assumed that the source of radiation is a small sample of a radioisotope, which can then be introduced directly into the counting volume of the detector. The great advantage of counting the source internally is the fact that no entrance window need come between the radiation source and active volume of the counter. Because window materials can

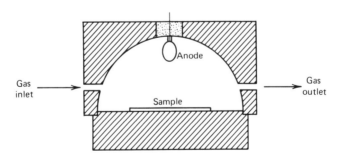

Figure 6-6   Diagram of a 2π gas flow proportional counter with a loop anode wire and hemispherical volume. The sample can often be inserted into the chamber by sliding a tray to minimize the amount of air introduced.

seriously attenuate soft radiations such as X-rays or alpha particles, the internal source configuration is most widely used for these applications.

The counting geometry can also be made very favorable for this type of detector. In the system shown in Fig. 6-6, virtually any quantum emerging from the surface of the source finds its way into the active volume of the counter and can generate an output pulse. The effective solid angle is therefore very close to 2π, and the detector can have an efficiency that is close to the maximum possible for sources in which the radiation emerges from one surface only.

Some means must be provided to introduce the source into the active volume of internal source counters. In some designs, the base of the counter consists of a rotating table in which several depressions or source wells are provided at regular angular intervals. The table can be rotated so that one well at a time is placed directly under the center of the tube active volume. At the same time, at least one other well is accessible from outside the chamber and can be loaded with a new source to be counted. By rotating the table, the new source can be brought into counting position and the prior sample removed. In other designs, the chamber may simply be opened to allow insertion of the source. In either case, some air will enter the chamber and must be eliminated. The counter is therefore purged by allowing a supply of proportional gas to flow through the chamber for some time and sweep away the residual air. Most proportional counters of this type are also operated as flow counters in which the gas continues to flow at a slower rate through the chamber during its normal use.

If the source is prepared on a backing that is thin compared with the range of the radiation of interest, particles (or photons) may emerge from either surface. In that case, the 4π geometry sketched in Fig. 6-7 can take advantage of the added counting efficiency

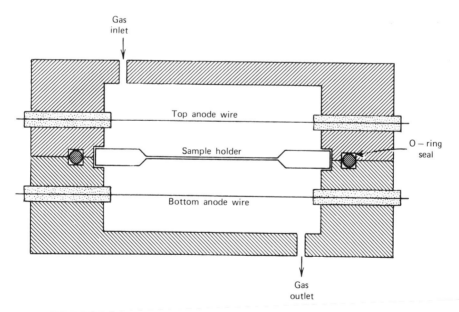

Figure 6-7 A 4π gas flow proportional counter used to detect radiations that emerge from both surfaces of the sample. The top and bottom halves are provided with separate anode wires and can be separated to introduce the sample that is mounted between them.

possible from such sources. The two halves of the chamber can be operated independently or in coincidence to carry out selection of the event to be counted.

C. Fill Gases

Because gas multiplication is critically dependent on the migration of free electrons rather than negative ions, the fill gas in proportional counters must be chosen from those species that do not exhibit an appreciable electron attachment coefficient. Because air is not one of these, proportional counters must be designed with provision to maintain the purity of the gas. The gas can be either permanently sealed within the counter or circulated slowly through the chamber volume in designs of the continuous flow type. Sealed counters are more convenient to use, but their lifetime is sometimes limited by microscopic leaks which lead to gradual contamination of the fill gas. Continuous flow counters require gas supply systems that can be cumbersome but bypass many potential problems involving gas purity. The continuous flow design also permits the flexibility of choosing a different fill gas for the counter when desired. These systems may be of the "once through" type, in which the exit gas is vented to the atmosphere, or the gas may be recycled after passage through a purifier. The purifier must remove traces of oxygen or other electronegative impurities, and in its most common form[8] consists of a heated porcelain tube filled with calcium turnings maintained at a temperature of 350°C. The influence of electronegative impurities is most pronounced in large-volume counters and for small values of the electron drift velocity. In carbon dioxide, a gas with relatively slow electron drift, it has been shown[9] that an oxygen concentration of 0.1% results in the loss of approximately 10% of the free electrons per centimeter of travel.

Gas multiplication in the proportional counter is based on the secondary ionization created in collisions between electrons and neutral gas molecules. In addition to ionization, these collisions may also produce simple excitation of the gas molecule without creation of a secondary electron. These excited molecules do not contribute directly to the avalanche but decay to their ground state through the emission of a visible or ultraviolet photon. Under the proper circumstances, these deexcitation photons could create additional ionization elsewhere in the fill gas through interaction with less tightly bound electron shells or could produce an electron by photoelectric interaction at the wall of the counter. Although such photon-induced events are important in the Geiger–Mueller region of operation, they are generally undesirable in proportional counters because they can lead to a loss of proportionality and/or spurious pulses. It has been found that the addition of a small amount of polyatomic gas, such as methane, to many of the common fill gases will suppress the photon-induced effects by preferentially absorbing the photons in a mode that does not lead to further ionization. Most monatomic counter gases operated at high values of gas multiplication require the use of such a polyatomic stabilizing additive. This component is often called the *quench gas*.

The noble gases, either pure or in binary mixtures, can be useful proportional gases provided the gas multiplication factor is kept below about 100 (Ref. 10). Beyond this point, adding a quench gas is helpful in reducing instabilities and proportionality loss caused by propagation of ultraviolet photons. Because of cost factors, argon is the most widely used of the inert gases, and a mixture of 90% argon and 10% methane, known as P-10 gas, is probably the most common general-purpose proportional gas. When applications require high efficiency for the detection of gamma-ray photons by absorption within the gas, the heavier inert gases (krypton or xenon) are sometimes substituted. Many hydrocarbon gases, such as methane, ethylene, and so on, are also suitable proportional gases and are widely applied where stopping power is not a major consideration. In applications where the signal is used for coincidence or fast timing purposes, gases with high electron drift velocities are preferred (see Fig. 6-13). Proportional counters used for thermal neutron detection are operated with BF_3 or ^3He as the proportional gas (see Chapter 14), whereas proportional counters applied in fast neutron spectroscopy are filled with hydrogen, methane, helium, or some other low-Z gas (see Chapter 15). In dosimetry studies, it is often convenient to choose a fill gas that has the approximate composition of biological tissue. For such purposes, a mixture consisting of 64.4% methane, 32.4% carbon dioxide, and 3.2% nitrogen is recommended.[11]

Even though oxygen is electronegative, it has been shown[12,13] that air can serve as an acceptable proportional gas under special circumstances. If the distance that electrons must travel is only 1–2 mm and the electric field is kept high in the drifting region, enough electrons escape attachment to form small but detectable avalanches. For stable operation, ambient air must generally be purged of water vapor and organic vapors[14] before introduction into the counter volume.

The basic properties of a fill gas can be changed significantly by small concentrations of a second gas whose ionization potential is less than that of the principal component. One mechanism, known as the *Penning effect*, is related to the existence of long-lived or metastable excited states in the principal gas. If the excitation energy is larger than the ionization energy of the added component, then a collision between the metastable excited atom and a neutral additive atom can ionize the additive. Because the excitation energy would otherwise be lost without the additive, a greater number of ion pairs will be formed per unit energy lost by the incident radiation. For example, the W-value for argon can be reduced from 26.2 to 20.3 eV through the addition of a small concentration of ethylene.[15] Furthermore, because a greater fraction of the incident radiation energy is converted into ions, the relative fluctuation in the total number of ions is decreased by as much as a factor of two.[15] Because of the corresponding improvement in energy resolution, fill gases that consist of Penning mixtures are commonly chosen for proportional counters applied in radiation spectroscopy.[16-18]

III. PROPORTIONAL COUNTER PERFORMANCE

A. Gas Multiplication Factor

A study of the multiplication process in gases is normally subdivided into two parts. The single-electron response of the counter is defined as the total charge that is developed by gas multiplication if the avalanche is initiated by a single electron originating outside the region of gas multiplication. This process can be studied experimentally by creating irradiation conditions in which only one electron is liberated per interaction (e.g., by the photoelectric interaction of incident ultraviolet photons). Multiplication factors of the order of 10^5 are adequate to allow direct detection of the resulting pulses, and their amplitude distribution can supply information about the gas multiplication mechanisms within the counter. Studies of this type are described in Refs. 19–22. Results of these experiments are used later in this chapter under the discussion of energy resolution of proportional counters.

If the single-electron response is known, the amplitude properties of pulses produced by many original ion pairs can be deduced. Provided that space charge effects (discussed later in this chapter) are not large enough to distort the electric field, each avalanche is independent, and the total charge Q generated by n_0 original ion pairs is

$$Q = n_0 eM \qquad (6\text{-}4)$$

where M is the average *gas multiplication factor* that characterizes the counter operation.

Analyses have been carried out[23-27] which attempt to derive a general expression for the expected factor M in terms of the tube parameters and applied voltage. Physical assumptions that are usually made for simplification are that the only multiplication process is through electron collision (any photoelectric effects are neglected), that no electrons are lost to negative ion formation, and that space charge effects are negligible. The solution to the Townsend equation [Eq. (6-1)] in cylindrical geometry must take into account the radial dependence of the Townsend coefficient α caused by the radial variation of the electric field strength. In general, the mean gas amplification factor M can then be written

$$\ln M = \int_a^{r_c} \alpha(r)\, dr \qquad (6\text{-}5)$$

where r represents the radius from the center of the anode wire. The integration is carried out over the entire range of radii over which gas multiplication is possible, or from the anode radius a to the critical radius r_c beyond which the field is too low to support further gas multiplication. α is a function of the gas type and the magnitude of the electric field, $\mathscr{E}(r)$, and data on its behavior can be found in Ref. 28. Equation (6-5) is normally rewritten explicitly in terms of the electric field as

$$\ln M = \int_{\mathscr{E}(a)}^{\mathscr{E}(r_c)} \alpha(\mathscr{E}) \frac{\partial r}{\partial \mathscr{E}} d\mathscr{E} \qquad (6\text{-}6)$$

Now by introducing Eq. (6-3) for the shape of the electric field in cylindrical proportional tubes, we can write

$$\ln M = \frac{V}{\ln(b/a)} \int_{\mathscr{E}(a)}^{\mathscr{E}(r_c)} \frac{\alpha(\mathscr{E})}{\mathscr{E}} \frac{d\mathscr{E}}{\mathscr{E}} \qquad (6\text{-}7)$$

Various analytic expressions that appear in the literature to relate the gas multiplication factor to experimental parameters differ from each other in the form of the expression used for $\alpha(\mathscr{E})$. A review of possible options is presented in Refs. 29 and 30.

By assuming linearity between α and $\mathscr{E}$, Diethorn[31] derived a widely used expression for M:

$$\ln M = \frac{V}{\ln(b/a)} \cdot \frac{\ln 2}{\Delta V} \left(\ln \frac{V}{pa \ln(b/a)} - \ln K \right) \qquad (6\text{-}8)$$

where
$$M = \text{gas multiplication factor}$$
$$V = \text{applied voltage}$$
$$a = \text{anode radius}$$
$$b = \text{cathode radius}$$
$$p = \text{gas pressure.}$$

According to the model used in the derivation, ΔV corresponds to the potential difference

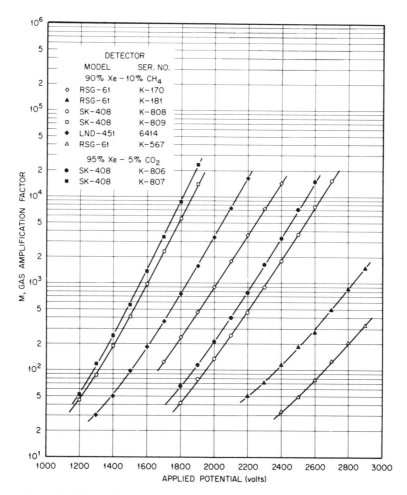

Figure 6-8 Variation of the gas multiplication factor M with voltage applied to various proportional counters. The tubes differ in their physical characteristics, but only the two indicated gases were used. (From Hendricks.[32])

through which an electron moves between successive ionizing events, and K represents the minimum value of $\mathscr{E}/p$ below which multiplication cannot occur. Both ΔV and K should be constants for any given fill gas.

Hendricks[32] studied the applicability of this expression to a variety of different counters in which the fill gas was either xenon–methane or xenon–CO_2 mixture. The gain versus voltage characteristics are shown in Fig. 6-8 and the same data are reduced to the form of a *Diethorn plot* shown in Fig. 6-9. Wolff[33] demonstrated the satisfactory applicability of Eq. (6-8) to a number of other fill gases listed in Table 6-1. The corresponding values for ΔV and K are listed in the table. The Diethorn approach treats the two constants ΔV and K as empirically fitted parameters. This procedure is normally sufficient to represent experimental data over limited ranges of voltage, pressure, and counter dimensions. Introducing a third fitted parameter[29,30] can be justified on theoretical grounds and leads to generalized models that can represent proportional counter behavior over much wider ranges of operation.

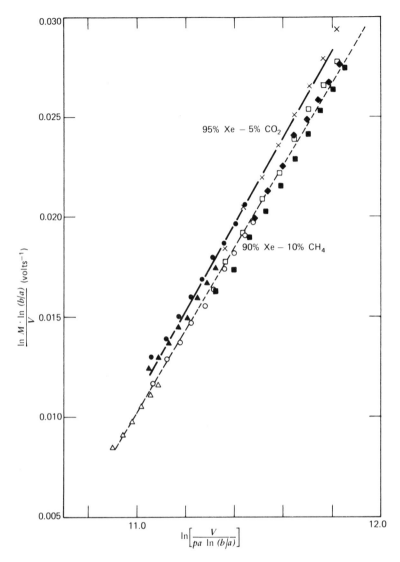

Figure 6-9 A Diethorn plot of the same data shown in Fig. 6-8. As predicted by Eq.
(6-8), a graph of $[\ln M \cdot \ln(b/a)/V]$ versus $\ln[V/pa \ln(b/a)]$ for all counters using the
same gas closely matches a straight line with slope of $(\ln 2/\Delta V)$ and intercept
$(-\ln 2)(\ln K)/\Delta V$. (From Hendricks.[32])

For a given proportional tube at constant gas pressure, Eq. (6-8) shows that the gas
multiplication increases rapidly with applied voltage V. Neglecting the slowly varying
logarithmic term, the multiplication M varies primarily as an exponential function of V.
Proportional counters must therefore be operated with extremely stable voltage supplies
to prevent changes in M over the course of the measurement.

B. Space Charge Effects

In the avalanche process on which proportional counters depend, both electrons and
positive ions are created. The electrons are collected relatively quickly at the anode,

TABLE 6-1   Diethorn Parameters for Proportional Gases[a]

Gas Mixture	K (10^4 V/cm · atm)	ΔV (V)	Reference
90% Ar, 10% CH_4 (P-10)	4.8	23.6	33
95% Ar, 5% CH_4 (P-5)	4.5	21.8	33
100% CH_4 (methane)	6.9	36.5	33
100% C_3H_8 (propane)	10.0	29.5	33
96% He, 4% isobutane	1.48	27.6	33
75% Ar, 15% Xe, 10% CO_2	5.1	20.2	33
69.4% Ar, 19.9% Xe, 10.7% CH_4	5.45	20.3	33
64.6% Ar, 24.7% Xe, 10.7% CO_2	6.0	18.3	33
90% Xe, 10% CH_4	3.62	33.9	32
95% Xe, 5% CO_2	3.66	31.4	32

[a]See Eq. (6-8)

leaving behind the positive ions that move much more slowly and gradually diffuse by drifting outward toward the tube cathode wall. The space charge represented by these net positive charges can, under some circumstances, appreciably distort the electric field from its value if the space charge were absent. Because the ions are formed preferentially near the anode wire where most gas multiplication takes place, the effect of the space charge will be to reduce the electric field at small radii below its normal value. If the magnitude of the effect is sufficiently large, it can have a measurable effect in reducing the size of the output pulse below that which one would ordinarily expect. Furthermore, because the amount of the reduction is likely to vary depending on the detailed geometry of each original ionizing event, the energy resolution of the proportional counter may also be adversely affected by the space charge nonlinearities.

There are two different categories of space charge effects. *Self-induced* effects arise when the gas gain is sufficiently high so that positive ions formed during a given avalanche can alter the field and reduce the number of electrons produced in further stages of the same avalanche. This effect depends on the magnitude of the gas multiplication and the geometry of the tube[34] but does not depend on the pulse rate. The *general* space charge effect includes the cumulative effect of positive ions created from many different avalanches. This effect can be important at lower values of the gas multiplication and becomes more serious as the rate of events within the tube is increased. The general problem of space charge in proportional tubes has been reviewed by Hendricks,[35] who presents a general analytic formulation to allow estimation of the resulting nonlinear effects under a variety of conditions. In general, most authors advocate using gas multiplication factors that are as low as possible consistent with signal-to-noise requirements to avoid potential loss of energy resolution caused by space charge effects within the tube. This recommendation is particularly significant for applications that involve high counting rates.

If the gas multiplication is perfectly linear, one should be able to record exactly the same pulse height spectrum from the detector by decreasing counter voltage and to offset the corresponding drop in gas multiplication by increasing the gain of the amplifier in the subsequent electronics. If space-charge-induced nonlinearities are significant, the large pulse amplitudes will be more severely limited at the higher voltage, and the recorded shape of the spectrum will change. This procedure can serve as a simple check to detect the presence of nonlinearities at the chosen operating voltage of the chamber.

C. Energy Resolution

1. STATISTICAL CONSIDERATIONS

The charge Q, which is developed in a pulse from a proportional counter in the absence of nonlinear effects, can be assumed to be the sum of the charges created in each individual avalanche. There will be n_0 of these avalanches, each triggered by a separate electron from the n_0 ion pairs created by the incident radiation. In the following discussion, we let A represent the electron multiplication factor for any one avalanche triggered by a single electron and M represent the average multiplication factor from all the avalanches which contribute to a given pulse:

$$M = \frac{1}{n_0} \sum_{i=1}^{n_0} A_i \equiv \bar{A} \tag{6-9}$$

Also, because eA_i is the charge contributed by the ith avalanche,

$$M = \frac{Q}{en_0}$$

or

$$Q = n_0 eM \tag{6-10}$$

From the discussions given in Chapter 4, the pulse amplitude from the detector normally is proportional to Q. This amplitude is subject to fluctuation from pulse to pulse even in the case of equal energy deposition by the incident radiation because both n_0 and M in Eq. (6-10) will show some inherent variation. Because these factors are assumed to be independent, we can use the error propagation formula [Eq. (3-41)] to predict the expected relative variance in Q

$$\left(\frac{\sigma_Q}{Q}\right)^2 = \left(\frac{\sigma_{n_0}}{n_0}\right)^2 + \left(\frac{\sigma_M}{M}\right)^2 \tag{6-11}$$

It is convenient to rewrite the second term in the above sum in terms of the variance of the *single-electron* multiplication factor A. Again, because each avalanche is assumed to be independent, we can apply the error propagation formula to Eq. (6-9) to obtain

$$\sigma_M^2 = \left(\frac{1}{n_0}\right)^2 \sum_{i=1}^{n_0} \sigma_A^2$$

$$\sigma_M^2 = \frac{1}{n_0} \sigma_A^2 \tag{6-12}$$

where A represents the typical avalanche magnitude. Combining Eqs. (6-11) and (6-12), we obtain

$$\left(\frac{\sigma_Q}{Q}\right)^2 = \left(\frac{\sigma_{n_0}}{n_0}\right)^2 + \frac{1}{n_0}\left(\frac{\sigma_A}{\bar{A}}\right)^2 \tag{6-13}$$

This expression now allows an analysis of the expected fractional variance of the pulse amplitude $(\sigma_Q/Q)^2$ in terms of the separate contributions of ion pair fluctuations $(\sigma_{n_0}/n_0)^2$ and single-electron multiplication variations $(\sigma_A/\overline{A})^2$.

a. Variations in the Number of Ion Pairs

The first term in Eq. (6-13) represents the relative variance in the original number n_0 of ion pairs. These fluctuations were discussed at the beginning of Chapter 5 and can be expressed in terms of the Fano factor F:

$$\sigma_{n_0}^2 = Fn_0$$

or

$$\left(\frac{\sigma_{n_0}}{n_0}\right)^2 = \frac{F}{n_0} \tag{6-14}$$

Estimates of the size of the Fano factor for proportional gases range from about 0.05 to 0.20, and some specific values are given later in Table 6-2. The lowest values of F are generally observed for binary gas mixtures in which the Penning effect is important (see prior discussion under fill gas properties).

b. Variations in Single-Electron Avalanches

The second term in Eq. (6-13) represents the contribution of the fluctuations in single-electron avalanche magnitude, a topic that has been the subject of considerable experimental and theoretical investigation. Some examples can be found in Refs. 22, 36, 37, and 38. Reference 22 also contains a fairly complete citation of the earlier literature on this topic.

A simple theoretical prediction for the single-electron avalanche distribution can be carried out with the assumption that the probability of ionization by an electron is dependent only on the electric field strength and is independent of its previous history. It can then be shown that the expected distribution in the number of electrons produced in a given avalanche should be predicted by the Furry distribution,

$$P(A) = \frac{\left(1 - 1/\overline{A}\right)^{A-1}}{\overline{A}} \tag{6-15}$$

where A = avalanche multiplication, or number of electrons in the avalanche

$\overline{A}$ = mean value of A $(= M)$.

If $\overline{A}$ is reasonably large, say greater than 50 or 100 as is almost always the case, then the Furry distribution reduces to a simple exponential form,

$$P(A) \cong \frac{e^{-A/\overline{A}}}{\overline{A}} \tag{6-16}$$

which predicts a relative variance

$$\left(\frac{\sigma_A}{\overline{A}}\right)^2 = 1$$

Experiments carried out at low values of the electric field tend to confirm the predicted exponential shape of the avalanche distribution. However, at higher fields, which are more typical of proportional counter operation, a somewhat more complex model must be used to represent experimental results adequately.

In strong electric fields, the probability of ionization by an electron can no longer be considered totally independent of its past history, and the conditions leading to the exponential distribution for single electrons are violated. Single-electron spectra measured under these conditions tend to show a peaked distribution in which the number of low-amplitude events is suppressed below that predicted by the exponential distribution. In a model proposed by Byrne,[39] the simple exponential distribution of Eq. (6-16) is replaced by a Polya distribution,

$$P(A) = \left(\frac{A(1 + \theta)}{\bar{A}} \right)^{\theta} \exp\left(\frac{-A(1 + \theta)}{\bar{A}} \right) \tag{6-17}$$

where θ is a parameter related to the fraction of electrons whose energy exceeds a threshold energy for ionization[36] and is in the range $0 < \theta < 1$. It can be shown that the relative variance predicted by the Polya distribution is

$$\left(\frac{\sigma_A}{\bar{A}} \right)^2 = \frac{1}{\bar{A}} + b \tag{6-18}$$

where $b \equiv (1 + \theta)^{-1}$ and is observed to have a value of about 0.5. For large values of the multiplication factor $\bar{A}$,

$$\left(\frac{\sigma_A}{\bar{A}} \right)^2 \cong b \tag{6-19}$$

c. Overall Statistical Limit

Regardless of the form assumed for the single-electron avalanche distribution, the distribution in pulse amplitude Q for large values of n_0 (say > 20) approaches a Gaussian distribution in shape. Because this is true in most applications, a symmetric Gaussian-shaped peak in the pulse amplitude distribution from monoenergetic radiation should be expected from fluctuations in n_0 and A.

In order to predict the relative variance of this distribution, we return to Eq. (6-13):

$$\left(\frac{\sigma_Q}{Q} \right)^2 = \left(\frac{\sigma_{n_0}}{n_0} \right)^2 + \frac{1}{n_0} \left(\frac{\sigma_A}{\bar{A}} \right)^2$$

Evaluating $(\sigma_{n_0}/n_0)^2$ from Eq. (6-14) and $(\sigma_A/\bar{A})^2$ from Eq. (6-19), we obtain

$$\left(\frac{\sigma_Q}{Q} \right)^2 = \frac{F}{n_0} + \frac{b}{n_0}$$

$$\left(\frac{\sigma_Q}{Q} \right)^2 = \frac{1}{n_0}(F + b) \tag{6-20}$$

where F is the Fano factor (typical value of 0.05–0.20) and b is the parameter from the Polya distribution that characterizes the avalanche statistics (typical value of 0.4–0.7). From the relative magnitudes of F and b, we see that the pulse amplitude variance is dominated by the fluctuations in avalanche size and that the fluctuations in the original number of ion pairs are typically a small contributing factor.

TABLE 6-2 Resolution-Related Constants for Proportional Gases

Gas	W (eV/ion pair)	Fano Factor F Calculated[a]	Measured	Multiplication Variance b	Energy Resolution at 5.9 keV Calculated[b]	Measured
Ne	36.2	0.17		0.45	14.5%	
Ar	26.2	0.17		0.50	12.8%	
Xe	21.5		≤ 0.17			
Ne + 0.5% Ar	25.3	0.05		0.38	10.1%	11.6%
Ar + 0.5% C_2H_2	20.3	0.075	≤ 0.09	0.43	9.8%	12.2%
Ar + 0.8% CH_4	26.0	0.17	≤ 0.19			
Ar + 10% CH_4	26 [c]			0.50	12.8%	13.2%

[a]From Alkhazov et al.[15]
[b]Given by $2.35[W(F + b)/5900 \text{ eV}]^{1/2}$ [see Eq. (6-22)].
[c]From Wolff.[33]

Source: Adapted from Sipila.[40]

The relative standard deviation of the pulse amplitude distribution is obtained by taking the square root of Eq. (6-20):

$$\frac{\sigma_Q}{Q} = \left(\frac{F + b}{n_0} \right)^{1/2} \tag{6-21}$$

Because $n_0 = E/W$, where E is the energy deposited by the incident radiation and W is the energy required to form one ion pair

$$\frac{\sigma_Q}{Q} = \left(\frac{W(F + b)}{E} \right)^{1/2} \tag{6-22}$$

$$\frac{\sigma_Q}{Q} = \left(\frac{C}{E} \right)^{1/2} \tag{6-23}$$

where $C = W(F + b)$ and is constant for a given fill gas. The statistical limit of the energy resolution of a proportional counter is thus expected to vary inversely with the square root of the energy deposited by the incident radiation. Using values of $W = 35$ eV/ion pair, $F = 0.20$, and $b = 0.61$ (the lower limit estimated by Byrne), we calculate a value for the constant in Eq. (6-23) of 0.0283 keV. Because the conventional definition of energy resolution (see Fig. 4-5) is given by 2.35 times the relative standard deviation, the expected energy resolution should be about 12.5% at 10 keV and 3.9% at 100 keV for our example.

Various proportional gases will have somewhat different values for W, F, and b, and some specific cases are listed in Table 6-2. The limiting resolution is proportional to $\sqrt{W(F + b)}$, so this parameter can serve as a guide when comparing the potential resolution obtainable from different gases. Recent measurements[40] indicate that the parameter b may be somewhat smaller than originally estimated by Byrne, and these values are included in the table.

There is some indication[37,40] that the relative variance of the gas multiplication is smallest at low values of the multiplication factor. This implies that the parameter b may be mildly dependent on the electric field, which is not an unreasonable assumption. Several authors therefore advocate the use of the smallest possible voltage and gas multiplication consistent with keeping electronic noise to a negligible level if the ultimate in proportional counter energy resolution is to be achieved. Low values of the multiplication also will minimize potential harmful effects that can arise due to nonlinearities caused by space charge effects (see previous discussion).

2. OTHER FACTORS AFFECTING ENERGY RESOLUTION

Under well-controlled circumstances, the energy resolution actually observed for proportional counters comes very close to the statistical limit predicted by Eq. (6-22). Some data gathered by Charles and Cooke[41] on the observed energy resolution of a proportional counter for low-energy photons are shown in Fig. 6-10. The data points are very well described by the square root energy dependence given by Eq. (6-23). Other data[42] tend to confirm this energy dependence at gas pressures of 1 atm or less, but the resolution was observed to worsen somewhat at higher pressures.

In order to achieve energy resolutions that approach this statistical limit, care must be taken to minimize potentially harmful effects of electronic noise, geometric nonuniformities in the chamber, and variations in the operating parameters of the detector. The noise

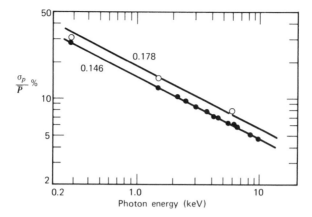

Figure 6-10 Fractional standard deviation of the peak (energy resolution/2.35) for a proportional counter filled with P-10 gas at 1 atm. Two sets of measurements are shown, one for a "standard" tungsten anode wire of 0.05 mm diameter (○) and the other for a similar wire of improved uniformity (●). The corresponding values for $\sqrt{C}$ (in keV$^{1/2}$) in Eq. (6-23) are indicated. (From Charles and Cooke.[41])

levels of modern solid-state preamplifiers are generally sufficiently low so that electronic noise is a negligible contributor to energy resolution, even for applications involving low-energy radiation.

The most critical of the geometric factors is the uniformity and smoothness of the anode wire used in the proportional counter. Variations in the wire diameter as small as 0.5% can be significant.[23] Other factors include possible eccentricity of the wire or nonuniformity of the cylindrical cathode, but these are substantially less critical factors. As discussed earlier in the section on design, the ends of the chamber present special problems. Unless care is taken in shaping the electric field through the use of field tubes or by other methods, some interactions may take place in regions of reduced field. The subsequent avalanches will be smaller than normal and can increase the width of the observed pulse amplitude distribution.

Significant operating parameters that can affect energy resolution are gas purity, gas pressure, and stability of the high voltage applied to the chamber. Trace quantities of electronegative gases can significantly reduce the gas multiplication and introduce additional fluctuations in the pulse amplitude. Changes in gas pressure of the order of a few tenths of a percent can be significant,[23,43] as reflected in the corresponding change in gas multiplication. This is seldom a problem for sealed tubes (except for window flexing caused by changes in barometric pressure) but can be an important cause of resolution loss for gas flow proportional counters in which the gas pressure is not well regulated. Finally, the extreme dependence of the gas multiplication on applied voltage is indicated by Eq. (6-8). Variations in the applied voltage of 0.1 or 0.2% can significantly change the gas multiplication factor under typical conditions, and thus voltage supplies must be well regulated and free from long-term drifts if the ultimate energy resolution of a proportional counter is to be preserved.

It is also commonly observed[44] that changes in the counting rate from proportional counters can influence the gas multiplication factor. If the irradiation rate varies during the course of a measurement, the associated gain change can adversely affect the energy resolution. While a full physical explanation for rate-dependent effects cannot always be identified, some potential factors[45] include the following: space charge buildup in the multiplication region due to slowly moving positive ions, enhanced volume recombination near the anode, possible accumulation on the anode wire of a loosely bound layer of polarizable molecules, and changes in the ballistic deficit (see p. 184) due to differences in the mean pulse rise time. Whatever the specific cause(s) may be, the *count rate effect* has proved to be a significant source of resolution loss in many applications of proportional counters.

When used over long periods of time, proportional counters can show "aging" effects that lead to a drop in multiplication factor and deterioration in energy resolution. Leakage of air into sealed tubes is one possible cause, but such changes can occur even in gas flow counters. One potential cause is the gradual buildup of solid deposits or "whiskers" on the anode wire. In gases that contain methane, the conditions created in an avalanche can cause the formation of polymers that collect on the wire surface. In argon–methane and xenon–methane mixtures, significant performance degradation[46] sets in after the accumulation of enough avalanches to represent a total collected charge of about 10^{-6}–10^{-5} C/mm of wire length. Trace contaminants in fill gases can also lead to the development of corrosion products on the wire that may cause spurious pulses from the counter.[47]

D. Time Characteristics of the Signal Pulse

The same type of analysis carried out in Chapter 5 for pulse-type ion chambers can be applied to derive the shape of the output pulse from proportional counters.[48] However, several major differences arising from the following physical considerations change the character of the output pulse rather dramatically.

1. Virtually all the charge generated within the proportional tube originates within the avalanche region, regardless of where the original ion pairs are formed. The time history of the output pulse is therefore conveniently divided into two stages: the *drift time* required for the free electrons created by the radiation to travel from their original position to the region near the anode wire where multiplication can take place, and the *multiplication time* required from the onset of the avalanche to its completion.

For any given original ion pair, the contribution to the output pulse during the drift time is negligible compared with the contribution of the much larger number of charge carriers formed in the subsequent avalanche. The effect of the drift time therefore is to introduce a delay between the time of formation of the ion pair and the start of the corresponding output pulse. The drift time (perhaps a microsecond or so) is normally much greater than the multiplication time and varies depending on the radial position of the original ion pair within the tube.

2. Because most of the ions and electrons are created very close to the anode wire, the bulk of the output pulse is attributable to drift of the positive ions rather than the movement of the electrons. At first, the positive ions are in a high-field region and move rapidly, leading to a fast-rising initial component of the pulse. Eventually, however, the ions reach regions of the tube at greater radii where the field is smaller, and their drift velocity decreases. The latter part of the pulse therefore rises very slowly and is often not observed in practice due to the finite shaping times of the subsequent electronic circuits.

A simplified analysis of the pulse shape due to the collection of electrons and positive ions formed in an avalanche follows, using arguments that extend the derivation beginning on p. 151 for parallel plate ion chambers. One major difference is that cylindrical

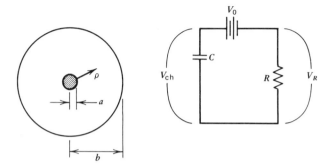

Figure 6-11 On the left is shown a cross-section of the cylindrical geometry used in deriving the pulse shape induced by a single avalanche. The equivalent circuit across which the signal is developed is at the right. Here C represents the capacitance of the detector and associated wiring, V_R is the signal voltage developed across the load resistance R, and V_{ch} is the voltage remaining across the detector.

geometry must now be used to represent the shape of the electric field around the anode wire. In general, the energy absorbed (dE) by the motion of a positive charge (Q) through a difference in electric potential ($d\varphi$) is given by

$$dE = -Q\,d\varphi \tag{6-24}$$

In terms of the electric field $\mathscr{E}(r) = -d\varphi(r)/dr$,

$$\frac{dE}{dr} = Q\mathscr{E}(r) = Q\frac{V_0}{r\ln(b/a)} \tag{6-25}$$

where Eq. (6-3) has been used for the value of $\mathscr{E}(r)$.

Let us first assume that n_0 electrons and positive ions are formed in the avalanche at a fixed distance ρ from the surface of the anode wire (see Fig. 6-11). Setting $Q = n_0 e$, the energy absorbed by the motion of the positive ions to the cathode is then

$$E^+ = \int_{a+\rho}^{b} \frac{dE}{dr}\,dr = \frac{QV_0}{\ln(b/a)}\int_{a+\rho}^{b}\frac{dr}{r}$$

$$= \frac{QV_0}{\ln(b/a)}\ln\frac{b}{a+\rho} \tag{6-26}$$

The energy absorbed by the motion of the negatively charged electrons inward to the cathode is

$$E^- = -\frac{QV_0}{\ln(b/a)}\int_{a+\rho}^{a}\frac{dr}{r} = \frac{QV_0}{\ln(b/a)}\ln\frac{a+\rho}{a} \tag{6-27}$$

The sum of the energy absorbed after both species have been collected is, from Eqs. (6-26) and (6-27),

$$\Delta E = E^+ + E^- = \frac{QV_0}{\ln(b/a)}\ln\left(\frac{b}{a+\rho}\cdot\frac{a+\rho}{a}\right)$$

$$= QV_o \tag{6-28}$$

As in the case of the parallel plate geometry, this energy must come at the expense of the energy stored on the detector capacitance:

$$\tfrac{1}{2}CV_{ch}^2 = \tfrac{1}{2}CV_0^2 - \Delta E$$

$$\tfrac{1}{2}C(V_{ch} + V_0)(V_{ch} - V_0) = -\Delta E$$

Assuming $V_{ch} + V_0 \cong 2V_0$ and substituting $V_R = V_0 - V_{ch}$, we obtain

$$V_R = \frac{\Delta E}{CV_0} = \frac{QV_0}{CV_0} = \frac{Q}{C} \tag{6-29}$$

which is the same result shown for parallel plate geometry. This value is the maximum pulse amplitude that would be developed if the time constant RC is long compared with the ion collection time. In practice, this condition almost never holds for proportional counters. The maximum pulse amplitude then depends on the shape of the voltage–time profile.

Most of the electrons and ions created in an avalanche are formed close to the anode wire surface. The exponential growth characterized by Eq. (6-2) predicts that half will be formed within one mean free path of the anode, typically only a few micrometers from the surface. From Eqs. (6-26) and (6-27), the ratio of the maximum signal amplitude from electron drift to that from ion drift is given by

$$\frac{E^-}{E^+} = \frac{\ln\left[(a + \rho)/a\right]}{\ln\left[b/(a + \rho)\right]} \tag{6-30}$$

Choosing values of $a = 25 \ \mu\text{m}$ and $b = 1 \ \text{cm}$ for the tube dimensions, and assuming that $\rho = 3 \ \mu\text{m}$, we find

$$\frac{E^-}{E^+} = 0.019$$

For this example, less than 2% of the maximum signal results from the motion of electrons, and it is the positive ion drift that dominates the pulse formation. We therefore proceed by neglecting the electron contribution and assuming that the entire signal pulse develops from drift of the ions that are created essentially at the anode wire surface.

From Eqs. (5-2) and (6-3), the drift velocity of the ions varies with radial position as

$$v^+(r) = \mu \frac{\mathscr{E}(r)}{p} = \frac{\mu}{p} \frac{V_0}{\ln(b/a)} \cdot \frac{1}{r} \tag{6-31}$$

By putting this expression into the law of motion,

$$\int_a^{r(t)} \frac{dr}{v^+(r)} = \int_0^t dt$$

and carrying out the integration, we obtain the following expression for the time-dependent position of the ions:

$$r(t) = \left(2 \frac{\mu}{p} \frac{V_0}{\ln(b/a)} t + a^2\right)^{1/2} \tag{6-32}$$

The time required to collect the ions can be found by substituting $r(t) = b$ in the above expression:

$$t^+ = \frac{(b^2 - a^2)p \ln(b/a)}{2\mu V_0} \tag{6-33}$$

Using typical values for the parameters, this collection time is very long, with a representative value of several hundred microseconds. However, a large fraction of the signal is developed during the very early phase of the ion drift. The energy absorbed by the motion of the ions as a function of time is

$$E^+(t) = \frac{QV_0}{\ln(b/a)} \int_a^{r(t)} \frac{dr}{r} = \frac{QV_0}{\ln(b/a)} \ln \frac{r(t)}{a} \tag{6-34}$$

Using Eq. (6-32) for $r(t)$ in the above equation and setting $V_R(t) = E^+(t)/CV_0$, we find the time profile of the signal pulse to be

$$V_R(t) = \frac{Q}{C} \frac{1}{\ln(b/a)} \ln \left(\frac{2\mu V_0}{a^2 p \ln(b/a)} t + 1 \right)^{1/2} \tag{6-35}$$

This equation predicts that the pulse will reach half its maximum amplitude within a time given by

$$t|_{\text{half amplitude}} = \frac{a}{a+b} t^+ \tag{6-36}$$

where t^+ is the full ion collection time given by Eq. (6-33). At this point, the radial position of the ions is given by $\sqrt{ab}$, where the value of the electric field has dropped to a fraction given by $\sqrt{a/b}$ of its value at the anode wire surface.

Again evaluating for $a = 25$ μm and $b = 1$ cm, Eq. (6-36) predicts that the half-amplitude point is reached after only 0.25% of the full ion drift time, typically a fraction of a microsecond. At that point the ions have moved 475 μm from the wire surface, where the electric field is down to 5% of its surface value. This fast leading edge of the pulse is followed by a much slower rise corresponding to the drift of the ions through the lower-field regions found at larger radial distance.

If all original ion pairs are formed at a fixed radius, the electron drift times will all be identical, and Eq. (6-35) will also describe the shape of the output pulse for these events. Most situations, however, involve the formation of ion pairs along the track of the incident radiation and thus cover a range in radii. The spread in electron drift times will introduce additional spread in the rise time of the output pulse. Figure 6-12 shows the shape of the expected leading edge of the output pulse under two conditions: ion pairs formed at a constant radius, and ion pairs uniformly distributed throughout the volume of the counter. The rise time of the output pulse is seen to be considerably greater for the latter case. To minimize this rise time and the associated timing uncertainties, a short electron drift time is helpful. This objective is served by keeping the electric field values as high as possible in the drift region and by choosing a gas with high electron drift velocities.[50] Some examples of both "slow" and "fast" fill gases are shown in Fig. 6-13.

When pulses from proportional counters are shaped using time constants of several microseconds, the slow component of the drift of the ions no longer contributes to the pulse amplitude. The shaped pulse therefore has an amplitude that is less than that

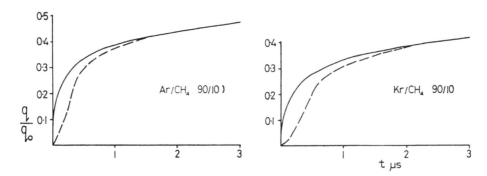

Figure 6-12 Shape of the output pulse leading edge calculated for a typical tube for two different proportional counter gases. In each case, the solid curve represents initial ionization formed at a single radius (constant drift time), whereas the dashed line assumes uniform ionization along a diameter. (From Gott and Charles.[49])

corresponding to an infinite time constant by an amount known as the *ballistic deficit*. If all ion pairs were formed at a constant radius, the shape of the pulse and the ballistic deficit would always be the same, and the net effect would simply be to reduce all pulse amplitudes by a constant factor. When the interactions are randomly distributed over a variety of radii, the pulse shapes will vary depending on the radial distribution of the original ion pairs. Unless shaping times of many microseconds are used, the ballistic deficit will also vary depending on this radial distribution, and the energy resolution of the detector may therefore suffer. The ballistic deficit caused by various pulse-shaping networks for pulses from proportional counters has been the subject of a number of investigations[49, 52–54] which allow estimation of its magnitude. In general, shaping times

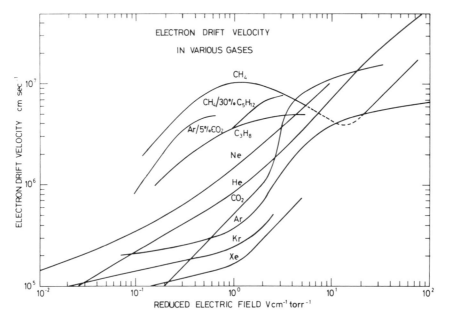

Figure 6-13 The electron drift velocities in various gases, as compiled by Jeavons et al.[51]

that are large compared with the variation in the rise time of the pulses (usually several microseconds) should be used to minimize the effect of variations in the ballistic deficit on energy resolution.

The variations in pulse rise time can sometimes be used to good advantage. Radiations with a short range (heavy charged particles or secondary particles from neutron interactions) will often create ions with a limited range of radii within the tube. Background radiation or undesirable events (such as fast electrons or secondary electrons created by gamma-ray interactions) may have much longer ranges and therefore tend to create ions with a greater spread of initial radii. These background pulses will then have a longer rise time than the desired signal pulses and can be eliminated using methods of rise time discrimination outlined in Chapter 17. This method can be applied advantageously for background reduction in the counting of soft X-ray or low-energy beta particles,[55-57] as well as in the suppression of gamma-ray-induced pulses for the neutron detectors described in Chapters 14 and 15.

E. Spurious Pulses

In some circumstances, satellite pulses may be generated following the primary pulse from a proportional counter. These secondary pulses have nothing to do with the incident radiation but are generated from secondary processes that arise from effects within the primary avalanche. These spurious pulses can lead to multiple counting, where only one pulse should be recorded, and are a potential cause of counter instability.

Spurious afterpulses are often very small, corresponding to the amplitude of an avalanche triggered by a single electron. Under many circumstances, they may therefore be eliminated by simple amplitude discrimination. At moderate values of the electric field, their rate of occurrence may not exceed a few hundredths of a percent of the primary rate in common counter gases. Needlessly high values of the gas multiplication should be avoided in those situations which are sensitive to these satellite pulses, because their probability of generation increases rapidly with the multiplication factor.

The various physical mechanisms leading to the production of spurious pulses have been reviewed by Campion.[11] One important mechanism originates with the optical photons emitted by excited atoms produced in the avalanche. These photons can reach the cathode wall with virtually no time delay and may produce a low-energy electron at the cathode surface through the photoelectric effect. This single electron will then drift toward the anode and create another avalanche, which will be delayed from the primary pulse by the time required for the electron to drift from the cathode to the anode. In counters of ordinary dimensions, this time will be of the order of several microseconds. In cylindrical counters, all drift times between the cathode and anode will be the same, and consequently, if more than one photoelectron is created, the resulting spurious pulse may have an amplitude considerably greater than that caused by a single primary electron.

Other mechanisms, which are not fully understood, can lead to spurious pulses that are delayed by as much as several hundred microseconds from the primary pulse.[58] Possible causes may involve interaction of the positive ion upon collection at the cathode, field emission of electrons at the cathode wall, metastable states of the counter gas, or electrons delayed by space charge effects from a previous avalanche. Spurious pulses with such long delays almost always have an amplitude corresponding to a single primary electron and therefore are detected only when the gas multiplication factor is rather large.

The end of the counting plateau discussed in Section IV is reached (under some circumstances) when the applied voltage is sufficiently high to permit counting of these afterpulses. The apparent count rate then abruptly increases beyond the rate at which true interactions are occurring within the counter. A sensitive method for determining the end of the plateau has been suggested by Campion[59] in which the number of pulses that occur within a fixed period of time (several tens of microseconds) is monitored following a primary pulse from the counter. Under normal circumstances, the number of these pulses will be very low and can accurately be predicted from counting statistics based on the observed counting rate. If spurious pulses begin to be detected, however, the rate of afterpulses will abruptly increase and will clearly signal the end of the counting plateau.

IV. DETECTION EFFICIENCY AND COUNTING CURVES

A. Selection of the Operating Voltage

For charged radiations such as alpha or beta particles, a signal pulse will be produced for every particle that deposits a significant amount of energy in the fill gas. At high values of the gas multiplication, a single ion pair can trigger an avalanche with enough secondary ionization to be detectable using preamplifiers of reasonable noise behavior. Unless required by the application, proportional counters are seldom operated in a mode that is sensitive to single avalanches because the measurement is then prone to nonlinearities for larger pulses due to space charge effects. Furthermore, many background and spurious pulses may also be counted, which often correspond to very few original ion pairs.

Instead, lower values of the gas multiplication are typically used, which require that the pulse originate from a finite number of ion pairs in order to have an amplitude large enough to exceed the discrimination level of the counting system. Because the gas multiplication factor varies with the voltage applied to the detector (see Fig. 6-8), a "counting curve" of the type described in Chapter 4 can be recorded to select an operating voltage appropriate to the specific application. In this measurement, the counting rate is recorded under constant source conditions as the detector voltage is varied. An operating point is then selected which normally corresponds to a flat region or "plateau" on the resulting rate versus voltage curve.

B. Alpha Counting

If almost all pulses from the detector are of the same size, the differential pulse height spectrum has a single isolated peak as shown at the left in the sketch below.

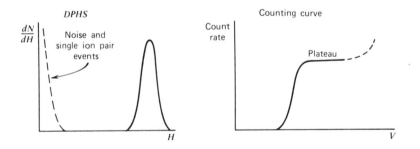

The corresponding counting curve shown at the right then has a simple plateau. An operating point on the counting plateau then assures that all the pulses of interest from the detector are counted. This simple behavior is characteristic of the G-M counter discussed in Chapter 7 but is observed in proportional counters only in special circumstances. One such case is for monoenergetic charged particles whose range in the counter gas is less than the dimensions of the chamber. Alpha particle sources often fall in this category, and proportional counters can therefore easily record each particle that enters the active volume with virtually 100% efficiency. To avoid energy loss in the entrance windows of sealed tubes, windowless flow counters are often used for alpha counting. Absolute measurements of alpha source activity are therefore relatively straightforward and involve evaluation of the effective solid angle subtended by the counter active volume (which can be close to 2π for internal source flow counters) and small corrections for alpha self-absorption and scattering in the source itself.

C. Beta Counting

For beta particles of typical energies, the particle range greatly exceeds the chamber dimensions. The number of ion pairs formed in the gas is then proportional to only that small fraction of the particle energy lost in the gas before reaching the opposite wall. Beta particle pulses are therefore normally much smaller than those induced by alpha particles of equivalent kinetic energy and will also cover a broader range of amplitude because of the spread in beta particle energies and the variations in possible paths through the gas. A differential pulse height spectrum, which might represent a typical source of mixed beta and alpha activity, is sketched at the left portion of the diagram below. The corresponding counting curve now shows two plateaus: the first at the point at which only alpha particles are counted, and the second where both alpha and beta particles are counted. Because the beta particle pulse height distribution is broader and less well separated from the low-amplitude noise, the "beta plateau" is generally shorter and shows a greater slope than the "alpha plateau."

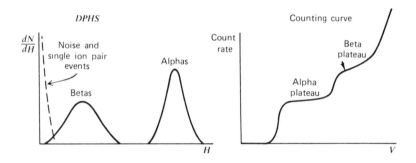

Absolute beta activity measurements are often carried out in a 4π flow counter of the type sketched in Fig. 6-7. The greater range of beta particles allows preparation of sources on backings that are sufficiently thin to allow most of the particles to emerge from either side of the backing, and only small corrections are required to account for those particles that do not reach either half of the chamber active volume.

D. Mixed Sources

It is often desired to count only the alpha particle component of a mixed alpha–beta emitting source. With a proportional counter, it is simple to use the inherent difference between the typical alpha and beta pulse amplitude to choose an operating voltage on the alpha plateau which automatically carries out this discrimination. This is the normal mode of operation, for example, when counting low-level samples of environmental alpha activity. As a bonus, the background rate under these circumstances can be extremely low. Most sources of background (see Chapter 20) such as cosmic rays or ambient gamma rays produce ionization of low density typical of beta particles and are therefore also eliminated by the inherent amplitude discrimination. The residual background is largely due to the natural alpha activity of the counter construction materials and can be as low as 1 or 2 counts per hour.

E. X-Ray and Gamma-Ray Sources

Proportional counters can be used for the detection and spectroscopy of soft X-rays or gamma rays whose energy is low enough to interact with reasonable efficiency in the counter gas. Figure 6-14 shows the interaction probability in a 5.08 cm thickness of the common high-Z proportional gases at STP. The discontinuities caused by the gas absorption edges (discussed in Chapter 2) are very evident. For xenon, the useful response can be extended appreciably above 100 keV photon energy. The spectroscopy of low-energy X-rays is one of the most important applications of proportional counters and is based on fully absorbing the photoelectrons formed by photon interactions within the gas. Because the photoelectron energy is directly related to the X-ray energy (see Chapter 2), photon energies can be identified from the position of corresponding "full-energy" peaks in the measured pulse height spectrum.

The response function for low-energy X-rays can be complicated by several effects related to characteristic X-rays generated by interaction of the primary radiation within the detector. The most significant involves the characteristic K X-rays which usually follow the photoelectric absorption of the primary radiation in the fill gas. Because the

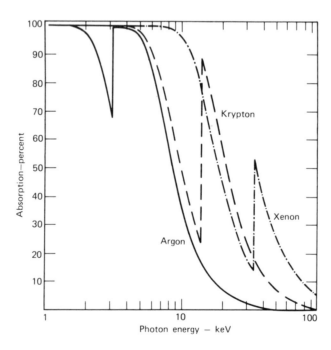

Figure 6-14 Fraction of incident photons absorbed in a 5.08 cm thick layer of several proportional gases at 1 atm pressure.

corresponding energy can be relatively large (K_α X-ray energies are 2.97, 12.6, and 29.7 keV in argon, krypton, and xenon, respectively), this X-ray may escape from the gas without further interaction. A corresponding "escape peak" will then appear in the response function which lies below the full-energy peak by an amount equal to the characteristic X-ray energy. This escape peak may be suppressed by choosing a gas whose characteristic energy lies above the incident X-ray energy. Other peaks in the response function may also arise from the absorption of characteristic X-rays generated by interaction of the primary radiation in the entrance window or walls of the counter. A low-Z window material such as beryllium is often chosen to minimize this contribution.

At higher photon energies more typical of gamma rays, proportional counters are no longer attractive as a detector. The counting efficiency becomes very small because the direct interaction probability of the photon in the gas drops rapidly with energy (see Fig. 6-14). The remaining gamma sensitivity then arises from photon interactions in the counter wall for which the resulting secondary electron reaches the counter gas. This process results in an intrinsic efficiency of only a percent or so (see Fig. 7-9) and cannot be used as the basis of energy measurements because of the variable loss of electron energy before reaching the gas. When gas counters are applied to gamma-ray detection, they are more commonly used in the Geiger–Mueller region to take advantage of the resulting simplicity of operation.

V. VARIANTS OF THE PROPORTIONAL COUNTER DESIGN

A. Parallel Plate Avalanche Counters

A variant of the traditional proportional counter has gained some attention[60,61] for applications in which the need for fast timing information is more important than good energy resolution. The *parallel plate avalanche detector* is particularly useful in applications involving heavy charged particles where the associated radiation damage in solid-state detectors (see p. 371) may prohibit their use. The counter consists of two parallel plate electrodes separated by a gap that is normally kept as small as possible for best timing information. Alternatively, fine mesh grids can be used as electrodes[62] instead of solid plates to reduce their thickness. The electrodes are enclosed in a container in which a proportional gas is introduced, and a homogeneous electric field is produced between the plates which, under conditions of relatively low gas pressure, can reach values as high as 4×10^6 V/m · atm (Ref. 61). A charged particle that traverses the gap between the plates leaves a trail of ions and electrons that are multiplied through the usual gas multiplication process. Electrons formed nearest the cathode are obviously subjected to more multiplication than those formed near the anode. Maximum gas gains of the order of 10^4 are possible.

Because typical applications involve particles that fully traverse the gap between the electrodes, the number of ions that are formed reflects the amount of energy lost by the particle in its transit. This type of detector is commonly called a ΔE *detector*, and other examples are given later in this text. Because any typical output pulse contains a mix of gas amplification factors, the energy resolution is seldom better than about 20%. Nonetheless, this performance is often adequate to separate different types of particles with widely different specific energy loss.

The output shape consists of a fast-rising component generated by the electron collection, and a slow component resulting from the motion of the positive ions. For timing purposes, only the fast component is used and the slow portion is eliminated by the shorter time constant of the collection circuit. For gaps of the order of 1 mm, the fast component rise time is about 2 ns under high field conditions.[59] If the energy deposited by the incident particle is relatively uniform, the time resolution (as defined in Chapter 17) can be as short as 160 ps.[61, 63]

B. Position-Sensitive Proportional Counters

Because the position of the avalanche in a proportional counter is limited to a small portion of the anode wire length, some schemes have been developed which are capable of sensing the position of an event taking place within the tube. In the common cylindrical geometry, electrons drift inwardly from their place of formation along radial field lines, so that the position of the avalanche is a good indicator of the axial position at which the original ion pair was formed. If the track of the incident radiation extends for some distance along the length of the tube, then many avalanches will also be distributed along the anode, and only an average position can be deduced.

The most common method of position sensing in proportional tubes is based on the principle of charge division, illustrated in Fig. 6-15. The anode wire is fabricated to have significant resistance per unit length, so that the collected charge is divided between the amplifiers placed at either end of the wire in a proportion that is simply related to the position of the interaction. By summing the output of the two amplifiers, a conventional output pulse is produced with an amplitude proportional to the total charge. A position signal is generated by dividing the output of a single amplifier by the summed signal to give a pulse that indicates the relative position along the anode wire length. This division can be carried out using either analog methods[64] or digital techniques.[65]

An alternative to the charge division method, which is capable of excellent spatial resolution, has been developed by Borkowski and Kopp.[66] The method is based on observing the relative rise time from preamplifiers placed at either end of a resistive anode wire. Events occurring far from one of the preamplifiers will exhibit a longer rise time than those near that preamplifier. A position signal can then be derived from the rise time difference between the pulses produced by the two preamplifiers. The spatial resolution reported for this type of detector for well-collimated alpha particles is 0.15 mm FWHM for a tube of 200 mm total length.[67]

Extensive reviews of position-sensitive proportional counters have been compiled describing their application as X-ray and neutron detectors[68] and as focal plane detectors for magnetic spectroscopy of charged particles.[69]

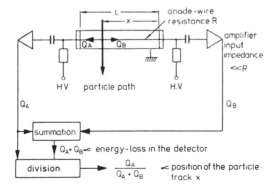

Figure 6-15 Position localization in proportional counters using the charge-division method. (From Fischer.[65])

C. Multiwire Proportional Counters

In some situations, it is advantageous to provide more than one anode wire within a proportional counter. For example, detectors with very large surface area can be constructed economically by placing a grid of anode wires between two large flat plates that serve as cathodes on either side of the counter. Figure 6-16 shows a sketch of this arrangement together with the configuration of the electric field that results. Electrons formed by ionization of the gas drift inward toward the plane of the anode wires, initially in a nearly uniform field. As they approach, they are accelerated toward the nearest wire into its surrounding high-field region where avalanches are formed. Because the signal then appears only on one anode wire, the event is automatically localized in the dimension perpendicular to the wires. One of the position-sensing techniques discussed earlier can then be used to localize the event along the length of the wire, providing two-dimensional localization. Alternatively, the second coordinate can be obtained by dividing one cathode plate into narrow strips perpendicular to the wires and sensing the induced charge on the nearest strip. Devices of similar design, often extending to several square meters in cross-sectional area, serve as a very common position-sensing detector used in high-energy particle research.

While many multiwire chambers are operated in the proportional mode, the relatively low signal amplitude can be a limiting factor in position-sensing applications. It has been demonstrated that the self-quenched streamer mode of operation (see p. 684), a hybrid between proportional and Geiger modes, may be preferable in some applications because of the larger signal size that is produced.

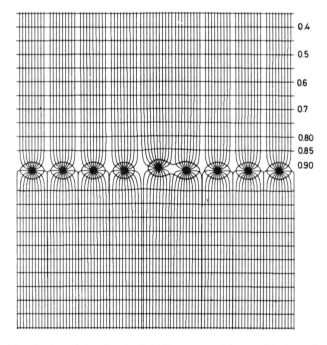

Figure 6-16 A plot of the electric field lines created by a grid of anode wires (running perpendicular to the page) that are placed equidistant between two parallel cathode plates at the top and bottom of the figure. Over much of the volume away from the grid, the field is nearly uniform. A high-field region is created in the immediate vicinity of each grid wire. (From Charpak and Sauli.[70])

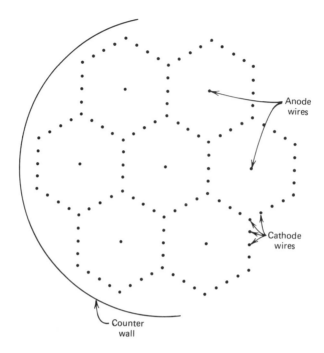

Figure 6-17 The cross-sectional view of a multicell proportional counter. A single anode wire is at the center of each cell, surrounded by a grid of cathode wires.

The use of multiple anode wires can also help in the design of fast response proportional counters. Because the speed of response is largely related to the drift time of the initial electron to the multiplying region, the response time can be minimized by providing many multiplying regions throughout the volume of the detector. For example, Breskin[71] describes a 40 mm × 40 mm planar detector in which anode wires are spaced at 1 mm intervals. When operated with ethylene fill gas at a pressure of 5 torr (666 Pa), a time resolution of 0.8 ns FWHM was achieved using 5.5 MeV alpha particle irradiation.

Multiple wires can also be oriented in a single gas enclosure to form layers or cells that function as independent proportional counters. For example, Fig. 6-17 illustrates a cylindrical arrangement in which a center cell is surrounded by an annulus of cells, each made up of an anode wire at the center of a hexagonal array of wires that act as the cell cathode. Independent signals are derived from each anode. This type of design provides multiple detector units that are in direct contact without intervening walls or dead layers. Through the use of coincidence or anticoincidence selection of events that involve more than one cell, those of a specific type can be preferentially accepted. For example, the background in the inner cell when used to count soft beta particles can be reduced[72] by rejecting longer-range particles that create a coincident signal in one of the outer cells. In another application, the complicating effects of particles that strike the wall of gas-filled detectors can be eliminated from measured energy spectra by restricting events to those taking place in the center cell alone.

D. Gas Proportional Scintillation Counters

An interesting hybrid detector has been developed which combines some of the properties of a proportional counter with those of the scintillation detector discussed in Chapter 8.

The *gas proportional scintillation (GPS) counter* is based on generating a signal pulse from the visible and ultraviolet photons emitted by excited gas atoms or molecules. In a conventional gas scintillator, these excited gas molecules are created by direct interaction of incident radiation as it passes through the gas. The lifetime of the excited states is normally quite short, so that the light appears within a few nanoseconds of the time of passage of the incident particle.

If an electric field is applied to the gas, the free electrons from ion pairs created along the particle track will drift as they do in an ionization chamber or proportional counter. If the field is sufficiently strong, inelastic collisions between these electrons and neutral gas molecules can elevate some of the molecules to excited states, which may then deexcite through the emission of a photon. This process also occurs in a conventional proportional counter, but the photons are treated as something of a nuisance because their subsequent interaction in other parts of the counter can lead to spurious pulses. In the gas proportional scintillation counter, however, the photons emitted by excited gas molecules are detected by a photomultiplier tube, which produces an electrical pulse proportional to the number of photons incident on the tube. The mechanisms involved in the generation of this "secondary scintillation" light in gases have been reviewed in Ref. 73, which also contains an extensive list of citations to early work involving the development of GPS detectors. Current designs and performance are reviewed in Ref. 74.

The light output from a GPS detector consists of prompt and delayed responses. The prompt light component is relatively weak and consists of the conventional scintillation light created along the track of the incident radiation. A delay on the order of a microsecond is then typically required for the electrons to drift from the particle track to the high-field region near the anode wire. No light is created during this drift time. The electron then enters the high-field region where excited molecules are created over times that are again of the order of a microsecond, giving rise to a relatively slow but strong secondary signal. Although the intensity of the prompt scintillation in typical gases will be only a few percent of the equivalent light output from the standard scintillator NaI(Tl), the secondary light yield can be several orders of magnitude greater. A single primary electron could, in principle, produce an unlimited number of secondary photons if the high-field region were to extend indefinitely. In most practical designs, the high field is created in the vicinity of anode wires so that the region ends at the wire surface. With voltages that are below the point of creating gas multiplication conditions, one electron can produce hundreds of photons before reaching the wire.

It has been estimated[75,76] that, in pure noble gases, as much as 75–97% of the energy gained by the electron from the electric field is converted into light. Noble gases have no vibrational or rotational states, so excitation requires the elevation of the atom to the lowest electronic excited state at around 10 eV. The electron typically undergoes thou-

TABLE 6-3 Spectral Properties of Light Emitted in Gas Proportional Scintillation Counters

Gas	Peak Wavelength of Emission Spectrum (nm)	FWHM of Emission Spectrum (nm)
Argon	128	10
Krypton	147	12
Xenon	173	14

Data from Suzuki and Kubota.[81]

sands of nonradiative elastic scattering collisions before gaining enough kinetic energy from the field to create an excited atom. The presence of molecular impurities with low-lying vibrational or rotational states will therefore strongly inhibit this excitation process, and purity of the gas is essential to good secondary scintillation yield. Deexcitation of the excited atom R* generally proceeds as a two-step process: collisions with neutral atoms first produce a molecular rare gas excimer, R_2^*. This species then deexcites by breaking up into two ground-state atoms plus an emitted photon ($R_2^* \rightarrow 2R + h\nu$). This light lies in the vacuum ultraviolet region of the spectrum for the commonly used noble gases (see Table 6-3). It can be detected directly using special UV-sensitive photomultiplier tubes or, alternatively, with standard tubes after shifting the wavelength to the visible range by absorption and reemission in an organic optical layer on the exit window. (Optical wavelength shifting is discussed more fully beginning on p. 246.) Another option is the use of photodiodes[77, 78] and photoionization devices (see p. 282)

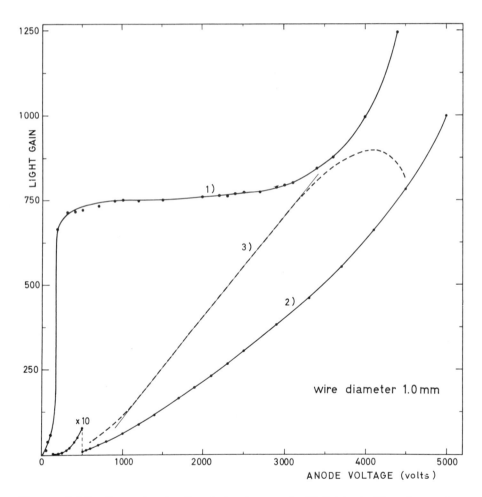

Figure 6-18 Curves showing the relative charge gain (1), light gain (2), and secondary light per collected electron (3), for a gas proportional scintillation counter. A numerical scale is shown for curve (2) only. Charge multiplication begins at about 3 kV, but secondary light production is observed at much lower anode voltages. (From Policarpo et al.[82])

that can detect light into the ultraviolet region. In particular, position-sensitive photoionization detectors have proved to be extremely useful in recording both the intensity and the spatial position of the secondary scintillation light in large-area GPS devices applied in radiation imaging.[79, 80]

The voltage characteristics of a typical GPS counter are plotted in Fig. 6-18. Increasing the chamber voltage increases the secondary signal amplitude, but an operating point that is short of the gas multiplication region is preferred from energy resolution considerations. By avoiding the formation of an avalanche, the fluctuations in electron multiplication that limit proportional counter resolution are not a factor, and the fluctuations in secondary scintillation yield are generally somewhat smaller. The energy resolution for low-energy radiations can be as much as a factor of 2 better than that for proportional counters,[82, 83] and this major advantage has been the basis of most successful applications of gas proportional scintillation detectors.

One disadvantage of GPS counters is that they are more susceptible to deleterious effects of gas impurities than are conventional proportional counters. As a result, most of these detectors have been operated using a continuous flow gas purification system to control the impurities. Published efforts,[84] however, have shown that the use of ultrahigh-vacuum techniques and close attention to gas purity can allow the use of sealed detectors over time periods of at least 6 months.

PROBLEMS

6-1. Assuming that $W = 26.2$ eV/ion pair and the Fano factor $F = 0.17$ for argon, find the mean and expected standard deviation in the number of ion pairs formed by the absorption of 1 MeV of radiation energy.

6-2. Explain why air is widely used as a fill gas for ionization chambers but is seldom used in proportional counters.

6-3. **(a)** A proportional counter with anode radius of 0.003 cm and cathode radius of 1.0 cm is filled with P-10 gas at 1 atm. Using the Diethorn parameters from Table 6-1, find the voltage required to achieve a gas multiplication factor of 1000.
(b) At the same voltage, by what factor would the multiplication change if the anode were twice as large?
(c) With the original anode radius, by what factor would the multiplication change if the cathode radius were doubled?

6-4. A windowless flow proportional counter is used to detect 5 MeV alpha particles that are totally stopped in the fill gas. The tube has an anode radius of 0.005 cm, a cathode radius of 5.0 cm, and is filled with P-10 gas at 1 atm. Using the data from Table 6-1, estimate the amplitude of the voltage pulses from the counter for an applied voltage of 2000 V and a collection capacitance of 500 pF.

6-5. Describe a measurement that can determine whether a proportional tube is operating in the true proportional region. Be explicit about which quantities are measured and how the data are used to make the determination. Assume that you have available any necessary type of radioisotope source.

6-6. Explain the following statement: In a windowless proportional counter, the output pulse height from an alpha particle source will increase with increasing alpha energy, whereas for beta particles the opposite is usually true.

6-7. A given voltage-sensitive preamplifier requires a minimum input pulse amplitude of 10 mV for good signal/noise performance. What gas multiplication factor is required in an argon-filled proportional counter with 200 pF capacitance if 50 keV X-rays are to be measured?

6-8. A cylindrical proportional tube has an anode wire radius of 0.003 cm and a cathode radius of 2.0 cm. It is operated with an applied voltage of 2000 V. If a minimum electric field of 1.0 MV/m is required to initiate gas multiplication, what fraction of the internal volume of the tube corresponds to the multiplication region?

6-9. Explain the function of the "field tube" found in some proportional tube designs.

6-10. Explain the sudden discontinuities in the curves for the absorption properties of various gases shown in Fig. 6-14.

6-11. Why are gas scintillation proportional detectors usually operated with a voltage below that required for charge multiplication?

REFERENCES

1. P. J. Campion, *Int. J. Appl. Radiat. Isotopes* **19**, 219 (1968).
2. W. Bambynek, *Nucl. Instrum. Meth.* **112**, 103 (1973).
3. M. Matoba et al., *IEEE Trans. Nucl. Sci.* **NS-32**(1), 541 (1985).
4. E. F. Bennett and T. J. Yule, ANL-7763, (1971).
5. A. Spernol and B. Denecke, *Int. J. Appl. Radiat. Isotopes* **15**, 195 (1963).
6. F. B. Riggs, Jr., *Rev. Sci. Instrum.* **34**, 392 (1963).
7. H. M. Horstman, G. Ventura, and G. R. Vespignani, *Nucl. Instrum. Meth.* **112**, 619 (1973).
8. J. Byrne and F. Shaikh, *Nucl. Instrum. Meth.* **79**, 286 (1970).
9. P. Povinec, *Nucl. Instrum. Meth.* **163**, 363 (1979).
10. H. Sipila, *Nucl. Instrum. Meth.*, **140**, 389 (1977).
11. P. J. Campion, *Nucl. Instrum. Meth.* **112**, 75 (1973).
12. T. Aoyama, H. Miyai, and T. Watanabe, *Nucl. Instrum. Meth.* **221**, 644 (1984).
13. T. Aoyama and T. Watanabe, *Health Phys.* **48**, 773 (1985).
14. T. Aoyama, M. Totogawa, and T. Watanabe, *Nucl. Instrum. Meth.* **A255**, 524 (1987).
15. G. D. Alkhazov, A. P. Komar, and A. A. Vorob'ev, *Nucl. Instrum. Meth.* **48**, 1 (1967).
16. M.-L. Järvinen and H. Sipilä, *IEEE Trans. Nucl. Sci.* **NS-31**(1), 356 (1984).
17. T. Z. Kowalski and J. Zajac, *Nucl. Instrum. Meth. Phys. Res.* **A249**, 426 (1986).
18. J. P. Sephton, M. J. L. Turner, and J. W. Leake, *Nucl. Instrum. Meth. Phys. Res.* **A256**, 561 (1987).
19. S. C. Curran, A. L. Cockroft, and J. Angus, *Philos. Mag.* **40**, 929 (1949).
20. H. Schlumbohm, *Z. Phys.* **182**, 306 (1965).
21. R. Gold and E. F. Bennett, *Phys. Rev.* **147**(1), 201 (1966).
22. H. Genz, *Nucl. Instrum. Meth.* **112**, 83 (1973).
23. A. Williams and R. I. Sara, *Int. J. Appl. Radiat. Isotopes* **13**, 229 (1962).
24. A. Zastawny, *J. Sci. Instrum.* **43**, 179 (1966).
25. G. D. Alkhazov, *Nucl. Instrum. Meth.* **75**, 161 (1969).
26. P. J. Campion, *Phys. Med. Biol.* **16**, 611 (1971).
27. P. Hopstone and S. Shalev, *Trans. Am. Nuc. Soc.* **21**, 132 (1975).
28. A. Von Engel, in *Handbuch der Physik*, Vol. 21 (S. Flugge, ed.), Springer-Verlag, Berlin, 1956, p. 504.

29. T. Aoyama, *Nucl. Instrum. Meth.* **A234**, 125 (1985).

30. T. Z. Kowalski, *Nucl. Instrum. Meth.* **A243**, 501 (1986).

31. W. Diethorn, NYO-6628 (1956).

32. R. W. Hendricks, *Nucl. Instrum. Meth.* **102**, 309 (1972).

33. R. S. Wolff, *Nucl. Instrum. Meth.* **115**, 461 (1974).

34. C. Mori, M. Uno, and T. Watanabe, *Nucl. Instrum. Meth.* **196**, 49 (1982).

35. R. W. Hendricks, *Rev. Sci. Instrum.* **40**(9), 1216 (1969).

36. J. Byrne, *Nucl. Instrum. Meth.* **74**, 291 (1969).

37. G. D. Alkhazov, *Nucl. Instrum. Meth.* **89**, 155 (1970).

38. B. Breyer, *Nucl. Instrum. Meth.* **112**, 91 (1973).

39. J. Byrne, *Proc. R. Soc. Edinburg Sect. A* **66**, 33 (1962).

40. H. Sipila, *Nucl. Instrum. Meth.* **133**, 251 (1976).

41. M. W. Charles and B. A. Cooke, *Nucl. Instrum. Meth.* **61**, 31 (1968).

42. F. Shaikh, J. Byrne, and J. Kyles, *Nucl. Instrum. Meth.* **88**, 317 (1970).

43. W. Hink, A. N. Scheit, and A. Ziegler, *Nucl. Instrum. Meth.* **87**, 137 (1970).

44. B. Bednarek, *Nucl. Instrum. Meth.* **175**, 431 (1980).

45. T. Z. Kowalski, *Nucl. Instrum. Meth.* **A239**, 551 (1985).

46. A. Smith and M. J. L. Turner, *Nucl. Instrum. Meth.* **192**, 475 (1982).

47. D. J. Grady and J. C. Robertson, *Nucl. Instrum. Meth.* **179**, 317 (1981).

48. D. H. Wilkinson, *Ionization Chambers and Counters*, Cambridge University Press, Cambridge, 1950.

49. R. Gott and M. W. Charles, *Nucl. Instrum. Meth.* **72**, 157 (1969).

50. L. G. Christophorov, D. L. McCorkle, D. V. Maxey and J. G. Carter, *Nucl. Instrum. Meth.* **163**, 141 (1979).

51. A. Jeavons et al., *Nucl. Instrum. Meth.* **176**, 89 (1980).

52. E. Mathieson and M. W. Charles, *Nucl. Instrum. Meth.* **72**, 155 (1969).

53. E. Mathieson, *Nucl. Instrum. Meth.* **72**, 355 (1969).

54. B. Breyer and M. Cimerman, *Nucl. Instrum. Meth.* **92**, 19 (1971).

55. P. Gorenstein and S. Mickiewicz, *Rev. Sci. Instrum.* **39**(6), 816 (1968).

56. S. Sudar, L. Vas, and T. Biro, *Nucl. Instrum. Meth.* **112**, 399 (1973).

57. R. C. Hochel and D. W. Hayes, *Nucl. Instrum. Meth.* **130**, 183 (1975).

58. H. Genz, D. S. Harmer, and R. W. Fink, *Nucl. Instrum. Meth.* **60**, 195 (1968).

59. P. J. Campion, *J. Phys. E* **3**, 920 (1970).

60. G. Hempel, F. Hopkins, and G. Schatz, *Nucl. Instrum. Meth.* **131**, 445 (1975).

61. H. Stelzer, *Nucl. Instrum. Meth.* **133**, 409 (1976).

62. D. Fabris et al., *Nucl. Instrum. Meth.* **216**, 167 (1983).

63. A. Breskin and N. Zwang, *Nucl. Instrum. Meth.* **144**, 609 (1977).

64. G. P. Westphal, *Nucl. Instrum. Meth.* **134**, 387 (1976).

65. B. E. Fischer, *Nucl. Instrum. Meth.* **141**, 173 (1977).

66. C. J. Borkowski and M. K. Kopp, *Rev. Sci. Instrum.* **39**, 1515 (1968).

67. C. J. Borkowski and M. K. Kopp, *IEEE Trans. Nucl. Sci.* **NS-17**(3), 340 (1970).

68. R. W. Hendricks, *Trans. Am. Crystallogr. Assoc.* **12**, 103 (1976).

69. J. L. C. Ford, Jr., *Nucl. Instrum. Meth.* **162**, 277 (1979).

70. G. Charpak and F. Sauli, *Nucl. Instrum. Meth.* **162**, 405 (1979).

71. A. Breskin, *Nucl. Instrum. Meth.* **141**, 505 (1977).

72. P. Povinec, J. Szarka, and S. Usacev, *Nucl. Instrum. Meth.* **163**, 369 (1979).

73. P. E. Thiess and G. H. Miley, *IEEE Trans. Nucl. Sci.* **NS-21**(1), 125 (1974).

74. A. J. P. L. Policarpo, *Space Sci. Instrum.* **3**, 77 (1977).

75. R. D. Andresen, E. A. Leimann, and A. Peacock, *Nucl. Instrum. Meth.* **140**, 371 (1977).

76. M. Alegria Feio et al., *Nucl. Instrum. Meth.* **176**, 473 (1980).

77. J. C. Van Staden et al., *Nucl. Instrum. Meth.* **157**, 301 (1978).

78. A. J. de Campos, *IEEE Trans. Nucl. Sci.* **NS-31**(1), 133 (1984).

79. A. J. P. L. Policarpo, *Nucl. Instrum. Meth.* **196**, 53 (1982).

80. D. F. Anderson, *IEEE Trans. Nucl. Sci.* **NS-32**(1), 495 (1985).

81. M. Suzuki and S. Kubota, *Nucl. Instrum. Meth.* **164**, 197 (1979).

82. A. J. P. L. Policarpo, M. A. F. Alves, M. C. M. Dos Santos, and M. J. T. Carvalho, *Nucl. Instrum. Meth.* **102**, 337 (1972).

83. M. Alice, F. Alves, A. J. P. L. Policarpo, and M. Salete, *IEEE Trans. Nucl. Sci.* **NS-22**, 109 (1975).

84. R. D. Andresen, L. Karlsson, and B. G. Taylor, *IEEE Trans. Nucl. Sci.* **NS-23**(1), 473 (1976).

CHAPTER · 7

Geiger–Mueller Counters

The Geiger–Mueller counter (commonly referred to as the G-M counter, or simply Geiger tube) is one of the oldest radiation detector types in existence, having been introduced by Geiger and Mueller in 1928. However, the simplicity, low cost, and ease of operation of these detectors have led to their continued use to the present time. General textbook references in the field of G-M counters include the reviews by Wilkinson,[1] Price,[2] Sharpe,[3] and Emery.[4] Construction details and specific tube designs beyond those given in the following discussion can be found in all these references.

As a complement to the ion chambers and proportional counters discussed in the two previous chapters, G-M counters comprise the third general category of gas-filled detectors based on ionization. In common with proportional counters, they employ gas multiplication to increase greatly the charge represented by the original ion pairs formed along the radiation track, but in a fundamentally different manner. In the proportional counter, each original electron leads to an avalanche that is basically independent of all other avalanches formed from other electrons associated with the original ionizing event. Because all avalanches are nearly identical, the collected charge remains proportional to the number of original electrons.

In the G-M tube, substantially higher electric fields are created which enhance the intensity of each avalanche. Under proper conditions, a situation is created in which one avalanche can itself trigger a second avalanche at a different position within the tube. At a critical value of the electric field, each avalanche can create, on the average, at least one more avalanche, and a self-propagating chain reaction results. At still greater values of the electric field, the process becomes rapidly divergent and, in principle, an exponentially growing number of avalanches could be created within a very short time. Once this *Geiger discharge* reaches a certain size, however, collective effects of all the individual avalanches come into play and ultimately terminate the chain reaction. Because this limiting point is always reached after about the same number of avalanches have been created, *all pulses from a Geiger tube are of the same amplitude regardless of the number of original ion pairs that initiated the process.* A Geiger tube can therefore function only as a simple counter of radiation-induced events and cannot be applied in direct radiation spectroscopy because all information on the amount of energy deposited by the incident radiation is lost.

A typical pulse from a Geiger tube represents an unusually large amount of collected charge, about 10^9–10^{10} ion pairs being formed in the discharge. Therefore, the output pulse amplitude is also large (typically of the order of volts). This high-level signal allows considerable simplification to be made in the associated electronics, often completely eliminating the need for a preamplifier stage. Because the tubes themselves are relatively inexpensive, a G-M counter is often the best choice when a simple and economical counting system is needed.

Besides the lack of energy information, a major disadvantage of G-M counters is their unusually large dead time, which greatly exceeds that of any other commonly used radiation detector. These detectors are therefore limited to relatively low counting rates, and a dead time correction must be applied to situations in which the counting rate would otherwise be regarded as moderate (a few hundred pulses per second). Some types of Geiger tube also have a limited lifetime and will begin to fail after a fixed number of total pulses have been recorded (see following discussion of the quench gas).

I. THE GEIGER DISCHARGE

In a typical Townsend avalanche created by a single original electron, many excited gas molecules are formed by electron collisions in addition to the secondary ions. In a time that is usually no more than a few nanoseconds, these excited molecules return to their ground state through the emission of photons whose wavelength may be in the visible or ultraviolet region. These photons are the key element in the propagation of the chain reaction that makes up the Geiger discharge. If one of these photons interacts by photoelectric absorption in some other region of the gas or at the cathode surface, a new electron is liberated which can subsequently migrate toward the anode and will trigger another avalanche (see Fig. 7-1). If the gas multiplication factor M as defined in Chapter 6 is relatively low (say, 10^2–10^4) as in a proportional tube, the number of excited molecules formed in a typical avalanche (n_0') is not very large. Also, because most gases are relatively transparent in the visible and UV wavelength regions, the probability p of photoelectric absorption of any given photon is also relatively low. Under these conditions, $n_0' p \ll 1$ and the situation is "subcritical" in that relatively few avalanches are formed in addition to those created by the original free electrons. Many proportional gases also contain an additive to absorb the photons preferentially without creation of free electrons, further suppressing the possibility of new avalanches.

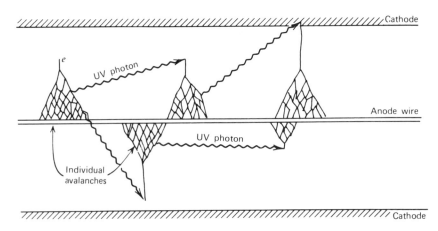

Figure 7-1 The mechanism by which additional avalanches are triggered in a Geiger discharge.

In a Geiger discharge, however, the multiplication represented by a single avalanche is much higher (10^6–10^8) and therefore n_0' is also much larger. Now the conditions of "criticality" can be achieved

$$n_0' p \geq 1$$

and an ever-increasing number of avalanches may potentially be created throughout the tube. The time required for the spread of these avalanches is relatively short. Because each avalanche begins only when the free electron has drifted to within a few mean free paths of the anode wire, the time required to produce all the ion pairs and excited molecules in a given avalanche is a small fraction of a microsecond. Because the lifetime of the excited molecules is short and the photons travel at the speed of light, a secondary free electron can be created elsewhere in the tube almost in coincidence with the first avalanche. This electron then need only drift to the multiplying region before creating a secondary avalanche. Because these drift times are also typically a fraction of a microsecond, the entire Geiger discharge process takes place within about a microsecond. This time is less than that required to fully develop the output pulse from a single avalanche, and thus the pulse amplitude simply represents the sum of all the charges created in the Geiger discharge before its termination.

In a proportional counter, each avalanche is formed at an axial position within the tube which corresponds to that of the original free electron created by the incident radiation. Furthermore, the free electron approaches the anode wire from one radial direction, and the secondary ions are preferentially formed on that side of the anode wire.[5,6] In a Geiger discharge, however, the rapid propagation of the chain reaction leads to many avalanches, which initiate at random radial and axial positions throughout the tube. Secondary ions are therefore formed throughout the cylindrical multiplying region which surrounds the anode wire. The Geiger discharge therefore grows to envelope the entire anode wire, regardless of the position at which the primary initiating event occurred.

The process that leads to the termination of a Geiger discharge has as its origin the positive ions that are created along with each electron in an avalanche. The mobility of these positive ions is much less than that of the free electrons, and they remain essentially motionless during the time necessary to collect all the free electrons from the multiplying region. When the concentration of these positive ions is sufficiently high, their presence begins to change the magnitude of the electric field in the vicinity of the anode wire. Because the ions represent a positive space charge, the region between the ions and the positive anode will have an electric field below that predicted by Eq. (6-3) in the absence of the space charge. Because further gas multiplication requires that an electric field above some minimum value be maintained, the buildup of positive ion space charge eventually terminates the Geiger discharge.

For a fixed applied voltage to the tube, the point at which the Geiger discharge is terminated will always be the same, in the sense that a given density of positive ions will be needed to reduce the electric field below the minimum value required for further multiplication. Thus, each Geiger discharge is terminated after developing about the same total charge, regardless of the number of original ion pairs created by the incident radiation. All output pulses are therefore about the same size, and their amplitude can provide no information about the properties of the incident radiation.

As the high voltage is raised in a Geiger tube, the magnitude of the Geiger discharge increases and the amplitude of the output pulse also increases in size. A higher voltage means a larger initial electric field before the discharge, which then requires a greater buildup of space charge before the field is reduced below the critical value. The pulse amplitude increases roughly in proportion to the *overvoltage*, defined as the difference between the applied voltage and the minimum voltage required to initiate a Geiger discharge.

II. FILL GASES

Gases used in a Geiger tube must meet some of the same requirements discussed in Chapter 6 for proportional counters. Because both types of detector are based on gas multiplication, even trace amounts of gases which form negative ions (such as oxygen) must be avoided in either case. The noble gases are widely used for the principal component of the fill gas in G-M tubes, with helium and argon the most popular choices. A second component is normally added to most Geiger gases for purposes of *quenching*, discussed in the next section.

The average energy attained by a free electron between collisions with gas molecules depends on the ratio of the electric field magnitude to the gas pressure, $\mathscr{E}/p$ [see Eq. (5-2)]. Because the number of excited molecules n_0' should also increase with this parameter, the development of a full Geiger discharge requires a certain minimum value of $\mathscr{E}/p$, which depends on the specific gas mixture used in the G-M tube. It is common practice to design sealed tubes with a fill gas kept at less than atmospheric pressure in order to minimize the value of $\mathscr{E}$ needed to achieve the necessary $\mathscr{E}/p$ ratio. For normal gases at pressures of several tenths of an atmosphere, operating voltages of about 500–2000 V are required to reach the necessary field strength using anode wires of about 0.1 mm diameter [see Eq. (6-3)].

Geiger tubes can also be designed to operate at atmospheric pressure. In such cases, the mechanical design can be simplified (especially for large tubes) because no pressure differential need be supported across the walls or window of the tube. Just as in proportional counters, either the gas can be permanently sealed within the tube or arrangements can be made to have the gas flow through the counter volume to replenish the gas continuously and to flush out any impurities.

III. QUENCHING

If a Geiger tube is filled with a single gas such as argon, then all the positive ions formed are ions of that same gas species. After the primary Geiger discharge is terminated, the positive ions slowly drift away from the anode wire and ultimately arrive at the cathode or outer wall of the counter. Here they are neutralized by combining with an electron from the cathode surface. In this process, an amount of energy equal to the ionization energy of the gas minus the energy required to extract the electron from the cathode surface (the *work function*) is liberated. If this liberated energy also exceeds the cathode work function, it is energetically possible for another free electron to emerge from the cathode surface. This will be the case if the gas ionization energy exceeds twice the value of the work function. The probability is always small that any given ion will liberate an electron in its neutralization, but if the total number of ions is large enough, there will likely be at least one such free electron generated. This electron will then drift toward the anode and will trigger another avalanche, leading to a second full Geiger discharge. The entire cycle will now be repeated, and under these circumstances the G-M counter, once initially triggered, would produce a continuous output of multiple pulses.

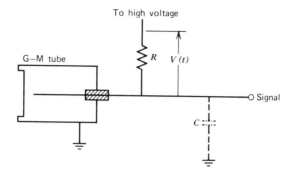

Figure 7-2 Equivalent counting circuit for a G-M tube. The product of resistance R and capacitance C (usually only inherent capacitance of the tube and electronics) determines the time constant of the restoration of the high voltage following a Geiger discharge.

The problem of multiple pulsing is potentially much more severe in Geiger tubes than in proportional counters. In the latter case, the number of positive ions per event is small enough so that only an occasional spurious pulse may result. The amplitude of the spurious pulse will correspond only to a single avalanche and will generally be considerably smaller than that produced by the initial event. With a Geiger tube, the increased number of positive ions reaching the cathode greatly enhances the free electron emission probability. Furthermore, the secondary discharge, even though caused by only a single electron, builds to an amplitude equal to that of the primary discharge.

For these reasons, special precautions must be taken in Geiger counters to prevent the possibility of excessive multiple pulsing. *External quenching* consists of some method for reducing the high voltage applied to the tube, for a fixed time after each pulse, to a value that is too low to support further gas multiplication. Then, secondary avalanches cannot be formed, and even if a free electron is liberated at the cathode, it cannot cause another Geiger discharge. The voltage must be suppressed for a time that is greater than the transit time of the positive ion from its place of formation to the cathode (generally a few hundred microseconds), plus the transit time of the free electron (generally about a microsecond).

One method of external quenching is simply to choose R in Fig. 7-2 to be a large enough value (typically 10^8 ohms) so that the time constant of the charge collection circuit is of the order of a millisecond. This "external resistor" method of quenching has the disadvantage of requiring several milliseconds for the anode to return to near its normal voltage, and therefore full Geiger discharges for each event are produced only at very low counting rates.

It is much more common to prevent the possibility of multiple pulsing through *internal quenching*, which is accomplished by adding a second component called the *quench gas* to the primary fill gas. It is chosen to have a lower ionization potential and a more complex molecular structure than the primary gas component and is present with a typical concentration of 5–10%. Although it is given the same name as the component added to proportional counter gases to absorb UV photons, the quench gas in a Geiger counter serves a different function. It prevents multiple pulsing through the mechanism of charge transfer collisions. The positive ions formed by the incident radiation are mostly of the primary component, and they subsequently make many collisions with neutral gas molecules as they drift toward the cathode. Some of these collisions will be with molecules of the quench gas and, because of the difference in ionization energies, there will be a tendency to transfer the positive charge to the quench gas molecule. The original positive ion is thus neutralized by transfer of an electron, and a positive ion of the quench gas begins to drift in its place. If the concentration of the quench gas is sufficiently high, these charge-transfer collisions assure that all the ions that eventually arrive at the cathode will be those of the quench gas. When they are neutralized, the excess energy may now go into disassociation of the more complex molecules in preference to liberating a free electron from the cathode surface. With the proper choice of quench gas, the probability of disassociation can be made much larger than that of electron emission, and therefore no additional avalanches are formed within the tube.

Many organic molecules possess the proper characteristics to serve as a quench gas. Of these, ethyl alcohol and ethyl formate have proved to be the most popular and are widely used in many commercial G-M tubes. Because the quench gas is disassociated in carrying out its function, it is gradually consumed during the lifetime of the tube. An organic quenched tube may typically have a limit of about 10^9 counts in its useful lifetime before multiple pulsing can no longer be prevented because of depletion of the quench gas.

To avoid the problem of limited lifetime, some tubes use halogens (chlorine or bromine) as the quench gas. Although the halogen molecules also disassociate when carrying out their quenching function, they may be replenished by spontaneous recombination at a later time. In principle, halogen-quenched tubes therefore have an infinite lifetime and are preferred in situations where the application involves extended use at high counting rates. However, there are other mechanisms besides the consumption of the quench gas that limit the lifetime of Geiger tubes. Two of the more important appear to be contamination of the gas by reaction products produced in the discharge, and changes on the anode surface caused by the deposition of polymerized reaction products.[7,8]

IV. TIME BEHAVIOR

A. Pulse Profile

In Chapter 6, the development of the pulse from a single avalanche was shown to derive almost entirely from the motion of the positive ions away from the anode wire. An initial fast component (with rise time less than a microsecond) results from the motion of the ions through the high-field region near the anode wire. A much slower rise follows which corresponds to the slower drift of the ions through the lower-field region that exists in most of the tube volume. In a Geiger discharge, the pulse corresponds to the cumulative effect of many avalanches that have been formed along the entire length of the anode wire. The time required for secondary electrons to reach the multiplying region to trigger these multiple avalanches will be variable, but differences will be limited by the relatively rapid motion of the electrons. The initial pulse rise will be somewhat slower than that for a single avalanche but still will develop over a time that is typically less than a microsecond. The situation is also complicated by the electric field distortions that take place from the buildup of space charge near the termination of the discharge.

The net effect is that the output pulse from a G-M tube still exhibits a fast rise on the order of a microsecond or less, followed by a much slower rise due to the slow drift of the ions. If the time constant of the collection circuit were infinite, it would take the full ion collection time for the output pulse to reach its theoretical maximum. From considerations of maximum counting rate, time constants are often chosen that are much less than

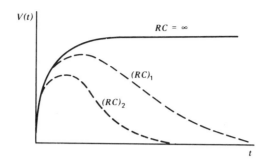

Figure 7-3 Shape of the G-M tube output pulse for different assumed time constants RC of the counting circuit. Here $(RC)_2 < (RC)_1 <$ infinite time constant. The signal voltage $V(t)$ is assumed to be measured as indicated in Fig. 7-2, giving a positive polarity. More conventionally, the pulse is measured with respect to ground within the amplifier, leading to inverted shape or negative polarity.

100 μs, which largely eliminate the slow-rising portion of the pulse and leave only the fast leading edge, as illustrated in Fig. 7-3. Even though a significant fraction of the potential output pulse amplitude may be lost, the large amount of charge generated in the Geiger discharge produces so large a pulse that some amplitude loss can easily be tolerated. Furthermore, because all Geiger discharges are approximately uniform in size and time profile, all pulses will be attenuated by the same fraction in the shaping process and the output pulses will remain almost of one amplitude.

As in proportional counters, there is a delay between the time of formation of the first ion pair within the gas and the initiation of the first avalanche. This delay corresponds to the drift time of the electron from its point of origin to the multiplying region and can be as much as a microsecond in large tubes. When G-M tubes are used for timing purposes, one should therefore expect a time uncertainty of this magnitude if radiation interactions occur randomly throughout the tube volume.

B. Dead Time

The building of the positive ion space charge that terminates the Geiger discharge also ensures that a considerable amount of time must pass before a second Geiger discharge can be generated in the tube. As the positive ions drift radially outward, the space charge becomes more diffuse and the electric field in the multiplying region begins to return to its original value. After the positive ions have traveled some of the distance, the field will have recovered sufficiently to permit another Geiger discharge. If the field has not been fully restored, however, the discharge will be less intense than the original because fewer positive ions will be required to shut down the discharge by again reducing the electric field below the critical point. The first pulses that appear are therefore reduced in amplitude compared with the original and may or may not be registered by the counting system, depending on its sensitivity. When the positive ions have eventually drifted all the way to the cathode, the electric field is fully restored and another ionizing event will trigger a second Geiger discharge of full amplitude. This behavior is sketched in Fig. 7-4, which is similar to an oscilloscope trace that can be experimentally observed from Geiger tubes at high counting rates.

Immediately following the Geiger discharge, the electric field has been reduced below the critical point by the positive space charge. If another ionizing event occurs under these conditions, a second pulse will not be observed because gas multiplication is prevented.

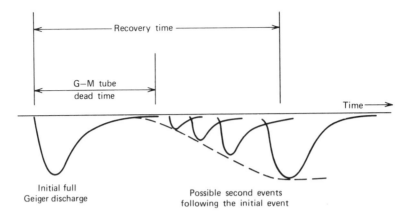

Figure 7-4   Illustration of the dead time of a G-M tube. Pulses of negative polarity conventionally observed from the detector are shown.

During this time the tube is therefore "dead" and any radiation interactions that occur in the tube during this time will be lost. Technically, the *dead time* of the Geiger tube is defined as the period between the initial pulse and the time at which a second Geiger discharge, regardless of its size, can be developed. In most Geiger tubes, this time is of the order of 50–100 μs. In any practical counting system, some finite pulse amplitude must be achieved before the second pulse is recorded, and the elapsed time required to develop a second discharge that exceeds this amplitude is sometimes called the *resolving time* of the system. In practice, these two terms are often used interchangeably and the term dead time may also be used to describe the combined behavior of the detector–counting system. The *recovery time* is the time interval required for the tube to return to its original state and become capable of producing a second pulse of full amplitude.

V. THE GEIGER COUNTING PLATEAU

Because the Geiger tube functions as a simple counter, its application requires only that operating conditions be established in which each pulse is registered by the counting system. In practice, this operating point is normally chosen by recording a *plateau curve* from the system under conditions in which a radiation source generates events at a constant rate within the tube. The counting rate is recorded as the high voltage applied to the tube is raised from an initially low value. The general properties of the resulting counting curve have been discussed in Chapter 4. The results are particularly simple for the case of the Geiger tube in which pulses with amplitudes centered about a single value are involved. The differential pulse height distribution then has the simple appearance sketched in Fig. 7-5a for a fixed value of the voltage applied to the tube. If the voltage is increased, the average pulse amplitude also increases, and the peak in the differential pulse height distribution shifts to the right. Distributions are shown for two hypothetical values of applied voltage in the sketch.

If the minimum amplitude required by the counting system is H_d as indicated on the figure, then no pulses should be counted when the applied voltage is 1000 V, but all the pulses should be counted at 1200 V. The corresponding counting rate therefore abruptly changes from zero to the maximum value between these two voltage settings. Increasing the voltage further only moves the peak more to the right in the differential distribution

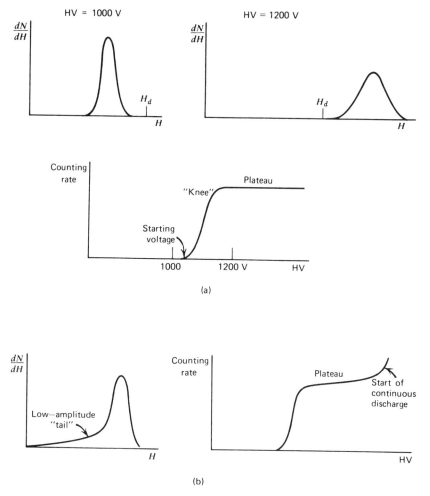

Figure 7-5 (a) The establishment of the counting plateau for a G-M tube. As the high voltage is varied in this example from 1000–1200 V, the output pulses change from falling below the counter threshold H_d to a situation in which all pulses are larger than H_d. (b) The low-amplitude tail on the pulse height spectrum at the left causes a finite slope of the plateau in the counting curve.

but does not increase the number of recorded events. Therefore, the counting curve shows a flat plateau at voltages above 1200 V, and in principle any choice of operating voltage above this value would assure stable operation. The minimum voltage at which pulses are first registered by the counting system is often called the *starting voltage*, whereas the transition between the rapid rise of the curve and the plateau is its *knee*.

If the voltage is raised sufficiently high, the plateau abruptly ends because of the onset of continuous discharge mechanisms within the tube. These can be corona discharges that initiate from any sharp irregularities on the anode wire, or multiple pulsing caused by failure of the quenching mechanism. The continuous discharge process is potentially harmful if sustained for any length of time, and one should therefore immediately decrease the applied voltage when the end of the plateau is observed. In the interests of long operating life, an operating point is normally chosen which is only sufficiently far up the plateau to ensure that the flat region has been reached. The Geiger discharges are thus kept near the minimum size compatible with good counting stability, and the quench gas is consumed at the minimum rate.

In real cases, the counting plateau always shows some finite slope, as shown in Fig. 7-5b. Any effect that adds a low-amplitude tail to the differential pulse height distribution can be a contributing cause of the slope. For example, some regions near the ends of the tube may have a lower than normal electric field strength, and the discharges originating in these regions may be smaller than normal. Also, any pulses that occur during the recovery time will also be abnormally small. In order to exclude this latter effect when measuring the plateau characteristics, the counting rate should not exceed a few percent of the inverse of the recovery time. For typical G-M tubes, this restriction means that the plateau measurement should be made at a rate below a few hundred counts per second.

Another cause of slope in the plateau of many G-M tubes is the occasional failure of the quenching mechanism which may lead to a satellite or spurious pulse in addition to the primary Geiger discharge. The probability of such an occurrence is normally low but will increase as the number of positive ions that ultimately reach the cathode also increases. The active volume of the tube may also increase slightly with voltage as the very low electric field near corners of the tube is increased and ion pair recombination is thereby prevented.

The plateau curve may also show a slight hysteresis effect, in that the counting curve recorded by increasing the voltage may differ somewhat from that obtained while decreasing the voltage. This hysteresis arises because the electrical charges on the insulators are slow to equilibrate and can influence the electric field configuration inside the tube.

Geiger tubes are often rated based on the slope shown by the linear portion of the plateau. Organic quench tubes tend to show the flattest plateau, with slopes of the order of 2–3% per 100 V change in the applied voltage. Halogen-quenched tubes have a plateau of greater typical slope, but their longer useful lifetime can sometimes offset this disadvantage. Halogen tubes can also be operated at lower applied voltage values. For example, argon–alcohol counters usually require a starting voltage of at least 1000 V, but neon-filled tubes quenched with bromine can show a starting voltage as low as 200–300 V (Ref. 9). The mechanisms that lead to this unusually low starting voltage strongly depend on the partial pressure of the halogen and the $\mathscr{E}/p$ value within the tube. The interaction of bromine atoms with metastable and excited neon atoms apparently plays an important role in propagation of the discharge and in determining the resulting tube properties.[10]

VI. DESIGN FEATURES

Geiger tubes share some general design features with proportional counters, but in many respects they are less demanding. Because gas multiplication and avalanche formation are again important, the use of a fine anode wire is nearly universal in order to produce the high local values of the electric field necessary to create a multiplying region. In the Geiger tube, less attention to uniformity of the wire is necessary because the Geiger discharge spreads over the entire length of the anode, and nonuniformities are therefore averaged out. Because the pulse amplitude carries no quantitative information, uniformity of response is less important, and therefore the field tubes used in proportional counters to shape the electric field are seldom found in Geiger tubes. Some designs can be used interchangeably as proportional or Geiger tubes, although a change in the fill gas is normally recommended for optimal performance.

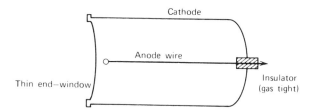

Figure 7-6 A cross-section of a typical end-window Geiger tube.

A typical design of Geiger tubes is the *end-window* type illustrated in Fig. 7-6. The anode wire is supported at one end only and is located along the axis of a cylindrical cathode made of metal or glass with a metallized inner coating. Radiation enters the tube through the entrance window, which may be made of mica or other material that can maintain its strength in thin sections. Because most Geiger tubes are operated below atmospheric pressure, the window may have to support a substantial differential pressure. The window should be as thin as possible when counting short-range particles, such as alphas, but may be made more robust for applications that involve beta particles or gamma rays.

Geiger tubes can also be manufactured in many other shapes and configurations. A simple wire loop anode inserted into an arbitrary volume enclosed by a conducting cathode will normally work as a Geiger counter, but some care must be taken to avoid low-field regions in corners where events may be lost. Geiger counters can also take the form of "needle counters," in which the anode consists of a sharply pointed needle supported within a gas enclosed by the cathode. The field in the vicinity of the point varies as $1/r^2$, rather than $1/r$ as in the common cylindrical geometry. Counters with small active volumes can be conveniently manufactured using the needle geometry, and representative designs are described in Refs. 11 and 12.

For applications involving very soft radiations such as low-energy heavy charged particles or soft beta particles, it may be preferable to introduce the source directly into the counting volume. Continuous flow Geiger counters, which are similar in design to the gas flow proportional counters discussed in Chapter 6, are often used for this purpose. By continuously flowing the counter gas through the chamber, the gas purity is maintained despite small amounts of air introduced with the sample.

Very high counting efficiencies for some soft radiations can be obtained by introducing the source in the form of a gas mixed directly with the counter gas. For example, low concentrations of ^{14}C are often counted by conversion of the carbon to CO_2 which can then be mixed with a conventional argon–alcohol fill gas. Reasonable properties as a Geiger gas are maintained provided the concentration of CO_2 is not too high. The soft beta particles from tritium can also be efficiently counted using a fill gas consisting of hydrogen in which a small amount of tritiated water vapor has been added.

Because no preamplifier is normally required, a counting system involving Geiger tubes can be as simple as that shown in Fig. 7-7. The series resistance R between the high-voltage supply and the anode of the tube is the load resistance across which the signal voltage is developed. The parallel combination of R with the capacitance of the tube and associated wiring (shown as C_s) determines the time constant of the charge collection circuit. As elaborated in the section on time response, this time constant is normally chosen to be a few microseconds so that only the fast-rising components of the pulse are preserved. Also shown is the coupling capacitor C_c, needed to block the tube

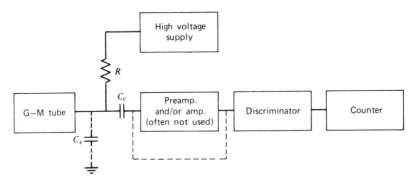

Figure 7-7 Block diagram of the counting electronics normally associated with a G-M tube. These functions are often combined on a single chassis traditionally called a *scaler* (see Chapter 17).

high voltage but transmit the signal pulse at ground potential to the following circuits. In order to carry out this function without attenuating the pulse amplitude, its value must be large enough so that RC_c is large compared with the pulse duration.

VII. COUNTING EFFICIENCY

A. Charged Particles

Because a single ion pair formed within the fill gas of the Geiger tube can trigger a full Geiger discharge, the counting efficiency for any charged particle that enters the active volume[†] of the tube is essentially 100%. In most practical situations, the effective counting efficiency is therefore determined by the probability that the incident radiation penetrates the window of the tube without absorption or backscattering. With alpha particles, absorption within the window is the major concern, and windows with thickness as small as 1.5 mg/cm^2 are commercially available. Thicker windows can generally be used for counting beta particles, but some will be backscattered from the window without penetrating if the window thickness is a significant fraction of the electron range.

B. Neutrons

There are several reasons why Geiger tubes are seldom used to count neutrons. For thermal neutrons, the conventional Geiger gases have low capture cross sections and therefore have an unacceptably low counting efficiency. Gases with a high capture cross section (such as BF_3) can be substituted, but then the detector is much more sensibly operated in the proportional region than in the Geiger region. In the proportional counter, neutron-induced events are of much larger amplitude than pulses generated by gamma-ray background and therefore are easily distinguished (see Chapter 14). In the Geiger region, all pulses are of the same amplitude and gamma-ray discrimination is not possible.

Fast neutrons can produce recoil nuclei (see Chapter 15) in a Geiger gas which generate ion pairs and a subsequent discharge. Therefore, Geiger tubes will, to some extent, respond to a fast neutron flux, particularly those that are helium filled. However,

[†]A small dead space is often encountered just behind the entrance window of end-window designs where the local electric field may be too small to prevent recombination or electron loss.

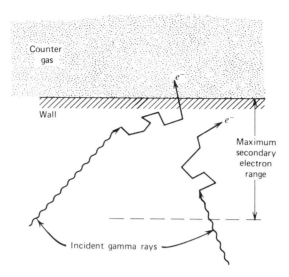

Figure 7-8 The principal mechanism by which gas-filled counters are sensitive to gamma rays involves creation of secondary electrons in the counter wall. Only those interactions that occur within an electron range of the wall surface can result in a pulse.

gas-filled fast neutron detectors are normally operated as proportional counters rather than as Geiger tubes to take advantage of the spectroscopic information provided only in the proportional region (see Chapter 15).

C. Gamma Rays

In any gas-filled counter, the response to gamma rays of normal energy comes about by way of gamma-ray interactions in the solid wall of the counter. If the interaction takes place close enough to the inner wall surface so that the secondary electron created in the interaction can reach the gas and create ions, a pulse will result. Because only a single ion pair is required, the secondary electron need only barely emerge from the wall near the end of its track in order to generate a pulse from a Geiger tube.

The efficiency for counting gamma rays therefore depends on two separate factors: (1) the probability that the incident gamma ray interacts in the wall and produces a secondary electron, and (2) the probability that the secondary electron reaches the fill gas before the end of its track. As shown in Fig. 7-8, only the innermost layer of the wall near the gas may contribute secondary electrons. This region has a thickness equal to the maximum range of the secondary electrons that are formed, or typically a millimeter or two. Making the wall thicker than this value contributes nothing to the efficiency because electrons formed in regions of the wall farther from the gas have no chance to initiate a discharge.

The probability of gamma-ray interaction within this critical layer generally increases with atomic number of the wall material. Therefore, the gamma-ray efficiency of Geiger tubes is highest for those tubes constructed with a cathode wall of high-Z material. G-M tubes with bismuth ($Z = 83$) cathodes have been widely used in the past for gamma-ray detection. As shown in Fig. 7-9, however, the probability of interaction within the sensitive layer remains small even for high-Z materials, and typical gamma-ray counting efficiencies are seldom higher than several percent.

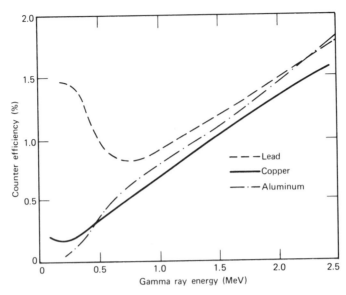

Figure 7-9 The efficiency of G-M tubes for gamma rays normally incident on the cathode. [From W. K. Sinclair, Chap. 5 in *Radiation Dosimetry* (G. J. Hine and G. L. Brownell, eds.). Copyright 1956 by Academic Press. Used with permission.]

If the photon energy is low enough, direct gamma-ray interaction in the fill gas of the tube may no longer be negligible. Therefore, the counting efficiency for low-energy gamma rays and X-rays can be enhanced by using gases of high atomic number at a pressure as high as possible. Xenon and krypton are common choices in these applications, and counting efficiencies near 100% can be achieved for photon energies below about 10 keV (see Fig. 6-14).

VIII. G-M SURVEY METERS

A common type of survey meter used in gamma-ray monitoring consists of a portable Geiger tube, high-voltage supply, and pulse counting rate meter. The pulse rate is then taken as an indication of the intensity of the gamma-ray exposure. The count rate meter scales are often calibrated in terms of exposure rate units, but under some circumstances these readings can be in error by a factor of 2 or 3 or more.

The difficulty arises from the fact that the count rate from a Geiger tube, in contrast to the measured current from an ion chamber, bears no fundamental relation to the gamma-ray exposure rate. For a given gamma-ray energy, the count rate obviously scales linearly with the intensity, but applications are likely to involve gamma rays of many different and variable energies. The exposure rate scale may be calibrated accurately at any fixed gamma-ray energy, but if the survey meter is applied to measurements involving other gamma-ray energies, the variation of the counter efficiency with energy must be considered. Plots of this type are given in Fig. 7-9 for different cathode materials. Ideally, one would like an efficiency versus energy curve that exactly matches a plot of exposure per gamma-ray photon versus energy. If this match were exact, then the G-M survey meter could generally be applied to all energies because the importance of each photon would be weighted correctly by the inherent efficiency of the detector. The ratio of the

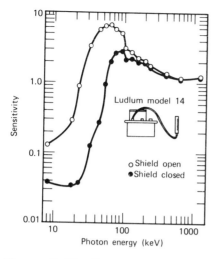
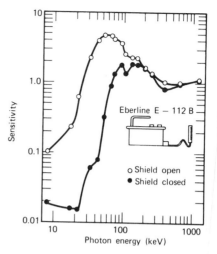

Figure 7-10 The sensitivity versus energy of two commercially manufactured Geiger survey meters. Sensitivity is defined as the indicated exposure rate divided by the true exposure rate. (From Storm et al.[13])

two curves at any given energy is a measure of the correction factor that should be applied to the survey meter readings. A plot of this correction factor versus gamma-ray energy for two commercial survey meters is shown in Fig. 7-10.

PROBLEMS

7-1. Why must the quench gas in a Geiger–Mueller tube have an ionization potential below that of the major fill gas?

7-2. Predict the effect each of the following changes would have on the observed starting voltage for a Geiger–Mueller tube if all other parameters remain the same:
   (a) Doubling the diameter of the anode wire.
   (b) Doubling the fill gas pressure.
   (c) Doubling the trace concentration of the quench gas.

7-3. Why does the pulse height from a Geiger tube continue to increase with applied voltage even after a full Geiger discharge is obtained?

7-4. Estimate the voltage at which the counter of Problem 6-3 enters the Geiger region of operation. Assume that three excited atoms are formed for every ion pair in a typical avalanche and that the subsequent deexcitation photons have a probability of 10^{-5} of creating an additional avalanche.

7-5. Both the proportional counter and Geiger tube are based on internal gas multiplication. Comment on each separately and contrast their behavior with regard to:
   (a) Variation of pulse height with applied voltage.
   (b) The need for a quench gas and its function.
   (c) Ability to differentiate heavy charged particle and electron radiations.
   (d) Ability to register high counting rates.
   (e) Typical counting efficiency for 1 MeV gamma rays.

76. The intrinsic detection efficiency for a gas-filled counter (proportional counter or G-M tube) when used with medium-energy gamma rays (say 1 MeV) will depend on the counter wall thickness in the following way:

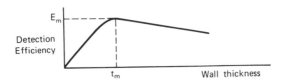

(a) Explain the general shape of this curve.

(b) Give an order-of-magnitude estimate for the optimum wall thickness t_m and relate it to basic physical properties.

7-7. A G-M tube is operated with a counting system whose threshold requires that a full Geiger discharge take place to register a count. Is its dead time behavior likely to be better described by the paralyzable or nonparalyzable model?

7-8. In a given counter gas operated at a pressure of 0.5 atm, the mobility of a free electron is 1.5×10^{-4} (m/s) · (m/V) · atm. The threshold electric field for the onset of avalanche formation is 2×10^6 V/m. If this gas is used in a cylindrical tube with anode radius of 0.005 cm and cathode radius of 2 cm, calculate the drift time of an electron from the cathode to the multiplying region for an applied voltage of 1500 V.

REFERENCES

1. D. H. Wilkinson, *Ionization Chambers and Counters*, Cambridge University Press, London, 1950.

2. W. J. Price, *Nuclear Radiation Detection*, 2nd ed., Chap. 5, McGraw-Hill, New York, 1964.

3. J. Sharpe, *Nuclear Radiation Detectors*, 2nd ed., Methuen and Co., London, 1964.

4. E. W. Emery, "Geiger–Mueller and Proportional Counters," in *Radiation Dosimetry*, Vol. II (F. H. Attix and W. C. Roesch, eds.), Academic Press, New York, 1966.

5. J. Fischer, H. Okuno, and A. H. Walenta, *IEEE Trans. Nucl. Sci.* **NS-25**(1), 794 (1978).

6. E. Mathieson, T. J. Harris, and G. C. Smith, *Nature* **272**, 709 (1978).

7. F. E. Lox and Z. Eeckhaut, *Rev. Sci. Instrum.* **40**(9), 1206 (1969).

8. A. Peeva and T. Karatoteva, *Nucl. Instrum. Meth.* **118**, 49 (1974).

9. S. Usacev and M. Seman, *Nucl. Instrum. Meth.* **73**, 41 (1969).

10. D. Srdoc, *Nucl. Instrum. Meth.* **99**, 321 (1972).

11. Y. Fujita, Y. Taguchi, M. Imamura, T. Inoue, and S. Tanaka, *Nucl. Instrum. Meth.* **128**, 523 (1975).

12. T. Aoyama and T. Watanabe, *Nucl. Instrum. Meth.* **197**, 357 (1982).

13. E. Storm, D. W. Lier, and H. Israel, *Health Phys.* **26**, 179 (1974).

CHAPTER · 8

Scintillation Detector Principles

The detection of ionizing radiation by the scintillation light produced in certain materials is one of the oldest techniques on record. The scintillation process remains one of the most useful methods available for the detection and spectroscopy of a wide assortment of radiations. In this chapter we discuss the various types of scintillators available and the important considerations in the efficient collection of the scintillation light. The following chapters cover modern light sensors—photomultiplier tubes and photodiodes—required to convert the light into an electrical pulse, and the application of scintillation detectors in radiation spectroscopy.

The ideal scintillation material should possess the following properties:

1. It should convert the kinetic energy of charged particles into detectable light with a high scintillation efficiency.
2. This conversion should be linear—the light yield should be proportional to deposited energy over as wide a range as possible.
3. The medium should be transparent to the wavelength of its own emission for good light collection.
4. The decay time of the induced luminescence should be short so that fast signal pulses can be generated.
5. The material should be of good optical quality and subject to manufacture in sizes large enough to be of interest as a practical detector.
6. Its index of refraction should be near that of glass ($\sim$ 1.5) to permit efficient coupling of the scintillation light to a photomultiplier tube.

No material simultaneously meets all these criteria, and the choice of a particular scintillator is always a compromise among these and other factors. The most widely applied scintillators include the inorganic alkali halide crystals, of which sodium iodide is the favorite, and organic-based liquids and plastics. The inorganics tend to have the best light output and linearity, but with several exceptions are relatively slow in their response time. Organic scintillators are generally faster but yield less light. The intended application also has a major influence on scintillator choice. The high Z-value of the constituents and high density of inorganic crystals favor their choice for gamma-ray spectroscopy, whereas organics are often preferred for beta spectroscopy and fast neutron detection.

The process of *fluorescence* is the prompt emission of visible radiation from a substance following its excitation by some means. It is conventional to distinguish several other processes that can also lead to the emission of visible light. *Phosphorescence* corresponds to the emission of longer wavelength light than fluorescence, and with a characteristic time that is generally much slower. *Delayed fluorescence* results in the same emission spectrum as prompt fluorescence but again is characterized by a much longer emission time following excitation. To be a good scintillator, a material should convert as large a fraction as possible of the incident radiation energy to prompt fluorescence, while minimizing the generally undesirable contributions for phosphorescence and delayed fluorescence.

A number of general reviews of the basic theory and application of scintillators have been published, with the exhaustive book by Birks[1] as an outstanding example. We discuss some of the basic mechanisms necessary for an understanding of the properties of the most common types of scintillator, together with some of their important properties as practical radiation detectors.

I. ORGANIC SCINTILLATORS

A. Scintillation Mechanism in Organics

The fluorescence process in organics arises from transitions in the energy level structure of a single molecule and therefore can be observed from a given molecular species independent of its physical state. For example, anthracene is observed to fluoresce as either a solid polycrystalline material, as a vapor, or as part of a multicomponent solution. This behavior is in marked contrast to crystalline inorganic scintillators such as sodium iodide, which require a regular crystalline lattice as a basis for the scintillation process.

A large category of practical organic scintillators is based on organic molecules with certain symmetry properties which give rise to what is known as a π-electron structure. The π-electronic energy levels of such a molecule are illustrated in Fig. 8-1. Energy can be absorbed by exciting the electron configuration into any one of a number of excited states. A series of singlet states (spin 0) are labeled as $S_0, S_1, S_2, \ldots$ in the figure. A similar set of triplet (spin 1) electronic levels are also shown as $T_1, T_2, T_3, \ldots$. For molecules of interest as organic scintillators, the energy spacing between S_0 and S_1 is 3 or 4 eV, whereas spacing between higher-lying states is usually somewhat smaller. Each of these electronic configurations is further subdivided into a series of levels with much finer spacing which correspond to various vibrational states of the molecule. Typical spacing of these levels is of the order of 0.15 eV. A second subscript is often added to distinguish these vibrational states, and the symbol S_{00} represents the lowest vibrational state of the ground electronic state.

Because the spacing between vibrational states is large compared with average thermal energies (0.025 eV), nearly all molecules at room temperature are in the S_{00} state. In Fig. 8-1 the absorption of energy by the molecule is represented by the arrows pointing upward. In the case of a scintillator, these processes represent the absorption of kinetic energy from a charged particle passing nearby. The higher singlet electronic states that are excited are quickly (on the order of picoseconds) deexcited to the S_1 electron state through radiationless internal conversion. Furthermore, any state with excess vibrational energy (such as S_{11} or S_{12}) is not in thermal equilibrium with its neighbors and again

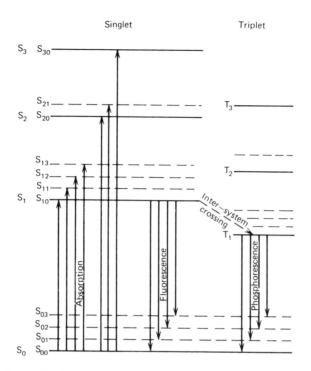

Figure 8-1 Energy levels of an organic molecule with π-electron structure. (From J. B. Birks, *The Theory and Practice of Scintillation Counting.* Copyright 1964 by Pergamon Press, Ltd. Used with permission.)

quickly loses that vibrational energy. Therefore, the net effect of the excitation process in a simple organic crystal is to produce, after a negligibly short time period, a population of excited molecules in the S_{10} state.

The principal scintillation light (or prompt fluorescence) is emitted in transitions between this S_{10} state and one of the vibrational states of the ground electronic state. These transitions are indicated by the downward arrows in Fig. 8-1. If τ represents the fluorescence decay time for the S_{10} level, then the prompt fluorescence intensity at a time t following excitation should simply be

$$I = I_0 e^{-t/\tau} \tag{8-1}$$

In most organic scintillators, τ is a few nanoseconds, and the prompt scintillation component is therefore relatively fast.

The lifetime for the first triplet state T_1 is characteristically much longer than that of the singlet state S_1. Through a transition called *intersystem crossing*, some excited singlet states may be converted into triplet states. The lifetime of T_1 may be as much as 10^{-3} s and the radiation emitted in a deexcitation from T_1 to S_0 is therefore a delayed light emission characterized as phosphorescence. Because T_1 lies below S_1, the wavelength of this phosphorescence spectrum will be longer than that for the fluorescence spectrum. While in the T_1 state, some molecules may be excited back to the S_1 state and subsequently decay through normal fluorescence. This process represents the origin of the delayed fluorescence sometimes observed for organics.

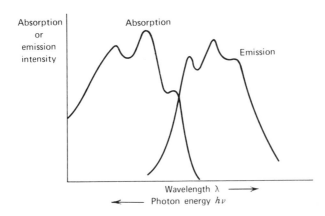

Figure 8-2   The optical absorption and emission spectra for a typical organic scintillator with the level structure shown in Fig. 8-1.

Figure 8-1 can also be used to explain why organic scintillators can be transparent to their own fluorescence emission. The length of the upward arrows corresponds to those photon energies that will be strongly absorbed in the material. Because all the fluorescence transitions represented by the downward arrows (with the exception of S_{10}–S_{00}) have a lower energy than the minimum required for excitation, there is very little overlap between the optical absorption and emission spectra, and consequently little self-absorption of the fluorescence. An example of these spectra for a typical organic scintillator is given in Fig. 8-2.

The *scintillation efficiency* of any scintillator is defined as the fraction of all incident particle energy which is converted into visible light. One would always prefer this efficiency to be as large as possible, but unfortunately there are alternate deexcitation modes available to the excited molecules which do not involve the emission of light and in which the excitation is degraded mainly to heat. All such radiationless deexcitation processes are grouped together under the term *quenching*. In the fabrication and use of organic scintillators, it is often important to eliminate impurities (such as dissolved oxygen in liquid scintillators), which degrade the light output by providing alternate quenching mechanisms for the excitation energy.

In almost all organic materials, the excitation energy undergoes substantial transfer from molecule to molecule before deexcitation occurs. This energy transfer process is especially important for the large category of organic scintillators which involves more than one species of molecules. If a small concentration of an efficient scintillator is added to a bulk solvent, the energy that is absorbed, primarily by the solvent, can eventually find its way to one of the efficient scintillation molecules and cause light emission at that point. These "binary" organic scintillators are widely used both as liquid and plastic solutions incorporating a variety of solvents and dissolved organic scintillants.

A third component is sometimes added to these mixtures to serve as a "waveshifter." Its function is to absorb the light produced by the primary scintillant and reradiate it at a longer wavelength. This shift in the emission spectrum can be useful for closer matching to the spectral sensitivity of a photomultiplier tube or to minimize bulk self-absorption in large liquid or plastic scintillators. Birks and Pringle[2] have reviewed the energy transfer mechanisms in binary and tertiary organic mixtures together with their influence on scintillation efficiency and pulse timing characteristics.

B. Types of Organic Scintillator

1. PURE ORGANIC CRYSTALS

Only two materials have achieved widespread popularity as pure organic crystalline scintillators. Anthracene is one of the oldest organic materials used for scintillation purposes and holds the distinction of having the highest scintillation efficiency (or greatest light output per unit energy) of any organic scintillator. Stilbene has a lower scintillation efficiency but is preferred in those situations in which pulse shape discrimination is to be used to distinguish among scintillations induced by charged particles and electrons (see the later discussion of this technique). Both materials are relatively fragile and difficult to obtain in large sizes. Also, the scintillation efficiency is known to depend on the orientation of an ionizing particle with respect to the crystal axis.[3] This directional variation, which can be as much as 20–30%, spoils the energy resolution obtainable with these crystals if the incident radiation will produce tracks in a variety of directions within the crystal.

2. LIQUID ORGANIC SOLUTIONS

A category of useful scintillators is produced by dissolving an organic scintillator in an appropriate solvent. Liquid scintillators can consist simply of these two components, or a third constituent is sometimes added as a wavelength shifter to tailor the emission spectrum to better match the spectral response of common photomultiplier tubes.

Liquid scintillators are often sold commercially in sealed glass containers and are handled in the same manner as solid scintillators. In certain applications, large-volume detectors with dimensions of several meters may be required. In these cases the liquid scintillator is often the only practical choice from a cost standpoint. In many liquids, the presence of dissolved oxygen can serve as a strong quenching agent and can lead to substantially reduced fluorescence efficiency. It is then necessary for the solution to be sealed in a closed volume from which most of the oxygen has been purged.

Liquid scintillators are also widely applied to count radioactive material which can be dissolved as part of the scintillator solution. In this case all radiations emitted by the source immediately pass through some portion of the scintillator and the counting efficiency can be almost 100%. The technique is widely used for counting low-level beta activity such as that from carbon-14 or tritium. This large field of *liquid scintillation counting* is described further in Chapter 10.

3. PLASTIC SCINTILLATORS

If an organic scintillator is dissolved in a solvent which can then be subsequently polymerized, the equivalent of a solid solution can be produced. A common example is a solvent consisting of styrene monomer in which an appropriate organic scintillator is dissolved. The styrene is then polymerized to form a solid plastic. Because of the ease with which they can be shaped and fabricated, plastics have become an extremely useful form of organic scintillator.

Plastic scintillators are available commercially with a good selection of standard sizes of rods, cylinders, and flat sheets. Because the material is relatively inexpensive, plastics are often the only practical choice if large-volume solid scintillators are needed. In these cases the self-absorption of the scintillator light may no longer be negligible, and some attention should be given to the attenuation properties of the material. The distance in which the light intensity will be attenuated by a factor of 2 can be as much as several meters, although much smaller attenuation lengths are observed for some plastics.[4]

4. THIN FILM SCINTILLATORS

Very thin films of plastic scintillator play a unique role in the field of radiation detectors. Because ultrathin films with a thickness as low as 20 $\mu g/cm^2$ can be fabricated, it is easy to provide a detector that is thin compared with the range of even weakly penetrating particles such as heavy ions. These films thus serve as transmission detectors, which respond to only the fraction of energy lost by the particle as it passes through the detector. The thickness can be as much as one or two orders of magnitude smaller than the minimum possible with other detector configurations, such as totally depleted silicon surface barriers. The films are available commercially[5] with thickness down to approximately 10 μm. Even thinner films can be produced by the user through techniques such as evaporation from a solution of plastic scintillator[6-9] or through the spin coating process.[10] The film can be deposited directly on the face of a photomultiplier tube,[7,11] or the light can be collected indirectly through a transparent light pipe in contact with the edges of the film.[12,13] Alternatively, the film can be placed within a reflecting cavity.[12,14,15] The response of these films does not follow directly from the expected energy loss of ions in the detector and is a more complex function of the ion velocity and atomic number.[9,11,16-19] The light yield per unit energy loss increases with decreasing atomic number of the ion, so that thin films can be useful transmission detectors for protons or alpha particles[8,14] even when the energy deposited is relatively small. In common with other organic scintillators, thin film detectors show scintillation decay times of only several ns, and they have proved very useful in fast timing measurements.[20]

5. LOADED ORGANIC SCINTILLATORS

Organic scintillators as a category are generally useful for the direct detection of beta particles (fast electrons) or alpha particles (positive ions). They also are readily adaptable to the detection of fast neutrons through the proton recoil process (see Chapter 15). Because of the low Z-value of their constituents (hydrogen, carbon, and oxygen), however, there is virtually no photoelectric cross section for gamma rays of typical energies. As a result, typical organic scintillators show no photopeak and will give rise only to a Compton continuum in their gamma-ray pulse height spectrum.

To provide some possibility for photoelectric conversion of gamma rays, attempts have been made to add high-Z elements to organic scintillators. The most common form is the addition of lead or tin to common plastic scintillators up to a concentration of 10% by weight. It has also been demonstrated[21] that tin can be added to liquid organic scintillator solutions in concentrations of up to 54% by weight while retaining a weak scintillation light output. At low gamma-ray energies, the photopeak efficiency of these materials can be made relatively high,[22] and they have the additional advantages of fast response and low cost compared with more conventional gamma-ray scintillators.[23,24] Unfortunately, the addition of these high-Z elements inevitably leads to a decreased light output, and the energy resolution that can be achieved is therefore considerably inferior to that of inorganic scintillators.

Another example of the loading of organic scintillators is the addition of gadolinium to liquid organic scintillators. The high capture cross section of gadolinium for neutrons induces beta and gamma activity, which is then detected directly in the organic scintillator. Large-volume neutron detectors can be constructed economically by using this material in a liquid-containing tank viewed by multiple photomultiplier tubes.

C. Response of Organic Scintillators

An overall compilation of the properties of organic scintillators is given in Table 8-1.

TABLE 8-1 Properties of Some Commercially Available Organic Scintillators

	Scintillator	Type	Density (g/cm³)	Refractive Index	Melting, Softening or Boiling Point (°)	Light Output, % Anthracene[a]	Decay Constant Main Component (ns)	Wavelength of Maximum Emission (nm)	Content of Loading Element (% by wt.)	Number of H Atoms/Number of C Atoms	Principal Applications[b]
Crystal	Anthracene	Crystal	1.25	1.62	217	100	30	447		0.715	γ, α, β fast neutrons
	Stilbene	Crystal	1.16	1.626	125	50	4.5	410		0.858	Fast neutrons (PSD), γ, etc.
Plastic	NE 102	Plastic	1.032	1.581	75	65	2.4	423		1.104	γ, α, β, fast n dosimetry
	NE 105	Plastic	1.037	1.58	75	46	~ 2.4	423			
	NE 110	Plastic	1.032	1.58	75	60	3.3	434		1.104	γ, α, β, fast neutrons, etc.
	NE 111	Plastic	1.032	1.58	75	55	1.7	375		1.096	Ultrafast timing
	NE 113	Plastic	1.032	1.58	75	60	3.3	434		1.108	Solvent bondable
	NE 115	Plastic		1.58		35	225	385			Long decay time
	NE 140	Plastic	1.045	1.58		58	~ 2	425	Sn 5%	1.100	
	Pilot B	Plastic	1.032	1.58	75	68	1.8	408		1.104	Fast counting
	Pilot F	Plastic	1.032	1.58	75	64	2.1	425		1.100	Fast neutrons, protons, electrons
	Pilot U	Plastic	1.032	1.58	75	67	1.36	391		1.100	Ultrafast timing
	Pilot Y	Plastic	1.032	1.58	75	60	3.1	432		1.102	Fast neutrons, protons, electrons, Large-area applications
Liquid	NE 213	Liquid	0.874	1.508	141	78	3.7	425		1.213	Fast neutrons (PSD)
	NE 216	Liquid	0.885	1.523	141	78	3.5	425		1.171	α, β (internal counting)
	NE 220	Liquid	1.036	1.442	104	65	3.8	425		1.669	α, β (internal, aqueous sample)
	NE 221	Gel	1.08	1.442	104	55	4	425		1.669	α, β (internal counting)
	NE 224	Liquid	0.877	1.505	169	80	2.6	425		1.330	γ, fast neutrons
	NE 226	Liquid	1.61	1.38	80	20	3.3	430		0	γ, insensitive to neutrons
	NE 228	Liquid	0.735	1.403	99	45		385		2.00	Neutrons
	NE 230	Deuterated liquid	0.945	1.50	81	60	3.0	425	D 14.2%	0.984	(D/C) special applications
	NE 231	Liquid	0.88	1.50	80	58	2.8	425		0.984	Special applications
	NE 232	Deuterated liquid	0.89	1.43	81	60	4	430	D 24.5%	1.96	(D/C) special applications
	NE 233	Liquid	0.874	1.506	117	74	3.7	425		1.118	α, β (internal counting)
	NE 235A (235H)	Liquid	0.858	1.47	350	40 (50)	4	420		2.0	α, fast neutrons, large tanks
	NE 250	Liquid	1.035	1.452	104	50	4	425		1.760	α, β (internal, aqueous sample)
	NE 260	Liquid				40		425			α, β (internal counting)
Loaded Liquid	NE 311 (311A)	B(^{10}B) loaded	0.91	1.411	85	65	3.8	425	B 5%	1.701	Neutrons, β
	NE 313	Gd loaded	0.88	1.506	136	62	4.0	425	Gd 0.5%	1.220	Neutrons
	NE 316	Sn loaded	0.93	1.496	148.5	35	4.0	425	Sn 10%	1.411	γ, X-rays
	NE 323	Gd loaded	0.879	1.50	161	60	3.8	425	Gd 0.5%	1.377	Neutrons

[a] NaI(Tl) is 230% on this scale.
[b] PSD represents neutron–gamma pulse shape discrimination.

Source: Table of Physical Constants of Scintillators, Nuclear Enterprises, Inc.

1. LIGHT OUTPUT

A small fraction of the kinetic energy lost by a charged particle in a scintillator is converted into fluorescent energy. The remainder is dissipated nonradiatively, primarily in the form of lattice vibrations or heat. The fraction of the particle energy which is converted (the scintillation efficiency) depends on both the particle type and its energy. In some cases, the scintillation efficiency may be independent of energy, leading to a linear dependence of light yield on initial energy.

For organic scintillators such as anthracene, stilbene, and many of the commercially available liquid and plastic scintillators, the response to electrons is linear for particle energies above about 125 keV.[25] The response to heavy charged particles such as protons or alpha particles is always less for equivalent energies and is nonlinear to much higher initial energies. As a typical example, Fig. 8-3 shows the scintillation response of a common, commercially available plastic scintillator. At energies of a few hundred keV, the response to protons is smaller by a factor of 10 compared with the light yield of equivalent energy electrons. At higher energies the discrepancy is less, but the proton response is always below the electron response.

The response of organic scintillators to charged particles can best be described by a relation between dL/dx, the fluorescent energy emitted per unit path length, and dE/dx, the specific energy loss for the charged particle. A widely used relation first suggested by Birks[1] is based on the assumption that a high ionization density along the track of the particle leads to quenching from damaged molecules and a consequent lowering of the scintillation efficiency. If we assume that the density of damaged molecules along the wake of the particle is directly proportional to the ionization density, we can represent their density by $B(dE/dx)$, where B is a proportionality constant. Birks assumes that some fraction k of these will lead to quenching. A further assumption is that, in the

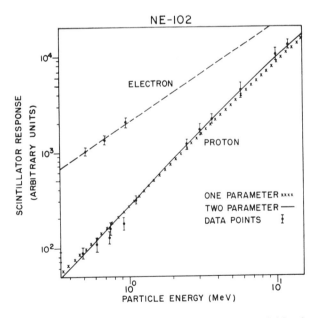

Figure 8-3 The scintillation light yield for a commercially available plastic scintillator (NE 102) when excited by electrons and protons. The data are fit by curves from Eq. (8-3) (one parameter) and Eq. (8-9) (two parameter). (From Craun and Smith.[26])

absence of quenching, the light yield is proportional to energy loss:

$$\frac{dL}{dx} = S\frac{dE}{dx} \tag{8-2}$$

where S is the normal scintillation efficiency. To account for the probability of quenching, Birks then writes

$$\frac{dL}{dx} = \frac{S\dfrac{dE}{dx}}{1 + kB\dfrac{dE}{dx}} \tag{8-3}$$

Equation (8-3) is commonly referred to as *Birks' formula*. As a practical matter the product kB is treated as an adjustable parameter to fit experimental data for a specific scintillator. In many cases, this single adjustable parameter can give very good fits to the shape of experimental data, with the value of S providing the absolute normalization.

When excited by fast electrons (either directly or from gamma-ray irradiation), dE/dx is small for sufficiently large values of E and Birks' formula then predicts

$$\frac{dL}{dx}\bigg|_e = S\frac{dE}{dx} \tag{8-4}$$

or the incremental light output per unit energy loss is a constant

$$\frac{dL}{dE}\bigg|_e = S \tag{8-5}$$

This is the regime in which the light output

$$L \equiv \int_0^E \frac{dL}{dE}\,dE = SE \tag{8-6}$$

is linearly related to the initial particle energy E.

On the other hand, for an alpha particle, dE/dx is very large so that saturation occurs along the track and Birks' formula becomes

$$\frac{dL}{dx}\bigg|_\alpha = \frac{S}{kB} \tag{8-7}$$

The appropriate value of kB can therefore be determined by taking the ratio of these two responses:

$$kB = \frac{dL}{dE}\bigg|_e \bigg/ \frac{dL}{dx}\bigg|_\alpha \tag{8-8}$$

In order to match experimental data more closely, other formulas for dL/dx have been proposed by a number of authors. These are in effect semiempirical formulas that introduce one or more additional fitting parameters. An extensive analysis of the response of a number of organic scintillators has been carried out by Craun and Smith.[26] Their analysis is based largely on the data of Smith et al.[27] and uses an extended version of

Birks' formula

$$\frac{dL}{dx} = \frac{S \dfrac{dE}{dx}}{1 + kB\dfrac{dE}{dx} + C\left(\dfrac{dE}{dx}\right)^2}$$ (8-9)

where C is again treated as an empirically fitted parameter. This expression approaches the simple Birks' formula Eq. (8-3) for small values of dE/dx. Parameters for Eq. (8-9) are derived and listed in Ref. 26 for a variety of organic scintillators and exciting particles. Also given is a useful table of dE/dx values for organic scintillators of different composition for protons and deuterons.

The *alpha-to-beta ratio* is a widely used parameter to describe the difference of light output for an organic scintillator for electrons and charged particles of the same energy. The light yield for electrons is always higher than that for a charged particle of the same kinetic energy, and therefore the alpha-to-beta ratio is always less than 1. This ratio will depend on the energy at which the comparison is made, and no fixed value is applicable over the entire energy range. Measurements have been reported by Czirr[28] for a variety of organic scintillators.

The light output of NE 213 liquid scintillator as measured by Maier and Nitschke[29] is shown in Fig. 8-4. Typical of many organic scintillators, the light output can be represented as proportional to $E^{3/2}$, for energies below about 5 MeV, and becomes approximately linear for higher energies.

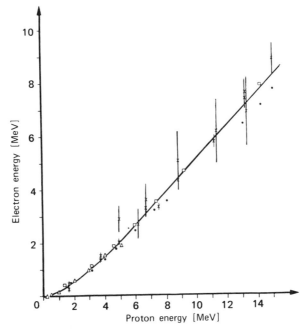

Figure 8-4   The light output (expressed in equivalent electron energy deposition) versus proton energy for liquid scintillator NE 213. (From Maier and Nitschke.[29])

Clark[30] has compiled a set of data on the absolute scintillation efficiency [S in Eq. (8-5)] of plastic scintillators for gamma-ray excitation. These data are a useful supplement to older tabulations which can be found in Ref. 1. Clark points out the variability of different measurements attempting to determine absolute efficiencies, which in some cases can be as much as a factor of 2 discrepant. This variability can be traced in part to differences in the purity and past history of the scintillation material, and absolute values should always be used with caution. In some organics, the partial overlap of the emission and absorption spectra leads to a size dependency of the apparent efficiency for scintillation. Also, the anisotropic response of anthracene and stilbene further complicates efficiency measurements.

Prolonged exposure to ionizing radiation leads to a general deterioration of the properties of most organic scintillators. Plastic scintillators exposed to light and oxygen have also been shown to undergo a long-term deterioration due to polymer degradation.[31] In addition to internal effects, the surface of plastics can often become crazed on exposure to extreme environments. The surface crazing leads to a substantial drop in observed light output from large scintillators because of the decreased efficiency of internal light reflection.

2. TIME RESPONSE

If it can be assumed that the luminescent states in an organic molecule are formed instantaneously and only prompt fluorescence is observed, then the time profile of the light pulse should be a very fast leading edge followed by a simple exponential decay [Eq. (8-1)]. Much of the published literature quotes a *decay time* that characterizes the prompt scintillation yield from various organic materials. Although this simple representation is often adequate for many descriptions of the scintillator behavior, a more detailed model of the time dependence of the scintillation yield must take into account two other effects: the finite time required to populate the luminescent states, and the slower components of the scintillation corresponding to delayed fluorescence and phosphorescence.

Times of approximately half a nanosecond are required to populate the levels from which the prompt fluorescence light arises. For the very fast scintillators, the decay time from these levels is only three or four times greater, and a full description of the expected pulse shape must take into account the finite rise time as well. One approach[32, 33] assumes that the population of the optical levels is also exponential and that the overall shape of the light pulse is given by

$$I = I_0 \left(e^{-t/\tau} - e^{-t/\tau_1} \right) \qquad (8\text{-}10)$$

where τ_1 is the time constant describing the population of the optical levels and τ is the time constant describing their decay. Values for these parameters for several plastic scintillators are given in Table 8-2. Other observations[34] have concluded that the population step is better represented by a Gaussian function $f(t)$ characterized by a standard deviation σ_{ET}. The overall light versus time profile is then described by

$$\frac{I}{I_0} = f(t) e^{-t/\tau} \qquad (8\text{-}11)$$

Best fit values for σ_{ET} are also shown in Table 8-2. Experimentally, the rise and fall of the light output can be characterized by the full width at half maximum (FWHM) of the resulting light versus time profile, which can be measured using very fast timing proce-

TABLE 8-2   Some Timing Properties of Fast Plastic Scintillators

	Parameters for Eq. (8-10)		Parameters for Eq. (8-11)		Measured FWHM
	τ_1 (rise)	τ (decay)	σ_{ET}	τ	
NE 111	0.2 ns	1.7 ns	0.2 ns	1.7 ns	1.54 ns
Naton 136	0.4 ns	1.6 ns	0.5 ns	1.87 ns	2.3 ns
NE 102A	0.6 ns	2.4 ns	0.7 ns	2.4 ns	3.3 ns

Data from Bengtson and Moszynski.[34]

dures. It is becoming increasingly popular to specify the performance of ultrafast organic scintillators by their FWHM time rather than the decay time alone.

Studies of the time response of organic scintillators have been reported in Refs. 35–38. In these measurements the emphasis is on conventional plastic scintillators, with the smallest FWHM reported to be 1.3 ns for NE 111. Lynch[33] has also reported results on light output and time response of a variety of liquid scintillators. Some attention[39–41] has focused on plastics to which a quenching agent has been deliberately added. Although the light output is reduced by an order of magnitude or more, the FWHM can be as small as a few hundred picoseconds. For these very fast scintillators, effects other than the scintillation mechanism can sometimes affect the observed time response. Among these is the finite flight time of the photons from the point of scintillation to the photomultiplier tube. Particularly in large scintillators, transit time fluctuation due to multiple light reflections at scintillator surfaces can amount to a sizable spread in the arrival time of the light at the photomultiplier tube photocathode.[42] Also, there is evidence[43,44] that self-absorption and reemission of the fluorescence plays an important role in causing an apparent worsening of the time resolution as the dimensions of a scintillator are increased.

Although fast decay time is an advantage in nearly all applications of scintillators, there is at least one application in which a slow decay is needed. In the phoswich detector (see p. 326) two different scintillation materials are employed, one of which must have an appreciably slower decay time than the other. For such applications, a plastic scintillator has been developed[45] with an unusually long decay of 225 ns, more typical of slower inorganic materials.

3. PULSE SHAPE DISCRIMINATION

For the vast majority of organic scintillators, the prompt fluorescence represents most of the observed scintillation light. A longer-lived component is also observed in many cases, however, corresponding to delayed fluorescence. The composite yield curve can often be represented adequately by the sum of two exponential decays—called the fast and slow components of the scintillation. Compared with the prompt decay time of a few nanoseconds, the slow component will typically have a characteristic decay time of several hundred nanoseconds. Because the majority of the light yield occurs in the prompt component, the long-lived tail would not be of great consequence except for one very useful property: *The fraction of light that appears in the slow component often depends on the nature of the exciting particle.* One can therefore make use of this dependence to differentiate between particles of different kinds which deposit the same energy in the detector. This process is often called *pulse shape discrimination* and is widely applied to eliminate gamma-ray-induced events when organic scintillators are used as neutron detectors.

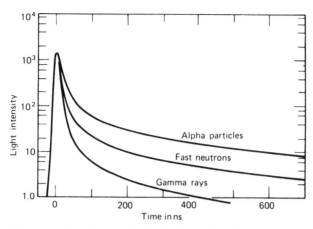

Figure 8-5 The time dependence of scintillation pulses in stilbene (equal intensity at time zero) when excited by radiations of different types. (From Bollinger and Thomas.[46])

There is strong evidence that the slow scintillation component originates with the excitation of long-lived triplet states (labeled T_1 in Fig. 8-1) along the track of the ionizing particle. Bimolecular interactions between two such excited molecules can lead to product molecules, one in the lowest singlet state (S_1) and the other in the ground state. The singlet state molecule can then deexcite in the normal way, leading to delayed fluorescence. The variation in the yield of the slow component can then be partially explained by the differences expected in the density of triplet states along the track of the particle, because the bimolecular reaction yield should depend on the square of the triplet concentration. Therefore, the slow component fraction should depend primarily on the rate of energy loss dE/dx of the exciting particle and should be greatest for particles with large dE/dx. These predictions are generally confirmed by measurements of the scintillation pulse shape from a wide variety of organics.

Certain organic scintillators, including stilbene crystals and a number of commercial liquid scintillators, are particularly favored for pulse shape discrimination because of the large differences in the relative slow component induced by different radiations. Figure 8-5 shows the differences observed in stilbene for alpha particles, fast neutrons (recoil protons), and gamma rays (fast electrons). In such scintillators, it is not only possible to differentiate radiations with large dE/dx differences (such as neutrons and gamma rays) but also to separate events arising from various species of heavy charged particles as well. Reviews of the pulse shape discrimination properties of different types of organic scintillators and examples of applications are given in Refs. 47–52. Electronic circuits designed to carry out this pulse shape discrimination are described in Chapter 17.

II. INORGANIC SCINTILLATORS

A. Scintillation Mechanism in Inorganic Crystals with Activators

The scintillation mechanism in inorganic materials depends on the energy states determined by the crystal lattice of the material. As shown in Fig. 8-6, electrons have available only discrete bands of energy in materials classified as insulators or semiconductors. The lower band, called the valence band, represents those electrons that are essentially bound at lattice sites, whereas the conduction band represents those electrons

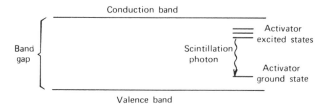

Figure 8-6   Energy band structure of an activated crystalline scintillator.

that have sufficient energy to be free to migrate throughout the crystal. There exists an intermediate band of energies, called the forbidden band, in which electrons can never be found in the pure crystal. Absorption of energy can result in the elevation of an electron from its normal position in the valence band across the gap into the conduction band, leaving a hole in the normally filled valence band. In the pure crystal, the return of the electron to the valence band with the emission of a photon is an inefficient process. Furthermore, typical gap widths are such that the resulting photon would be of too high an energy to lie in the visible range.

To enhance the probability of visible photon emission during the deexcitation process, small amounts of an impurity are commonly added to inorganic scintillators. These impurities, called *activators*, create special sites in the lattice at which the normal energy band structure is modified from that of the pure crystal. As a result, there will be energy states created within the forbidden gap through which the electron can deexcite back to the valence band. Because the energy is less than that of the full forbidden gap, this transition can now give rise to a visible photon and therefore serve as the basis of the scintillation process.

A charged particle passing through the detection medium will form a large number of electron–hole pairs created by the elevation of electrons from the valence to the conduction band. The positive hole will quickly drift to the location of an activator site and ionize it, because the ionization energy of the impurity will be less than that of a typical lattice site. Meanwhile, the electron is free to migrate through the crystal and will do so until it encounters such an ionized activator. At this point the electron can drop into the impurity site, creating a neutral impurity configuration which can have its own set of excited energy states. These states are illustrated in Fig. 8-6 as lines within the forbidden gap. If the activator state that is formed is an excited configuration with an allowed transition to the ground state, its deexcitation will occur very quickly and with high probability for the emission of a corresponding photon. If the activator is properly chosen, this transition can be in the visible energy range. Typical half-lives for such excited states are on the order of 10^{-7} s. Because the migration time for the electron is much shorter, all the excited impurity configurations are formed essentially at once and will subsequently deexcite with the half-life characteristic of the excited state. It is the decay time of these states that therefore determines the time characteristics of the emitted scintillation light. Many inorganic scintillators can adequately be characterized by a single decay time or a simple exponential, although more complex time behavior is sometimes observed.

There are processes that compete with the one just described. For example, the electron upon arriving at the impurity site can create an excited configuration whose transition to the ground state is forbidden. Such states then require an additional increment of energy to raise them to a higher-lying state from which deexcitation to the ground state is possible. One source of this energy is thermal excitation and the resulting slow component of light is called *phosphorescence*. It can often be a significant source of background light or "afterglow" in scintillators.

A third possibility exists when an electron is captured at an activator site. Certain radiationless transitions are possible between some excited states formed by electron capture and the ground state, in which case no visible photon results. Such processes are called *quenching* and represent loss mechanisms in the conversion of the particle energy to scintillation light.

As an alternative to the independent migration of the electron and hole described above, the pair may instead migrate together in a loosely associated configuration known as an *exciton*. In this case the electron and hole remain associated with each other but are free to drift through the crystal until reaching the site of an activator atom. Similar excited activator configurations can again be formed and give rise to scintillation light in their deexcitation to the ground configuration.

A measure of the efficiency of the scintillation process follows from a simple energy calculation. For a wide category of materials, it takes on the average about three times the bandgap energy to create one electron–hole pair. In sodium iodide, this means about 20 eV of charged particle energy must be lost to create one electron–hole pair. For 1 MeV of particle energy deposited in the scintillator, about 5×10^4 electron–hole pairs are thus created. Various experimental determinations have shown that the absolute scintillation efficiency of thallium-activated sodium iodide is about 12%. Absorption of 1 MeV of energy should therefore yield about 1.2×10^5 eV in total light energy, or 4×10^4 photons with an average energy of 3 eV. The yield is thus very close to 1 photon per electron–hole pair originally formed, and the energy transfer to activator sites must be extremely efficient. The processes of energy transfer in alkali halide scintillators have been reviewed by Murray[53] and by Kaufman et al.[54]

One important consequence of luminescence through activator sites is the fact that the crystal can be transparent to the scintillation light. In the pure crystal, roughly the same energy would be required to excite an electron–hole pair as that liberated when that pair recombines. As a result the emission and absorption spectra will overlap and there will be substantial self-absorption. As we have seen, however, the emission from an activated crystal occurs at an activator site where the energy transition is less than that represented by the creation of the electron–hole pair. As a result the emission spectrum is shifted to longer wavelengths and will not be influenced by the optical absorption band of the bulk of the crystal.

The emission spectrum of the light produced by a number of inorganic scintillators is shown in Fig. 8-7. To make full use of the scintillation light, the spectrum should fall near the wavelength region of maximum sensitivity for the device used to detect the light. For reference, the responses of several common photocathodes are also plotted in Fig. 8-7 (other photocathode responses are plotted in Fig. 9-2).

The scintillation properties of a collection of inorganic scintillators are compiled in Table 8-3. The decay constants shown in the fourth column are only approximate, and in many cases more than one decay component contributes to the overall light yield. The next column gives the rise time of a typical output pulse from a PM tube that collects the light with a long time constant measuring circuit. The total light yield values in the sixth column are based on measurements[57] using a photodiode with broad spectral response. The values shown are averages derived from many samples, and a variation of $\pm 20\%$ is not unusual in different crystals of the same material. The seventh column lists the calculated absolute scintillation efficiency (fraction of particle energy converted into light)

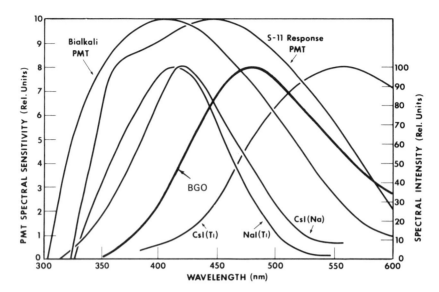

Figure 8-7   The emission spectra of several common inorganic scintillators. Also shown are the response curves for two widely used photocathodes. (Primarily from *Scintillation Phosphor Catalog*, The Harshaw Chemical Company. The emission spectrum for BGO is from Ref. 55.)

for fast secondary electrons produced by gamma ray interactions in the material. The last column shows the relative pulse amplitude to be expected from secondary electrons of the same energy when the light is detected using a PM tube with bialkali photocathode. The variability of photocathode spectral response may also introduce additional variations in these values.

B. Characteristics of Alkali Halide Scintillators

1. NaI(Tl)

In 1948, Robert Hofstadter[59] first demonstrated that crystalline sodium iodide, in which a trace of thallium iodide had been added in the melt, produced an exceptionally large scintillation light output compared with the organic materials that had previously received primary attention. This discovery, more than any other single event, ushered in the era of modern scintillation spectrometry of gamma radiation. It is remarkable that the same material remains preeminent in the field despite four decades of subsequent research into other scintillation materials. Large ingots can be grown from high-purity sodium iodide to which about 10^{-3} mole fraction of thallium has been added as an activator. Scintillators of unusual size or shape can also be fabricated by pressing small crystallites together. NaI(Tl) is hygroscopic and will deteriorate due to water absorption if exposed to the atmosphere for any length of time. Crystals must therefore be "canned" in an air-tight container for normal use.

The most notable property of NaI(Tl) is its excellent light yield. Its response to electrons (and gamma rays) is close to linear over most of the significant energy range (see Fig. 8-8). It has come to be accepted as the standard scintillation material for routine gamma-ray spectroscopy and can be machined into a wide assortment of sizes and shapes. The crystal is somewhat fragile and can easily be damaged by mechanical or thermal shock.

TABLE 8-3 Properties of Common Inorganic Scintillators

Material	Specific Gravity	Wavelength of Maximum Emission (nm) λ_{max}	Index of Refraction at λ_{max}	Principal Decay Constant (μs)	Pulse 10–90% Rise Time (μs)	Total Light Yield in Photons/MeV	Absolute Scintillation Efficiency for Fast Electrons	Relative γ-Ray Pulse Height with Bialkali PM Tube
NaI(Tl)	3.67	415	1.85	0.23	0.5	38000	11.3%	1.00
CsI(Tl)	4.51	540	1.80	1.0	4	52000	11.9	0.49
CsI(Na)	4.51	420	1.84	0.63	4	39000	11.4	1.11
LiI(Eu)	4.08	470	1.96	1.4	—	11000	2.8	0.23
BGO	7.13	505	2.15	0.30	0.8	8200	2.1	0.13
BaF$_2$ slow component	4.89	310	1.49	0.62	3	10000	4.5	0.13
BaF$_2$ fast component	4.89	220	—	0.0006	—	—	—	0.03[a]
ZnS(Ag) (polycrystalline)	4.09	450	2.36	0.2	—	—	—	1.30[b]
CaF$_2$(Eu)	3.19	435	1.44	0.9	4	24000	6.7	0.78
CsF	4.11	390	1.48	0.004	—	—	—	0.05
Li glass[c]	2.5	395	1.55	0.075	—	—	1.5	0.10

For comparison, a typical organic (plastic) scintillator:

Material	Specific Gravity	Wavelength of Maximum Emission (nm) λ_{max}	Index of Refraction at λ_{max}	Principal Decay Constant (μs)	Pulse 10–90% Rise Time (μs)	Total Light Yield in Photons/MeV	Absolute Scintillation Efficiency for Fast Electrons	Relative γ-Ray Pulse Height with Bialkali PM Tube
NE 102A	1.03	423	1.58	0.002	—	10000	3.0	0.25

[a] Using UV-sensitive PM tube.
[b] For alpha particles.
[c] Properties vary with exact formulation. Also see Table 15-1.

Source: Data derived primarily from Refs. 56–58.

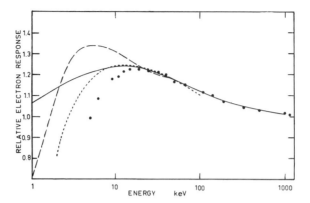

Figure 8-8   The light output of NaI(Tl) for totally absorbed electrons, expressed as total light per unit initial energy. A completely linear response would therefore be a horizontal line. The various lines and symbols represent different experimental data (normalized at 45 keV) as compiled by Prescott and Narayan.[60]

The dominant decay time of the scintillation pulse is 230 ns, uncomfortably long for some fast timing or high counting rate applications. In addition to this prompt yield, a phosphorescence with characteristic 0.15 s decay time has also been measured,[61] which contributes about 9% to the overall light yield. Other longer-lived phosphorescence components have also been measured.[62] Because the anode time constant of photomultiplier tubes is usually set much shorter than these decay times, each photoelectron associated with the phosphorescence is normally resolved individually. At low counting rates, the result is then a series of single-electron pulses that follow the main scintillation pulse and usually are well below the amplitude of interest in the measurement. There are applications, however, in which single-electron sensitivity is needed and the influence of these phosphorescence pulses can be significant. At high counting rates, the phosphorescence will tend to build up due to the multiple overlap from many preceding pulses. This afterglow is often an undesirable characteristic of sodium iodide used in high-rate applications.

The properties summarized above are for thallium-activated sodium iodide operated at room temperature. In some applications, the scintillator must be operated at either lower or higher ambient temperatures. Figure 8-9 shows the dependence of the total light yield in thallium-activated sodium iodide as a function of temperature. The dropoff in scintillation yield with increasing temperature, typical of most scintillation materials, results in generally poorer energy resolution when the scintillator must be used at elevated temperatures. The scintillation decay time in NaI(Tl) is also a function of temperature (see Fig. 8-10), with somewhat faster response at higher temperatures.

2. CsI(Tl) and CsI(Na)

Cesium iodide is another alkali halide that has gained substantial popularity as a scintillation material. It is commercially available with either thallium or sodium as the activator material, and very different scintillator properties are produced in the two cases. Cesium iodide has a somewhat larger gamma-ray absorption coefficient per unit size compared to sodium iodide. This advantage is of primary importance for applications such as space instrumentation where size and weight are at a premium. Because it is less brittle than sodium iodide, it can be subjected to more severe conditions of shock and vibration. When cut into thin sheets, cesium iodide may be bent into various shapes

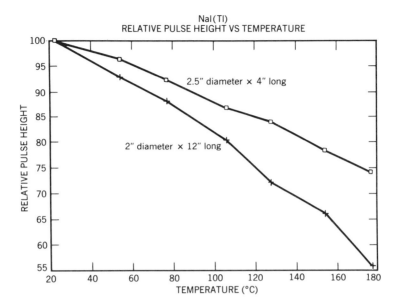

Figure 8-9 The temperature dependence of the light yield measured from two NaI(Tl) crystals. The measurements were made using an oven equipped with a light pipe, and the temperature of the photomultiplier tube was held constant. The difference in behavior between the two crystals is probably due to changes in surface reflectivity. (Data courtesy R. Dayton, Bicron Corporation, Newbury, Ohio.)

without fracturing, and it is reasonably soft and malleable. An extensive bibliography of the properties of cesium iodide (both sodium and thallium activated) can be found in Ref. 64.

A most useful property of CsI(Tl) is its variable decay time for various exciting particles. Pulse shape discrimination techniques can therefore be used to differentiate among various types of radiation. Particularly clean separations can be achieved between

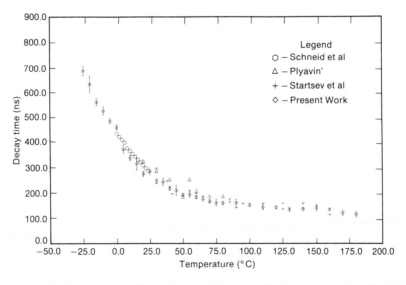

Figure 8-10 Temperature dependence of the scintillation decay time in NaI(Tl). (From Schweitzer and Ziehl.[63])

charged particles such as protons or alpha particles on one hand, and electron events on the other hand. The material is less hygroscopic than NaI(Tl) but will deteriorate if exposed to water or high humidity.

The emission spectrum of CsI(Tl) is peaked at much longer wavelengths than that for NaI(Tl) (see Fig. 8-7) and is poorly matched to the response of PM tubes with S-11 or bialkali photocathodes. For that reason, the light output is often quoted as being substantially lower in CsI(Tl). However, when measurements are made using photodiodes with extended response into the red region of the spectrum, the scintillation yield is actually higher[57] than that of any other scintillator.

CsI(Na) has an emission spectrum that is similar to that of NaI(Tl) with a light yield that is slightly greater, but its relatively slow decay is a disadvantage. The decay is reported to consist of two components with mean lives of 0.46 and 4.18 μs, and a dependence of the slow to fast component ratio has been shown for various exciting particles.[65] Long-lived phosphorescence components in cesium iodide have also been reported.[66] CsI(Na) is hygroscopic and must be sealed against exposure to ambient atmospheres.

3. LiI(Eu)

Lithium iodide (europium activated) is an alkali halide of special interest in neutron detection. As discussed in Chapter 14, crystals prepared using lithium enriched in ^{6}Li provide for the efficient detection of low-energy neutrons through the ^{6}Li(n, α) reaction. Scintillation properties of this material are included in Table 8-3.

C. Other Inorganic Crystals

1. BISMUTH GERMANATE (or BGO)

In the late 1970s, a scintillation material, $Bi_4Ge_3O_{12}$, became commercially available, and crystals of this material (commonly abbreviated as BGO) have been applied in a rapidly growing number of applications. A major advantage of BGO is its high density (7.3 g/cm^3) and the large atomic number (83) of the bismuth component. These properties result in the largest probability per unit volume of any commonly available scintillation material for the photoelectric absorption of gamma rays. Its mechanical and chemical properties make it easy to handle and use, and detectors using BGO can be made more rugged than those employing the more fragile and hygroscopic sodium iodide. Unfortunately, the light yield from BGO is relatively low, being variously reported at 10–20% of that of NaI(Tl). It is therefore of primary interest when the need for high gamma-ray counting efficiency outweighs considerations of energy resolution. Some comparative gamma-ray spectra for BGO and NaI(Tl) are shown in Chapter 10.

Figure 8-11 shows the time profile of the light emitted in a scintillation event in both BGO and NaI(Tl). In BGO, the principal decay time of 300 ns is preceded by a faster component of the light with 60 ns decay time that represents about 10% of the total light yield. On the other hand, the initial decay of the NaI(Tl) light pulse is slightly slower than the 230 ns principal decay. These differences, coupled with the much lower light yield from BGO, result in an overall timing resolution for BGO that is about a factor of 2 worse than that for NaI(Tl).[67] In BGO there are almost none of the long decay components that lead to afterglow in sodium iodide and some other scintillators. BGO has therefore found widespread application in X-ray computerized tomography scanners where scintillators operated in current mode must accurately follow rapid changes in X-ray intensity.

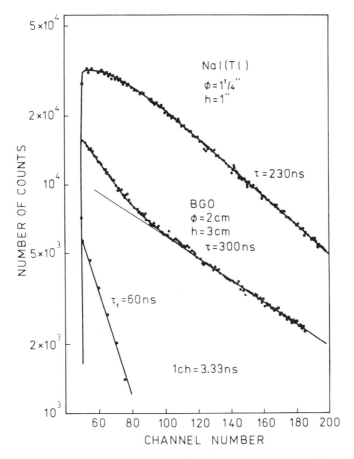

Figure 8-11 Measurements of the light pulse shapes from BGO and NaI(Tl). The abscissa represents time, the ordinate the relative light output. The BGO yield is represented as the sum of separate decay components with 60 and 300 ns decay times. (From Moszynski et al.[67])

BGO is an example of a "pure" inorganic scintillator that does not require the presence of a trace activator element to promote the scintillation process. Instead, the luminescence is associated[68] with an optical transition of the Bi^{3+} ion that is a major constituent of the crystal. There is a relatively large shift between the optical absorption and emission spectra (called the *Stokes shift*) of the Bi^{3+} states. Therefore, relatively little self-absorption of the scintillation light occurs, and the crystal remains transparent to its own emission over dimensions of many centimeters. The scintillation efficiency depends strongly on the purity of the crystal, and some of the variability in the light yield reported from BGO in the past can be attributed to using crystals with different residual levels of impurities.[69] The crystals are typically grown by the Czochralski method in which a crystal boule is pulled from a molten mixture of bismuth oxide and germanium oxide at a rate of a few millimeters per hour. The boule can then be cut and polished using conventional methods. BGO remains two to three times more costly than NaI(Tl) and is currently available only in limited sizes.

In common with many other scintillators, the light output from BGO decreases with increasing temperature. Figure 8-12 shows the relative light output and the scintillation

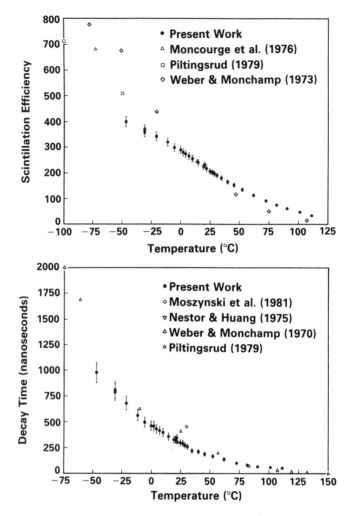

Figure 8-12 Measurements of the temperature dependence of the scintillation light yield (top) and the scintillation decay time (bottom) in BGO. (From Melcher et al.[55])

decay time as a function of temperature. Since the light yield is already low at room temperature, its rapid dropoff severely limits the usefulness of BGO in high-temperature applications.

2. BARIUM FLUORIDE (BaF$_2$)

Another pure inorganic scintillator, barium fluoride, has attracted recent attention[70-73] because of its potential application in fast timing measurements. It is currently the only high-Z scintillation material that exhibits a scintillation component with decay time of less than 1 ns. This combination of properties therefore makes the material attractive for scintillation detectors in which both high detection efficiency per unit volume and a fast response are required.

Unactivated BaF$_2$ has been recognized as a scintillation material for a number of years.[74] However, it has only recently been shown[70] that the scintillation light actually consists of two components: a fast component with decay time of 0.6 ns emitted in the

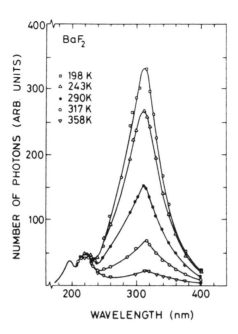

Figure 8-13 The scintillation emission spectra from BaF_2 measured at various temperatures. The fast component (corresponding to the two small peaks at the left) does not display the strong temperature dependence of the slow component. (From Schotanus et al.[75, 76])

short-wavelength region of the spectrum, and a slower component with 630 ns decay time at somewhat longer wavelengths. The two components are identified in the emission spectrum shown in Fig. 8-13. The fast component went unobserved for many years because many photomultiplier tubes are not sensitive to this short-wavelength region of the spectrum. However, if tubes or photodiodes are used that are sensitive in the near ultraviolet, about 20% of the total scintillation light at room temperature is measured in the fast component. The total light yield is only about 20% of that typically observed in NaI(Tl), so the energy resolution of BaF_2 scintillation detectors is considerably poorer. Nonetheless, applications have been demonstrated[77] in which the fast timing characteristics coupled with the high density and atomic number of the material have made BaF_2 the scintillation material of choice.

3. ZnS(Ag)

Silver-activated zinc sulfide is one of the older inorganic scintillators. It has a very high scintillation efficiency comparable to that of NaI(Tl) but is available only as a polycrystalline powder. As a result, its use is limited to thin screens used primarily for alpha particle or other heavy ion detection. Thicknesses greater than about 25 mg/cm^2 become unusable because of the opacity of the multicrystalline layer to its own luminescence. These scintillation screens played a key role in the early experiments of Rutherford, in which alpha particle interactions were visually observed through a low-power microscope.

4. CaF_2(Eu)

Europium-activated calcium fluoride is notable as a nonhygroscopic and inert inorganic scintillator that can often be used where severe environmental conditions preclude other choices. It is relatively fracture resistant and nonreactive, with a vapor pressure low enough to be usable under vacuum conditions. Its scintillation properties are included in Table 8-3.

5. CsF

Cesium fluoride is another inorganic scintillator with unusually fast decay time. For excitation by incident gamma rays, essentially all the scintillation light yield occurs with a decay time of 4.4 ns.[78] The light yield is low, variously reported as from 5 to 10% of that of NaI(Tl). Consequently, the energy resolution to be expected using this scintillator is poor. However, the fast decay time permits its application to timing measurements[79] with excellent timing resolution.

D. Glass Scintillators

Silicate glasses containing lithium and activated with cerium are widely used as neutron detectors, and a collection of their physical and scintillation properties may be found in Table 15-1. Bollinger et al.[80] investigated the properties of a variety of lithium and boron glasses as scintillation materials. The light output observed for the lithium formulations averaged over a factor of 10 larger than that for the boron-containing glasses. The lithium formulations have therefore predominated. In silicate glasses, cerium is the only activator that produces a useful amount of scintillation light. The emission is peaked in the blue region of the spectrum and is associated with the Ce^{3+} sites within the glass. Modern scintillation glasses[81] are made from various mixtures of SiO_2, LiO_2, Al_2O_3, MgO, and Ce_2O_3, with BaO sometimes added to increase the glass density.

The absolute scintillation efficiency for converting fast electron energy to light has been measured[82] in the range from 1 to 2%, generally varying inversely with lithium concentration of the glass formulation. The response to charged particles is nonlinear and always considerably below that to electrons of the same energy. For example, the light output for 1 MeV protons, deuterons, and alpha particles is lower than that for 1 MeV electrons by factors of 2.1, 2.8, and 9.5, respectively.[83]

Despite the fact that their relative light output is quite low (typically 20–30% of that of anthracene) glass scintillators are sometimes applied in beta or gamma-ray counting when severe environmental conditions prevent the use of more conventional scintillators. Examples include conditions in which the scintillator must be exposed to corrosive chemical environments or operated at high temperatures. The decay time of glass scintillators (typically 50–75 ns) is intermediate between faster organics and slower crystalline inorganics. Because the material can be fabricated in much the same manner as other glasses, it is widely available in physical forms other than the more common cylindrical shapes. For example, filaments of diameter from 0.1 to 1 mm with lengths up to 20 cm are commercially available and lend themselves to detector designs with specific directional properties. The material is also available as a powder or small-diameter spheres for use in flow cells in which the radioactive material is allowed to pass through a chamber filled with the porous scintillator.

Because glasses may contain naturally radioactive thorium or potassium, precautions must be taken if scintillators are to be applied in low-level counting systems where a minimum background is required. Ordinary glass scintillators will show a spontaneous background rate of about 100–200 disintegrations per minute for every 100 g of material. Glasses are now commercially available which are manufactured from low-thorium materials and which are processed to minimize contamination by any other alpha-active materials. These low-background glasses will have a background activity of less than 20 disintegrations per minute per 100 g.

When used as a neutron counter, the lithium content of the glasses is normally enriched to 95% or more in ^{6}Li. If applied as a beta or gamma counter, this enrichment is not required and can actually be a hindrance if the detector is operated with any appreciable neutron background. Therefore, one of the commercially available forms fabricated using either natural lithium or lithium depleted in ^{6}Li is normally chosen for these applications.

E. Scintillator Gases

Certain high-purity gases can serve as useful scintillation detection media. Considerable experience has been gained for the noble gases, with xenon and helium receiving the most attention. The mechanism giving rise to the scintillation photons is relatively simple: The incident radiation or charged particle leaves a population of excited gas molecules in its wake as it passes through the scintillator. As these excited molecules return to their ground state, through a variety of different mechanisms, photons will be emitted. The emission spectra correspond to the *second continuum* spectra observed in rare gas discharges. The photons are emitted during the transition from the two lowest molecular excited states to the ground state and much of the scintillation emission lies in the ultraviolet region of the spectrum rather than in the visible.[†] As a result, photomultiplier tubes or photodiodes that are sensitive in the near-ultraviolet must be employed if the scintillations are to be detected directly. Alternatively, a small concentration of a second gas such as nitrogen may be added to shift the emission spectrum by absorbing the ultraviolet and reradiating the energy at a longer wavelength. Due to a variety of competing parallel modes of deexcitation, such as intermolecular collisions or internal quenching, the overall scintillation efficiency of gases is characteristically quite low. However, the transitions take place in a very short time, typically a few nanoseconds or less, and gas scintillators can be among the fastest of all radiation detectors.

In addition to their fast response, gas scintillators have the relative advantage of easily variable size, shape, and stopping power for incident radiations. They also tend to be unusually linear over wide ranges of particle energy and dE/dx. Their major disadvantage is the low light yield, which is, at best, over an order of magnitude below that of NaI(Tl) for equivalent particle energy loss. They have been widely applied in the spectroscopy of heavy ions, often with a pressurized active volume. Some scintillation properties of several noble gases are given in Table 8-4. Extensive reviews of gas scintillators can be found in Refs. 1 and 86, and some measurements of specific properties in Refs. 87–90.

Some noble gases when condensed as a cryogenic liquid or solid have also been observed to be quite efficient scintillators. Results have been reported[91,92] on liquid and solid argon, krypton, and xenon and also on liquid helium.[93] There are obvious operational difficulties in efficiently collecting the scintillation light while the condensed gas must be maintained at cryogenic temperature. Xenon in liquid or solid form has the most appeal as a potential gamma-ray detector because of its high atomic number of 54.

[†] There is recent evidence[84] that Ne, Ar, Kr, and Xe also emit a significant scintillation component in the near-infrared region of the spectrum. By using a PM tube with extended long-wavelength response, this component in argon can be converted into pulses that are large enough to carry out alpha particle spectroscopy with a 5.1% energy resolution.[85]

TABLE 8-4   Properties of Gas Scintillators at Atmospheric Pressure

Gas	Mean Wavelength of Emission	Number of Photons with $\lambda > 200$ nm per 4.7 MeV Alpha Particle
Xenon	325 nm	3,700
Krypton	318 nm	2,100
Argon	250 nm	1,100
Helium	390 nm	1,100
Nitrogen (for comparison)	390 nm	800
NaI(Tl)	415 nm	41,000

Source: J. B. Birks, *The Theory and Practice of Scintillation Counting.* Copyright 1964 by Pergamon Press, Ltd. Used with permission.

Kubota et al.[91] present measurements of the decay time characteristics of these condensed gases, with all showing both fast and slow components with relative intensities that depend on the type of exciting particle. Values for the principal decay component range from 6 ns in condensed argon to about 30 ns in condensed xenon.

III. LIGHT COLLECTION AND SCINTILLATOR MOUNTING

A. Uniformity of Light Collection

In any scintillation detector, one would like to collect the largest possible fraction of the light emitted isotropically from the track of the ionizing particle. Two effects arise in practical cases which lead to less than perfect light collection: optical self-absorption within the scintillator and losses at the scintillator surfaces. With the exception of very large scintillators (many centimeters in dimension) or rarely used scintillation materials (e.g., ZnS), self-absorption is usually not a significant loss mechanism. Therefore, the uniformity of light collection normally depends primarily on the conditions that exist at the interface between the scintillator and the container in which it is mounted.

The light collection conditions affect the energy resolution of a scintillator in two distinct ways. First, the statistical broadening of the response function discussed in Chapter 10 will worsen as the number of scintillation photons that contribute to the measured pulse is reduced. The best resolution can therefore be achieved only by collecting the maximum possible fraction of all photons emitted in the scintillation event. Second, the *uniformity* of the light collection will determine the variation in signal pulse amplitude as the position of the radiation interaction is varied throughout the scintillator. Perfect uniformity would assure that all events depositing the same energy, regardless of where they occur in the scintillator, would give rise to the same mean pulse amplitude. With ordinary scintillators of a few centimeters in dimension, uniformity of light collection is seldom a significant contributor to the overall energy resolution. In larger scintillators, particularly those that are viewed along a thin edge, variations in light collection efficiency can often dominate the energy resolution.

Because the scintillation light is emitted in all directions, only a limited fraction can travel directly to the surface at which the photomultiplier tube or other sensor is located. The remainder, if it is to be collected, must be reflected one or more times at the scintillator surfaces. Two situations may prevail when the light photon reaches the surface, as illustrated in Fig. 8-14. If the angle of incidence θ is greater than the critical

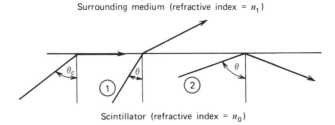

Figure 8-14 Conditions at the interface of dissimilar optical media ($n_0 > n_1$). Ray ① may escape, but ray ② will be internally reflected at the surface.

angle θ_c, total internal reflection will occur. If θ is less than θ_c, partial reflection (called *Fresnel* reflection) and partial transmission through the surface will occur. The fraction of reflected light drops to only a few percent when the angle of incidence is near zero. The critical angle θ_c is determined by the indices of refraction for the scintillation medium n_0 and the surrounding medium (often air) n_1:

$$\theta_c = \sin^{-1} \frac{n_1}{n_0} \tag{8-12}$$

To recapture the light that does escape from the surface, the scintillator is normally surrounded by a reflector at all surfaces except that at which the photomultiplier tube is mounted. Reflectors can be either *specular* or *diffuse*. A polished metallic surface will act as a specular reflector for which the angle of reflection equals the angle of incidence. Better results are usually obtained, however, with a diffuse reflector such as magnesium oxide or aluminum oxide. Here the angle of reflection is approximately independent of the angle of incidence and follows a distribution given by Lambert's law:

$$\frac{dI(\psi)}{dI_0} = \cos \psi \tag{8-13}$$

where ψ is the reflection angle with respect to the direction perpendicular to the surface. Commercially prepared sodium iodide crystals are normally reflected with dry MgO powder packed around surfaces of the crystal which have been slightly roughened with an abrasive. Plastic scintillators are usually left with a polished surface, especially when the light must be collected over large dimensions involving multiple reflections. White reflecting paint is easy to apply to scintillators but is an inferior reflector to dry powder.

Although total internal reflection is desirable at reflecting surfaces, it must be minimized at the surface from which the scintillator is viewed to prevent internal trapping of the light. Ideally, one would like to optically couple the scintillator to the photocathode of the PM tube through a transparent medium of the same index of refraction as the scintillator. Then Eq. (8-12) would predict no internal reflection and all light incident on the surface would be conducted to the photocathode. The refractive index of plastic, liquid, and glass scintillators is normally fairly close to that of the glass end window (about 1.5) on which the photocathode is deposited. In these cases, near-perfect coupling will be achieved if the interface between the scintillator and PM tube is filled with an optical coupling fluid of the same refractive index. Some internal reflection will inevitably occur for scintillators with a higher index (such as NaI or LiI), and a light photon will be reflected back into the scintillator if its angle of incidence is too large. The usefulness of scintillators with a very high refractive index is seriously hampered because of excessive internal trapping of the scintillation light. High-viscosity silicone oil is almost universally used as the coupling agent between the scintillator and PM tube. Design of the scintillation detector must provide a means to prevent the long-term flow of the fluid away from the interface. If the detector will be subjected to severe temperature cycling or vibration, it is sometimes preferable to couple the scintillator with a transparent epoxy cement.

A simple test can be made of light-collecting uniformity by allowing a narrowly collimated gamma-ray beam to strike selected portions of a scintillation crystal. With a uniform surface treatment it is often found that those areas closest to the photocathode give rise to a larger pulse height because of more favorable light collection. In order to compensate for this natural tendency, the surface is often preferentially treated to enhance light collection from the points in the crystal farthest from the photocathode.

Good light collection in the scintillator also requires that either a single crystal or single piece of plastic be used for the scintillating medium, or that special precautionary techniques be used in the event that separate pieces are cemented together. If separate pieces are used, the additional surfaces will often introduce added reflection and can seriously affect the overall uniformity of light collection.

Any scintillation counter must be shielded from ambient room light. For temporary arrangements not involving hygroscopic crystals, a simple wrapping of the reflected scintillator and photomultiplier tube with black paper, followed by a layer of the ubiquitous black tape, will often suffice. Commercial sodium iodide scintillation crystals are usually canned in a metallic container and must be hermetically sealed. The surface through which the light is to be collected is provided with a glass or quartz window. The opposite surface is covered with an opaque but thin metallic sheet to provide an entrance window for soft radiations. In cases in which the scintillation counter must be used in low background counting, special care must be given to the proper choice of materials used in contact with the crystal. Some types of glass, for example, are very high in potassium and as a result will generate a large background contribution from the natural ^{40}K activity.

Light collection in large scintillators can often be enhanced by the use of more than one photomultiplier tube. Although this is not usually an attractive option for routine use because of the added complexity, the gains that can be achieved are, in some cases, substantial. The average number of reflections required for a typical event to reach a photomultiplier tube will obviously be less if more than one escape surface is provided. The fewer the reflections, the greater the light collection efficiency, and consequently the greater uniformity of pulse height response. Because of the importance of self-absorption in large scintillators, there often is no substitute for multiple PM tubes in these cases.

More complete discussions of scintillator mounting and reflection are given by Birks[1] and Bell.[94] A useful review of light collection methods and surface properties has also been published by Keil.[95]

B. Light Pipes

It is often unadvisable or even impossible to couple a photomultiplier tube directly to one face of a scintillator. For example, the size or shape of the scintillator may not conveniently match the circular photocathode area of commercially available PM tubes. One solution is to place the unreflected scintillator near the center of a large box, whose interior surfaces are coated with a diffuse reflector. One or more PM tubes can then view the interior of the box to record some fraction of the light that eventually escapes from the scintillator. If the fraction of the box surface replaced by PM tubes is small and the reflectivity of the coating is high,[96] the light is thoroughly randomized before being detected. Very uniform light collection can therefore be achieved even for scintillators with complex or unusual shape. Because of inevitable light losses, however, the total fraction of light collected is typically small. Better light collection efficiency usually can be achieved by using a transparent solid, known as a *light pipe*, to physically couple the scintillator to the PM tube and to act as a guide for the scintillation light.

Light pipes also serve a useful purpose in other situations. If scintillation measurements are to be made in a strong magnetic field, the PM tube must be shielded from the field and this often implies its removal to a location some distance away from the scintillator. Very thin scintillators should not be mounted directly on the PM tube end window to avoid the pulse height variations that can arise due to photocathode nonuniformities. A light pipe between the thin scintillator and the PM tube will spread the light from each scintillation event over the entire photocathode to average out these nonuniformities and improve the pulse height resolution.

Light pipes operate on the principle of total internal reflection outlined in the previous section. They are generally optically transparent solids with a relatively high index of refraction to minimize the critical angle for total internal reflection. Surfaces are highly polished and are often surrounded by a reflective wrapping to direct back some of the light that escapes at angles less than the critical angle. Lucite, with an index of refraction of 1.49–1.51, is the most widely used material and can easily be formed into complex shapes. A study of the effects of various surface and outer reflector arrangements with Lucite light pipes has been reported by Kilvington et al.,[97] and the data are summarized in Fig. 8-15.

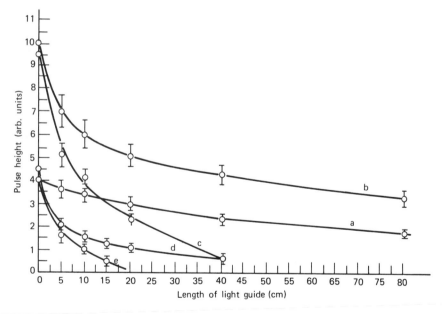

Figure 8-15 Variation of pulse height with length of light guide for various reflective wrappings. (a) Total internal reflection only, (b) total internal reflection with reflective covering, (c) surface of light guide painted with NE 560 reflector paint, (d) specular reflector without light guide, (e) diffuse reflector without light guide. (From Kilvington et al.[97])

For an isotropic radiator within a cylindrical rod, the following approach can be followed to find the fraction of light which will be conducted in one direction down the length of the rod by successive internal reflections. In the sketch below, only the light emitted within the cone angle ϕ_c will be incident on the rod surface at the critical angle θ_c or greater and therefore undergoes total internal reflection. Because the angle of reflection equals the angle of incidence, subsequent arrivals of the reflected light at the rod surface will also be above the critical angle, and this light is therefore "piped" along the rod length as in an optical fiber. The fractional solid angle subtended by this angle ϕ_c is calculated as

$$F = \frac{\Omega}{4\pi} = \frac{1}{4\pi}\int_{\phi=0}^{\phi=\phi_c}d\Omega = \frac{1}{4\pi}\int_0^{\phi_c}2\pi\sin\phi\,d\phi$$

$$= \tfrac{1}{2}(1 - \cos\phi_c) = \tfrac{1}{2}(1 - \sin\theta_c)$$

$$= \tfrac{1}{2}(1 - n_1/n_0) \tag{8-14}$$

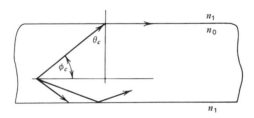

In a cylindrical rod of refractive index $n_0 = 1.5$ surrounded by air ($n_1 \cong 1$), Eq. (8-14) predicts that 16.7% of the isotropically generated light is piped in one direction. Another 16.7% is piped in the opposite direction and may be reflected back depending on the surface conditions at the end of the cylinder.

In slab geometry, the light that reaches a surface with incident angle θ greater than the critical angle θ_c will be trapped and conducted to the slab edges by total internal reflection. For rays with $\theta < \theta_c$, there is some possibility that Fresnel reflection will prevent escape at the first surface. However, then the ray is reflected back into the slab at an angle equal to its angle of incidence, and it arrives at the opposite surface again with $\theta < \theta_c$. Even if several Fresnel reflections happen to occur, eventually the ray will escape. Thus all rays that escape the slab are confined to a double ended cone with apex at the point of origin of the light, axis perpendicular to the slab surfaces, and a vertex angle of θ_c. The total escaping fraction of light is thus given by

$$E = 2\cdot\frac{\Omega}{4\pi} = \frac{1}{2\pi}\int_{\theta=0}^{\theta=\theta_c}d\Omega = \frac{1}{2\pi}\int_0^{\theta_c}2\pi\sin\theta\,d\theta$$

$$= 1 - \cos\theta_c = 1 - \sqrt{1 - \left(\frac{n_1}{n_0}\right)^2}$$

The fraction of light trapped in the slab is therefore

$$F = 1 - E = \sqrt{1 - \left(\frac{n_1}{n_0}\right)^2} \tag{8-15}$$

For a slab with refractive index of 1.5 surrounded by air, this result predicts that 75% of the light is piped to the edges.

To maximize the fraction of light that is piped in either geometry, one wants the refractive index of the light pipe n_0 to be as large as possible. In reality, however, the light is generated inside the scintillator, not the light pipe, and it is usually the scintillator refractive index that determines the fraction of light collected. This is particularly true when the scintillator is long in the direction perpendicular to the viewing surface, and a typical scintillation photon is multiply reflected before collection. The scintillator then acts as its own light pipe between the point of scintillation and the exit surface.

For the simple case of a cylindrical scintillation crystal and tube of equal diameter, the light pipe can be a simple cylinder of the same diameter. More often, however, the light pipe cross section shape must vary along its length in order to serve as a smooth transition between the scintillator exit surface and the PM tube end window. No matter how complex the shape of a conventional light pipe may be, the flux of light photons per unit area per unit solid angle can never be greater at any point inside the pipe than at its input. Any light pipe whose cross-sectional area decreases from scintillator to PM tube will therefore result in some light loss. If the cross-sectional area is maintained constant and sharp bends are avoided, however, the pipe can theoretically transmit all the light that enters within the acceptance angle at the input end. Light pipes with this property are called *adiabatic*[98] and obviously require the use of a PM tube with a photocathode area at least as large as the scintillator exit surface.

In cases in which the edge of a thin but large-area scintillator is to be viewed, a unique arrangement known as the strip light guide has found widespread application. As shown in Fig. 8-16, the coupling is accomplished through a number of twisted strips that

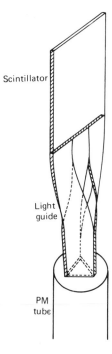

Figure 8-16 A strip light guide can be used to couple the edge of a large, flat scintillator to a PM tube.

are aligned at the scintillator edge but converge to a more compact pattern at the photomultiplier end. The unit can easily be made from flat Lucite strips, which are then bent following heating and formed to the required shape. Practical guides for these procedures are given by Dougan et al.[99] and Piroue.[100]

C. Wavelength Shifters

Light collection from large scintillators or complex geometries can sometimes be aided through the use of optical elements that employ *wavelength shifting* techniques. Many liquid or plastic scintillators routinely incorporate an organic additive whose function is to absorb the primary scintillation light and reradiate the energy at a longer wavelength. In that case, the objective is to better match the emission spectrum to the response peak of a photomultiplier tube. The same process can be used to help light collection by exploiting the fact that the reradiated light is emitted isotropically so that its direction bears no relation to the direction of the absorbed light. This feature allows the light to "turn corners" that would otherwise be impossible.

To illustrate, suppose that the primary scintillation light is generated within a large slab scintillator shown in Fig. 8-17. Rather than collect light from the edges of the slab, there are circumstances in which it may be preferable instead to couple the light to one or more cylindrical rods or optical fibers that run perpendicular to the slab surface. (Such could be the case, for example, if a large number of slabs were involved, and a few PM tubes were used at the ends of the rods to record light from all the slabs.) The rods would then be made of optically transparent material to act as light pipes. If the scintillator and rods have similar indices of refraction, it is very difficult to couple light efficiently from the slab to a perpendicular rod in optical contact. Light that is conducted by total internal reflection along the slab arrives at the rod at angles that are not favorable for subsequent internal reflection within the rod. Also, if other slabs are in optical contact with the rod, the light piping properties of the rod are no longer preserved.

A solution to both difficulties follows if the rod is doped with a wavelength shifter[102-105] and passed through the slab in an air-filled hole of slightly larger diameter (see Fig. 8-17). Now the index change is preserved over the entire surface of the rod, and it will therefore act as a near-ideal light pipe. Some fraction of the light from the slab that arrives at a hole may pass across the air gap and enter the rod. The doping level of the wavelength shifter is adjusted so that there is a good probability of absorption of the primary light within the rod. The reradiated light is now isotropic, and one-third or more will typically be piped along the length of the rod. The shifted wavelength is now away

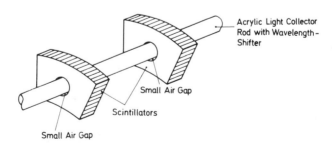

Figure 8-17 Sections of two slab scintillators are shown coupled to a common light pipe loaded with wavelength shifter. Light passing across the air gap and entering the pipe can be absorbed, reradiated, and guided to the ends of the rod. (From Eckardt et al.[101])

from the strong absorption bands of the dopant, so that further loss along the rod length can largely be avoided.

Since there are several inefficient steps in this example, the overall light collection efficiency of such a scheme cannot compare with that obtainable with a more direct coupling of the photomultiplier tube to the scintillator. In some applications, however, the scintillation light yield per event may be large enough to allow one to tolerate considerable loss in the collection process. In such cases, the added flexibility afforded by wavelength shifting techniques has led to successful applications in complex detector geometries[106] or in the compact readout of large-area scintillators.[107]

PROBLEMS

8-1. Calculate the scintillation efficiency of anthracene if 1 MeV of particle energy loss creates 20,300 photons with average wavelength of 447 nm.

8-2. Assuming an inverse decay constant of 230 ns, how much time is required for a NaI(Tl) scintillation event to emit 99% of the total light yield?

8-3. Assuming that the scintillation light pulse in each case is a pure exponential, find the ratio of the maximum brightness (rate of photon emission) of pulses generated by equal electron energy deposition in NaI(Tl) and anthracene.

8-4. Make a selection between a typical inorganic scintillator [say, NaI(Tl)] and a typical organic (say, a plastic scintillator) on the basis of the following properties:

(a) Speed of response.
(b) Light output.
(c) Linearity of light with deposited energy.
(d) Detection efficiency for high-energy gamma rays.
(e) Cost.

8-5. Explain the function of the activator added in trace quantities to many inorganic scintillators. Why are they not needed in organic scintillators?

8-6. Which scintillation material is most efficient at converting the energy of a 2 MeV electron into light?

8-7. Explain the following statement: Organic crystalline scintillators remain good energy-to-light converters when dissolved in a solvent, while inorganics no longer function as scintillators if dissolved.

8-8. Scintillation light is emitted isotropically within a slab of plastic scintillator (see Table 8-1). If the other dimensions of the slab are assumed to be infinite, calculate the fraction of the light that escapes from either slab surface.

8-9. The dark-adapted human eye may be able to detect as few as 10 visible photons as a single flash. Will an observer with pupil diameter of 3 mm be able to see individual scintillation events caused by a 1 MeV beta particle in NaI(Tl) while viewing the surface of the scintillator at a distance of 10 cm?

REFERENCES

1. J. B. Birks, *The Theory and Practice of Scintillation Counting*, Pergamon Press, Oxford, 1964.
2. J. B. Birks and R. W. Pringle, *Proc. R. Soc. Edinburgh Sect. A* **70**, 233 (1972).
3. D. B. Oliver and G. F. Knoll, *IEEE Trans. Nucl. Sci.* **NS-15**(3), 122 (1968).
4. J. K. Walker, *Nucl. Instrum. Meth.* **68**, 131 (1969).

5. G. Bendiscioli, V. Filippini, G. Fumagalli, E. L. Rizzini, C. Marciano, C. Milani, A. Rotondi, and A. Venaglioni, *Nucl. Instrum. Meth.* **206**, 471 (1983).

6. M. L. Muga, D. J. Burnsed, and W. E. Steeger, *Nucl. Instrum. Meth.* **104**, 605 (1972).

7. N. N. Ajitanand and K. N. Iyengar, *Nucl. Instrum. Meth.* **133**, 71 (1976).

8. H. Geissel et al., *Nucl. Instrum. Meth.* **144**, 465 (1977).

9. C. Manduchi, M. T. Russo-Manduchi, and G. F. Segato, *Nucl. Instrum. Meth. Phys. Res.* **A243**, 453 (1986).

10. E. Norbeck, T. P. Dubbs, and L. G. Sobotka, *Nucl. Instrum. Meth. Phys. Res.* **A262**, 546 (1987).

11. F. D. Brooks, W. A. Cilliers, and M. S. Allie, *Nucl. Instrum. Meth. Phys. Res.* **A240**, 338 (1985).

12. T. Batsch and M. Moszynski, *Nucl. Instrum. Meth.* **125**, 231 (1975).

13. J. W. Kohl, *Nucl. Instrum. Meth.* **125**, 413 (1975).

14. K. Ettling and W. Von Witsch, *Nucl. Instrum. Meth.* **148**, 299 (1978).

15. G. Bendiscioli, V. Filippini, C. Marciano, A. Rotondi, and A. Zenoni, *Nucl. Instrum. Meth. Phys. Res.* **227**, 478 (1984).

16. L. Muga, *Nucl. Instrum. Meth.* **124**, 541 (1975).

17. M. L. Muga and J. D. Bridges, *Nucl. Instrum. Meth.* **134**, 143 (1976).

18. I. Kanno and T. Nakagome, *Nucl. Instrum. Meth. Phys. Res.* **A251**, 108 (1986).

19. I. Kanno and T. Nakagome, *Nucl. Instrum. Meth. Phys. Res.* **A244**, 551 (1986).

20. T. Batsch and M. Moszynski, *Nucl. Instrum. Meth.* **123**, 341 (1975).

21. C. B. Ashford, I. B. Berlman, J. M. Flournoy, L. A. Franks, S. G. Iversen, and S. S. Lutz, *Nucl. Instrum. Meth. Phys. Res.* **A243**, 131 (1986).

22. Z. H. Cho, I. Ahn, and C. M. Tsai, *IEEE Trans. Nucl. Sci.* **NS-21**(1), 218 (1974).

23. L. A. Eriksson, C. M. Tsai, Z. H. Cho, and C. R. Hurlbut, *Nucl. Instrum. Meth.* **122**, 373 (1974).

24. Z. H. Cho, C. M. Tsai, and L. A. Eriksson, *IEEE Trans. Nucl. Sci.* **NS-22**(1), 72 (1975).

25. E. Brannon and G. L. Olde, *Radiat. Res.* **16**, 1 (1962).

26. R. L. Craun and D. L. Smith, *Nucl. Instrum. Meth.* **80**, 239 (1970).

27. D. L. Smith, R. G. Polk, and T. G. Miller, *Nucl. Instrum. Meth.* **64**, 157 (1968).

28. J. B. Czirr, *Nucl. Instrum. Meth.* **25**, 106 (1963).

29. K. H. Maier and J. Nitschke, *Nucl. Instrum. Meth.* **59**, 227 (1968).

30. D. Clark, *Nucl. Instrum. Meth.* **117**, 295 (1974).

31. N. A. Weir, *Int. J. Appl. Radiat. Isotopes* **23**, 371 (1972).

32. A. Raviart and V. Koechlin, *Nucl. Instrum. Meth.* **29**, 45 (1964).

33. F. J. Lynch, *IEEE Trans. Nucl. Sci.* **NS-15**(3), 102 (1968).

34. B. Bengtson and M. Moszynski, *Nucl. Instrum. Meth.* **117**, 227 (1974).

35. T. M. Kelly, J. A. Merrigan, and R. M. Lambrecht, *Nucl. Instrum. Meth.* **109**, 233 (1973).

36. P. B. Lyons and J. Stevens, *Nucl. Instrum. Meth.* **114**, 313 (1974).

37. F. J. Lynch, *IEEE Trans. Nucl. Sci.* **NS-22**(1), 58 (1975).

38. S. Sanyal, S. C. Pancholi, and S. L. Gupta, *Nucl. Instrum. Meth.* **136**, 157 (1976).

39. P. B. Lyons, C. R. Hurlbut, and L. P. Hocker, *Nucl. Instrum. Meth.* **133**, 175 (1976).

40. P. B. Lyons et al., *IEEE Trans. Nucl. Sci.* **NS-24**(1), 177 (1977).

41. K. G. Tirsell et al., *IEEE Trans. Nucl. Sci.* **NS-24**(1), 250 (1977).

42. R. E. Pixley and H. von Fellenberg, *Nucl. Instrum. Meth.* **129**, 487 (1975).

43. M. Moszynski and B. Bengtson, *Nucl. Instrum. Meth.* **142**, 417 (1977).

44. B. Sipp and J. A. Miehe, *Nucl. Instrum. Meth.* **114**, 255 (1974).

45. M. Bantel et al., *Nucl. Instrum. Meth.* **226**, 394 (1984).

46. L. M. Bollinger and G. E. Thomas, *Rev. Sci. Instrum.* **32**, 1044 (1961).

47. T. G. Miller, *Nucl. Instrum. Meth.* **63**, 121 (1968).

48. J. B. Czirr, *Nucl. Instrum. Meth.* **88**, 321 (1970).

49. R. A. Winyard and G. W. McBeth, *Nucl. Instrum. Meth.* **98**, 525 (1972).

50. D. B. C. B. Syme and G. I. Crawford, *Nucl. Instrum. Meth.* **104**, 245 (1972).

51. I. B. Berlman and O. J. Steingraber, *Nucl. Instrum. Meth.* **108**, 587 (1973).

52. M. Ahmed, *Nucl. Instrum. Meth.* **143**, 255 (1977).

53. R. B. Murray, *IEEE Trans. Nucl. Sci.* **NS-22**(1), 54 (1975).

54. R. G. Kaufman, W. B. Hadley, and H. N. Hersh, *IEEE Trans. Nucl. Sci.* **NS-17**(3), 82 (1970).

55. C. L. Melcher et al., *IEEE Trans. Nucl. Sci.* **NS-32**(1), 529 (1985).

56. E. Sakai, *IEEE Trans. Nucl. Sci.* **NS-34**(1), 418 (1987).

57. I. Holl, E. Lorenz, and G. Mageras, *IEEE Trans. Nucl. Sci.* **35**(1), 105 (1988).

58. *Scintillation Phosphor Catalog*, The Harshaw Chemical Company.

59. R. Hofstadter, *Phys. Rev.* **74**, 100 (1948).

60. J. R. Prescott and G. H. Narayan, *Nucl. Instrum. Meth.* **75**, 51 (1969).

61. S. Koicki, A. Koicki, and V. Ajdacic, *Nucl. Instrum. Meth.* **108**, 297 (1973).

62. C. F. G. Delaney and A. M. Lamki, *Int. J. Appl. Radiat. Isotopes* **19**, 169 (1968).

63. J. S. Schweitzer and W. Ziehl, *IEEE Trans. Nucl. Sci.* **NS-30**(1), 380 (1983).

64. C. J. Crannell, R. J. Kurz, and W. Viehmann, *Nucl. Instrum. Meth.* **115**, 253 (1974).

65. S. Keszthelyi-Landori and G. Hrehuss, *Nucl. Instrum. Meth.* **68**, 9 (1969).

66. P. E. Francois and D. T. Martin, *Int. J. Appl. Radiat. Isotopes* **21**, 687 (1970).

67. M. Moszynski et al., *Nucl. Instrum. Meth.* **188**, 403 (1981).

68. M. J. Weber and R. R. Monchamp, *J. Appl. Phys.* **44**, 5495 (1973).

69. R. G. L. Barnes et al., *IEEE Trans. Nucl. Sci.* **NS-31**(1), 249 (1984).

70. M. Laval et al., *Nucl. Instrum. Meth.* **206**, 169 (1983).

71. K. Wisshak and F. Kaeppeler, *Nucl. Instrum. Meth.* **227**, 91 (1984).

72. Y. C. Zhu et al., *Nucl. Instrum. Meth.* **A244**, 577 (1986).

73. H. J. Karwowski et al., *Nucl. Instrum. Meth.* **A245**, 207 (1986).

74. M. R. Farukhi and C. F. Swinehart, *IEEE Trans. Nucl. Sci.* **NS-18**(1), 200 (1971).

75. P. Schotanus et al., *Nucl. Instrum. Meth.* **A238**, 564 (1985).

76. P. Schotanus et al., *Nucl. Instrum. Meth.* **A259**, 586 (1987).

77. M. Moszynski et al., *Nucl. Instrum. Meth.* **226**, 534 (1984).

78. M. Moszynski et al., *Nucl. Instrum. Meth.* **179**, 271 (1981).

79. M. Moszynski et al., *Nucl. Instrum. Meth.* **205**, 239 (1983).

80. L. M. Bollinger, G. E. Thomas, and R. J. Ginther, *Nucl. Instrum. Meth.* **17**, 97 (1962).

81. A. D. Bross, *Nucl. Instrum. Meth.* **A247**, 319 (1986).

82. A. W. Dalton, *Nucl. Instrum. Meth. Phys. Res.* **A259**, 545 (1987).

83. A. W. Dalton, *Nucl. Instrum. Meth. Phys. Res.* **A254**, 361 (1987).

84. P. Lindblom and O. Solin, *Nucl. Instrum. Meth. Phys. Res.* **A268**, 204 (1988).

85. P. Lindblom and O. Solin, *Nucl. Instrum. Meth. Phys. Res.* **A268**, 212 (1988).

86. M. Mutterer, *Nucl. Instrum. Meth.* **196**, 73 (1982).

87. M. Mutterer, J. Pannicke, K. Scheele, W. Spreng, J. P. Theobald, and P. Wastyn, *IEEE Trans. Nucl. Sci.* **NS-27**(1), 184 (1980).

88. M. Suzuki, J. Ruangen, and S. Kubota, *Nucl. Instrum. Meth.* **192**, 565 (1982).

89. M. Suzuki, *Nucl. Instrum. Meth.* **215**, 345 (1983).

90. P. Grimm, F.-J. Hambsch, M. Mutterer, J. P. Theobald, and S. Kubota, *Nucl. Instrum. Meth. Phys. Res.* **A262**, 394 (1987).

91. S. Kubota et al., *Nucl. Instrum. Meth.* **196**, 101 (1982).

92. W. Baum, S. Götz, H. Heckwolf, P. Heeg, M. Mutterer, and J. P. Theobald, *IEEE Trans. Nucl. Sci.* **35**(1), 102 (1988).

93. A. Helaly et al., *Nucl. Instrum. Meth.* **A241**, 169 (1985).

94. P. R. Bell, "The Scintillation Method," in *Beta- and Gamma-Ray Spectroscopy* (K. Siegbahn, ed.), Elsevier–North Holland, Amsterdam, 1955.

95. G. Keil, *Nucl. Instrum. Meth.* **87**, 111 (1970).

96. S. P. Ahlen, B. G. Cartwright, and G. Tarle, *Nucl. Instrum. Meth.* **143**, 513 (1977).

97. A. I. Kilvington, C. A. Baker, and P. Illinesi, *Nucl. Instrum. Meth.* **80**, 177 (1970).

98. R. L. Garwin, *Rev. Sci. Instrum.* **23**, 755 (1952).

99. P. Dougan, T. Kivikas, K. Lugner, W. Ramsay, and W. Stiefler, *Nucl. Instrum. Meth.* **78**, 317 (1970).

100. P. A. Piroue, Conference on Instrumentation Techniques in Nuclear Pulse Analysis, National Academy of Sciences, National Research Council Publication 1184 (Nuclear Science Series Report #40) 1964.

101. V. Eckardt et al., *Nucl. Instrum. Meth.* **155**, 389 (1978).

102. W. Selove, W. Kononenko, and B. Wilsker, *Nucl. Instrum. Meth.* **161**, 233 (1979).

103. W. Viehmann and R. L. Frost, *Nucl. Instrum. Meth.* **167**, 405 (1979).

104. W. Kononenko, W. Selove, and G. E. Theodosiou, *Nucl. Instrum. Meth.* **206**, 91 (1983).

105. C. Aurouet et al., *Nucl. Instrum. Meth.* **211**, 309 (1983).

106. V. I. Kryshkin and A. T. Ronzhin, *Nucl. Instrum. Meth.* **A247**, 583 (1986).

107. G. Keil, *Nucl. Instrum. Meth.* **83**, 145 (1970).

CHAPTER · 9

Photomultiplier Tubes and Photodiodes

I. INTRODUCTION

The widespread use of scintillation counting in radiation detection and spectroscopy would be impossible without the availability of devices to convert the extremely weak light output of a scintillation pulse into a corresponding electrical signal. The photomultiplier (PM) tube accomplishes this task remarkably well, converting light signals that typically consist of no more than a few hundred photons into a usable current pulse without adding a large amount of random noise to the signal. Although there has been some recent progress (described later in this chapter) in the development of semiconductor photodiodes for use with scintillators, the PM tube remains the most widely used device for this purpose. A great variety of commercially available PM tubes are sensitive to radiant energy in the ultraviolet, visible, and near-infrared regions of the electromagnetic spectrum. They find many applications in optical spectroscopy, laser measurements, and astronomy. A useful review of PM tube properties and design characteristics can be found in Ref. 1. In this chapter the discussion is limited to those designs of primary interest for scintillation counting. Morton has published[2] a very readable historical account of the development of tubes for this purpose. Useful guides and standards for the testing of PM tubes for scintillation counting have been developed as part of a series of such standards published[3] by the IEEE.

The simplified structure of a typical photomultiplier tube is illustrated in Fig. 9-1. The two major elements consist of a photosensitive layer, called the *photocathode*, coupled to an electron multiplier structure. The photocathode serves to convert as many of the incident light photons as possible into low-energy electrons. If the light consists of a pulse from a scintillation crystal, the photoelectrons produced will also be a pulse of similar time duration. Because only a few hundred photoelectrons may be involved in a typical pulse, their charge is too small at this point to serve as a convenient electrical signal. The electron multiplier section in a PM tube provides an efficient collection geometry for the photoelectrons as well as serving as a near-ideal amplifier to greatly increase their number. After amplification through the multiplier structure, a typical scintillation pulse will give rise to 10^7–10^{10} electrons, sufficient to serve as the charge signal for the original scintillation event. This charge is conventionally collected at the anode or output stage of the multiplier structure.

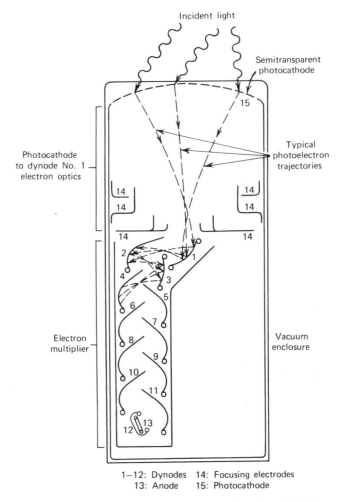

Incident light

Semitransparent
photocathode

15

Photocathode
to dynode No. 1
electron optics

Typical
photoelectron
trajectories

14   14
14   14
14   14

2   1
4   3
5
6
7
8
9
10
11
12 13

Electron
multiplier

Vacuum
enclosure

1–12: Dynodes   14: Focusing electrodes
13: Anode   15: Photocathode

Figure 9-1 Basic elements of a PM tube. (From Ref. 1.)

Most photomultipliers perform this charge amplification in a very linear manner, producing an output pulse that remains proportional to the number of original photoelectrons over a wide range of amplitude. Much of the timing information of the original light pulse is also retained. Typical tubes, when illuminated by a very short duration light pulse, will produce an electron pulse with a time width of a few nanoseconds after a delay time of 20–50 ns.

PM tubes are commercially available in a wide variety of sizes and properties. We begin our discussion with the important elements of PM tube design and their influence on overall performance.

II. THE PHOTOCATHODE

A. The Photoemission Process

The first step to be performed by the PM tube is the conversion of incident light photons into electrons. This process of photoemission can be thought of as occurring in three sequential stages: (1) the absorption of the incident photon and transfer of energy to an electron within the photoemissive material, (2) the migration of that electron to the surface, and (3) the escape of the electron from the surface of the photocathode.

The energy that can be transferred from the photon to an electron in the first step is given by the quantum energy of the photon $h\nu$. For blue light typical of that emitted by many scintillators, the quantum energy is about 3 eV. In step 2, some of that energy will be lost through electron–electron collisions in the migration process. Finally, in step 3, there must be sufficient energy left for the electron to overcome the inherent potential barrier that always exists at any interface between material and vacuum. This potential barrier (often called the *work function*) is normally greater than 3 or 4 eV for most metals but can be as low as 1.5–2 eV for suitably prepared semiconductors.

From these energy considerations, some general comments can be made regarding photocathodes. First, the finite potential barrier in step 3 imposes a minimum energy on the incoming light photon even if all other energy losses are zero. All photocathodes therefore have a long-wavelength (small ν) cutoff that is usually in the red or near-infrared portion of the spectrum.[†] Even for higher-energy light photons, the surface barrier should be as low as possible to maximize the number of escaping electrons. The rate of energy loss as the electron migrates to the surface should be kept small in order to maximize the depth in the material (called the *escape depth*) at which electrons may originate and still reach the surface with sufficient energy to overcome the potential barrier. The rate of energy loss in metals is relatively high, and an electron can travel no more than a few nanometers before its energy drops below the potential barrier. Therefore, only the very thin layer of material lying within a few nanometers of the surface will contribute any photoelectrons from common metals. In semiconductors, the rate of energy loss is much lower and the escape depth can extend to about 25 nm. This, however, is still a very small thickness even with respect to stopping visible light. Photocathodes of this thickness are semitransparent and will cause less than half the visible light to interact within the photosensitive layer. Therefore, such photocathodes cannot come close to converting all the visible light photons into electrons, no matter how low the potential barrier may be.

In order for an incident light photon to be absorbed in a semiconductor, its energy must exceed the bandgap energy E_g. (For a discussion of the band structure in semiconductors, see Chapter 11.) The absorption process consists simply of elevating an electron from the normally populated valence band to the conduction band. Within about a picosecond, these electrons rapidly lose energy through phonon interactions with the crystal until their energy is at the bottom of the conduction band. In normal semiconductors, the electron potential outside the surface is higher than the bottom of the conduction band by an amount called the *electron affinity*. If an electron is to escape, it must reach the surface in the short time before phonon interactions have reduced its energy to the bottom of the conduction band. The electron, however, will remain at the bottom of the conduction band for perhaps another 100 ps before recombining with a hole and dropping to the valence band. The use of *negative electron affinity* materials, discussed more fully later in this chapter, leads to a much greater escape depth by allowing electrons that have dropped to the bottom of the conduction band to also escape if they reach the surface (see Fig. 9-4).

B. Spontaneous Electron Emission

The surface potential barrier influences another important property of photocathodes: *thermionic noise*. Normal conduction electrons within the photocathode material will

---

[†] The presence of low concentrations of impurity states in some photocathode materials can result in a small but measurable sensitivity beyond the normal cutoff wavelength into the infrared region of the spectrum.

always have some thermal kinetic energy which, at room temperature, will average about 0.025 eV. There is a spread in this distribution, however, and those electrons at the extreme upper end of the distribution can occasionally have an energy that exceeds the potential barrier. If that electron is close enough to the surface, it may escape and give rise to a spontaneous thermally induced signal. In metals, the thermal emission rate is low ($\sim 100/\text{m}^2 \cdot \text{s}$) because of the relatively high potential barrier. In semiconductors, the lower potential barrier leads to thermal emission rates as high as 10^6–$10^8/\text{m}^2 \cdot \text{s}$. Their superior photosensitivity is therefore achieved only at the price of a higher noise rate from thermally stimulated electron emission.

On theoretical grounds, the rate of thermionic emission should rise exponentially with photocathode temperature. The observed rate of spontaneous electron emission does increase with temperature, but the dependence is generally much milder than this prediction, indicating the influence of nonthermal effects in determining the overall emission rate.

C. Fabrication of Photocathodes

Photocathodes can be constructed as either opaque or semitransparent layers. Each type is used in a somewhat different geometric arrangement. An opaque photocathode is normally fabricated with a thickness somewhat greater than the maximum escape depth and is supported by a thick backing material. Photoelectrons are collected from the same surface on which the light is incident. Semitransparent photocathodes generally are no thicker than the escape depth and are deposited on a transparent backing (often the glass end window of the PM tube). Light first passes through the transparent backing and subsequently into the photocathode layer, and photoelectrons are collected from the opposite surface. Because they are more readily adaptable to tube designs that use a flat end window, semitransparent photocathodes are more common in PM tubes designed for scintillation counters.

An important practical property of photocathodes is the uniformity to which their thickness can be held over the entire area of the photocathode. Variations in thickness give rise to corresponding changes in the sensitivity of the photocathode and can be one source of resolution loss in scintillation counters. This problem is especially serious for large-diameter PM tubes.

D. Quantum Efficiency and Spectral Response

The sensitivity of photocathodes can be quoted in several ways. When applied to dc light measurements, it is traditional to quote an overall photocathode efficiency in terms of current produced per unit light flux on its surface (amperes per lumen). A unit of greater significance in scintillation counting is the *quantum efficiency* (*QE*) of the photocathode. The quantum efficiency is simply defined as

$$QE = \frac{\text{number of photoelectrons emitted}}{\text{number of incident photons}} \qquad (9\text{-}1)$$

The quantum efficiency would be 100% for an ideal photocathode. Because of the limitations mentioned earlier, practical photocathodes show maximum quantum efficiencies of 20–30%.

The quantum efficiency of any photocathode will be a strong function of wavelength or quantum energy of the incident light, as shown in Fig. 9-2. To estimate the effective

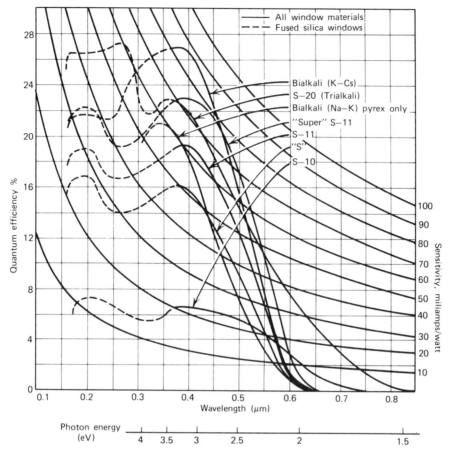

Figure 9-2 The spectral sensitivity of a number of photocathode materials of interest in PM tubes. The use of silica or quartz windows is necessary to extend the response into the ultraviolet region. (Courtesy of EMI GENCOM Inc., Plainview, NY.)

quantum efficiency when used with a particular scintillator, these curves must be averaged over the emission spectrum of the scintillator. One consideration in selecting a photocathode is to match the spectral response curve as closely as possible to that of the emission spectrum of the scintillator being used.

The sensitivity at the long-wavelength or low photon energy end of the scale is largely limited by the reduced absorption of light in the photocathode and the low energy imparted to the photoelectron. At a sufficiently high λ this electron no longer has sufficient energy to escape the surface of the photocathode and the response drops to zero. The response at the opposite end of the scale is normally not a function of the photocathode itself but rather of the window through which the light must enter to reach the photoemissive layer. For normal glass, the cutoff will be at a wavelength in the region of 350 nm, which is usually adequate for most scintillation materials. For some scintillators (e.g., noble gases), however, a significant part of the emission spectrum can be in the ultraviolet region with shorter wavelength. For such applications, special PM tubes with entrance windows made from fused silica or quartz can be used to extend the sensitivity to wavelengths as short as about 160 nm.

An alternative measure of quantum efficiency is sometimes quoted for PM tubes used in scintillation counting. The number of photoelectrons produced per unit energy lost in a scintillator mounted with close optical coupling to the photocathode is proportional to the quantum efficiency averaged over the emission spectrum of the scintillator. Because of the widespread use of thallium-activated sodium iodide as a scintillation crystal, the standard for quotation is the number of photoelectrons produced from a given photocathode per keV of energy loss by fast electrons in a NaI(Tl) crystal from which nearly all the light is collected. For photocathodes with peak quantum efficiency of 25–30%, measurements give about 8–10 photoelectrons per keV energy loss.[4,5] The reciprocal of this value, or the average energy loss required to create one photoelectron, is therefore approximately 100–120 eV. As emphasized elsewhere in this text, this value for the energy loss required to produce one basic information carrier in a typical scintillation detector is much larger than the equivalent value in gas-filled or semiconductor detectors.

Presently available materials for photocathodes include a *multialkali* material based on the compound Na_2KSb. Prepared by activation with a small amount of cesium, this material was the first to show a relatively high quantum efficiency of up to 30% in the blue region of the spectrum. A later formulation based on K_2CsSb activated with oxygen and cesium is given the name *bialkali* and can show an even higher efficiency in the blue. Furthermore, thermionic emission from bialkali photocathodes tends to be significantly lower than that from the multialkali materials, leading to lower spontaneous noise rates from tubes with this photocathode. Negative electron affinity materials are under active development as photocathodes and hold promise for high quantum efficiency over a wide spectral range. Results reported on one of these materials, GaAs, show excellent efficiency well into the infrared region.[6]

III. ELECTRON MULTIPLICATION

A. Secondary Electron Emission

The multiplier portion of a PM tube is based on the phenomenon of secondary electron emission. Electrons from the photocathode are accelerated and caused to strike the surface of an electrode, called a *dynode*. If the dynode material is properly chosen, the energy deposited by the incident electron can result in the reemission of more than one electron from the same surface. The process of secondary electron emission is similar to that of photoemission discussed in the previous section. In this case, however, electrons within the dynode material are excited by the passage of the energetic electron originally incident on the surface rather than by an optical photon.

Electrons leaving the photocathode have a kinetic energy on the order of 1 eV or less. Therefore, if the first dynode is held at a positive potential of several hundred volts, the kinetic energy of electrons on arrival at the dynode is determined almost entirely by the magnitude of the accelerating voltage. The creation of an excited electron within the dynode material requires an energy at least equal to the bandgap, which typically may be of the order of 2–3 eV. Therefore, it is theoretically possible for one incident electron to create on the order of 30 excited electrons per 100 V of accelerating voltage. Because the direction of motion of these electrons is essentially random, many will not reach the surface before their deexcitation. Others that do arrive at the surface will have lost sufficient energy so that they cannot overcome the potential barrier at the surface and are therefore incapable of escaping. Therefore, only a small fraction of the excited electrons ultimately contribute to the secondary electron yield from the dynode surface.

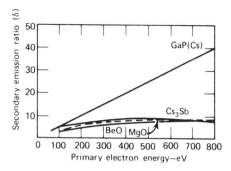

Figure 9-3 Variation of the secondary emission yield with primary electron energy for standard dynode materials (lower three curves) and an NEA material (GaP(Cs)). (From Krall et al.[7])

The secondary electron yield is a sensitive function of incident electron energy. If a relatively low-energy electron strikes the dynode surface, little energy is available for transfer to electrons in the dynode material, and relatively few electrons will be excited across the gap between the valence and conduction bands. At the same time, because the distance of penetration is not large, most of these excited electrons will be formed near the surface. For incident electrons of higher energy, more excited electrons will be created within the dynode but at greater average depth. Because the probability of escape will diminish with increasing depth, the observed electron yield will be maximized at an optimum incident electron energy.

The overall multiplication factor for a single dynode is given by

$$\delta = \frac{\text{number of secondary electrons emitted}}{\text{primary incident electron}} \quad (9\text{-}2)$$

and should be as large as possible for maximum amplification per stage in the photomultiplier tube. A plot of δ versus incident electron energy is given in Fig. 9-3 for several dynode materials. For the conventional dynode materials of BeO, MgO, and Cs_3Sb, the maximum multiplication factor reaches about 10 for an optimum incident energy near 1 keV, although values of 4–6 are more typical at conventional interdynode voltages of a few hundred volts.

B. Negative Electron Affinity Materials

The secondary emission yield of dynodes can be increased significantly through the use of *negative electron affinity* (NEA) materials[7-11] developed in the 1970s. The most successful of these materials has been gallium phosphide (GaP), heavily doped to a concentration of about 10^{19} atoms/cm^3 with a p-type material such as zinc. The zinc creates acceptor sites within the bulk of the gallium phosphide. A thin, nearly monatomic layer of an electropositive material such as cesium is then applied to one surface. The acceptors at the surface attract an electron from the electropositive cesium, and each cesium atom becomes ionized and is held to the surface by electrostatic forces.

The effect of this surface treatment can best be illustrated through the band structure diagram shown in Fig. 9-4. At the left is shown a conventional band diagram for undoped gallium phosphide, which is also representative of conventional dynode materials. The series of arrows on the left shows a typical history of an electron that does not escape.

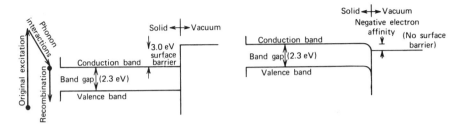

Figure 9-4   Band structure near the surface for conventional semiconductors (left) and NEA materials (right). (Adapted from Krall et al.[7])

The original excitation causes a number of electrons to be elevated from the valence band to some point well up into the conduction band. As these electrons diffuse, they lose energy primarily through phonon interactions, such that within about a picosecond, these "hot" electrons have come into thermal equilibrium with their local environment and their energy has relaxed to near the bottom of the conduction band. If this electron is to escape, it must reach the surface with an energy greater than the potential that exists on the vacuum side of the surface. Once the electron drops to the bottom of the conduction band, its energy is normally below the vacuum potential and is too low to permit escape. In this case only a short time is available for the electron to escape and it cannot travel large distances from its point of origin. Therefore, relatively small escape depths of only a few nanometers are possible. However, the electron will tend to diffuse for a substantially longer time (typically 100 ps) before dropping across the gap to rejoin the valence band.

On the right is shown the band bending created by the filling of acceptor sites at the surface by the thin cesium layer. The effect of the bending is to bring the vacuum potential below that of the bottom of the conduction band in the interior of the material. Therefore, electrons that have already dropped to the bottom of the conduction band still have sufficient energy to escape if they happen to diffuse to the surface. Because the thickness of the bent-band region is very small, it can be less than a mean free path and the electron may escape without further energy loss in the surface region. The net effect is that the electrons that have already reached the bottom of the conduction band are still candidates for escape and remain so for a period of time that is about 100 times greater than in the previous case. The average escape depth will therefore tend to be much greater and can reach tens or hundreds of nanometers.

The effect of this change on the secondary electron yield is profound. The increased time over which electrons may escape enhances the escape probability for any typical electron. Furthermore, excited electrons created deep within the dynode material remain candidates for escape. Therefore, the secondary electron yield will continue to increase with increasing primary electron energy until the distance of penetration of the primary is very large. These effects are reflected in the yield curve for a GaP(Cs) dynode shown in Fig. 9-3. Secondary electron yields of 50 or 60 are readily achieved with an interstage voltage of 1000 V, and much higher values are possible in principle if even larger voltages are permitted by the PM tube design.

A secondary advantage of NEA materials is evident in PM tubes used for ultrafast timing applications. Because almost all escaping electrons have previously dropped to the bottom of the conduction band, their average energy is lower and much more uniform than secondary electron energies from conventional materials. Because variations in initial energy contribute to the time spread in the multiplier section, a narrower distribution leads to lower time broadening. Furthermore, the higher gain per stage permits a reduction in the number of stages required for a given total gain, also reducing the overall time spread.

C. Multiple Stage Multiplication

To achieve electron gains on the order of 10^6, all PM tubes employ multiple stages. Electrons leaving the photocathode are attracted to the first dynode and produce δ electrons for each incident photoelectron. The secondaries that are produced at the surface of the first dynode again have very low energies, typically a few eV. Thus, they are quite easily guided by another electrostatic field established between the first dynode and a second similar dynode. This process can be repeated many times, with low-energy secondary electrons from each dynode accelerated toward the following dynode. If N stages are provided in the multiplier section, the overall gain for the PM tube should be given simply by

$$\text{overall gain} = \alpha \delta^N \qquad (9\text{-}3)$$

where α is the fraction of all photoelectrons collected by the multiplier structure. Conventional dynode materials are characterized by a typical value of $\delta = 5$, and α is near unity for well-designed tubes. Ten stages will therefore result in an overall tube gain of 5^{10}, or about 10^7. If high-yield NEA dynodes are used with $\delta = 55$, the same gain can be achieved with only four stages.

The overall gain of a PM tube is a sensitive function of applied voltage V. If δ were a linear function of interdynode voltage, then the overall gain of a 10-stage tube would vary as V^{10}. As shown in Fig. 9-3, however, δ for conventional dynodes varies as some fractional power of interdynode voltage so that the overall gain is more typically proportional to V^6–V^9.

D. Statistics of Electron Multiplication

If δ were strictly a constant, each photoelectron would be subject to exactly the same multiplication factor. Under fixed operating conditions, all output pulses that originate from a single photoelectron would then have the same amplitude. In actuality, the emission of secondary electrons is a statistical process, and therefore the specific value of δ at a given dynode will fluctuate from event to event about its mean value. The shape of the single photoelectron pulse-height spectrum observed from a real PM tube is an indirect measure of the degree of fluctuation in δ and has thus been the subject of extensive investigation.

In the most simple model, the production of secondary electrons at a dynode can be assumed to follow a Poisson distribution about the average yield. For a single photoelectron incident on the first dynode, the number of secondaries produced has a mean value of δ and standard deviation σ of $\sqrt{\delta}$ (see Chapter 3). The relative variance, defined as $(\sigma/\delta)^2$, is thus equal to $1/\delta$. When this process is now compounded over N identical stages of the PM tube, the mean number of electrons collected at the anode (and hence the pulse amplitude) is given by δ^N. It can be demonstrated from the properties of Poisson statistics that the relative variance in this number is now

$$\frac{1}{\delta} + \frac{1}{\delta^2} + \frac{1}{\delta^3} + \cdots + \frac{1}{\delta^N} \cong \frac{1}{\delta - 1} \qquad (9\text{-}4)$$

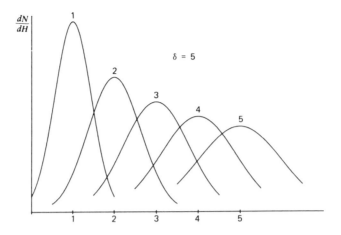

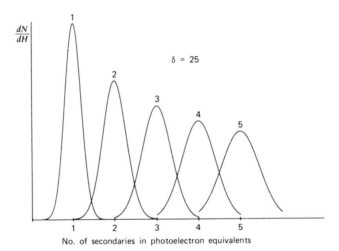

Figure 9-5 Statistical broadening of the secondary electron yield from the first dynode of a PM tube. Numbers identify the number of incident photoelectrons. Two cases are shown representing conventional dynodes ($\delta = 5$) and NEA materials ($\delta = 25$).

Thus, if $\delta \gg 1$, the relative variance or spread in the output pulse amplitude is dominated by fluctuations in the yield from the first dynode where the absolute number of electrons is smallest.

In many applications of scintillators, hundreds or thousands of photoelectrons contribute to each pulse, and they are therefore much larger than single photoelectron pulses. When poor light collection or low-energy radiations are involved, however, signal pulses corresponding to only a few photoelectrons may be involved. Then the fluctuations in electron multiplication may interfere with the ability to discriminate against noise events, many of which correspond to single photoelectrons. Figure 9-5 shows the expected distribution in the number of secondaries produced by the first dynode when struck by different numbers of photoelectrons. If the value of δ is small, it is impossible to separate cleanly the events caused by one photoelectron from those in which more photoelectrons are involved. If the dynodes are characterized by a larger value of δ, however, the

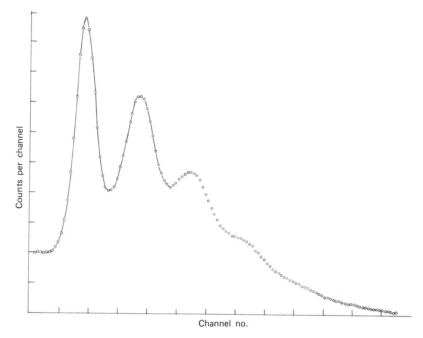

Figure 9-6 The measured pulse height spectrum for weak scintillation events obtained from a RCA 8850 photomultiplier tube. The high-gain first dynode results in distinguishable peaks in the spectrum corresponding to 1, 2, and 3 photoelectrons per pulse. (From Houtermans.[12])

separation is much more distinct and it is possible to distinguish peaks in the distribution corresponding to discrete numbers of photoelectrons up to about 4 or 5. This behavior is demonstrated in Fig. 9-6 in the pulse height spectrum observed from a PM tube with a high-δ first dynode made from a NEA material.

Experimental measurements of the single photoelectron pulse height spectra from PM tubes generally show a peaked distribution,[13,14] but with a larger relative variance than that predicted by the Poisson model. In fact, observations made under some conditions show no peak at all, but rather an exponential-like distribution.[15] These discrepancies have led to alternate models of the multiplication statistics in which a Polya distribution[16] or compound Poisson[17,18] is substituted for the simple Poisson description of electron multiplication. No universal descriptions have as yet emerged which can accommodate all experimental measurements, and it is possible that differences in specific electron trajectories and dynode properties may preclude a general model applicable to all PM tubes.

IV. PHOTOMULTIPLIER TUBE CHARACTERISTICS

A. Structural Differences

Figure 9-7 shows some representative construction details of PM tubes of various designs. All consist of a semitransparent photocathode, a photoelectron collection region between the photocathode and the first dynode, a multistage electron multiplier section, and an anode for collection of the amplified charge. These structures are enclosed in a glass vacuum envelope, through which electrical leads are conducted at the base. Tubes with

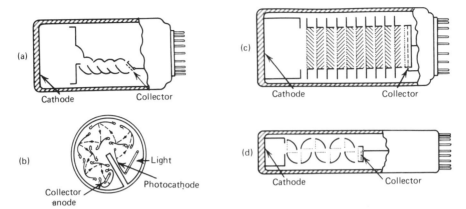

Figure 9-7 Configurations of some common types of PM tubes. (a) Focused linear structure. (b) Circular grid. (c) Venetian blind. (d) Box-and-grid. (Courtesy of EMI GENCOM Inc., Plainview, NY.)

flat end plate windows are the only types now in general use for scintillation counting, so cylindrical scintillation crystals can easily be mounted directly on the end window adjacent to the photocathode. Tubes are available commercially with circular photo-cathodes ranging in diameter from a few millimeters to over 20 cm. A nominal 2-in. (5 cm) diameter is one of the common choices in many scintillation applications, and the widest selection is generally available in this size. PM tubes with square and hexagonal photocathodes have also been introduced for specialized applications in which an array of tubes must be closely packed.

The "venetian blind" type of construction is one of the oldest used for photomulti-plier tubes and is now virtually obsolete. It is readily adaptable to tubes of varying numbers of stages but suffers from a relatively slow response time due to low electric fields at the surface of the dynodes. The "box and grid" structure is also fairly old and slow, but it is still standard in many tubes of large diameter. The circular grid and linear multiplier structures were introduced to speed up the electron transit time through the multiplier structure and are used in those PM tubes with the fastest response time.

Photomultiplier tubes should be protected from excessive mechanical shock or vibration to avoid physical damage to internal components. Also, some modulation of the output signal can be induced by vibration of the multiplier structures. Most tube manufacturers offer a rugged version of some of their photomultiplier tubes (often originally designed for space applications). These are provided with electron multiplier structures and other internal components mechanically arranged to resist vibration and shock. Standard tubes are quite adequate in this respect for most scintillation counting applications, but rugged versions can be helpful in counters designed for field use where they may be subjected to rough handling.

Another type of specialized electron multiplier is the continuous channel, illustrated in Fig. 9-8. This device is extremely simple, consisting of a hollow glass tube whose inner surface acts as a secondary electron emitter. A potential difference is applied across the length of the tube, thereby attracting electrons from the entrance end to the exit end. Electrons entering the tube will eventually strike the wall giving rise to a pulse of secondary electrons. These will be further accelerated along the length of the tube until they in turn also strike the wall, giving rise to further secondaries. The device acts much

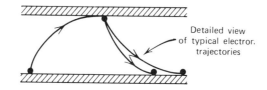

Detailed view
of typical electron
trajectories

Figure 9-8 Continuous channel electron multiplier.

like a photomultiplier tube with continuous dynodes available along its entire length. The number of times an entering electron and its subsequent secondaries strike the wall is an accident of past orientation and individual trajectories and will vary considerably for different entering electrons. When operated at high voltages, the electron gain may be large enough so that the resulting space charge near the exit of the channel limits the total charge per pulse to about 10^6 or 10^7 electrons at saturation.

Channel multipliers must be shaped to prevent feedback problems that can arise when positive ions occasionally formed within the channel are accelerated in the reverse direction from the electrons. By forming the channel as a curve or chevron, these ions can be made to strike a wall before their energy is high enough to create secondary electrons.

Clusters of many thousands of tubes can be fabricated to form a *microchannel plate*. Each channel is of very small diameter (typically 15–50 μm) and acts as an independent

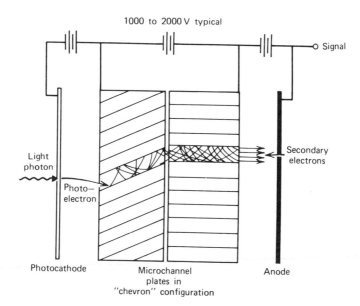

Figure 9-9 Elements of a PM tube based on microchannel plate electron multiplication.

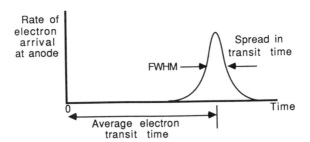

Figure 9-10   The response of a PM tube to a short pulse of light on the photocathode.

electron multiplier. Their application as the multiplier element in PM tubes is reviewed in Refs. 19–22. In the configuration shown in Fig. 9-9, each photoelectron generally enters a separate channel. If all channels are operated in charge saturation and are identical, then the total collected charge is proportional to the number of original photoelectrons.

The primary advantage of PM tubes based on microchannel plates lies in their excellent timing properties.[23-25] The total electron transit time through a channel is a few nanoseconds, compared with 20–50 ns for conventional multiplier structures (see the following section). The spread in transit time, which determines timing performance, is about 100 ps, a factor of 2 or 3 better than the fastest currently available, conventional PM tube.

B. Pulse Timing Properties

Because the time required for photoemission in the photocathode or secondary emission from dynodes is very short (0.1 ns or less), the time characteristics of the PM tube are determined exclusively by the electron trajectories. The *electron transit time* of a PM tube is defined as the average time difference between the arrival of a photon at the photocathode and the collection of the subsequent electron burst at the anode. In PM tubes of various designs, electron transit times range from 20 to 80 ns. In most timing applications, however, the transit time itself is not of primary importance because if it were always a constant, it would introduce only a fixed delay in the derived signal. Instead, the *spread in transit time* is a more important quantity because it determines the time width of the pulse of electrons arriving at the anode of the tube. The timing response of a typical PM tube is illustrated in Fig. 9-10.

The region between the photocathode and first dynode is critical in determining the timing properties. To allow uniform collection over large photocathodes, this distance is kept fairly large compared with interdynode distances (see Fig. 9-11). The difference in paths between a photoelectron leaving the center of the photocathode and one at the edge is often a dominant factor in the observed spread in transit time. The photocathode is often curved to minimize the transit time spread across its diameter. It is convenient to have a flat outer surface for scintillator mounting, so an end window with a plano-concave shape is frequently used with the photocathode deposited on the inner curved surface. A second source of transit time spread arises from the distribution in initial velocities of photoelectrons leaving the photocathode. This effect can be minimized by using a large voltage difference between the photocathode and first dynode.

The amount of transit time spread observed for a specific pulse also depends on the number of initial photoelectrons per pulse. To simplify the analysis and comparison of different photomultipliers, many of the measurements reported in the literature con-

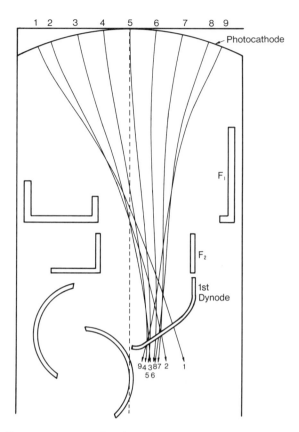

Figure 9-11 Computer-generated trajectories of electrons accelerated from the photo-cathode to the first dynode in a PM tube. In this design, the photocathode is curved to minimize the spread in transit time. Structures labeled F_1 and F_2 are electrodes carrying adjustable voltage that electrostatically focuses the electrons for optimum performance. (From Kume et al.[26])

centrate on the transit time spread due to a *single* photoelectron. This parameter does not include the important contribution of the cathode to first dynode space,[27,28] and is only a measure of the time spreads introduced subsequent to the first dynode. As the number of photoelectrons increases, a larger number of samplings of various possible electron trajectories through the tube are made. If the distribution in the various possible transit times is assumed to be Gaussian, then statistical theory predicts that the relative spread in transit times should vary inversely with the square root of the number of photoelectrons. In Fig. 9-12, this behavior is verified for a typical PM tube. Thus, high light yield from a scintillator is important in timing applications as well as in pulse height measurements. The time spread attributable to the multiplier section also decreases with increasing interdynode voltage, and the best timing performance is normally obtained by operating the tube at maximum voltage permitted by the ratings.

Leskovar and Lo[29] experimentally investigated the timing properties of a number of fast PM tubes and present a comprehensive survey of the literature on this topic. Fast pulse performance remains an active area of PM tube design innovation and the timing properties listed later in Table 9-1 are likely to be improved with further development. When used with slow inorganic scintillators, most current PM tubes are fast enough so

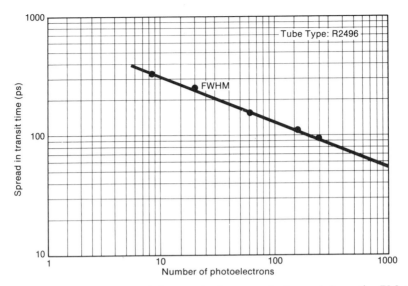

Figure 9-12 Measurements of the transit time spread observed from the PM tube shown in Fig. 9-11 as a function of the average number of photoelectrons generated per pulse. (From Kume et al.[26])

that their contribution to the overall time response usually is not an important factor. It is only when scintillators with shorter decay time are employed to derive a fast timing signal that the PM tube can become a significant element in determining the resultant timing properties.

C. Maximum Ratings

All commercial photomultipliers are supplied with a set of maximum voltage and current ratings that are not to be exceeded during routine use. Detailed specifications will often give individual values for maximum photocathode to first dynode, dynode to dynode, last dynode to anode, and cathode to anode maximum voltages as well. These limits are set by a variety of conditions including ohmic leakage and dark current considerations. The user is often more interested in a single figure for the maximum applied voltage when using the manufacturer's recommended voltage divider string. Because virtually every PM tube will show an increase in gain as the voltage is increased, the maximum value for applied voltage practically determines the maximum gain obtainable from the tube.

Maximum values are also often given for photocathode current and anode current. The first limit is usually set by consideration of photocathode resistivity, which can give rise to distorted electrostatic fields at high currents, whereas the anode limit is set by heat dissipation considerations. In normal pulse-mode scintillation counting, current limits seldom are of concern because the maximum counting rate permitted by pile up and pulse resolution considerations nearly always corresponds to an average current far less than the maximum ratings.

D. Photomultiplier Tube Specifications

Manufacturers will conventionally quote performance of PM tubes in terms of certain characteristics, which are defined here.

    1. *Overall Luminous Sensitivity.* Defined as the ratio of the measured anode current at operating voltage to the luminous flux from a tungsten light source of specified

TABLE 9-1 Properties of Some Commercially Available Photomultiplier Tubes

A	B	C	D	E	F	H	J	K	L	M	P	Q	R	S
Amp	58AVP	3000	130	IL	110	S-11	70	56	100(2400)	131	269	2	45	2
EMI	9256B	1500	51	IL	45	bia.	110	12	0.45(800)	51.5	115	2.5	—	0.2
EMI	9758B	2100	75	VB	67	bia.	90	12	0.6(1000)	78	129	10	65	0.5
EMI	9815B	2500	50	IL	46	bia.	75	11	0.67(1650)	53	149	1.8	34	0.3
EMI	9954B	2800	53	IL	46	bia.	90	10.5	0.56(1750)	53.3	149	2	41	2
EMR	541N-01-14	3000	35	VB	25	bia.	54	—	1.0(2000)	35	108	—	—	0.1
HM	R268	1500	28	BX	25	bia.	95	11.5	2.1(1000)	30	127	12	60	2
HM	R331-05	2500	51	IL	46	bia.	80	10	1.5(1500)	53.5	141	2.6	48	18
HM	R647-01	1250	13	IL	10	bia.	80	10.5	1.0(1000)	14.0	86	2.3	24	5
HM	R1288	1800	25	CC	21	bia.	40	6.0	0.38(1500)	25.9	100	2.2	14	0.1
HM	R1306	1300	51	BX	46	bia.	95	11.5	0.32(1000)	51.5	137	12	57	5
HM	R1307	1500	76	BX	70	bia.	95	11.5	0.32(1000)	76.8	150	11	64	5
HM	R1332	2700	51	IL	46	bia.	70	9.0	1.0(1500)	53.3	147	3.0	43	10
RCA	4516	1800	19	IL	12.7	bia.	66	9.2	0.52(1500)	19.3	100	1.8	19	0.1
RCA	6199	1250	39	CC	31.5	S-11	70	6.2	3.6(1000)	39.6	116	2.8	32	3.2
RCA	6342A	1500	52	CC	42.7	S-11	80	8.0	0.35(1250)	52.3	148	—	—	15
RCA	8575	3000	53	IL	45.7	bia.	60	9.2	27(2000)	53.3	148	2.8	37	1.0
RCA	8850	3000	53	IL	45.7	bia.	64	9.0	21(2000)	53.3	148	2.8	37	1.0
RCA	8852	2500	51	IL	45.7	mult.	180	—	0.6(1500)	53.3	148	3.3	42	5.5
RCA	8854	3000	127	IL	114	bia.	67	10	50(2000)	129	299	4.0	78	100

A = manufacturer: Amp = Amperex, HM = Hamamatsu.
B = model number.
C = maximum voltage overall.
D = nominal diameter (mm).
E = dynode structure: CC = circular, VB = venetian blind, IL = in-line or linear, BX = box.
F = photocathode diameter (mm).
H = spectral class: bia. = bialkali, mult. = multialkali.
J = cathode luminous sensitivity (microamperes per lumen).
K = filtered luminous sensitivity (Corning Cs 5-58 (blue) polished to 1/2 stock thickness).
L = gain $\times 10^6$ at given voltage (in parentheses).
M = maximum diameter (mm).
P = maximum length (mm).
Q = rise time at voltage in L (ns).
R = transit time at voltage in L (ns).
S = dark current (a very approximate number because of large variations in the method of measurement between different manufacturers) (nanoamperes).

temperature incident on the photocathode. This quantity is an overall measure of the expected current from the PM tube per unit incident light from a broad-band source. The units are amperes per lumen (A/lm).

2. *Cathode Luminous Sensitivity.* Defined as above, except that the current of photo-electrons leaving the photocathode is substituted in the numerator for the anode current. This quantity is again measured in amperes per lumen, is a characteristic only of the photocathode, and is independent of the electron multiplier structure.

3. *Overall Radiant Sensitivity.* This parameter is defined as the ratio of anode current to radiant power at a given wavelength incident on the photocathode. Units are amperes per watt (A/W).

4. *Cathode Radiant Sensitivity.* Defined as above, except that the photocathode current is substituted for the anode current.

5. *Dark Current.* Normally specified in terms of anode current measured without photocathode illumination when the tube is operated to provide a given overall luminous sensitivity.

6. *Anode Pulse Rise Time.* Quoted as the time taken for the output pulse to rise from 10 to 90% of the peak when the photocathode is illuminated by a flash of light of very short duration.

7. *Anode Pulse Width.* Normally quoted as the time width of the output pulse measured at half maximum amplitude, again for short-duration illumination of the photocathode.

Table 9-1 lists specifications for a number of commercial photomultiplier tubes of significant current interest in scintillation counting. Because PM tube design remains a rapidly developing area, potential users should always seek up-to-date information directly from the tube manufacturers.

E. Linearity

The electron multiplication factor in nearly all PM tubes remains constant for pulses that range in size from a single photoelectron to many thousands. Under these conditions the amplitude of the pulse collected at the anode is linearly related to the number of photoelectrons, and consequently to the intensity of the scintillation light flash. Nonlinearities can arise for very large pulses due to space charge effects between the last dynode and anode where the number of electrons is greatest. The buildup of space charge affects the trajectories of electrons in this region and causes some to be lost which would otherwise be collected. Another factor that can cause nonlinearities at high pulse amplitudes is any deviation of dynode voltages from their equilibrium value during the course of the pulse. Under normal circumstances in scintillation pulse counting, these effects are seldom important with an adequately designed tube base (see later discussion) and the photomultiplier tube remains in the linear range.

F. Noise and Spurious Pulses

Usually the most significant source of random noise from a photomultiplier tube results from thermionic electrons which are spontaneously emitted by the photocathode. The pulses that result from this process correspond to a single photoelectron, so their amplitude is limited to the lowest end of the scale. Because most scintillation counting is done under conditions in which a scintillation pulse corresponds to many photoelectrons, amplitude discrimination is usually sufficient to eliminate all contributions of this thermal noise. In applications in which very-low-energy radiation is measured or in which single photons are detected, valid signal events may also correspond to a single photoelectron and therefore may be indistinguishable from thermionic noise. In that event there is no alternative but to try to reduce the noise contribution as much as possible. The rate at which these pulses are observed is proportional to the area of the photocathode, and therefore one should select a tube of the smallest diameter required for a specific application in order to minimize these dark pulses. The rate at which thermionic electrons are emitted per unit area varies greatly between photocathode materials, and bialkali photocathodes are among the most quiet. The dark current specifications given in Table 9-1 largely reflect the contribution of thermionic electrons, and the tubes with lowest noise rates are those with the lowest dark currents for equivalent luminous sensitivity.

For some photocathodes the rate at which thermionic electrons are emitted can be drastically reduced by cooling the tube. Reductions by a factor of 100 or more may be observed with proper temperature reduction. Dry ice or liquid nitrogen are often used for cooling, but self-contained refrigerators are also commercially available for this purpose. Problems that can arise in connection with PM tube cooling include water vapor condensation on exposed cold surfaces and increased photocathode electrical resistance at lower temperatures. Large photocathode resistance can distort the electrostatic field between the photocathode and first dynode and may lead to a loss in photoelectron collection efficiency.

Photomultiplier tubes should be stored in the dark when not in use. Exposure to room light is disastrous while voltage is supplied to the tube because very high illumination levels lead to anode currents that greatly exceed the maximum ratings and that can quickly damage the multiplier structures. Incident light, especially from fluorescent tubes, is to be avoided even when no voltage is applied to the tube because of the temporary increase in dark pulses which will result. It is not unusual to observe an increase of 100 or more in the rate of dark pulses immediately after exposure to intense room light, and a measurable increase can persist for several hours. It appears that the major cause of this increased noise rate is the emission of light from phosphorescence states in the glass envelope of the PM tube.

Another source of dark pulses originates with natural radioactivity in the structure of the tube itself. The most important components are usually potassium-40 and thorium contained in the glass envelope. A beta particle produced in radioactive decay will give rise to a flash of Cerenkov radiation, which can liberate photoelectrons from the photocathode in much the same way as normal scintillation events. For applications in which the ultimate in low background is a necessity, special tubes with low-activity glass to minimize these events are available. Scintillation or Cerenkov light produced in the glass by external radiation can also be a significant source of dark pulses.[30] One such source is cosmic radiation (see p. 719), which generally results in dark pulses of small amplitude from Cerenkov light produced in the thin end window of the tube. Because of the low specific energy loss of the secondary cosmic radiations and the low light yield from the Cerenkov process, the corresponding dark pulses correspond to only a few photoelectrons and are therefore usually discarded by amplitude discrimination in typical scintillation applications. However, if very weak scintillation events must be recorded, these dark pulses can be of the same size as the signal. In such cases, the rate from cosmic radiation can be minimized by operating the tube with its major axis oriented horizontally so that the end window presents minimum cross section to the cosmic secondaries that are directed preferentially in a downward vertical direction.

Afterpulses are another type of noise sometimes observed in PM tubes.[31] These are satellite pulses that will sometimes follow a true signal pulse after a short delay period. One mechanism that can give rise to afterpulsing is the emission of light from the latter stages of the multiplier structure, which finds its way back to the photocathode. Such afterpulses will be delayed by a time characteristic of the electron transit time through the tube, or roughly 20–50 ns. Because these pulses often correspond to a single photon, their amplitude is usually quite small. Another cause of afterpulsing can be an imperfect vacuum within the tube. Traces of residual gas can be ionized by the passage of electrons through the multiplier structure. The positive ions that are formed will drift in the reverse direction and some may find a path back to the photocathode. Typical ions will liberate tens or hundreds of photoelectrons when they strike the photocathode, and the resulting pulse will be of rather large size. Because the velocity of the positive ion is relatively low, the time it takes to drift back to the photocathode can range from hundreds of nanoseconds to a microsecond or more. The time spacing between the primary pulse and the afterpulse from this mechanism is therefore relatively large. Because the amount of residual gas can vary considerably between tubes of identical design, problems due to afterpulsing in some applications may often be eliminated by simple substitution of another tube. The probability of production of an afterpulse increases linearly with primary pulse amplitude in at least one tube design but is usually no more than a few percent. Afterpulsing is therefore of little consequence in ordinary scintillation spectroscopy but can become a serious perturbing effect in timing measurements where the pulses of interest follow a preceding high-intensity burst of radiation. Discussions of the afterpulsing characteristics of some PM tubes, together with electronic means of afterpulse suppression, may be found in Refs. 32–35.

G. Photocathode Nonuniformities

Direct measurement (e.g., see Ref. 10, 36, or 37) has shown that the sensitivity of photocathodes, especially those with large diameters, is far from uniform across the entire photocathode area. This problem is further compounded by the difficulties of achieving uniform photoelectron collection to the first dynode from the entire photocathode area. The combination of these two effects can lead to situations in which the anode pulse observed for a given flash of light may vary by as much as 30–40% as the position of the illumination is moved across the photocathode area. This nonuniformity is a potentially serious problem in scintillation counting because response variations will tend to spoil the energy resolution of the system. One means of reducing the problem is to place a light pipe between the scintillator and end window of the PM tube. This is especially important for thin scintillators and tends to spread light from any scintillation event over the entire photocathode, thereby averaging out much of the nonuniformity.

H. Gain Variations with Counting Rate

Another nonideality of PM tubes of which the user should be aware is the possibility of gain changes during the course of a measurement. The most common situation is one in which the counting rate changes by a large factor. If the divider string current is too low (see p. 272), changes in count rate can lead to gain changes due to resulting variations in the dynode potentials.[38] Even if this effect is totally eliminated, the tube may require several hours to stabilize at a constant count rate as thermal and space charge effects created by the electron current through the multiplier structure of the tube reach equilibrium.[39] The gain changes may not be fully reversible, and hysteresis effects have been observed experimentally.[40] Photomultiplier structures can be designed to minimize these drifts, and a good tube will not change its gain by more than 1% as the counting rate is varied from 10^3 to 10^4 per second.[3] Specifications of this type can often be found for many PM tube designs. The gradual drift in tube gain that often follows a large change in tube current or counting rate is called *fatigue*[41,42] and can be a serious problem if the tube current changes by orders of magnitude during the course of the measurement.

V. ANCILLARY EQUIPMENT REQUIRED WITH PHOTOMULTIPLIER TUBES

A. High-Voltage Supply and Voltage Divider

An external voltage source must be connected to the PM tubes in such a way that the photocathode and each succeeding multiplier stage are correctly biased with respect to

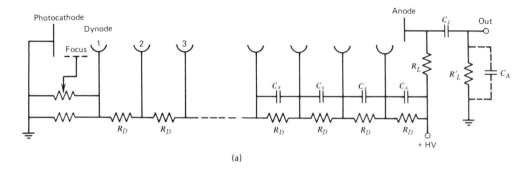

(a)

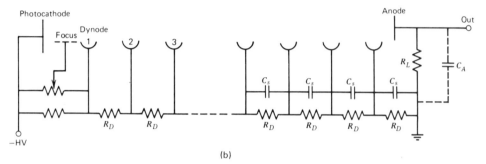

(b)

Figure 9-13 Typical wiring diagrams for the base of a PM tube. Scheme (a) utilizes positive high voltage and a grounded photocathode. Scheme (b) uses negative high voltage, and the photocathode must be isolated from ground. Values of the divider string resistors R_D are chosen using criteria given in the text. The equivalent anode load resistance is R_L in (b), and the parallel combination of R_L and R'_L in (a). Other identifications are given in the text.

one another. Because electrons must be attracted, the first dynode must be held at a voltage that is positive with respect to the photocathode, and each succeeding dynode must be held at a positive voltage with respect to the preceding dynode. For efficient photoelectron collection, the voltage between photocathode and first dynode is often several times as great as the dynode-to-dynode voltage differences.

The interstage voltage requirements of a PM tube may, in principle, be supplied by individual voltage sources, such as a multicell battery. Battery supplies are practical in some applications where counting rates are low, but often are unattractive because of the rapid discharge rates due to the current demands of the latter stages of the PM tube. The internal current in the PM tube is at a maximum between the last dynode and the anode, so the battery cell that supplies the voltage to the last dynode must be capable of a current drain equal to the average dc current leaving this dynode.

In the vast majority of cases, the voltage differences are provided instead by a resistive voltage divider and a single source of high voltage. Figure 9-13a shows a typical wiring diagram for the base of a PM tube in which a positive polarity high voltage is used. In this case the cathode of the PM tube is grounded, and the divider string supplies successively increasing positive voltages to each dynode down the multiplying string. The anode is at a dc potential equal to the supply voltage, and signal pulses must therefore be capacitively coupled from the anode to allow the pulse component to be passed on at ground potential to succeeding electronic devices. The load resistor R_L can be chosen by the experimenter so that the resulting anode circuit time constant is of proper magnitude.

The anode capacitance C_A usually is not a physical capacitor, but only the stray capacitance associated with the anode structure and connecting cables. An analysis of the pulse shape expected from this anode circuit is given later in this chapter.

The direct current through the divider string is determined simply by the ratio of the applied high voltage and the summed resistance of all the resistors in the divider string. In order to use the simplest and least expensive voltage supply, one would like to keep this current at a minimum. Small currents also minimize problems due to heat dissipation in the divider resistors. However, the divider string current should in principle be kept large compared with the internal PM tube current represented by the pulse of electrons flowing from dynode to dynode. If the internal current at the peak of a pulse becomes comparable with the divider current, the voltages of the dynodes normally will begin to deviate from their equilibrium values, leading to drift of the PM tube gain. This problem is especially serious for the last few stages of the PM tube where peak currents are at a maximum. To suppress this effect, it is commonplace to provide *stabilizing capacitors* (labeled as C_s in Fig. 9-13) to the stages of the divider string near the anode to help hold these latter dynode voltages at a constant value throughout the pulse. The stabilizing capacitor momentarily supplies the current lost from the dynode during the pulse and is recharged by the divider string current during the period between pulses. To prevent a more than 1% interdynode voltage change, the charge stored on the stabilizing capacitor (given by the product of capacitance value and the interdynode voltage) must be 100 times greater than the charge emitted by that dynode during the pulse.

A numerical example is instructive at this point. Assume we start with a scintillation event that liberates 1000 photoelectrons from the photocathode of the PM tube. Further assume that the PM tube provides an overall gain of 10^6, so that 10^9 electrons per pulse leave the last dynode and are collected by the anode. If these scintillation pulses are occurring at a rate of 10^5 per second, then the average dc anode current can easily be calculated as

$$I_{\text{avg}} = 10^9 \frac{\text{electrons}}{\text{pulse}} \times 1.6 \times 10^{-19} \frac{\text{coulomb}}{\text{electron}} \times 10^5 \frac{\text{pulses}}{\text{second}}$$

$$= 1.6 \times 10^{-5}\,\text{A} = 0.016\,\text{mA} \qquad (9\text{-}5)$$

Because this current appears in discrete pulses, however, the peak current during a pulse is substantially higher. As an extreme case, assume we have a very fast organic scintillator whose decay time combined with the transit time spread in the PM tube produces a narrow electron pulse of 5 ns width. The peak pulse current is then approximately

$$I_{\text{peak}} = 10^9\,\text{electrons} \times 1.6 \times 10^{-19} \frac{\text{coulomb}}{\text{electron}} \times \frac{1}{5 \times 10^{-9}\,\text{s}}$$

$$= 0.032\,\text{A} = 32\,\text{mA} \qquad (9\text{-}6)$$

For photomultipliers operated in either current or pulse mode, the current through the voltage divider string should be kept large compared with the average dc signal current. In the example shown, this criterion can easily be met with a voltage supply capable of a few tenths of a milliampere. In pulse mode, however, the instantaneous demands in the last few stages of the PM tube can be many times greater. In the case of the example above, it would be impractical to attempt to design the divider string with a dc current much larger than the maximum pulse current as well. In that case, a divider string current of 10 or 20 mA might be used and stabilizing capacitors provided for the last few stages in which the peak pulse current is greater than 1–2 mA. Alternate schemes which employ zener diodes or transistors to stabilize the dynode voltages are described in Refs. 43 and 44.

The polarity (+ or −) of the high voltage used with PM tubes is in some sense an arbitrary choice. In the example given previously, the photocathode end of the divider string was grounded, which then required the application of positive high voltage to the anode end of the string. Exactly the same interdynode voltages can be achieved by grounding the opposite (or anode) end of the divider string and applying negative high voltage to the photocathode end. This latter arrangement is shown in Fig. 9-13b. The designer of the tube base circuit must choose one of these alternatives, and both are in common use. It is therefore important that users be aware of which convention has been chosen by the manufacturer of their own tube base before initial use of the equipment with a PM tube. Mistakenly applying the wrong polarity to a PM tube is not usually fatal to the tube, but electrons will refuse to swim uphill and the PM tube will not work. Designing the tube base for positive high voltage is in some ways simpler because it allows the photocathode of the PM tube to be grounded. Because this end is often in contact with a scintillation crystal whose cover must be exposed, it is convenient to keep the crystal at ground potential as well. Operating the tube with negative high voltage means that the photocathode will be at the full high voltage supplied to the tube, and care must be taken to prevent spurious pulses due to high-voltage leakage through the glass tube envelope to nearby grounded structures. A high electric field across the tube end window can also induce spurious pulses in some tubes due to electroluminescence in the glass. An advantage that stems from use of negative high voltage is the elimination of the coupling capacitor (C_c in Fig. 9-13a) required if positive polarity is used. The anode is now at ground potential, and signal pulses can be direct-coupled into subsequent measuring circuits. This advantage is particularly important for fast pulse applications in which it is often desirable to couple the anode directly into a 50 ohm transmission line structure.

Because the gain of a PM tube is extremely sensitive to changes in voltage, it is very important that sources of high voltage be well regulated and free of ripple. Drifts in the high voltage will show up as corresponding changes in tube gain and can, if sufficiently large, deteriorate the energy resolution of a scintillation counter. Any ripple is likely to be superimposed on the output signal, especially if the tube is operated with the photo-cathode grounded. In that case, any fluctuations in the supply voltage can be capacitively coupled directly onto the signal output. If the tube is operated with negative high voltage, the anode end of the tube is at ground potential, and ripples in the high voltage supply are important only to the extent that they modulate the overall gain of the tube.

The very large changes in gain that accompany changes in voltage with a PM tube are often a great convenience in setting up a counting system. Although the timing properties of the tube are optimized only when operated near the recommended voltage, other general properties such as linearity and relative signal-to-noise are not seriously changed over wide ranges of voltage. Therefore, it is often possible to operate a PM tube well below its recommended voltage without appreciably hurting its performance. The experimenter can therefore conveniently change the gain of the PM tube by orders of magnitude to suit the needs of the remainder of the signal chain simply by changing the applied voltage. This procedure is often followed in simple scintillation counters to determine counting plateaus and optimum operating conditions.

In those cases in which the tube voltage is likely to be varied over wide regions, the performance sometimes can suffer at the lower voltages due to reduced gain at the critical first dynode. It is here that the statistical properties of the PM tube are largely determined. In these cases a zener diode is sometimes used in conjunction with the divider string to hold the voltage between the photocathode and first dynode at a constant value, whereas the remainder of the divider string is varied.

In the usual case, the electrical signal from a PM tube is derived from the anode. The burst of electrons in the last stage of the PM tube is simply collected, resulting in a charge Q of negative polarity. As shown in Section VII of this chapter, a corresponding voltage pulse is then developed across the load resistance in the anode circuit. In some specialized applications, there may be some advantage in deriving an alternative signal from a preceding dynode. At each dynode stage, a net positive charge is induced during a pulse since more electrons leave the stage than were originally collected. This positive charge becomes progressively larger for each succeeding dynode stage due to the electron multiplication produced by each dynode. If a load resistor is placed between a given dynode and the voltage divider chain shown in Fig. 9-13, then a positive voltage pulse of smaller amplitude than the negative anode pulse can be observed. There has been some demonstration[25, 45, 46] that a dynode signal taken several stages before the anode has some advantage in fast timing measurements. However, the vast majority of scintillation pulses are taken from the PM tube anode, and the analysis of pulse shape beginning on p. 278 applies equally well to dynode or anode pulses. For simplicity, the analysis assumes that the anode pulse also corresponds to a positive charge Q, even though in practice the polarity of the anode pulse will be negative.

B. Magnetic Shielding

The electron optics within a PM tube are particularly sensitive to stray magnetic fields because of the low average energy (on the order of 100 eV) of the electrons traveling from stage to stage. Even the influence of the earth's magnetic field is sufficient to have an appreciable effect on the trajectories of these electrons. In situations in which the tube is likely to be physically moved or brought near equipment with stray magnetic fields, it is essential that a magnetic shield be provided to prevent gain shifts of the PM tube. The most common form consists of a thin cylinder of mu-metal which fits closely around the outside glass envelope of the PM tube. For most tube designs, this shield must be held at photocathode potential in order to avoid noise due to electroluminescence in the glass envelope.

VI. PHOTODIODES AS SUBSTITUTES FOR PHOTOMULTIPLIER TUBES

A. Potential Advantages

Photomultiplier tubes are the most common light amplifiers used with scintillators, both in pulse and current mode operation. However, recent advances in the development of semiconductor photodiodes have led to the substitution of newly available solid-state devices for PM tubes in some applications. In general, photodiodes offer the advantages of higher quantum efficiency (and therefore the potential for better energy resolution), lower power consumption, more compact size, and improved ruggedness compared with PM tubes used in scintillation counting. Photodiodes are also virtually insensitive to magnetic fields and therefore can sometimes be substituted in experiments where magnetic fields prevent the use of PM tubes. Because of the relatively small dimensions over which charges must move in these devices, their time response is comparable to that of conventional PM tubes, and they can be used to good advantage in coincidence and other timing applications.

There are two general designs that have received attention as possible substitutes for PM tubes. *Conventional photodiodes* have no internal gain and operate by directly converting the optical photons from the scintillation detector to electron–hole pairs that are simply collected. *Avalanche photodiodes* incorporate internal gain through the use of higher electric fields that increase the number of charge carriers that are collected.

B. Conventional Photodiodes

When light is incident on a semiconductor, electron–hole pairs are generated in a manner similar to that detailed in Chapter 11 for incident ionizing radiation. Photons corresponding to typical scintillation light carry about 3–4 eV of energy, sufficient to create electron–hole pairs in a semiconductor with bandgap of approximately 1–2 eV. (This process of conversion of visible light to electrical carriers is the basis of the common solar cell.) The conversion is not limited by the need for charge carriers to escape from a surface as in a conventional photocathode, so the quantum efficiency of the process can be as high as 60–80%, several times larger than in a PM tube. However, the absence of internal gain results in a very small signal size. In a typical scintillation event, only a few

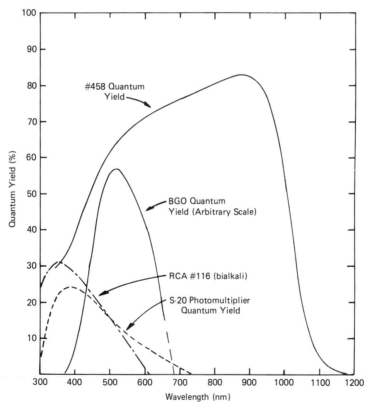

Figure 9-14 A comparison of the quantum efficiency of a silicon photodiode (labeled #458) with representative bialkali and S-20 photocathode quantum efficiencies. The emission spectrum from a BGO scintillator is shown for reference. (From Groom.[53])

thousand visible photons are produced, so the size of the charge pulse that can be developed is limited to no more than the same number of electronic charges. Because of the small signal amplitude, noise from fluctuations in thermally generated charge carriers (the dark current) is a major problem in pulse mode operation, especially for large-area detectors and low-energy radiations. In an extensive study of the performance of many different scintillators with a silicon photodiode, Sakai[5] measured energy resolution values at 662 keV that were always much poorer (often by a factor of 2) than in equivalent measurements with a PM tube. Differences were smaller for higher-energy gamma rays, but the PM tube performance was always superior. Successful applications to date have largely been limited to high-energy radiations[47-49] and/or small-diameter diodes[50,51] for which the associated dark current and capacitance are also small. In current mode, the cumulative effect of many scintillation events at high rates can override the inherent noise, resulting in excellent operational characteristics. For example, photodiodes have become the light detector of choice for current mode scintillators used in X-ray CT scanners for medical imaging.[52]

The spectral response of a typical silicon photodiode is plotted in Fig. 9-14. The quantum efficiency reaches higher values and extends much farther into the long wavelength region than that of typical photocathodes. This extended spectral response is particularly important for scintillators (such as CsI(Tl) or BGO) with emission spectra that have significant yield at longer wavelengths (see Fig. 8-7). The relative light yields for various scintillation materials given previously in Table 8-3 were measured using a typical PM tube photocathode response, and the values change significantly when measured with a photodiode. For example, the pulse amplitude from CsI(Tl) is smaller than the

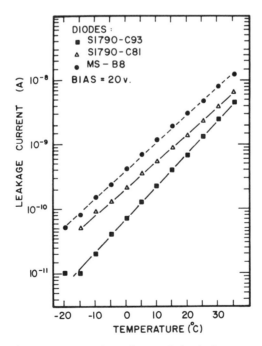

Figure 9-15 The temperature dependence of the leakage current for some typical silicon photodiodes. (From Bian et al.[57])

equivalent pulse from NaI(Tl) when measured with PM tubes but becomes over twice as large if the measurement is made with a silicon photodiode.[5]

Since there is no internal gain, photodiodes are very stable and rugged compared with PM tubes. Applied voltages of only tens or hundreds of volts are adequate for good charge collection, contrasted with voltages of 1000 V or more typically needed for PM tubes. A number of different semiconductor materials can be used for conventional photodiodes, including silicon, germanium, and, more recently, mercuric iodide.[54,55] The last is an example of a wide bandgap semiconductor in which very low dark currents are attainable at room temperature. The spectral response is also very broad, and a quantum efficiency approaching 90% is realizable between 380 and 550 nm. The energy resolution can therefore be excellent, and a value of just under 5% at 662 keV has been demonstrated[56] with a 2.54 cm diameter HgI_2 photodiode coupled to a CsI(Tl) scintillator. This performance is significantly better than typically seen with PM tubes and represents something of a world's record in the energy resolution of any type of scintillation spectrometer.

For materials with narrower bandgap, such as silicon or germanium, the dark current at room temperature is large enough to result in a noise figure that is generally an order of magnitude larger than that in an equivalent PM tube. Because the dark current is due to thermally generated charge carriers, there is a strong dependence on temperature as shown in Fig. 9-15. The noise level from a photodiode can therefore be strongly reduced by cooling the device. The rapid rise in dark current above room temperature has generally prevented the use of silicon photodiodes in applications requiring operation at elevated temperatures.

C. Avalanche Photodiodes

The small amount of charge that is produced in a conventional photodiode by a typical scintillation event can be increased through an *avalanche* process that occurs in a semiconductor at high values of the applied voltage. The charge carriers are accelerated sufficiently between collisions to create additional electron–hole pairs along the collection path, in much the same way that gas multiplication occurs in a proportional counter. (This same process is described in Chapter 13 in connection with *avalanche diode* detectors for ionizing radiations.) The internal gain helps pull the signal up from the noise level and permits good energy resolution in pulse mode at lower radiation energy than possible using conventional photodiodes. Because the gain factor is very sensitive to applied voltage, avalanche photodiodes require well-regulated high-voltage supplies for stable operation. For current mode applications, the inherent stability of conventional photodiodes (without gain) is usually preferred.

Silicon avalanche photodiodes have recently become commercially available[58] in relatively large diameter (up to 4.4 cm). In the past, they have been plagued by nonuniformity of the gain across the active area of the device. Recent improvements have been made possible by the use of silicon that has been doped with a very high degree of uniformity through neutron transmutation doping. Excellent spectroscopic results have been obtained[59,60] from scintillators of practical size coupled to avalanche photodiodes that offer promise for the future substitution of these devices for PM tubes in routine scintillation counting.

Both conventional and avalanche photodiodes respond directly to ionizing radiation in a manner similar to that described in Chapter 11 for silicon junction detectors. In scintillation spectroscopy, such interactions are often an undesirable background that can interfere with interpretation of the information recorded from the scintillator. This background can be minimized by avoiding direct irradiation of the photodiode through shielding and by choosing diode-sensitive thickness (the *depletion depth*) no larger than needed. Because the capacitance of the diode (and therefore the noise level of the diode–preamplifier system) decreases with increasing depletion depth, some compromise must then be struck in choosing this parameter.

VII. SCINTILLATION PULSE SHAPE ANALYSIS

The shape of the voltage pulse produced at the anode of a PM tube following a scintillation event depends on the time constant of the anode circuit. As discussed in Chapter 4, two extremes can be identified, both of which are commonly used in connection with scintillation counting. The first corresponds to those situations in which the time constant is chosen to be large compared with the decay time of the scintillator. This is the situation usually chosen if good pulse height resolution is a major objective and pulse rates are not excessively high. Then each pulse of electrons is integrated by the anode circuit to produce a voltage pulse whose amplitude is equal to Q/C, the ratio of the collected electron charge to the anode circuit capacitance. The second extreme is obtained by setting the anode circuit time constant to be much smaller than the scintillator decay time. As the following analysis will show, a much faster pulse results, which can often be an advantage in fast timing applications or when high pulse rates are encountered. At the same time, a sacrifice is then made in pulse amplitude and resolution.

The anode circuit can be idealized as shown in Fig. 9-16. C represents the capacitance of the anode itself, plus capacitance of the connecting cable and input capacitance of the circuit to which the anode is connected. The load resistance R may be a physical resistor wired into the tube base (see Fig. 9-13) or, if none is provided, the input impedance of the connected circuit. The current flowing into the anode $i(t)$ is simply the current of electrons from a single pulse, assumed to begin at $t = 0$. The shape of $i(t)$ will obviously influence the shape of the anode voltage pulse, and we choose for analysis a simplified representation of a typical electron pulse following a scintillation event. The principal component of emitted light from most scintillators can be adequately represented as a simple exponential decay. If the spread in transit time of the PM tube is small compared with this decay time, then a realistic model of the electron current arriving at the PM tube anode is simply

$$i(t) = i_0 e^{-\lambda t} \tag{9-7}$$

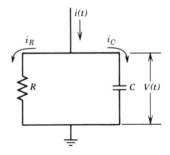

Figure 9-16 Simple parallel *RC* circuit representing a PM tube anode circuit.

where λ is the scintillator decay constant. The initial current i_0 can be expressed in terms of the total charge Q collected over the entire pulse by noting

$$Q = \int_0^\infty i(t)\, dt = i_0 \int_0^\infty e^{-\lambda t}\, dt = \frac{i_0}{\lambda} \tag{9-8}$$

Therefore

$$i_0 = \lambda Q \tag{9-9}$$

and

$$i(t) = \lambda Q e^{-\lambda t} \tag{9-10}$$

To derive the voltage pulse $V(t)$ expected at the anode, we first note that the current flowing into the parallel RC circuit must be the sum of the current flowing into the capacitance i_c and the current through the resistance i_R

$$i(t) = i_C + i_R \tag{9-11}$$

$$i(t) = C\frac{dV(t)}{dt} + \frac{V(t)}{R} \tag{9-12}$$

Inserting Eq. (9-10) for $i(t)$ and dividing by C, we obtain

$$\frac{dV(t)}{dt} + \frac{1}{RC}V(t) = \frac{\lambda Q}{C}e^{-\lambda t} \tag{9-13}$$

The solution to this first-order inhomogeneous differential equation with initial condition $V(0) = 0$ can be shown to be

$$V(t) = \frac{1}{\lambda - \theta} \cdot \frac{\lambda Q}{C}(e^{-\theta t} - e^{-\lambda t}) \tag{9-14}$$

where $\theta \equiv 1/RC$ is the reciprocal of the anode time constant.

Case 1. Large Time Constant

If the anode time constant is made large compared with the scintillator decay time, then $\theta \ll \lambda$ and Eq. (9-14) can be approximated by

$$V(t) \cong \frac{Q}{C}(e^{-\theta t} - e^{-\lambda t}) \tag{9-15}$$

A plot of this pulse form is shown in Fig. 9-17. Because $\theta \ll \lambda$, the first exponential in Eq. (9-15) decays slowly and the short time behavior is approximately

$$V(t) \cong \frac{Q}{C}(1 - e^{-\lambda t}) \qquad \left(t \ll \frac{1}{\theta}\right) \tag{9-16}$$

After a sufficiently long time, the second exponential decays to zero, and the long-time behavior is determined by the first exponential:

$$V(t) \cong \frac{Q}{C}e^{-\theta t} \qquad \left(t \gg \frac{1}{\lambda}\right) \tag{9-17}$$

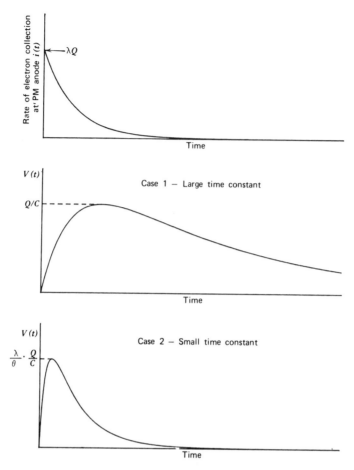

Figure 9-17   For the assumed exponential light pulse shown at the top, plots are given of the anode pulse $V(t)$ for the two extremes of large and small anode time constant. The duration of the pulse is shorter for Case 2, but the maximum amplitude is much smaller.

The following important observations can now be made:

1. The leading edge of the pulse has the time behavior $(1 - e^{-\lambda t})$ and its rise time therefore is determined by the scintillator decay constant λ. Fast scintillators have large λ values that lead to fast-rising pulses.

2. The tail of the pulse has the time behavior $e^{-\theta t}$ and therefore decays away at a rate determined by the anode circuit time constant $RC \equiv 1/\theta$.

3. The amplitude of the pulse is given simply by Q/C, but this value is reached only if $\theta \ll \lambda$. Restated, the anode circuit time constant must be large compared with the scintillator decay time.

Most scintillation counting is carried out in this mode because the pulse height is maximized and subsequent sources of noise will have minimum degrading effect on pulse height resolution. Furthermore, the pulse amplitude achieved is not sensitive to changes in load resistance or to small changes in time characteristics of the electron pulse.

The experimenter must then choose a time constant that is at least 5–10 times greater than the scintillator decay time but that is not excessively long to prevent needless pulse pileup with the tail from a preceding pulse at high rates. The time constant is determined by the product RC, and either the anode capacitance or load resistance can be varied to change its value. In most applications, however, it is the resistance that must be tailored to achieve the desired time constant because the capacitance is intentionally held at its minimum value to maximize the pulse amplitude (Q/C).

Case 2. Small Time Constant

In the opposite extreme, the anode time constant is set at a small value compared with the scintillator decay time, or $\theta \gg \lambda$. Now Eq. (9-14) becomes

$$V(t) = \frac{\lambda}{\theta} \cdot \frac{Q}{C}(e^{-\lambda t} - e^{-\theta t}) \tag{9-15a}$$

This pulse shape is also graphed in Fig. 9-17. The behavior at small values of t is now

$$V(t) \cong \frac{\lambda}{\theta} \cdot \frac{Q}{C}(1 - e^{-\theta t}) \qquad \left(t \ll \frac{1}{\lambda}\right) \tag{9-16a}$$

whereas for large t

$$V(t) \cong \frac{\lambda}{\theta} \cdot \frac{Q}{C}e^{-\lambda t} \qquad \left(t \gg \frac{1}{\theta}\right) \tag{9-17a}$$

The following general conclusions now apply:

1. The leading edge of the pulse has the time behavior $(1 - e^{-\theta t})$, which is determined by the anode time constant $RC \equiv 1/\theta$.
2. The tail of the pulse has the time behavior $e^{-\lambda t}$, which is identical to that of the scintillator light.
3. The maximum amplitude of the pulse is now $(\lambda Q/\theta C)$, a great deal smaller than the Case 1 maximum (Q/C) because, by definition of Case 2, $\lambda \ll \theta$.

The voltage pulse is now of much shorter duration than in Case 1, and its shape approaches that of the scintillator-produced PM tube current as the time constant is made smaller and smaller. This short duration comes at the price of a much reduced pulse amplitude, which varies linearly with the time constant and inversely with the scintillator decay time. Under fixed conditions, however, the pulse amplitude still is a linear measure of the charge Q collected by the anode, although it is more likely to be subject to fluctuations due to noise and component instabilities.

The simplified model we have used assumes a continuous and smooth current $i(t)$, which does not fully represent the discreteness or "clumped" nature of the anode current that ultimately arises from discrete photoelectrons. In Case 1, effects of the discreteness are largely smoothed out by the current integration process that takes place. In Case 2, however, no integration is carried out and the pulse form is much more sensitive to fluctuations that originate from the statistical nature of the photoelectron production. These fluctuations in pulse shape and amplitude are most serious for weak scintillation events, which produce only a small number of photoelectrons.

VIII. POSITION-SENSING PHOTOMULTIPLIER TUBES

In ordinary scintillation counting, the light from a typical scintillation event is spread over the entire PM tube photocathode, and therefore the point of interaction within the photocathode is of no particular interest. However, there are some specialized applications, for example, in particle physics or in nuclear medicine, in which the position of the arriving photon on the photocathode carries some information. For these applications, there has been some recent interest in developing PM tubes that can provide some position information regarding the incident light.

A first requirement in any such design is to use an electron-multiplying structure that maintains spatial separation between the multiplied electron clouds originating from photoelectrons generated at separate locations on the photocathode. This requirement rules out the common multiplier structures such as those illustrated in Fig. 9-7 in which electrons from all areas of the photocathode are focused onto a common dynode. However, the multichannel plate configuration discussed earlier in this chapter does preserve the spatial position of the multiplied charge and can be used in position-sensitive tubes. Alternatively, newer electron multipliers have been developed for this purpose,[61–65] which consist of multiple layers of a fine mesh made from secondary-electron-emitting material. Each layer serves the purpose of a separate dynode stage, but the trajectories of electrons are confined to a small area of the mesh as they travel from the photocathode to the anode.

At the anode, several options are available to encode the position information. The simplest method is to segment the anode into separate elements, each segment collecting only those events generated in its vicinity.[63,66] Alternatively, other position-sensing methods similar to those used in gas-filled detectors can be employed, such as the use of resistive anodes[67] or orthogonal layers of crossed anode wires.[62] Position-sensing PM tubes of various designs are now becoming commercially available and will likely play an important part in future designs of scintillation instruments used for radiation imaging.

IX. PHOTOIONIZATION DETECTORS

There is another alternative to PM tubes that is undergoing active development for applications in which the light to be detected is in the ultraviolet portion of the spectrum. Certain organic compounds in the gaseous phase can be ionized by UV photons to form ion pairs. If the organic vapor is incorporated as a component of the fill gas of a conventional detector that is sensitive to ionization (such as a proportional counter), then the signal pulse amplitude will reflect the number of incoming photons that have undergone conversion to ions. Furthermore, if position-sensitive detectors such as multiwire proportional counters are used, the spatial position of the conversion point can also be determined.

Some organic compounds that are of interest as photoionization agents are listed in Table 9-2. Of these, TMAE and TEA have received the most attention. It has been shown that each can be successfully incorporated into standard proportional gases with typical concentration of up to 3%. In the case of TMAE, its vapor pressure of 0.35 torr at 20°C (Ref. 69) is a limit on its room temperature absolute concentration in the gas phase. At that partial pressure, its quantum efficiency is approximately 2% per mm of path length for the fast component of the emission spectrum (see Fig. 8-13) from BaF_2 (Ref. 70). The vapor reacts with oxygen and organic materials such as vacuum grease and O-rings, so that some care is needed in chamber design and gas handling procedures. TEA is

TABLE 9-2 Some Organics Potentially Useful in Photoionization Detectors

Acronym	Chemical Name	Vapor Phase Ionization Potential (eV)	Long-Wavelength Cutoff (nm)
TMAE	Tetrakis(dimethylamino)ethylene	5.36	231
TMBI	1,1′,3,3′-Tetramethyl-2,2′-biimidazolidinylidene	5.41	229
TMAB	1,1,4,4-Tetrakis(dimethylamino)butadiene	5.60	221
TMPD	N,N,N',N'-tetramethyl-p-phenylenediamine	6.20	200
TEA	Triethylamine	7.5	165

Data as tabulated by Policarpo.[68]

somewhat easier to handle and has a higher vapor pressure, but its higher photoionization potential limits its response to shorter wavelengths.

Photoionization detectors are as yet far from being commercially available, but they can offer interesting properties for specialized applications. They can be made in almost unlimited size with excellent uniformity of response over the large area entrance window. Good spatial resolution can also be achieved (of interest in imaging applications) by incorporating the position-sensing techniques discussed in Chapter 5 for large-area proportional detectors. They have found useful application in the detection of UV emissions from scintillators,[70,71] Cerenkov detectors,[72] and gas scintillation proportional counters.[68,72]

PROBLEMS

9-1. Calculate the long-wavelength limit of the sensitivity of a photocathode layer with work function of 1.5 eV.

9-2. Find the transit time for an electron between typical dynodes in a PM tube if the interdynode spacing is 12 mm and the potential difference is 150 V per stage. For simplicity, assume a uniform electric field.

9-3. Using the data plotted in Fig. 9-3, find the total applied voltage necessary for a PM tube with a six-stage multiplier using GaP(Cs) dynodes to achieve an electron gain factor of 10^6.

9-4. The dark current from a PM tube with electron gain of 10^6 is measured to be 2 nA. What is the corresponding electron emission rate from the photocathode?

9-5. The gain per dynode δ of a 10-stage PM tube varies as $V^{0.6}$, where V is the interdynode voltage. If the tube is operated at an overall voltage of 1000 V, how much voltage fluctuation can be tolerated if the gain is not to change by more than 1%?

9-6. The decay time (inverse of the decay constant) for scintillations in NaI(Tl) is 230 ns. Neglecting any time spread introduced by the PM tube, find the maximum value of the voltage pulse amplitude for anode circuit time constants of 10, 100, and 1000 ns. Express your answer as a ratio to the amplitude that would be observed for an infinite time constant. What is the minimum value of the time constant if this ratio is to be at least 0.9?

9-7. A current pulse of the following shape flows into a parallel *RC* circuit:

$$i(t) = I \quad 0 \le t < T$$
$$i(t) = 0 \quad T \le t$$

Find the general solution $V(t)$ for the voltage appearing across the circuit, assuming $V(0) = 0$. Sketch the solution in the two limits:

(a) $RC \gg T$.
(b) $RC \ll T$.

9-8. Calculate the amplitude of the signal pulse expected from a NaI(Tl)–PM tube combination under the following circumstances:

Radiation energy loss: 1.2 MeV.

Light collection efficiency: 70%.

Photocathode quantum efficiency: 20%.

PM tube electron gain: 100,000.

Anode capacitance: 100 pF.

Anode load resistance: 10^5 ohms.

Any other physical parameters you may need can be found in the text.

9-9. What is the principal advantage of microchannel plate PM tubes compared with more conventional designs?

9-10. The bandgap energy in silicon at room temperature is 1.11 eV. Calculate the longest wavelength of light that is energetically capable of exciting an electron across this gap to create an electron–hole pair in a photodiode.

REFERENCES

1. *RCA Photomultiplier Manual*, Technical Series PT-61, RCA Solid State Division, Electro-Optics and Devices, Lancaster, PA, 1970.
2. G. A. Morton, *IEEE Trans. Nucl. Sci.* **NS-22**(1), 26 (1975).
3. "Test Procedures for Photomultipliers for Scintillation Counting and Glossary for Scintillation Counting Field," IEEE Standard 398-1972, (1972).
4. M. Miyajima, S. Sasaki, and E. Shibamura, *Nucl. Instrum. Meth.* **224**, 331 (1984).
5. E. Sakai, *IEEE Trans. Nucl. Sci.* **NS-34**(1), 418 (1987).
6. B. F. Williams, *IEEE Trans. Nucl. Sci.* **NS-19**(3), 39 (1972).
7. H. R. Krall, F. A. Helvy, and D. E. Persyk, *IEEE Trans. Nucl. Sci.* **NS-17**(3), 71 (1970).
8. R. E. Simon and B. F. Williams, *IEEE Trans. Nucl. Sci.* **NS-15**(3), 167 (1968).
9. G. A. Morton, H. M. Smith, Jr., and H. R. Krall, *IEEE Trans. Nucl. Sci.* **NS-16**(1), 92 (1969).
10. B. Leskovar and C. C. Lo, *IEEE Trans. Nucl. Sci.* **NS-19**(3), 50 (1972).
11. H. R. Krall and D. E. Persyk, *IEEE Trans. Nucl. Sci.* **NS-19**(3), 45 (1972).
12. H. Houtermans, *Nucl. Instrum. Meth.* **112**, 121 (1973).
13. T. Sandor, *Nucl. Instrum. Meth.* **78**, 8 (1970).
14. R. Bosshard, R. Rausch, M. Sauce, C. Zajde, and G. Amsel, *IEEE Trans. Nucl. Sci.* **NS-19**(3), 107 (1972).
15. F. W. Inman and J. J. Muray, *IEEE Trans. Nucl. Sci.* **NS-16**(2), 62 (1969).
16. J. R. Prescott, *Nucl. Instrum. Meth.* **39**, 173 (1966).

17. J. P. Ballini, *Nucl. Instrum. Meth.* **116**, 109 (1974).

18. J. Ballini, P. Cazes, and P. Turpin, *Nucl. Instrum. Meth.* **134**, 319 (1976).

19. B. Leskovar, *Phys. Today* **30**(11), 42 (1977).

20. G. Pietri, *IEEE Trans. Nucl. Sci.* **NS-22**(5), 2084 (1975).

21. S. Dhawan and R. Majka, *IEEE Trans. Nucl. Sci.* **NS-24**(1), 270 (1977).

22. C. C. Lo, P. Lecomte, and B. Leskovar, *IEEE Trans. Nucl. Sci.* **NS-24**(1), 302 (1977).

23. G. Pietri, *IEEE Trans. Nucl. Sci.* **NS-24**(1), 228 (1977).

24. J. Uyttenhove, J. Demuynck, and A. Deruytter, *IEEE Trans. Nucl. Sci.* **NS-25**(1), 566 (1978).

25. B. Leskovar and C. C. Lo, *IEEE Trans. Nucl. Sci.* **NS-25**(1), 582 (1978).

26. H. Kume et al., *IEEE Trans. Nucl. Sci.* **NS-33**(1), 364 (1986).

27. F. de la Barre, *Nucl. Instrum. Meth.* **102**, 77 (1972).

28. B. Sipp and J. A. Miehe, *Nucl. Instrum. Meth.* **114**, 249 (1974).

29. B. Leskovar and C. C. Lo, *Nucl. Instrum. Meth.* **123**, 145 (1975).

30. F. A. Johnson, *Nucl. Instrum. Meth.* **87**, 215 (1970).

31. G. A. Morton, H. M. Smith, and R. Wasserman, *IEEE Trans. Nucl. Sci.* **NS-14**(1), 443 (1967).

32. S. J. Hall and J. McKeown, *Nucl. Instrum. Meth.* **112**, 545 (1973).

33. R. Staubert, E. Bohm, K. Hein, K. Sauerland, and J. Trumper, *Nucl. Instrum. Meth.* **84**, 297 (1970).

34. S. S. Stevens and J. W. Longworth, *IEEE Trans. Nucl. Sci.* **NS-19**(1), 356 (1972).

35. G. P. Lamaze, J. K. Whittaker, R. A. Schrack, and O. A. Wasson, *Nucl. Instrum. Meth.* **123**, 403 (1975).

36. J. M. Paul, *Nucl. Instrum. Meth.* **89**, 285 (1970).

37. H. J. Kellermann, H. O. Klages, R. Langkau, U. Paschen, and J. Remen, *Nucl. Instrum. Meth.* **115**, 301 (1974).

38. A. K. Gupta and N. Nath, *Nucl. Instrum. Meth.* **53**, 352 (1967).

39. C. Weitkamp, G. G. Slaughter, W. Michaelis, and H. Schmidt, *Nucl. Instrum. Meth.* **61**, 122 (1968).

40. M. Yamashita, *Nucl. Instrum. Meth.* **142**, 435 (1977).

41. L. Cathy, "Control of Fatigue in Photomultipliers," DP-642 (1961).

42. R. D. Conner and M. K. Husain, *Nucl. Instrum. Meth.* **6**, 337 (1960).

43. R. D. Hiebert, H. A. Thiessen, and A. W. Obst, *Nucl. Instrum. Meth.* **142**, 467 (1977).

44. C. R. Kerns, *IEEE Trans. Nucl. Sci.* **NS-24**(1), 353 (1977).

45. J. De Vries and F. E. T. Kelling, *Nucl. Instrum. Meth. Phys. Res.* **A262**, 385 (1987).

46. B. Bengtson and M. Moszynski, *Nucl. Instrum. Meth.* **204**, 129 (1982).

47. R. Glasow et al., *Nucl. Instrum. Meth.* **228**, 354 (1985).

48. G. Hall et al., *IEEE Trans. Nucl. Sci.* **NS-33**(1), 310 (1986).

49. W. G. Gong, Y. D. Kim, G. Poggi, Z. Chen, C. K. Gelbke, W. G. Lynch, M. R. Maier, T. Murakami, M. B. Tsang, H. M. Xu, and K. Kwiatkowski, *Nucl. Instrum. Meth. Phys. Rev.* **A268**, 190 (1988).

50. S. E. Derenzo, *Nucl. Instrum. Meth.* **219**, 117 (1984).

51. H. Grassmann et al., *Nucl. Instrum. Meth.* **A234**, 122 (1985).

52. D. P. Boyd, *IEEE Trans. Nucl. Sci.* **NS-26**(2), 2836 (1979).

53. D. E. Groom, *Nucl. Instrum. Meth.* **219**, 141 (1984).

54. J. S. Iwanczyk et al., *Nucl. Instrum. Meth.* **213**, 123 (1983).

55. J. Markakis et al., *IEEE Trans. Nucl. Sci.* **NS-32**(1), 559 (1985).

56. J. Markakis, *IEEE Trans. Nucl. Sci.* **NS-35**(1), 356 (1988).

57. Z. Bian, J. Dobbins, and N. Mistry, *Nucl. Instrum. Meth.* **A239**, 518 (1985).

58. M. R. Squillante, G. Reiff, and G. Entine, *IEEE Trans. Nucl. Sci.* **NS-32**(1), 563 (1985).

59. R. Lecomte et al., *IEEE Trans. Nucl. Sci.* **NS-32**(1), 482 (1985).

60. A. W. Lightstone et al., *IEEE Trans. Nucl. Sci.* **33**(1), 456 (1986).

61. M. Salomon and S. S. A. Williams, *Nucl. Instrum. Meth.* **A241**, 210 (1985).

62. H. Kume, S. Muramatsu, and M. Iida, *IEEE Trans. Nucl. Sci.* **NS-33**(1), 359 (1986).

63. S. Suzuki, T. Matsushita, T. Suzuki, S. Kimura, and H. Kume, *IEEE Trans. Nucl. Sci.* **NS-35**(1), 382 (1988).

64. J. Vallerga, J. Hull, and M. Lampton, *IEEE Trans. Nucl. Sci.* **NS-35**(1), 539 (1988).

65. H. Uchida et al., *IEEE Trans. Nucl. Sci.* **NS-33**(1), 464 (1986).

66. M. Salomon and J. Marans, *IEEE Trans. Nucl. Sci.* **NS-33**(1), 254 (1986).

67. M. Lampton and C. W. Carlson, *Rev. Sci. Instrum.* **50**(9), 91 (1979).

68. A. J. P. L. Policarpo, *Nucl. Instrum. Meth.* **196**, 53 (1982).

69. D. F. Anderson, *IEEE Trans. Nucl. Sci.* **NS-28**(1), 842 (1981).

70. C. L. Woody, C. I. Petridou, and G. C. Smith, *IEEE Trans. Nucl. Sci.* **NS-33**(1), 136 (1986).

71. D. F. Anderson et al., *Nucl. Instrum. Meth.* **225**, 8 (1984).

72. D. F. Anderson, *IEEE Trans. Nucl. Sci.* **NS-32**(1), 495 (1985).

CHAPTER · 10

Radiation Spectroscopy with Scintillators

In the early 1950s, the availability of thallium-activated sodium iodide as a scintillation material ushered in the modern era of gamma-ray spectroscopy. With its introduction, a practical detector was available which could provide a high efficiency for the detection of gamma rays and, at the same time, was capable of sufficiently good energy resolution to be useful in separating the contributions of polyenergetic gamma-ray sources. Gamma-ray spectroscopy using scintillators has since developed into a mature science with applications in an impressive array of technical fields.

Despite the fact that it was virtually the first practical solid detection medium used for gamma-ray spectroscopy, NaI(Tl) remains the most popular scintillation material for this application. This extraordinary success stems from its extremely good light yield, excellent linearity, and the high atomic number of its iodine constituent. Other scintillators mentioned in Chapter 8, notably cesium iodide and BGO, have also achieved some success in gamma-ray spectroscopy, but the combined use of all other materials extends to only a small fraction of the cases in which sodium iodide is found. In the sections that follow, we concentrate on NaI(Tl), with the understanding that most of the discussion and general conclusions can be extended to other scintillation materials by taking into account the differences in their gamma-ray interaction probabilities and scintillation properties.

Useful textbook reviews of scintillation spectroscopy of gamma radiation have been published by Birks,[1] Shafroth,[2] and Siegbahn.[3]

I. GENERAL CONSIDERATIONS IN GAMMA-RAY SPECTROSCOPY

An X-ray or gamma-ray photon is uncharged and creates no direct ionization or excitation of the material through which it passes. The detection of gamma rays is therefore critically dependent on causing the gamma-ray photon to undergo an interaction that transfers all or part of the photon energy to an electron in the absorbing material. These interaction processes are detailed in Chapter 2 and represent sudden and major alterations of the photon properties, as opposed to the continuous slowing down of charged particles or electrons through many simultaneous interactions.

Because the primary gamma-ray photons are "invisible" to the detector, it is only the fast electrons created in gamma-ray interactions that provide any clue to the nature of the incident gamma rays. These electrons have a maximum energy equal to the energy of the incident gamma-ray photon and will slow down and lose their energy in the same manner as any other fast electron, such as a beta particle. Energy loss is therefore through ionization and excitation of atoms within the absorber material and through bremsstrahlung emission (see Chapter 2).

In order for a detector to serve as a gamma-ray spectrometer, it must carry out two distinct functions. First, it must act as a conversion medium in which incident gamma rays have a reasonable probability of interacting to yield one or more fast electrons; second, it must function as a conventional detector for these secondary electrons. In the discussion that follows, we first assume that the detector is sufficiently large so that the escape of secondary electrons (and any bremsstrahlung created along their track) is not significant. For incident gamma rays of a few MeV, the most penetrating secondary electrons will also be created with a few MeV kinetic energy. The corresponding range in typical solid detector media is a few millimeters (most bremsstrahlung photons generated along the electron track will be considerably less penetrating). The assumption of complete electron absorption therefore implies a detector whose minimum dimension is at least about a centimeter. Then only a small fraction of the secondary electrons, which are created more or less randomly throughout the volume of the detector, lie within one range value of the surface and could possibly escape. Later in this chapter we discuss the complicating effects of electron and bremsstrahlung escape in small detectors as a perturbation on the simpler model that follows.

The following discussions are kept relatively general so that they apply not only to other scintillation materials but also to other solid or liquid detection media used in gamma-ray spectroscopy. Chapters 12 and 13 discuss semiconductor detectors, which have been widely applied to gamma-ray spectroscopy over the past decade. The following section also serves as a general introduction to these chapters, because the basic modes of gamma-ray interactions are identical for all detector types.

The requirement of full absorption of the secondary electron rules out gas-filled detectors for the spectroscopy of gamma rays, other than those with very low energy. The penetration distance of a 1 MeV electron in STP gases is several meters, so that detectors of any practical size can never come close to absorbing all the secondary electron energy. To complicate the situation further, most gamma-ray-induced pulses from a gas-filled counter arise from gamma-ray interactions taking place in the solid counter wall, following which the secondary electron finds its way to the gas. Under these conditions, the electron loses a variable and an indeterminant amount of energy in the wall, which does not contribute to the detector output pulse, and therefore virtually all hope of relating the electron to incident gamma-ray energy is lost.

II. GAMMA-RAY INTERACTIONS

Of the various ways gamma rays can interact in matter, only three interaction mechanisms have any real significance in gamma-ray spectroscopy: photoelectric absorption, Compton scattering, and pair production. As detailed in Chapter 2, photoelectric absorption predominates for low-energy gamma rays (up to several hundred keV), pair production predominates for high-energy gamma rays (above 5–10 MeV), and Compton scattering is the most probable process over the range of energies between these extremes. The atomic number of the interaction medium has a strong influence on the relative probabilities of these three interactions, as can be seen from the formulas and plots given in Chapter 2. The most striking of these variations involves the cross section for photoelectric absorption, which varies approximately as $Z^{4.5}$. As we shall see from the following discussion, because photoelectric absorption is the preferred mode of interaction, there is a premium on choosing detectors for gamma-ray spectroscopy from materials that incorporate elements with high atomic number.

A. Photoelectric Absorption

Photoelectric absorption is an interaction in which the incident gamma-ray photon disappears. In its place, a photoelectron is produced from one of the electron shells of the absorber atom with a kinetic energy given by the incident photon energy $h\nu$ minus the binding energy of the electron in its original shell (E_b). This process is shown in the diagram below. For typical gamma-ray energies, the photoelectron is most likely to emerge from the K shell, for which typical binding energies range from a few keV for low-Z materials to tens of keV for materials with higher atomic number.

The vacancy that is created in the electron shell as a result of the photoelectron emission is quickly filled by electron rearrangement. In the process, the binding energy is liberated either in the form of a characteristic X-ray or Auger electron. In iodine, a characteristic X-ray is emitted in about 88% of the cases.[4] The Auger electrons have extremely short range because of their low energy. The characteristic X-rays may travel some distance (typically a millimeter or less) before being reabsorbed through photoelectric interactions with less tightly bound electron shells of the absorber atoms. Although escape of these X-rays can at times be significant, for now we assume that they are also fully absorbed in keeping with our simplified model.

Thus, the effect of photoelectric absorption is the liberation of a photoelectron, which carries off most of the gamma-ray energy, together with one or more low-energy electrons corresponding to absorption of the original binding energy of the photoelectron. If nothing escapes from the detector, then the sum of the kinetic energies of the electrons that are created must equal the original energy of the gamma-ray photon.

Photoelectric absorption is therefore an ideal process if one is interested in measuring the energy of the original gamma ray. The total electron kinetic energy equals the incident gamma-ray energy and will always be the same if monoenergetic gamma rays are involved. Under these conditions, the differential distribution of electron kinetic energy for a series of photoelectric absorption events would be a simple delta function as shown below. The single peak appears at a total electron energy corresponding to the energy of the incident gamma rays.

B. Compton Scattering

The result of a Compton scattering interaction is the creation of a recoil electron and scattered gamma-ray photon, with the division of energy between the two dependent on the scattering angle. A sketch of the interaction is given below.

Before   After

The energy of the scattered gamma ray $h\nu'$ in terms of its scattering angle θ is given by

$$h\nu' = \frac{h\nu}{1 + \left(h\nu/m_0c^2\right)(1 - \cos\theta)} \tag{10-1}$$

where m_0c^2 is the rest mass energy of the electron (0.511 MeV). The kinetic energy of the recoil electron is therefore

$$E_{e^-} = h\nu - h\nu' = h\nu\left(\frac{\left(h\nu/m_0c^2\right)(1 - \cos\theta)}{1 + \left(h\nu/m_0c^2\right)(1 - \cos\theta)}\right) \tag{10-2}$$

Two extreme cases can be identified:

1. A grazing angle scattering, or one in which $\theta \cong 0$. In this case, Eqs. (10-1) and (10-2) predict that $h\nu' \cong h\nu$ and $E_{e^-} \cong 0$. In this extreme, the recoil Compton electron has very little energy and the scattered gamma ray has nearly the same energy as the incident gamma ray.

2. A head-on collision in which $\theta = \pi$. In this extreme, the incident gamma ray is backscattered toward its direction of origin, whereas the electron recoils along the direction of incidence. This extreme represents the maximum energy that can be transferred to an electron in a single Compton interaction. Equations (10-1) and (10-2) yield for this case

$$h\nu'|_{\theta=\pi} = \frac{h\nu}{1 + 2h\nu/m_0c^2} \tag{10-3}$$

$$E_{e^-}|_{\theta=\pi} = h\nu\left(\frac{2h\nu/m_0c^2}{1 + 2h\nu/m_0c^2}\right) \tag{10-4}$$

In normal circumstances, all scattering angles will occur in the detector. Therefore, a continuum of energies can be transferred to the electron, ranging from zero up to the maximum predicted by Eq. (10-4). Figure 10-1 shows the shape of the distribution of Compton recoil electrons predicted by the Klein–Nishina cross section (Chapter 2) for several different values of the incident gamma-ray energy. For any one specific gamma-ray

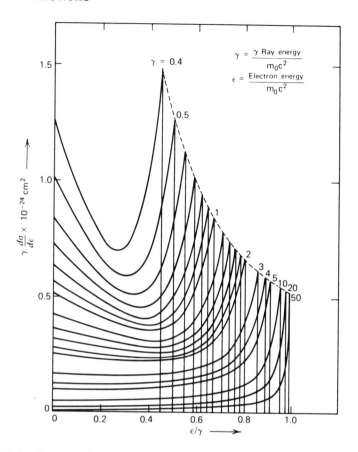

Figure 10-1 Shape of the Compton continuum for various gamma-ray energies. (From S. M. Shafroth (ed.), *Scintillation Spectroscopy of Gamma Radiation*. Copyright 1964 by Gordon & Breach, Inc. By permission of the publisher.

energy, the electron energy distribution has the general shape shown in the sketch below.

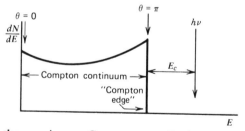

The gap between the maximum Compton recoil electron energy and the incident gamma-ray energy is given by

$$E_C \equiv h\nu - E_{e^-}|_{\theta = \pi} = \frac{h\nu}{1 + 2h\nu/m_0c^2} \qquad (10\text{-}5)$$

In the limit that the incident gamma-ray energy is large, or $h\nu \gg m_0c^2/2$, this energy difference tends toward a constant value given by

$$E_C \cong \frac{m_0c^2}{2}(= 0.256 \text{ MeV}) \qquad (10\text{-}6)$$

The preceding analysis is based on the assumption that Compton scattering involves electrons that are initially free or unbound. In actual detector materials, the binding energy of the electron prior to the scattering process can have a measurable effect on the shape of the Compton continuum. These effects will be particularly noticeable for low incident gamma-ray energy. They involve a rounding-off of the rise in the continuum near its upper extreme and the introduction of a finite slope to the abrupt drop of the Compton edge. These effects are often masked by the finite energy resolution of the detector but can be evident in the spectra from detectors with high inherent resolution (see Fig. 13-9).

C. Pair Production

The third significant gamma-ray interaction is pair production. The process occurs in the field of a nucleus of the absorbing material and corresponds to the creation of an electron–positron pair at the point of complete disappearance of the incident gamma-ray photon. Because an energy of $2m_0c^2$ is required to create the electron–positron pair, a minimum gamma-ray energy of 1.02 MeV is required to make the process energetically possible. If the incident gamma-ray energy exceeds this value, the excess energy appears in the form of kinetic energy shared by the electron–positron pair. Therefore, the process consists of converting the incident gamma-ray photon into electron and positron kinetic energies, which total

$$E_{e^-} + E_{e^+} = h\nu - 2m_0c^2 \tag{10-7}$$

For typical energies, both the electron and positron travel a few millimeters at most before losing all their kinetic energy to the absorbing medium. A plot of the total (electron + positron) charged particle kinetic energy created by the incident gamma ray is again a simple delta function, but it is now located $2m_0c^2$ below the incident gamma-ray energy, as illustrated in the sketch below. In our simple model, this amount of energy will be deposited each time a pair production interaction occurs within the detector. As introduced in the next section, this energy corresponds to the position of the *double escape peak* in actual gamma-ray pulse height spectra.

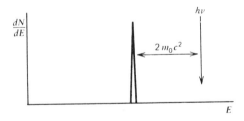

The pair production process is complicated by the fact that the positron is not a stable particle. Once its kinetic energy becomes very low (comparable to the thermal energy of normal electrons in the absorbing material), the positron will annihilate or combine with a normal electron in the absorbing medium. At this point both disappear, and they are replaced by two annihilation photons of energy m_0c^2 (0.511 MeV) each. The time required for the positron to slow down and annihilate is small, so that the annihilation radiation appears in virtual coincidence with the original pair production interaction.

III. PREDICTED RESPONSE FUNCTIONS

A. "Small" Detectors

As an example of one extreme in gamma-ray detector behavior, we first examine the expected response of detectors whose size is small compared with the mean free path of the *secondary gamma radiations* produced in interactions of the original gamma rays. These secondary radiations consist of Compton scattered gamma rays, together with annihilation photons formed at the end of the tracks of positrons created in pair production. Because the mean free path of the secondary gamma rays is typically of the order of several centimeters, the condition of "smallness" is met if the detector dimensions do not exceed 1 or 2 cm. At the same time, we retain our original simplifying assumption that all charged particle energy (photoelectron, Compton electron, pair electron, and positron) is completely absorbed within the detector volume.

The predicted electron energy deposition spectra under these conditions are illustrated in Fig. 10-2. If the incident gamma-ray energy is below the value at which pair production is significant, the spectrum results only from the combined effect of Compton scattering and photoelectric absorption. The continuum of energies corresponding to

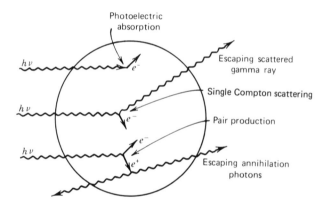

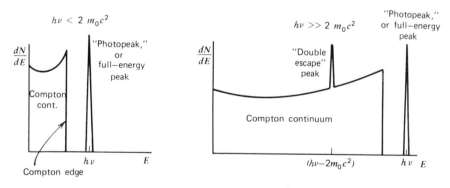

Figure 10-2 The "small detector" extreme in gamma-ray spectroscopy. The processes of photoelectric absorption and single Compton scattering give rise to the low-energy spectrum at the left. At higher energies, the pair production process adds a double escape peak shown in the spectrum at the right.

Compton scattered electrons is called the *Compton continuum*, whereas the narrow peak corresponding to photoelectrons is designated as the *photopeak*. For the "small" detector, only single interactions take place, and the ratio of the area under the photopeak to the area under the Compton continuum is the same as the ratio of the photoelectric cross section to the Compton cross section in the detector material.

If the incident gamma-ray energy is sufficiently high (several MeV), the results of pair production are also evident in the electron energy spectrum. For a small detector, only the electron and positron kinetic energies are deposited, and the annihilation radiation escapes. The net effect is to add a *double escape peak* to the spectrum located at an energy of $2m_0c^2$ ($\sim$ 1.02 MeV) below the photopeak. The term double escape refers to the fact that both annihilation photons escape from the detector without further interaction.

B. Very Large Detectors

As an opposite extreme case, imagine that gamma rays could be introduced near the center of a very large detector, perhaps in an arrangement resembling that of Fig. 10-3. The detector dimensions are now assumed to be sufficiently large so that all secondary radiations, including Compton scattered gamma rays and annihilation photons, also interact within the detector active volume and none escapes from the surface. For typical gamma-ray energies, this condition would translate into requiring detector dimensions on the order of many tens of centimeters, unrealistically large for most practical cases.

Nonetheless, it is helpful to see how increasing the detector size greatly simplifies its response function. Some typical histories, obtained by following a particular source gamma ray and all subsequent secondary radiation, are sketched in Fig. 10-3. If the initial interaction is a Compton scattering event, the scattered gamma ray will subsequently interact at some other location within the detector. This second interaction may also be a Compton scattering event, in which case a scattered photon of still lower energy is produced. Eventually, a photoelectric absorption will occur and the history is terminated at that point.

It is important to appreciate the small amount of time required for the entire history to take place. The primary and secondary gamma rays travel at the speed of light in the detector medium. If the average migration distance of the secondary gamma rays is of the order of 10 cm, the total elapsed time from start to finish of the history will be less than a nanosecond. This time is substantially less than the inherent response time of virtually all practical detectors used in gamma-ray spectroscopy. Therefore, the net effect is to create the Compton electrons at each scattering point and the final photoelectron in time coincidence. The pulse produced by the detector will therefore be the sum of the responses due to each individual electron. If the detector responds linearly to electron energy, then a pulse is produced which is proportional to the *total* energy of all the electrons produced along the history. Because nothing escapes from the detector, this total electron energy must simply be the original energy of the gamma-ray photon, no matter how complex any specific history may be. *The detector response is therefore the same as if the original gamma-ray photon had undergone a simple photoelectric absorption in a single step.*

The same type of argument can be used if the history involves a pair production event. The annihilation photons formed when the positron is stopped are now assumed to interact through Compton scattering or photoelectric absorption elsewhere in the detector. Again, if the detector is large enough to prevent any secondary radiation from escaping, the sum of the kinetic energies of the electron–positron pair and subsequent

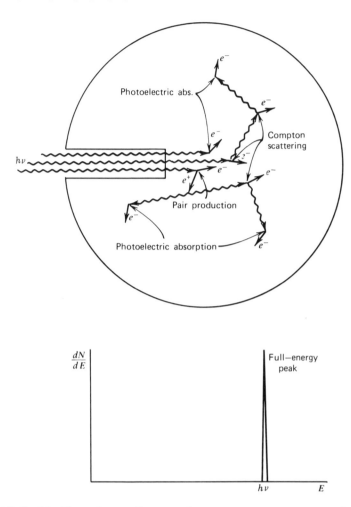

Figure 10-3 The "large detector" extreme in gamma-ray spectroscopy. All gamma-ray photons, no matter how complex their mode of interaction, ultimately deposit all their energy in the detector. Some representative histories are shown at the top.

Compton and photoelectrons produced by interaction of the annihilation radiation must equal the original gamma-ray photon energy. Therefore, the detector response is again simply proportional to the original gamma-ray energy.

The conclusion to be reached is therefore very simple: If the detector is sufficiently large and its response linearly dependent on electron kinetic energy, then the signal pulse is identical for all gamma-ray photons of the same energy, regardless of the details of each individual history. This circumstance is very fortunate because the detector response function now consists of the single peak shown in Fig. 10-3 rather than the more complex function shown in Fig. 10-2. The ability to interpret complex gamma-ray spectra involving many different energies is obviously enhanced when the response function consists of a single peak.

By common usage, the corresponding peak in the response function is called the *photopeak*, just as in the case of the small detector. It should be realized, however, that in addition to simple photoelectric events, much more complex histories involving multiple

Compton scattering or pair production also contribute pulses that fall within this peak. A better name is the *full-energy peak* because it represents all histories in which all of the original gamma-ray energy is fully converted to electron kinetic energy.

C. Intermediate Size Detectors

Real detectors of the sizes in common use for gamma-ray spectroscopy are neither small nor large by the standards given above. For usual geometries in which the gamma rays are incident externally on the surface of the detector, even large-volume detectors appear finite because some interactions will take place near the entrance surface. Normal detector response functions therefore combine some of the properties discussed for the two previous cases, as well as additional features related to *partial* recovery of the secondary gamma-ray energy. Some representative histories that illustrate these added possibilities are shown in Fig. 10-4, together with corresponding features in the response function.

The spectrum for low to medium gamma-ray energies (where pair production is not significant) again consists of a Compton continuum and photopeak. Now, however, the ratio of the area under the photopeak to that under the Compton continuum is significantly enhanced over that for the very small detector due to the added contribution of multiple events to the photopeak. The lower the incident gamma-ray energy, the lower will be the average energy of a Compton scattered photon and the corresponding average distance of migration. Thus, even detectors of moderate size will appear to be large, and the relative area under the photopeak increases with decreasing incident photon energy. At very low energies (say, < 100 keV) the Compton continuum may effectively disappear.

At medium energies, the possibility of multiple Compton scattering followed by escape of the final scattered photon can lead to a total energy deposition that is greater than the maximum predicted by Eq. (10-4) for single scattering. These multiple events can thus partially fill in the gap between the Compton edge and the photopeak, as well as distort the shape of the continuum predicted for single scattering.

If the gamma-ray energy is high enough to make pair production significant, a more complicated situation prevails. The annihilation photons now may either escape or undergo further interaction within the detector. These additional interactions may lead to either partial or full-energy absorption of either one or both of the annihilation photons.

If both annihilation photons escape without interaction, events occur that contribute to the double escape peak discussed previously. Another relatively frequent occurrence is a history in which one annihilation photon escapes but the other is totally absorbed. These events contribute to a *single escape peak*, which now appears in the spectrum at an energy of m_0c^2 (0.511 MeV) below the photopeak. A continuous range of other possibilities exists in which one or both of the annihilation photons are partially converted to electron energy through Compton scattering and subsequent escape of the scattered photon. Such events accumulate in a broad continuum in the pulse height spectrum lying between the double escape peak and the photopeak.

The response function to be expected for a real gamma-ray detector will depend on the size, shape, and composition of the detector, and also the geometric details of the irradiation conditions. For example, the response function will change somewhat if a point gamma-ray source is moved from a position close to the detector to one that is far away. The variation is related to the differences in the spatial distribution of the primary interactions that occur within the detector as the source geometry is changed. In general, the response function is too complicated to predict in detail other than through the use of Monte Carlo calculations, which simulate the histories actually taking place in a detector of the same size and composition.

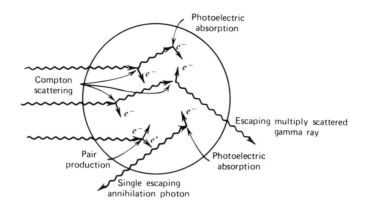

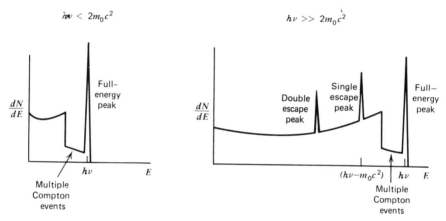

Figure 10-4 The case of intermediate detector size in gamma-ray spectroscopy. In addition to the continuum from single Compton scattering and the full-energy peak, the spectrum at the left shows the influence of multiple Compton events followed by photon escape. The full-energy peak also contains some histories that began with Compton scattering. At the right, the single escape peak corresponds to initial pair production interactions in which only one annihilation photon leaves the detector without further interaction. A double escape peak as illustrated in Fig. 10-2 will also be present due to those pair production events in which both annihilation photons escape.

Some properties of the response function are of general interest in gamma-ray spectroscopy. The *photofraction* is defined as the ratio of the area under the photopeak (or full-energy peak) to that under the entire response function. It is a direct measure of the probability that a gamma ray that undergoes interaction of any kind within the detector ultimately deposits its full energy. Large values of the photofraction are obviously desirable to minimize the complicating effects of Compton continua and escape peaks in the spectrum.

At high gamma-ray energies, the single and double escape peaks are quite prominent parts of the response function and can, under some circumstances, become larger than the photopeak. The ratio of the area under the single or double escape peak to the area under the photopeak is also a widely quoted property of the response function which can help in the interpretation of complex spectra.

D. Complications in the Response Function

1. SECONDARY ELECTRON ESCAPE

If the detector is not large compared with typical secondary electron ranges, a significant fraction of the electrons may leak from the detector surface and their energy will not be fully collected. This effect is enhanced for high gamma-ray energies for which the average secondary electron energy is also high. Electron leakage will tend to distort the response function by moving some events to a lower amplitude from that which would be observed if the entire electron energy were collected. The shape of the Compton continuum will therefore be altered somewhat to favor lower amplitudes. Because some events will be lost from the photopeak, the photofraction will be reduced as compared with the situation in which electron leakage is not important.

2. BREMSSTRAHLUNG ESCAPE

One of the mechanisms by which secondary electrons lose energy is by the radiation of bremsstrahlung photons. The fraction lost by this process increases sharply with electron energy and becomes the dominant process for electrons with energy over a few MeV (see Chapter 2). Even though the electron itself may be fully stopped within the detector, there is a possibility that some fraction of the bremsstrahlung photons may escape without being reabsorbed. The effects on the response function are similar to those described in the previous paragraph for electron escape and are again most important when the incident gamma-ray energy is large. For both secondary electron or bremsstrahlung escape, the effects are to change the shape of the response function somewhat, but peaks or sharp features are not introduced.

3. CHARACTERISTIC X-RAY ESCAPE

In the photoelectric absorption process, a characteristic X-ray is emitted by the absorber atom. In the majority of cases this X-ray energy is reabsorbed fairly near the original interaction site. If the photoelectric absorption occurs near a surface of the detector, however, the X-ray photon may escape. In this event, the energy deposited in the detector is decreased by an amount equal to the X-ray photon energy. Without the X-ray escape, the original gamma ray would have been fully absorbed and the resulting pulse would have contributed to the photopeak. With escape, a new category of events is created in which an amount of energy equal to the original gamma-ray energy minus the characteristic X-ray energy is repeatedly deposited in the detector. Therefore, a new peak will appear in the response function and will be located at a distance equal to the energy of the characteristic X-ray below the photopeak. These peaks are generally labeled "X-ray escape peaks" and tend to be most prominent at low incident gamma-ray energies and for detectors whose surface-to-volume ratio is large. Examples are shown in the spectra of Figs. 10-10 and 13-7.

4. SECONDARY RADIATIONS CREATED NEAR THE SOURCE

a. Annihilation Radiation

If the gamma-ray source consists of an isotope that decays by positron emission, an additional peak in the spectrum at 0.511 MeV is to be expected from the annihilation photons created when the positron is stopped. Most standard gamma-ray sources are encapsulated in a covering sufficiently thick to fully stop all the positrons, and thus they undergo annihilation in the region immediately surrounding the source. This region therefore acts as a source of 0.511 MeV annihilation radiation, which is superimposed on the gamma-ray spectrum expected from decay of the source itself.

b. Bremsstrahlung

Most common gamma-ray sources decay by beta-minus emission, and the source encapsulation is usually also thick enough to stop these beta particles. In other cases, an external absorber may be used to prevent the beta particles from reaching the detector where their energy deposition would needlessly complicate the gamma-ray spectrum. In the absorption process, however, some secondary radiation in the form of bremsstrahlung will be generated and may reach the detector and contribute to the measured spectrum. In principle, the bremsstrahlung spectrum may extend to an energy equal to the maximum beta particle energy, but significant yields are confined to energies that are much lower than this value. Some examples of bremsstrahlung energy spectra are given in Fig. 10-5, which illustrates the shape of the spectrum favoring low-energy bremsstrahlung photon emission. Because these spectra are continua, they do not lead to peaks in the recorded spectra but rather can add a significant continuum on which all other features of the gamma-ray spectra are superimposed. Because the bremsstrahlung contribution cannot

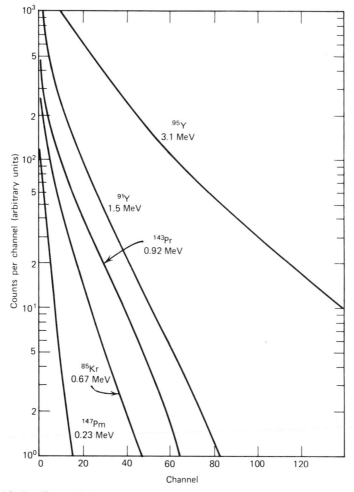

Figure 10-5 Shape of the bremsstrahlung spectra produced by beta particles with the indicated endpoint energies. (From Heath.[5])

simply be subtracted as a background, its inclusion can lead to errors in quantitative measurements of areas under peaks in the gamma-ray spectrum. To minimize the generation of bremsstrahlung, the use of beta absorbers made from low atomic number materials, such as beryllium, is often preferred.

5. EFFECTS OF SURROUNDING MATERIALS

In any practical application, a detector used for gamma-ray spectroscopy is surrounded by other materials that can have a measurable influence on its response. At a very minimum, the detector is encapsulated to provide a barrier against moisture and light or is mounted within a vacuum enclosure. To reduce natural background, most gamma-ray detectors are also operated within a shielded enclosure. The gamma-ray source itself is often part of a larger sample of the material or is contained within some type of encapsulation. All these materials are potential sources of secondary radiations that can be produced by interactions of the primary gamma rays emitted by the source. If the secondary radiations reach the detector, they can influence the shape of the recorded spectrum to a noticeable extent. Some possibilities are illustrated in Fig. 10-6.

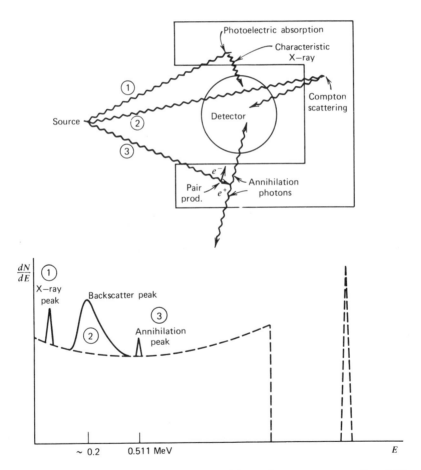

Figure 10-6 Influence of surrounding materials on detector response. In addition to the expected spectrum (shown as a dashed line), the representative histories shown at the top lead to the indicated corresponding features in the response function.

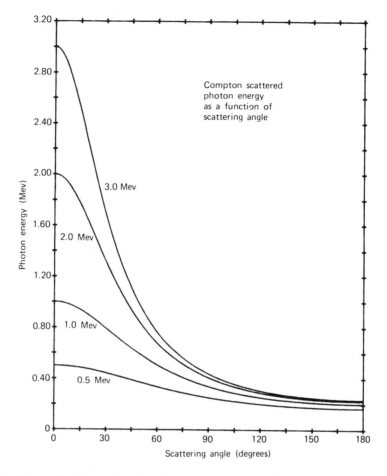

Figure 10-7 Variation of scattered gamma-ray energy with scattering angle.

a. Backscattered Gamma Rays

Pulse height spectra from gamma-ray detectors often show a peak in the vicinity of 0.2–0.25 MeV, called the *backscatter peak*. The peak is caused by gamma rays from the source which have first interacted by Compton scattering in one of the materials surrounding the detector. Figure 10-7 shows the energy dependence of these scattered gamma rays as a function of the scattering angle. From the shape of these curves, it can be seen that any scattering angle greater than about 110–120° results in scattered photons of nearly identical energy. Therefore, a monoenergetic source will give rise to many scattered gamma rays whose energy is near this minimum value, and a peak will appear in the recorded spectrum. The energy of the backscatter peak will correspond to Eq. (10-3):

$$hv'|_{\theta=\pi} = \frac{hv}{1 + 2hv/m_0c^2}$$

In the limit that the primary gamma-ray energy is large ($hv \gg m_0c^2/2$), this expression reduces to

$$h\nu'|_{\theta=\pi} \cong \frac{m_o c^2}{2} \qquad (10\text{-}8)$$

Thus, the backscatter peak always occurs at an energy of 0.25 MeV or less.

b. Other Secondary Radiations

In addition to Compton scattering, other interactions of the primary gamma rays in the surrounding materials can give noticeable peaks in the recorded spectrum. For example, photoelectric absorption in the materials immediately surrounding the detector can lead to generation of a characteristic X-ray that may reach the detector. If the atomic number of the material is high, the X-ray photon will be relatively energetic and can penetrate significant thicknesses of intervening material. Therefore, high-Z materials should be avoided in the immediate vicinity of the detector. On the other hand, the most effective shielding materials are those with high atomic numbers such as lead. A *graded shield* is one in which the bulk of the shield is made from high-Z materials, but the inner surface is lined with a material with lower atomic number. This inner lining serves to absorb the characteristic X-ray emitted by the bulk of the shield, at the same time emitting only low-energy or weakly penetrating X-rays of its own.

If the energy of the primary gamma rays is high, pair production within surrounding materials can give a significant yield of annihilation radiation. A peak can therefore appear at 0.511 MeV in the spectrum from the detection of these secondary photons. There is a danger of confusing this peak with that expected from annihilation radiation produced by radioactive sources that are positron emitters, and care must therefore be exercised in identifying the source of these annihilation photons.

E. Summation Effects

Additional peaks caused by the coincident detection of two (or more) gamma-ray photons may also appear in the recorded pulse height spectrum. The most common situation occurs in applications involving an isotope that emits multiple cascade gamma rays in its decay, as illustrated in Fig. 10-8. If we assume that no isomeric states are involved, the lifetime of the intermediate state is generally so short that the two gamma rays are, in effect, emitted in coincidence. It is then quite possible for both gamma-ray photons from a single decay to interact and deposit all their energy within a time that is short compared with the response time of the detector or the resolving time of the following electronics. If enough of these events occur, a *sum peak* will be observable in the spectrum which occurs at a pulse height that corresponds to the sum of the two individual gamma-ray energies. A continuum of sum events will also occur at lower amplitudes due to the summation of partial energy loss interactions.

The relative number of events expected in the sum peak depends on the branching ratio of the two gamma rays, the angular correlation that may exist between them, and the solid angle subtended by the detector. A complete analysis is often quite complex, but the following simplified derivation illustrates the general approach that can be applied.

Let ϵ_1 be the intrinsic peak efficiency of the detector for gamma ray ①, and let Ω be the fractional solid angle (steradians/4π) subtended by the detector. Then the full-energy

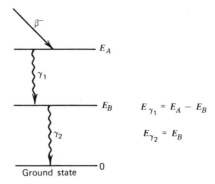

Figure 10-8 Simplified nuclear decay scheme which can lead to summation effects in gamma spectra. Provided the intermediate state (at E_B) is short-lived, γ_1 and γ_2 are emitted in virtual coincidence.

peak area for gamma ray ① in the absence of summing effects is

$$N_1 = \epsilon_1 \Omega S y_1 \tag{10-9}$$

where S is the number of source decays over the observation period and y_1 is the yield of gamma ray ① per disintegration. Applying the same definitions to gamma ray ②, we obtain

$$N_2 = \epsilon_2 \Omega S y_2 \tag{10-10}$$

The probability of simultaneous detection of both gamma rays is the product of both individual detection probabilities, multiplied by a factor $W(0°)$ to account for any angular correlation between the gamma-ray photons. $W(0°)$ is defined as the relative yield of γ_2 per unit solid angle about the $0°$ direction defined by the detector position, given that γ_1 is emitted in the same direction. Then the sum peak area should be

$$N_{12} = S(\epsilon_1 \Omega y_1)(\epsilon_2 \Omega y_2) W(0°)$$

$$= S \epsilon_1 \epsilon_2 y_1 y_2 \Omega^2 W(0°) \tag{10-11}$$

The summation process not only creates the sum peak but also removes events that would otherwise fall within individual gamma-ray full-energy peaks. The remaining number of full-energy events for γ_1 is [from Eqs. (10-9) and (10-11)]

$$N_1\big|_{\substack{\text{with} \\ \text{summation}}} = N_1 - N_{12}$$

$$= \epsilon_1 \Omega S y_1 \left[1 - \epsilon_2 \Omega y_2 W(0°)\right] \tag{10-12}$$

Because a coincident event of any kind from γ_2 (not just a photopeak event) will remove a count from N_1, the detection efficiency ϵ_2 should now be interpreted as the intrinsic *total* efficiency. For these losses to remain small, the fractional solid angle Ω is often restricted to small values to keep the second term in the above equation much smaller than the first.

If the solid angle is too large, quantitative measurements based on determination of the area under full-energy peaks can be in error unless an accounting of the second term is provided.[6]

The summation process described above involves multiple radiations from the same nuclear decay event and therefore is classified as a *true coincidence* by the definitions given in Chapter 17. Another process can also lead to summed pulses due to the accidental combination of two separate events from independent decays that occur closely spaced in time. Because the time intervals separating adjacent events are randomly distributed, some will be less than the inherent resolving time of the detector or pulse-processing system. These *chance coincidences* increase rapidly with increasing counting rate and will occur even in the absence of true coincidences. A corresponding sum peak can therefore appear in spectra from isotopes that emit only a single radiation per decay.

Chance coincidences will occur if a second pulse arrives within the resolving time t_r following a typical signal pulse. For a random pulse rate of r_s and $r_s t_r \ll 1$, the rate at which coincidences occur should be the fraction of all time that lies within t_r of a preceding pulse (given by $r_s t_r$) multiplied by the rate of pulse arrival (r_s), or

$$r_{\text{ch}} = r_s^2 t_r \qquad\qquad (10\text{-}13)$$

Therefore, the accidental sum peak will have an intensity that is proportional to the square of the counting rate, whereas both the true sum peak or normal photopeaks will be linearly related to the counting rate. When multiple radiations are involved, accidental sum peaks may potentially occur at all possible combinations of any two single energies. At normal rates and typical detector solid angles, however, sum peaks are usually lost in fluctuations in the continua and background present from other energies, except at the upper energy extremes of the spectrum where such backgrounds are low.

As a practical matter, the resolving time t_r is normally set by the shaping time constants of the linear amplifier used in the pulse-processing chain from the detector. The chance coincidences therefore take the form of *pulse pileup* in the amplifier, which is further detailed in the discussions of Chapter 17.

F. Coincidence Methods in Gamma-Ray Spectrometers

1. CONTINUUM REDUCTION

For an ideal gamma-ray detector, the response function would simply be a single well-resolved peak with no associated continuum. Then the pulse height spectrum from a complex gamma-ray source could be most easily interpreted, and the presence of high-energy gamma rays would not hinder the detection of weak radiations at lower energies.

At the price of added complexity, some steps can be taken to approach this ideal more closely, even for gamma-ray detectors with response functions that are inherently more complicated. These methods involve placing other detectors around the primary detector and employ coincidence techniques to select preferentially those events that are most likely to correspond to full-energy absorption. For the case of sodium iodide spectrometers, the most common methods involve the use of an annular detector surrounding the primary crystal for Compton suppression by anticoincidence, or the use of two or more adjacent crystals in the *sum-coincidence* mode. Representative descriptions of sodium iodide spectrometers in which one or both of these methods of continuum suppression have been applied are given in Refs. 7–11.

An explanation of these techniques is postponed until Chapter 12, where their use with germanium detectors is detailed. Although significant improvements in peak-to-continuum ratios can be achieved by applying these methods to NaI(Tl) spectrometers, current attention has focused on their application to germanium systems where continua are much more prominent and greater gains can be achieved through their suppression.

2. THE COMPTON SPECTROMETER

The combination of two separated gamma-ray detectors operated in coincidence, as shown in Fig. 10-9, is another configuration that can simplify the response function at the expense of detection efficiency. A collimated beam of gamma rays is allowed to strike the first detector in which the desired mode of interaction is now Compton scattering. Some fraction of the scattered gamma rays will travel to the second detector where they may also interact to give a second pulse. Because the separation distance is normally no greater than a few tens of centimeters, the pulses are essentially in time coincidence. By selectively recording only those pulses from the first crystal that are in coincidence with a pulse from the second crystal, the recorded spectrum largely reflects only single Compton scattering events. Because the angle of scattering is fixed, a constant amount of energy is

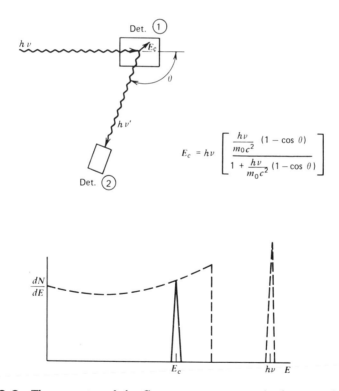

$$E_c = h\nu \left[\frac{\dfrac{h\nu}{m_0 c^2}(1 - \cos\theta)}{1 + \dfrac{h\nu}{m_0 c^2}(1 - \cos\theta)} \right]$$

Figure 10-9 The geometry of the Compton spectrometer is shown at the top. The spectrum of those events from detector ① that are in coincidence with pulses from detector ② is shown as the solid curve at the bottom. The normal spectrum from detector ① is shown as the dashed curve.

deposited for each scattering interaction involving monoenergetic incident gamma rays. Photoelectric absorption and all other events that do not lead to coincidence between the two detectors are excluded. The response function is thus reduced to a single peak, which appears at a position within the original Compton continuum determined by the scattering angle.

IV. PROPERTIES OF SCINTILLATION GAMMA-RAY SPECTROMETERS

A. Response Function

Sodium iodide gained much of its early popularity because the relatively high atomic number ($Z = 53$) of its iodine constituent assures that photoelectric absorption will be a relatively important process. The corresponding high intrinsic detection efficiency and large photofraction have contributed to the success of sodium iodide scintillation spectrometers. Other materials, such as cesium iodide or BGO, have even higher density or effective atomic number, and therefore the response function for these materials shows an even greater detection efficiency and photofraction. However, the relatively high light output and smaller decay time have led to the dominance of NaI(Tl) in spectroscopy with scintillators.

The importance of many of the factors that influence the shape of the response function for NaI(Tl) scintillators is detailed by Mueller and Maeder.[12] An extensive catalog of experimentally measured gamma-ray spectra for nearly 300 radionuclides as recorded by a 3 in.× 3 in. NaI(Tl) spectrometer has been published by Heath.[5] A later compilation by Adams and Dams[6] contains spectra for both 3 in.× 3 in. and 4 in.× 4 in. cylindrical sodium iodide crystals. These published data can be of considerable help in

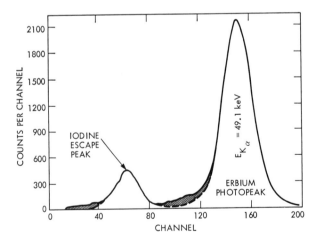

Figure 10-10 A low-energy spectrum from a NaI(Tl) scintillator for incident 49.1 keV X-rays from erbium. The iodine characteristic X-ray escape peak lies 25 keV below the photopeak. (From Dell and Ebert.[13])

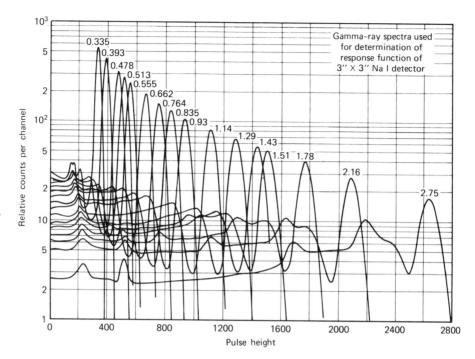

Figure 10-11 Response functions for a 7.62 cm × 7.62 cm cylindrical NaI(Tl) scintillator for gamma rays from 0.335 to 2.75 MeV. (From Heath.[5])

predicting the response function to be expected from a scintillation spectrometer when applied to gamma-ray-emitting isotopes. Examples of gamma-ray spectra taken from these compilations are given in Figs. 10-10 through 10-13. Many of the features described in the previous discussion can be observed in these spectra.

A direct comparison of gamma-ray spectra observed under similar conditions from equal size crystals of sodium iodide and BGO is shown in Fig. 10-14. The considerably better energy resolution of sodium iodide is evidenced by the smaller widths of the peaks in the bottom spectrum. The higher efficiency of BGO is apparent, especially at the higher energy, from the larger area under the photopeaks in the upper spectrum. The larger photofraction of BGO also results in less prominent Compton continua and escape peaks.

Because of the lack of suitable sources, it is often difficult experimentally to measure the response function for all gamma-ray energies of potential interest. It is then necessary to resort to calculations to derive the response function. Because of the complexity of the situation in which multiple interactions play an important role, virtually all practical calculations are done using the Monte Carlo method. The work of Berger and Seltzer[15] is a good example of such a calculation for sodium iodide cylindrical detectors. This publication also contains an extensive list of references to prior experimental and calculational efforts. Figure 10-15 shows the good match to experimental data which can be obtained when the computational model contains sufficient detail to provide adequate representation of all important interactions taking place within the detector. Figure 10-16 illustrates a set of calculated response functions for a sodium iodide scintillator extending to relatively high gamma-ray energies. The gradual disappearance of the photopeak and the broadening of the escape peaks as the gamma-ray energy increases is clearly

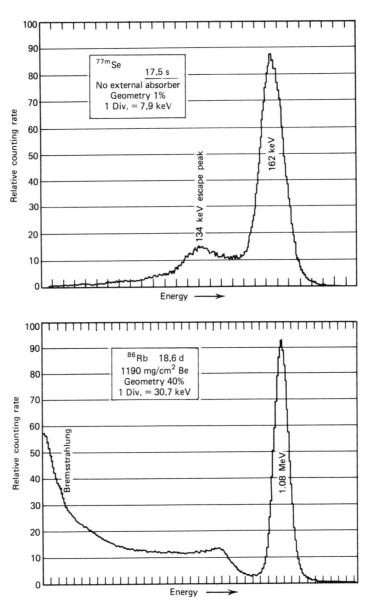

Figure 10-12 Experimentally measured pulse height spectra in NaI(Tl) scintillators for two radioisotopes emitting monoenergetic gamma rays. The top spectrum shows the iodine X-ray escape peak, which can be evident for gamma rays of relatively low energy. The bottom spectrum, for a higher-energy gamma ray, shows the bremsstrahlung generated by stopping the beta particles within the source, in addition to the Compton continuum and photopeak. (From F. Adams and R. Dams, *Applied Gamma-Ray Spectrometry*, 2nd ed. Copyright 1970 by Pergamon Press, Ltd. Used with permission.)

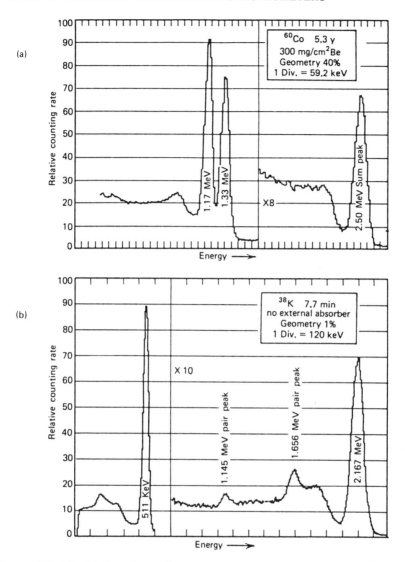

Figure 10-13   (a) Experimentally measured pulse height spectrum for ^{60}Co (1.17 and 1.33 MeV gamma rays) from a 10.16 cm × 10.16 cm cylindrical NaI(Tl) scintillator. The large fractional solid angle of 40% subtended by the detector enhances the sum peak intensity. (b) Spectrum from ^{38}K (2.167 MeV gamma ray plus positron annihilation photons) from a 7.62 cm × 7.62 cm NaI(Tl) scintillator. The single and double escape peaks for the high-energy gamma ray are evident, as well as a backscatter peak at ~ 200 keV. (From F. Adams and R. Dams, *Applied Gamma-Ray Spectrometry*, 2nd ed. Copyright 1970 by Pergamon Press, Ltd. Used with permission.)

illustrated by this series of plots. The photopeak disappears because pair production becomes the dominant mode of interaction, and the high-energy pair that is produced is difficult to absorb fully because of losses from bremsstrahlung emission and leakage from the surface of the crystal. The broadening of the escape peak is due to the additional statistical fluctuations in the number of photoelectrons produced as their number increases (see Section B.2 below).

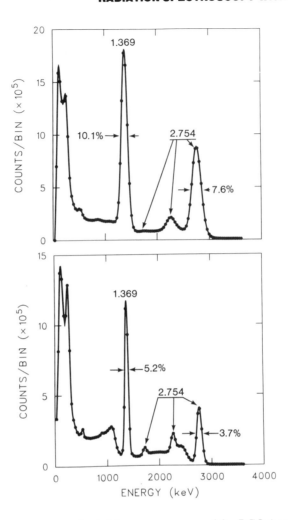

Figure 10-14   Comparative pulse height spectra measured for BGO (top) and NaI(Tl) (bottom) scintillators of equal 7.62 cm × 7.62 cm size for gamma rays from ^{24}Na. (From Moss et al.[14])

B. Energy Resolution

In contrast to the theoretical energy deposition spectra shown in Figs. 10-2 to 10-4, the measured response functions shown above contain the "blurring" effects due to the finite energy resolution of the detector. The most striking difference is the fact that all peaks now have some finite width rather than appearing as narrow, sharp lines.

As introduced in Chapter 4, the *energy resolution R* is defined as

$$R = \frac{\text{FWHM}}{H_0} \tag{10-14}$$

where   FWHM = full width at half maximum of the full-energy peak

H_0 = mean pulse height corresponding to the same peak

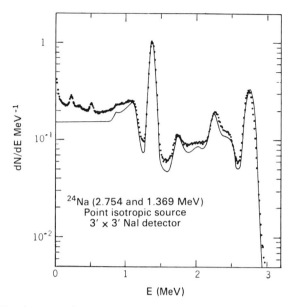

Figure 10-15 A comparison of a measured pulse height spectrum (points) with a theoretical spectrum calculated by Monte Carlo methods. (From Berger and Seltzer.[15])

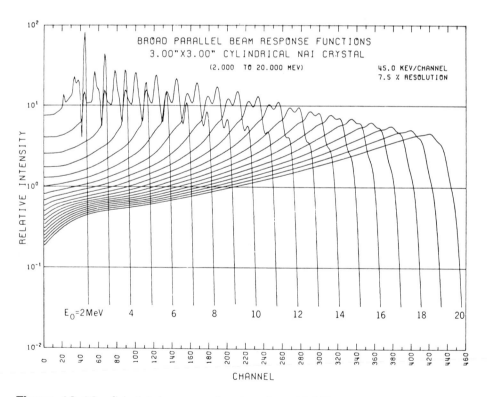

Figure 10-16 Calculated response functions for a NaI(Tl) scintillation detector extending to 20 MeV. (From Berger and Seltzer.[15])

The energy resolution of scintillators is the poorest of any commonly used detector, and therefore the spectra have relatively broad peaks.

1. ORIGINS OF RESOLUTION LOSS

From the arguments of Chapter 4, the finite energy resolution of any detector may contain contributions due to the separate effects of charge collection statistics, electronic noise, variations in the detector response over its active volume, and drifts in operating parameters over the course of the measurement. For scintillation detectors, the fluctuations in PM tube gain from event to event (see p. 259) can also add to the measured resolution. For the majority of applications of scintillators to gamma-ray spectroscopy, the first of these sources is normally the most significant. Contributions of electronic noise are usually negligible, so that preamplifiers and other electronic components used with scintillators need not include elaborate schemes for noise reduction. Variations in the light collected from scintillation events over the volume of the crystal can be a significant problem, so the techniques discussed in Chapter 8 for promoting uniform light collection are quite important. In commercially prepared crystals of typical sizes, these nonuniformities normally are a small part of the total peak width. Drifts in the operating parameters are usually associated with the PM tube and can be severe if the detector is subject to large changes in counting rate or temperature. Some methods of spectrum stabilization discussed in Chapter 18 can be applied in these cases but usually are necessary only under extreme conditions.

2. PHOTOELECTRON STATISTICS

Statistical spreads are therefore left as the single most important cause of peak broadening in scintillators. The statistical fluctuations will be most significant at the point in the signal chain at which the numbers of information carriers are at a minimum. For scintillation counters, this point is reached after conversion of the scintillation light to electrons by the photocathode of the PM tube.

A representative numerical example is helpful in illustrating this point. In the signal chain illustrated below, it is assumed that exactly 0.5 MeV of electron energy is deposited by the gamma-ray photon in the scintillation crystal. In thallium-activated sodium iodide, the scintillation efficiency is about 12%. In the example, 60 keV is thus converted by the scintillator into visible light with an average photon energy of about 3 eV. Therefore, about 20,000 scintillation photons are produced per pulse. With allowance for some light loss at the crystal surface and at the crystal–phototube interface, perhaps 15,000 of these light photons ultimately reach the photocathode. If we assume that the average quantum efficiency of the photocathode over the scintillation spectrum is 20%, then 3000 photoelectrons are produced. This number is now the minimum in the signal chain because subsequent stages in the PM tube multiply the number of electrons.

The amplitude of the signal produced by the PM tube is proportional to this number of photoelectrons. For repeated events in which exactly the same energy is deposited in

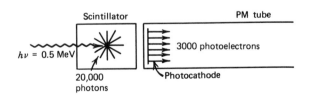

the crystal, the pulse amplitude will fluctuate due to the inherent statistical variation in the number of photoelectrons collected for each event. If we assume that Poisson statistics hold, then the standard deviation of the number of photoelectrons should be the square root of the mean number produced. Therefore, in our example we would expect a standard deviation of $\sqrt{3000}$, or about 1.8% of the mean value. Energy resolution is formally defined in terms of the full width at half maximum of the peak rather than its standard deviation. For an assumed Gaussian shape, the FWHM is 2.35 values of the standard deviation, and therefore the statistical contribution to the energy resolution for the example should be 4.3%.

In many practical cases, the statistical broadening of the peak predominates over other potential sources of resolution loss. In that event, the variation of the resolution with gamma-ray energy can be predicted simply by noting that the FWHM of the peak is proportional to the square root of the gamma-ray photon energy. The average pulse height produced is directly proportional to the gamma-ray energy. Therefore, from the definition of energy resolution,

$$R \equiv \frac{\text{FWHM}}{H_0} = K \frac{\sqrt{E}}{E} = \frac{K}{\sqrt{E}} \tag{10-15}$$

The energy resolution should thus be inversely proportional to the square root of the gamma-ray energy. If we take the logarithm of both sides of Eq. (10-15), we derive

$$\ln R = \ln K - \tfrac{1}{2} \ln E \tag{10-16}$$

Therefore, a plot of $\ln R$ versus $\ln E$ should be a straight line with slope of $-\tfrac{1}{2}$.

Figure 10-17 shows a plot of experimentally determined resolution values for a sodium iodide scintillator as a function of the gamma-ray energy at which they were

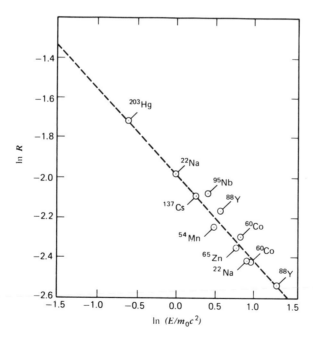

Figure 10-17 Experimentally measured resolution R from a NaI(Tl) scintillation detector for various gamma-ray energies E. (From Beattie and Byrne.[16])

measured. The data adhere fairly closely to a straight line, but the slope is not as steep as predicted, indicating the influence of nonstatistical sources of the peak broadening. A more adequate representation of measured data can take the form

$$R = \frac{(\alpha + \beta E)^{1/2}}{E} \tag{10-17}$$

where α and β are constants particular to any specific scintillator–PM combination.

3. OTHER FACTORS IN ENERGY RESOLUTION

Other sources of resolution loss in scintillation spectrometers are conveniently categorized into three groups: those that are characteristic of the crystal itself (the *intrinsic crystal resolution*), those effects that are characteristic of the PM tube, and the variable probability that a visible photon generated by a scintillation event in the crystal produces a photoelectron that is collected by the first dynode of the PM tube (the *transfer variance*).

The intrinsic crystal resolution includes any variation caused by local fluctuations in the scintillation efficiency of the crystal. With modern fabrication techniques, the uniformity of sodium iodide crystals is generally sufficiently good so that the line broadening due to this effect alone would be very small (less than 2%). More significant fluctuations arise because of the less-than-perfect reflection conditions that exist at the surface of the crystal. The consequent nonuniform light collection efficiency can introduce significant line broadening, especially in crystals of large size.

The variance introduced by the photomultiplier tube can be a significant contribution. Uniformity of photoelectron collection from the photocathode is an important factor, as is the statistical fluctuation in the electron multiplication. There is considerable variation in the performance of different photomultipliers in this regard, even among different samples of the same design. For example, in a study of several hundred PM tubes sampled from a few standard types, Persyk and Moi[17] observed an average NaI scintillator energy resolution of 10–11% for [57]Co radiation (122 keV). The best PM tube included in the sample, however, gave a corresponding value of 8.5%.

A more subtle component of the intrinsic crystal resolution arises from the slight nonlinearity of sodium iodide scintillation response (see Fig. 8-8). If all incident gamma rays underwent an interaction in which their entire energy were converted to a single electron, this nonlinearity would not be a source of resolution loss. However, the incident gamma-ray energy may be subdivided among two or more secondary electrons through single or multiple Compton scattering followed by photoelectric absorption. Furthermore, even if simple photoelectric absorption occurs, the excited atom that remains may convert its excitation energy in a number of ways which lead to varying electron energy spectra, primarily in the form of Auger electrons. Consequently, even a monoenergetic flux of incident gamma rays will lead to a wide distribution of electron energies within the crystal. If the response of the crystal is not linear with electron energy, the total light yield will be different from event to event, depending on details of the energy subdivision between the various electrons that are produced. These effects appear to be significant for gamma-ray energies above a few hundred keV, for which multiple interactions are predominant, and have been studied in detail by Narayan and Prescott.[18]

Because the energy resolution varies with energy, values are usually specified at a fixed gamma-ray energy for comparison purposes. It is conventional to quote the energy resolution for gamma rays from [137]Cs (0.662 MeV) as a standard. The energy resolution for other gamma-ray energies can then be estimated through the use of Eq. (10-15).

Good quality solid sodium iodide scintillators coupled to modern PM tubes can achieve an energy resolution of about 6–7% at 0.662 MeV. If the shape of the crystal is more complicated than a simple right cylinder, the added difficulty of uniformity in light collection will often make the resolution somewhat worse. For example, the energy resolution for crystals with a center well is generally 1–3% larger than the equivalent solid cylinder.

4. PREVENTION OF RESOLUTION LOSS DUE TO LONG-TERM DRIFT

In scintillator measurements that must extend over many hours or days, some resolution loss can be experienced due to drifts in the gain of the PM tube and other circuit components. Some electronic methods that can be used to minimize or completely eliminate these effects are outlined in Chapter 18. These techniques work best when there is a single isolated peak in the spectrum from which an error signal can be derived to adjust a variable gain component in the signal chain. If a strong isolated peak does not exist in the measured spectrum, or if the counting rates are low, these methods may be impractical.

For scintillation counters, an alternative method can be used, which is based on providing a reference light source within the scintillation package to produce an artificial peak in the spectrum. If the light pulses are of constant intensity, a feedback signal can be generated to adjust the system gain to hold the resulting peak at a constant position in the measured spectrum. Light sources used for this purpose fall into two general categories: those that consist of a combination of a radioactive isotope with a suitable phosphor, and those that are basically electronic in design. A common requirement is that their yield be extremely stable over long periods of time because any change in light output will be interpreted as a drift in the gain of the counting system. Discussions of the design of light sources used for this purpose are given in Refs. 19–23.

C. Linearity

For all scintillators, the *scintillation efficiency* or amount of light generated per unit energy loss (dL/dE) depends both on the particle type and its kinetic energy. For an ideal spectrometer, dL/dE would be a constant independent of particle energy. The total light yield would then be directly proportional to the incident particle energy, and the response of the scintillator would be perfectly linear.

For electrons in NaI(Tl), the scintillation efficiency does vary mildly with electron energy (see Fig. 8-8). For direct electron spectroscopy, some significant nonlinearity should thus be expected. In gamma-ray measurements, monoenergetic gamma rays create varied combinations of secondary electrons with widely different energies. Figure 8-8 is therefore sampled at many points over the energy scale even for a single photon energy, and some of the nonlinearity is averaged out. The average pulse height observed for gamma rays is therefore closer to being linear with photon energy, but measurable nonlinearities remain.

Figure 10-18 shows some experimental results for linearity measurements with gamma rays on NaI(Tl). This nonlinearity must be taken into account when relating the pulse height scale to gamma-ray energy. In practice, a calibration is usually carried out in which peak positions are plotted versus known gamma-ray energies. Because Fig. 10-18 represents the slope of this calibration line, a small degree of curvature or nonlinearity should be expected. For interpolation between narrowly spaced peaks of known energy, however, assumption of linearity normally leads to a negligible error.

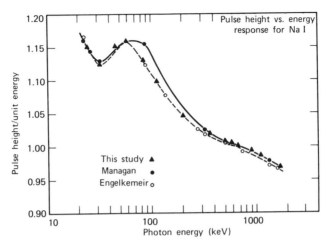

Figure 10-18 The differential linearity measured for a NaI(Tl) scintillator. (From Heath.[5])

D. Detection Efficiency

One of the marked advantages in using scintillation crystals for gamma-ray measurements is the fact that many standard size detectors can be mass produced with virtually identical properties. Because the physical dimensions of these crystals can be controlled to within very small tolerances, the interaction probability for gamma rays will be identical for all crystals of the same size and shape.

One of the common applications of sodium iodide scintillators is to measure the absolute intensity of a given source of gamma rays. From the discussion in Chapter 4, such a measurement based on simple counting data requires a prior knowledge of the efficiency of the radiation detector. Undoubtedly, there are more published data available on the detection efficiency of sodium iodide scintillators for gamma rays than for any other detector type or application. The number of different sizes and shapes of NaI(Tl) crystals in routine use is relatively limited so that reasonably complete data can be compiled on each of the common configurations.

1. CRYSTAL SHAPES

Two general crystal shapes are in widespread use for applications in gamma-ray detection. The solid right circular cylinder is simple to manufacture and encapsulate and can be mounted directly to the circular face plate of most PM tubes. If the height-to-diameter ratio of the cylinder does not greatly exceed unity, the light collection properties are quite favorable in this geometry. If the height-to-diameter ratio is much less than 1, the pulse height resolution can often be improved by interposing a light pipe between the crystal and photomultiplier tube to spread the light more uniformly from each scintillation event over the entire photocathode, and thereby average out spatial variations of its quantum efficiency.

A *well crystal* is a right circular cylinder into which a cylindrical well has been machined, usually along the cylindrical axis. A significant advantage of this geometry is the very high counting efficiency that can be achieved by placing the samples to be counted at the bottom of the well. In this position, almost all the gamma rays that are emitted isotropically from the source are intercepted by at least a portion of the crystal.

For low-energy gamma rays, the counting efficiency in this geometry can therefore approach 100%. At higher energies, some of the advantage is lost because the average path length through the crystal is somewhat less than if the gamma rays were externally incident on a solid crystal. Because the efficiency for sources near the bottom of the well is not a sensitive function of position, well counters can also simplify the counting of multiple samples with different physical properties while providing nearly identical counting efficiency.

2. EFFICIENCY DATA

Data on detector efficiency are commonly presented in the form of a graph of the efficiency value as a function of gamma-ray energy. Some examples are shown in Figs. 10-19 through 10-22. It is important to point out those parameters that must be specified before using this type of data:

1. The specific category of efficiency which is being tabulated must be clearly identified. As defined in Chapter 4, detector efficiencies are classified as either absolute or intrinsic, with the latter being the more common choice. An additional specification must be made that deals with the type of event accepted by the counting system. Here the most common choices are either peak or total efficiencies, with the distinction hinging on whether only full-energy events or all events are accepted (see Chapter 4).

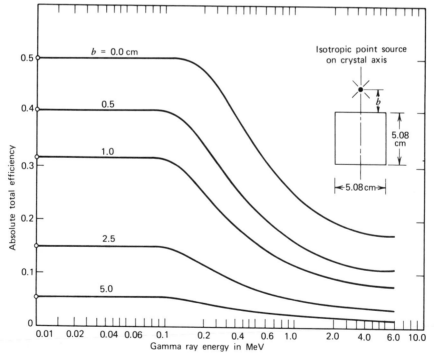

Figure 10-19 The absolute total efficiency calculated for a 5.08 cm × 5.08 cm solid cylindrical NaI(Tl) scintillator. Different values of the source location are shown. (From Snyder.[24])

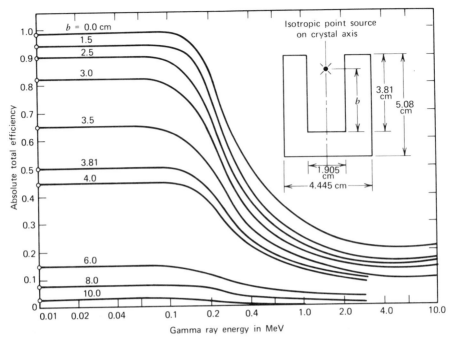

Figure 10-20 The absolute total efficiency calculated for point gamma-ray source and a NaI(Tl) well-type scintillator with the dimensions shown. The parameter b is the source height above the well bottom. (From Snyder.[24])

2. The size and shape of the scintillation crystal have a strong influence on the counting efficiency. Although the major influence on the intrinsic efficiency is the thickness of the crystal in the direction of the incident gamma radiation, mild variation with other detector dimensions should also be expected.

3. The size and physical nature of the source also influence the counting efficiency. Data are widely available for the relatively simple case of an isotropic point

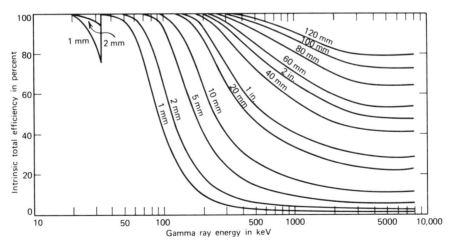

Figure 10-21 The intrinsic total efficiency of various thicknesses of NaI(Tl) for gamma rays perpendicular to its surface.

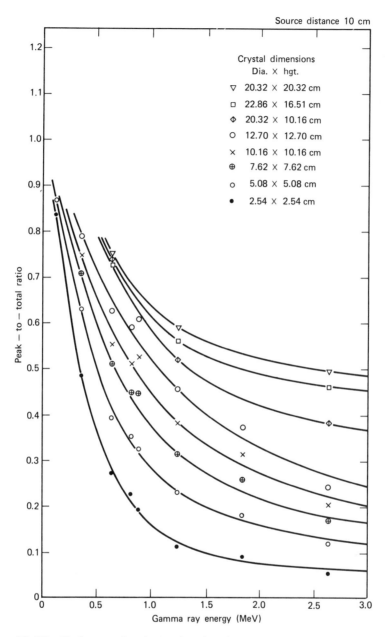

Figure 10-22 Peak-to-total ratio (or the photofraction) for various solid cylinders of NaI(Tl) for a point gamma-ray source 10 cm from the scintillator surface. (Courtesy of Harshaw Chemical Company.)

gamma-ray source located a specified distance from the detector face along its axis. Although absolute efficiencies are quite sensitive to the source–detector spacing, this dependence is much milder for intrinsic efficiencies and vanishes entirely if the source is sufficiently far from the detector. Other common source conditions for which data can be found include the case of a parallel beam of gamma rays uniformly irradiating one surface of the detector (equivalent to a point source at an infinite distance) and a narrowly collimated beam (a "pencil beam") incident at a specified point on the detector surface. Limited data also are available for distributed sources consisting of disks or volumetric sources under specified source–detector geometries.

4. Any absorption taking place between the point of gamma-ray emission and the scintillation crystal will also influence the detection efficiency. Published data normally neglect the effects of the crystal housing and other material between the source and detector, but some data can be found that account for self-absorption within large-volume gamma-ray sources.

There are two general methods by which efficiency data may be generated. The most straightforward is simply to measure the appropriate counting rate induced by a gamma-ray source of known activity. Sets of "standard" gamma-ray sources can be purchased whose absolute activity can be quoted to about 1% precision. If the experimenter has a set of such sources available covering the energy range of interest, then an efficiency curve can be determined experimentally for the specific detector in use. In many cases, however, a set of absolutely calibrated sources is not available or they do not adequately represent the geometric irradiation conditions of the actual experiment. Then the experimenter must turn to published efficiency data, which can include a greater variety of energies and experimental conditions.

A second means of obtaining efficiency data is through calculation based on an assumed knowledge of the various gamma-ray interaction probabilities. The simplest case is the total efficiency, which is completely determined by the total linear absorption coefficient μ that characterizes the detector material. The intrinsic total efficiency is just the value of the gamma-ray interaction probability $(1 - e^{-\mu l})$ integrated over all path lengths l taken by those gamma rays that strike the detector. For simple geometries, this

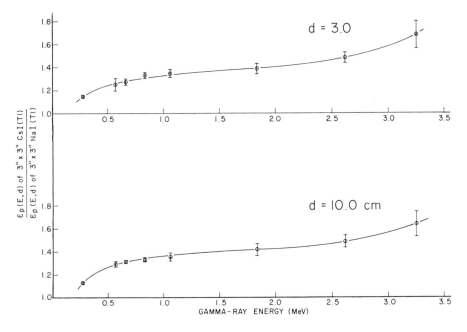

Figure 10-23 Relative intrinsic peak efficiencies of 7.62 cm $\times$ 7.62 cm cylindrical crystals of CsI(Tl) and NaI(Tl); d is the crystal–source distance. (From Irfan and Prasad.[34])

integration can be carried out analytically. In most cases, however, the peak efficiency is of greater interest. Because more complex processes involving multiple interactions contribute to full-energy events, the peak efficiency, in general, no longer can be calculated analytically. Instead, recourse must be made to Monte Carlo calculations, which attempt to simulate the behavior of gamma-ray photons based on knowledge of the individual probabilities for photoelectric absorption, Compton scattering, and pair production. Because each Monte Carlo calculation is, in effect, a computer experiment, the results obtained are subject to statistical uncertainties determined by the number of histories that have been calculated. Furthermore, the results are specific to the detector geometry and gamma-ray energy assumed and cannot be generalized further.

Extensive tables and graphs of sodium iodide detector efficiencies can be found in Refs. 5, 6, and 25. A good review of both experimentally determined and calculated efficiencies for solid cylindrical sodium iodide crystals has been published by Grosswendt and Waibel.[26] Extensive references are also tabulated in Refs. 27 and 28 for calculations and measurements of efficiencies for well-type crystals, and in Ref. 29 for other crystal shapes.

When the gamma-ray-emitting sample is not negligibly small, self-absorption effects can substantially reduce the efficiency that would be obtained for point sources. The usual procedure is to apply a multiplicative correction factor to standard efficiencies to account for absorption within the sample itself. This approach gives adequate representation for a wide range of physical situations.[30] Reviews of scintillator efficiencies for absorbing disk sources are given in Refs. 31 and 32.

Most of the available efficiency data concern sources that are located along the axis of symmetry of the detector, and only limited data are available regarding the directional response of scintillation crystals to off-axis sources. One such analysis for cylindrical

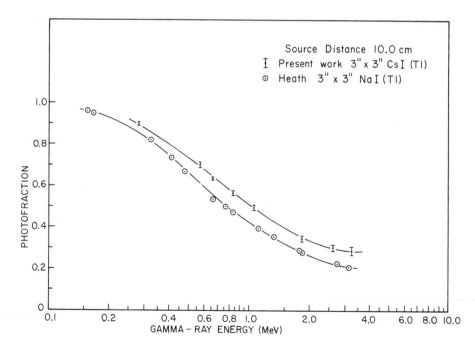

Figure 10-24 Experimental photofractions of 7.62 cm $\times$ 7.62 cm cylindrical crystals of CsI(Tl) and NaI(Tl). (From Irfan and Prasad.[34])

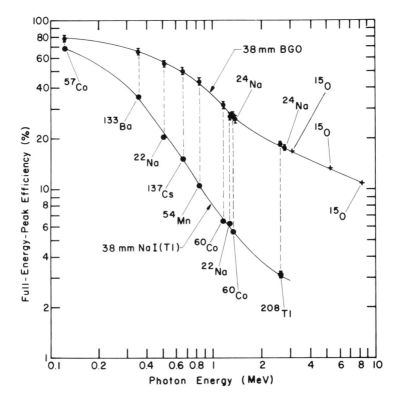

Figure 10-25 Intrinsic peak efficiencies for BGO and NaI(Tl) scintillators of equal 38 mm × 38 mm size. Radioisotope sources used for various photon energies are indicated. (From Evans and Orndoff.[35])

scintillators is given in Ref. 33, which also contains references to other previously published data on directional characteristics.

Efficiency data can also be found for scintillation materials other than sodium iodide. Some measurements comparing cesium iodide with sodium iodide are shown in Figs. 10-23 and 10-24. CsI(Tl) exhibits both a higher efficiency and a greater photofraction, but its somewhat lower light output (as measured using standard PM tubes) leads to a poorer energy resolution. Corresponding differences are even more extreme in the case of BGO. Figure 10-25 shows that the intrinsic peak efficiency for an equivalent size BGO crystal can be almost a factor of 10 greater than for sodium iodide at high gamma-ray energies, but the energy resolution is even poorer than in cesium iodide. Gamma-ray detection efficiencies for NE-213 organic liquid scintillators are given in Refs. 36 and 37, and for lead-doped plastic scintillators in Ref. 38.

3. PEAK AREA DETERMINATION

To apply the peak efficiency data for any detector, the area under the full-energy peaks that appear in its spectrum must be determined. Even after subtraction of the normal background, nearly all such peaks will be superimposed on a continuum caused by many of the complicating effects described earlier in this chapter. It is therefore not always a simple task to determine the number of events that contribute to a given full-energy peak.

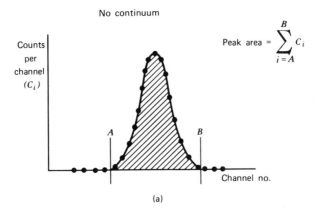

No continuum

Counts per channel (C_i)

$$\text{Peak area} = \sum_{i=A}^{B} C_i$$

A   B

Channel no.

(a)

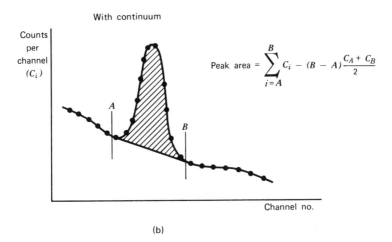

With continuum

Counts per channel (C_i)

$$\text{Peak area} = \sum_{i=A}^{B} C_i - (B - A)\frac{C_A + C_B}{2}$$

A

B

Channel no.

(b)

Figure 10-26   Methods of obtaining peak areas from multichannel spectra.

If the peak were a simple isolated one without any superimposed continuum, as shown in Fig. 10-26a, its area could be determined by simple integration between the limits shown. When the spectrum is recorded in a multichannel analyzer, the equivalent process is a simple addition of the content of each channel between the indicated limits. If a continuum is also present, as in Fig. 10-26b, some additional unwanted counts are included in this process and must be subtracted. Some shape must therefore be assumed for the continuum within the region under the peak, and a number of fitting procedures of varying degrees of complexity can be applied. A linear interpolation between the continuum values on either side of the peak is the easiest approach and will give sufficient accuracy for many purposes.

At times, closely spaced or overlapping peaks do not allow the straightforward summation method to be applied. More complex methods must then be used to separate the individual contributions of each of the closely lying peaks. These methods normally involve fitting an analytic shape to that portion of the peak which can be clearly resolved, and assuming that the remainder of the peak is described by the fitted function. A Gaussian curve fitted to the points that lie within one standard deviation on either side of

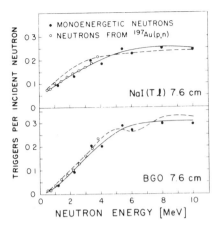

Figure 10-27 The prompt pulse response to fast neutrons in 7.6 cm $\times$ 7.6 cm scintillation detectors of NaI(Tl) and BGO. (From Hausser et al.[40])

the peak value adequately represents the shape of the measured photopeak from a NaI(Tl) scintillator over an assortment of source geometry and counting conditions.[39] More complex shapes are sometimes necessary for spectra recorded at high rates or under nonideal circumstances. Because a good deal of complexity is involved in such fitting routines, nearly all are carried out by computer programs, which are described in further detail in Chapter 18.

V. RESPONSE OF SCINTILLATION DETECTORS TO NEUTRONS

Some types of scintillators specifically designed to be used as neutron detectors are described in Chapters 14 and 15. However, virtually any scintillation material will respond to some extent if exposed to a neutron flux. Fast neutrons are sometimes unavoidably present in gamma-ray measurements made around accelerators or reactors, and the pulses they produce can be an undesirable background. These pulses fall into two general categories: prompt pulses that are produced within a few nanoseconds of the time the neutron enters the scintillator, and pulses that occur after some delay.

In NaI(Tl) and BGO, the prompt pulses are principally due to the detection of gamma rays produced in inelastic scattering interactions of the neutron with the scintillator.[40] Figure 10-27 shows the detection efficiency corresponding to the prompt pulses in these two materials. BGO has a better gamma ray to neutron sensitivity ratio compared with NaI(Tl).

The delayed pulses can be triggered by two categories of events. A neutron first may be moderated (requiring perhaps 100 μs) and then captured in the detector as a thermal neutron. The resulting capture gamma rays may lead to a detected pulse. Pulses that occur after a much longer delay may result if the neutron capture produces a radioactive species that subsequently decays. Examples with their corresponding half-lives are ^{24}Na (15 h) and ^{128}I (25 min) in sodium iodide, and ^{75}Ge (83 min) and ^{77}Ge (11.3 h) in BGO.

VI. ELECTRON SPECTROSCOPY WITH SCINTILLATORS

Scintillators can also be applied to the measurement of fast electrons (such as beta particles) that are incident on one surface of the crystal. Although it has become more common to use lithium-drifted silicon detectors for this purpose (see Chapter 13), applications sometimes arise in which the size limitation of silicon detectors or other considerations dictate the use of scintillators.

The nature of the electron response function depends on the scintillation material, its physical thickness, and the angle of incidence of the electrons. Electrons from an external source normally must pass through some protective covering and/or light reflector before reaching the surface of the scintillator itself. In the discussion that follows, the energy loss that may occur in these intervening materials is not explicitly considered but may be important if the electron energy is small. We also assume that the scintillators under consideration are thicker than the maximum range of the incident electrons. Even so, the detector may not be totally opaque to the secondary bremsstrahlung photons that will be generated along the path of the electron.

In general, the response functions show a pronounced full-energy peak corresponding to the total absorption of the incident electron energy, together with a tail extending to lower energies. The major cause of such partial energy absorption is backscattering, in which the electron reemerges from the surface through which it entered after having undergone only partial energy loss. Other events that contribute to the tail are those electrons which are fully stopped within the scintillator but which generate bremsstrahlung photons that escape from the front or back surface of the detector.

Both the probability of backscattering and the fraction of the electron energy loss to bremsstrahlung increase markedly with the atomic number of the scintillator. Because both processes detract from the full-energy peak and add the unwanted tail to the response function, scintillators with low atomic number are generally preferred for electron spectroscopy (just the opposite criterion than that desired for gamma-ray spectroscopy). Therefore, organic scintillators such as anthracene or plastics are most commonly applied in electron measurements. Table 10-1 lists the probability for backscattering of normally incident electrons from some common scintillator materials. The much lower backscattering probabilities for the low-Z materials are evident.

Figure 10-28 shows measured pulse height spectra in both cesium iodide and a plastic scintillator for normally incident 1 MeV electrons. The low-energy tail is more pronounced for cesium iodide because of its higher atomic number. Additional data for other scintillation materials, electron energies, and varying angles of incidence are given in Ref. 41.

TABLE 10-1 Fraction of Normally Incident Electrons Backscattered from Various Detector Surfaces

	Electron Energy (MeV)				
Scintillator	0.25	0.50	0.75	1.0	1.25
Plastic	0.08 ± 0.02	0.053 ± 0.010	0.040 ± 0.007	0.032 ± 0.003	0.030 ± 0.005
Antracene	0.09 ± 0.02	0.051 ± 0.010	0.038 ± 0.004	0.029 ± 0.003	0.026 ± 0.004
NaI(Tl)	0.450 ± 0.045	0.410 ± 0.010	0.391 ± 0.014	0.375 ± 0.008	0.364 ± 0.007
CsI(Tl)	0.49 ± 0.06	0.455 ± 0.023	0.430 ± 0.013	0.419 ± 0.018	0.404 ± 0.016

Source: Titus.[41]

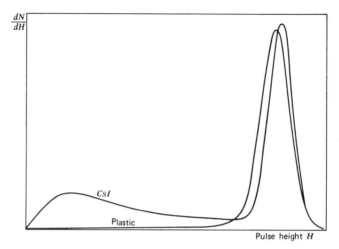

Figure 10-28 Experimental pulse height spectra from CsI(Tl) and plastic scintillators for 1.0 MeV electrons at normal incidence. The spectra are normalized to the same maximum pulse height. (From Titus.[41])

VII. SPECIALIZED DETECTOR CONFIGURATIONS BASED ON SCINTILLATION

A. The Phoswich Detector

The combination of two dissimilar scintillators optically coupled to a single PM tube is often called a *phoswich* (or phosphor sandwich) detector. The scintillators are chosen to have different decay times so that the shape of the output pulse from the PM tube is dependent on the relative contribution of scintillation light from the two scintillators. Most applications involve the use of this pulse shape difference to distinguish events that have occurred in only one scintillator from those that occur in both. For example, lightly penetrating radiations can be made to stop fully in the first scintillator, but more penetrating particles may generate light in both. Sodium iodide and cesium iodide are often chosen as the two materials because their decay times are quite different (0.23 and 1.0 μs), and pulses arising from only one decay are easily distinguished from those with both components, using the pulse shape discrimination methods of Chapter 17. Other common scintillator combinations include BGO and cesium iodide[42,43] or two different plastic scintillators,[44,45] one with a fast decay time and the other with a slow decay. Phoswich detectors employing pulse shape discrimination have proved to be useful in suppressing background in the counting of X-rays and beta particles.[46,47]

Alternatively, separate electronic pulses may be derived from the fast and slow components of the PM tube signal.[45,48] Independent measurements of the energy deposited in each scintillator can then be obtained without the need for a second PM tube. Using a thin fast scintillator in front of a thick slow scintillator allows simultaneous measurements of dE/dx and E for particles that penetrate the thin detector. The particle identification techniques described on p. 380 can then be applied to distinguish one type of heavy charged particle from another.

B. The Moxon–Rae Detector

Another type of gamma-ray detector can be built based on the detection of secondary electrons produced by gamma-ray interactions in a converter that is external to the detector. Although the resulting detection efficiency is typically quite low, the efficiency versus energy dependence can be tailored in certain ways to suit the need of specialized applications. For example, Moxon and Rae[49] first described the properties of a thick low-Z converter used to produce secondary electrons that were then detected by a thin plastic scintillator. By choosing a converter that consisted of graphite with a small component of bismuth, they found that the detection efficiency could be made nearly proportional to the incident gamma-ray energy. This proportionality allows a simplified analysis to be carried out for a class of experiments involving neutron capture because the detection probability of any given radiative capture event becomes independent of the specific deexcitation cascade mechanisms following the capture.[49] The proportionality of efficiency with gamma-ray energy results from the combined probabilities of electron production within the converter and of their escape from the surface facing the electron detector. Detailed analyses of the theory and performance of Moxon–Rae detectors can be found in Refs. 49–51.

C. Liquid Scintillation Counters

The liquid scintillation media discussed in Chapter 8 can be applied to avoid some of the difficulties that arise when measuring low-energy beta particles or alpha particles using conventional methods. The approach, sometimes called *internal source liquid scintillation counting*, involves dissolving the sample to be counted directly in the liquid scintillator. Under these conditions, problems relating to sample self-absorption, attenuation of particles by detector windows, and beta backscattering from the detector are completely avoided. These advantages are particularly important for low-energy radiations such as the beta particles emitted by tritium and ^{14}C. The endpoint energies for these beta spectra are 18 and 160 keV, respectively, and average beta energies are about one-third these values. Because these isotopes are particularly important in chemical and biomedical applications, much of the development of the liquid scintillation technique has taken place in connection with these sciences. Several texts have been published[52-54] which thoroughly review the fundamental principles of liquid scintillation counting. Proceedings of conferences dedicated to various applications in this field can be found in Refs. 55–57.

The first step in the technique involves incorporation of the sample within a suitable liquid scintillation solution. Problems can often arise in this step because most liquid scintillation solutions are based on toluene or other organic solvent, whereas many samples are often more conveniently prepared in an aqueous solution. Detailed discussions of various methods for obtaining stable solutions through the use of solubilizing or complexing agents are given in Refs. 58 and 59. A common problem is that the introduction of the sample tends to reduce the scintillation light output compared with the pure scintillator. This phenomenon, commonly called *quenching*, often limits the amount of sample material that can effectively be incorporated within the scintillator solution. The quenching can be due to either alteration of the optical properties of the solution by the sample (*color quenching*) or interference with the energy transfer process within the scintillator itself. Insoluble materials can sometimes be introduced as a suspension of fine particles. In those cases where rapid settling of the suspension is a problem, some methods have been developed that involve converting the solution to a gel immediately after preparation of these suspensions.

After the sample has been prepared, the solution is normally loaded into a glass vial and placed in a light-tight enclosure viewed by one or more PM tubes. Because of the effects of quenching and the fact that typical applications involve radiations of low energy, pulses produced in the PM tube often correspond to no more than a few photoelectrons. Under these circumstances, the measurements are potentially very sensitive to sources of noise that may interfere with accurate and reproducible counting of the sample. Significant sources of noise are thermally generated electrons from the photocathode of the PM tube, long-lived phosphorescence in the scintillator, and chemiluminescence (light generated by chemical reactions within the sample–scintillator solution). Virtually all these noise sources correspond to the generation of only a single electron per pulse, so they can normally be rejected by placing a discrimination level in the signal chain to eliminate those pulses whose amplitude corresponds to a single photoelectron. However, because the signal also consists of only a few photoelectrons, there is a risk that this discrimination process will also eliminate some of the signal.

Because all beta particles emitted by the sample pass through some portion of the scintillator, and the great majority are fully stopped within the solution, the counting efficiency can potentially be close to 100%. The degree to which the few-photoelectron signal can be distinguished from single-electron noise determines the practical counting efficiency. As a gauge of the development of the technique, it is interesting to note the improvement in the counting efficiency for tritium as improvements in PM tube design and other techniques have been implemented. A counting efficiency of about 20% in 1960 was improved to about 60% in 1970 through the use of low-noise bialkali PM tubes, and to a present value of about 90% through the application of PM tubes with gallium phosphide high-gain dynodes.[60]

One method of eliminating the PM tube noise is to use two PM tubes to view the scintillator from opposite sides. Only those pulses that are observed in coincidence between the two tubes are counted. Because the noise generated in each tube will be uncorrelated, a true coincidence will not be observed for these events, and the recorded counts will correspond only to events generated within the liquid scintillator. The summed output from both PM tubes can then be used to record the pulse height spectrum from the sample.

Although most commonly applied to samples emitting beta particles or conversion electrons, liquid scintillation counting has also been used to count samples that are alpha active.[61–63] Although the best reported energy resolution[62] of 5–8% is much inferior to that attainable with semiconductor diode detectors, the advantages of high counting efficiency and uniform counting geometry offer some attraction for applications such as counting low-level environmental samples.[63] Because typical alpha energies are several MeV, the light output is much greater than for low-energy beta particle counting, and therefore a counting efficiency that approaches 100% is relatively easy to achieve. Pulse shape discrimination can be applied to eliminate backgrounds due to beta particles or gamma-ray-induced events.[64] By using these techniques, an exceptionally low background level of 0.01 count/min has been reported[62] while maintaining essentially 100% counting efficiency for alpha particles within the sample.

D. Position-Sensitive Scintillators

1. ONE-DIMENSIONAL POSITION SENSING

Because the light from a scintillator is generated along the track of the ionizing particle, it is possible to sense the position of interaction by localizing the source of the scintillation light. For sensing position in one dimension, a long rod or bar of scintillation material can be used with PM tubes or photodiodes positioned at either end as in the sketch below:

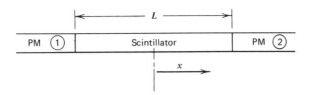

In this type of geometry, it is generally observed that the intensity of the light measured at one end of the rod drops off exponentially with the distance at which the scintillation light is generated. Thus, we can write for the signal from PM tube ①

$$E_1 = \frac{E_\gamma P}{E_0} \exp\left[-\alpha(L/2 + x)\right] \tag{10-18}$$

where E_γ = energy deposited by gamma ray

$P =$ probability that light quantum produced at one end will generate a photoelectron in adjacent tube

$E_0 =$ energy deposited per light photon created in scintillator

α = light attenuation coefficient.

Similarly, for PM tube ②,

$$E_2 = \frac{E_\gamma P}{E_0} \exp\left[-\alpha\left(\frac{L}{2} - x\right)\right] \tag{10-19}$$

By dividing one signal by the other, we obtain

$$\frac{E_2}{E_1} = \frac{\exp\left[-\alpha(L/2 - x)\right]}{\exp\left[-\alpha(L/2 + x)\right]} = \exp\left(+2\alpha x\right)$$

$$\ln \frac{E_2}{E_1} = 2\alpha x$$

$$x = \frac{1}{2\alpha} \ln \frac{E_2}{E_1} \tag{10-20}$$

Therefore, by electronically deriving the logarithm of the ratio of the two PM tube signals, we obtain a linear indication of the position at which the scintillation occurs.

By multiplying Eq. (10-18) and (10-19), we obtain

$$E_\gamma^2 = E_1 E_2 \left(\frac{E_0}{P}\right)^2 e^{\alpha L}$$

Now taking the square root of both sides, we obtain

$$E_\gamma = \sqrt{E_1 E_2}\, \frac{E_0}{P} e^{\alpha L/2} \tag{10-21}$$

Thus, the square root of the product of the two PM tube signals serves as a measure of the total scintillation light, independent of the position within the bar.

For best spatial resolution,[65] the light attenuation coefficient α in the expressions above should have an optimal value of $2.9/L$. This light attenuation can be accomplished either by adjusting the reflection conditions at the surface of the scintillator or by allowing some absorption of the light to occur in the bulk of the scintillator. In a typical application of this technique,[65] a spatial resolution of 10 mm was obtained in a 500 mm long sodium iodide rod, using gamma rays of 660 keV energy.

An alternative method of sensing the scintillation position in one dimension is to use the same geometry described above but to exploit the fact that there will be a slight difference in the time of arrival of the scintillation light at the two ends of the bar. This time difference is maximum for an event that occurs at either end of the bar and decreases linearly to zero for an event at the center. The time differences are small, since the velocity of propagation of the light traveling directly from the scintillation site is c/n, where c is the velocity of light and n is the index of refraction of the scintillation medium. For typical scintillators, this corresponds to a flight time of about 5 ps/mm. However, most of the light is reflected many times from the surface of the scintillator as it travels along the length of the bar and therefore the actual flight path and propagation time are substantially extended. In one application of this method using a 250 mm long plastic scintillator rod,[66] a spatial resolution of between 17 and 23 mm was obtained for 511 keV gamma rays.

2. TWO-DIMENSIONAL POSITION SENSING (IMAGING DETECTORS)

In nuclear medicine, it is often necessary to form the image of the distribution of gamma-ray-emitting isotopes distributed throughout the patient. The *gamma-ray camera* is a device that senses the two-dimensional coordinates of a gamma-ray photon as it interacts in a large-area detector and forms an image through the accumulation of many

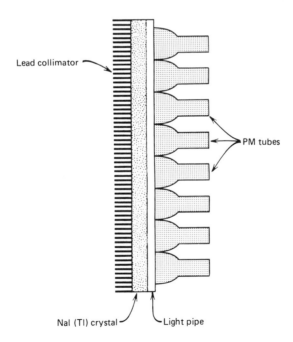

Figure 10-29 Elements of a two-dimensional position-sensitive scintillation detector, commonly called a gamma camera.

such events over the exposure time. A lead pinhole or parallel hole collimator is used to restrict the gamma rays that strike the detector so that the image can be directly interpreted as the spatial distribution of the emitting isotope.

The most common type of gamma-ray camera is based on an original design by Anger.[67] Its basic elements are diagrammed in Fig. 10-29. The detection medium consists of a flat single scintillation crystal (generally sodium iodide) with diameter of 25–50 cm and thickness of about 1 cm. The light generated by gamma-ray interactions in this crystal is sensed by an array of PM tubes that completely cover one of its flat faces. The two-dimensional position of each event across the area of the crystal is deduced from the relative size of the signals produced from these tubes.

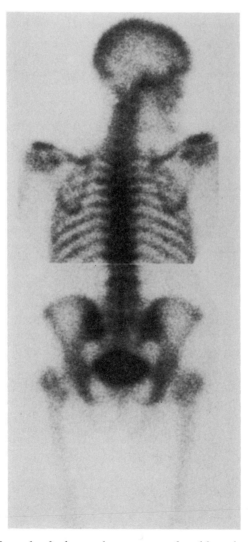

Figure 10-30 Example of a human bone scan produced by using a gamma camera to image the distribution of methylene diphosphonate labeled with ^{99m}Tc. The upper and lower halves of the image each were produced by recording approximately 10^6 counts over a 3 min period. (Courtesy of W. L. Rogers, University of Michigan Medical Center.)

Each scintillation event will generate output pulses of significant amplitude from all the PM tubes that are near the location of the interaction. The largest signal will generally be from the tube nearest the position, with smaller pulses from tubes at a greater distance. The "center of gravity" of the light is interpolated from these signals using either a charge division process in which a portion of each signal is coupled to X and Y output lines, or through the use of readout schemes based on delay line encoding of the position.[68] The resulting analog X and Y position signals are typically digitized using fast analog-to-digital converters and accumulated in a two-dimensional digital memory to form the image. The summed output from all the tubes is a good measure of the total energy deposited in the crystal and is normally passed through a single-channel analyzer to record selectively only those events that correspond to the full source energy. This selection eliminates loss of contrast in the image due to gamma rays that have initially scattered in the patient or elsewhere before reaching the camera and would therefore give a false position indication.

An example of an image from a typical gamma-ray camera is given in Fig. 10-30. The intrinsic spatial resolution of the camera when recording the most common type of image generated by ^{99m}Tc (140 keV) is limited to about 3 mm FWHM. In practice, additional spatial resolution loss normally takes place due to the geometric uncertainties in the source position when it is some distance from the face of the collimator.

PROBLEMS

10-1. A gamma-ray photon after Compton scattering through an angle of 90° has an energy of 0.5 MeV. Find its energy before the scattering.

10-2. A 2 MeV gamma-ray photon is incident on a detector, undergoes two sequential Compton scatterings, and then escapes. If the angles of scattering are 30° and 60°, respectively, how much total recoil electron energy is deposited in the detector? Does the answer change if the sequence of the scattering angles is reversed?

10-3. Find the maximum energy that can be deposited by a 1 MeV gamma-ray photon if it undergoes two successive Compton scattering events and then escapes the detector.

10-4. Estimate the time that separates two successive gamma-ray scattering interactions that are 3 cm apart in sodium iodide. Compare with the characteristic decay time for the light that is generated in the same material.

10-5. The cross sections for photoelectric, Compton, and pair production interactions in sodium iodide at 2 MeV are in the ratio $1:20:2$, respectively. Will the pulse height spectrum from 2 MeV gamma rays incident on a sodium iodide scintillator give a peak-to-total ratio of less than, more than, or about equal to $1/23$?

10-6. If the energy resolution of a particular NaI(Tl) scintillation detector is 7% for ^{137}Cs gamma rays (0.662 MeV), estimate its energy resolution for the 1.28 MeV gamma rays from ^{22}Na.

10-7. The mass attenuation coefficient of NaI at 0.5 MeV is 0.955 cm^2/g. Find the intrinsic total efficiency of a slab detector 0.50 cm thick at this energy. If the photofraction is 40% at the same energy, what is the intrinsic peak efficiency?

10-8. (a) Find the energy of the Compton edge for the 1.17 MeV gamma rays from ^{60}Co.

(b) Calculate the backscatter peak energies corresponding to incident gamma rays of 1, 2, and 3 MeV.

10-9. Listed below are a number of parameters of interest in gamma-ray spectroscopy using scintillation detectors:

  (a) Density of the detector medium.
  (b) Kinetic energy required to create a scintillation photon in the crystal.
  (c) Average atomic number (Z-value) of the detector medium.
  (d) Geometry of the source–detector system.
  (e) Gain of the photomultiplier tube.
  (f) Quantum efficiency of the photocathode in the photomultiplier.
  (g) Gain of the amplifier used between the detector and pulse analysis system.
  (h) Fraction of light generated in the crystal that reaches the photocathode of the PM tube (light collection efficiency).

  Identify those parameters from this list that have a major influence on the detector *intrinsic peak efficiency*. Repeat, but now identify those that have a major influence on *energy resolution*.

10-10. Calculate the pulse amplitude from the anode of a PM tube used with a NaI(Tl) scintillator under the following conditions: A 1 MeV electron loses all its energy in the scintillator, the light collection efficiency to the photocathode is 50%, the average quantum efficiency of the photocathode is 20%, and 80% of the photoelectrons are collected at the first dynode. Assume that the PM tube has 10 stages with a multiplication factor $\delta = 2.5$ per stage. The anode load resistance is 100 kΩ, and the anode capacitance is 100 pF.

10-11. A particular radioisotope emits two coincident gamma rays, each with 100% yield per decay, with no angular correlation between the photon directions. A sample is placed 10 cm from the surface of a 5 cm radius cylindrical detector along its axis. The intrinsic peak efficiency of the detector for γ_1 is 50%, and for γ_2 it is 30%.

  (a) If the sample activity is low enough so that chance coincidences are negligible, calculate the ratio of the counts under the sum peak in the recorded pulse height spectrum to the counts under the γ_1 full-energy peak.
  (b) Calculate the rate at which events are recorded in the sum peak if the source activity is 100 kBq. For a detector resolving time of 3 μs, what additional rate should be expected from chance coincidences between γ_1 and γ_2?

10-12. From Fig. 2-14, estimate the range of a 1MeV electron in sodium iodide. From your answer, calculate the percentage of the total volume of a 5.08 cm $\times$ 5.08 cm cylindrical crystal that lies near enough to the surface so that electron escape is possible.

10-13. A radioisotope source is known not to emit any gamma-ray photons with energy of 511 keV, but a peak is observed at this position in the recorded gamma-ray spectrum. Give two possible origins for this peak.

10-14. If the energy resolution of a scintillator is 8.5% at 662 keV, find the standard deviation (in energy units) of the Gaussian curve that would be a fit to the photopeak at that energy.

10-15. Why are materials with low atomic number often preferred as scintillators for electron spectroscopy, while the opposite is true for gamma-ray spectroscopy?

10-16. Explain the major advantage of liquid scintillation counting when applied to low-energy beta emitters compared with conventional solid scintillation detectors.

REFERENCES

1. J. B. Birks, *The Theory and Practice of Scintillation Counting*, Pergamon Press, Oxford, 1964.

2. S. M. Shafroth (ed.), *Scintillation Spectroscopy of Gamma Radiation*, Gordon & Breach, London, 1967.

3. K. Siegbahn (ed.), *Alpha-, Beta-, and Gamma-Ray Spectroscopy*, North Holland, Amsterdam, 1968.

4. C. D. Broyles, D. A. Thomas, and S. K. Haynes, *Phys. Rev.* **89**, 715 (1953).

5. R. L. Heath, *Scintillation Spectrometry Gamma-Ray Spectrum Catalogue*, IDO-16880, Vols. 1 and 2, 1964.

6. F. Adams and R. Dams, *Applied Gamma-Ray Spectrometry*, 2nd ed. and revision of original publication by C. E. Crouthamel, Pergamon Press, Oxford, 1970.

7. N. A. Wogman, D. E. Robertson, and R. W. Perkins, *Nucl. Instrum. Meth.* **50**, 1 (1967).

8. N. A. Wogman, R. W. Perkins, and J. H. Kaye, *Nucl. Instrum. Meth.* **74**, 197 (1969).

9. B. A. Euler, D. F. Covell, and S. Yamamoto, *Nucl. Instrum. Meth.* **72**, 143 (1969).

10. B. Bengtson and M. Moszynski, *Nucl. Instrum. Meth.* **85**, 133 (1970).

11. M. D. Hasinoff, S. T. Lim, D. F. Measday, and T. J. Mulligan, *Nucl. Instrum. Meth.* **117**, 375 (1974).

12. R. Mueller and D. Maeder, "Single Crystal Spectroscopy," Chap. VII in *Scintillation Spectroscopy of Gamma Radiation* (S. M. Shafroth, ed.), Gordon & Breach, New York, 1967.

13. J. R. Dell and P. J. Ebert, *Nucl. Instrum. Meth.* **68**, 335 (1969).

14. C. E. Moss et al., *Nucl. Instrum. Meth.* **219**, 558 (1984).

15. M. J. Berger and S. M. Seltzer, *Nucl. Instrum. Meth.* **104**, 317 (1972).

16. R. J. D. Beattie and J. Byrne, *Nucl. Instrum. Meth.* **104**, 163 (1972).

17. D. E. Persyk and T. E. Moi, *IEEE Trans. Nucl. Sci.* **NS-25**(1), 615 (1978).

18. G. H. Narayan and J. R. Prescott, *IEEE Trans. Nucl. Sci.* **NS-15**(3), 162 (1968).

19. R. D. Bolton, H. W. Baer, J. D. Bowman, and L. Gordon, *Nucl. Instrum. Meth.* **174**, 411 (1980).

20. J. S. Kapustinsky, R. M. De Vries, N. J. DiGiacomo, W. E. Sondheim, J. W. Sunier, and H. Coombes, *Nucl. Instrum. Meth. Phys. Res.* **A241**, 612 (1985).

21. C. A. Benulis and W. K. McFarlane, *Nucl. Instrum. Meth. Phys. Res.* **A240**, 135 (1985).

22. L. Holm, H. W. Fielding, and G. C. Neilson, *Nucl. Instrum. Meth. Phys. Res.* **A234**, 517 (1985).

23. B. K. Utts, *Nucl. Instrum. Meth. Phys. Res.* **A242**, 516 (1986).

24. B. J. Snyder, "Calculation of Gamma Ray Scintillation Detector Efficiencies and Photofraction by Monte Carlo Methods," Ph.D. Dissertation, The University of Michigan, 1965.

25. C. C. Grosjean and W. Bossaert, *Table of Absolute Detection Efficiencies of Cylindrical Scintillation Gamma-Ray Detectors*, Computing Laboratory, University of Ghent, 1965.

26. B. Grosswendt and E. Waibel, *Nucl. Instrum. Meth.* **133**, 25 (1976).

27. P. Holmberg and R. Rieppo, *Int. J. Appl. Radiat. Isotopes* **24**, 99 (1973).

28. R. Rieppo, *Int. J. Appl. Radiat. Isotopes* **27**, 453 (1976).

29. R. Rieppo, *Nucl. Instrum. Meth.* **115**, 541 (1974).

30. Y. S. Horowitz, S. Mordechai, and A. Dubi, *Nucl. Instrum. Meth.* **122**, 399 (1974).

31. T. Nakamura, *Nucl. Instrum. Meth.* **105**, 77 (1972).

32. M. Belluscio, R. De Leo, A. Pantaleo, and A. Vox, *Nucl. Instrum. Meth.* **118**, 553 (1974).

33. I. Petr, A. Adams, and J. B. Birks, *Nucl. Instrum. Meth.* **95**, 253 (1971).

34. M. Irfan and R. D. G. Prasad, *Nucl. Instrum. Meth.* **107**, 583 (1973).

35. A. E. Evans and J. D. Orndoff, NUREG/CR-1398 (LA-8301), 1980.

36. N. A. Lurie, L. Harris, Jr., and J. P. Wondra, *Nucl. Instrum. Meth.* **129**, 619 (1975).

37. R. De Leo, G. D'Erasmo, and A. Pantaleo, *Nucl. Instrum. Meth.* **129**, 501 (1975).

38. A. Schaarschmidt and H. Durner, *Nucl. Instrum. Meth.* **105**, 504 (1972).

39. T. S. Mudhole and N. Umakantha, *Nucl. Instrum. Meth.* **116**, 401 (1974).

40. O. Hausser et al., *Nucl. Instrum. Meth.* **213**, 301 (1983).

41. F. Titus, *Nucl. Instrum. Meth.* **89**, 93 (1970).

42. E. Costa, E. Massaro, and L. Piro, *Nucl. Instrum. Meth. Phys. Res.* **A243**, 572 (1986).

43. L. Piro, E. Massaro, S. Fiacconi, G. E. Gigante, and E. Costa, *Nucl. Instrum. Meth. Phys. Res.* **A257**, 429 (1987).

44. M. Bantel et al., *Nucl. Instrum. Meth.* **226**, 394 (1984).

45. F. Lidén, J. Nyberg, A. Johnson, and A. Kerek, *Nucl. Instrum. Meth. Phys. Res.* **A253**, 305 (1987).

46. M. R. Mayhugh, A. C. Lucas, and B. K. Utts, *IEEE Trans. Nucl. Sci.* **NS-25**(1), 569 (1978).

47. B. H. Erkkila, M. A. Wolf, Y. Eisen, W. P. Unruh, and R. J. Brake, *IEEE Trans. Nucl. Sci.* **NS-32**(1), 969 (1985).

48. J. Pouliot, Y. Chan, A. Dacal, A. Harmon, R. Knop, M. E. Ortiz, E. Plagnol, and R. G. Stokstad, Lawrence Berkeley Laboratory Report LBL-24396, Dec. 1987.

49. M. C. Moxon and E. R. Rae, *Nucl. Instrum. Meth.* **24**, 445 (1963).

50. S. S. Malik, *Nucl. Instrum. Meth.* **125**, 45 (1975).

51. S. S. Malik and C. F. Majkrzak, *Nucl. Instrum. Meth.* **130**, 443 (1975).

52. Y. Kobayashi and D. V. Maudsley, *Biological Applications of Liquid Scintillation Counting*, Academic Press, New York, 1974.

53. D. L. Horrocks, *Applications of Liquid Scintillation Counting*, Academic Press, New York, 1974.

54. M. Crook and P. Johnson (eds.), *Liquid Scintillation Counting*, Vol. 4, Heyden and Son, London, 1977.

55. P. E. Stanley and B. A. Scoggins, *Liquid Scintillation Counting: Recent Developments*, Academic Press, New York, 1974.

56. A. A. Noujaim, C. Ediss, and L. I. Wiebe, *Liquid Scintillation Science and Technology*, Academic Press, New York, 1976.

57. D. L. Horrocks and C. T. Peng (eds.), *Organic Scintillators and Liquid Scintillation Counting*, Academic Press, New York, 1971.

58. J. H. Parmentier and F. E. L. Ten Haaf, *Int. J. Appl. Radiat. Isotopes* **20**, 305 (1969).

59. B. W. Fox, *Techniques of Sample Preparation for Liquid Scintillation Counting*, North-Holland, Amsterdam, 1976.

60. R. Vaninbroukx and I. Stanef, *Nucl. Instrum. Meth.* **112**, 111 (1973).

61. J. W. McKlveen, H. W. Berk, and W. R. Johnson, *Int. J. Appl. Radiat. Isotopes* **23**, 337 (1972).

62. J. W. McKlveen and W. J. McDowell, *Trans. Am. Nuc. Soc.* **22**, 149 (1975).

63. W. J. McDowell, *IEEE Trans. Nucl. Sci.* **NS-22**(1), 649 (1975).

64. P. Cross and G. W. McBeth, *Health Phys.* **30**, 303 (1976).

65. J. N. Carter et al., *Nucl. Instrum. Meth.* **196**, 477 (1982).

66. R. Myllyla, H. Heusala, and M. Karras, *IEEE Trans. Nucl. Sci.* **NS-28**(1), 167 (1981).

67. H. O. Anger, *Rev. Sci. Instrum.* **29**, 27 (1958).

68. T. Hiramoto, E. Tanaka, and N. Nohara, *J. Nucl. Med.* **12**, 160 (1971).

CHAPTER · 11

Semiconductor Diode Detectors

In many radiation detection applications, the use of a solid detection medium is of great advantage. For the measurement of high-energy electrons or gamma rays, detector dimensions can be kept much smaller than the equivalent gas-filled detector because solid densities are some 1000 times greater than that for a gas. Scintillation detectors offer one possibility of providing a solid detection medium, and their application to the detection and measurement of various radiations has been described in Chapter 10.

One of the major limitations of scintillation counters is their relatively poor energy resolution. The chain of events that must take place in converting the incident radiation energy to light and the subsequent generation of an electrical signal involves many inefficient steps. Therefore, the energy required to produce one information carrier (a photoelectron) is of the order of 100 eV or more, and the number of carriers created in a typical radiation interaction is usually no more than a few thousand. The statistical fluctuations in so small a number place an inherent limitation on the energy resolution that can be achieved under the best of circumstances, and nothing can be done about improving the energy resolution beyond this point. As detailed in Chapter 10, the energy resolution for sodium iodide scintillators is limited to about 6% when detecting 0.662 MeV gamma rays and is largely determined by the photoelectron statistical fluctuations.

The only way to reduce the statistical limit on energy resolution is to increase the number of information carriers per pulse. As we show in this chapter, the use of semiconductor materials as radiation detectors can result in a much larger number of carriers for a given incident radiation event than is possible with any other detector type. Consequently, the best energy resolution achievable today is realized through the use of such detectors. The basic information carriers are *electron–hole pairs* created along the path taken by the charged particle (primary radiation or secondary particle) through the detector. The electron–hole pair is somewhat analogous to the ion pair created in gas-filled detectors. Their motion in an applied electric field generates the basic electrical signal from the detector.

Devices employing semiconductors as the basic detection medium became practically available in the early 1960s. Early versions were called *crystal counters*, but modern detectors are referred to as *semiconductor diode detectors* or simply *solid-state detectors*. Although the latter term is somewhat ambiguous in the sense that technically scintillation counters can also be thought of as solid detectors, it has come into widespread use to characterize only those devices that are based on electron–hole pair collection from semiconductor media.

In addition to superior energy resolution, solid-state detectors can also have a number of other desirable features. Among these are compact size, relatively fast timing characteristics, and an effective thickness that can be varied to match the requirements of the application. Drawbacks may include the limitation to small sizes and the relatively high susceptibility of these devices to performance degradation from radiation-induced damage.

Of the available semiconductor materials, silicon predominates in the diode detectors used primarily for charged particle spectroscopy and discussed in this chapter. Germanium is more widely used in the gamma-ray measurements described in Chapter 12, whereas devices that use other semiconductor materials are covered in Chapter 13.

Several comprehensive books are available on the topic of solid-state detectors, including Refs. 1–6. Each of these contains a rather complete citation of the literature up to the time of publication, and the other references in these chapters are largely limited to those that have appeared more recently.

I. SEMICONDUCTOR PROPERTIES

A. Band Structure in Solids

The periodic lattice of crystalline materials establishes allowed energy bands for electrons that exist within that solid. The energy of any electron within the pure material must be confined to one of these energy bands, which may be separated by gaps or ranges of forbidden energies. A simplified representation of the bands of interest in insulators or semiconductors is shown in Fig. 11-1. The lower band, called the *valence band*, corresponds to those electrons that are bound to specific lattice sites within the crystal. In the case of silicon or germanium, they are parts of the covalent bonding which constitute the interatomic forces within the crystal. The next higher-lying band is called the *conduction band* and represents electrons that are free to migrate through the crystal. Electrons in this band contribute to the electrical conductivity of the material. The two bands are separated by the *bandgap*, the size of which determines whether the material is classified as a semiconductor or an insulator. The number of electrons within the crystal is just adequate to fill completely all available sites within the valence band. In the absence of

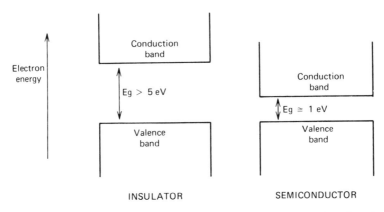

Figure 11-1 Band structure for electron energies in insulators and semiconductors.

thermal excitation, both insulators and semiconductors would therefore have a configuration in which the valence band is completely full and the conduction band completely empty. Under these circumstances, neither would theoretically show any electrical conductivity.

In a metal, the highest occupied energy band is not completely full. Therefore, electrons can easily migrate throughout the material because they need achieve only small incremental energy to be above the occupied states. Metals are therefore always characterized by very high electrical conductivity. In insulators or semiconductors, on the other hand, the electron must first cross the bandgap to reach the conduction band and the conductivity is therefore many orders of magnitude lower. For insulators, the bandgap is usually 5 eV or more, whereas for semiconductors, the bandgap is considerably less.

B. Charge Carriers

At any nonzero temperature, some thermal energy is shared by the electrons in the crystal. It is possible for a valence electron to gain sufficient thermal energy to be elevated across the bandgap into the conduction band. Physically, this process simply represents the excitation of an electron that is normally part of a covalent bond such that it can leave the specific bonding site and drift throughout the crystal. The excitation process not only creates an electron in the otherwise empty conduction band, but it also leaves a vacancy (called a *hole*) in the otherwise full valence band. The combination of the two is called an *electron–hole pair* and is roughly the solid-state analogue of the ion pair in gases. The electron in the conduction band can be made to move under the influence of an applied electric field. The hole, representing a net positive charge, will also tend to move in an electric field, but in a direction opposite that of the electron. The motion of both of these charges contributes to the observed conductivity of the material.

The probability per unit time that an electron–hole pair is thermally generated is given by

$$p(T) = CT^{3/2}\exp\left(-\frac{E_g}{2kT}\right) \tag{11-1}$$

where

T = absolute temperature

E_g = bandgap energy

k = Boltzmann constant

C = proportionality constant characteristic of the material.

As reflected in the exponential term, the probability of thermal excitation is critically dependent on the ratio of the bandgap energy to the absolute temperature. Materials with a large bandgap will have a low probability of thermal excitation and consequently will show the very low electrical conductivity characteristic of insulators. If the bandgap is as low as several electron volts, sufficient thermal excitation will cause a conductivity high enough for the material to be classified as a semiconductor. In the absence of an applied electric field, the thermally created electron–hole pairs ultimately recombine, and an equilibrium is established in which the concentration of electron–hole pairs observed at any given time is proportional to the rate of formation. From Eq. (11-1), this equilibrium concentration is a strong function of temperature and will decrease drastically if the material is cooled.[†]

[†] Because the ionization potential for gases is typically 15 eV or more, the probability of a thermally generated ion pair is negligibly small in gas ionization chambers, even at room temperature.

TABLE 11-1   Properties of Intrinsic Silicon and Germanium

	Si	Ge
Atomic number	14	32
Atomic weight	28.09	72.60
Stable isotope mass numbers	28-29-30	70-72-73-74-76
Density (300 K); g/cm^3	2.33	5.32
Atoms/cm^3	4.96×10^{22}	4.41×10^{22}
Dielectric constant	12	16
Forbidden energy gap (300 K); eV	1.115	0.665
Forbidden energy gap (0 K); eV	1.165	0.746
Intrinsic carrier density (300 K); cm^{-3}	1.5×10^{10}	2.4×10^{13}
Intrinsic resistivity (300 K); $\Omega \cdot$ cm	2.3×10^5	47
Electron mobility (300 K); cm^2/V $\cdot$ s	1350	3900
Hole mobility (300 K); cm^2/V $\cdot$ s	480	1900
Electron mobility (77 K); cm^2/V $\cdot$ s	2.1×10^4	3.6×10^4
Hole mobility (77 K); cm^2/V $\cdot$ s	1.1×10^4	4.2×10^4
Energy per electron–hole pair (300 K); eV	3.62	
Energy per electron–hole pair (77 K); eV	3.76	2.96
Fano factor (77 K)	0.143 (Ref. 7)	0.129 (Ref. 9)
	0.084 (Ref. 8)	0.08 (Ref. 10)
	0.085 ⎫	< 0.11 (Ref. 11)
	to ⎬ (Ref. 12)	0.057 ⎫ (Ref. 12)
	0.137 ⎭	0.064 ⎭
	0.16 (Ref. 13)	0.058 (Ref. 14)

Source: G. Bertolini and A. Coche (eds.), *Semiconductor Detectors*, Elsevier-North Holland, Amsterdam, 1968, except where noted.

C. Migration of Charge Carriers in an Electric Field

If an electric field is applied to the semiconductor material, both the electrons and holes will undergo a net migration. The motion will be the combination of a random thermal velocity and a net *drift velocity* parallel to the direction of the applied field. The motion of the conduction electrons is a relatively easy process to visualize, but the fact that holes also contribute to conductivity is less obvious. A hole moves from one position to another if an electron leaves a normal valence site to fill an existing hole. The vacancy left behind by the electron then represents the new position of the hole. Because electrons will always be drawn preferentially in an opposite direction to the electric field vector, holes move in the same direction as the electric field. This behavior is consistent with that expected of a point positive charge, because the hole actually represents the absence of a negatively charged electron.

At low-to-moderate values of the electric field intensity, the drift velocity v is proportional to the applied field. Then a *mobility* μ for both electrons and holes can be defined by

$$v_h = \mu_h \mathscr{E} \tag{11-2}$$

$$v_e = \mu_e \mathscr{E} \tag{11-3}$$

where $\mathscr{E}$ is the electric field magnitude. In gases, the mobility of the free electron is much larger than that of the positive ion, but in semiconductor materials the mobility of the

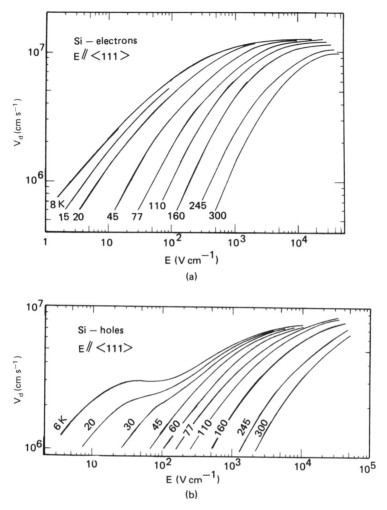

Figure 11-2 Drift velocity as a function of electric field applied parallel to the ⟨111⟩ crystallographic direction. Absolute temperature is the parameter for the different curves. (a) Electrons in silicon; (b) holes in silicon; (c) electrons in germanium; (d) holes in germanium. (From Ottaviani et al.[15])

electron and hole are roughly of the same order. Numerical values for common semiconductor materials are given in Table 11-1.

At higher electric field values, the drift velocity increases more slowly with the field. Eventually, a *saturation velocity* is reached which becomes independent of further increases in the electric field. Figure 11-2 shows the dependence of the drift velocity on field magnitude for silicon and germanium.

Many semiconductor detectors are operated with electric field values sufficiently high to result in saturated drift velocity for the charge carriers. Because these saturated velocities are of the order of 10^7 cm/s, the time required to collect the carriers over typical dimensions of 0.1 cm or less will be under 10 ns. Semiconductor detectors can therefore be among the fastest-responding of all radiation detector types.

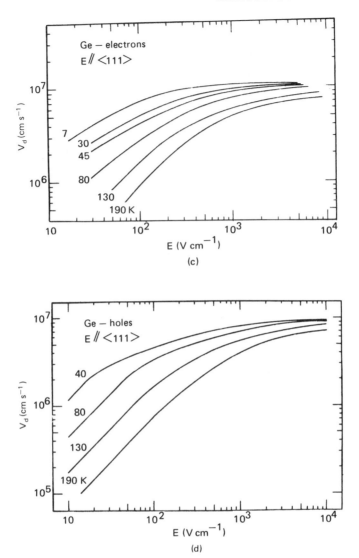

Figure 11-2 (*Continued*)

D. Effect of Impurities or Dopants

1. INTRINSIC SEMICONDUCTORS

In a completely pure semiconductor, all the electrons in the conduction band and all the holes in the valence band would be caused by thermal excitation (in the absence of ionizing radiation). Because under these conditions each electron must leave a hole behind, the number of electrons in the conduction band must exactly equal the number of holes in the valence band. Such material is called an *intrinsic* semiconductor. Its properties can be described theoretically, but in practice it is virtually impossible to achieve. The electrical properties of real materials tend to be dominated by the very small levels of residual impurities; this is true even for silicon and germanium, which are the semiconductors available in the highest practical purities.

In the discussions that follow, we let n represent the concentration (number per unit volume) of electrons in the conduction band. Also, p represents the concentration of holes in the valence band. In the intrinsic material (subscript i), the equilibrium established by the thermal excitation of electrons from the valence to conduction band and their subsequent recombination leads to equal numbers of electrons and holes, or

$$n_i = p_i \tag{11-4}$$

The quantities n_i and p_i are known as the intrinsic carrier densities. From Eq. (11-1), it is clear that these densities will be lowest for materials with large bandgap energy and when the material is used at low temperature. Intrinsic hole or electron densities at room temperature are 1.5×10^{10} cm^{-3} in silicon, and 2.4×10^{13} cm^{-3} in germanium.

2. n-TYPE SEMICONDUCTORS

To illustrate the effect of doping on semiconductor properties, we use crystalline silicon as an example. Germanium and other semiconductor materials behave in a similar way. Silicon is tetravalent and in the normal crystalline structure forms covalent bonds with the four nearest silicon atoms. A sketch of this situation is shown in Fig. 11-3a, where each of the dashes represents a normal valence electron involved in a covalent bond. Thermal excitation in the intrinsic material consists of breaking loose one of these covalent electrons, leaving behind an unsaturated bond or hole.

We now consider the effect of the small concentration of impurity which may be present in the semiconductor either as a residual amount after the best purification processes, or as a small amount intentionally added to the material to tailor its properties. We first assume that the impurity is pentavalent or is found in group V of the periodic table. When present in small concentrations (of the order of a few parts per million or less) the impurity atom will occupy a substitutional site within the lattice, taking the place of a normal silicon atom. Because there are five valence electrons surrounding the impurity atom, there is one left over after all covalent bonds have been formed. This extra electron is somewhat of an orphan and remains only very lightly bound to the original impurity site. It therefore takes very little energy to dislodge it to form a conduction electron without a corresponding hole. Impurities of this type are referred to as *donor impurities* because they readily contribute electrons to the conduction band. Because they are not part of the regular lattice, the extra electrons associated with donor impurities can occupy a position within the normally forbidden gap. These very loosely bound electrons will have an energy near the top of the gap as shown in Fig. 11-3b. The energy spacing between these donor levels and the bottom of the conduction band is sufficiently small so that the probability of thermal excitation given by Eq. (11-1) is high enough to ensure that a large fraction of all the donor impurities are ionized. In nearly all cases, the concentration of impurity N_D is large compared with the concentration of electrons expected in the conduction band for the intrinsic material. Therefore, the number of conduction electrons becomes completely dominated by the contribution from the donor impurities, and we can write

$$n \cong N_D \tag{11-5}$$

The added concentration of electrons in the conduction band compared with the intrinsic value increases the rate of recombination, shifting the equilibrium between electrons and holes. As a result, the equilibrium concentration of holes is decreased by an amount such that the equilibrium constant given by the product of n and p is the same as for the

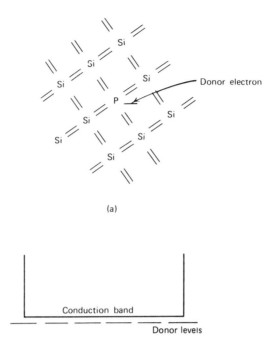

(a)

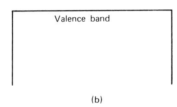

(b)

Figure 11-3 (a) Representation of a donor impurity (phosphorus) occupying a substitutional site in a silicon crystal. (b) Corresponding donor levels created in the silicon bandgap.

intrinsic material:

$$np = n_i p_i \tag{11-6}$$

For example, in room-temperature silicon, the intrinsic carrier densities are about 10^{10} cm^{-3}. If a donor impurity is present at a concentration of 10^{17} atoms/cm^3 (about 2 parts per million), the density of conduction electrons n will be 10^{17} cm^{-3} and the concentration of holes p will be 10^3 cm^{-3}. Because the *total* number of charge carriers of both types is now much greater (10^{17} cm^{-3} versus 2×10^{10} cm^{-3}), the electrical conductivity of a doped semiconductor is always much larger than that of the corresponding pure material.

Even though conduction electrons now greatly outnumber the holes, charge neutrality is maintained because of the presence of ionized donor impurities. These sites represent net positive charges which exactly balance the excess electron charges. They are not, however, to be confused with holes because the ionized donors are fixed in the lattice and cannot migrate.

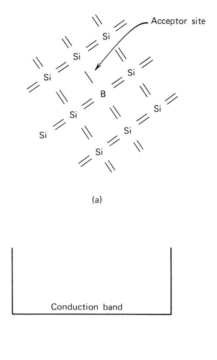

(a)

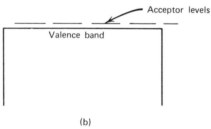

(b)

Figure 11-4 (a) Representation of an acceptor impurity (boron) occupying a substitutional site in a silicon crystal. (b) Corresponding acceptor levels created in the silicon bandgap.

The net effect in *n*-type material is therefore to create a situation in which the number of conduction electrons is much greater and the number of holes much smaller than in the pure material. The electrical conductivity is then determined almost exclusively by the flow of electrons, and holes play a very small role. In this case, the electrons are called the *majority carriers* and holes the *minority carriers*.

3. *p*-TYPE SEMICONDUCTORS

The addition of a trivalent impurity such as an element from group III of the periodic table to a silicon lattice results in a situation sketched in Fig. 11-4a. If the impurity occupies a substitutional site, it has one fewer valence electron than the surrounding silicon atoms and therefore one covalent bond is left unsaturated. This vacancy represents a hole similar to that left behind when a normal valence electron is excited to the conduction band, but its energy characteristics are slightly different. If an electron is captured to fill this vacancy, it participates in a covalent bond that is not identical to the bulk of the crystal because one of the two participating atoms is a trivalent impurity. An electron filling this hole, although still bound to a specific location, is slightly less firmly attached than a typical valence electron. Therefore, these *acceptor impurities* also create electron sites within the normally forbidden energy gap. In this case, the acceptor levels lie near the bottom of the gap because their properties are quite close to sites occupied by normal valence electrons.

Normal thermal excitation in the crystal assures that there will always be some electrons available to fill the vacancies created by the acceptor impurities or to occupy the acceptor sites shown in Fig. 11-4b. Because the energy difference between typical acceptor sites and the top of the valence band is small, a large fraction of all the acceptor sites are filled by such thermally excited electrons. These electrons come from other normal covalent bonds throughout the crystal and therefore leave holes behind in the valence band. To a good approximation, an extra hole is created in the valence band for every acceptor impurity that is added. If the concentration N_A of acceptor impurities is made to be large compared with the intrinsic concentration of holes p_i, then the number of holes is completely dominated by the concentration of acceptors, or

$$p \cong N_A \qquad (11\text{-}7)$$

The increased availability of holes enhances the recombination probability between conduction electrons and holes and therefore decreases the equilibrium number of conduction electrons. Again, the same equilibrium constant discussed earlier holds, and $np = n_i p_i$. In *p*-type material, holes are the majority carrier and dominate the electrical conductivity. The filled acceptor sites represent fixed negative charges that balance the positive charge of the majority holes.

One measure of the impurity level of semiconductor materials is the electrical conductivity, or its inverse, the resistivity. The resistivity of the intrinsic material, in which all charge carriers are created by thermal excitation, can be predicted from Eq. (11-1). The corresponding values for silicon and germanium are listed in Table 11-1. In practice, these theoretical values of resistivity are never observed because of the unavoidable residual impurities. Using the most advanced purification methods available at this writing, silicon resistivity of about 50,000 Ω-cm can be achieved, compared with a theoretical value of over 200,000 Ω-cm.

At room temperature, the influence of impurities on resistivity is much more pronounced for silicon than for germanium. Thermally excited electron–hole pairs are much more common in germanium because of its lower bandgap energy, and therefore a greater concentration of impurity is required to change the intrinsic resistivity significantly. For example, in silicon an impurity concentration of 10^{13} atoms/cm³ for *p*-type material corresponds to a resistivity of about 500 Ω-cm (Ref. 3), markedly below the intrinsic resistivity of over 10^5 Ω-cm. The same impurity concentration in germanium results in resistivities of 50 Ω-cm for *p*-type material and 15 Ω-cm for *n*-type germanium, not greatly reduced from the intrinsic value of 50 Ω-cm. On the other hand, if germanium is cooled to liquid nitrogen temperature, the increased intrinsic resistivity is changed by a much greater factor by the same impurity concentration.

4. COMPENSATED MATERIAL

If donor and acceptor impurities are present in a semiconductor in equal concentration, the material is said to be *compensated*. Such material has some of the properties of an intrinsic semiconductor because electrons contributed by donor impurities are removed to some extent by their capture at the site of acceptor impurities. Despite the potential confusion with purified intrinsic material, compensated regions in semiconductors are commonly given the designation *i* because of their near intrinsic properties.

In practice, it is impossible to achieve exact compensation at the time of fabrication of the doped material because any small imbalance in the acceptor or donor concentration quickly leads to *n*-type or *p*-type behavior. At present, the only practical means for achieving compensation over large volumes in silicon or germanium is through the lithium ion drifting process after the crystal has been fabricated. This procedure is discussed in Chapter 13.

5. HEAVILY DOPED MATERIAL

Thin layers of semiconductor material that have an unusually high concentration of impurity are often given a special notation. Thus, n^+ and p^+ designate heavily doped *n*- and *p*-type layers that, as a result, have very high conductivity. These layers are often used in making electrical contact with semiconductor devices, because the very low minority carrier density allows their application as "blocking" contacts described later in this chapter.

E. Trapping and Recombination

Once electrons and holes are formed in a semiconductor, they will tend to migrate either spontaneously or under the influence of an applied electric field until they are either collected at an electrode or recombination takes place. There are theoretical predictions[16] that the average lifetime of charge carriers before recombination in perfectly pure semiconductors could be as large as a second. In practice, lifetimes at least three or four orders of magnitude smaller than a second are actually observed which are dominated entirely by the very low level of impurities remaining in the material. Some of these impurities, such as gold, zinc, cadmium, or other metallic atoms occupying substitutional lattice positions, introduce energy levels near the middle of the forbidden gap. They are therefore classified as "deep impurities" (as opposed to acceptor or donor impurities which, because the corresponding energy levels lie near the edges of the forbidden band, are called "shallow impurities"). These deep impurities can act as *traps* for charge carriers in the sense that if a hole or electron is captured, it will be immobilized for a relatively long period of time. Although the trapping center ultimately may release the carrier back to the band from which it came, the time delay is often sufficiently long to prevent that carrier from contributing to the measured pulse.

Other types of deep impurities can act as *recombination centers*. These impurities are capable of capturing both majority and minority carriers, causing them to annihilate. An impurity level near the center of the forbidden gap might, for example, first capture a conduction electron. At a slightly later time, a hole from the valence band might also be captured, with the electron then filling the hole. The impurity site is thus returned to its original state and is capable of causing another recombination event. In most crystals, recombination through such centers is far more common than direct recombination of electrons and holes across the full bandgap.

Both trapping and recombination contribute to the loss of charge carriers and tend to reduce their average lifetime in the crystal. For the material to serve as a good radiation detector, a large fraction (preferably 100%) of all the carriers created by the passage of the incident radiation should be collected. This condition will hold provided the collection time for the carriers is short compared with their mean lifetime. Collection times of the order of 10^{-7}–10^{-8} s are fairly common, so that carrier lifetimes of the order of 10^{-5} s or longer are usually sufficient.

Another widely quoted specification is the *trapping length* within the material. This quantity is simply the mean distance traveled by a carrier before trapping or recombination and is given by the product of the mean lifetime and the average drift velocity. In order to have an acceptable detector, the trapping length should be long compared with the physical dimensions over which the charge must be collected.

In addition to impurities, structural defects within the crystal lattice can also lead to trapping and charge carrier loss. These imperfections include point defects such as vacancies or interstitials which tend to behave as acceptors or donors, respectively. Carrier loss may also occur at line defects or dislocations that may be produced in stressed crystals. A dislocation represents the slippage of one crystal plane with respect to another, and its intersection with the surface of the crystal leads to a pit upon chemical etching. The density of these etched pits is often quoted as a measure of the crystalline perfection of a semiconductor sample.

II. THE ACTION OF IONIZING RADIATION IN SEMICONDUCTORS

A. The Ionization Energy

When a charged particle passes through a semiconductor with the band structure shown in Fig. 11-1, the overall significant effect is the production of many electron–hole pairs along the track of the particle. The production process may be either direct or indirect, in that the particle produces high-energy electrons (or *delta rays*) that subsequently lose their energy in producing more electron–hole pairs. Regardless of the detailed mechanisms involved, the quantity of practical interest for detector applications is the average energy expended by the primary charged particle to produce one electron–hole pair. This quantity, often loosely called the *ionization energy* and given the symbol ϵ, is experimentally observed to be largely independent of both the energy and type of the incident radiation. This important simplification allows interpretation of the number of electron–hole pairs produced in terms of the incident energy of the radiation, provided the particle is fully stopped within the active volume of the detector.

The dominant advantage of semiconductor detectors lies in the smallness of the ionization energy. The value of ϵ for either silicon or germanium is about 3 eV (see Table 11-1), compared with about 30 eV required to create an ion pair in typical gas-filled detectors. Thus, the number of charge carriers is 10 times greater for the semiconductor case, for a given energy deposited in the detector. The increased number of charge carriers has two beneficial effects on the attainable energy resolution. The statistical fluctuation in the number of carriers per pulse becomes a smaller fraction of the total as the number is increased. This factor often is predominant in determining the limiting energy resolution of a detector for medium to high radiation energy. At low energies, the resolution may be limited by electronic noise in the preamplifier, and the greater amount of charge per pulse leads to a better signal/noise ratio.

More detailed examination shows that ϵ depends on the nature of the incident radiation. Most detector calibrations are carried out using alpha particles, and the values for ϵ shown in Table 11-1 are based on this mode of excitation. All experimental values obtained using other light ions or fast electrons seem to be fairly close,[17-19] but differences as large as 2.2% have been reported[20] between proton and alpha particle excitation in silicon. These observed differences point up the need to carry out detector calibration using a radiation type that is identical to that involved in the measurement itself if precise energy values are required.

A much larger difference is measured for ϵ when heavy ions or fission fragments are involved. The value of ϵ is significantly higher than for alpha particle excitation, leading to a lower than anticipated number of charge carriers. The physical origins of this *pulse height defect* are discussed later in this chapter.

The ionization energy is also temperature dependent. For the most significant detector materials, the value of ϵ increases with decreasing temperature. As shown in Table 11-1, ϵ in silicon is about 3% greater at liquid nitrogen temperature compared with room temperature.[18]

B. The Fano Factor

In addition to the mean number, the fluctuation or variance in the number of charge carriers is also of primary interest because of the close connection of this parameter with energy resolution of the detector. As in gas counters, the observed statistical fluctuations in semiconductors are smaller than expected if the formation of the charge carriers were a Poisson process. The Poisson model would hold if all events along the track of the ionizing particle were independent and would predict that the variance in the total number of electron–hole pairs should be equal to the total number produced, or E/ϵ. The Fano factor F is introduced as an adjustment factor to relate the observed variance to the Poisson predicted variance:

$$F \equiv \frac{\text{observed statistical variance}}{E/\epsilon} \tag{11-8}$$

For good energy resolution, one would like the Fano factor to be as small as possible. Although a complete understanding of all the factors that lead to a nonunity value for F does not yet exist, rather sophisticated models have been developed[21] which at least qualitatively account for experimental observations. Some numerical values for silicon and germanium are given in Table 11-1.

There is considerable variation in reported experimental values, particularly for silicon. The Fano factor is usually measured by observing the energy resolution from a given detector under conditions in which all other factors that can broaden the full-energy peak (such as noise or drift) can be eliminated. The assumption is then made that the residual width can be attributed to statistical effects only. If nonstatistical residual factors remain, however, the Fano factor will appear to be larger than it actually is. This may explain the historical trend toward lower values as measurement procedures are refined. It has also been postulated[13] that the value of the Fano factor may depend on the nature of the particle that deposits the energy.

III. SEMICONDUCTORS AS RADIATION DETECTORS

A. Electrical Contacts

In order to construct a practical radiation detector, some means must be provided to collect the electrical charges created by the radiation at either boundary of the semiconductor material. An *ohmic* contact is a nonrectifying electrode through which charges of either sign can flow freely. If two ohmic contacts are fitted on opposite faces of a slab of semiconductor and connected to a detection circuit, the equilibrium charge carrier concentrations in the semiconductor will be maintained. If an electron or hole is collected at one electrode, the same carrier species is injected at the opposite electrode to maintain the equilibrium concentrations in the semiconductor.

The steady-state leakage currents that are observed using ohmic contacts are too high, even with the highest resistivity material available (see following section), to permit their general application to semiconductor detectors. Instead, *noninjecting* or *blocking* electrodes are universally employed to reduce the magnitude of the current through the bulk of the semiconductor. If blocking electrodes are used, charge carriers initially removed by the application of an electric field are not replaced at the opposite electrode, and their overall concentration within the semiconductor will drop after application of an electric field. The leakage current can thus be reduced to a sufficiently low value to allow the detection of the added current pulse created by the electron–hole pairs produced along the track of an ionizing particle.

The most appropriate type of blocking contacts are the two sides of a *p-n* semiconductor junction. It is very difficult to inject electrons from the *p* side of this junction because holes are the majority carrier and free electrons are relatively scarce. At the opposite side, electrons are the majority carrier and holes cannot readily be injected. In this chapter, we discuss detectors that are created by placing the *p*- and *n*-type materials in direct contact, forming a *p-n* junction. In Chapters 12 and 13, detectors in which the *p* and *n* regions are separated by an intrinsic or compensated region (the *i* region) are described.

B. Leakage Current

In order to create an electric field large enough to achieve an efficient collection of the charge carriers from any semiconductor detector, an applied voltage of typically hundreds or thousands of volts must be imposed across the active volume. Even in the absence of ionizing radiation, all detectors will show some finite conductivity and therefore a steady-state *leakage current* will be observed. Random fluctuations that inevitably occur in the leakage current will tend to obscure the small signal current that momentarily flows following an ionizing event and will represent a significant source of noise in many situations. Methods of reducing the leakage current are therefore an important consideration in the design of semiconductor detectors.

The resistivity of the highest purity silicon currently available is about 50,000 Ω-cm. If a 1 mm thick slab of this silicon were cut with 1 cm^2 surface area and fitted with ohmic contacts, the electrical resistance between faces would be 5000 Ω. An applied voltage of 500 V would therefore cause a leakage current through the silicon of 0.1 A. In contrast, the peak current generated by a pulse of 10^5 radiation-induced charge carriers would only be about 10^{-6} A. It is therefore essential to reduce this bulk leakage current greatly through the use of blocking contacts. In critical applications, the leakage current must not exceed about 10^{-9} A to avoid significant resolution degradation.

At these levels, leakage across the surface of the semiconductor can often become more significant than bulk leakage. Great care is taken in the fabrication of semiconductor detectors to avoid contamination of the surfaces, which could create leakage paths. Some configurations may also use grooves in the surface or guard rings to help suppress surface leakage (see Chapter 5).

C. The Semiconductor Junction

1. BASIC JUNCTION PROPERTIES

The radiation detectors described in this chapter are based on the favorable properties that are created near the junction between *n*- and *p*-type semiconductor materials. Charge carriers are able to migrate across the junction if the regions are brought together in good thermodynamic contact. Simply pressing together two pieces of the material will not suffice because gaps will inevitably be left which will be large compared with the interatomic lattice spacing. In practice, the junction is therefore normally formed in a single crystal by causing a change in the impurity content from one side of the junction to the other.

As an illustration, assume that the process begins with a *p*-type crystal that has been doped with a uniform concentration of acceptor impurity. In the concentration profile at the top of Fig. 11-5, this original acceptor concentration N_A is shown as a horizontal line.

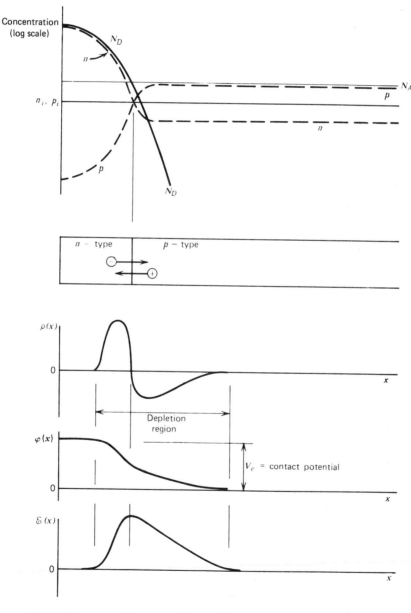

Figure 11-5 The assumed concentration profiles for the *n-p* junction shown at the top are explained in the text. The effects of carrier diffusion across the junction give rise to the illustrated profiles for space charge $\rho(x)$, electric potential $\varphi(x)$, and electric field $\mathscr{E}(x)$.

We now assume that the surface of the crystal on the left is exposed to a vapor of an n-type impurity that diffuses some distance into the crystal. The resulting donor impurity profile is labeled N_D on the figure and falls off as a function of distance from the surface. Near the surface, the donor impurities can be made to outnumber the acceptors, converting the left portion of the crystal to n-type material.

The approximate variation of equilibrium charge carrier concentration is also plotted at the top of Fig. 11-5 and labeled as p (hole concentration) and n (conduction electron concentration). These profiles are subsequently altered in the vicinity of the p-n junction due to the effects of charge carrier diffusion. In the n-type region at the left, the density of conduction electrons is much higher than in p-type. The junction between the two regions therefore represents a discontinuity in the conduction electron density. Wherever such a sharp gradient exists for any carrier that is free to migrate, a net diffusion from regions of high concentration to those of low concentration must take place. Thus, there will be some net diffusion of conduction electrons into the p-type material, where they will quickly combine with holes. In effect, this annihilation represents the capture of the conduction electron by one of the vacancies existing in the covalent bonds in the p-type material. The diffusion of conduction electrons out of the n-type material leaves behind immobile positive charges in the form of ionized donor impurities. A similar and symmetric argument leads to the conclusion that holes (the majority in the p-type material) must also diffuse across the junction because they also see an abrupt density gradient. Each hole that is removed from the p side of the junction leaves behind an acceptor site that has picked up an extra electron and therefore represents a fixed and immobile negative charge. The combined effect is to build up a net negative space charge on the p side and a positive space charge on the n side of the junction.

The accumulated space charge creates an electric field that diminishes the tendency for further diffusion. At equilibrium, the field is just adequate to prevent additional net diffusion across the junction, and a steady-state charge distribution is therefore established.

The region over which the charge imbalance exists is called the *depletion region* and extends into both the p and n sides of the junction. If the concentrations of donors on the n side and acceptors on the p side are equal, the diffusion conditions are approximately the same for both holes and electrons, and the depletion region extends equal distances into both sides. Usually, however, there is a marked difference in the doping levels on one side of the junction compared with the other. For example, if the donor concentration in the n-type material is higher than that of acceptor atoms in the p-type, the electrons diffusing across the junction will tend to travel a greater distance into the p-type material before all have recombined with holes. In this case, the depletion region would extend farther into the p side.

The buildup of net charge within the region of the junction leads to the establishment of an electric potential difference across the junction. The value of the potential φ at any point can be found by solution of Poisson's equation

$$\nabla^2\varphi = -\frac{\rho}{\epsilon} \tag{11-9}$$

where ϵ is the dielectric constant of the medium, and ρ is the net charge density. In one dimension, Eq. (11-9) takes the form

$$\frac{d^2\varphi}{dx^2} = -\frac{\rho(x)}{\epsilon} \tag{11-10}$$

so that the shape of the potential across the junction can be obtained by twice integrating the charge distribution profile $\rho(x)$. Graphical examples are shown in Fig. 11-5. At equilibrium, the potential difference across the junction (called the *contact potential*) amounts to nearly the full bandgap value of the semiconductor material. The direction of this potential difference is such that it opposes the further diffusion of electrons from left to right and holes from right to left in Fig. 11-5.

Where a difference in electrical potential exists, there must also be an electric field $\mathscr{E}$. Its magnitude is found by taking the gradient of the potential

$$\mathscr{E} = -\operatorname{grad} \varphi \qquad (11\text{-}11)$$

which, in one dimension, is simply

$$\mathscr{E}(x) = -\frac{d\varphi}{dx} \qquad (11\text{-}12)$$

The electric field will extend over the width of the depletion region, in which charge imbalance is significant and the potential has some gradient. Its variation is also sketched in Fig. 11-5.

The depletion region exhibits some very attractive properties as a medium for the detection of radiation. The electric field which exists causes any electrons created in or near the junction to be swept back toward the n-type material, and any holes are similarly swept toward the p-type side. The region is thus "depleted" in that the concentration of holes and electrons is greatly suppressed.[†] The only significant charges remaining in the depletion region are the immobile ionized donor sites and filled acceptor sites. Because these latter charges do not contribute to conductivity, the depletion region exhibits a very high resistivity compared with the n- and p-type materials on either side of the junction. Electron–hole pairs, which are created within the depletion region by the passage of radiation, will be swept out of the depletion region by the electric field, and their motion constitutes a basic electrical signal.

2. REVERSE BIASING

Thus far, we have discussed a semiconductor diode junction to which no external voltage is applied. Such an unbiased junction will function as a detector, but only with very poor performance. The contact potential of about 1 V which is formed spontaneously across the junction is inadequate to generate a large enough electric field to make the charge carriers move very rapidly. Therefore, charges can be readily lost due to trapping and recombination, and incomplete charge collection often results. The thickness of the depletion region is quite small, and the capacitance of an unbiased junction is high. Therefore, the noise properties of an unbiased junction connected to the input stage of a preamplifier are quite poor. For these reasons, unbiased junctions are not used as practical radiation detectors, but instead, an external voltage is applied in the direction to cause the semiconductor diode to be reverse biased.

The p-n junction is most familiar in its role as a diode. The properties of the junction are such that it will readily conduct current when voltage is applied in the "forward" direction, but it will conduct very little current when biased in the "reverse" direction. In the configuration of Fig. 11-5, first assume that a positive voltage is applied to the p side

---

[†] The carrier density remaining in the depletion region of practical silicon detectors is as low as 100 electrons or holes per cm^3, compared with typical carrier densities of about 10^{10} in high-resistivity material.

of the junction with respect to the *n* side. The potential will tend to attract conduction electrons from the *n* side as well as holes from the *p* side across the junction. Because, in both cases, these are the majority carriers, conductivity through the junction is greatly enhanced. The contact potential shown in Fig. 11-5 is reduced by the amount of the bias voltage that is applied, which tends to lessen the potential difference seen by an electron from one side of the junction to the other. This is the direction of forward biasing, and only small values of the forward bias voltage are needed to cause the junction to conduct large currents.

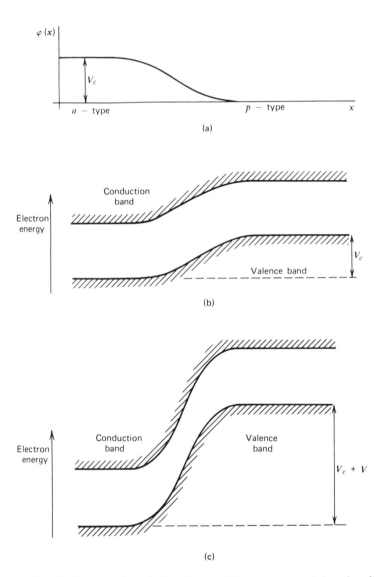

Figure 11-6 (a) The variation of electric potential across an *n-p* junction from Fig. 11-5. (b) The resulting variation in electron energy bands across the junction. The curvature is reversed because an increase in electron energy corresponds to a decrease in conventional electric potential $\varphi(x)$ defined for a positive charge. (c) The added displacement of the bands caused by application of a reverse bias V across the junction.

If the situation is reversed, and the p side of the junction is made negative with respect to the n side, the junction is reverse biased. Now the natural potential difference from one side of the junction to the other is enhanced, as shown in Fig. 11-6c. Under these circumstances, it is the minority carriers (holes on the n side and electrons on the p side) which are attracted across the junction and, because their concentration is relatively low, the reverse current across the diode is quite small. Therefore, the p-n junction serves as a rectifying element, allowing relatively free flow of current in one direction while presenting a large resistance to its flow in the opposite direction. If the reverse bias is made very large, a sudden breakdown in the diode will occur and the reverse current will abruptly increase, often with destructive effects.

3. PROPERTIES OF THE REVERSE BIAS JUNCTION

When a reverse bias is applied to the junction, virtually all the applied voltage will appear across the depletion region, because its resistivity is much higher than that of the normal n- and p-type material. Because the effect of the reverse bias is to accentuate the potential difference across the junction, Poisson's equation [Eq. (11-9)] demands that space charge must also increase and extend a greater distance on either side of the junction. Thus, the thickness of the depletion region is also increased, extending the volume over which radiation-produced charge carriers will be collected. Practical detectors are almost always operated with a bias voltage that is very large compared with the contact potential, so that the applied voltage completely dominates the magnitude of the potential difference across the junction.

In the analysis that follows, we first assume that the semiconductor wafer in which the junction is formed is sufficiently thick so that the depletion region does not reach either surface and is contained within the interior volume of the wafer. This condition holds for *partially depleted* detectors in which some portion of the wafer thickness remains undepleted. Many semiconductor detectors are operated with sufficient reverse bias voltage so that the depletion region extends through the full wafer thickness, creating a *fully depleted* (or *totally depleted*) detector. These configurations share many of the properties derived below, except that the depletion region is obviously limited by the physical thickness of the wafer.

Some properties of the reverse bias junction can be derived if we represent the charge distribution sketched in Fig. 11-5 by the idealized distribution shown below:

$$\rho(x) = \begin{cases} eN_D & (-a < x \le 0) \\ -eN_A & (0 < x \le b) \end{cases}$$

Here the electron diffusion is assumed to result in a uniform positive space charge (the ionized donor sites) over the region $-a < x \le 0$ on the n side of the junction. A corresponding negative space charge (the filled acceptor sites) resulting from hole diffusion is assumed to extend over the region $0 < x \le b$ on the p side. Because the net charge must be zero, $N_D a = N_A b$.

Equation (11-10) applied to this case takes the form

$$
\frac{d^2\varphi}{dx^2} =
\begin{cases}
-\dfrac{eN_D}{\epsilon} & (-a < x \le 0) \\[2ex]
+\dfrac{eN_A}{\epsilon} & (0 < x \le b)
\end{cases}
$$

We now carry out an integration and apply the boundary conditions that the electric field $\mathscr{E} = -d\varphi/dx$ must vanish at both edges of the charge distribution:

$$
\frac{d\varphi}{dx}(-a) = 0 \quad \text{and} \quad \frac{d\varphi}{dx}(b) = 0
$$

The result is then

$$
\frac{d\varphi}{dx} =
\begin{cases}
-\dfrac{eN_D}{\epsilon}(x + a) & (-a < x \le 0) \\[2ex]
+\dfrac{eN_A}{\epsilon}(x - b) & (0 < x \le b)
\end{cases}
$$

The corresponding shape of the electric field $\mathscr{E} = -d\varphi/dx$ is sketched below:

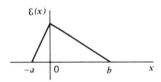

Another integration will now yield the electric potential $\varphi(x)$. The difference in potential from the n side to the p side of the junction, if we neglect the relatively small contact potential, is just the value of the applied reverse bias V. We can therefore apply the boundary conditions

$$
\varphi(-a) = V \quad \text{and} \quad \varphi(b) = 0
$$

The solution then takes the form

$$
\varphi(x) =
\begin{cases}
-\dfrac{eN_D}{2\epsilon}(x + a)^2 + V & (-a < x \le 0) \\[2ex]
+\dfrac{eN_A}{2\epsilon}(x - b)^2 & (0 < x \le b)
\end{cases}
$$

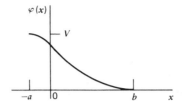

Since the solutions for either side of the junction must match at $x = 0$, we can write

$$V - \frac{eN_Da^2}{2\epsilon} = \frac{eN_Ab^2}{2\epsilon}$$

or

$$N_Ab^2 + N_Da^2 = \frac{2\epsilon V}{e}$$

Now since $N_Da = N_Ab$, the expression above can be rewritten:

$$(a + b)b = \frac{2\epsilon V}{eN_A}$$

The total width of the depletion region d is the entire distance over which the space charge extends, or $d = a + b$.

For purposes of the present example, we have assumed that the n-side doping level is much higher than on the p side, so that $N_D \gg N_A$. Because $N_Da = N_Ab$, it follows that $b \gg a$, and therefore the space charge extends much farther into the p side than the n side. Then $d \cong b$ and we can write

$$d \cong \left(\frac{2\epsilon V}{eN_A} \right)^{1/2}$$

If we had started from the opposite assumption that the p-side doping level was predominant, a similar result would be obtained except that N_A in the above expression would be replaced by N_D. A generalized solution for the thickness of the depletion region is therefore

$$d \cong \left(\frac{2\epsilon V}{eN} \right)^{1/2} \tag{11-13}$$

In this expression, N now represents the dopant concentration (either donors or acceptors) on the side of the junction that has the lower dopant level. (For surface barriers described later in this chapter, N is the dopant concentration in the bulk of the crystal.)

The resistivity ρ_d of the doped semiconductor is given by $1/e\mu N$, where μ is the mobility of the majority carrier. Equation (11-13) may thus be written

$$d \cong \left(2\epsilon V\mu\rho_d \right)^{1/2} \tag{11-14}$$

Because one often would like the largest depletion width possible for a given applied voltage, it is advantageous to have the resistivity as high as possible. This resistivity is limited by the purity of the semiconductor material before the doping process, because enough dopant must be added to override the nonuniform effects of the residual impurities. A premium is therefore placed on obtaining detectors fabricated from the highest purity material possible.

Because of the fixed charges that are built up on either side of the junction, the depletion region exhibits some properties of a charged capacitor. If the reverse bias is increased, the depletion region grows thicker and the capacitance represented by the separated charges therefore decreases. The value of the capacitance per unit area is

$$C = \frac{\epsilon}{d} \cong \left(\frac{e\epsilon N}{2V} \right)^{1/2} \tag{11-15}$$

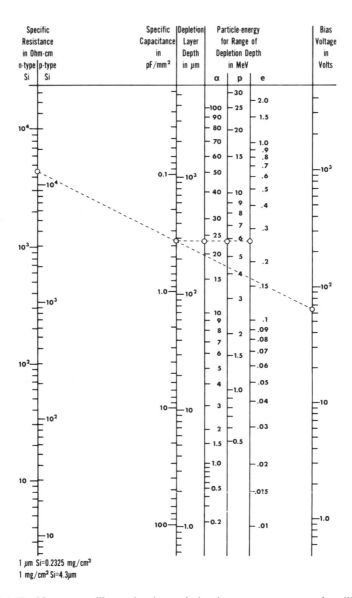

Figure 11-7 Nomogram illustrating interrelation between parameters for silicon junction detectors. (Similar to nomogram originally published by Blankenship.[22])

Good energy resolution under conditions in which electronic noise is dominant depends on achieving a small detector capacitance and is thus promoted by using the largest possible applied voltage, up to the point that the detector becomes fully depleted.

The maximum electric field will occur at the point of transition between the n- and p-type material. Its magnitude is given by

$$\mathcal{E}_{max} \cong \frac{2V}{d} = \left(\frac{2VNe}{\epsilon} \right)^{1/2} \qquad (11\text{-}16)$$

and can easily reach 10^6–10^7 V/m under typical conditions. For partially depleted junctions, the depletion layer thickness d is proportional to $\sqrt{V}$ so that the value of $\mathcal{E}_{max}$ increases with applied voltage as $\sqrt{V}$.

The interrelation between these parameters is illustrated in the nomogram for silicon detectors given in Fig. 11-7. Also shown are scales corresponding to the ranges of various charged particles to allow selection of conditions required to produce a depletion depth that exceeds the range.

The maximum operating voltage for any diode detector must be kept below the breakdown voltage to avoid a catastrophic deterioration of detector properties. Commercially manufactured detectors are supplied with a maximum voltage rating that should always be strictly observed. Additional protection can be provided by monitoring the leakage current during application of the voltage (see the discussion later in this chapter).

To summarize, the reverse biased p-n junction makes an attractive radiation detector because charge carriers created within the depletion region can be quickly and efficiently collected. The width of the depletion region represents the active volume of the detector and is changed in partially depleted detectors by varying the reverse bias. The variable active volume of semiconductor junctions is unique among radiation detectors and sometimes is used to good advantage. The capacitance of a partially depleted detector also varies with applied voltage, and stable operation therefore requires the use of charge sensitive preamplifiers (see Chapter 17).

IV. SEMICONDUCTOR DETECTOR CONFIGURATIONS

A. Diffused Junction Detectors

One common fabrication method for semiconductor diode detectors starts with a homogeneous crystal of p-type material. One surface is treated by exposing it to a vapor of n-type impurity (typically phosphorus), which then converts a region of the crystal near the surface from p-type to n-type material. A junction is therefore formed some distance from the surface at the point at which the n- and p-type impurities reverse their relative concentration. Typical depths of the diffused n-type layer range from 0.1 to 2.0 μm. Because the n-type surface layer is heavily doped compared with the p-type original crystal, the depletion region extends primarily into the p side of the junction. Therefore, much of the surface layer remains outside the depletion region and represents a *dead layer* or window through which the incident radiation must pass before reaching the depletion region. In charged particle spectroscopy, this dead layer can be a real disadvantage because a portion of the particle energy will be lost before the active region of the detector is reached. Methods for experimentally determining its thickness are given later in this chapter.

To avoid the disadvantages of the dead layer, diffused junction detectors have been replaced in many applications by surface barrier detectors, described in the following section. Diffused junction detectors are still commercially manufactured, however, and offer some advantage over surface barrier detectors. They are somewhat more rugged and less prone to the problems that can arise due to the accumulation of oil or other foreign matter on the surface of the detector.

B. Surface Barrier Detectors

The role of the *p*-type material in forming the junction can be assumed by a high density of electron traps formed at the surface of an *n*-type crystal. The resulting depletion region behaves in much the same way as discussed earlier for a diffused junction detector. Formation of the surface states is carried out using recipes that have evolved somewhat empirically. One such set of typical procedures is described in Ref. 23. The usual treatment is etching of the surface, followed by evaporation of a thin gold layer for electrical contact. Best results are obtained if the evaporation is carried out under conditions that promote slight oxidation of the surface; the resulting oxide layer between the gold and silicon apparently plays an important role in the resulting properties of the surface barrier. Surface barriers can also be produced by starting with a *p*-type crystal and evaporating aluminum to form an equivalent *n*-type contact. The very thin dead layers that characterize surface barrier detectors are further discussed later in this chapter.

One potential disadvantage of surface barriers is their sensitivity to light. The thin entrance windows are optically transparent, and photons striking the detector surface can reach the active volume. The energy of visible light photons of 2–4 eV is greater than the

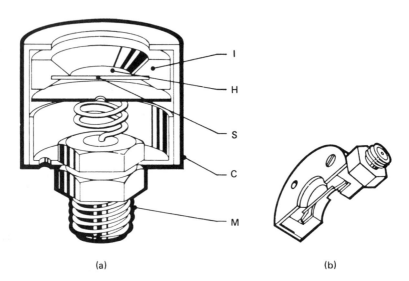

(a)   (b)

Figure 11-8 Construction and mounting of silicon junction detectors shown in cross-sectional view. (a) Surface barrier mount with coaxial connector (M) at rear. The silicon wafer (S) is mounted in a ceramic ring (I) with electrical contact made between either side of the junction and opposite metalized surfaces of the ring. The front surface is connected to the outer case (C) and grounded, whereas the back surface is connected to the center conductor of the coaxial connector. (b) Cutaway view of a transmission mount, in which both surfaces of the silicon wafer are accessible. The coaxial connector is placed at the edge of the ceramic ring. (Courtesy of EG & G ORTEC, Oak Ridge, TN.)

bandgap energy of most semiconductors, and electron–hole pairs can therefore be produced by photon interactions. A very high noise level is produced by normal room lighting, but the vacuum enclosure required for most charged particle applications usually reduces light-induced noise to insignificant levels. The thin entrance window also makes the detector sensitive to damage from exposure to vapors, and the front surface must never be directly handled.

A cross-sectional diagram of a typical mounting arrangement for a surface barrier detector is shown in Fig. 11-8a. The outer housing and front surface are normally grounded, and an electrical lead from the back surface of the semiconductor wafer attaches to the center electrode of the coaxial connector at the rear. Because normal surface barriers are usually created on n-type crystals, a positive polarity voltage is required to reverse bias the junction.

C. Ion Implanted Layers

An alternative method of introducing doping impurities at the surface of the semiconductor is to expose that surface to a beam of ions produced by an accelerator. This method is known as *ion implantation* and can be used to form n^+ or p^+ layers by accelerating, for example, either phosphorus or boron ions, respectively. At a fixed accelerator voltage (typically about 10 kV) monoenergetic ions are produced which have a well-defined range in the semiconductor material. By changing the energy of the incident ions, the concentration profile of the added impurity can be closely controlled. Following exposure to the ion beam, an annealing step is normally carried out to reduce the effects of radiation damage caused by the incident ions. One of the advantages of ion implantation is that the annealing temperature required (less than 500°C) is considerably lower than that needed for the thermal diffusion of dopants to form a diffused junction. The structure of the crystal is therefore less disturbed, and carrier lifetimes are not unnecessarily reduced. Compared with surface barriers, ion-implanted detectors tend to be more stable and less subject to ambient conditions. Also, they can be formed with entrance windows as thin as 34 nm silicon equivalent[24,25] and they are now available commercially. A review of the use of ion implantation to form radiation detectors can be found in Ref. 26.

D. Fully Depleted Detectors

As shown by Eq. (11-13), the width of the depletion region associated with a p-n junction increases as the reverse bias voltage is increased. If the voltage can be increased far enough, the depletion region eventually extends across virtually the entire thickness of the silicon wafer, resulting in a fully depleted (or totally depleted) detector. Because of the several advantages this configuration presents over partially depleted detectors, the fully depleted configuration is becoming the preferred type in a growing number of applications.

In the usual case, one side of the junction is made up of a heavily doped n^+ or p^+ layer or, alternatively, a surface barrier. The opposite side of the junction generally consists of high-purity semiconductor material that is only mildly n or p type. (Such material is often designated ν or π, respectively.) The reason that high-purity material is important is reflected in Eq. (11-13). For a given applied voltage, the depletion depth is maximized by minimizing the concentration of doping impurities on the higher-purity side of the junction. Thick depletion regions can therefore only be obtained by starting from semiconductor material with the lowest possible impurity concentration. Also, with a large difference in the doping levels, the depletion layer essentially extends only into the

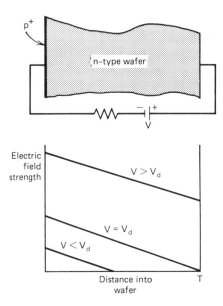

Figure 11-9 The electric field shape in a reverse bias semiconductor detector. Three plots are shown for bias voltages that are below, equal to, and above the depletion voltage V_d.

high-purity side of the junction. The heavily doped layer can then be very thin, providing an *entrance window* for weakly penetrating radiations.

In Fig. 11-9, we assume that we have such a junction formed between a heavily doped p^+ surface layer and a high-purity n-type silicon wafer. As the reverse bias voltage applied to the detector is raised from zero, the depletion region extends further from the p^+ surface into the bulk of the wafer. For low values of the voltage, the wafer is only partially depleted and the electric field goes to zero at the far edge of the depletion region. Between this point and the back surface of the wafer, a region of undepleted silicon exists in which there is no electric field. This region then represents a very thick dead layer from which charge carriers are not collected. For all practical purposes, partially depleted detectors are therefore only sensitive to charged particles incident on the front surface.

If the applied voltage is increased further, the depletion region may be made to extend all the way to the back surface of the wafer. The voltage required to achieve this condition is sometimes called the *depletion voltage*. Its value is found by setting the depletion depth d in Eq. (11-13) equal to the wafer thickness T:

$$V_d = \frac{eNT^2}{2\epsilon}$$

When this stage is reached, a finite electric field exists all the way through the wafer, and the back dead layer thickness is reduced to that of the surface electrical contact that is employed. This condition is represented by the middle plot in Fig. 11-9. Once the wafer is fully depleted, raising the applied voltage further simply results in a constant increase in the electric field everywhere in the wafer. At voltages much larger than the depletion voltage, the electric field profile therefore tends to become more nearly uniform across the entire wafer thickness. Under these conditions, the detector is sometimes said to be *over-*

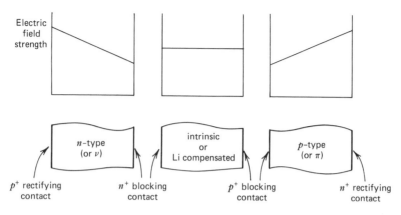

Figure 11-10 The electric field shapes for fully depleted planar semiconductor detectors of different configurations.

depleted. Because of the advantages of having a high electric field everywhere within the detector active volume, virtually all totally depleted detectors are operated at sufficient voltage to achieve this condition.

Figure 11-10 shows several configurations that are typical of fully depleted detectors, together with the corresponding electric field profile through the wafer. In order to deplete the wafer fully at as low a voltage as possible, one normally starts with material with the highest available purity, either n or p type. The junction is then formed by providing a heavily doped surface layer of the opposite type. This is often called the *rectifying contact.* Because of its high doping level, it also serves as an excellent blocking contact in which the minority carrier concentration is very low. In the nearly pure bulk of the wafer, however, the minority carriers are not highly suppressed and an additional blocking contact is normally provided at the opposite face of the wafer. If the high-purity silicon is mildly n type, then a thin n^+ layer is provided at this back surface. Since both materials are n type, no semiconductor junction exists at this surface. Instead, the n^+ layer provides the noninjecting conditions necessary to suppress leakage current due to minority carrier motion across the junction.

As shown on the right in Fig. 11-10, the roles of the n^+ and p^+ surface layers are reversed if one starts with high-purity material that is mildly p type. In both cases, the electric field is a maximum at the rectifying contact and decreases linearly to a minimum at the blocking contact. Shown in the center in Fig. 11-10 is the case in which intrinsic or perfectly compensated material is used for the wafer. In this case, the distinction between the two contacts disappears and the electric field is uniform throughout the entire wafer. The detector is fully depleted even for very low values of applied voltage. This *p-i-n* configuration is discussed in connection with lithium-drifted detectors in the next chapter.

Fully depleted silicon detectors are very useful as *transmission detectors* for incident particles that have sufficient energy to pass completely through the wafer. The pulse amplitude then indicates the energy lost by the incident radiation during its transit through the device. Totally depleted silicon detectors are commercially available in thicknesses from about 50 to 2000 μm. Transmission mounts of the type shown in Fig. 11-8b allow access to both surfaces of the wafer.

Several properties of totally depleted detectors are of primary importance. The dead layers must be as small as possible at both the front and rear surfaces of the detector if the pulse is to indicate accurately the energy loss of the particle during its transit. An empirical test is often carried out to determine the minimum bias voltage at which these detectors are totally depleted. The pulse height from a monoenergetic source of charged particles is recorded for the particles incident on both the front and back face of the detector. When the detector is totally depleted, the pulse height should be approximately the same for either orientation. In interpreting such measurements, allowance must be made for the fact that the inherent window thicknesses of the front and back contacts of these detectors are often not the same.

In partially depleted detectors, the thickness uniformity of the crystal from which the detector is fabricated is not critical because the active volume of the detector is determined by the limited depletion depth. In fully depleted detectors, however, the wafer thickness must be kept quite uniform to avoid energy loss variations across the surface of the detector. Consequently, considerable effort is taken to provide uniform crystal wafers when totally depleted configurations are produced.

Fully depleted detectors have other advantages over partially depleted configurations in which there is an undepleted back dead layer. The finite electrical resistance of this dead layer is a source of Johnson noise that can contribute to the degradation of energy resolution. It is eliminated in a fully depleted detector by extending the depletion region all the way to the back contact. Timing properties also tend to be superior in fully depleted configurations. In a partially depleted detector, the electric field drops to zero at the edge of the depletion region. Charge carrier velocities therefore become very low in these low-field regions, slowing the rise of the signal pulse. In a totally depleted detector, the electric field can be maintained at a high value everywhere within the detector volume. Finally, some added stability results from the fact that the active volume and capacitance of a fully depleted detector are no longer functions of the applied voltage as they are in a partially depleted configuration.

The thickness of wafer that can be fully depleted using voltages short of catastrophic breakdown depends on the purity of the semiconductor. In this respect, there is a significant difference between silicon and germanium. Using ultrapure germanium described in the following chapter, depletion of over 1 cm thickness can be achieved. The impurity levels in currently available silicon are somewhat higher, and depletion thicknesses are generally limited to no more than several millimeters. Greater thicknesses in silicon are currently possible only through the use of material compensated by the lithium-drifting process described in Chapter 13.

E. Passivated Planar Detectors

The newest method of fabricating silicon junction detectors combines the techniques of ion implantation and photolithography to produce detectors with very low leakage currents and excellent operational characteristics.[27-29] Methods that were first developed in the semiconductor industry to produce integrated circuits have now been adapted successfully[30,31] to the fabrication of detectors. The techniques described below lend themselves to the batch production of multiple detectors simultaneously starting with a large-area silicon wafer, thus providing potential cost savings. The techniques can also accommodate the type of complex electrode geometry required, for example, in the silicon microstrip detectors described in Chapter 13.

The planar fabrication process generally begins with high-purity silicon that is mildly n type due to residual donor impurities. The steps in the fabrication process are shown in Fig. 11-11. After the wafer has been polished and cleaned, the surface is "passified" through the creation of an oxide layer at elevated temperature. Next, the techniques of

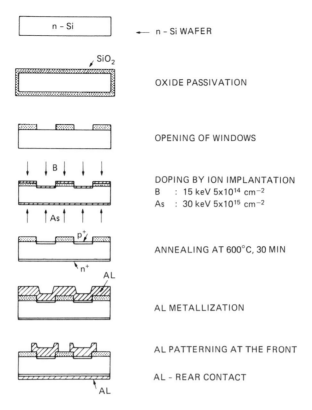

n – Si   ← n – Si WAFER

SiO₂

OXIDE PASSIVATION

OPENING OF WINDOWS

B   DOPING BY ION IMPLANTATION
           B : 15 keV 5x10¹⁴ cm⁻²
           As : 30 keV 5x10¹⁵ cm⁻²
As

p⁺   ANNEALING AT 600°C, 30 MIN

n⁺
    AL
           AL METALLIZATION

           AL PATTERNING AT THE FRONT

           AL – REAR CONTACT
    AL

Figure 11-11 Steps in the fabrication of passivated planar silicon diode detectors. (From Kemmer.[30])

photolithography are used to remove selectively areas of the oxide where the entrance windows of the finished detectors are to be located. The junction is then formed by converting a very thin layer of silicon within the windows into p-type material through the implanting of acceptor ions (boron) using an accelerator. To serve as a blocking electrical contact, the rear surface of the wafer is converted into n^+ material through implantation of donor (As) ions. The radiation damage in the implanted layers is next removed through annealing at elevated temperature. Finally, aluminum is evaporated and patterned by photolithography to provide thin ohmic electrical contacts at the front and rear surfaces. The individual detectors are then separated and encapsulated.

One advantage of this planar fabrication process is that the junction edges are defined by the ion implantation pattern and can be kept within the bulk of the wafer. The oxide-passivated surface keeps leakage currents much lower than in surface barrier detectors, where the junction edge extends all the way to the edge of the wafer. Much of the leakage current in these designs then occurs where these edges are encapsulated in epoxy or similar material. Formation of the p^+ layer through ion implantation also provides planar detectors with a very thin and uniform entrance window or dead layer, an important consideration in preserving good energy resolution for the detector. The aluminized front surface is more rugged and less subject to damage compared with the gold front surface used in surface barrier fabrication.

This incorporation of integrated circuit techniques in the fabrication of silicon detectors opens the possibility of integrating some detector electronics on the same wafer used to produce the detector. A number of research laboratories are currently carrying out development work with this goal in mind. (For example, see Ref. 32.) The resulting hybrid would have potential cost advantages and would offer a much smaller and rugged system compared with separate detector and preamplifier components.

V. OPERATIONAL CHARACTERISTICS

A. Leakage Current

When voltage is applied to a junction detector in the normal fashion, that is, to reverse bias the junction, a small current of the order of a microampere is normally observed. The origins of this *leakage current* are related both to the bulk volume and surface of the detector. Bulk leakage currents arising internally within the volume of the detector can be caused by either of two mechanisms.

The direction of the electric field across the depletion region is such that any majority carriers that diffuse from the normal p and n regions of the detector to the edges of the depletion region will be repelled away from the junction. However, the minority carriers in either case are attracted and will therefore be conducted across the junction. Because the minority carriers are generated continuously on both sides of the junction and are free to diffuse, a steady-state current will result which will be roughly proportional to the area of the junction. In most cases, the minority carrier current is small and is seldom an important leakage source.

A second source of bulk leakage is the thermal generation of electron–hole pairs within the depletion region. This rate will obviously increase with the volume of the depletion region and can be reduced only by cooling the material. Silicon detectors of usual dimensions have a sufficiently low thermally generated current to allow their use at room temperature, but germanium detectors, because of the lower gap energy, must always be operated at reduced temperatures.

Surface leakage effects take place at the edges of the junction where relatively large voltage gradients must be supported over small distances. The amount of surface leakage can vary greatly, depending on such factors as the type of detector encapsulation used, humidity, and any contamination of the detector surface by fingerprints, vacuum pump oil, or other condensable vapors. Guard rings analogous to those described in Chapter 5 are sometimes incorporated into the design of semiconductor diode detectors[33,34] to reduce the surface leakage, but in commercial detectors the normal approach is to rely on clean encapsulation techniques to keep the surface leakage within tolerable levels. The recent introduction of the planar fabrication process has allowed the production of detectors in which the junction edges are buried within the silicon wafer. As a result, leakage current is reduced to a small fraction of that typically observed in either diffused junction or surface barrier devices.

In addition to the effects on energy resolution discussed in the following section, the leakage current has another practical influence on detector operation. The bias voltage to the detector is always supplied through a series resistor (R_L in Fig. 17-5) for signal isolation purposes. Therefore, the true bias voltage applied to the junction is reduced from that of the voltage source by the product of the leakage current and the series resistance. If the leakage current is large enough, the drop across the resistor can appreciably diminish the actual voltage applied to the detector, and the supply voltage must then be raised to compensate for this loss. It is therefore a fairly common practice to monitor the leakage current with an ammeter in series with the voltage supply.

Monitoring the leakage current can also detect the onset of abnormal detector behavior. During steady operation, the leakage should normally maintain a steady value, and any abrupt change or increase in the leakage current can indicate a change in detector performance, which may degrade the energy resolution. Also, it is useful to monitor the leakage current as the bias voltage is first applied to the detector. Normally, the leakage current will increase as the bias voltage is raised. However, any sudden increases can signal the approach of the breakdown of the diode, and the voltage should therefore be reduced to a lower value. Finally, the long-term behavior of the leakage current is often a useful monitor on the degree of radiation damage suffered by a given detector when used under conditions in which such damage is significant.

B. Detector Noise and Energy Resolution

Sources of noise attributable to the detector can be categorized into three groups:

1. Fluctuations in the bulk generated leakage current, commonly known as *shot noise*.

2. Fluctuations in surface leakage current.

3. Johnson noise associated with series resistance or poor electrical contacts to the detector.

The noise contribution of the first source can be reduced only through temperature reduction of the crystal. Surface leakage is much more variable and depends on fabrication methods and past history of the detector. The third source includes the series resistance of the undepleted region of partially depleted junction detectors and surface barrier detectors. This contribution can be eliminated if totally depleted detectors are used.

These noise sources add in quadrature, so that

$$\left(\Delta E_{\text{noise}}\right)^2 = \left(\Delta E_{\text{bulk}}\right)^2 + \left(\Delta E_{\text{surface}}\right)^2 + \left(\Delta E_{\text{Johnson}}\right)^2 \qquad (11\text{-}17)$$

where ΔE represents the peak broadening due to each independent mechanism. The noise width combines in quadrature with broadening due to the statistical fluctuation in charge carrier generation to give the overall width characteristic of the detector itself.

If trapping effects become significant in the detector, it is usually evidenced by the appearance of low-energy tails on the peaks observed from monoenergetic sources of radiation. These tails correspond to pulses in which less than the total amount of charge generated by the radiation has been collected. Because the amount of trapping varies according to the distance traveled by the carriers before reaching the collection electrodes, the amount of energy loss is variable and the full-energy peaks are spread only to the low-energy side.

Examples of the energy resolution attainable with semiconductor diode detectors are given in the section on alpha particle spectroscopy later in this chapter.

C. Changes with Detector Bias Voltage

When the bias voltage and electric field are low, the pulse height from radiations that are fully stopped within the depletion layer continues to rise with applied voltage. This variation is caused by the incomplete collection of charge carriers because of trapping or recombination along the track of the incident particle. The fraction that escape collection will decrease as the electric field is increased. Similar losses to recombination are observed in a gas-filled ion chamber at low values of the electric field. Once the electric field is sufficiently high, charge collection becomes complete and the pulse height no longer changes with further increases in the detector bias voltage. This region of operation is called the *saturation region* and corresponds to the region of ion saturation in a gas-filled ion chamber.

If radiations of a single energy and type are involved, it is sometimes possible to operate the detector at a bias voltage that is short of true saturation without significant deterioration in the energy resolution, because the fraction of charge lost for each event is likely to be nearly constant. When measuring radiations of diverse energy and specific ionization, however, it is quite important to ensure that the detector is operating in the region of true saturation to avoid significant deterioration in the energy resolution. To reach the saturation region, somewhat higher electric fields are generally required for diffused junction detectors than for surface barriers, and the necessary field for both types tends to increase as the detector undergoes radiation damage.[35]

If the electric field is made sufficiently high, multiplication effects can be induced in a semiconductor detector which are analogous to gas multiplication in proportional counters or Geiger–Mueller tubes. The multiplication arises when electrons liberated in the initial radiation interaction gain sufficient energy from the field to create further electron–hole pairs as they drift toward the collecting electrodes. These multiplication effects are discussed in some detail in Ref. 36 and differ significantly in diffused junction detectors compared with surface barriers. In diffused junctions, the multiplication is evidenced by a more or less uniform increase in gain which is discernible as a general shift of the recorded pulse height spectrum to larger amplitude. The minimum field required is quite high (about 10^7 V/m), which approaches the bulk breakdown field in silicon. In surface barriers, however, multiplication can be observed with fields that are an order of magnitude lower. Here the effects are often nonuniform and normally can be detected as the appearance of a tail at the high-energy side of peaks recorded from monoenergetic sources.

D. Pulse Rise Time

Semiconductor diode detectors are typically among the fastest of all commonly used radiation detectors. A general review of theoretical and experimental work on their timing properties is given in Ref. 37. Under normal conditions, the observed pulse rise time is of the order of 10 ns or less. The detector contribution to this rise time is composed of the *charge transit time* and the *plasma time*.

The charge transit time corresponds to the migration of the electrons and holes formed by the incident radiation across the region of high electric field in the depletion region. The rise time of the output pulse is therefore limited by the time required for complete migration of these charges from their point of formation to the opposite extremes of the depletion region. These times are minimized in detectors with high electric fields and small depletion widths. In totally depleted detectors, the depletion width is fixed by the physical thickness of the silicon wafer, and therefore the transit time is decreased as the bias voltage is increased. In partially depleted detectors, however, the depletion width increases with increasing bias [see Eq. (11-13)], and therefore the effect of a larger bias voltage is to increase both the electric field and the distance over which charges must be collected. Furthermore, because the electric field is not uniform, the drift velocity of electrons and holes will vary as they move across the depletion region. The dependence of the charge transit time on bias voltage in these detectors is therefore somewhat more complicated, but it can be shown to be independent of the voltage if certain simplifying assumptions are made.[38] A derivation is given in Chapter 12 (see p. 403) of the time profile of the signal pulse attributable to charge migration in solid-state detectors in which the electric field is uniform.

For the case of a particle range that is much less than the width of the depletion region, all the charge carriers are created near one boundary. The collection time of one type of carrier corresponds to its migration across the entire depletion region and is therefore much longer than that for the other carrier. For a surface barrier on an *n*-type crystal, it is thus the electron collection time that dominates the time response for weakly penetrating particles.

With the assumption of a constant mobility value for the electrons, the transit time under these conditions can be shown[39] to be

$$t_c = \frac{0.53d^2}{\mu V} \tag{11-18}$$

where t_c is in units of seconds, d is the depletion width in cm, μ is the electron mobility in $cm^2/V \cdot s$, and V is the applied bias in volts. Experimental observations, which show that the rise time for heavily ionizing particles is considerably greater than that predicted by Eq. (11-18), have led to the realization that an additional component of the rise time is involved for these cases.

A second component called the *plasma time* is observed when heavy charged particles, such as alpha particles or fission fragments, comprise the incident radiation. For these radiations, the density of electron–hole pairs along the track of the particle is sufficiently high to form a plasma-like cloud of charge that shields the interior from the influence of the electric field. Only those charge carriers at the outer edge of the cloud are subject to the influence of the field, and they begin to migrate immediately. The outer regions are gradually eroded away until the charges at the interior are finally subject to the applied field and also begin to drift. The plasma time is roughly defined as the time required for the charge cloud to disperse to the point where normal charge collection proceeds.

A number of theoretical models have been developed to describe the plasma erosion process;[40-44] it is predicted[40] that the plasma time should vary inversely with the electric field strength at the position of the track and increase as the cube root of the linear carrier density along the track. The effects of the plasma formation are observed to be a fixed delay of several nanoseconds between the time of track formation and the onset of the rise of the output pulse together with a slowing of the rise time of the output pulse. Measurements of the delay time with silicon surface barrier detectors,[45-50] give typical values of 1–3 ns for alpha particles, and 2–5 ns for heavy ions and fission fragments.

The actual rise time observed from a detector–preamplifier combination may also be influenced by the preamplifier properties. The time constant of the equivalent input circuit must be short if the rise time is to be held to that determined by the detector charge collection and plasma time properties only. One contributor to the input time constant is the series resistance of the undepleted region in partially depleted detectors. Therefore, fully depleted detectors in which the series resistance is largely eliminated are often favored in fast-timing situations.

E. Entrance Window or Dead Layer

When heavy charged particles or other weakly penetrating radiations are involved, the energy loss that may take place before the particle reaches the active volume of the detector can be significant. Because the thickness of the dead layer includes not only the metallic electrode but also an indeterminant thickness of silicon immediately beneath the electrode in which charge collection is inefficient, the dead layer can be a function of the applied voltage. Its effective thickness must often be measured directly by the user if accurate compensation is to be made.

The simplest and most frequently used technique is to vary the angle of incidence of a monoenergetic charged particle radiation. When the angle of incidence is zero (i.e., perpendicular to the detector surface), the energy loss in the dead layer is given by

$$\Delta E_0 = \frac{dE_0}{dx} t \qquad (11\text{-}19)$$

where t is the thickness of the dead layer. The energy loss for an angle of incidence of θ is given by

$$\Delta E(\theta) = \frac{\Delta E_0}{\cos \theta} \qquad (11\text{-}20)$$

Therefore, the difference between the measured pulse height for angles of incidence of zero and θ is given by

$$E' = (E_0 - \Delta E_0) - (E_0 - \Delta E(\theta))$$

$$E' = \Delta E_0 \left(\frac{1}{\cos \theta} - 1 \right) \qquad (11\text{-}21)$$

If a series of measurements are made as the angle of incidence is varied, a plot of E' as a function of $(1/\cos \theta - 1)$ should be a straight line whose slope is equal to ΔE_0. Using tabular data for dE_0/dx for the incident radiation, we can calculate the dead layer thickness from Eq. (11-19).

One possible flaw in this method involves the assumption that the energy loss through the dead layer depends only on the total path length traversed and not on the relative orientation of the particle path with respect to the detector axis. There is some evidence that recombination should be more severe for particle paths parallel to the direction of the electric field in the detector compared with paths perpendicular to the field. This recombination would tend to cause a lower than expected response for paths near normal incidence and should be evidenced by a curvature in the plot described above.

The thinnest dead layers are produced in semiconductor detectors of the ion implanted or surface barrier types. Typical values of 100 nm of silicon equivalent correspond to an energy loss of 4 keV for 1 MeV protons, 14 keV for 5 MeV alpha particles, and several hundred keV for fission fragments. Because variations in this energy loss due to straggling or variable angle of incidence will potentially detract from energy resolution, thin dead layers are quite important in high-resolution charged particle spectroscopy. Using special techniques, dead layers of less than 30 nm have been successfully fabricated.[51]

F. Channeling

In crystalline materials, the rate of energy loss of a charged particle can depend on the orientation of its path with respect to the crystal axes.[52] Particles that travel parallel to crystal planes can, on the average, show a rate of energy loss that is lower than that for particles directed in some arbitrary direction. Therefore, these "channeled" particles can penetrate significantly farther through the crystal. The effects are particularly significant for thin totally depleted detectors because the amount of energy deposited is then dependent on the orientation of the crystal planes with respect to the particle direction. To minimize the tendency for incident particles to channel, detectors are normally fabricated from silicon cut so that the $\langle 111 \rangle$ crystal orientation is perpendicular to the wafer surface.

Channeling can affect the recorded pulse height even in situations in which the particle is fully stopped within the active volume. Nuclear collisions are less probable for channeled particles and therefore the pulse height defect for heavy ions (discussed later in this chapter) may be reduced.

G. Radiation Damage

The proper operation of any semiconductor detector depends on the near perfection of the crystalline lattice to prevent defects that can trap charge carriers and lead to incomplete charge collection. Any extensive use of these detectors, however, ensures that some damage to the lattice will take place due to the disruptive effects of the radiation being measured as it passes through the crystal. These effects tend to be relatively minor for lightly ionizing radiations (beta particles or gamma rays) but can become quite significant under typical conditions of use for heavy charged particles. For example, prolonged exposure of silicon surface barrier detectors to fission fragments will lead to a measurable increase in the leakage current and a significant loss in energy resolution of the detector. With extreme radiation damage, multiple peaks may appear in the pulse height spectrum recorded for monoenergetic particles. Furthermore, the time characteristics of the detector may be degraded even at doses that are too low to show measurable spectral effects.[53]

The form of the most common type of radiation damage is the *Frenkel defect*, produced by the displacement of an atom of the semiconductor material from its normal lattice site. The vacancy left behind, together with the original atom now at an interstitial position, constitutes a trapping site for normal charge carriers. When enough of these defects have been formed, carrier lifetime is reduced and the energy resolution of the detector is degraded due to fluctuations in the amount of charge lost. The increase in leakage current appears to be more directly related to edge effects[54] and also contributes to a loss of detector resolution from the corresponding increase in leakage current fluctuation. Some minor annealing of the radiation damage can occur over long periods of time (see Fig. 11-14), but for all intents and purposes, the damage is permanent.

The number of Frenkel defects produced by a fission fragment is estimated to be about 100–1000 times greater than that produced by an alpha particle.[54] At the other extreme, an incident electron or beta particle requires a minimum of about 145 keV to produce a defect, and very little damage is observed for electrons whose energy is much below 250 keV.[55] The severity of damage to be anticipated is therefore a strong function of the nature of the radiation involved.

For silicon surface barriers irradiated on the gold or front surface, various data have been published on the integrated flux of charged particles required to produce a significant deterioration in detector performance. Although subject to a great deal of variability, depending on the specifics of each experiment, serious changes appear to take place for fast electron irradiations of about $10^{14}/cm^2$ (Ref. 55), 10^{12} to 10^{13} protons/cm^2 (Refs. 56 and 57), 10^{11} alpha particles/cm^2 (Ref. 58), and about 3×10^8 fission fragments/cm^2 (Ref. 54). Exposure to fast neutron fluxes of about 3×10^{11} neutrons/cm^2 (Ref. 59) and gamma-ray doses of about 10^6 R (Ref. 60) are also sufficient to lead to significant performance degradation.

For penetrating radiations such as gamma rays or neutrons, the damage is generally distributed throughout the detector and the direction of incidence of the radiation has little effect. For electrons or charged particles, however, the orientation with respect to the detector is important. Irradiation of the front (or gold) surface of totally depleted detectors requires exposures that are several orders of magnitude less than those needed to produce the same effects by irradiation of the back (or aluminum) contact.[56]

In general, diffused junction detectors are somewhat less susceptible to radiation damage effects than surface barriers. Also, fully depleted detectors are less sensitive than partially depleted devices because the average electric field throughout the detector is somewhat higher. The reduced carrier lifetime is then less important.

H. Energy Calibration

When applied to the measurement of fast electrons or light ions such as protons or alpha particles, semiconductor diode radiation detectors respond very linearly, and the energy calibration obtained for one particle type is very close to that obtained using a different radiation type. Some observations show that there is a small difference in the pulse height observed for protons and alpha particles of the same energy,[61, 62] but such differences are usually on the order of 1% or less. It is not clear whether these differences reflect a fundamental difference in the ionization energy ϵ for various particles, or whether other factors related to the modes of energy loss may be responsible. For any application in which an absolute energy calibration of about 1% or less is required, it is always best to calibrate the detector using the same type of particle that is involved in the measurement itself.

The most common calibration source is the alpha-emitting isotope ^{241}Am. This isotope emits alpha particles of 5.486 MeV (85%) and 5.443 MeV (13%), and a representative pulse height spectrum is illustrated in Fig. 11-12. An accurate calibration of the energy scale requires that account be taken of the energy loss of these alpha particles in the source itself, in any intervening material between the source and detector, and in the window or dead layer of the detector.

I. Pulse Height Defect

The response of semiconductor detectors to heavy ions such as fission fragments is less straightforward. There is abundant evidence that the pulse height observed for heavy ions is substantially less than that observed for a light ion of the same energy. The *pulse height defect* is defined in units of energy as the difference between the true energy of the heavy ion and its apparent energy, as determined from an energy calibration of the detector obtained using alpha particles.

A plot of the effects of the pulse height defect is given in Fig. 11-13. Measurements of the pulse height defect for surface barriers when irradiated by fission fragments[65] show that a value as large as 15 MeV for the defect is possible, compared with an average

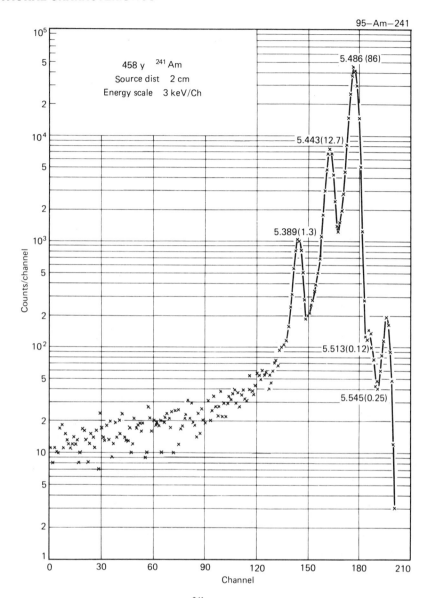

Figure 11-12 Upper portion of the ^{241}Am alpha spectrum as recorded by a high-resolution surface barrier detector. (From Chanda and Deal.[63])

energy of about 80 MeV. Other measurements[66-68] using fragments that were separated in mass and energy obtained smaller defect values of 3–7 MeV.

Analysis has shown[64,69-71] that three separate phenomena contribute to the observed pulse height defect. The first and simplest contributor is the energy loss of the ion in the entrance window and dead layer of the detector. The magnitude of the contribution to the pulse height defect can be calculated from the stopping power of the ion and measurements of the dead layer thickness using methods discussed earlier. Because heavy ions such as fission fragments show maximum dE/dx at the start of their range, whereas light

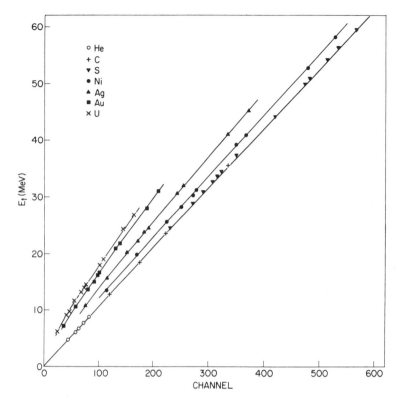

Figure 11-13   The true energy of various ions versus the pulse height channel number from a silicon surface barrier detector. (From Wilkins et al.[64])

ions such as alpha particles show the reverse behavior, the fractional energy loss in the dead layer will be much more significant for the heavy ions.

A second contributor to the pulse height defect involves the tendency for heavy ions to lose energy by means other than simple electronic collisions. As the velocity of the ion decreases, nuclear collisions become important, and recoil nuclei are the direct result of such interactions. Because of the low velocity of these recoil nuclei, the probability of electronic interactions is reduced and a net decrease in the efficiency with which electron–hole pairs are produced is observed. The contribution of nuclear collisions increases with the effective charge on the ion and is therefore most significant for heavy ions. A calculation of the magnitude of this effect has been carried out by Haines and Whitehead[72] from basic theories of ion slowing down in solids. An experimental study of the effect in surface barriers is described in Ref. 73.

A third factor in the pulse height defect involves the high rate of electron–hole recombination expected in the dense plasma created along the ion track, particularly near its end.[65, 74] The magnitude of the recombination would be expected to decrease with increasing bias voltage and may also depend on the relative orientation of the particle path with respect to the electric field within the detector. A practical method of reducing the pulse height defect is to minimize the effect of recombination by creating the largest possible electric field within the detector.

Because the effects of trapping and recombination are influenced by radiation damage within the detector, it should be anticipated that the pulse height defect may

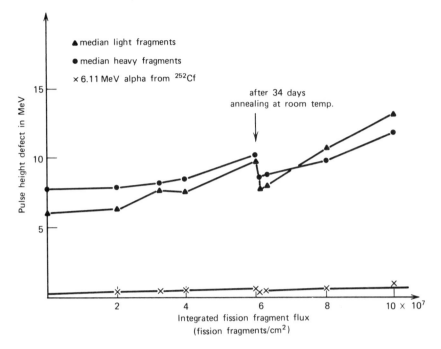

Figure 11-14 Dependence of measured pulse height defect versus fission fragment exposure for a heavy ion silicon detector. (From Bozorgmanesh.[75])

increase with normal use. Figure 11-14 shows the measured pulse height defect for a silicon heavy ion detector as a function of integrated fission fragment flux. The general trend toward increasing pulse height defect is evident, together with the effects of partial annealing of the radiation damage over several weeks following removal of the detector from the flux.

VI. APPLICATIONS OF SILICON DIODE DETECTORS

A. General Charged Particle Spectroscopy

Since their development as practical detectors in the early 1960s, silicon diodes have become the detectors of choice for the majority of applications in which heavy charged particles are involved. Some of the more common applications involving the spectroscopy of alpha particles and fission fragments and the measurement of energy loss of charged particles in transmission detectors are discussed in the following section. Most of the conclusions carry over into other charged particle spectroscopy involving protons, deuterons, or other heavy ions.

Silicon diode detectors are sometimes used for the measurement of beta particles and fast electrons, particularly as thin totally depleted transmission detectors. However, because it is more common to use ion-drifted detectors for this purpose, a discussion of the response of semiconductor detectors to fast electrons will be postponed until Chapter 13.

Compared with competing techniques, the use of semiconductor detectors often provides advantages in a number of key areas. These include exceptionally good energy resolution, good stability and freedom from drift, excellent timing characteristics, very thin entrance windows, and simplicity of operation. The relatively small size can be an advantage in some situations but is also a limitation in those applications in which a detector with a large surface area is required. Silicon diodes are commercially available with surface area up to 20 cm^2, but the corresponding large capacitance results in a poorer energy resolution than is attainable with smaller detectors. More usual sizes range from 1 to 5 cm^2. Depletion depths up to 5 mm can be obtained commercially in some special configurations, but the more common detectors are limited to a depletion depth of 1 mm or less.

In the event that the detector depletion depth is greater than the range of the incident heavy ions, the response of the detector is very simple. For monoenergetic incident particles, only a single full-energy peak is observed because there are no competing processes that can significantly scatter out the ion or otherwise lead to partial energy deposition.[†] For fully depleted detectors, the depletion depth is given simply by the thickness of the silicon wafer. In partially depleted detectors, the depletion depth increases with applied bias and is therefore limited by the maximum detector bias that can be applied without risking detector breakdown. The maximum bias and the corresponding depletion depth are normally provided as a specification by the detector manufacturer. If a measurement of the radiation energy is not required, simple counting of charged particle radiation can be carried out with semiconductor detectors whose depletion depth is less than the range, provided the energy deposited within the depletion region is sufficiently high to generate a pulse that lies above the noise level of the instrumentation system.

B. Alpha Particle Spectroscopy

Silicon diodes operated at room temperature are near-ideal detectors for alpha particles and other light ions. Because of the wide availability of convenient monoenergetic sources of alpha particles, the performance of semiconductor detectors conventionally is tested by recording the pulse height spectrum from such sources. The most common of these is ^{241}Am, and the corresponding alpha spectrum is widely used for comparison of the energy resolution of solid-state detectors. A representative spectrum taken with a detector of good resolution is shown in Fig. 11-12.

With alpha particles of this energy (5.486 MeV), the noise contribution of the preamplifier and other electronic components is normally smaller than the inherent energy resolution of the detector itself. Silicon detectors of small size are currently available with resolutions of 10–11 keV. A more typical resolution for larger detectors is 15–20 keV. This performance is somewhat disappointing in that the limiting resolution from charge carrier statistics using estimated values of the Fano factor should be only 3–4 keV. The difference represents the additional contributions to peak broadening that are still significant in these detector systems.

[†]For incident ions with relatively high energies, some complications in the response can be observed due to nuclear reactions induced by the incident particle in the material of the detector. One example occurs in silicon detectors due to the inelastic scattering of light ions from ^{28}Si nuclei in the detector that are followed by the emission of a 1.78 MeV gamma ray. Since the probability is high that this gamma ray escapes from the detector, a satellite peak located 1.78 MeV below the full-energy peak can sometimes be observed.[76] In the case of incident protons, the probability for this process remains lower than 0.1% per incident particle for energies below 10 MeV.

One of these contributions is due to the fact that a small portion of the alpha particle energy is transferred to recoil nuclei rather than to electrons. These low-energy recoil nuclei lose their energy in quasielastic collisions with surrounding atoms and form almost no additional electron–hole pairs. If the fraction of energy lost in this manner were constant for each alpha particle, there would be no effect on energy resolution. However, this energy loss is subject to large fluctuation since it is influenced by a few relatively large events. It has been estimated[77] that the FWHM contribution due to these fluctuations in silicon amount to about 3.5 keV for 6 MeV alpha particles.

Other significant contributions to the peak broadening are the effects of incomplete charge collection, and variations in the energy lost by the particle in dead layers at the detector surface. For lower charged particle energies and detectors of large capacitance, the electronic noise can also be a significant contribution. The noise level from the detector–preamplifier–amplifier combination is dominated by fluctuations in the detector leakage current, the inherent preamplifier noise, and the characteristics of the FET used in the input stage of the preamplifier.

Since all these sources of peak broadening are normally independent, the square of each FWHM that would theoretically be observed for each source alone can be summed together to give the square of the overall FWHM.

C. Heavy Ion and Fission Fragment Spectroscopy

The energy measurement of fission fragments or other ions of large mass involves several special concerns. Most stem from the high density of charge carriers that are created along the track of these ions. Recombination of electron–hole pairs is accentuated, and the detector may require a higher bias voltage than that required to saturate the signal from alpha particles. The pulse height defect discussed previously is also accentuated by this high carrier density, complicating the energy calibration procedures. Also, the prolonged exposure of detectors to heavy ions or fission fragments creates rapid performance deterioration due to radiation damage in the detector.

Detector manufacturers generally offer some silicon surface barrier detectors especially tailored for heavy ion spectroscopy. They are designed to minimize the problems of slow rise time and pulse height defect caused by the high specific energy loss along the particle track. The most effective step is to ensure that the electric field is as high as possible. One approach is to use thin slices of low-resistivity silicon to prepare totally depleted detectors that require a large value of the reverse bias to become fully depleted. Special methods of preparing contacts can also be used[78] to allow extreme overbiasing (beyond the point of full depletion), which can raise the field magnitude by as much as a factor of 20 over its value at the depletion voltage.[79] The silicon used for heavy ion detectors should also have large carrier lifetimes to help reduce the pulse height defect due to recombination.

Typical measurements[80-82] of the response function of silicon diodes to monoenergetic heavy ions show an asymmetric peak with significant tailing toward the low energy side. The cause of the tailing is likely related to fluctuations in the energy loss due to the entrance window or in the charge lost to recombination along the particle track. Representative energy resolution figures range from about 100 keV FWHM for 35–50 MeV oxygen ions, to about 300 keV for 50 MeV sulfur ions.[81]

Recording the fission fragment spectrum from the spontaneously fissioning isotope ^{252}Cf serves as a standard test of heavy ion detector performance. Thin sources of this isotope are readily available and are widely used to monitor detector properties. Figure

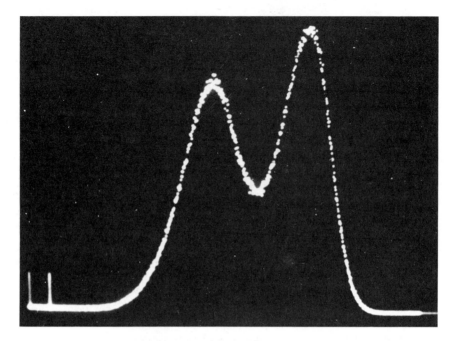

Figure 11-15 ^{252}Cf fission fragment pulse height spectrum. The spectrum parameters defined on the diagram can be used for energy calibration and detector evaluation (see text). (From Bozorgmanesh[75] and Schmitt and Pleasonton.[83])

11-15 shows a ^{252}Cf fission fragment spectrum taken with a good quality silicon surface barrier detector. Schmitt et al. have shown that a careful analysis of this spectrum can give an accurate energy calibration of the detector. The spectrum also provides a check on detector properties, such as resolution, low-energy tailing, and any possible effects due to internal multiplication in the detector.

In the first work reported by these investigators,[35,84] a series of surface barrier and diffused junction silicon detectors were carefully calibrated by recording the pulse height spectrum from monoenergetic heavy ions produced in an accelerator. These results show that for a given heavy ion species, the fragment energy versus pulse height relation is of the form $E = ax + b$, where E is the ion energy and x is the measured pulse height. This relation was found to hold over a wide energy range and illustrates the constancy of the pulse height defect for a given particle type. By studying a number of different ions, a general relation was demonstrated for an ion of mass m of the form

$$E(x, m) = (a + a'm)x + b + b'm \qquad (11\text{-}22)$$

where the constants a, a', b, and b' assume different values depending on the specific detector.

The californium fission product spectrum was also recorded for the same detectors. A scheme was then developed by which the values of the constants in Eq. (11-22) can be extracted from properties of the ^{252}Cf spectrum. A more recent study by Weissenberger et al.[85] has confirmed this general approach to energy calibration of the detector, with the numerical results shown below. Referring to spectrum features defined in Fig. 11-15, the

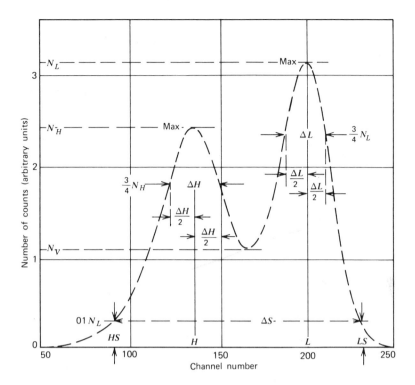

following values were obtained for each of the constants:

$$a = \frac{24.300}{P_L - P_H} \qquad \left(\frac{\text{MeV}}{\text{channel}}\right)$$

$$a' = \frac{0.0283}{P_L - P_H} \qquad \left(\frac{\text{MeV}}{\text{channel} \cdot \text{amu}}\right)$$

$$b = 90.397 - aP_L \quad (\text{MeV})$$

$$b' = 0.1150 - a'P_L \quad \left(\frac{\text{MeV}}{\text{amu}}\right)$$

where P_L and P_H are the channel numbers corresponding to L and H on the figure.

TABLE 11-2   Parameters of the ^{252}Cf Fission Fragment Spectrum

Spectrum Parameter[a]	Reasonable Limit	Expected Value
N_L/N_V	> 2.85	~ 2.9
N_H/N_V	~ 2.2	~ 2.2
N_L/N_H	—	~ 1.30
$\Delta L/(L - H)$	< 0.38	~ 0.36
$\Delta H/(L - H)$	≲ 0.45	≲ 0.44
$(H - HS)/(L - H)$	< 0.70	≲ 0.69
$(LS - L)/(L - H)$	≲ 0.49	≲ 0.48
$(LS - HS)/(L - H)$	≲ 2.18	~ 2.17

[a] Definitions are given in Fig. 11-15.

Source: Schmitt and Pleasonton.[83]

Schmitt and Pleasonton[83] have published a set of acceptance criteria based on the ^{252}Cf spectrum that should be displayed by a good quality heavy ion detector. These are shown in Table 11-2 and refer to the same spectrum shape parameters. The physical significance of each of these terms is detailed in Ref. 83. Deviations from these criteria can indicate poor detector resolution or other undesirable performance characteristics. Periodic monitoring of the californium spectrum can, for example, indicate the onset of performance degradation caused by radiation damage over prolonged periods.

D. Energy Loss Measurements — Particle Identification

Previously, we gave examples of charged particle spectroscopy which involved totally stopping these particles within the depletion region of the semiconductor detector. Neglecting the pulse height defect, the number of charge carriers created is proportional to the total energy of the incident radiation. Applications sometimes arise in which the specific energy loss dE/dx of the incident radiation is of interest, rather than its total energy. For these applications, detectors that are thin compared with the particle range are chosen. The number of charge carriers created within a detector of small thickness Δt will simply be $(dE/dx)\Delta t/\epsilon$. The particle passes completely through the detector, retaining most of its initial energy, and a signal proportional to dE/dx is observed. In such applications, the device is often called a ΔE detector.

A number of detector types are used for such measurements. Thin film scintillators discussed in Chapter 8 can be made in uniform thicknesses which are the smallest of any solid detector, but they do not offer good energy resolution. A totally depleted detector, on the other hand, can be manufactured from a semiconductor wafer as thin as 10 μm and will provide the excellent energy resolution generally observed for all solid-state detectors. The thickness uniformity of transmission detectors is very important if the inherently good energy resolution is to be preserved. For example, a variation of only 1 μm in a wafer of 20 μm thickness will introduce a 5% variation in the signal—considerably larger than a typical energy resolution figure.

For low-energy particles, the range may be too small to allow the use of even the thinnest available silicon detector as a ΔE detector. A number of applications[86-89] therefore use a gas-filled ionization or proportional counter for this purpose. Their advantage is that their thickness can be made very uniform and adjustable by changing the gas pressure, but the limiting energy resolution is theoretically poorer. In many applications, however, the ΔE signal fluctuations are dominated by energy straggling of the incident particles, and the energy resolution of gas-filled detectors may be adequate.

Transmission detectors are commonly used in conjunction with a normal surface barrier or other "thick" detector in the *particle identifier telescope* arrangement diagrammed in Fig. 11-16. By accepting only those events that occur in coincidence between the two detectors, a simultaneous measurement of dE/dx and E is carried out for each incident particle.

For nonrelativistic charged particles of mass m and charge ze, Bethe's formula [Eq. (2-2)] predicts that

$$\frac{dE}{dx} = C_1 \frac{mz^2}{E} \ln C_2 \frac{E}{m} \tag{11-23}$$

where C_1 and C_2 are constants. If we form the product $E(dE/dx)$, the result is only mildly dependent on the particle energy but is a sensitive indicator of the mz^2 value that characterizes the particle involved. If the incident radiation consists of a mixture of different particles whose energies do not differ by large factors, the product of the pulse amplitudes from both detectors will therefore be a nearly unique parameter for each

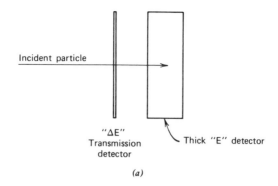

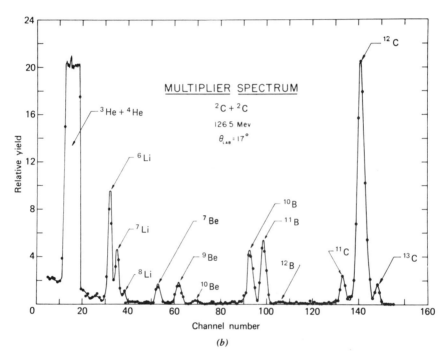

Figure 11-16 (a) A particle identifier arrangement consisting of tandem ΔE and E detectors operated in coincidence. (b) Experimental spectrum obtained for the $\Delta E \cdot E$ signal product for a mixture of different ions. (From Bromley.[90])

different particle type. Because the incident energy can be obtained by summing the pulse amplitudes from the ΔE and E detectors, simultaneous determination of both mass and energy of each incident particle is therefore possible. Figure 11-16 shows a distribution of the $E \Delta E$ product for a particle identifier telescope in which a large number of individual particle types can be distinguished.

An alternative approach described by Goulding and Landis[91] is based on the observation that the range R and energy E for a wide variety of charged particles can be related by a power-law approximation (see Fig. 2-7)

$$R(E) = aE^b \qquad (11\text{-}24)$$

where a and b are constants. If the incident particle deposits an energy ΔE in the transmission detector of thickness Δt, and the remainder of its energy E_r in the thick detector, then

$$\Delta t = R(E_r + \Delta E) - R(E_r)$$

or

$$\frac{\Delta t}{a} = (E_r + \Delta E)^b - E_r^b \tag{11-25}$$

The value of a is a constant for a given particle type roughly proportional to $1/mz^2$. The value of b does not change greatly if ions of similar mass are involved ($b = 1.73$ for protons and $b = 1.65$ for carbon ions). Therefore, choosing a reasonable value for b and using a signal-processing element to carry out the operations on ΔE and E_r indicated in Eq. (11-25) will yield a parameter that is characteristic of the particle type and independent of its energy.

Particle identification schemes based on the above methods and their subsequent modifications are compared in the review article by Goulding and Harvey.[92] Their applicability is always limited by the approximations made to the actual energy loss or range behavior of the particles that are involved. An alternative "brute force" approach,[93] which can avoid any approximations, is to make use of a table of actual energy loss data stored in a computer memory and to carry out on-line digital comparisons with these data. The fundamental limit to the ability of any scheme based on ΔE and E measurements to distinguish particle types generally is set by the fluctuations in the ΔE signal produced by energy straggling.[94]

E. Current Mode Operation

The overwhelming majority of applications of silicon junction detectors involve pulse mode operation to take advantage of their excellent pulse height resolution. However, it has been demonstrated[28] that these detectors can also be useful for monitoring high radiation fields when operated in current mode. In these applications, the output current is directly measured from the detector and has been demonstrated to have excellent proportionality to the intensity of the incident radiation up to output currents of several amperes. In any such application, one must bear in mind the susceptibility of semiconductor detectors to radiation damage and limit the integrated particle flux below those listed on p. 372 as sufficient to cause appreciable internal detector damage.

It is possible to measure radiation-induced current from a semiconductor detector, even in the absence of an applied voltage.[95, 96] This *photovoltaic* mode of operation is similar to that of a common solar cell, except that the charge carriers are created by the ionizing radiation rather than by incident light. As can be seen from Fig. 11-5, the simple junction of n- and p-type semiconductors creates a contact potential and a depletion region. Electrons and holes formed within the depletion region will drift under the influence of the corresponding electric field, and current will be measured in an external circuit connected across the junction. This photovoltaic mode is not widely used because the contact potential is small, amounting to less than 1 V in silicon. The thickness of the depletion region is also small, usually less than the range of charged particles of interest. Junction detectors are therefore normally operated with a large reverse bias voltage to improve their charge collection efficiency and increase the thickness of their active volume. However, the small depletion region present without applied voltage may be adequate to result in a useful current when the incident radiation intensity is sufficiently high.

PROBLEMS

11-1. From the values for intrinsic carrier densities given in Table 11-1, estimate the impurity levels (in parts per billion) that begin to change intrinsic silicon and germanium into doped materials at room temperature.

11-2. Find the ratio of the number of charge carriers created in silicon by a 1 MeV proton to the number created by the same energy deposition in air.

11-3. From the data in Table 11-1, calculate the mean value and variance in the number of electron–hole pairs created by the loss of 100 keV of particle energy in silicon.

11-4. By what factor is the rate of thermal generation of electron–hole pairs in germanium reduced by cooling from room temperature to liquid nitrogen temperature (77 K)?

11-5. In order to reverse bias the junction, should the positive terminal of the voltage supply be connected to the p or n side of the junction? Justify your answer.

11-6. Indicate the functional dependence of the following properties of a p-n junction on the magnitude of the applied reverse bias:

(a) Depletion width.
(b) Capacitance.
(c) Maximum electric field.

11-7. There is often a premium on fabricating detectors with the largest possible depletion width for a given applied voltage. Explain why starting with semiconductor material of the highest available purity enhances this objective.

11-8. Use the nomogram of Fig. 11-7 to find the bias voltage necessary to create a depletion depth of 0.1 mm in a junction detector prepared from n-type silicon with 1000 Ω-cm resistivity.

11-9. The primary alpha peak from a ^{241}Am calibration source was centered in channel number 461 of a multichannel analyzer when the alpha particles were collimated to be perpendicular to the surface of a silicon junction detector. The geometry was then changed to cause the alpha particles to be incident at an angle of 35° to the normal, and the observed peak shifted to channel number 455. Assuming no zero offset in the MCA, find the dead layer thickness in units of alpha energy loss.

11-10. Why does the typical energy resolution for surface barrier detectors worsen as the surface area of the detector increases?

11-11. A totally depleted silicon detector with 0.1 mm thickness is operated with large overbias so as to saturate the carrier velocities everywhere within the wafer. Estimate the maximum electron and hole collection times.

11-12. A partially depleted silicon surface barrier is operated with sufficient bias voltage to ensure that the depletion depth greatly exceeds the range of incident 5 MeV alpha particles. If used with a voltage-sensitive preamplifier, how much will the pulse amplitude change if the bias voltage changes by 5%?

11-13. A 10 MBq source of alpha particles is located 10 cm in front of a silicon surface barrier detector. After what length of exposure time is radiation damage to the detector likely to become significant?

11-14. In a given heavy ion detector in which zero energy deposition corresponds to channel zero in an associated multichannel pulse height analyzer, the 5.486 MeV alpha particles from ^{241}Am are recorded in channel number 116. If heavy ions of 21.0 MeV energy are recorded in channel 402, what is their pulse height defect?

11-15. Sketch the expected differential pulse height spectra from a silicon surface barrier detector under the following conditions:

(a) Incident 5 MeV alpha particles, depletion depth of the detector greater than the alpha range.

(b) On the same graph, the corresponding spectrum for a depletion depth of one-half the alpha range.

(c) Again on the same graph, the spectrum with depletion depth as in part (a), but for the same alpha particles after they have passed through an absorber whose thickness is equal to one-half the alpha range in the absorber material.

REFERENCES

1. G. Dearnaley and D. C. Northrop, *Semiconductor Counters for Nuclear Radiations*, 2nd ed., Wiley, New York, 1966.
2. G. Bertolini and A. Coche (eds.), *Semiconductor Detectors*, Elsevier-North Holland, Amsterdam, 1968.
3. G. T. Ewan, "Semiconductor Spectrometers," in *Progress in Nuclear Techniques and Instrumentation*, Vol. 3, F. J. M. Farley (ed.), Elsevier-North Holland, New York, 1968.
4. A. Kuhn, *Halbleiter und Kristallzahler*, Akademische Verlagsgesellschaft, Geest and Portig, Leipzig, 1969.
5. W. L. Brown, W. A. Higinbotham, G. L. Miller, and R. L. Chase (eds.), *Semiconductor Nuclear-Particle Detectors and Circuits*, Publication 1593, National Academy of Sciences, Washington, DC, 1969.
6. H. Büker, *Theorie und Praxis der Halbleiterdetektoren für Kernstrahlung*, Springer-Verlag, Berlin, 1971.
7. J.M. Palms, P. V. Rao, and R. E. Wood, *Nucl. Instrum. Meth.* **76**, 59 (1969).
8. J. E. Eberhardt, *Nucl. Instrum. Meth.* **80**, 291 (1970).
9. H. R. Bilger, *Phys. Rev.* **163**, 238 (1967).
10. R. H. Pehl and F. S. Goulding, *UCRL* **19530**, 333 (1969).
11. A. H. Sher and W. J. Keery, *IEEE Trans. Nucl. Sci.* **NS-17**(1), 39 (1970).
12. H. R. Zulliger and D. W. Aikten, *IEEE Trans. Nucl. Sci.* **NS-17**(3), 187 (1970).
13. T. Yamaya, R. Asano, H. Endo, and K. Umeda, *Nucl. Instrum. Meth.* **159**, 181 (1979).
14. N. Stroken, V. Ajdacic, and B. Lalovic, *Nucl. Instrum. Meth.* **94**, 147 (1971).
15. G. Ottaviani, C. Canali, and A. Alberigi Quaranta, *IEEE Trans. Nucl. Sci.* **NS-22**(1), 192 (1975).
16. W. Van Roosbroeck and W. Shockley, *Phys. Rev.* **94**, 1558 (1954).
17. R. H. Pehl, F. S. Goulding, D. A. Landis, and M. Lenzlinger, *Nucl. Instrum. Meth.* **59**, 45 (1968).
18. C. Canali, M. Martini, G. Ottaviani, and A. Alberigi Quaranta, *IEEE Trans. Nucl. Sci.* **NS-19**(4), 9 (1972).
19. R. D. Ryan, *IEEE Trans. Nucl. Sci.* **NS-20**(1), 473 (1973).
20. R. A. Langley, *Nucl. Instrum. Meth.* **113**, 109 (1973).
21. C. A. Klein, *IEEE Trans. Nucl. Sci.* **NS-15**(3), 214 (1968).
22. J. L. Blankenship, *IRE Trans. Nucl. Sci.* **NS-7**(2–3), 190 (1960).
23. N. J. Hansen, R. G. Scott, and D. J. Henderson, *Nucl. Instrum. Meth.* **104**, 333 (1972).
24. H. R. Zulliger, W. E. Drummond, and L. M. Middleman, *IEEE Trans. Nucl. Sci.* **NS-19**(3), 306 (1972).
25. H. Grahmann and S. Kalbitzer, *Nucl. Instrum. Meth.* **136**, 145 (1976).
26. J. W. Mayer, *Nucl. Instrum. Meth.* **63**, 141 (1968).

27. J. Von Borany, G. Mende, and B. Schmidt, *Nucl. Instrum. Meth.* **212**, 489 (1983).

28. P. Burger, M. O. Lampert, R. Henck, and J. Kemmer, *IEEE Trans. Nucl. Sci.* **NS-31**(1), 344 (1984).

29. I. Ahmad, *Nucl. Instrum. Meth. Phys. Res.* **A242**, 395 (1986).

30. J. Kemmer, *Nucl. Instrum. Meth.* **226**, 89 (1984).

31. P. Burger and Y. Beroud, *Nucl. Instrum. Meth.* **226**, 45 (1984).

32. B. J. Hosticka and G. Zimmer, *IEEE Trans. Nucl. Sci.* **NS-32**(1), 402 (1985).

33. J. M. Jaklevic and F. S. Goulding, *IEEE Trans. Nucl. Sci.* **NS-19**(1), 384 (1972).

34. E. Belcarz, J. Chwaszczewska, M. Slapa, M. Szymczak, and J. Tys, *Nucl. Instrum. Meth.* **77**, 21 (1970).

35. H. W. Schmitt, W. M. Gibson, J. H. Neiler, F. J. Walter, and T. D. Thomas, *Proceedings of the IAEA Conference on The Physics and Chemistry of Fission*, Salzburg, 1965, p. 531.

36. F. J. Walter, *IEEE Trans. Nucl. Sci.* **NS-11**(3), 232 (1964).

37. A. Alberigi Quaranta, M. Martini, and G. Ottaviani, *IEEE Trans. Nucl. Sci.* **NS-16**(2), 35 (1969).

38. H. M. Mann, J. W. Haslett, and G. P. Lietz, *IRE Trans. Nucl. Sci.* **NS-8**(1), 157 (1961).

39. R. L. Williams and P. P. Webb, *R.C.A. Rev.*, 23 (Mar. 1962).

40. W. Seibt, K. E. Sundstrom, and P. A. Tove, *Nucl. Instrum. Meth.* **113**, 317 (1973).

41. E. C. Finch, *Nucl. Instrum. Meth.* **121**, 431 (1974).

42. E. C. Finch, *Nucl. Instrum. Meth.* **129**, 617 (1975).

43. C. F. G. Delaney and E. C. Finch, *Nucl. Instrum. Meth.* **215**, 219 (1983).

44. I. Kanno, *Rev. Sci. Instrum.* **58**(10), 1926 (1987).

45. R. N. Williams and E. M. Lawson, *Nucl. Instrum. Meth.* **120**, 261 (1974).

46. H. Henschel, H. Hipp, A. Kohnle, and F. Gonnenwein, *Nucl. Instrum. Meth.* **125**, 365 (1975).

47. E. C. Finch, C. F. G. Delaney, and M. Asghar, *IEEE Trans. Nucl. Sci.* **NS-27**(1), 286 (1980).

48. E. C. Finch, A. A. Cafolla, and M. Asghar, *Nucl. Instrum. Meth.* **198**, 547 (1982).

49. H.-O. Neidel, H. Henschel, H. Geissel, and Y. Laichter, *Nucl. Instrum. Meth.* **212**, 299 (1983).

50. R. Butsch, J. Pochodzalla, and B. Heck, *Nucl. Instrum. Meth. Phys. Res.* **228**, 586 (1985).

51. C. Inskeep, E. Elad, and R. A. Sareen, *IEEE Trans. Nucl. Sci.* **NS-21**(1), 379 (1974).

52. G. Dearnaley, *IEEE Trans. Nucl. Sci.* **NS-11**(3), 249 (1964).

53. P. Mulas and E. L. Haines, *Rev. Sci. Instrum.* **40**, 507 (1969).

54. F. Shiraishi, *Nucl. Instrum. Meth.* **69**, 316 (1969).

55. Y. M. Liu and J. A. Coleman, *IEEE Trans. Nucl. Sci.* **NS-18**(1), 192 (1971).

56. J. A. Coleman, D. P. Love, J. H. Trainor, and D. J. Williams, *IEEE Trans. Nucl. Sci.* **NS-15**(3), 363 (1968).

57. K. Ohba, T. Shoji, S. Ito, and J. Hiratate, *IEEE Trans. Nucl. Sci.* **NS-30**(1), 371 (1983).

58. F. A. Hanser and B. Sellers, *Rev. Sci. Instrum.* **41**, 780 (1970).

59. G. Dearnaley and A. B. Whitehead, *Nucl. Instrum. Meth.* **12**, 205 (1961).

60. R. W. Klingensmith, *IRE Trans. Nucl. Sci.* **NS-8**(1), 112 (1961).

61. K. W. Kemper and J. D. Fox, *Nucl. Instrum. Meth.* **105**, 333 (1972).

62. W. N. Lennard, H. Geissel, K. B. Winterbon, D. Phillips, T. K. Alexander, and J. S. Forster, *Nucl. Instrum. Meth. Phys. Res.* **A248**, 454 (1986).

63. R. N. Chanda and R. A. Deal, "Catalogue of Semiconductor Alpha-Particle Spectra," IN-1261 (1970).

64. B. D. Wilkins, M. J. Fluss, S. B. Kaufman, C. E. Gross, and E. P. Steinberg, *Nucl. Instrum. Meth.* **92**, 381 (1971).

65. E. C. Finch and A. L. Rodgers, *Nucl. Instrum. Meth.* **113**, 29 (1973).

66. H. Wohlfarth, W. Lang, H. Dann, H. G. Clerc, K. H. Schmidt, and H. Schrader, *Nucl. Instrum. Meth.* **140**, 189 (1977).

67. E. C. Finch et al., *Nucl. Instrum. Meth.* **142**, 539 (1977).

68. E. C. Finch, F. Gönnenwein, P. Geltenbort, A. Oed, E. Weissenberger, *Nucl. Instrum. Meth. Phys. Res.* **228**, 402 (1985).

69. E. P. Steinberg, S. B. Kaufman, B. D. Wilkins, and C. E. Gross, *Nucl. Instrum. Meth.* **99**, 309 (1972).

70. M. Ogihara, J. Nagashima, W. Galster, and T. Mikumo, *Nucl. Instrum. Meth. Phys. Res.* **A251**, 313 (1986).

71. E. C. Finch, *Nucl. Instrum. Meth. Phys. Res.* **A257**, 381 (1987).

72. E. L. Haines and A. B. Whitehead, *Rev. Sci. Instrum.* **37**, 190 (1966).

73. G. Forcinal, P. Siffert, and A. Coche, *IEEE Trans. Nucl. Sci.* **NS-15**(1), 475 (1968).

74. E. C. Finch, M. Asghar, and M. Forte, *Nucl. Instrum. Meth.* **163**, 467 (1979).

75. H. Bozorgmanesh, Ph.D. dissertation, The University of Michigan, 1976.

76. T. H. Zabel et al., *Nucl. Instrum. Meth.* **174**, 459 (1980).

77. G. D. Alkhazov, A. P. Komar, and A. A. Vorob'ev, *Nucl. Instrum. Meth.* **48**, 1 (1967).

78. J. B. A. England and V. W. Hammer, *Nucl. Instrum. Meth.* **96**, 81 (1971).

79. J. B. A. England, *Nucl. Instrum. Meth.* **102**, 365 (1972).

80. E. D. Klema, J. X. Saladin, J. G. Alessi, and H. W. Schmitt, *Nucl. Instrum. Meth.* **178**, 383 (1980)

81. E. D. Klema, F. J. Camelio, and T. K. Saylor, *Nucl. Instrum. Meth. Phys. Res.* **225**, 72 (1984).

82. J. C. Overley and H. W. Lefevre, *Nucl. Instrum. Meth. Phys. Res.* **B10/11**, 237 (1985).

83. H. W. Schmitt and F. Pleasonton, *Nucl. Instrum. Meth.* **40**, 204 (1966).

84. H. W. Schmitt, W. E. Kiker, and C. W. Williams, *Phys. Rev.* **137**, B837 (1965).

85. E. Weissenberger, P. Geltenbort, A. Oed, F. Gönnenwein, and H. Faust, *Nucl. Instrum. Meth. Phys. Res.* **A248**, 506 (1986).

86. J. Barrette, P. Braun-Munzinger, and C. K. Gelbke, *Nucl. Instrum. Meth.* **126**, 181 (1975).

87. D. R. Maxson, D. C. Palmer, and J. P. Bading, *Nucl. Instrum. Meth.* **142**, 479 (1977).

88. B. Sundqvist et al., *IEEE Trans. Nucl. Sci.* **NS-24**(1), 652 (1977).

89. E. Rosario-Garcia and R. E. Benenson, *Nucl. Instrum. Meth.* **143**, 245 (1977).

90. D. A. Bromley, *IRE Trans. Nucl. Sci.* **NS-9**(3), 135 (1962).

91. F. S. Goulding and D. A. Landis, "Recent Advances in Particle Identifiers at Berkeley," in *Semiconductor Nuclear-Particle Detectors and Circuits*, Publication 1593, National Academy of Sciences, Washington, DC, 1969, p. 757.

92. F. S. Goulding and B. G. Harvey, *Ann. Rev. Nucl. Sci.* **25**, 167 (1975).

93. D. G. Perry and L. P. Remsberg, *Nucl. Instrum. Meth.* **135**, 103 (1976).

94. A. G. Seamster, R. E. L. Green, and R. G. Korteling, *Nucl. Instrum. Meth.* **145**, 583 (1977).

95. S. C. Klevenhagen, *Phys. Med. Biol.* **22**, 353 (1977).

96. A. Maruhashi, *Nucl. Instrum. Meth.* **141**, 87 (1977).

CHAPTER · 12

Germanium Gamma-Ray Detectors

I. GENERAL CONSIDERATIONS

The simple junction and surface barrier detectors discussed in Chapter 11 find widespread application for the detection of alpha particles and other short-range radiations but are not easily adaptable for applications that involve more penetrating radiations. Their major limitation is the maximum depletion depth or active volume that can be created. Using silicon or germanium of normal semiconductor purity, depletion depths beyond 2 or 3 mm are difficult to achieve despite applying bias voltages that are near the breakdown level. Much greater thicknesses are required for the detectors intended for gamma-ray spectroscopy, which are the topic of this chapter. From Eq. (11–13) the thickness of the depletion region is given by

$$d = \left(\frac{2\epsilon V}{eN} \right)^{1/2} \tag{12-1}$$

where V is the reverse bias voltage and N is the net impurity concentration in the bulk semiconductor material. At a given applied voltage, greater depletion depths can only be achieved by lowering the value of N through further reductions in the net impurity concentration.

There are two general approaches that can be taken to accomplish this goal. The first is to seek further refining techniques capable of reducing the impurity concentration to approximately 10^{10} atoms/cm^3. At this impurity level in germanium, Eq. (12-1) predicts that a depletion depth of 10 mm can be reached using a reverse bias voltage of less than 1000 V. However, such a low impurity concentration corresponds to levels that are less than 1 part in 10^{12}, a virtually unprecedented degree of material purity. Techniques have been developed to achieve this goal in germanium, but not in silicon.[†] Detectors that are manufactured from this ultrapure germanium are usually called *intrinsic germanium* or *high-purity germanium* (HPGe) detectors, and they have become available with depletion depths of 1 cm or more.

The second approach to reducing net impurity concentration is to create a compensated material in which the residual impurities are balanced by an equal concentration

---

[†]One reason for this difference is that the higher melting point for silicon (1410°C versus 959°C for germanium) makes the exclusion of impurities in the refining process more difficult.

of dopant atoms of the opposite type. This compensation cannot be carried out simply by adding the appropriate amount of dopant to the semiconductor before the crystal is grown, because the balance between acceptors and donors will never be exactly right. Therefore, the material always turns out to be n or p type depending on which impurity shows a predominance, no matter how slight. Instead, the process of *lithium ion drifting* has been applied in both silicon and germanium crystals to compensate the material after the crystal has been grown. Residual acceptor impurities are exactly balanced over a thickness of up to 2 cm by the addition of interstitial lithium donor atoms. The resulting compensated material has many of the properties of intrinsic or pure material. Even if the compensation is not perfect, the residual net impurity level may be low enough so that the drifted region can easily be depleted over its entire thickness.

Germanium detectors produced by the lithium drifting process are given the designation Ge(Li). They became commercially available in the early 1960s and served as the common type of large-volume germanium detector for two decades. The widespread availability of high-purity germanium in the early 1980s provided an alternative to lithium drifting, and most manufacturers have now discontinued production of Ge(Li) detectors in favor of the HPGe type. A major reason for this evolution, discussed later in this chapter, is the much greater operational convenience afforded by HPGe detectors. Whereas Ge(Li) detectors must be continuously maintained at low temperature, HPGe detectors can be allowed to warm to room temperature between uses. Nonetheless, some Ge(Li) detectors remain in use, and so we include their characteristics as part of the general discussion that follows. In general, most important performance characteristics such as detection efficiency and energy resolution are nearly identical for Ge(Li) and HPGe detectors of the same size.

In silicon, large depletion thicknesses can only be achieved by lithium drifting because of the present limit on available purity. A detailed discussion of the lithium-drifting process is therefore postponed until Chapter 13 in which lithium-drifted silicon detectors are described.

II. CONFIGURATIONS OF GERMANIUM DETECTORS

A. High-Purity Germanium (HPGe) Detector Fabrication

Techniques for the production of ultrapure germanium with impurity levels as low as 10^{10} atoms/cm^3 first were developed in the mid 1970s. The starting material is bulk germanium intended for use in the semiconductor industry. This material, already of high purity, is further processed using techniques of zone refining. The impurity levels are progressively reduced by locally heating the material and slowly passing a melted zone from one end of the sample to the other. Since impurities tend to be more soluble in the molten germanium than in the solid, impurities are preferentially transferred to the molten zone and are swept from the sample. After many repetitions of this process, impurity levels as low as 10^9 atoms/cm^3 have been achieved. The resulting germanium is perhaps the most highly purified and completely analyzed material of any kind that has ever been produced in commercial volume. Large single crystals of germanium are then slowly grown from this purified feedstock. If the remaining low-level impurities are acceptors (such as aluminum) the electrical properties of the semiconductor crystal grown from the material is mildly p type. (The designation π type is often used to represent this high-purity p-type material.) Alternatively, if donor impurities remain, high purity n type (designated ν type) is the result.

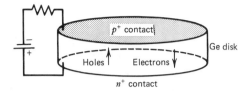

Figure 12-1 Configuration of a planar HPGe detector. The Ge semiconductor may be ν type (p^+ contact is rectifying), π type (n^+ contact is rectifying), or lithium drifted. See Fig. 11-10 for sketches of the electric field shape for these three cases.

B. Planar Configuration

A representative configuration for a *planar* HPGe detector fabricated from high-purity p-type (or π-type) germanium is shown in Fig. 12-1. In a planar configuration, the electrical contacts are provided on the two flat surfaces of a germanium disk. The n^+ contact can be formed either by lithium evaporation and diffusion onto one surface of the wafer, or by direct implantation of donor atoms using an accelerator. The detector depletion region is formed by reverse biasing this n^+-p junction. The contact at the opposite face of the crystal must be a noninjecting contact for a majority carrier. It may consist of a p^+ contact produced by ion implantation of acceptor atoms, or a metal–semiconductor surface barrier that acts as the electrical equivalent. Ion implantation techniques have the advantage that the contact layers can be made very thin to serve as entrance windows for weakly penetrating radiations such as low-energy X-rays.

High-purity germanium detectors are generally operated as fully depleted detectors (see Chapter 11). Reverse biasing requires that a positive voltage be applied to the n^+ contact with respect to the p^+ surface. The depletion region effectively begins at the n^+ edge of the central region and extends further into the π region as the voltage is raised until the detector becomes fully depleted. Further increases in the voltage then serve to increase the electric field everywhere in the detector by a uniform amount (see Fig. 11-9). It is preferable to apply sufficient overvoltage so that the minimum electric field is high enough to impart saturated drift velocity to the charge carriers, minimizing the collection time and the detrimental effects due to carrier recombination and trapping. In germanium at the normal operating temperature of 77 K, saturated electron velocities are reached at a minimum field of about 10^5 V/m, but field strengths three to five times larger are required to fully saturate the hole velocity (see Fig. 11-2). Practical problems related to breakdown and surface leakage often limit the maximum voltage to values at which electrons but not holes will reach saturated drift velocity.

Germanium detectors may also be fabricated starting with high-purity n-type material (also known as ν-type) rather than with p-type material as described above. In this case, n^+ and p^+ contacts are again provided at each surface of the wafer. Reverse biasing still requires the application of a positive voltage to the n^+ contact with respect to the p^+ surface. However, the roles of the two contacts are reversed. The p^+ layer now serves as the rectifying contact and the depletion region extends from this contact as the voltage is raised. The n^+ contact now functions as a *blocking* or noninjecting contact in which the population of holes is very low.

A third type of diode configuration represents the case of lithium-drifted detectors. In this case, the lithium-drifting process (see p. 445) is used to convert the bulk of the wafer to compensated or *intrinsic* material. Here the concentrations of donors and acceptors exactly balance, and the material (often designated *i type*) is neither n nor p type. Again,

n^+ and p^+ contacts are provided on each face of the wafer. In this case, there is no distinction between the rectifying contact and the blocking contact (see Fig. 11-10).

When fully depleted and operated with a large overvoltage, the electric field in planar detectors is almost uniform from one contact to the other. Holes or electrons therefore drift under the influence of a nearly constant field throughout the detector active volume. Under these circumstances, both the n^+-π-p^+ configuration and the n^+-ν-p^+ configuration described above show very similar properties as planar geometry detectors. In contrast, the electric field is quite nonuniform in the coaxial geometry detectors discussed below, and therefore the configuration choice has a more significant influence on the details of the charge carrier motion.

C. Coaxial Configuration

For the planar detectors outlined above, the diameter of the cylindrical crystal from which the wafer is cut is typically no more than a few centimeters. The maximum depletion depth (or thickness of the lithium-drifted region) is limited to less than 1 or 2 cm. The total active volume available in planar detectors therefore does not exceed about 10–30 cm³. To produce a detector with a larger active volume as needed in gamma-ray spectroscopy, a different approach is used. The detector is then constructed in cylindrical or *coaxial* geometry as illustrated in Fig. 12-2. In this case, one electrode is fabricated at the outer cylindrical surface of a long germanium cylindrical crystal. A second cylindrical contact is provided by removing the core of the crystal and placing a contact over the inner cylindrical surface. Because the crystal can be made long in the axial direction, much larger active volumes can be produced (up to 400 cm³ at the time of this writing).

A *closed-ended coaxial* configuration is one in which only part of the central core is removed and the outer electrode is extended over one flat end of the cylindrical crystal (see Fig. 12-2). Most commercial fabricators of HPGe detectors choose the closed-ended configuration over a true coaxial geometry to avoid the complications of dealing with leakage currents at the front surface. Also, the closed-ended configuration provides a planar front surface that can serve as an entrance window for weakly penetrating radiations if fabricated with a thin electrical contact. In the closed-ended configuration, the electric field lines are no longer completely radial as they are in the true coaxial case, and there is some tendency to produce regions of reduced field strength near the corners of the crystal where carrier velocities may be lower than normal. Extending the central hole to reach close to the front surface helps keep the field lines as nearly radial as possible, and some manufacturers round the front corners of the crystal and hole (called *bulletizing*) to help eliminate the low-field regions.

Coaxial HPGe detectors are also available in *well* configurations in which the housing is shaped to allow external access to the central hole. Small radioisotope sources can then be placed within this well for measurements in which the source is nearly surrounded by germanium and the detection efficiency can be unusually high.

In coaxial geometry, the rectifying contact that forms the semiconductor junction can in principle be placed either at the inner or outer surface of the crystal. The electric field conditions that result are quite different. If the rectifying contact is on the outer surface, the depletion layer grows inward as the voltage is raised until reaching the inner hole surface at the depletion voltage. If the inner surface has the rectifying contact, the depletion region grows outward, and a much larger voltage is required to fully deplete the

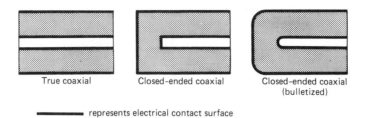

True coaxial   Closed-ended coaxial   Closed-ended coaxial
(bulletized)

━━━━━ represents electrical contact surface

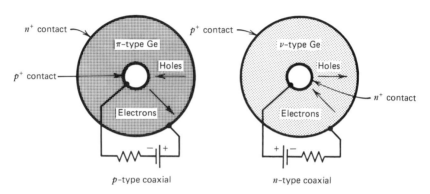

p-type coaxial   n-type coaxial

Figure 12-2 At the top are shown the three common shapes of large-volume coaxial detectors. Each represents a cross-sectional view through the axis of a cylindrical crystal. The outer electrode is extended over the flat front (left) surface in both closed-ended cases. Cross sections perpendicular to the cylindrical axis of the crystal are shown at the bottom. The HPGe material may be either high-purity p or n type. The corresponding electrode configurations are shown for each type.

detector volume. The first choice also maintains a higher electric field value in the outer regions of the cylindrical crystal (where most of the volume lies) as the voltage is raised above full depletion. For these reasons, manufacturers universally choose to locate the rectifying contact at the outer surface. Specifically, the outer contact will be n^+ for p-type HPGe and p^+ for n-type HPGe. The inner surface is provided with a blocking contact of the opposite type. The derivations given in the next section of this chapter assume this configuration for coaxial detectors.

D. Electric Field and Capacitance

The electric field in germanium detectors determines the drift velocity of the charge carriers, so its configuration is important in considerations of pulse shape, timing behavior, and completeness of the charge collection process. The spatial variation of the field strength across the detector active volume is markedly different in planar and coaxial geometry, and the two cases are considered separately in the following discussion.

In each geometry, we solve Poisson's equation

$$\nabla^2\varphi = -\frac{\rho}{\epsilon} \tag{12-2}$$

to find the electric potential φ in the presence of the charge density ρ (ϵ is the dielectric constant). The charge density is dependent on the material from which the detector is fabricated. For π-type germanium, $\rho = -eN_A$, where e is the electronic charge and N_A is the density of acceptor impurities. This negative charge represents the filled acceptor sites on the π side of the junction. Similarly, the ionized donor sites in ν-type germanium represent a positive charge density given by $\rho = eN_D$. In lithium-drifted germanium, impurities of each type are exactly compensated, and $\rho = 0$. In the derivations that follow, we state the results explicitly for the first case of π-type germanium. The corresponding results for ν-type and lithium-drifted detectors can be obtained by substituting the appropriate value for ρ.

1. PLANAR GEOMETRY

From the results derived in Chapter 11 for planar diodes [Eq. (11-13)], the detector depletion depth is

$$d = \left(\frac{2\epsilon V}{\rho} \right)^{1/2} \tag{12-3}$$

Full depletion requires a minimum applied voltage V_d (the *depletion voltage*) at which the depletion depth extends entirely across the slab thickness T:

$$V_d = \frac{\rho T^2}{2\epsilon} \tag{12-4}$$

In one-dimensional slab geometry, Poisson's equation [Eq. (12-2)] becomes

$$\frac{d^2\varphi}{dx^2} = -\frac{\rho}{\epsilon} \tag{12-5}$$

For an applied voltage less than that required for full depletion, the electric field $\mathscr{E} = -\text{grad } \varphi = -d\varphi/dx$ is obtained by solution of Eq. (12-5) with the boundary condition

$$\varphi(d) - \varphi(0) = V$$

The result is

$$-\mathscr{E}(x) = \frac{V}{d} + \frac{\rho}{\epsilon}\left(\frac{d}{2} - x \right) \tag{12-6a}$$

or

$$|\mathscr{E}(x)| = \frac{V}{d} + \frac{eN_A}{\epsilon}\left(x - \frac{d}{2} \right) \tag{12-6b}$$

where x is the distance from the p^+ contact. For $V < V_d$, the portion of this solution corresponding to the undepleted region of the detector is not applicable, and the field is zero. Equation (12-6) also holds for $V > V_d$, because the effect of the overvoltage $(V - V_d)$ is to increase the field by a constant amount $(V - V_d)/T$ everywhere within the detector. Plots of some electric field configurations predicted by Eq. (12-6) are shown in Fig. 12-3.

As in surface barrier or junction detectors, the capacitance varies with the applied bias voltage up to the value at which the detector is fully depleted. In planar geometry,

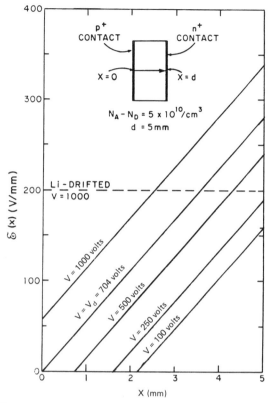

Figure 12-3 The variation of the electric field strength throughout the active volume of a planar HPGe detector for different values of the reverse bias V. The bias value at which full depletion is achieved is labeled V_d. The uniform field present in a Ge(Li) detector of the same dimension is shown for comparison. (From Llacer.[1])

the detector capacitance per unit area prior to the point of full depletion is given by

$$C = \left(\frac{\epsilon \rho}{2V} \right)^{1/2} \tag{12-7}$$

For $V > V_d$, the detector capacitance is a constant obtained by setting $V = V_d$ in Eq. (12-7). The independence of detector capacitance on applied bias is often taken as an indication of full depletion within the detector.

2. COAXIAL GEOMETRY

Poisson's equation [Eq. (12-2)] in cylindrical coordinates becomes

$$\frac{d^2\varphi}{dr^2} + \frac{1}{r}\frac{d\varphi}{dr} = -\frac{\rho}{\epsilon} \tag{12-8}$$

We treat the case of a true coaxial detector with inner and outer radii of r_1 and r_2. A boundary condition is that the potential difference between these radii is given by the applied voltage V, or $\varphi(r_2) - \varphi(r_1) = V$. Solving Eq. (12-8) for $\mathscr{E}(r) = -d\varphi/dr$, we find

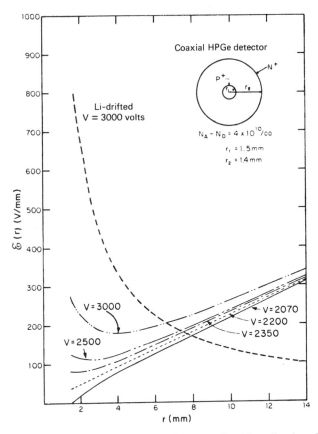

Figure 12-4 The variation of the electric field strength with radius in a HPGe coaxial detector. For the example shown, a minimum value of the reverse bias voltage V of 2070 V is required to fully deplete the detector. The $1/r$ field variation in a coaxial Ge(Li) detector is shown for comparison. (From Llacer.[1])

that the resulting electric field configuration is

$$-\mathscr{E}(r) = -\frac{\rho}{2\epsilon}r + \frac{V + (\rho/4\epsilon)(r_2^2 - r_1^2)}{r\ln(r_2/r_1)} \qquad (12\text{-}9)$$

or

$$|\mathscr{E}(r)| = \frac{eN_A}{2\epsilon}r + \frac{V - (eN_A/4\epsilon)(r_2^2 - r_1^2)}{r\ln(r_2/r_1)} \qquad (12\text{-}10)$$

provided N_A (the acceptor concentration) is constant over the detector volume. Some plots of this field shape for different values of V are shown in Fig. 12-4. More detailed plots are presented in Ref. 2 for different detector dimensions and acceptor concentrations.

The absolute value of the electric field profile remains the same for both the p-type and n-type coaxial configurations shown in Fig. 12-2, provided the same voltage and dopant concentrations are involved. Only the polarity of the field changes, reversing the drift directions of the holes and electrons. Note that both cases maintain the rectifying contact at the outer surface.

The voltage V_d needed to fully deplete the detector can be found by setting $\mathscr{E}(r_1) = 0$ in Eq. (12-9) above. The result is

$$V_d = \frac{\rho}{2\epsilon}\left[r_1^2 \ln\left(\frac{r_2}{r_1}\right) - \frac{1}{2}\left(r_2^2 - r_1^2\right)\right]$$

The depletion voltage decreases linearly with ρ (or dopant level) and is theoretically zero for perfectly compensated germanium.

The capacitance per unit length of a fully depleted true coaxial detector is

$$C = \frac{2\pi\epsilon}{\ln\left(r_2/r_1\right)} \qquad (12\text{-}11)$$

Because there is generally an advantage in minimizing detector capacitance, the radius of the central core r_1 is normally kept to a minimum consistent with allowing the fabrication of an appropriate electrical contact.

E. Surface Dead Layer

To a first approximation, the active volume of a germanium detector is simply the region between the n^+ and p^+ contacts. However, these contacts may have appreciable thickness and can represent a dead layer on the surface of the crystal through which the incident radiation must pass. For example, some traditional methods of fabricating contacts, such as evaporating and diffusing lithium into the surface to form an n^+ layer, can produce thicknesses of several hundred micrometers. For gamma rays of about 200 keV energy or greater, the attenuation in such layers is generally negligible, and the gamma-ray detection efficiency is not appreciably affected by the presence of the dead layer. However, if gamma rays or X-rays of lower energy are to be measured, the presence of layers of this thickness must be avoided to prevent attenuation.

For such applications, the ion implantation technique has become more popular for the formation of the contacts. For example, the p^+ contact can be produced by implanting boron ions that have been accelerated to about 20 keV into the crystal surface. Such contacts are only a few tenths of a micrometer in thickness and can serve as suitable entrance windows for soft X-rays. A large-volume germanium detector fabricated with a thin ion-implanted surface contact will therefore be useful for the measurement of photons over a wide energy range, from X-rays of a few keV through gamma rays of many MeV energy.

III. GERMANIUM DETECTOR OPERATIONAL CHARACTERISTICS

A. Detector Cryostat and Dewar

Because of the small bandgap (0.7 eV), room-temperature operation of germanium detectors of any type is impossible because of the large thermally induced leakage current that would result. Instead, germanium detectors must be cooled to reduce the leakage current to the point that the associated noise does not spoil their excellent energy resolution. Normally, the temperature is reduced to 77 K through the use of an insulated dewar in which a reservoir of liquid nitrogen is kept in thermal contact with the detector.

For Ge(Li) detectors, the low temperature must be maintained continuously to prevent a catastrophic redistribution of the drifted lithium that will rapidly take place at

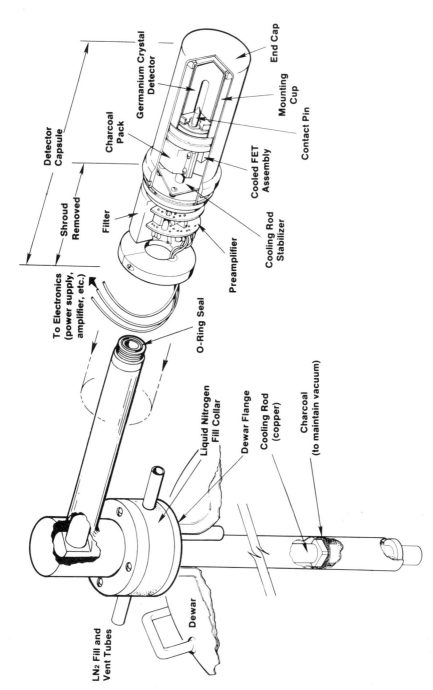

Figure 12-5 Diagram showing the location of a HPGe detector within its vacuum capsule. In this design, the capsule can be connected, without using vacuum pumps, to a variety of cryostats or cryostat–dewar assemblies. The example shown is the common horizontal configuration of the cryostat above the liquid nitrogen dewar (see inset). (Courtesy of M. Martini; EG & G ORTEC, Oak Ridge, TN.)

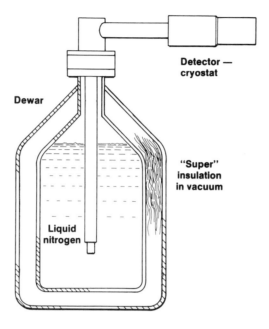

Figure 12-5   (*Continued*)

room temperature. Lithium drifting is eliminated in HPGe detectors, and they can be allowed to warm to room temperature between uses. Fabrication techniques have been developed to the point that modern detectors will withstand indefinite cycling of the temperature. This operational advantage is the major reason that HPGe detectors have supplanted Ge(Li) configurations.

The detector must be housed in a vacuum-tight cryostat to inhibit thermal conductivity between the crystal and the surrounding air. The cryostat is normally evacuated and sealed by the manufacturer, although a pumping port may be provided to facilitate unusual operations such as high-temperature annealing (see p. 440). A thin end window is usually located near the crystal to minimize attenuation of gamma rays before they enter the germanium. The cryostat can be mounted on the liquid nitrogen dewar in a number of orientations, including horizontal or vertical extensions to suit specific shielding or other application requirements. An example is shown in Fig. 12-5. To avoid the necessity of filling the dewar more frequently than weekly, its capacity must be about 30 liters. The corresponding size and weight of the dewar are much greater than those of the detector itself, and portable operation is difficult. For such applications, some manufacturers provide a smaller dewar with a more limited holding time. Sealed detector capsules are also available so that the same germanium crystal can easily be attached to many different cryostat–dewar configurations (horizontal, vertical, portable, etc.).

Although it is conventional to operate HPGe detectors at liquid nitrogen temperature, in some applications it may be more convenient to allow the detector temperature to rise above this nominal 77 K value. Several studies[3-5] have shown that the performance of coaxial detectors does not suffer until the absolute temperature rises to about 130 K. There are several alternatives to liquid nitrogen cooling that may be used. Mechanical closed-cycle refrigerators have been demonstrated[6,7] that are capable of cooling detectors to as low as 50–60 K, but care must be exercised to prevent electronic noise due to microphonics from mechanical vibrations. Commercial systems based on the Solvay cycle are sufficiently free of vibration to allow their use while the detector is in operation. Coolers based on the Joule–Thomson effect have also been used in detector applications.[8] Here a supply of high-pressure gas, typically nitrogen at 100 atm, is expanded in a capillary tube to provide the cooling. These systems provide alternatives in situations where liquid nitrogen is either not available or impractical.

The preamplifier is normally incorporated as part of the cryostat package in modern HPGe systems. It is always advantageous to locate the preamplifier as near the detector as possible to minimize capacitance. The input stages of the preamplifier normally are also cooled along with the detector to reduce electronic noise. There is also some advantage in keeping the exterior of the cryostat free of electronic "boxes" to facilitate close-fitting shielding and/or Compton suppression systems (see p. 418).

B. Energy Resolution

The dominant characteristic of germanium detectors is their excellent energy resolution when applied to gamma-ray spectroscopy. In Fig. 12-6, comparative pulse height spectra are shown for a NaI(Tl) scintillator and a germanium detector for identical incident gamma-ray spectra. The great superiority of the germanium system in energy resolution allows the separation of many closely spaced gamma-ray energies, which remain unresolved in the NaI(Tl) spectrum. Consequently, virtually all gamma-ray spectroscopy that involves complex energy spectra is now carried out with germanium detectors.

The overall energy resolution achieved in a germanium system is normally determined by a combination of three factors: the inherent statistical spread in the number of charge carriers, variations in the charge collection efficiency, and contributions of electronic noise. Which of these factors dominate depends on the energy of the radiation and the size and inherent quality of the detector in use. The full width at half maximum W_T of a typical peak in the spectrum due to the detection of a monoenergetic gamma ray can be synthesized as follows

$$W_T^2 = W_D^2 + W_X^2 + W_E^2 \qquad (12\text{-}12)$$

where the W values on the right-hand side are the peak widths that would be observed due only to effects of carrier statistics, charge carrier collection, and electronic noise.

The first of these factors, W_D^2, represents the inherent statistical fluctuation in the number of charge carriers created and is given by

$$W_D^2 = (2.35)^2 F \epsilon E \qquad (12\text{-}13)$$

where F is the Fano factor, ϵ is the energy necessary to create one electron–hole pair, and E is the gamma-ray energy. The variation of the statistical contribution to the observed FWHM of a small Ge(Li) detector is shown in Fig. 12-7.

The contribution of the second term, W_X^2, is due to incomplete charge collection and is most significant in detectors of large volume and low average electric field. Its magnitude can often be experimentally estimated by carrying out a series of FWHM measurements as the applied voltage is varied. The necessary assumption is that, if the electric field could be made infinitely large, the effects of incomplete charge collection could be reduced to an insignificant level. Therefore, a plot of the observed FWHM versus the reciprocal of the average electric field within the detector, such as shown in Fig. 12-8, allows an extrapolation to infinite field conditions. The residual value of FWHM given by this extrapolation is then assumed to arise only from the remaining two factors

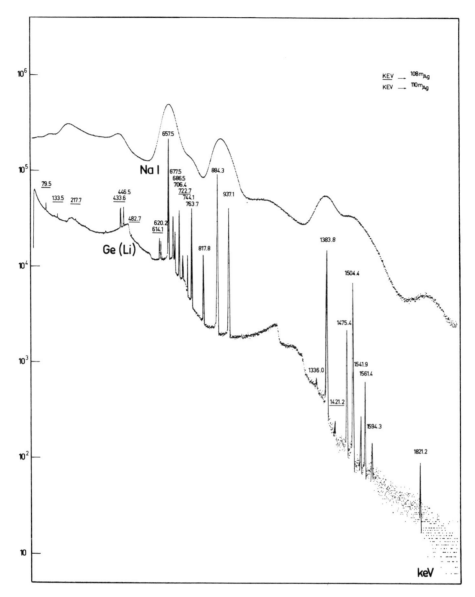

Figure 12-6 Comparative pulse height spectra recorded using a sodium iodide scintillator and a Ge(Li) detector. The source was gamma radiation from the decay of [108m]Ag and [110m]Ag. Energies of peaks are labeled in keV. (From Philippot.[9])

in Eq. (12-12). One should use this type of analysis with caution, however, since the drift velocities of the carriers eventually saturate at high electric field values. Data taken under these conditions would no longer be expected to show a continuing reduction in the effects of incomplete charge collection with further increases in voltage.

The third factor, W_E^2, represents the broadening effects of all electronic components following the detector. Its magnitude can be conveniently measured by supplying the output of a precision pulser with a highly stable amplitude to the preamplifier and recording the corresponding peak in the pulse height spectrum. These measurements

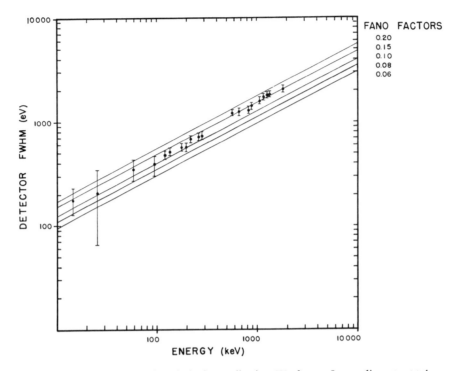

Figure 12-7 The measured statistical contribution W_D for an 8 mm diameter $\times$ 4 mm planar Ge(Li) detector. Lines are shown corresponding to various assumed values for the Fano factor. (From Palms et al.[10])

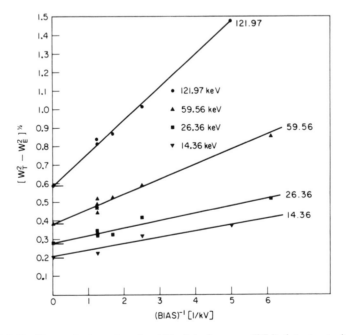

Figure 12-8 Extrapolation of peak width data from an Si(Li) detector to infinite field conditions. The electronic broadening (W_E) has already been subtracted quadratically from the measured FWHM, so that the remaining contribution given by the extrapolation should only be due to statistical effects. (From Zulliger et al.[11])

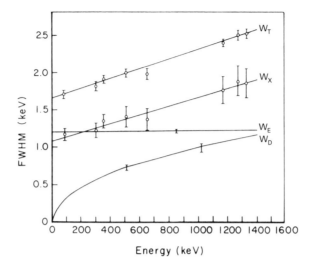

Figure 12-9 Example of the variation of the FWHM of the full-energy peak of a 86 cm³ HPGe detector with gamma-ray energy. Each of the three terms in Eq. (12-12) is shown separately, together with the total FWHM. (From Owens.[12])

should be made while the detector is normally connected to the preamplifier so that capacitive loading of the preamplifier input is typical of conditions under actual use. A parallel test pulse input to the preamplifier is normally provided for this purpose (see Fig. 17-6).

As an illustration of the relative importance of the three major contributions to the overall energy resolution, Fig. 12-9 presents the variation with gamma-ray energy of each of the three terms in Eq. (12-12) for a typical HPGe detector. At low energies, the contributions from electronic noise and charge collection dominate, while the additional broadening due to carrier statistics becomes significant at higher energies. Small-volume detectors will usually show a better overall energy resolution than larger detectors of the same quality due to two factors. Small detectors will generally have lower capacitance values, and the electronic noise of the system increases with detector capacitance. Also, the effects of carrier loss due to trapping are magnified in large-volume detectors with relatively large charge collection distances. The best energy resolution figures are therefore obtained using detectors that are tailored to the energy range and application of interest. For example, small planar detectors give the best performance for X-rays and low-energy gamma rays where a large active volume may not be necessary.

Energy resolution figures for germanium detectors are often specified at 5.9 keV (^{55}Fe), 122 keV (^{57}Co), 662 keV (^{137}Cs), or 1333 keV (^{60}Co). Representative FWHM values for commercially available systems with a small planar germanium detector are about 150–250 eV at 5.9 keV, increasing to 400–600 eV at 122 keV. Larger coaxial detectors will produce FWHM values of 800–1200 eV at 122 keV, rising to 1.7–2.3 keV at 1333 keV.

C. Pulse Shape and Timing Properties

1. THE CHARGE COLLECTION PROCESS
The detailed shape of the signal pulse developed by a germanium detector is important on several counts. In standard pulse height spectroscopy, it is necessary for the shaping times of the pulse-processing electronics to be substantially larger than the longest rise time

likely to be encountered from the detector if resolution loss due to ballistic deficit is to be avoided. In applications where timing information must be obtained from the pulse, both the rise time and the detailed shape of the leading edge of the pulse become important when considering various time pick-off methods. The ultimate time resolution obtainable from germanium detectors is critically dependent both on the overall average rise time and the significant variation in the pulse shape from event to event.

Assuming that the equivalent circuit of the measuring electronics presents a large time constant compared with the largest rise time produced by the detector, the leading edge of the signal pulse is almost entirely determined by the details of the charge collection process within the detector. Because it is always advantageous to have the smallest possible rise time, conditions are normally sought in which charge collection occurs in the minimum possible time. In Fig. 11-2, drift velocity plots were shown for electrons and holes in germanium as a function of the applied electric field. For low values of the electric field, the velocity increases linearly with the field, which implies a constant value for the electron or hole mobility. At sufficiently high electric fields, however, the velocity ceases to increase and approaches a constant saturation value. For electrons in germanium at 77 K, this saturation drift velocity is approximately 10^5 m/s and is achieved at a field value of about 10^5 V/m. The saturated velocity for holes is similar but requires a minimum field of approximately 3×10^5 V/m. Pulses of minimum rise time are therefore obtained by operating the detector with sufficient applied voltage so that an electric field of at least this magnitude is present everywhere within the active volume. Of the various detector configurations, the planar and true coaxial geometries are those in which regions of low electric fields are most easily avoided. Therefore, these types have predominated when the detector must be used in critical timing situations.

Two factors limit the ultimate time resolution that can be achieved with germanium detectors. One is that the charge collection process is inherently slow. Even at the saturated drift velocity, the time required for a charge carrier to travel a distance of 1 cm (which may be of the order of the detector thickness) is approximately 100 ns. Typical pulse rise times will therefore be of the same order of magnitude. These times are much longer than the output from fast detectors (such as organic scintillators) and therefore the time performance can never be as good. A second factor, however, makes the timing properties even worse. As will be illustrated below, the detailed shape of the pulse rise from germanium detectors can change substantially from event to event, depending on the position at which the electron–hole pairs are created within the active volume. When the volume of the detector is uniformly irradiated, these positions are more or less randomly distributed, and therefore the output pulses show a great variation in the shape of their leading edge. This variable pulse shape creates difficulties with many methods of deriving timing signals, and special time pick-off techniques are required to minimize the accompanying difficulties (see Chapter 17).

A general review of the pulse shape and timing properties of germanium detectors is given in Ref. 13. More detailed experimental and analytic investigations can be found in Refs. 14–19.

2. MODELS FOR THE PULSE SHAPE

Just as in other detectors in which signal carriers must be collected over appreciable distances, the shape of the leading edge of pulses from germanium detectors depends on the position at which the charge carriers are formed within the active volume. The simplest case is that of a very short-range particle, which, at first approximation, creates all the electron–hole pairs at one location within the detector. If that location lies within the interior of the active volume, there will be unique and separate collection times for holes and electrons because each species must travel a fixed distance before being collected. If the point of interaction is near either edge of the active volume, the observed pulse rise will then be due primarily to the motion of only one type of charge carrier. For charged radiations whose range is not small compared with the active volume, a distribution of collection times will result from the corresponding spatial distribution of the points at which holes and electrons are formed. If the orientation of the particle track can change significantly from event to event, an additional variation in the pulse rise time will be introduced.

The shape of the leading edge of the output pulse is subject to much the same analysis as was given for ion chambers in Chapter 5. One major difference from gas-filled counters is that the mobilities of the positive and the negative charge carriers (holes and electrons) are somewhat similar, whereas in a gas, the positive ion mobility is much less than that of free electrons. If a sufficient number of simplifying assumptions are made, analytic expressions can be derived to describe the shape of the pulse expected in simple germanium detector configurations. The common assumptions are:

1. All charge carriers are created at a fixed position within the detector active volume. This assumption ignores the finite range of secondary electrons produced by gamma-ray interactions, as well as the fact that many gamma-ray pulses arise from multiple interactions (e.g., Compton scattering followed by photoelectric absorption) that occur at widely separated points within the detector.

2. Trapping and detrapping of the charge carriers are ignored. The capture of charges in shallow traps followed by a rapid release can significantly alter the transit time of charge carriers and can be important when the concentration of such shallow traps is not negligible.

3. All charge carriers are assumed to be generated entirely within the active volume of the detector where the electric field has its full expectation value.

4. The electric field within the active volume of the detector is sufficiently high to cause saturation of the drift velocity of both electrons and holes.

With this set of simplifying assumptions, analytic expressions can be obtained for the expected pulse shape for two relatively simple geometries: the planar detector in which the electric field is approximately uniform, and the true coaxial configuration in which the electric field varies inversely with the radial distance from the detector axis.

a. Planar Geometry

In general, the energy dE absorbed by the motion of a positive charge q_0 through a potential difference $d\varphi$ is given by

$$dE = -q_0 \, d\varphi \tag{12-14}$$

In terms of the electric field $\mathscr{E}(x) = -d\varphi(x)/dx$

$$\frac{dE}{dx} = q_0 \mathscr{E}(x) = q_0 \frac{V_0}{d} \tag{12-15}$$

where V_0 is the applied voltage and d is the detector thickness. The energy absorbed by the motion from x_0 to x is

$$\Delta E = \int_{x_0}^{x} dE = \frac{q_0 V_0}{d} \int_{x_0}^{x} dx = \frac{q_0 V_0}{d}(x - x_0) \tag{12-16}$$

Using the same arguments introduced for parallel plate ion chambers beginning on p. 151, we obtain the corresponding signal voltage:

$$\Delta V_R = \frac{\Delta E}{C V_0} = \frac{q_0}{C} \frac{(x - x_0)}{d} \tag{12-17}$$

This voltage corresponds to an *induced charge* of

$$\Delta Q = C \Delta V_R = q_0 \frac{(x - x_0)}{d} \tag{12-18}$$

The development of the signal pulse can thus be represented by the growth of a time-dependent induced charge $Q(t)$. There are two separate components of the induced charge, one corresponding to the motion of the electrons and the other to the motion of the holes. From Eq. (12-18), we can write the combination as

$$Q(t) = \frac{q_0}{d} \left(\begin{array}{c} \text{electron drift} \\ \text{distance} \end{array} + \begin{array}{c} \text{hole drift} \\ \text{distance} \end{array} \right) \tag{12-19}$$

This induced charge starts at zero when the electrons and holes are first formed by the ionizing particle and reaches its maximum of q_0 when both species have been collected. As in the ion chamber case, $q_0 = n_0 e$, where n_0 is now the number of electron–hole pairs and e is the electronic charge.

We now assume that these charges are formed at a distance x from the n^+ contact of the planar detector (see Fig. 12-10). The following definitions are used:

$t_e \equiv$ electron collection time

     $= x/v_e$,   where v_e is the saturation electron velocity

$t_h \equiv$ hole collection time

     $= \dfrac{d - x}{v_h}$,   where v_h is the saturation hole velocity

Equation (12-19) can be divided into four possible time domains as follows. While both holes and electrons are drifting ($t < t_h$ and $t < t_e$):

$$Q(t) = q_0 \left(\frac{v_e}{d} t + \frac{v_h}{d} t \right) \tag{12-20a}$$

If electrons have been collected, but holes are still drifting ($t_e < t < t_h$):

$$Q(t) = q_0 \left(\frac{x}{d} + \frac{v_h}{d} t \right) \tag{12-20b}$$

If holes have been collected, but electrons are still drifting ($t_h < t < t_e$):

$$Q(t) = q_0 \left(\frac{v_e}{d} t + \frac{(d - x)}{d} \right) \tag{12-20c}$$

After both holes and electrons have been collected ($t > t_h$ and $t > t_e$):

$$Q(t) = q_0 \tag{12-20d}$$

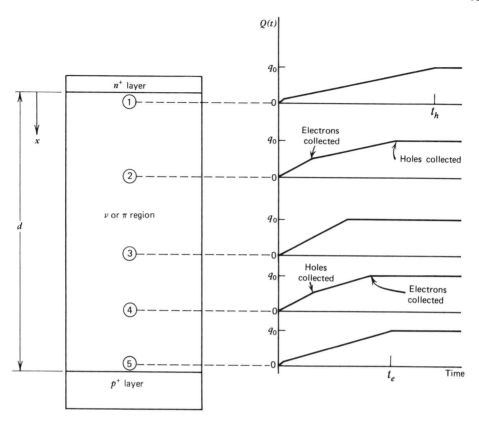

Figure 12-10 Shape of the leading edge of the output pulse $Q(t)$ for various interaction points within the detector [from Eq. (12-20)]. The initial slope of all pulses is the same, corresponding to the period when both electrons and holes are drifting. For point ①, electrons are collected very quickly and most of the rise corresponds to the drift of holes to the p^+ layer. The opposite is true for point ⑤. t_h and t_e are the assumed drift times for holes and electrons, respectively, across the entire thickness d.

Figure 12-10 shows several such pulse shapes for varying values of x. The variation in effective rise time of the pulses amounts to approximately a factor of 2.

b. Coaxial Geometry

The pulse shape generated by the collection of electrons and holes in a coaxial detector shares many of the general features derived above for a planar type, but the linear portions of the pulse leading edge take on a curvature because of the different geometry. The equivalent to Eq. (12-15) for cylindrical geometry is

$$\frac{dE}{dr} = q_0 \mathscr{E}(r) \tag{12-21}$$

The absolute value of the electric field $\mathscr{E}(r)$ in p-type HPGe true coaxial detectors [Eq. (12-10)] can be written

$$\mathscr{E}(r) = 2\alpha r + \frac{\beta}{r} \tag{12-22}$$

where

$$\alpha \equiv \frac{eN_A}{4\epsilon} \quad \text{and} \quad \beta \equiv \frac{V_0 - \alpha(r_2^2 - r_1^2)}{\ln(r_2/r_1)}$$

Putting this expression into Eq. (12-21) and integrating, we obtain

$$\Delta E = q_0 \int_{r_0}^r \mathscr{E}(r)\, dr$$

$$= q_0\alpha(r^2 - r_0^2) + q_0\beta \ln\frac{r}{r_0} \tag{12-23}$$

The time-dependent charge induced by the motion of the electrons outward toward the n^+ contact can then be expressed in terms of the radial position of the electrons $r_e(t)$:

$$Q^-(t) = \frac{\Delta E^-}{V_0} = \frac{q_0\alpha}{V_0}\left[r_e^2(t) - r_0^2\right] + \frac{q_0\beta}{V_0}\ln\frac{r_e(t)}{r_0} \tag{12-24}$$

Similarly, the induced charge due to the motion of the holes inward to the p^+ contact is

$$Q^+(t) = \frac{q_0\alpha}{V_0}\left[r_0^2 - r_h^2(t)\right] + \frac{q_0\beta}{V_0}\ln\frac{r_0}{r_h(t)} \tag{12-25}$$

and the total induced charge is

$$Q(t) = Q^-(t) + Q^+(t) \tag{12-26}$$

The shape of $Q(t)$ depends on assumptions made about the motion of the electrons and holes as reflected in $r_e(t)$ and $r_h(t)$. Because of the radial variation of the electric field strength, the drift velocities may change as the carriers are collected. One approach[20] is to assume an empirical dependence of the drift velocities with electric field strength $\mathscr{E}$ of the form

$$v = \frac{\mu_0\mathscr{E}}{\left[1 + (\mathscr{E}/\mathscr{E}_0)^\gamma\right]^{1/\gamma}}$$

where μ_0 is the mobility at low $\mathscr{E}$ values. Both γ and $\mathscr{E}_0$ are treated as adjustable parameters fitted to experimental drift velocity measurements. The radial position of each species can then be solved numerically. Plots of $Q(t)$ derived in this way are given in Ref. 20 for typical p-type HPGe coaxial detectors.

If the assumption of constant drift velocities is retained, the radial position of each carrier species is

$$r_e(t) = r_0 + v_e t \quad \text{and} \quad r_h(t) = r_0 - v_h t$$

and solutions to Eq. (12-26) are straightforward. For the simple case of Ge(Li) configurations where no net fixed charge is present in the intrinsic region, $\alpha = 0$ and $\beta = V_0/\ln(r_2/r_1)$. The result is then given by

$$Q(t) = \frac{q_0}{\ln(r_2/r_1)}\left[\ln\left(1 + \frac{v_e t}{r_0}\right) - \ln\left(1 - \frac{v_h t}{r_0}\right)\right] \tag{12-27}$$

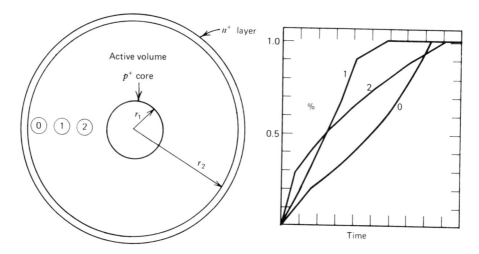

Figure 12-11 Leading edge of the output pulse for a coaxial drifted detector. Three different interaction points are indicated as 0, 1, and 2. (From Gadeken and Robertson.[17])

Equation (12-27) holds only for the first time domain, during which both holes and electrons are drifting. As in the previous case for planar geometry, the first and second terms in the square brackets become constants [equal to $\ln(r_2/r_0)$ and $\ln(r_1/r_0)$] when the electrons or holes are collected, and a slope change occurs in the waveform at the corresponding times. After both are collected, $Q(t) = q_0$. Plots of this pulse shape for various assumed interaction radii are given in Fig. 12-11. Here the variation in effective rise time can be even greater than for the planar case.

3. EFFECTS OF TRAPPING AND DETRAPPING

Martini and McMath[21] have made an analysis of the effects of trapping and detrapping on the expected pulse shape. To simplify the following discussion, we assume that all the electron–hole pairs are created near one boundary of the active volume. The signal pulse then corresponds to the motion of only that carrier species which is collected at the opposite boundary. If we assume that the carriers are formed at the p^+ contact of a planar detector, we can set $x = d$ in Eq. (12-20), which then reduces to

$$Q(t) = \begin{cases} q_0 \dfrac{t}{t_e} & (t \leq t_e) \\[2mm] q_0 & (t > t_e) \end{cases} \tag{12-28}$$

and the simple linear pulse rise corresponds to the drift of electrons across the active volume.

If we now assume that a uniform concentration of electron traps exists in this region, some of the charge may be lost, at least temporarily, from the signal pulse. If no detrapping takes place, the loss is permanent and the resulting pulse takes the form

$$Q(t) = \begin{cases} \dfrac{Q_0 \tau_T}{t_e}(1 - e^{-t/\tau_T}) & (t \leq t_e) \\[2mm] \dfrac{Q_0 \tau_T}{t_e}(1 - e^{-t_e/\tau_T}) & (t > t_e) \end{cases} \tag{12-29}$$

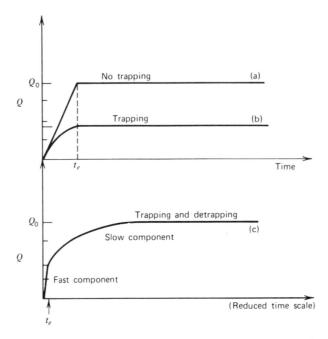

Figure 12-12 Sketches illustrating the pulse leading edge for interactions in a planar detector at one extreme of the active volume: (a) no trapping, (b) permanent trapping, and (c) trapping with slow detrapping. (From Martini and McMath.[21])

where τ_T is the *mean free drift time* of the electrons before trapping occurs. Permanent trapping adds curvature to the rise of the pulse and attenuates its amplitude, as shown in Fig. 12-12.

Ultimately, the trapped electrons may be released by thermal excitation if the trap is shallow. If τ_D is the mean lifetime of a trapped electron before detrapping occurs, the resulting pulse shape is composed of "fast" and "slow" components. The fast component or leading edge is characterized by

$$Q(t) = Q_0 \frac{\tau_e}{t_e}\left(\frac{t}{\tau_D} + \frac{\tau_e}{\tau_T}(1 - e^{-t/\tau_e}) \right) \qquad (t \le t_e) \qquad (12\text{-}30)$$

where

$$\tau_e \equiv \frac{\tau_T \tau_D}{\tau_T + \tau_D}$$

The slow component does not lend itself to a simple analytic expression but serves to build up $Q(t)$ to an asymptotic value of q_0 with a characteristic time on the order of τ_D. A sketch of this behavior is also illustrated in Fig. 12-12.

Trapping can thus lead to a reduction in expected pulse amplitude or, if detrapping occurs, to a slowing of a portion of the pulse's leading edge. The latter can have a significant effect on the timing properties of germanium detectors in which trapping is important. Furthermore, trapping can cause a deterioration in energy resolution due to the variable amount of charge lost per pulse. Detrapping is helpful in this respect only if it occurs on a short time scale compared with pulse-shaping times in the subsequent electronics.

In coaxial detectors, the cylindrical geometry adds further complication to the prediction of the expected pulse amplitude in the presence of trapping. In Ref. 22, a detailed analysis is presented that is shown to model accurately the behavior of coaxial detectors in which hole trapping has been made more severe by radiation damage.

4. SLOW OR DEFECTIVE PULSES

A fraction of the pulses observed from some germanium detectors have rise times substantially longer than predicted by the above analysis. Furthermore, these slow pulses are also "defective" in amplitude and can contribute to a loss of energy resolution or low-energy tails on peaks recorded in the pulse height spectrum. There is some evidence[23,24] that these defective pulses correspond to events that originate near the edges of the active volume, or in other regions where the local electric field is below normal.[25] Charge collection is then slowed, and losses due to trapping are more severe than for normal pulses. These effects are magnified in older detectors with a high concentration of defects or trapping centers, often the result of radiation damage from prolonged exposure of the detector to fast neutrons (see p. 437).

Because the slow-rising pulses are also deficient in amplitude, gains can sometimes be made in both the energy resolution and peak-to-Compton ratio (see p. 411) in gamma-ray spectra by discarding such pulses based on sensing their long rise times in an external circuit.[26] However, this approach will also reduce the detection efficiency if a substantial fraction of all pulses are involved. A more elaborate pulse height correction technique[27] can instead be used that avoids this efficiency loss. The principle of this technique is based on making a measurement of each pulse rise time and recording a set of independent pulse height spectra, each corresponding to pulses with nearly equal rise times. If a correlation between pulse height and rise time can be found, these multiple spectra can then be reassembled into a single spectrum by applying an appropriate correction factor to each.

IV. GAMMA-RAY SPECTROSCOPY WITH GERMANIUM DETECTORS

In the measurement of gamma-ray energies above several hundred keV, there are, at present, only two detector categories of major importance: inorganic scintillators, of which NaI(Tl) is the most popular, and germanium semiconductor detectors. Although there are many other potential factors, the choice in a given application most often revolves about a trade-off between counting efficiency and energy resolution. Sodium iodide scintillators have the advantage of availability in large sizes, which, together with the high density of the material, can result in very high interaction probabilities for gamma rays. The relatively high atomic number of iodine also assures that a large fraction of all interactions will result in complete absorption of the gamma-ray energy, and therefore the photofraction (or fraction of events lying under the full-energy peak in the pulse height spectrum) will also be relatively high.

The energy resolution of scintillators is poor. The comparative spectra shown in Fig. 12-6 illustrate the clear superiority of germanium detectors in situations where many closely spaced gamma-ray energies must be separated. Good germanium systems will have a typical energy resolution of a few tenths of a percent compared with 5–10% for sodium iodide. This improved energy resolution does not come without a price, however. The smaller sizes available and lower atomic number of germanium combine to give photopeak efficiencies an order of magnitude lower in typical cases. Furthermore, the

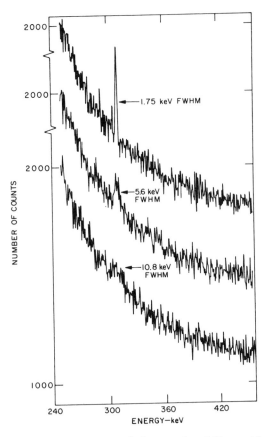

Figure 12-13 The effect of energy resolution on the ability to identify a weak peak superimposed on a statistically uncertain continuum. The area under the peak is the same in all three cases. (From Armantrout et al.[28])

photofractions are low and thus the continua in the pulse height spectra are more of a problem.

This latter disadvantage is somewhat offset by the superior energy resolution of germanium. Not only does good resolution help separate closely spaced peaks, but it also aids in the detection of weak sources of discrete energies when superimposed on a broad continuum. Detectors with equal efficiency will result in equal areas under the peak, but those with good energy resolution produce a narrow but tall peak that may then rise above the statistical noise of the continuum. This effect is illustrated in Fig. 12-13 in which the only variation between spectra is the resolution.

Germanium detectors are clearly preferred for the analysis of complex gamma-ray spectra involving many peaks. The choice is less obvious when only a few gamma-ray energies are involved, particularly if a measurement of their intensity rather than an accurate energy determination is the prime objective. Then the greater efficiency, larger photofraction, and lower cost of scintillators may well tip the balance in their favor.

Because the concentration of dopants is very small, the gamma-ray interaction characteristics are identical for lithium-drifted or high-purity germanium. The detector properties of response function and detection efficiency discussed in the remainder of this chapter are therefore the same for germanium detectors of similar size and shape, regardless of whether they are of the Ge(Li) or HPGe type. The extensive experience gained with Ge(Li) detectors over the several decades since their introduction is therefore still applicable in many ways to the more modern HPGe devices. We show examples of spectra and other detection properties published for both types, with the understanding that they are broadly representative of the behavior of germanium detectors in general. Detailed discussions of the use of germanium detectors in gamma-ray spectroscopy can be found in Ref. 29.

A. Response Function

1. GENERAL CHARACTERISTICS

Examples of pulse height spectra measured with a germanium detector for monoenergetic incident gamma rays are given in Fig. 12-14. The processes of photoelectric absorption, Compton scattering, and pair production all may contribute to the observed response in the manner described in Chapter 10 for the case of sodium iodide scintillators. The lower atomic number of germanium and the smaller typical active volume of germanium detectors compared with NaI scintillators lead to significant differences in the importance of various features of the pulse height spectrum.

The lower atomic number of germanium results in a photoelectric cross section that is smaller by a factor of 10–20 compared with sodium iodide, and an even greater discrepancy exists in comparison with high-Z scintillators such as BGO. As a direct result, the intrinsic peak efficiency of a germanium detector is many factors smaller than that of a sodium iodide scintillator of equivalent active volume. Even though the area under the full-energy peak is much smaller, these peaks still are a prominent and obvious part of the spectrum because their width is so small due to the superior energy resolution of semiconductor detectors. Events contributing to the full-energy peak are now much more likely to consist of multiple interactions, such as Compton scattering followed by photoelectric absorption of the scattered photon, and absorption of the full photon energy in a single photoelectric event is relatively rare. In Fig. 12-15, it can be seen that multiple events are the dominant contributor to the full-energy peak over all but the lowest range of gamma-ray energies.

The Compton continuum, as described in Chapter 10, is also a prominent part of germanium detector spectra. Because the ratio of the Compton to photoelectric cross section is much larger in Ge than in NaI, a much greater fraction of all detected events lie within this continuum rather than under the photopeak. The improved energy resolution leads to a more faithful reproduction of the shape of the distribution, including the rise in the continuum near the Compton edge. Because Compton scattering is an interaction with electrons, the position of the Compton edge is at the same energy for all detector types [see Eq. (10-4)].

The *peak-to-Compton ratio* is sometimes quoted as one feature of germanium detector gamma-ray spectra. This index of detector performance is defined[31] as the ratio of the count in the highest photopeak channel to the count in a typical channel of the Compton continuum associated with that peak. This sample of the continuum is to be taken in the relatively flat portion[†] of the distribution lying just to the left of the rise toward the

---

[†]Officially defined[31] as the interval from 358 to 382 keV for the 661 keV gamma rays from ^{137}Cs, and from 1040 to 1096 keV for the 1333 keV gamma rays from [60]Co.

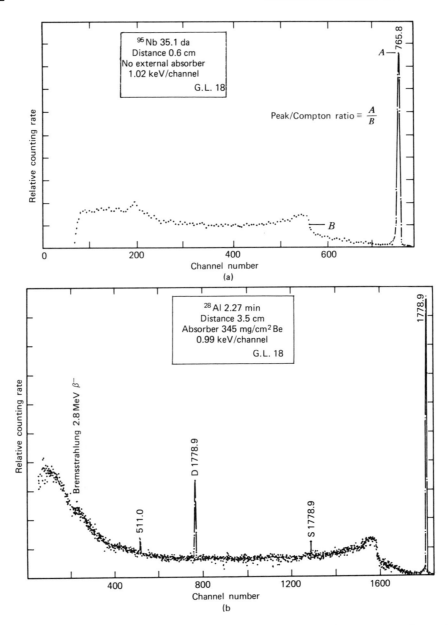

Figure 12-14 Measured pulse height spectra for monoenergetic gamma rays from a 18 cm^3 Ge(Li) detector. Counts in the lowest channels have been suppressed. (a) Spectrum from 764 keV gamma rays from [95]Nb. The full-energy peak, Compton continuum, and backscatter peak are evident. (b) Spectrum from 1779 keV gamma rays from [28]Al. Single and double escape peaks are evident, as well as a small peak at 511 keV from annihilation photons created in surrounding materials. (From F. Adams and R. Dams, *Applied Gamma-Ray Spectrometry*, 2nd ed. Copyright 1970 by Pergamon Press, Ltd. Used with permission.)

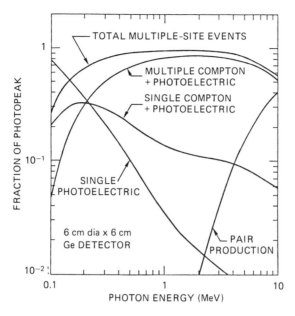

Figure 12-15 Fractions of the full-energy peak contributed by different energy loss mechanisms in a 6 cm × 6 cm coaxial HPGe detector, as predicted by Monte Carlo simulation. Absorption of the gamma-ray photon in a single photoelectric interaction predominates only for energies below about 140 keV. (From Roth et al.[30])

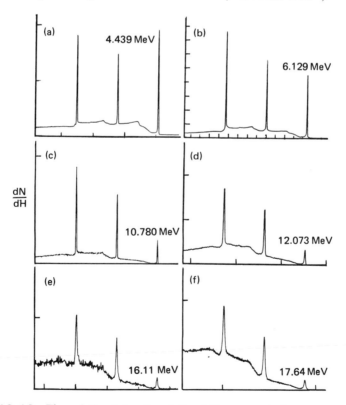

Figure 12-16 The relative intensity of the full-energy, single escape, and double escape peaks for a 100 cm³ Ge(Li) detector for gamma-ray energies from 4.439 to 17.64 MeV. (From Berg et al.[32])

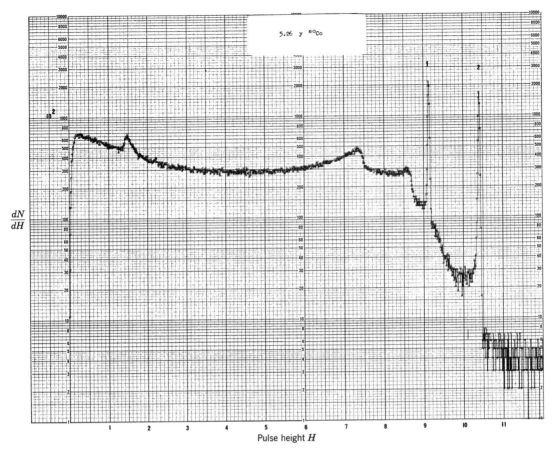

$\dfrac{dN}{dH}$

Pulse height H

Figure 12-17 Pulse height spectrum recorded from a 16 cm³ Ge(Li) detector for ^{60}Co gamma rays (1.174 and 1.333 MeV). A backscatter peak at ~ 0.2 MeV is evident. (From Balagna and Helmik.[37])

Compton edge (see Fig. 12-14). The ratio is conventionally quoted for the 1333 keV gamma ray from ^{60}Co and is a measure of the combined effects of detector energy resolution and photofraction. The ratio is adversely affected by any scattered gamma rays that enter the detector and it therefore also is influenced by structural materials near the detector or source. Typical values of the ratio range from 30 to 60 for representative coaxial germanium detectors measured under low-scatter conditions. For detectors with equal photofraction, the ratio will vary inversely with the FWHM value of the full-energy peak. For detectors with equal energy resolution, the ratio will be approximately proportional to the photofraction.

Because of the greater transparency of the detector to secondary gamma rays, escape peaks play a much more important role than in NaI scintillators. Escape peaks arise whenever a fixed amount of energy may be lost from the detector with a significant probability. The escape of the characteristic X-rays from germanium following photoelectric absorption can be significant, especially for small detectors with a large surface-to-volume ratio. A small peak will then be found in the spectrum at 11 keV below the photopeak, with the energy difference corresponding to the characteristic K X-ray energy for germanium. These X-ray escape peaks will be most prominent for incident low-energy gamma rays, because photoelectric absorption is then most probable, and interactions will tend to occur near the detector surface.

At high gamma-ray energies, the escape of annihilation radiation following pair production within the detector is very significant. The pair production process, described in Chapter 10, consists of creating an electron–positron pair at the site of the original gamma-ray interaction. The kinetic energy of both particles is expended in creating charge carriers, but the positron near the end of its track will annihilate and create two 0.511 MeV annihilation photons. In germanium detectors, there is a high probability that one or both of these photons will escape. Escape peaks will therefore appear in the spectrum corresponding to events in which one or both of the annihilation photons carry away a portion of the original gamma-ray energy. If both annihilation photons escape, a *double escape peak* appears at an energy of 1.022 MeV less than the full-energy peak. Events in which one annihilation photon escapes but the other is fully absorbed lead to a *single escape peak* appearing at an energy of 0.511 MeV below the full-energy peak. If the incident gamma-ray energy is sufficiently high, the probability of full-energy absorption may be very small, and the escape peaks often are the most prominent peaks that appear in the response function of the detector. Their location is illustrated in the spectra shown in Fig. 12-16. Because of their prominence, the escape peaks are often used in place of the full-energy peak to derive the primary energy of incident gamma rays. The primary energy can be obtained from the position of the single and double escape peaks by adding one or two units, respectively, of the annihilation photon energy (511.003 keV).

Careful measurements have demonstrated that the annihilation single escape peak can display subtle effects arising from the finite momentum of the positron–electron pair immediately before annihilation. This momentum causes one annihilation photon to have a slightly higher energy than its corresponding partner. Therefore, if only one photon is

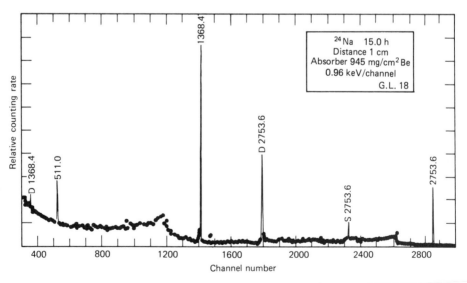

Figure 12-18 Pulse height spectrum recorded from a 18 cm³ Ge(Li) detector for ^{24}Na gamma rays (1.368 and 2.753 MeV). Single and double escape peaks are labeled S and D together with the gamma-ray energy. The peak at 0.511 MeV is from annihilation photons created by pair production in materials surrounding the detector. (From F. Adams and R. Dams, *Applied Gamma-Ray Spectrometry*, 2nd ed. Copyright 1970 by Pergamon Press, Ltd. Used with permission.)

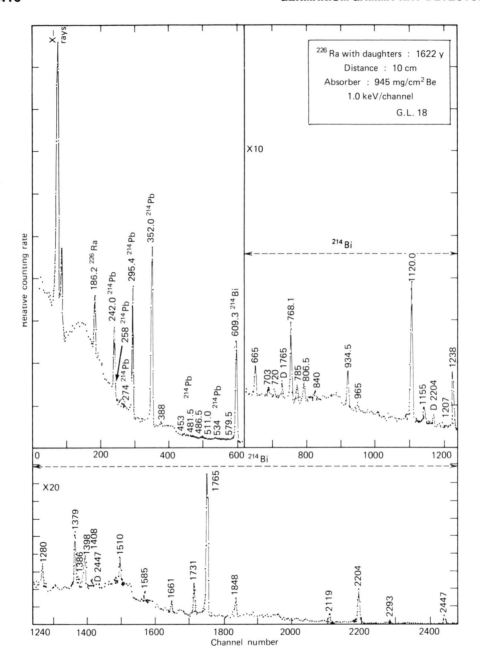

Figure 12-19 Pulse height spectrum recorded from a 18 cm³ Ge(Li) detector for gamma rays emitted by ²²⁶Ra in equilibrium with its daughters. Gamma-ray energies and yields are given in Table 12-3. (From F. Adams and R. Dams, *Applied Gamma-Ray Spectrometry*, 2nd ed. Copyright 1970 by Pergamon Press, Ltd. Used with permission.)

reabsorbed, a variation is introduced in the escaping energy. This variation results in a slight broadening of the single escape peak compared with either the full-energy peak or the double escape peak. It has also been demonstrated[33] that the centroid of the single escape peak in a coaxial detector can be slightly shifted (on the order of 50 eV) to higher energies because of effects due to the positron mobility in the electric field of the detector. One should therefore use single escape peak positions and widths with caution in precision measurements where accuracies of the order of 100 eV or less are to be achieved.

2. TYPICAL GERMANIUM PULSE HEIGHT SPECTRA

Because of the diversity of sizes and shapes of germanium detectors, catalogs of standard spectra are not as common as those published for sodium iodide scintillators. Some compilations of collected experimental spectra for specific detectors are given in Refs. 34–37. Representative spectra from some of these compilations are shown in Figs. 12-17 through 12-19.

Good results also have been achieved in calculating the response function of germanium detectors based on a knowledge of the cross section for the primary gamma rays and assumed behavior of the secondary radiations. As an example, some results from the Monte Carlo program of Meixner[38] are shown in Fig. 12-20. The excellent agreement

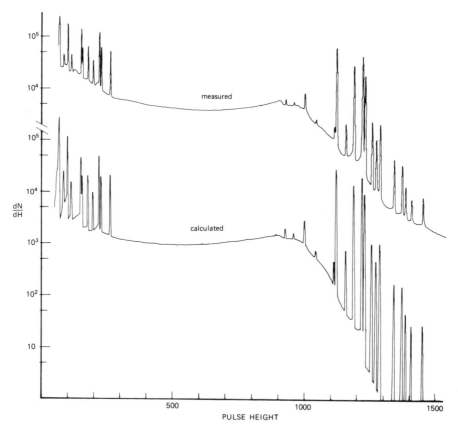

Figure 12-20 A comparison of the measured Ge(Li) spectrum for gamma rays from ^{182}Ta with a Monte Carlo calculated spectrum. (From Meixner.[38])

between calculation and measurement lends credence to the assumptions used in the Monte Carlo code, which can then be applied to a wide range of experimental situations. The plot shown in Fig. 12-21 shows a perspective diagram of pulse height spectra calculated by the same code as the gamma-ray energy is varied. The enhancement of the escape peaks as the gamma-ray energy increases is quite evident from this diagram.

3. PEAK SHAPE

The detailed shape of the peaks observed in germanium spectra is important if the area under the peak is to be accurately measured. Some details of possible features of the peak shape are shown in Fig. 12-22. Most empirical fits to the shape use modifications of a Gaussian distribution allowing for tailing on the low-energy side of the distribution.[40, 41] Tailing can arise from several physical effects, including imperfect charge collection in some regions of the detector, or secondary electron and bremsstrahlung escape from the active volume. The distinction between short-term and long-term tails in the figure is that the short-term tailing has a more serious effect on the shape of the peak near its base, whereas long-term tailing often can be treated as an additional portion of the background.

One method of specifying the severity of tailing for a specific detector is to quote the full width at one-tenth maximum (FW.1M) of the full-energy peak, in addition to the more conventional full width at half maximum (FWHM). For good quality germanium detectors with minimal tailing, the FW.1M should be less than double the FWHM (The FW.1M/FWHM ratio for a pure Gaussian peak is 1.823.) Another index even more sensitive to tailing is the ratio of the full width at 1/50 maximum (FWFM) to the FWHM, conventionally measured at 1.333 MeV. Good germanium detectors give values for this ratio between 2.5 and 3.0, compared with 2.376 for a pure Gaussian.

B. Methods for Continuum Reduction

As outlined in Chapter 10, the ideal gamma-ray spectrometer should have a response function consisting of a single peak only, with no associated continuum. As can be seen from the examples given earlier in this chapter, germanium spectra are characterized by very prominent continua, which can obscure low-intensity peaks from other gamma-ray energies. The photofraction values are typically much poorer than those of NaI(Tl) scintillators, and therefore the methods of continuum reduction outlined in Chapter 10 can potentially provide even greater benefits when applied to germanium detectors.

1. COMPTON REJECTION BY ANTICOINCIDENCE

The Compton continuum in gamma-ray spectra from germanium detectors is generated primarily by gamma rays that undergo one or more scatterings in the detector followed by escape of the scattered photon. In contrast, the full-energy absorption events result in no escaping photons. Therefore, coincident detection of the escaping photons in a surrounding annular detector can serve as a means to reject preferentially those events that only add to the continuum, without affecting the full-energy events. The rejection is carried out by passing the pulses from the germanium detector through an electronic gate that is closed if a coincident pulse is detected from the surrounding detector (called *anticoincidence mode*).

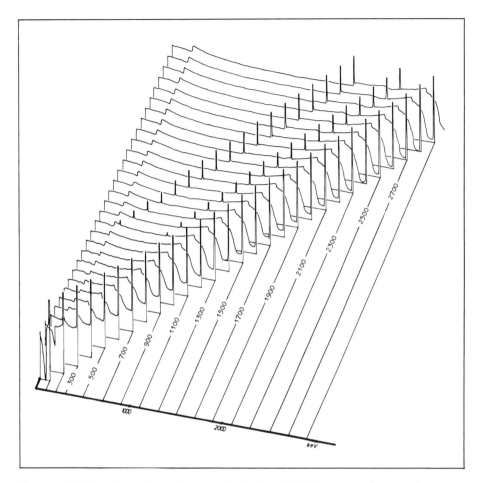

Figure 12-21 Perspective diagram of calculated Ge(Li) spectra for energies up to 3 MeV. (From Meixner.[38])

To be effective, the surrounding detector must be large enough to intercept most of the escaping photons and should have a good efficiency for their detection. Large scintillation detectors most readily meet these requirements, and both NaI(Tl) and BGO have been used for this purpose. BGO has the strong advantage that its high density and atomic number allow a more compact configuration compared with a sodium iodide detector of the same detection efficiency.

Good Compton suppression also requires that there be a minimum amount of absorbing material between the inner and outer detectors that could attenuate the escaping scattered photons. The relatively thick outer contact on p-type coaxial detectors is one example of such an absorber, and n-type germanium detectors with thinner contacts are generally preferred in this application.

Figure 12-23 shows one example of a Compton suppression system that combines BGO and NaI(Tl) scintillators in an annular geometry surrounding a germanium detector of the HPGe type. The reduction in the Compton continuum from the germanium detector by operating it and the scintillation detectors in anticoincidence mode is

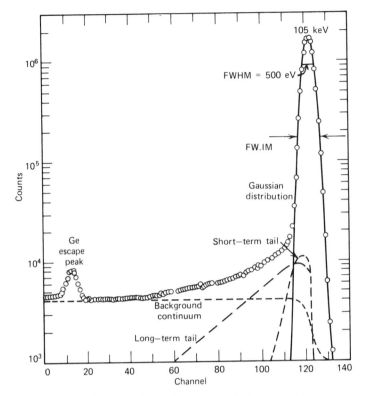

Figure 12-22 Detailed shape of an observed peak from a Ge(Li) detector. (From Meyer et al.[39])

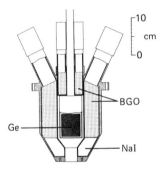

Figure 12-23 An example of a germanium detector surrounded by a Compton suppression system made up of NaI(Tl) and BGO scintillators. (From Nolan et al.[42])

illustrated in Fig. 12-24. Other examples of the design and performance of BGO-based Compton suppression systems are described in Refs. 43–46.

Because they also result in some coincident events in both detectors, pair production interactions of the incident gamma rays followed by escape of one or both of the annihilation photons are also suppressed by these systems. Single and double escape peaks in the recorded spectrum are therefore less prominent than in the original spectrum from the germanium detector.

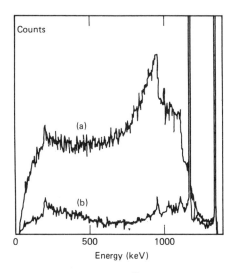

Figure 12-24 Pulse height spectra from a ^{60}Co gamma-ray source using the system shown in Fig. 12-23. A collimator shielded the scintillators from direct irradiation. (a) Normal spectrum recorded from the central germanium detector. (b) Spectrum recorded in anticoincidence with the scintillation detectors, showing the suppression of the Compton continuum. The full-energy peak areas were unaffected to within 1%. (From Nolan et al.[42])

One potential disadvantage of Compton suppression systems becomes apparent if a radioisotope source with a complex decay scheme is being measured. Many gamma rays can then be emitted in coincidence, and it is possible for independent gamma rays from the same disintegration to interact in both detectors. These events are therefore rejected, leading to unwanted suppression of some full-energy peaks.

2. SUM–COINCIDENCE MODE

Most of the Compton continuum consists of single Compton scatterings followed by escape of the scattered gamma ray, whereas full-energy events at typical gamma-ray energies are primarily comprised of multiple scattering sequences followed by a photo-electric absorption. The peak-to-Compton ratio can therefore be enhanced by requiring a recorded event to correspond to more than one interaction within the detector before its acceptance. In germanium detectors, this selection is usually accomplished by subdividing the detector into several segments (or providing several adjacent independent detectors) and seeking coincident pulses from two or more of the independent segments. When coincidences are found, the output from all detector segments is summed and recorded. The resulting spectrum is made up only of the full-energy peak lying above a featureless continuum that is greatly suppressed and has no abrupt Compton edges.

As an example, Palms et al.[47] describe a sum–coincidence spectrometer consisting of two concentric coaxial germanium detectors. For gamma-ray energies between 300 and 1800 keV, the ratio of the full-energy peak to the average Compton continuum was increased by a factor of 4 or 5 compared with a single germanium detector of the same volume. Compton edges and double escape peaks were essentially eliminated. These improvements are accomplished at the expense of a reduction in the full-energy peak

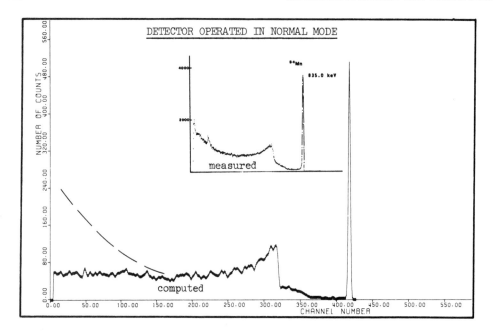

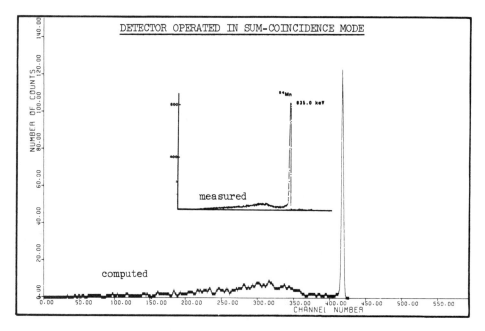

Figure 12-25 Pulse height spectra from a Ge(Li) detector in normal operation (top) compared with spectra from the same detector operated in sum–coincidence mode. (From Walker and Palms.[48])

efficiency, however, because events that correspond to full-energy absorption confined within a single detector segment are inevitably lost. The absolute reduction in the full-energy peak efficiency over the same energy range was approximately a factor of 5. Another example of the effectiveness of the sum-coincidence mode of operation is shown in Fig. 12-25.

Larsen and Strauss[49] describe the performance of separately housed germanium detectors in the sum–coincidence mode. A severe loss of efficiency takes place because the two detectors must then be separated significantly, and most success with sum–coincidence operation has been achieved by using segmented or split coaxial detectors mounted in close proximity within the same cryostat.[50]

3. THE PAIR SPECTROMETER

A different approach to simplifying the recorded germanium detector spectrum is to attempt to select only the double escape peak. If the gamma-ray energy is sufficiently high, a significant fraction of all interactions will correspond to pair production in which both photons produced by positron annihilation escape from the primary detector. Because these annihilation photons always are emitted in opposite directions, two additional detectors placed on opposite sides of the primary detector can intercept them

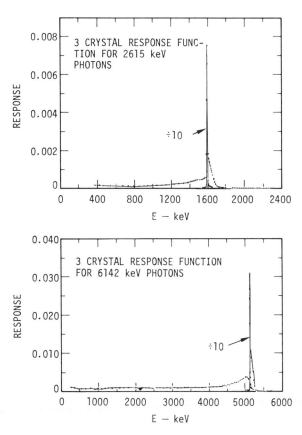

Figure 12-26 Response functions of a Ge(Li) crystal operated as a pair spectrometer in conjunction with two NaI(Tl) detectors. (From Stelts and Browne.[52])

with a reasonable efficiency. If coincidence is demanded among all three detectors, the selection of double escape events will be highly specific, and most of the continuum (as well as the full-energy peak) will be suppressed. A considerable sacrifice in counting efficiency must always be made, but the isolation of the double escape peak and suppression of other backgrounds can be very effective. Systems consisting of a central germanium detector and two surrounding NaI(Tl) scintillators are a common configuration whose performance has been well documented.[51, 52]

Figure 12-26 shows the very simple response function that can be obtained from such a pair spectrometer. Although the majority of the recorded events correspond to the isolated double escape peak, a small continuum still persists due to several nonidealities. Pulses above the peak can be recorded from events in which an annihilation photon undergoes small-angle scattering before leaving the central detector but still falls within the accepted pulse height window from the outer detector. Events below the peak can be generated from gamma-ray interactions that occur near a boundary of the central detector and for which either the positron or pair electron escapes without depositing all its energy. Robertson et al.,[51] have shown that this latter continuum can be suppressed significantly by using pulse shape discrimination to discard slow-rising pulses, which are most likely to be generated from interactions near the edges of the germanium central detector (see p. 409).

C. Energy Calibration

1. CALIBRATION SOURCES

In gamma-ray spectroscopy with germanium detectors, the pulse height scale must be calibrated in terms of absolute gamma-ray energy if various peaks in the spectrum are to be properly identified. In many routine applications, the gamma rays expected to appear in the spectrum are well known in advance and the corresponding peaks can readily be identified by inspection. In other applications, unknown gamma-ray spectra may be encountered which do not provide an unambiguous calibration of the energy scale. In such cases, a separate calibration gamma-ray source is conventionally used to supply peaks of known energy in the spectrum. Accurate calibration should involve a standard source with gamma-ray energies that are not widely different from those to be measured in the unknown spectrum. Because even the best spectrometer systems often show nonlinearities of a channel or two over a full range of several thousand channels, it is also useful to have multiple calibration peaks at various points along the measured energy range to account for these nonlinearities.

The precision to which the centroid of a peak in a pulse height spectrum can be localized is dependent on the spectrometer system resolution and its stability over the period of the measurement. With high-quality germanium systems, the uncertainty in the peak position can approach one part in 10^5, which is of the same order as the uncertainty in the calibration energy standards. Therefore, an important goal is to define closely the energy of the standards so that their energy uncertainty does not contribute unnecessarily to the overall imprecision of a gamma-ray measurement.

Criteria that enter into the selection of standards to be used for germanium spectrometer calibration have been reviewed by Kern.[53] When the ultimate in precision is required, use of one of two primary standards is recommended: the tungsten K-α_1 X-ray (near 59 keV) or the ^{198}Au gamma ray (near 411 keV). Precise values and estimated uncertainties for these energies are listed in Table 12-1. Some authors have also treated annihilation radiation (near 511 keV) as a primary standard, but its use in high-precision calibrations should be avoided. The observed line is always several keV wide due to finite

TABLE 12-1 Gamma Rays Used as Energy Calibration Standards[a]

Source	Energy (keV)	Source	Energy (keV)
[241]Am	59.536 ± 0.001	[192]Ir	468.060 ± 0.010
[109]Cd	88.034 ± 0.010	Annihilation	511.003 ± 0.002
[182]Ta	100.106 ± 0.001	[207]Bi	569.690 ± 0.030
[57]Co	122.046 ± 0.020	[208]Tl	583.139 ± 0.023
[144]Ce	133.503 ± 0.020	[192]Ir	604.378 ± 0.020
[57]Co	136.465 ± 0.020	[192]Ir	612.430 ± 0.020
[141]Ce	145.442 ± 0.010	[137]Cs	661.615 ± 0.030
[182]Ta	152.435 ± 0.004	[54]Mn	834.840 ± 0.050
[139]Ce	165.852 ± 0.010	[88]Y	898.023 ± 0.065
[182]Ta	179.393 ± 0.003	[207]Bi	1063.655 ± 0.040
[182]Ta	222.110 ± 0.003	[60]Co	1173.231 ± 0.030
[212]Pb	238.624 ± 0.008	[22]Na	1274.550 ± 0.040
[203]Hg	279.179 ± 0.010	[60]Co	1332.508 ± 0.015
[192]Ir	295.938 ± 0.010	[140]La	1596.200 ± 0.040
[192]Ir	308.440 ± 0.010	[124]Sb	1691.022 ± 0.040
[192]Ir	316.490 ± 0.010	[88]Y	1836.127 ± 0.050
[131]I	364.491 ± 0.015	[208]Tl	2614.708 ± 0.050
[198]Au	411.792 ± 0.008	[24]Na	2754.142 ± 0.060

[a] The primary X-ray standard: Tungsten $K - \alpha_1$ = 59.31918 ± 0.00035 keV (from Greenwood et al.).[54]

Source: Values from Gunnink et al.[55]

momentum effects at the annihilation point and will also be shifted below the electron rest-mass energy (511.003 keV) by up to 10 eV.[56,57] This shift reflects the effects of the finite binding energy of the electron involved in the annihilation process and therefore will be dependent on specific materials around the source. A detailed investigation of the annihilation peak[57] also shows that errors of 40–50 eV can easily be made in its centroid location unless account is taken of its slightly asymmetric shape.

Secondary gamma-ray standards more widely separated over the energy range are also listed in Table 12-1. Other energy standards between 26 and 100 keV are discussed in Ref. 58, between 50 and 420 keV in Ref. 54, and between 400 and 1300 keV in Ref. 59. For higher-energy gamma rays, the position of the double escape peak often will lie within the energy range of one of the listed standards, and the energy scale can thus be "bootstrapped" to higher energies by noting the position of the corresponding full-energy peak.

Another method of extrapolation to higher energies (often called the *cascade-cross-over* method) can be carried out if gamma rays from a cascade such as that illustrated in the sketch below are involved.

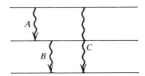

Here, the gamma ray labeled *C* should have an energy equal to that of the sum of the lower-energy gamma rays labeled *A* and *B*, after a small correction is made for the difference in nuclear recoil energy. If the energies of *A* and *B* lie within the calibration range, a new calibration point at the energy of *C* is thereby obtained.

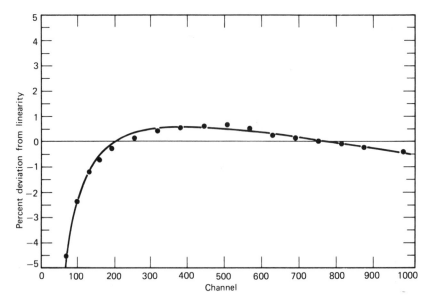

Figure 12-27 A typical differential linearity plot for a germanium detector system. (From Zulliger et al.[60])

2. CALIBRATION CURVE

Once energy calibration points have been established over the entire energy range of interest, a calibration curve relating energy to channel number is normally derived. Common techniques involve the least-square fitting of a polynomial of the form $E_i = \sum_{n=0}^{N} a_n C_i^n$, where E_i is the energy corresponding to the channel number C_i. A polynomial of order $N = 4$ or 5 normally is adequate for typical germanium spectrometers, depending on the severity of nonlinearity that is present.

Because the degree of nonlinearity is often quite small, the calibration curve is sometimes presented as a plot of the deviation from perfect linearity versus channel number. A representative plot of this type is given in Fig. 12-27.

3. DIRECTIONAL DEPENDENCE OF PEAK POSITIONS

When carrying out gamma-ray energy measurements of very high precision, some care must be taken that the unknown source and calibration source are placed so that the emitted gamma rays strike the detector from the same direction in either case. Shifts up to 110 eV have been reported in apparent peak position as the source of gamma rays was moved around the detector.[61,62] Such shifts are large enough to be of significance in high-precision measurements.

The peak shift with direction is thought to arise from two effects.[62] One involves the tendency for the secondary electrons produced in gamma-ray interactions to gain a slight amount of energy in the electric field of the detector. Because there is some correlation between the secondary electron direction and the incoming gamma-ray direction, the amount of energy gained from the field will differ slightly, depending on orientation. A second potential source of peak shift is any difference in charge collection efficiency in the detector. These differences often are a source of peak broadening when the detector is uniformly irradiated over its entire volume. However, gamma rays incident from a specific direction will tend to interact preferentially in certain regions of the detector volume. If the charge collection efficiency varies appreciably between regions, then the average peak position can also shift with incident gamma-ray direction.

D. Detection Efficiency

1. EFFICIENCY CALIBRATION

Any measurement of absolute emission rates of gamma rays (i.e., not relative to a similar source of known activity) requires knowledge of the detector efficiency. The emission rate for a point source can then be calculated from Eq. (4-19) by measuring the full-energy peak area over a fixed time and by determining the detector solid angle from its dimensions and the source–detector spacing. In germanium gamma-ray spectroscopy, an efficiency based on the area under the single or double escape peak is sometimes used in place of that based on the full-energy peak. Although efficiencies of germanium detectors can be estimated from published measurements or calculations for detectors of similar

TABLE 12-2 Decay Data for Radionuclides Used as Efficiency Standards

Nuclide	$T_{1/2}$		E (keV)	I^a (%)	$\Delta I/I^b$ (%)
^{22}Na	2.60	y	1274.5	99.95	0.0
^{24}Na	15.0	h	1368.5	100.0	0.0
			2754.0	99.85	0.0
^{46}Sc	83.7	d	889.2	99.98	0.0
			1120.5	99.99	0.0
^{54}Mn	312.5	d	834.8	99.98	0.0
^{57}Co	272	d	14.4	9.6	1.0
			122.1	85.6	0.3
^{60}Co	5.27	y	1173.2	99.88	0.0
			1332.5	99.98	0.0
^{85}Sr	64.8	d	13.4	50.7	1.5
			514.0	99.28	0.0
^{88}Y	106.6	d	14.2	52.5	1.5
			1836.1	99.4	0.2
^{95}Nb	35.15	d	765.8	99.80	0.0
^{113}Sn	115.2	d	24.1	79.5	2.0
^{131}I	8.02	d	364.5	82.4	0.5
^{134}Cs	2.06	y	604.6	97.5	0.2
^{137}Cs	30.0	y	31.8/32.2	5.64	2.0
			661.6	85.3	0.4
^{139}Ce	137.6	d	33.0/33.4	64.1	2.0
			165.8	80.0	0.4
^{141}Ce	32.5	d	35.6/36.0	12.6	2.0
			145.5	48.4	0.9
^{140}La	40.27	h	1596.6	95.6	0.3
^{198}Au	2.696	d	411.8	95.53	0.1
^{203}Hg	46.6	d	70.8/72.9	10.1	1.5
			279.2	81.3	0.2
^{241}Am	432	y	59.5	36.0	1.0

$^a I$: Gamma-ray photon yield per disintegration.

$^b \Delta I/I$: Uncertainty in yield figure.

Note: Only those gamma-ray lines are listed for which the yield uncertainty is 2.0% or less. Some of these nuclides emit other gamma rays in addition to those shown.

Source: Debertin et al.[63]

size, the accuracy of results based on these values will not be much better than 10–20%. One major difficulty is that the dimensions of these detectors are not standardized to any degree, and it is very difficult to determine precisely their active volume. Furthermore, long-term changes in charge collection efficiency and/or window thickness can lead to drifts in the detector efficiency over periods of time.

For these reasons, users will normally carry out their own periodic efficiency calibrations of their germanium detectors using sources calibrated by some other means. Any error in assumed detector dimensions will then apply both to the calibration and actual measurements and will not affect the accuracy of activity measurements. The source–detector distance still must be accurately measured and reproduced to avoid errors in the relative solid angle. The calibration is normally carried out for an assortment of gamma-ray energies covering the range of interest to allow construction of an empirical efficiency versus energy curve.

Many of the national standards laboratories will provide isotopes whose gamma ray emission rates have been calibrated to a precision ranging from 0.5 to 2%. Single isotope sources provide a few well-separated gamma-ray peaks whose area can be determined by simple methods to calibrate the detector efficiency. If the energy scale must extend over a relatively wide range, multiple sources must be used, either sequentially or in combination. Table 12-2 lists radionuclides used for efficiency calibrations, together with decay data necessary to compute gamma-ray yields from absolute activity.

Standard calibration sources are available as small deposits on thin backing material, so that they may closely approximate nonabsorbing point sources. However, radioactive samples to be measured often have non-negligible volume and mass, and gamma rays can be attenuated by self-absorption within the sample material itself. In such cases, correction must be made for this attenuation if an accurate determination of the gamma ray emission rate from the entire sample is needed. To aid in the calibration of large-volume samples, typically of water solutions or soil, a standard sample container called a

TABLE 12-3   **Gamma Rays Emitted by ^{226}Ra in Equilibrium with its Daughters**[a]

Isotope	Gamma-Ray Energy (keV)	Relative Intensity
^{226}Ra	186.211 ± 0.010	9.00 ± 0.10
^{214}Pb	241.981 ± 0.008	16.06 ± 0.19
^{214}Pb	295.213 ± 0.008	42.01 ± 0.53
^{214}Pb	351.921 ± 0.008	80.42 ± 0.81
^{214}Bi	609.312 ± 0.007	100 ± 0.92
^{214}Bi	768.356 ± 0.010	10.90 ± 0.15
^{214}Bi	934.061 ± 0.012	6.93 ± 0.10
^{214}Bi	1120.287 ± 0.010	32.72 ± 0.39
^{214}Bi	1238.110 ± 0.012	12.94 ± 0.17
^{214}Bi	1377.669 ± 0.012	8.87 ± 0.15
^{214}Bi	1509.228 ± 0.015	4.78 ± 0.09
^{214}Bi	1729.595 ± 0.015	6.29 ± 0.10
^{214}Bi	1764.494 ± 0.014	34.23 ± 0.44
^{214}Bi	1847.420 ± 0.025	4.52 ± 0.09
^{214}Bi	2118.551 ± 0.030	2.53 ± 0.05
^{214}Bi	2204.215 ± 0.040	10.77 ± 0.20
^{214}Bi	2447.860 ± 0.100	3.32 ± 0.08

[a]Only the strongest transitions are shown. Energies are measured relative to an assumed 411.794 keV gamma ray from ^{198}Au. Quoted errors do not include any error contribution from this reference standard.

Source: Zobel et al.[41]

Marinelli beaker has been specified[31] that closely fits over the endcap of the detector cryostat. Several standard sizes of Marinelli beaker are specified to accommodate samples of various volumes. With a standard geometry, the self-absorption in samples of similar composition and volume will be comparable. Some calibrated standard sources are available in Marinelli beaker form to simplify the efficiency determination for such bulk samples.

Recording of separate spectra for each source provides results that are simple to interpret, but the calibration process is then time consuming and tedious. The substitution of a single source that emits many different gamma-ray energies is a tempting alternative because only a single calibration spectrum need be recorded and analyzed. The problems of interference between multiple responses become much more severe, however, and, if precise results are to be obtained, more sophisticated methods of determining peak area must be used which take into account these interference effects. Several authors[41,64,65] suggest the use of ^{226}Ra in equilibrium with its decay products and provide tables of gamma-ray intensities per disintegration (see Table 12-3). These cover the energy range from 188 to 2446 keV and are particularly useful because a source of ^{226}Ra is often available in radioisotope laboratories. References 39 and 66 provide similar lists of gamma-ray intensities per disintegration for a number of other isotopes proposed for detector efficiency calibration. Of these, ^{152}Eu has gained recent popularity because of its convenient half-life (13 y) and the wide range of gamma-ray energies produced in its decay (see Table 12-4).

TABLE 12-4   Multiple Gamma Rays
Emitted in the Decay of ^{152}Eu

Energy (keV)	Relative Intensity	
121.8	141.	$\pm 4.$[a]
244.7	36.6	± 1.1
344.3	127.2	± 1.3
367.8	4.19	± 0.04
411.1	10.71	± 0.11
444.0	15.00	± 0.15
488.7	1.984	± 0.023
586.3	2.24	± 0.05
678.6	2.296	± 0.028
688.7	4.12	± 0.04
778.9	62.6	± 0.6
867.4	20.54	± 0.21
964.0	70.4	± 0.7
1005.1	3.57	± 0.07
1085.8	48.7	± 0.5
1089.7	8.26	± 0.09[b]
1112.1	65.0	± 0.7
1212.9	6.67	± 0.07
1299.1	7.76	± 0.08
1408.0	100.0	± 1.0
1457.6	2.52	± 0.09

[a] In order to use this line, no ^{154}Eu should be present.

[b] Not intended for use in calibrations because of the proximity to the more intense nearby energy.

Source: Data taken from Gehrke et al.[66]

2. SUM–COINCIDENCE EFFECTS

In employing any source that emits more than one radiation in coincidence, some care must be taken to ensure that measured peak intensities are not affected by the sum–coincidence effects previously described beginning on p. 302. If two gamma rays are emitted in coincidence from the source, they may interact simultaneously in the detector, and the resulting pulse will not, in general, lie within the full-energy peak corresponding to either one. The problem is particularly severe for those sources that involve many cascade gamma rays and, in the specific examples given above, can lead to errors when using reported peak intensities from ^{226}Ra (Ref. 67) or ^{152}Eu (Ref. 66). Not only must coincident detection of gamma rays be avoided, but any other coincident radiation, such as characteristic X-rays or bremsstrahlung, can also lead to summing effects that may reduce the apparent peak intensities. Depending on the decay scheme, coincidence summing may also increase the apparent intensity of some peaks whose energy corresponds to the sum of two lower energy peaks. Quantitative assessments of the changes that may occur in the full-energy peak intensity due to summing effects can be found in Refs. 66 and 68, together with suggested calculation methods to correct for such errors.

Because the summing effects will depend on the square of the detector solid angle [see Eq. (10-11)], whereas the simple peaks vary linearly, the relative effect of summing can be reduced by reducing the solid angle. Meyer[39] has presented data to show the variation of apparent peak intensities with source–detector spacing and recommends that for typical germanium detector sizes, the source–detector spacing be maintained to be at least 10 cm, and preferably 30–40 cm. Of course, there is a limit as to how far away one can practically place a source of low activity, and a compromise between summing inaccuracies and statistical errors in the peak must ultimately be made.

Other potential errors in the peak area measurement can arise from inaccurate treatment of system dead time and from pulse pile-up effects (see Chapter 17). Both phenomena are most important when pulse rates are relatively high, and their effect on absolute detector calibration has been reviewed by Mueller.[69]

3. ENERGY INTERPOLATION

Once the efficiency of a detector has been measured at several energies using calibrated sources, it is useful to fit a curve to these points which describes the detector efficiency over the entire energy range. A number of different empirical formulas have been recommended in the literature, and Singh[70] has tested the validity of a number of these formulas for both planar and coaxial germanium detectors. For the planar detectors studied, a formula first suggested by Mowatt[71] was found to give a very good representation of the full-energy peak efficiency of several different detectors between 60 and 1836 keV. This formula is written

$$\epsilon = \frac{K\left[\tau + \sigma Q \exp\left(-RE\right)\right]}{\tau + \sigma}\left\{1 - \exp\left[-P(\tau + \sigma)\right]\right\} \qquad (12\text{-}31)$$

where τ = photoelectric absorption coefficient in germanium at energy E

σ = Compton absorption coefficient at energy E.

K, Q, R, and P are parameters fitted to the experimental points.

A different empirical relation is needed to fit the response of coaxial detectors. An eight-parameter function, first suggested by McNelles and Campbell,[72] takes the form below, with the a's representing the fitted parameters:

$$\epsilon = \left(\frac{a_1}{E}\right)^{a_2} + a_3 \exp\left(-a_4 E\right) + a_5 \exp\left(-a_6 E\right) + a_7 \exp\left(-a_8 E\right) \quad (12\text{-}32)$$

In some cases, a satisfactory fit can be obtained by dropping the last term and using only six empirical parameters.[70] Equation (12-32) has been tested for various coaxial detectors over an energy range from 160 to 2598 keV and appears to fit experimental data with deviations of no more than 1 or 2%.[72]

Figure 12-28 illustrates such a fit for experimental measurements of the full-energy peak efficiency of a 38 cm³ coaxial Ge(Li) detector. The smooth curve was generated with

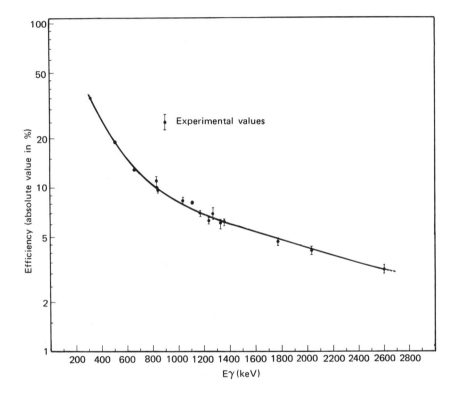

$\varepsilon = (a_1/E)^{a_2} + a_3 \exp(-a_4 E) + a_5 \exp(-a_6 E) +$
$\quad + a_7 \exp(-a_8 E)$

$a_1 = \quad 4.6181$	$a_5 = 11.3828$
$a_2 = \quad 0.4166$	$a_6 = \quad 0.6795$
$a_3 = 117.351$	$a_7 = \quad 0.0285$
$a_4 = \quad 5.1213$	$a_8 = \quad 0.9187$

$\varepsilon = (a_1/E)^{a_2} + a_3 \exp(-a_4 E) + a_5 \exp(-a_6 E)$

$a_1 = \quad 4.6258$	$a_4 = \quad 5.1213$
$a_2 = \quad 0.4166$	$a_5 = 11.3828$
$a_3 = 117.351$	$a_6 = \quad 0.6795$

Figure 12-28 Measured intrinsic full-energy peak efficiency for a coaxial 38 cm³ Ge(Li) detector. The smooth curve is Eq. (12-32) with the parameters shown. A six-parameter fit omitting the last term in Eq. (12-32) is indistinguishable from the full eight-parameter fit for this case. (From Singh.[70])

Eq. (12-32), using the parameter values also shown in the figure. In this case, a fit with six parameters is indistinguishable from that using the full eight parameters. If only an approximate fit is needed over a limited energy range above about 100 keV, a simple two-parameter relation of the form $\ln \epsilon = a + b \ln E$ often may be substituted with acceptable accuracy for the more complex fitting functions shown.

4. CONVENTIONS IN QUOTING DETECTION EFFICIENCY

The full-energy or peak efficiency of germanium detectors can be quoted in several different ways. The quantity most often directly measured is the absolute efficiency, defined as the ratio of counts in the full-energy peak divided by the number of gamma rays emitted by the source. A widely accepted standard energy at which this measurement is made is the 1.333 MeV gamma ray emitted by ^{60}Co. The absolute efficiency is obviously a strong function of both gamma-ray energy and the exact source–detector spacing used for the measurement. This dependence on spacing is greatly diminished if we instead use the intrinsic efficiency, defined as the number of full-energy peak counts divided by the number of gamma rays incident on the detector. At gamma-ray energies sufficiently high so that the detector is relatively transparent (interaction probabilities less than 20 or 30%), the intrinsic efficiency is approximately proportional to the total active volume of the detector. Consequently, most commercial detectors are specified by their total active volume measured in cm^3. For detectors with nonsymmetrical shapes, the intrinsic efficiency can vary substantially depending on orientation of the source relative to the detector axis, because the average thickness of the detector along the direction of the incident radiation is most critical in determining the intrinsic efficiency. For coaxial or closed-ended detectors in which the geometry is complex, the intrinsic efficiency is often difficult to calculate exactly from the observed count rate because of uncertainties in the effective detector solid angle.

In an effort to eliminate some of these difficulties in quoting detector efficiency, many commercial manufacturers specify the photopeak efficiency relative to that of a standard 3 in. × 3 in. (7.62 cm × 7.62 cm) cylindrical NaI(Tl) scintillation crystal. A source–detector spacing of 25 cm is assumed in both cases for standardization. The value is normally specified for the 1.333 MeV gamma-ray photopeak from ^{60}Co. The efficiency ratio can be directly measured by simply determining the photopeak areas from both detectors, using an uncalibrated ^{60}Co source. Alternatively, only the germanium photopeak area may be directly measured, and the NaI(Tl) value calculated by using a calibrated source and by assuming an absolute peak efficiency value of 1.2×10^{-3} (Ref. 73). An approximate rule of thumb for germanium coaxial detectors is that the efficiency ratio in percent is given by the detector volume in cm^3 divided by a factor of 5. The efficiency ratio of high-purity germanium detectors when they were first introduced in the 1960s was a few percent but has reached 80% for the largest detectors currently available. Further increases are likely in the future as crystal fabrication techniques continue to improve.

5. PUBLISHED EFFICIENCY DATA

If a set of calibrated sources is not available to the experimenter, an estimate of the detector efficiency may be made by referring to published data. In the case of sodium iodide scintillators discussed in Chapter 10, relatively few standard sizes of crystals have predominated, so it is often possible to find published data for a detector that is identical to that of the user. In the case of germanium detectors, there is much less standardization of sizes, and the variety of geometries (planar, coaxial, closed-ended) adds to the complexity of the situation. It is therefore much less likely that published data can be found for any specific detector geometry, and considerable interpolation between available values is often needed. The variation of efficiency with energy will be quite similar for detectors of roughly the same size and shape, even if the absolute efficiency values may differ.

Another approach is to derive efficiency data from calculations based on knowledge of the probability of each of the primary gamma-ray interactions within the detector. Monte Carlo codes provide the most accurate calculational methods for efficiency

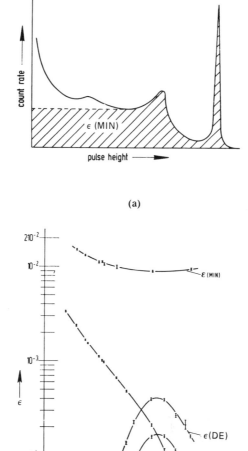

(a)

(b)

Figure 12-29 Absolute efficiency values for a 25 cm³ Ge(Li) detector for a point source 4.83 cm from the crystal face. ϵ(PP) is the full-energy peak efficiency, ϵ(SE) is the single escape peak efficiency, and ϵ(DE) is the double escape peak efficiency. ϵ(MIN) is the efficiency based on all the events indicated by the cross-hatched area on the pulse height spectrum. (From Waibel and Grosswendt.[80])

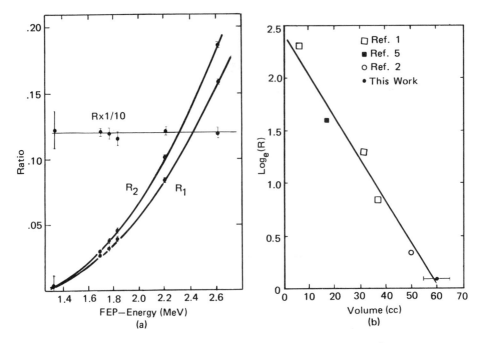

Figure 12-30 (a) The escape peak ratios R_1, R_2, and R for a 60 cm^3 Ge(Li) detector. (b) Plot of ln R versus volume for different Ge(Li) detectors. (From Johnson and Mann.[81])

determination, and examples of the Monte Carlo approach applied to germanium detectors can be found in Refs. 74–79. Compared with similar calculations of detection efficiency in NaI(Tl), the added importance of the escape of secondary radiations in germanium detectors (secondary electrons, bremsstrahlung, etc.) makes accurate modeling more difficult.

6. TYPICAL EFFICIENCY CURVES

A representative curve of germanium detector efficiency as a function of gamma-ray energy is plotted in Fig. 12-29. The full-energy peak efficiency drops continuously with energy and above an energy of several MeV becomes lower than the double escape peak efficiency. The single escape peak efficiency is always smaller than the double escape peak efficiency in the example shown, and both show a broad maximum at an energy of about 5 MeV. The total efficiency (including most of the Compton continuum) is represented at the top of the plot and shows a much milder energy dependence.

Several ratios of peak efficiencies can be defined for any specific detector:

$$R_1 = \text{single escape peak efficiency/full-energy peak efficiency}$$

$$R_2 = \text{double escape peak efficiency/full-energy peak efficiency}$$

$$R = R_2/R_1$$

A plot of these ratios for a specific detector over the energy range 1.33–2.60 MeV is shown in Fig. 12-30a. Over this range, both R_1 and R_2 increase with gamma-ray energy, indicating the increased importance of the escape peaks. The ratio R, however, is approximately energy independent and depends only on detector geometry. Figure 12-30b

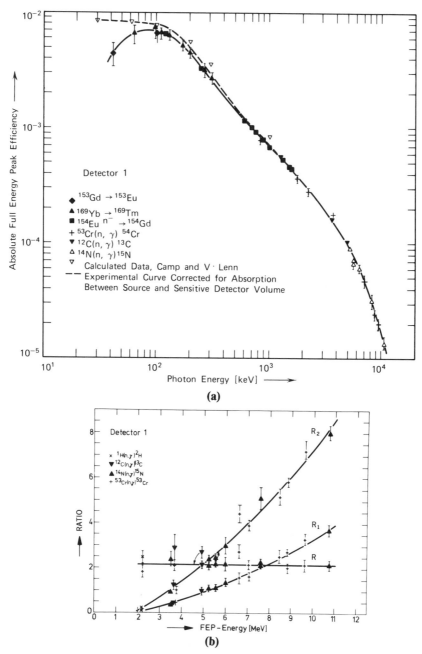

Figure 12-31 (a) Absolute full-energy peak efficiency for a point source 83 mm from the face of a 38 cm^3 true coaxial Ge(Li) detector. (b) Escape peak ratios (as defined in text) for the same detector. (From Seyfarth et al.[82])

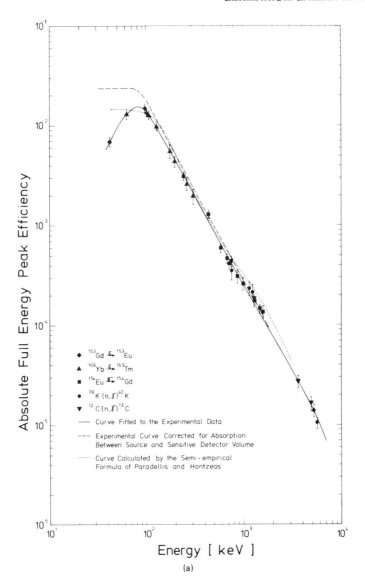

Figure 12-32 (a) As in Fig. 12-31, but for a point source 54 mm from the face of a 33 mm diameter × 6.8 mm thick (5.8 cm³ active volume) planar Ge(Li) detector. (b) Escape peak ratios for the planar Ge(Li) detector of part (a). (From Seyfarth.[83])

shows the dependence of R on detector volume. As the detector size increases, the probability of double escape approaches zero. Because the annihilation quanta are oppositely directed, one must always pass through a relatively thick portion of large-volume detectors before escaping.

Measured full-energy peak efficiencies for typical coaxial and planar germanium detectors are shown in Figs. 12-31 and 12-32. The escape peak ratios defined above are also shown for the same detectors.

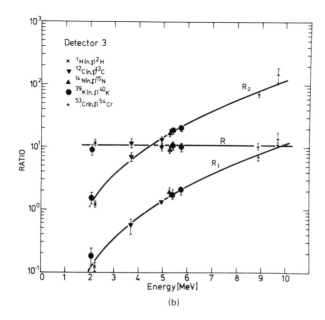

Figure 12-32 *(Continued)*

E. Miscellaneous Effects on Detector Performance

1. RADIATION DAMAGE

As discussed in Chapter 11, semiconductor detectors are relatively sensitive to performance degradation caused by damage created within the detector by incident radiation. The large volume and long charge collection paths in germanium gamma-ray detectors make them more susceptible to such degradation compared with the much thinner silicon diode detectors typically applied to charged particle detection. Because the amount of damage created by fast neutrons of a given fluence is large compared with the damage from an equivalent fluence of gamma rays, the most significant effects often arise in reactor or accelerator laboratories where fast neutrons may be present. An extensive review of the effects of this damage in both Ge(Li) and HPGe detectors is given in Ref. 84.

The principal consequence of radiation damage is to increase the amount of hole trapping within the active volume of the detector. In a damaged detector, the amount of charge collected is subject to a loss due to this trapping which will vary from pulse to pulse depending on the position of the interaction. Measured peaks in the pulse height spectrum will then show a tailing toward the low-energy side. The spectra shown in Fig. 12-33 illustrate a gradual broadening of gamma-ray peaks from a Ge(Li) detector as the fast neutron exposure is increased. Exposure of thick planar germanium detectors of either the Ge(Li) or HPGe types to a fast neutron fluence of about 10^9 n/cm^2 is sufficient to risk measurable changes in the detector resolution[86,87] and many become totally unusable after exposure to a fluence of 10^{10} n/cm^2.

In coaxial detectors, the specific detector configuration can have a strong influence on the measured spectral effects. In Ge(Li) detectors and HPGe coaxials fabricated from high-purity p-type germanium, holes are the carrier type that are drawn inward to the p^+

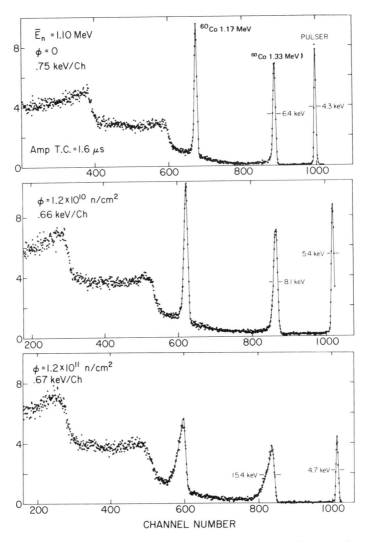

Figure 12-33 The deterioration of a Ge(Li) γ-spectrum with increasing neutron fluence. Only the upper part of the spectrum is shown. (From Kraner et al.[85])

contact near the cylindrical axis. For HPGe coaxials made from high-purity n-type germanium, the electrode polarity is reversed, and holes are instead collected at the p^+ contact now fabricated on the outer cylindrical surface (see Fig. 12-2). It is found that these n-type HPGe detectors (sometimes also called *reverse electrode* configurations) show much less performance degradation from radiation damage when compared with the more common p-type detectors. In one study[88] of similar coaxial detectors of both types, the n-type detector was able to withstand 28 times the neutron fluence before showing the same peak broadening as that observed from the p-type configuration. The explanation for this difference lies in the fact that the damage sites preferentially trap holes rather than electrons. Because of the cylindrical geometry and attenuation of the incident gamma rays, more interactions occur at large values of the detector radius than at small values. For such locations, the rise of the signal pulse is dominated by the species that

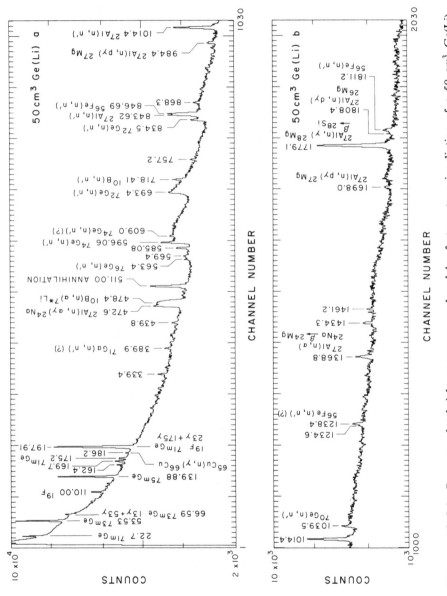

Figure 12-34 Prompt pulse height spectrum produced by fast neutron irradiation of a 50 cm³ Ge(Li) detector. (From Bunting and Kraushaar.[91])

drifts inward toward the detector center (see Fig. 12-2). If these are holes, the corresponding pulse is much more likely to be affected by hole trapping than if the holes must only travel the shorter distance to the outer electrode. As a result of their superior performance in the presence of damage, n-type HPGe coaxial detectors have become the configuration of choice for applications in which exposure to fast neutrons may occur over time.

One of the advantages of HPGe detectors over the older Ge(Li) types is that, should radiation damage occur, it can be repaired by annealing the detector. It has been shown[89] that raising the temperature to 120°C for 72 h can restore lightly damaged ($< 4 \times 10^9$ n/cm^2) n-type HPGe detectors to their original resolution performance. During the heating process, an external vacuum pump must be connected to preserve the cryostat vacuum. The process is simple enough so that many users are able to carry out this step in their own laboratories.

2. NEUTRON-INDUCED PULSES

When detectors are used under conditions in which fast neutrons may be present as a background, account must be taken of their effect on the recorded pulse height spectrum. The most obvious contribution is the appearance of spurious peaks that arise mainly due to excitation of germanium nuclei by inelastic neutron scattering, followed by the emission of deexcitation gamma rays, internal conversion electrons, or X-rays. The neutron-induced peaks are usually identifiable in the spectrum because their width is normally larger than that for gamma-ray-induced events. The peak broadening takes place because a fraction of the excitation energy goes into the recoiling germanium nucleus, which subsequently contributes a variable yield of electron–hole pairs, adding to those created by the deexcitation radiation.

Chasman et al.[90] investigated the response of a Ge(Li) detector to monoenergetic fast neutrons ranging from 1.2 to 16.3 MeV. Lines were observed that resulted from the inelastic excitation of four germanium isotopes, as well as from inelastic neutron excitation of other materials in the detector vacuum enclosure and cryostat. At neutron energies above 3 MeV, additional peaks were detected due to neutron-induced charged particles created in reactions with germanium atoms. Other peaks below 200 keV have been observed[91] which arise from isomers produced by neutron reactions on several germanium isotopes. These lower-energy peaks tend not to be broadened and can easily be confused with peaks from gamma rays. A spectrum showing many of the neutron-induced peaks is given in Fig. 12-34.

The radiation damage that may result from a given exposure to neutrons can be appraised from the previous section if an estimate of the fast neutron flux is obtained. The neutron-induced pulse height spectrum can be used for such an estimate, as suggested by Bell.[92] The neutron fluence is approximately given by

$$\frac{\text{neutrons}}{\text{cm}^2} = \frac{300 \times (\text{counts in 693.4 keV peak})}{\text{detector volume in cm}^3}$$

so that a quick determination of the spectrum can indicate whether further exposure of the detector to the neutron environment is advisable.

PROBLEMS

12-1. Assume that the Fano factor in germanium were half the currently assumed value. What quantitative effect would this change have on the theoretically obtainable energy resolution?

12-2. A planar Ge(Li) detector with a 10 mm thick intrinsic region is operated with sufficient applied voltage to saturate the carrier velocities. What is the approximate value of the required voltage? What must be the minimum charge carrier lifetimes if no more than 0.1% of either holes or electrons are to be lost from any pulse?

12-3. The detector described above is operated with a pulse-processing system that produces a peak with 1.2 keV equivalent FWHM from a pulser input. Estimate the energy resolution of the detector–electronics system for incident 140 keV gamma rays.

12-4. The Compton edge in a gamma-ray spectrum recorded with a germanium detector for a given isotope lies at an energy of 1.16 MeV. Find the energy of the incident gamma rays and the equivalent energy of the Compton edge in a sodium iodide detector.

12-5. Why are escape peaks generally more prominent in germanium detector gamma-ray spectra compared with sodium iodide detectors?

12-6. A germanium detector has a photopeak efficiency of 40% relative to that of a standard 7.62 cm × 7.62 cm NaI(Tl) scintillator. Find the counting rate for the 1.333 MeV peak for a ^{60}Co point source of 150 kBq activity at a distance of 40 cm.

12-7. Assuming that charge collection is complete and that electronic noise is negligible, find the expected energy resolution (in percent) of a germanium detector for the 0.662 MeV gamma rays from ^{137}Cs.

12-8. For the germanium detector of Fig. 12-31, find the expected number of counts under the full-energy, single escape, and double escape peaks for a 5 min measurement using a 0.25 MBq source of ^{137}Cs gamma rays located 10 cm from the detector face.

12-9. For incident 2.10 MeV gamma rays, at what energy does the peak appear in the spectrum from a pair spectrometer?

REFERENCES

1. J. Llacer, *Nucl. Instrum. Meth.* **98**, 259 (1972).
2. J. M. Marler and P. V. Hewka, *IEEE Trans. Nucl. Sci.* **NS-21**(1), 287 (1974).
3. G. A. Armantrout, *IEEE Trans. Nucl. Sci.* **NS-19**(3), 289 (1972).
4. G. H. Nakano, D. A. Simpson, and W. L. Imhof, *IEEE Trans. Nucl. Sci.* **NS-24**(1), 68 (1977).
5. R. H. Pehl, E. E. Haller, and R. C. Cordi, *IEEE Trans. Nucl. Sci.* **NS-20**(1), 494 (1973).
6. E. Sakai, Y. Murakami, and H. Nakatani, *IEEE Trans. Nucl. Sci.* **NS-29**(1), 760 (1982).
7. R. E. Stone, V. A. Barkley, and J. A. Fleming, *IEEE Trans. Nucl. Sci.* **NS-33**(1), 299 (1986).
8. G. Alberti, R. Clerici, and A. Zambra, *Nucl. Instrum. Meth.* **158**, 425 (1979).
9. J. Cl. Philippot, *IEEE Trans. Nucl. Sci.* **NS-17**(3), 446 (1970).
10. J. M. Palms et al., *Nucl. Instrum. Meth.* **64**, 310 (1968).
11. H. R. Zulliger, L. M. Middleman, and D. W. Aitken, *IEEE Trans. Nucl. Sci.* **NS-16**(1), 47 (1969).
12. A. Owens, *Nucl. Instrum. Meth.* **A238**, 473 (1985).
13. A. Alberigi Quaranta, M. Martini, and G. Ottaviani, *IEEE Trans. Nucl. Sci.* **NS-16**(2), 35 (1969).
14. T. W. Raudorf, T. J. Paulus, M. O. Bedwell, and M. Martini, *IEEE Trans. Nucl. Sci.* **NS-24**(1), 78 (1977).
15. B. C. Robertson and H. L. Malm, *Nucl. Instrum. Meth.* **150**, 401 (1978).

16. D. C. S. White and W. J. McDonald, *Nucl. Instrum. Meth.* **115**, 1 (1974).

17. L. L. Gadeken and B. C. Robertson, *Nucl. Instrum. Meth.* **136**, 255 (1976).

18. R. Chun, L. L. Gadeken, and B. C. Robertson, *Nucl. Instrum. Meth.* **137**, 295 (1976).

19. H. Engel, H. Schneider, and R. Spitz, *Nucl. Instrum. Meth.* **142**, 525 (1977).

20. T. W. Raudorf, M. O. Bedwell, and T. J. Paulus, *IEEE Trans. Nucl. Sci.* **NS-29**(1), 764 (1982).

21. M. Martini and T. A. McMath, *Nucl. Instrum. Meth.* **79**, 259 (1970).

22. T. W. Raudorf and R. H. Pehl, *Nucl. Instrum. Meth.* **A255**, 538 (1987).

23. E. Sakai, *IEEE Trans. Nucl. Sci.* **NS-18**(1), 208 (1971).

24. L. Karlsson, *Nucl. Instrum. Meth.* **109**, 101 (1973).

25. G. White, F. A. Smith, and C. F. Coleman, *Nucl. Instrum. Meth.* **A234**, 535 (1985).

26. M. Moszynski and B. Bengtson, *Nucl. Instrum. Meth.* **80**, 233 (1970).

27. N. Matsushita, J. Kasagi, and W. C. McHarris, *Nucl. Instrum. Meth.* **201**, 433 (1982).

28. G. A. Armantrout, A. E. Bradley, and P. L. Phelps, *IEEE Trans. Nucl. Sci.* **NS-19**(1), 107 (1972).

29. K. Debertin and R. G. Helmer, *Gamma- and X-Ray Spectrometry with Semiconductor Detectors*, North Holland Publishers, Amsterdam, 1988.

30. J. Roth, J. H. Primbsch, and R. P. Lin, *IEEE Trans. Nucl. Sci.* **NS-31**(1), 367 (1984).

31. ANSI/IEEE Standard 325-1986, *Test Procedures for Germanium Gamma Ray Detectors* (1986).

32. U. E. P. Berg, H. Wolf, B. Schaeffer, and K. Wienhard, *Nucl. Instrum. Meth.* **129**, 155 (1975).

33. I. K. MacKenzie and J. L. Campbell, *Nucl. Instrum. Meth.* **101**, 149 (1972).

34. R. L. Heath and J. E. Cline, *Gamma-Ray Spectrometry of Neutron Deficient Isotopes*, IDO-17188 (1966).

35. J. E. Cline and R. L. Heath, *Gamma Ray Spectrometry of Neutron Deficient Isotopes*, IDO-17222 (1967).

36. F. Adams and R. Dams, *Applied Gamma-Ray Spectrometry*, 2nd ed. and revision of original publication by C. E. Crouthamel, Pergamon Press, Oxford, 1970.

37. J. P. Balagna and S. B. Helmik, *An Atlas of Gamma-Ray Spectra*, LA-4312 (1970).

38. Ch. Meixner, *Nucl. Instrum. Meth.* **119**, 521 (1974).

39. R. A. Meyer, K. G. Tirsell, and G. A. Armantrout, *Proceedings, ERDA X- and Gamma-Ray Symposium*, Ann Arbor, MI (CONF-760539), p. 40 (1976).

40. H. H. Jorch and J. L. Campbell, *Nucl. Instrum. Meth.* **143**, 551 (1977).

41. V. Zobel, J. Eberth, U. Eberth, and E. Eube, *Nucl. Instrum. Meth.* **141**, 329 (1977).

42. P. J. Nolan, D. W. Gifford, and P. J. Twin, *Nucl. Instrum. Meth.* **A236**, 95 (1985).

43. M. Mohsen, *Nucl. Instrum. Meth.* **212**, 241 (1983).

44. R. M. Lieder et al., *Nucl. Instrum. Meth.* **220**, 363 (1984).

45. J. Verplancke et al., *IEEE Trans. Nucl. Sci.* **NS-33**(1), 340 (1986).

46. L. Hildingsson et al., *Nucl. Instrum. Meth.* **A252**, 91 (1986).

47. J. M. Palms, R. E. Wood, and O. H. Puckett, *IEEE Trans. Nucl. Sci.* **NS-15**(3), 397 (1968).

48. D. M. Walker and J. M. Palms, *IEEE Trans. Nucl. Sci.* **NS-17**(3), 296 (1970).

49. R. N. Larsen and M. G. Strauss, *IEEE Trans. Nucl. Sci.* **NS-17**(3), 254 (1970).

50. A. R. Sayres and J. A. Baicker, *IEEE Trans. Nucl. Sci.* **NS-15**(3), 393 (1968).

51. A. Robertson, G. C. Cormick, T. J. Kennett, and W. V. Prestwich, *Nucl. Instrum. Meth.* **127**, 373 (1975).

52. M. L. Steits and J. C. Browne, *Nucl. Instrum. Meth.* **133**, 35 (1976).

53. J. Kern, *Gamma Ray Standards*, CONF-7210117, p. 345 (1974).

54. R. C. Greenwood, R. G. Helmer, and R. J. Gehrke, *Nucl. Instrum. Meth.* **77**, 141 (1970).

55. R. Gunnink, J. B. Niday, R. P. Anderson, and R. A. Meyer, *Gamma-Ray Energies and Intensities*, UCID-15439 (1969).

56. K. Shizuma, H. Inoue, and Y. Yoshizawa, *Nucl. Instrum. Meth.* **137**, 599 (1976).

57. K. Fransson, A. Nilsson, J. DeRaedt, and K. G. Rensfelt, *Nucl. Instrum. Meth.* **138**, 479 (1976).

58. G. C. Nelson and B. G. Saunders, *Nucl. Instrum. Meth.* **84**, 90 (1970).

59. R. G. Helmer, R. C. Greenwood, and R. J. Gehrke, *Nucl. Instrum. Meth.* **96**, 173 (1971).

60. H. R. Zulliger, L. M. Middleman, and D. W. Aitken, *IEEE Trans. Nucl. Sci.* **NS-16**(1), 47 (1969).

61. P. C. Lichtenberger and I. K. MacKenzie, *Nucl. Instrum. Meth.* **116**, 177 (1974).

62. R. G. Helmer et al., *Nucl. Instrum. Meth.* **123**, 51 (1975).

63. K. Debertin et al., *Proceedings, ERDA X- and Gamma-Ray Symposium*, Ann Arbor, MI (CONF-760539), p. 59 (1976).

64. G. Walford and C. E. Doust, *Nucl. Instrum. Meth.* **62**, 353 (1968).

65. G. Wallace and G. E. Coote, *Nucl. Instrum. Meth.* **74**, 353 (1969).

66. R. J. Gehrke et al., *Nucl. Instrum. Meth.* **147**, 405 (1977).

67. A. Dolev et al., *Nucl. Instrum. Meth.* **68**, 176 (1969).

68. G. J. McCallum and G. E. Coote, *Nucl. Instrum. Meth.* **130**, 189 (1975).

69. J. W. Mueller, *Nucl. Instrum. Meth.* **112**, 47 (1973).

70. R. Singh, *Nucl. Instrum. Meth.* **136**, 543 (1976).

71. R. S. Mowatt, *Nucl. Instrum. Meth.* **70**, 237 (1969).

72. L. A. McNelles and J. L. Campbell, *Nucl. Instrum. Meth.* **109**, 241 (1973).

73. R. L. Heath, *Scintillation Spectrometry, Gamma-Ray Spectrum Catalog*, 2nd ed., Vol. 1, IDO-16880-1 (1964).

74. G. Aubin et al., *Nucl. Instrum. Meth.* **76**, 85 (1969).

75. B. Lal and K. V. K. Iyengar, *Nucl. Instrum. Meth.* **79**, 19 (1970).

76. D. E. Raeside and M. A. Ludington, *Nucl. Instrum. Meth.* **93**, 389 (1971).

77. B. F. Peterman et al., *Nucl. Instrum. Meth.* **104**, 461 (1972).

78. B. Grosswendt and E. Waibel, *Nucl. Instrum. Meth.* **131**, 143 (1975).

79. T. Nakamura, *Nucl. Instrum. Meth.* **131**, 521 (1975).

80. E. Waibel and B. Grosswendt, *Nucl. Instrum. Meth.* **131**, 133 (1975).

81. J. R. Johnson and K. C. Mann, *Nucl. Instrum. Meth.* **112**, 601 (1973).

82. H. Seyfarth et al., *Nucl. Instrum. Meth.* **105**, 301 (1972).

83. H. Seyfarth, *Nucl. Instrum. Meth.* **114**, 125 (1974).

84. H. W. Kraner, *IEEE Trans. Nucl. Sci.* **NS-27**(1), 218 (1980).

85. H. W. Kraner, C. Chasman, and K. W. Jones, *Nucl. Instrum. Meth.* **62**, 173 (1968).

86. P. H. Stelson, J. K. Dickens, S. Raman, and R. C. Trammell, *Nucl. Instrum. Meth.* **98**, 481 (1972).

87. H. W. Kraner, R. H. Pehl, and E. E. Haller, *IEEE Trans. Nucl. Sci.* **NS-22**(1), 149 (1975).

88. R. H. Pehl et al., *IEEE Trans. Nucl. Sci.* **NS-26**(1), 321 (1979).

89. T. W. Raudorf et al., *IEEE Trans. Nucl. Sci.* **NS-31**(1), 253 (1984).

90. C. Chasman, K. Jones, and R. Ristinen, *Nucl. Instrum. Meth.* **37**, 1 (1965).

91. R. L. Bunting and J. J. Kraushaar, *Nucl. Instrum. Meth.* **118**, 565 (1974).

92. R. A. I. Bell, "Tables for Calibration of Radiation Detectors," Australian National University Report ANU-P/606 (1974).

CHAPTER · 13

Other Solid-State Detectors

I. LITHIUM-DRIFTED SILICON DETECTORS

Using silicon of the highest currently available purity, the depletion depth that can be achieved by reverse biasing a conventional silicon diode detector of the type discussed in Chapter 11 is limited to 1–2 mm. If thicker silicon detectors are required, a different approach must be used to fabricate the device. The process of *lithium drifting* can instead be applied to create a region of compensated or "intrinsic" silicon in which the concentrations of acceptor and donor impurities are exactly balanced over thicknesses up to 5–10 mm. When provided with noninjecting electrodes (often called *blocking contacts*, see p. 350), this region then serves as the active volume of a detector. Such detectors are called *lithium-drifted silicon detectors*, and they are given the designation Si(Li). Their thickness is limited only by the distance over which the lithium drifting can be carried out successfully.

The process of compensating impurities by slowly drifting lithium into a semiconductor crystal was originally developed for producing detectors from either germanium or silicon. For two decades, drifted germanium [or Ge(Li)] detectors represented the only configuration in which a relatively large active volume could be provided in germanium. More recently, detectors of equivalent volume have become available in the high-purity germanium (or HPGe) configuration, and lithium drifting in germanium is no longer popular. However, the process remains important in silicon because its available material purity has not equaled that of germanium.

The lower atomic number of silicon ($Z = 14$) compared with that of germanium ($Z = 32$) means that for typical gamma-ray energies its photoelectric cross section is lower by about a factor of 50 (see Chapter 2). The gamma-ray peak efficiency for silicon detectors is therefore very low, and consequently they are not widely used in general gamma-ray spectroscopy. However, there are two main application areas in which the lower atomic number of silicon is not a hindrance but a help. One area involves the detection of very-low-energy gamma rays or X-rays, where the probability for photoelectric absorption can be reasonably high, even for silicon detectors of a few millimeters in thickness. Because of some advantages of silicon over germanium in this application, Si(Li) detectors have become the most common choice for low-energy photon spectrometry (LEPS) systems. These advantages include less prominent X-ray escape peaks from silicon, and the fact that the greater transparency of silicon for high energy gamma rays is actually helpful when low-energy X-rays must be measured in the presence of a gamma ray background. The second common application area is the detection and spectroscopy of beta particles or other externally incident electrons. Here the lower atomic number of silicon is a distinct advantage because fewer electrons will backscatter from the detector without depositing their full energy. Both of these application areas are more fully described later in this chapter.

The semiconductor properties of the two materials favor silicon over germanium to some extent. Its larger bandgap assures that the thermally generated leakage current at any given temperature will be smaller per unit volume in silicon than in germanium. The energy required to create an electron–hole pair and the Fano factor are not greatly different in the two materials, so the statistical contribution to energy resolution is nearly the same. If charge trapping is negligible, the energy resolution using equivalent electronic components may be better in silicon because of its lower leakage current.

In thin silicon surface barrier detectors, this bulk generated leakage current is not normally a significant contributor to the noise. With Si(Li) detectors, however, the compensated region is sufficiently thick so that at room temperature the fluctuations in this leakage current can be a significant noise source. As a result, nearly all low-noise applications are carried out by cooling the detector to liquid nitrogen temperature in much the same manner as described in Chapter 12 for germanium detectors. Under these conditions, overall noise contributions of less than 200 eV are quite common in commercially available Si(Li) detectors. Manufacturers generally recommend that these systems be continuously maintained at low temperature to prevent unnecessary thermal stresses due to temperature gradients, and to avoid the possibility of gradual redistribution of the drifted lithium at room temperature. In contrast with Ge(Li) detectors, the lithium migration at room temperature in Si(Li) detectors is slow enough that they can usually survive room temperature storage (without applied voltage) for days or weeks without performance degradation.

A. The Ion Drift Process

In both silicon and germanium, the material with highest available purity tends to be *p*-type, in which the best refining processes have left a predominance of acceptor impurities. Donor atoms must therefore be added to the material to accomplish the desired compensation. The alkali metals such as lithium, sodium, and potassium tend to form interstitial donors in crystals of silicon or germanium. The ionized donor atoms that are created when the donated electron is excited into the conduction band are sufficiently mobile at elevated temperatures so that they can be made to drift over periods of days under the influence of a strong electric field. Of the examples mentioned above, only lithium can be introduced into silicon or germanium in sufficient concentration to serve as a practical compensating dopant.

The fabrication process begins by diffusing an excess of lithium through one surface of the *p*-type crystal so that the lithium donors greatly outnumber the original acceptors, creating an *n*-type region near the exposed surface. The resulting *p-n* junction is then reverse biased while the temperature of the crystal is elevated to enhance the mobility of the ionized lithium donors. The lithium ions are slowly drawn by the electric field into the *p*-type region where their concentration will increase and approach that of the original acceptor impurities. At a typical drifting temperature of 40°C, several days or weeks are required for adequate results. A remarkable feature of the drifting process is that a nearly exact compensation must automatically take place because the lithium distribution in the drifted region tends toward a state in which the total space charge is zero at every point.

An equilibrium is thereby established in which the lithium ion continues to drift an increasing distance into the *p*-type region, but in which the only net change in lithium concentration occurs at the boundaries. It has been demonstrated[1] that any departure from exact compensation is unstable as the ions are drifting, and any imbalance in lithium concentration is quickly eroded away until a uniform concentration is reached. Through this drifting procedure, it is possible to achieve compensated regions that extend up to 5–10 mm in silicon and 10–15 mm in germanium.

Detailed analyses of the lithium-drifting process[2-4] demonstrate that the presence of thermally excited electron–hole pairs present during the drift contribute to the net space charge and can upset the exact compensation between lithium donor and acceptor impurities. The fabrication therefore normally takes place in two steps, in which a relatively long duration "clean-up" drift follows the primary drifting process. The temperature of the second step is kept much lower than that of the first so that the thermally excited carriers exert less influence, and a gradual redistribution of the lithium can take place to again approach near-perfect compensation. However, because the operating temperature of the detector is always considerably lower than the clean-up temperature, the potential for some residual charge imbalance and imperfect compensation remains. The resulting effects can be deleterious to the energy resolution of the detector, in that the electric field in some parts of the crystal may be weaker than would otherwise be expected.[3] With careful fabrication techniques, the level of uncompensated impurities in the drifted region can be kept below $10^9/cm^3$.

Lithium ions can be drifted into both silicon and germanium crystals to produce detectors. The lithium ion mobility is much greater in germanium and remains high enough at room temperature to permit an undesirable redistribution of the lithium from the compensated situation achieved during the drift. The lithium profile must therefore be preserved immediately after the drift in germanium by drastically reducing the crystal temperature, typically to that of liquid nitrogen (77 K). In silicon, the ion mobility is low enough at room temperature to permit temporary storage of lithium-drifted silicon detectors without cooling.

B. The *p-i-n* Configuration

Once the drifting process is completed, the resulting detector has the simplified configuration shown in Fig. 13-1. The excess lithium at the surface from which the drift was started serves to convert that surface into an n^+ layer that can be used as an electrical contact. The noncompensated *p* region at the opposite side is often given a metallic coating that acts as an ohmic contact.

The lifetime of charge carriers created within the compensated region (often simply called the intrinsic or *i* region) can be substantially greater than the time required to collect them at either boundary, and therefore good charge collection properties can result. It is necessary to collect the charges quickly, however, and thus substantial voltages are normally applied to assure that few electrons or holes are lost before collection is complete. Typical bias voltages are 500–4000 V across intrinsic regions of 5–10 mm thickness.

Because ideally, no net charge exists in the *i* region, the resulting electric potential predicted by Eq. (12-6a) for $\rho = 0$ varies linearly across the simple planar configuration illustrated in Fig. 13-1. The electric field will therefore be uniform across the *i* region. Because the resistivity of the compensated material is much higher than either the *p* or n^+ regions, virtually all the applied voltage appears across the *i* region, and the electric

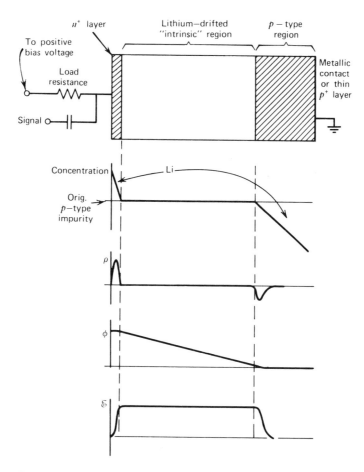

Figure 13-1 Basic configuration of a lithium-drifted *p-i-n* junction detector. Also shown are the corresponding profiles for impurity concentration, charge density ρ, electric potential φ, and electric field ℰ.

field drops sharply to zero at its boundaries. Therefore, the *i*-region dimensions determine the active volume of the detector, and the migration of charge carriers to the *p-i* and *i-n* boundaries gives rise to the basic signal pulse.

C. Electric Field and Pulse Shape

Most Si(Li) detectors are fabricated in planar geometry in which the lithium is drifted in from the flat surface of a thick silicon wafer. Although some large-volume coaxial configurations have been produced, the common applications of Si(Li) detectors to the measurement of soft X-rays or electrons benefit more from the large flat entrance surface that the planar configuration provides. In planar geometry, the magnitude of the constant electric field is

$$\mathscr{E} = \frac{V}{d} \tag{13-1}$$

where V is the applied voltage and d is the thickness of the i region. This simple result is

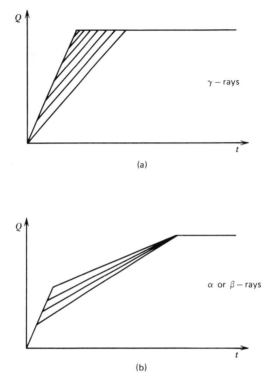

(a)

(b)

Figure 13-2 Simplified representation of the signal pulse leading edge for a Si(Li) detector. Variations arise from differences in the position of interaction within the detector active volume. (From Moszynski et al.[5])

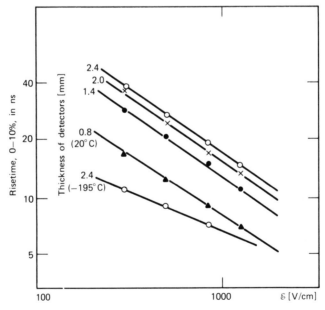

Figure 13-3 Observed rise time (see text for definition) for several Si(Li) planar detectors of different thicknesses. Bottom curve is for 77 K temperature; others are for 293 K. (From Moszynski et al.[5])

predicated on the assumption that the lithium compensation is perfect, so that no net charge exists in the i region. In practical Si(Li) detectors, departure from ideal compensation may introduce sufficient space charge to distort significantly the shape of the electric field from that described by this simple model (see below).

As in germanium detectors, the time profile $Q(t)$ of the induced charge at the detector electrodes depends on the position in the i region at which the charge carriers are formed. These variations are directly reflected in the shape of the leading edge of the pulses shown in Fig. 13-2. The variations in pulse shape are most severe for X-ray or gamma-ray irradiation of the detector, because then the interaction positions are randomly distributed throughout the detector volume. When a charged particle is incident on one face of the detector, the variation in pulse shape is usually somewhat less severe.

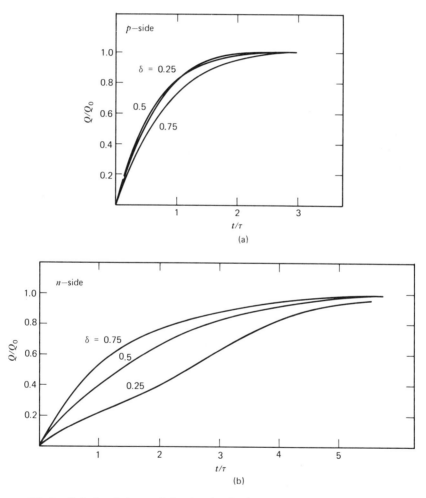

Figure 13-4 Calculated shape of the signal pulse leading edge for electrons incident on a Si(Li) detector. Various assumed electron ranges are shown, defined as $\delta \equiv$ electron range/detector thickness. The time scale is in units of $\tau \equiv d/\mu_e \mathscr{E}_0$, where $d =$ detector thickness, μ_e is the electron mobility, and $\mathscr{E}_0$ is the maximum electric field whose shape is assumed to be given by Eq. (13-2). (a) Electrons incident on the p side (high-field side) of the detector. (b) Electrons incident on the n side (low-field side) of the detector. (From Moroz and Moszynski.[7])

Figure 13-3 shows measured rise times for several different Si(Li) detectors using beta particle excitation. The rise time was measured for the initial portion of the pulse and was defined as the time required for the pulse to change by 10% of the final amplitude. The time resolution (see Chapter 17) achievable with Si(Li) detectors depends both on the average rise time of the pulses and the spread in this rise time due to the effects mentioned above. In this same study,[5] time resolutions of a few nanoseconds are quoted for beta particle excitation. Somewhat poorer time resolution should be expected for pulses from X-rays or gamma rays.

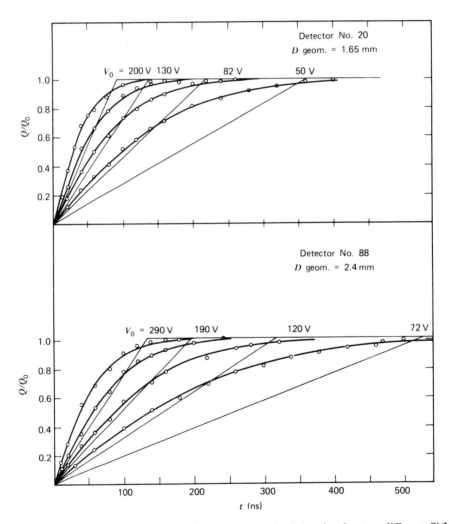

Figure 13-5 Observed shape of the signal pulse leading edge for two different Si(Li) detectors. In both cases, short-range alpha particles were incident on the p side of the planar detector. If the electric field were constant, the predicted pulse shape for various applied voltage values would be the linear waveforms shown. The actual data points are shown fitted by a curve of the form $Q/Q_0 = tgh(t/\tau)$, where τ is treated as an adjustable parameter. Since this behavior is predicted for a nonuniform electric field with parabolic shape, the data give further evidence for the validity of Eq. (13-2). (From Moszynski and Przyborski.[8])

The output pulse shape can also change if the electric field is not uniform throughout the active volume of the detector. The electric field is expected to be uniform in Si(Li) planar detectors only if the compensation of donors and acceptors within the active volume is perfect. It has been shown,[5,6] however, that inhomogeneities in the compensation of impurities can lead to an electric field that decreases strongly near the n side of the detector. If the residual space charge due to uncompensated impurities varies linearly across the active volume, then Poisson's equation [Eq. (12-2)] predicts a quadratic shape for the electric field. Moroz and Moszynski[7] showed that a quadratic variation of the form

$$\mathscr{E}(x) = \mathscr{E}_0\left(1 - \frac{x^2}{d^2}\right) \tag{13-2}$$

where

$\mathscr{E}_0$ = electric field strength at the negative electrode

x = distance from the negative electrode

d = thickness of the detector active volume

adequately predicted pulse shapes observed for monoenergetic electrons incident on both sides of the detector. Some of the resulting pulse shapes are shown in Fig. 13-4. Pulse shapes for alpha particle excitation from the p side of another Si(Li) detector are shown in Fig. 13-5.

As in the case of germanium detectors, "defective" pulses are sometimes observed in Si(Li) detectors from regions of poor charge collection (often near the boundaries of the

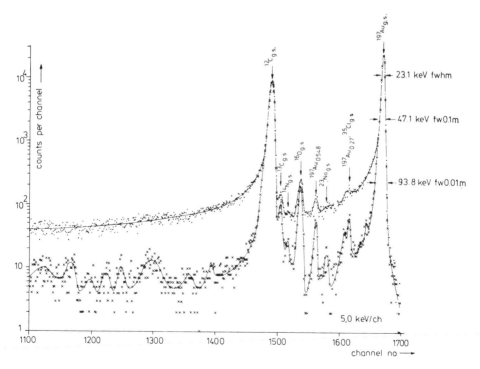

Figure 13-6 Comparison of pulse height spectra from an Si(Li) detector with (×) and without (●) rejection of pulses with rise times greater than 70 ns. The source was a proton beam scattered from several target nuclides. (From Lindstrom and Lisdat.[9])

active volume). These defective pulses have both a reduced amplitude and an unusually long rise time, because regions in which the effective charge collection is slow will also exhibit a high trapping probability. Figure 13-6 illustrates that many of these undesirable pulses may be rejected from an amplitude spectrum by excluding all signal pulses whose rise time exceeds a certain upper limit.

D. Low-Energy Photon Spectroscopy

1. RESPONSE FUNCTION

In silicon, the photoelectric process is more probable than Compton scattering for photon energies below about 55 keV (see Chapter 2). For energies below about 30 keV, photoelectric absorption is predominant, so that the response function of an Si(Li) detector to gamma rays below this energy is dominated by a single full-energy peak caused by absorption of the resulting photoelectron. If the electric field in the detector is sufficiently high, charge collection is complete and the shape of the full-energy peak can often be represented adequately by a Gaussian, although slight modifications of the Gaussian shape have been shown to be beneficial in exacting cases.[10]

Some photoelectric absorptions cause the prompt emission of a characteristic X-ray as the resulting vacancy in the electron shell is filled. At low incident gamma-ray energies, most of the absorptions occur near the surface of the detector where escape of the resulting X-ray can be significant. The result will be the appearance of a small X-ray escape peak in the response function, which in silicon will be located 1.8 keV below the full-energy peak (see Fig. 13-7). Figure 13-8 shows the intensity of the escape peak for a planar Si(Li) detector. The corresponding escape peak in a germanium detector spectrum will be much more prominent because of the smaller average penetration distance of the incident radiation and the greater escape probability of the Ge characteristic X-rays due

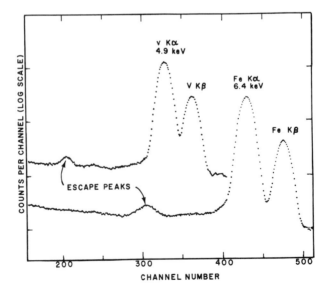

Figure 13-7 Upper portion of pulse height spectra of K X-rays from the decay of ^{51}Cr and ^{57}Co. Peaks caused by the escape of the Si X-ray from the detector are evident in both spectra. (From Wood et al.[11])

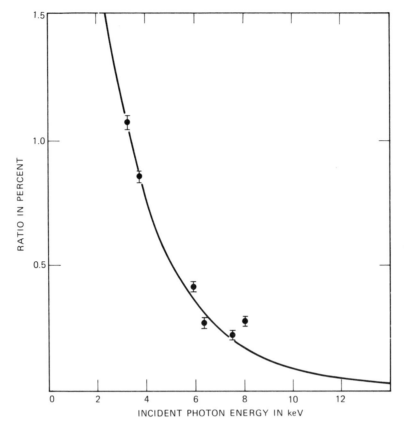

Figure 13-8 The ratio of the area under the X-ray escape peak to the area under the full-energy peak for a planar Si(Li) detector. Experimental points are shown together with a curve representing a theoretical prediction for a parallel beam incident on a semi-infinite slab of silicon. (From Dyson.[12])

to their higher energy (11 keV). The fluorescent yield in Ge (50%) is also much larger than that in Si (5%), so that the probability of emission of a characteristic X-ray is higher.

At higher gamma-ray energies, a significant contribution of Compton scattering adds a continuum to the spectrum. Figure 13-9 shows the shape of the Compton continuum for an incident photon of 140.5 keV energy. Also shown are the computed shapes of the electron energy distributions from single Compton scattering assuming both free electrons and bound electrons in silicon. It is evident that the binding effects must be taken into account if the shape of this continuum is to be described adequately. Because of the low photoelectric cross section in silicon, Si(Li) detectors are seldom used to measure gamma rays with energy above about 150 keV.

2. LINEARITY AND ENERGY RESOLUTION
The response of Si(Li) detectors to low-energy X-rays and gamma rays has been shown to be very linear, provided the applied voltage is high enough (approximately 250 V/mm) to avoid significant loss of charge due to trapping and recombination. In Fig. 13-10, the observed channel number for the photopeak is plotted versus X-ray energy for a number of characteristic X-rays produced by proton excitation.[14] The maximum possible nonlinearity is quoted as 1% over this low-energy range, where many potential causes of nonlinearity are most significant.

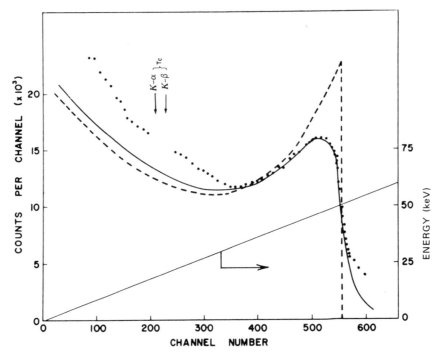

Figure 13-9 Calculated and observed shape of the Compton continuum for a ^{99m}Tc source (140.5 keV). The dashed curve is predicted based on scattering of free electrons, whereas the solid curve is calculated taking into account the finite electron binding energy in silicon. The points are experimental data. (From Felsteiner et al.[13])

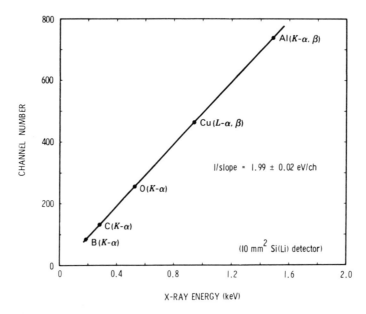

Figure 13-10 Experimentally determined linearity between pulse height and X-ray energy for an Si(Li) detector. (From Musket.[14])

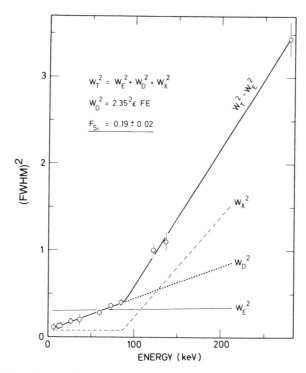

Figure 13-11 Contribution of various noise sources to the total width of the full-energy peak from an Si(Li) photon spectrometer. For definitions of the various terms, see the text. (From Hollstein.[15])

The resolution and noise behavior of a high-resolution Si(Li) detector have been studied by Hollstein[15] for energies between 5 and 280 keV. The results are shown in Fig. 13-11. As in germanium systems, the observed full width at half maximum of the full-energy peak W_T can be expressed as the quadrature sum of a number of independent components:

W_E—the FWHM-equivalent noise that can be attributed to the electronic components of the signal-processing chain.

W_D—the FWHM-equivalent spread due to the charge generation statistics within the detector. It is assumed that $W_D^2 = 2.35^2 \epsilon FE$, where ϵ is the energy necessary to create one electron–hole pair, F is the Fano factor, and E is the photon energy.

W_X—the FWHM equivalent attributable to detector leakage current and any charge collection problems within the detector.

Improvements in detector resolution have allowed FWHM values as low as 100 eV for state-of-the-art detectors of small surface area. Commercially available Si(Li) detectors with surface areas as large as 300 mm² can be obtained with an energy width of 255 eV measured at the 5.9 keV X-rays from ^{55}Fe (Ref. 16). Improvements in energy resolution have come about through the use of a double guard ring structure to reduce surface leakage currents, reduction of preamplifier noise by cooling the input FET, and the use of low-noise pulsed optical feedback within the preamplifier (see Chapter 17). Careful design of the detector–preamplifier first-stage coupling is also required to prevent additional broadening from microphonics (see p. 609) caused by mechanical vibrations

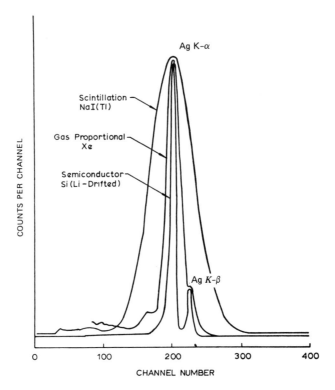

Figure 13-12 A comparison of the pulse height spectra from incident silver K-series X-rays as recorded by three different detectors with varying energy resolution. The K-α and K-β energies are approximately 21 and 25 keV. (From Muggleton.[18])

and/or acoustical noise. Loss of resolution from this cause has been shown[17] to be significant for some commercial systems.

The comparative energy resolution of three different detector types commonly applied to X-ray spectroscopy is shown graphically in Fig. 13-12. The generally superior energy resolution of Si(Li) detectors in this application allows a much better separation of the X-ray energy components compared with either a proportional counter or scintillation detector.

3. DETECTION EFFICIENCY

In principle, the full-energy peak efficiency of any Si(Li) detector can be determined from knowledge of its size and shape, together with the appropriate gamma-ray interaction cross sections. In practice, however, manufacturers often give only nominal values for detector dimensions, and these may be somewhat uncertain due to the effects of incomplete charge collection near the edges of the active volume. Furthermore, the detector itself is usually mounted inside a vacuum cryostat to permit its cooling to liquid nitrogen temperature. The exact spacing between source and detector is then sometimes difficult to determine. For these reasons, it is almost always necessary to carry out an experimental determination of the detector efficiency using calibrated sources, if reasonably accurate data are required. The task is complicated by the fact that there are relatively few standard sources available in the energy range of 5–50 keV, which is of primary interest in Si(Li) detectors.

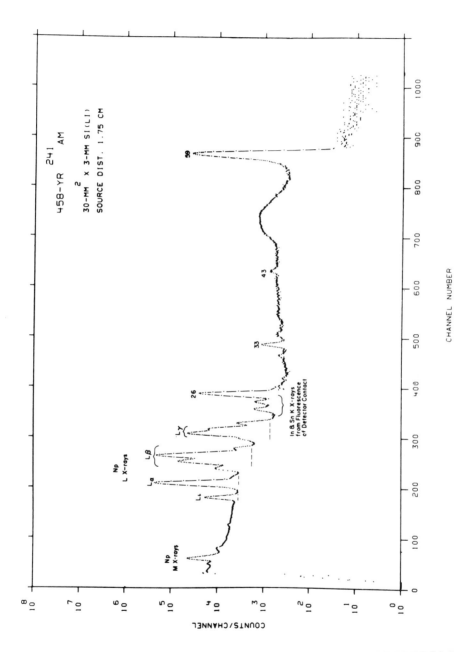

Figure 13-13 Photon spectrum from ^{241}Am obtained using a 30 mm^2 × 3 mm Si(Li) detector. (From Gehrke and Lokken.[20])

TABLE 13-1   Photon Intensities per Disintegration of ^{241}Am

Line	Energy (keV)	Percentage per Disintegration
L-α	13.9	13.3 ± 0.4
L-$\eta\beta$	17.8	19.4 ± 0.6
L-γ	20.8	4.9 ± 0.2
γ	26.35	2.4 ± 0.1
γ	59.54	35.82 ± 0.12

Data from Campbell and McGhee.[19]

One suitable calibration source is the alpha-active nuclide ^{241}Am, with a 458 year half-life. Two relatively intense gamma rays are emitted with energies of 26.35 and 59.54 keV, together with a number of characteristic L X-rays from the Np decay product. Data on relative yields of these photons are listed in Table 13-1, and an Si(Li) spectrum is shown in Fig. 13-13. A calibrated ^{241}Am source will therefore provide efficiency calibration points over the very important energy range of about 10–60 keV.

If absolutely calibrated sources are unavailable, an alternative approach can be used. The *X/gamma method*[21] makes use of radioisotopes that emit a dominant high-energy gamma ray, together with K X-rays from internal conversion or electron capture. The X-ray intensity can be determined absolutely from decay data if the gamma-ray intensity is somehow measured in an absolute manner. The gamma-ray measurement can be carried out to a high degree of accuracy at many laboratories, using standard NaI(Tl) scintillators or germanium detectors. With the data given in Table 13-2, these sources then provide X-ray calibration points with a similar absolute accuracy. The presence of the high-energy gamma ray is seldom a problem with Si(Li) detectors of typical dimensions because their probability of interaction is negligibly small.

As with other solid-state detectors, there is a need to repeat efficiency calibrations periodically because there is good evidence that the efficiency may change substantially over periods of weeks or months. These efficiency variations, although not fully understood, probably arise due to variations in charge collection efficiency and changes in dead layer thicknesses.

The energy-dependent efficiency of a representative Si(Li) detector is plotted in Fig. 13-14. Over a substantial range of incident energy (from about 7 to 20 keV in the example shown) the intrinsic peak efficiency of the detector is very close to 100%. This is the energy interval in which virtually all incident photons are absorbed in the detector by simple photoelectric absorption. The absolute counting efficiency then depends primarily on the surface area of the detector. As a result, essentially all low-energy photon detectors of this type are fabricated in planar geometry to maximize the surface area. At higher energies, the peak efficiency begins to fall off rapidly, reflecting the similar fall-off of the photoelectric cross section in silicon. Thicker detectors will help somewhat, but for all practical purposes, planar Si(Li) detectors are limited to energies below about 100 keV.

The efficiency at low energies is critically dependent on the thickness of window materials or dead layers associated with the detector and cryostat. (For very soft radiations, even a few centimeters of air between the source and detector can lead to significant attenuation.) In Fig. 13-14, the individual transmission efficiencies of typical window and dead layer values are shown separately, together with the resulting composite efficiency for the detector. The discontinuity at about 1.8 keV is a result of the K absorption edge of the silicon dead layer assumed for this detector. In practice, the dead

TABLE 13-2 Nuclides Suitable for Use in the X/γ Calibration Method

Nuclide	Half-Life (days)	X-Ray or Low-Energy Gamma	Energy (keV)	Energy of High-Energy Gammas (keV)	Intensity Ratio X/γ	
^{51}Cr	27.710	K-α	4.95	320.11	2.018	$\pm$ 0.021
		K-β	5.43		0.274	$\pm$ 0.0056
^{54}Mn	312.14	K-α	5.412	834.83	0.2234	$\pm$ 0.0011
		K-β	5.95		0.0305	$\pm$ 0.00034
^{57}Co	271.80	K-α	6.397	122.063	0.5863	$\pm$ 0.016
		K-β	7.06		0.0807	$\pm$ 0.0024
		γ	14.41		0.1086	$\pm$ 0.0023
^{65}Zn	244.0	K-α	8.041	1115.55	0.6790	$\pm$ 0.0044
		K-β	8.91		0.0946	$\pm$ 0.00077
^{75}Se	119.76	K-α	10.327	264.65	0.8273	$\pm$ 0.013
		K-β	11.72		0.1278	$\pm$ 0.0021
^{85}Sr	64.851	K-α	13.375	514.00	0.5035	$\pm$ 0.0035
		K-β	15.86		0.089	$\pm$ 0.00077
^{88}Y	106.62	K-α	14.142	898.03	0.5425	$\pm$ 0.0035
		K-β	16.765		0.0986	$\pm$ 0.00081
^{109}Cd	462.7	K-α	22.10	88.036	22.47	$\pm$ 0.58
		K-β'_1	24.93		4.022	$\pm$ 0.107
		K-β'_2	25.46		0.706	$\pm$ 0.023
^{113}Sn	115.09	K-α	24.14	391.69	1.225	$\pm$ 0.009
		K-β'_1	27.27		0.220	$\pm$ 0.005
		K-β'_2	27.86		0.0414	$\pm$ 0.00087
^{137}Cs	10964	K-α	32.06	661.63	0.0663	$\pm$ 0.0008
		K-β'_1	36.4		0.0127	$\pm$ 0.00016
		K-β'_2	37.3		0.00322	$\pm$ 0.00008
^{133}Ba	3841	K-α	30.85	356.0	1.604	$\pm$ 0.023
		K-β'_1	35.0		0.374	$\pm$ 0.006
		K-β'_2	35.8			
		γ	53.2		0.0352	$\pm$ 0.006
^{139}Ce	139.69	K-α	33.30	165.85	0.792	$\pm$ 0.011
		K-β'_1	37.8		0.151	$\pm$ 0.004
		K-β'_2	38.7		0.0388	$\pm$ 0.0009

Data from Campbell and McGhee.[19]

layer thickness can sometimes be determined by measuring the relative intensities of two or more X-rays emitted by a single source for which the emission yields are well known[23] or from separate sources with known absolute yields.[24]

A number of semiempirical relations have been used to serve as a functional fit to interpolate between measured efficiency points. One of these of the form

$$\epsilon = \Omega \exp(\alpha E^\beta)\left[1 - \exp(\gamma E^\delta)\right] \qquad (13\text{-}3)$$

was suggested by Gallagher and Cipolla[22] in which the five parameters α, β, γ, δ, and Ω are fit to the experimentally measured efficiency points. The applicability of Eq. (13-3) has been demonstrated for several different irradiation conditions,[22,25] and a least-square fit can normally be made to within the uncertainty of the data points themselves.

The intrinsic peak efficiency for four common types of detector used in low-energy photon spectroscopy is plotted in Fig. 13-15. The relative radiation transparency of the xenon proportional counter makes its efficiency drop off most rapidly with energy. Furthermore, above its K-edge energy of 34 keV, photoelectron absorption in the xenon

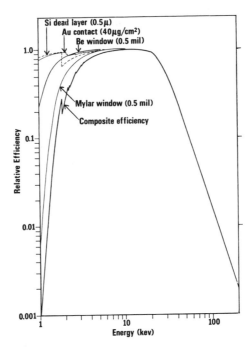

Figure 13-14 Calculated full-energy peak efficiency for a 3 mm thick Si(Li) detector. (From Gallagher and Cipolla.[22])

results in a characteristic X-ray with a fluorescent yield of 81%. This X-ray stands a very good chance of escaping from the detector without further interaction, and the pulse does not fall within the full-energy peak in the spectrum. Therefore, the photopeak efficiency of the detector does not follow the abrupt increase in the attenuation coefficient at this energy and actually shows a slight drop.

The efficiency for silicon is comparatively good over most of the X-ray range, and Si(Li) detectors are widely applied to this energy region. Characteristic escape effects in silicon are not serious because of its low K-edge energy and small fluorescent yield of 5%. The corresponding slight dip in the photopeak efficiency at 1.8 keV is barely perceptible in Fig. 13-15. Both germanium and sodium iodide, in reasonable thicknesses, are essentially opaque to electromagnetic radiation below 100 keV. The photopeak efficiencies also show the effect of characteristic X-ray escape, and the fluorescent yields in germanium and iodine are 50 and 88%.

4. APPLICATION IN FLUORESCENCE SPECTROSCOPY

Commercially available low-energy photon spectrometers (LEPS) consisting of a cooled Si(Li) planar detector, preamplifier, and liquid nitrogen dewar have come into widespread application for the measurement of X-ray spectra in the 1–50 keV energy region. A typical system is shown in Fig. 13-16. For the ultimate in low-noise performance, the input stages of the preamplifier will also be cooled to reduce electronic noise.

These systems are widely applied for the analysis of materials by means of the characteristic X-rays emitted following excitation by one of several means. If the sample is irradiated by X-rays with a photon energy above the K-shell binding energy of a specific element, photoelectric absorption will induce characteristic X-ray emission. This

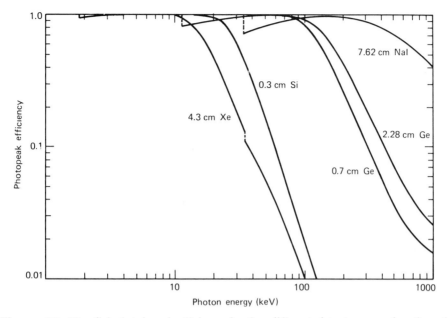

Figure 13-15 Calculated peak efficiency for five different detectors as a function of incident X- or gamma-ray energy. The thicknesses in the direction of the incident radiation are indicated on the figure. The detectors are a xenon-filled proportional counter, an Si(Li) detector, two different germanium detectors, and an NaI(Tl) scintillator. (From Israel et al.[26])

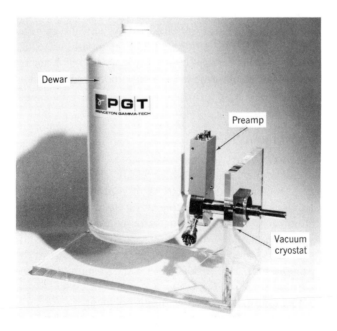

Figure 13-16 A typical low-energy photon spectrometer system consisting of a Si(Li) detector within a vacuum cryostat, cooled by a liquid nitrogen dewar. (Courtesy of Princeton Gamma-Tech, Princeton, NJ.)

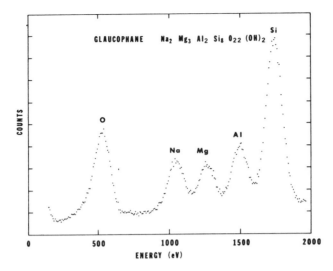

Figure 13-17 X-ray spectrum produced by electron excitation of the mineral glaucophane. All the elemental constituents (except hydrogen) are identifiable as peaks in the spectrum. (From Jaklevic and Goulding.[27])

process is called *X-ray fluorescence*. Other means of excitation can include alpha, electron, or other charged particle bombardment. Separate elements within the sample then can be identified by observing peaks in the pulse height spectrum corresponding to the characteristic X-rays of that particular element.

As an example, Fig. 13-17 shows the X-ray spectrum of a sample that has been excited by electron bombardment. Because the characteristic X-ray energies of light elements are quite small, a premium is placed on system energy resolution in order to resolve separate contributions from adjacent low-Z elements. Detection of characteristic X-rays for elements as light as boron (183 eV K-α energy) has been reported.[14] Many of the instrumental problems of importance in X-ray fluorescence analysis are discussed in the review by Goulding.[28]

E. Electron Spectroscopy

Lithium-drifted silicon detectors are also well suited to the measurement of electron energies. Thicknesses can be obtained which easily exceed the maximum distance of penetration of beta particles or other electrons of routine interest. The relatively low Z value of silicon also assures that backscattering of electrons incident on the face of the detector will be minimized. Because the detector is usually enclosed within a vacuum cryostat, electrons from external sources must penetrate the cryostat entrance window and may lose a significant amount of energy in the process. For low-energy electron spectroscopy, the sample must be introduced into the vacuum envelope to avoid window losses and air absorption (e.g., see Ref. 29).

The response of silicon detectors to normally incident monoenergetic electrons with energies between 0.15 and 5.0 MeV has been investigated both experimentally and through the use of a computational model by Berger et al.[30] A typical set of pulse height spectra is shown in Fig. 13-18. When the thickness of the detector is considerably larger

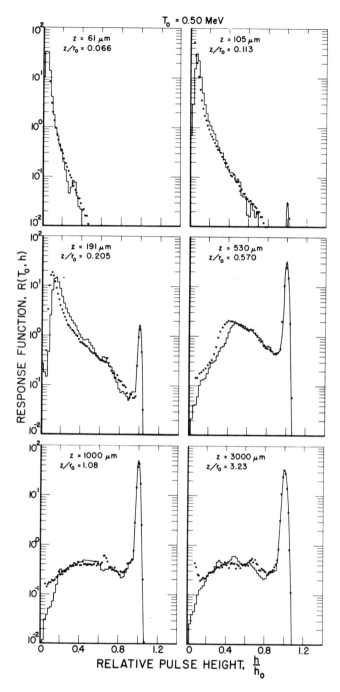

Figure 13-18 Experimental and calculated response functions for silicon detectors for 0.50 MeV electrons. The experimental data are shown as points, whereas the continuous plot is the result of a Monte Carlo calculation. z is the detector thickness, and z/r_0 is the ratio of this thickness to the mean range of the electrons. (From Berger et al.[30])

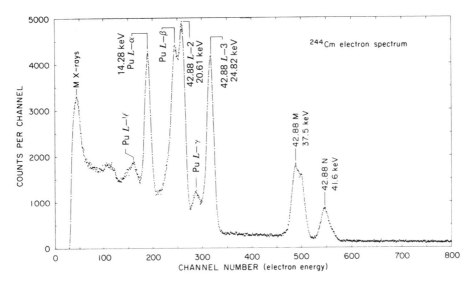

Figure 13-19 Conversion electron spectrum of a ^{244}Cm sample determined using a 80 mm^2 × 3 mm Si(Li) detector. (From Ahmad and Wagner.[29])

than the maximum penetration distance of the electron, the response function consists of a full-energy peak plus a continuum of lower amplitude events. The full-energy peak corresponds to all those electrons that are fully stopped within the active volume of the detector, and for which bremsstrahlung photons generated along its path are fully absorbed within the detector. The continuum[31] represents partial energy loss events, which can correspond to either the backscattering of electrons that reemerge from the incident surface without having lost all their energy, or the escape of bremsstrahlung photons generated along the track of an electron that otherwise may be fully stopped. As the detector is made thinner, some electrons begin to penetrate fully through the detector without losing their entire energy. The corresponding pulses shift from the full-energy peak to lower amplitudes in the spectrum. For very thin detectors, few electrons are fully absorbed and the pulse height spectrum corresponds to the energy loss distribution of electrons before they escape from the active volume. Numerical tables of response functions of the type shown in Fig. 13-18 can be found in Ref. 32 for detector thicknesses from 0.05 to 10.0 mm and electron energies from 0.15 to 5 MeV.

A Si(Li) detector whose intrinsic thickness exceeds that of the maximum penetration distance of electrons of interest makes a very suitable electron spectrometer. Figure 13-19 shows a recorded spectrum in which peaks appear due to monoenergetic electrons from the internal conversion process.

For electrons that are incident at oblique angles to the surface, a larger fraction are scattered back out of the detector before losing all their energy. As a result, the continuum in the response function is enhanced relative to the case of normal incidence. In one measurement[33] in which many angles were sampled by placing the detector close to the source to subtend a solid angle of 25% of 2π, the fraction of all events falling under the continuum rose to 65% compared with about 20% for small solid angles. If the source intensity permits, it is obviously of advantage to restrict the incident electrons to near-normal incidence only.

II. SEMICONDUCTOR MATERIALS OTHER THAN SILICON OR GERMANIUM

The large majority of semiconductor radiation detectors in current use are manufactured either from silicon or germanium. The widespread popularity of these materials is attributable to their excellent charge transport properties, which allow the use of large crystals without excessive carrier losses due to trapping or recombination. Detectors of practical size can therefore be manufactured in which virtually all electron–hole pairs created by the incident radiation are collected to make up the basic signal pulse.

Neither silicon nor germanium is ideal from certain standpoints. Germanium detectors must always be operated at low temperatures to reduce thermally generated leakage current. In low-noise applications, such as X-ray spectroscopy, silicon detectors must also be cooled for the same reason. In principle, a different semiconductor material with a wider bandgap (e.g., greater than 1.5 eV) could reduce the bulk-generated leakage current so that use at room temperature would be possible. In many applications, the convenience of room temperature operation would likely outweigh the disadvantages of a wider bandgap, such as the greater energy required to create an electron–hole pair (see Fig. 13-20).

In gamma-ray spectroscopy, detectors with a high Z value are at a premium. In this regard, germanium ($Z = 32$) is a great improvement over silicon ($Z = 14$), but many other elements would obviously be even better. A great deal of attention has therefore

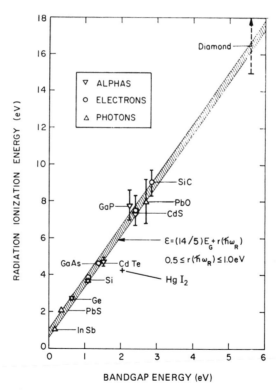

Figure 13-20   The average energy required to form one electron–hole pair (ϵ) versus bandgap energy for a number of semiconductor materials. (From Klein.[34])

TABLE 13-3   Properties of Semiconductor Materials

Material	Z	Density (g/cm^3)	Bandgap (eV)^a	Energy per Electron–Hole Pair (eV)^a	Best Gamma-Ray Energy Resolution (FWHM)^b
Si (300 K)	14	2.33	1.12	3.61	—
Ge (77 K)	32	5.32	0.74	2.98	420 eV at 100 keV 920 eV at 661 keV 1300 eV at 1330 keV
CdTe (300 K)	48–52	6.06	1.47	4.43	3500 eV at 122 keV 8000 eV at 661 keV
HgI$_2$ (300 K)	80–53	6.30	2.13	4.22	650 eV at 5.9 keV 2500 eV at 122 keV

^{a} Data from Cuzin.[35]

^{b} Representative resolution figures as tabulated in Refs. 36 and 37.

been focused on seeking other suitable semiconductor materials that incorporate at least one element of high atomic number.

To date, two compound semiconductors have received the most attention as potential room temperature radiation detectors: cadmium telluride (CdTe) and mercuric iodide (HgI$_2$). Some properties of these materials are compared with those of silicon and germanium in Table 13-3.

In order for useful detectors to be fabricated from these semiconductors, some practical requirements must also be met. Foremost is the ability to grow crystals of sufficient size to be interesting as detectors, while maintaining adequate purity. The density of trapping impurities determines the mean charge carrier lifetime, which sets a practical limit on the distance over which charges may be collected with good efficiency. A low net impurity concentration is also required if large depletion depths are to be created within the material [see Eq. (11-13)]. Compound semiconductors are also sensitive to stoichiometric imbalance, which can upset their theoretical behavior.

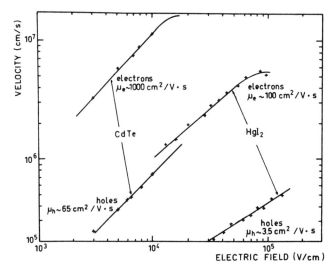

Figure 13-21   Drift velocity as a function of electric field strength for charge carriers in CdTe and HgI$_2$. In both materials, holes are much less mobile than electrons. (From Scharager et al.[38])

Most high-Z semiconductor materials with wide bandgaps that are suitable for operation at room temperature unfortunately tend to have low mobility (particularly for holes) compared with silicon or germanium. In Fig. 13-21, carrier velocities are plotted as a function of electric field for the two most popular such materials, CdTe and HgI_2. Even for large values of the electric field, the drift velocity for holes remains low in both materials. As a result, the effects of trapping and recombination are enhanced, and it is very difficult to achieve complete charge collection over distances greater than 1 mm. The signal pulse shapes and pulse height spectra then become strong functions of detector geometry and/or irradiation conditions.

These difficulties are at least partially offset by the excellent detection efficiency these materials exhibit for X-rays and gamma rays. Figure 13-22 plots their linear absorption

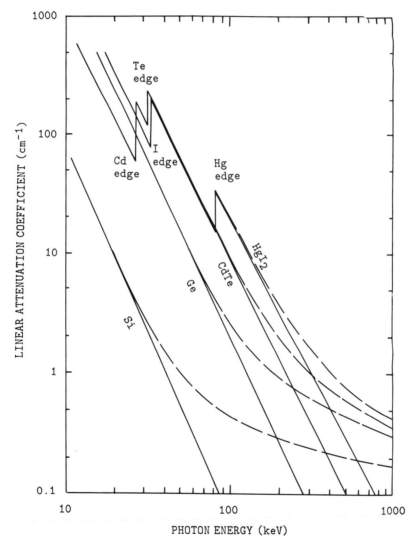

Figure 13-22 Photoelectric (solid lines) and total (dashed lines) linear attenuation coefficients for Si, Ge, CdTe, and HgI_2. K-shell absorption edges are shown. (Prepared by M. R. Squillante, Radiation Monitoring Devices, Inc.)

coefficients together with that of germanium and silicon for comparison. The much higher atomic numbers of cadmium telluride and mercuric iodide allow the use of much thinner detectors for equivalent detection efficiency, and charge transport over large distances can often be avoided.

A. Cadmium Telluride

Cadmium telluride (CdTe) combines relatively high atomic numbers (48 and 52) with a large enough bandgap energy (1.47 eV) to permit room temperature operation. The probability of photoelectric absorption per unit pathlength is roughly a factor of 4–5 times higher in cadmium telluride than in germanium, and 100 times larger than in silicon for typical gamma-ray energies. Applications of this material therefore most often involve situations in which a high gamma-ray detection efficiency per unit volume is at a premium.

High-purity CdTe crystals can be grown using a number of different fabrication techniques. These include growth from solution and, more recently, techniques based on zone refining in which a traveling heater is slowly passed through the stock material. Chlorine is often used as a compensating dopant, resulting in high-resistivity (up to 10^9 ohms · cm) p-type crystals. After slicing into a wafer of the appropriate thickness, metallic or carbon contacts are fabricated on opposite surfaces, and the crystal operated as a surface barrier diode detector. Another configuration that has been demonstrated[39, 40] consists of high-resistivity n-type CdTe with heavily doped p^+ and n^+ surface layers grown by liquid phase epitaxy techniques. This configuration then resembles the p^+-v-n^+ configuration common in high-purity germanium detectors (see Fig. 12-1).

Because of the rather poor collection efficiency for holes, energy resolution achievable in CdTe detectors is generally not comparable with that obtainable in silicon or germanium. Some representative figures are given in Table 13-3. The best energy resolution is obtained for low-energy gamma rays and X-rays where good detection efficiency can be obtained with a detector thickness as small as a few tenths of a millimeter. One example of such a spectrum is shown in Fig. 13-23. The larger thicknesses required for higher gamma-ray energies generally result in poorer spectroscopic performance. Results have been reported[40] for a 2 mm thick detector in which the FWHM for gamma rays from ^{137}Cs and ^{60}Co was 25 keV. This energy resolution was achieved by rejecting a large fraction of the pulses with incomplete charge collection by sensing their longer rise time in a supplementary pulse-processing circuit.

If spectroscopic information is not required, CdTe detectors can be applied effectively to simple pulse counting in a variety of applications in which the high gamma-ray efficiency allows for a very compact detector size. CdTe can also be operated in current mode in high gamma-ray fluxes. This includes the *photovoltaic* mode (see p. 382) in which no external voltage is applied to the detector. In this case, the charge carriers created by gamma-ray interactions are collected through the internal field formed by the contact potential of the junction, much the same as in a solar cell.[42, 43] CdTe detectors have also been used in the pulse mode spectroscopy of energetic charged particles,[44] but with an energy resolution inferior to that of silicon detectors.

A persistent problem in the use of some cadmium telluride detectors is the phenomenon of polarization, which under certain conditions of operation leads to a time-dependent decrease in the counting rate and charge collection efficiency.[45-47] This polarization is related to the capture of electrons by deep acceptors within the material. The resulting

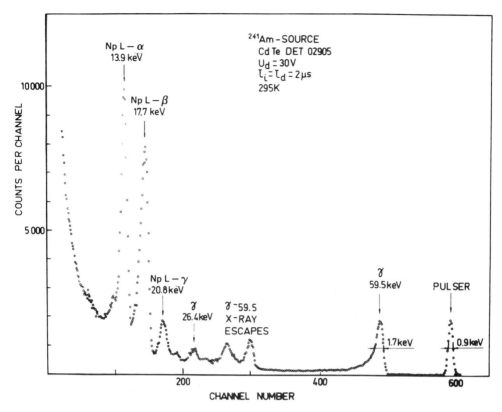

Figure 13-23 Photon spectrum from ^{241}Am obtained using a 1–2 mm diameter $\times 0.1$ mm thick CdTe detector. Emission intensities are given in Table 13-1. A similar spectrum taken with a larger Si(Li) detector is shown in Fig. 13-13. Note the peaks due to the escape of the characteristic K X-rays from Cd (23 keV) and Te (27 keV) following photoelectric absorption of the 59.5 keV gamma ray. (From Dabrowski et al.[41])

buildup of space charge not only interferes with carrier collection but also leads to a gradual decrease in the effective thickness of the depletion region.

Commercially available CdTe detectors range in size from 1 mm to over 1 cm in diameter. They are relatively rugged and stable in field use and can routinely be operated at temperatures up to 30°C without excessive thermal noise. Current mode operation is possible to temperatures up to 70°C. Some reviews of their performance in a variety of applications are given in Refs. 35, 37, 48–51.

B. Mercuric Iodide

Another material that has attracted considerable attention as a potential detector for X-rays and gamma rays is mercuric iodide. HgI_2 is a semiconducting material with a bandgap width of 2.1 eV. This unusually large value permits room temperature operation with very small thermally generated leakage current. Because of the high photoelectric cross section of Hg ($Z = 80$), low-energy gamma-ray interaction probabilities are as much as a factor of 50 larger than those for germanium. In this regard, the material is an

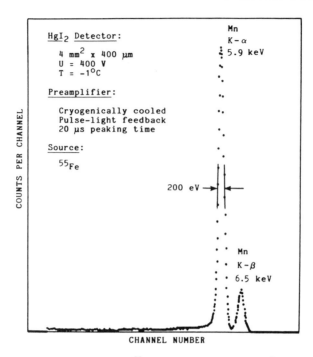

Figure 13-24   X-ray spectrum from ^{55}Fe obtained using a 4 mm^2 × 0.4 mm thick HgI$_2$ detector. (From Dabrowski et al.[55])

even more efficient photon detector than CdTe (see Fig. 13-22). For example, 85% of 100 keV photons are absorbed within a 1 mm thickness.

Several problems have restricted widespread application of HgI$_2$, and the development of detectors from this material remains largely at the research stage. A major problem is the low hole mobility that interferes with efficient charge collection over distances greater than about 1 mm. The buildup of trapped charges also can lead to polarization effects[52] in which the electric field is distorted and efficient charge collection is further disturbed. Also, the material must be encapsulated to prevent deterioration of the crystal surfaces on exposure to air over periods of weeks to months.

Despite these operational difficulties, there has been considerable progress[53-57] toward the development of practical HgI$_2$ detectors. Crystals with a thickness less than 1 mm can show good spectral qualities and have been applied successfully in the measurement of X-rays and low-energy gamma rays. Figure 13-24 shows one example of a low-energy X-ray spectrum taken with a 0.4 mm thick detector with 4 mm^2 area, typical of the sizes currently under evaluation as X-ray spectrometers. The best energy resolution is obtained by cooling the input FET of the preamplifier (to about 140 K), and a FWHM of 175 eV at 5.9 keV has been reported[58] from a 1 mm^2 room temperature detector under these conditions. Room temperature operation of the preamplifier resulted in an increase in the FWHM recorded from the same detector to 380 eV because of the added contribution of thermally generated electronic noise.

For good detection efficiency at higher gamma-ray energies, thicker detectors are needed. It has been possible to produce usable HgI$_2$ crystals with a thickness up to 1.2 cm and a volume of about 10 cm^3. Such detectors can be used directly as simple gamma-ray counters, but the poor hole transport makes the amplitude of the signal dependent on the position of the interaction within the crystal. Peaks resulting from monoenergetic radiations are therefore greatly broadened in the recorded spectrum. Efforts are underway[59-61] to restore the energy resolution in thick detectors through pulse-processing techniques. These methods are based on the principle of separately determining the contribution of electrons and holes to the initial rise of the pulse and then artificially adding a component to the pulse which is proportional to the amount of hole trapping expected for that event. Since the electrons are collected much more rapidly than holes, an estimate of the separate contribution of both carrier types can be made based on the slope of the pulse rise measured before and after the electrons have been collected. These electronic compensation techniques are currently capable of producing energy resolution figures of a few percent for detectors up to 5 mm thick.[60] While this energy resolution is still far inferior to that of germanium detectors, the very high detection efficiency per unit volume of HgI_2 could be an advantage in some applications.

The large bandgap value (2.10 eV) and resulting low leakage current allow such detectors to be operated as simple conductivity counters in which the creation of a depletion region is no longer necessary as in silicon or germanium detectors. Therefore either ohmic or blocking contacts can be used, and the detector no longer has the rectifying properties of a conventional junction or surface barrier detector. (The junction properties are necessary only in these latter materials to reduce the leakage current, which is already inherently low in wide bandgap materials such as HgI_2.) Consequently, these detectors can be operated with either positive or negative polarity voltage applied to either surface contact. The choice of polarity, however, is extremely important if soft radiation is involved that does not penetrate deeply into the detector. In that event, much better performance can be obtained by arranging the polarity so that holes are collected over the smaller distance (toward the entrance window), leaving the much more mobile electrons to traverse the full thickness of the detector.

Although mercuric iodide is primarily of interest as an X-ray or gamma-ray detector, measurements have also been made[62] of its response to charged particles. Because of its high density, HgI_2 has a stopping power for energetic protons of about twice that of silicon, the usual semiconductor material used in charged particle detectors. Nearly linear energy response was observed, with an energy resolution of 5–15% for H and He ions fully stopped in the detectors. While this energy resolution is more characteristic of scintillators than semiconductor detectors, some improvements might be expected by employing techniques mentioned above to compensate for charge loss due to hole trapping.

C. Miscellaneous Materials

Gallium arsenide (GaAs) is another semiconductor material with a bandgap sufficiently wide (1.42 eV) to permit room temperature operation. The atomic numbers of the constituents (31 and 33) bracket that of germanium, so the expected detection efficiency for gamma rays would resemble that of germanium detectors. After a series of investigations[63,64] in the 1970s, active development of GaAs as a detector material has not continued. Problems include the limitation of available high-purity crystals to less than 1 mm thickness and experimental problems with abnormal leakage currents and intermittent burst noise. However, the recent intense development of GaAs for fast microelectronic components could stimulate new interest in detector development with this material.

Table 13-4 lists a number of other compound semiconductor materials with wide bandgaps that have been evaluated as potential detector materials. Although several show

TABLE 13-4 Some Alternative Compound Semiconductor Materials

Material	Atomic Numbers	Bandgap (eV)	Energy per Electron–Hole Pair (eV)	Density (g/cm^3)	References
GaAs	31, 33	1.42	4.2	5.35	63, 64
Bi$_2$S$_3$	83, 16	1.3		6.73	65
PbI$_2$	82, 53	2.6	7.68	6.16	66
GaSe	31, 34	2.03	6.3	4.55	67, 68
AlSb	13, 51	1.62	5.05	4.26	69
CdSe	48, 34	1.75		5.74	70, 71

promise for further development, none of these has reached the point of commercial utilization. Most are limited to very small sizes, and there are often persistent problems with efficient charge collection. Development work continues, however, in the hope of producing an alternative to silicon and germanium capable of room temperature operation. Armantrout et al.[72] have carried out a preliminary screening of many additional compound semiconductor possibilities, none of which has received much attention as a potential detector material because of the lack of availability in practical crystal sizes.

There has also been some recent interest[73] in the use of amorphous silicon layers as radiation detectors. Silicon in amorphous form offers the potential of much lower cost and larger surface area compared with the single crystals conventionally used in silicon detector fabrication. Both the mobility and lifetime values for electrons and holes in amorphous material are many orders of magnitude below those observed in silicon crystals. As a result only very thin layers (up to 15 μm) have been used successfully as detectors, and then only with fractional charge collection. Layers of this type, although not generally useful in energy spectroscopy, can function as simple charged particle counters for applications in which energy resolution is not important. Pulses from alpha particles have been measured[73] for layers as thin as 5 μm.

Some limited utilization of diamond as a detector material has been made in specialized applications. Diamond is an insulating material with a very large bandgap (5.6 eV) and can be operated as a simple conduction counter by applying ohmic contacts to opposite faces of the crystal. Detectors of this type, although obviously limited to small sizes, are characterized by a fast response time of several nanoseconds and superior long-term stability. They can be operated in either conventional pulse mode or in current mode in high radiation fields.[74] Measurements have demonstrated an energy resolution of 82 keV for room temperature spectroscopy of alpha particles.[75] Because of their wide bandgap, diamond detectors are well suited for operation at elevated temperatures, and their use up to 250–300°C has been reported.[76, 77]

III. AVALANCHE DETECTORS

The normal semiconductor detector is the solid-state analogue of a conventional gas-filled ionization chamber. The objective in either case is simply to collect all charge carriers that are created by the incident radiation within the active volume. Under certain conditions, it is also possible to achieve charge multiplication in solid-state detectors. The resulting device is then the analogue of the proportional counter and is called an *avalanche detector*. Although not nearly as widely applied as gas proportional counters, they have achieved some degree of popularity for the detection of low-energy radiations.

Gain is achieved within the semiconductor material (normally silicon) by raising the electric field sufficiently high to enable the migrating electrons to create secondary ionization during the collection process. Avalanche detector designs must incorporate a device geometry that allows the generation of high fields within the volume of the detector without allowing the field at the surface to become so large as to create surface breakdown. This effect is sometimes accomplished by beveling or *contouring* the detector shape,[78] or through use of the *reach through* configuration.[79,80] Gains of up to several hundred in the total collected charge are possible, but there has been some difficulty in achieving uniform multiplication across the entire entrance window of such detectors.

Charge carriers created within the high-field region will be multiplied by a variable gain, depending on their position relative to the boundaries of the multiplying region. The internal configuration of avalanche detectors usually provides a normal *drift* region adjacent to the multiplying region where the electric field is too low to create multiplication. If all incident radiations interact within the drift region, a more uniform multiplication of the charge can be expected because all electrons will then be subject to the same degree of multiplication.

Avalanche detectors offer some attractive properties both from the standpoint of signal-to-noise ratio and in speed of response. Because the gain is provided directly within the detector itself prior to almost all noise sources, there is a considerable theoretical improvement in the signal-to-noise ratio compared with providing the equivalent gain externally in a preamplifier. The very high fields ensure that the transit time through the multiplying region is very short, and typical pulse rise times are of the order of 3 ns.[81] This combination of properties permits use of avalanche detectors with good signal-to-background characteristics, even for the detection of very soft radiation or under conditions in which the detector temperature is elevated. For example, Huth[81] reports that 1.5 keV X-rays can be detected with a 30–40% intrinsic efficiency at temperatures between 85 and 100°C while maintaining very low background levels.

Much of the noise reduction comes about because the electronic shaping times can be made very short due to the extremely fast rise time of pulses from these detectors. In many applications, the output signal of the avalanche detector is fed through a tunnel diode that is set with a discrimination point sufficiently high so as not to trigger on noise pulses. Because of the internal gain of the detector, X-rays with an energy as low as 0.6 keV can generate pulses that are above this discrimination point.

Avalanche detectors have found useful applications in the biomedical field for the in vivo measurement of plutonium and other low-energy X-ray emitters.[82,83] Other applications in portable gamma-ray monitors have also been described.[84]

IV. POSITION-SENSITIVE SEMICONDUCTOR DETECTORS

Detectors in which the position of interaction of the incident radiation is sensed together with its energy have application in a number of different areas. The two major types of detector used for position sensing for charged particles are gas-filled proportional tubes (as described in Chapter 6) and silicon or germanium semiconductor diode detectors. The latter types are sometimes preferred because of their compactness and low bias voltage compared with gas-filled detectors. Semiconductor detectors, because of their greater stopping power, are also better suited for applications involving long-range radiations.

In its simplest form, the semiconductor position-sensitive detector consists of a one-dimensional strip of silicon or germanium for which one contact is made to have a significant series resistance. As shown in Fig. 13-25, one end of this resistive contact is

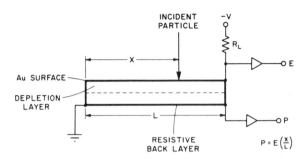

Figure 13-25 Basic configuration of a position-sensitive semiconductor detector. The position signal (labeled P) is obtained by resistive charge division from the back layer. The E signal is proportional to the energy deposited by the particle in the detector active volume. (From Kaufman et al.[85])

grounded whereas the other leads to an amplifier for the derivation of the position signal. The resistive contact acts as a charge divider, and the amount of charge delivered to the position signal amplifier is proportional to x/L, where x is the distance of the interaction from the grounded end, and L is the strip length. A second signal (the E signal) proportional to the total charge deposited in the detector is derived from the normal conductive front electrode. If the position signal is divided by the E signal, a pulse is produced whose amplitude reflects the position of the interaction, independent of the amount of charge actually deposited. A review and analysis of pulse-processing methods for position sensing are given in Ref. 86.

The performance of one such detector is illustrated in Fig. 13-26. The contributions of nine different locations are cleanly separated for incident fission fragments with a wide spread in energy. Position resolution can be of the order of 0.5–1.0% FWHM at room temperature over active lengths of 50–100 mm.[87] The position resolution normally deteriorates as the energy of the incident particle decreases, because one of the determining factors is the statistical variation in the number of charge carriers collected at either end of the resistive strip.

Under some conditions, the undepleted portion of a normal silicon surface barrier can be used as the resistive back layer. More typically, the resistive contact is formed by metal evaporation or by ion implantation techniques. Ion-implanted layers are usually superior because they can be made uniform in resistance.[87, 88] In other arrangements, the charge division is accomplished in an external resistive network.[89, 90]

An alternative method of position sensing is to subdivide one of the electrodes into a number of independent segments or strips. Since electron–hole pairs created within the volume of the detector will travel along field lines to the corresponding electrode segment, a signal will be derived only from those segments that have collected appreciable charge carriers. If short-range particles are involved, then only one such signal will be generated. For longer-range particles, the "center of gravity" of the position of the track can be determined by interpolation of the signals from nearby segments. A silicon *microstrip* detector is one in which a series of narrow parallel strip electrodes have been fabricated on one surface using ion implantation and/or photolithography techniques.[91, 92] For fine spatial resolution, such detectors have been produced with strip widths as small as 10 μm. To avoid the multiplicity of independent electronic channels that would be needed to measure the signal from each strip independently, schemes have been developed to find the strip nearest the interaction using one-dimensional charge division along a line that

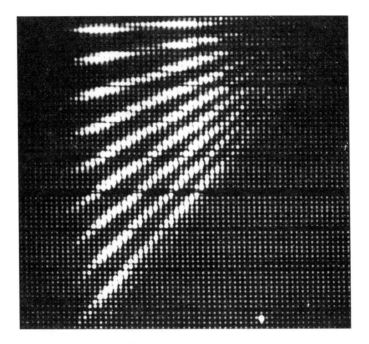

Figure 13-26 A two-parameter display of events recorded from a position-sensitive detector of the type shown in Fig. 13-25. The E signal amplitude is displayed along the horizontal axis and the P signal along the vertical axis. A collimator with nine equidistant slits was placed in front of the detector. Incident particles were fission fragments from ^{252}Cf. (From Kaufman et al.[85])

interconnects all the strips. Using this approach, one can obtain a spatial resolution approaching 10 μm in one dimension for incident alpha particles.

Position sensing in more than one dimension can also be carried out with certain detector configurations. Figure 13-27 illustrates one arrangement in which a germanium slab has been provided with orthogonal strip electrodes on both surfaces. Independent X and Y signals are then derived from each detector face to provide two-dimensional position sensing, with a spatial resolution determined by the strip width.

It has recently been demonstrated[93] that the drift time of charge carriers can also be used to deduce the position of their formation in a silicon detector. In the *semiconductor drift detector*, a unique electrode configuration is used as illustrated in Fig. 13-28. Semiconductor junctions are formed at *both* faces of a large-area wafer, and each is reverse biased until the detector is fully depleted. Electrons created by ionizing radiation within the semiconductor are confined within an electric potential well and caused to drift in a direction parallel rather than perpendicular to the wafer surface. A collecting anode is fabricated near the edge of the wafer. The time required for these electrons to drift to the anode is then a linear measure of the distance between the anode and the position of the interaction. Tests of prototype drift detectors[94] indicate that a spatial resolution of 4 μm may be achieved over a drift distance of several millimeters. Detectors with active area as large as 2.5 cm $\times$ 2.5 cm have been operated[95] with spatial resolution of less than 0.5 mm.

This type of configuration has an additional advantage that can be exploited[96,97] to improve energy resolution. Because the electrons can be drifted over large distances and

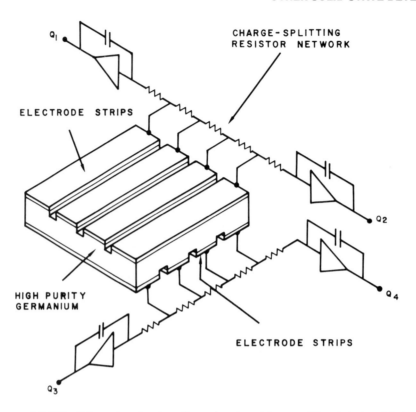

Figure 13-27 Schematic of a two-dimensional position-sensitive detector fabricated from intrinsic germanium. Position sensing is carried out independently in X and Y directions using external resistor networks with charge-sensitive preamplifiers at each end. The difference between the preamplifier signals gives the position information, whereas their sum gives the energy signal. (From Gerber et al.[90])

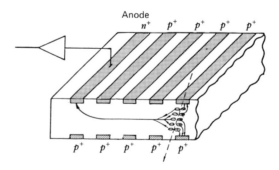

Figure 13-28 Configuration of a semiconductor drift detector. Minority carriers (holes) are collected at the anode strip nearest the interaction. Majority carriers (electrons) are transported parallel to the detector surface and collected on a single anode. (From Rehak et al.[94])

collected on an electrode of very small size, the capacitance of the detector can be much smaller than that of an equivalent semiconductor diode of conventional design. The detector capacitance is always an important factor in determining the overall energy resolution of a spectroscopic system, and low capacitance is a significant advantage. Such devices are still in the development stage but could represent an important step in the search for better energy resolution in semiconductor detectors.

PROBLEMS

13-1. Give two reasons why the X-ray escape peak is less intense in silicon detectors compared with germanium detectors.

13-2. Estimate the maximum charge collection time (using data given in Chapter 11) for a 4 mm thick Si(Li) detector operated at 2000 V.

13-3. What must be the energy resolution (in percent) for a Si(Li) detector if it is to resolve separately the K-characteristic X-rays from copper and zinc?

13-4. What physical effects cause the detection efficiency of Si(Li) detectors to drop off at low (less than 5 keV) incident X-ray energies?

13-5. By using a NaI(Tl) well counter with absolute peak efficiency of 83%, a net of 146,835 counts was recorded under the 122 keV photopeak from a ^{57}Co source over a 15 min live time. The same source was then placed 10 cm from the face of a Si(Li) detector with 300 mm^2 surface area and a spectrum recorded over a 60 min counting period. If 730 counts were recorded under the 6.4 keV X-ray peak, what is the efficiency of the Si(Li) detector at this energy (see Table 13-2)?

REFERENCES

1. E. M. Pell, *J. Appl. Phys.* **31**, 291 (1960).
2. A. Coche and P. Siffert, "Lithium Drifted Silicon and Germanium Detectors," in *Semiconductor Detectors* (G. Bertolini and A. Coche, eds.). Elsevier-North Holland, Amsterdam, 1968.
3. P. E. Gibbons and P. Iredale, *Nucl. Instrum. Meth.* **53**, 1 (1967).
4. A. Lauber, *Nucl. Instrum. Meth.* **75**, 297 (1969).
5. M. Moszynski, W. Kurcewicz, and W. Przyborski, *Nucl. Instrum. Meth.* **61**, 173 (1968).
6. A. S. Antonov, *Solid State Phys.* **8**, 1325 (1966).
7. Z. Moroz and M. Moszynski, *Nucl. Instrum. Meth.* **68**, 261 (1969).
8. M. Moszynski and W. Przyborski, *Nucl. Instrum. Meth.* **64**, 244 (1968).
9. G. Lindstrom and R. Lisdat, *Nucl. Instrum. Meth.* **116**, 181 (1974).
10. P. Van Espen, H. Nullens, and F.Adams, *Nucl. Instrum. Meth.* **145**, 579 (1977).
11. R. E. Wood, P. Venugopala Rao, O. H. Puckett, and J. M. Palms, *Nucl. Instrum. Meth.* **94**, 245 (1971).
12. N. A. Dyson, *Nucl.Instrum. Meth.* **114**, 131 (1974).
13. J. Felsteiner, S. Kahane, and B. Rosner, *Nucl. Instrum. Meth.* **118**, 253 (1974).
14. R. G. Musket, *Nucl. Instrum. Meth.* **117**, 385 (1974).
15. M. Hollstein, *Nucl. Instrum. Meth.* **82**, 249 (1970).
16. G. Bertolini, F. Cappellani, and G. Restelli, *Nucl. Instrum. Meth.* **112**, 219 (1973).
17. M. R. Khan, A. S. Lodhi, and D. Crumpton, *Nucl. Instrum. Meth.* **160**, 127 (1979).
18. A. H. F. Muggleton, *Nucl. Instrum. Meth.* **101**, 113 (1972).

19. J. L. Campbell and P. L. McGhee, *Nucl. Instrum. Meth.* **A248**, 393 (1986).

20. R. J. Gehrke and R. A. Lokken, *Nucl. Instrum. Meth.* **97**, 219 (1971).

21. J. L. Campbell and L. A. McNelles, *Nucl. Instrum. Meth.* **125**, 205 (1975).

22. W. J. Gallagher and S. J. Cipolla, *Nucl. Instrum. Meth.* **122**, 405 (1974).

23. B. Rosner, D. Gur, and L. Shabason, *Nucl. Instrum. Meth.* **131**, 81 (1975).

24. R. G. Musket and W. Bauer, *Nucl. Instrum. Meth.* **109**, 593 (1973).

25. J. L. Campbell, H. H. Jorch, and J. A. Thompson, *Nucl. Instrum. Meth.* **140**, 167 (1977).

26. H. I. Israel, D. W. Lier, and E. Storm, *Nucl. Instrum. Meth.* **91**, 141 (1971).

27. J. M. Jaklevic and F. S. Goulding, *IEEE Trans. Nucl. Sci.* **NS-18**(1), 187 (1971).

28. F. S. Goulding, *Nucl. Instrum. Meth.* **142**, 213 (1977).

29. I. Ahmad and F. Wagner, *Nucl. Instrum. Meth.* **116**, 465 (1974).

30. M. J. Berger, S. M. Seltzer, S. E. Chappell, J. C. Humphreys, and J. W. Motz, *Nucl. Instrum. Meth.* **69**, 181 (1969).

31. A. Damkjaer, *Nucl. Instrum. Meth.* **200**, 377 (1982).

32. M. J. Berger, S. M. Seltzer, S. E. Chappell, J. C. Humphreys, and J. W. Motz, *NBS Tech. Note* **489** (1969).

33. R. D. Von Dincklage and J. Gerl, *Nucl. Instrum. Meth.* **A235**, 198 (1985).

34. C. A. Klein, *J. Appl. Phys.* **39**, 2029 (1968).

35. M. Cuzin, *Nucl. Instrum. Meth.* **A253**, 407 (1987).

36. P. Siffert, A. Cornet, R. Stuck, R. Triboulet, and Y. Marfaing, *IEEE Trans. Nucl. Sci.* **NS-22**(1), 211 (1975).

37. P. Siffert, *Mat. Res. Soc. Symp. Proc.* **16**, 87 (1983).

38. C. Scharager, P. Siffert, A. Holtzer, and M. Schieber, *IEEE Trans. Nucl. Sci.* **NS-27**(1), 276 (1980).

39. S. H. Shin et al., *IEEE Trans. Nucl. Sci.* **NS-32**(1), 487 (1985).

40. T. Hazlett et al., *IEEE Trans. Nucl. Sci.* **NS-33**(1), 332 (1986).

41. A. J. Dabrowski et al., *Nucl. Instrum. Meth.* **150**, 25 (1978).

42. R. J. Fox and D. C. Agouridis, *Nucl. Instrum. Meth.* **157**, 65 (1978).

43. G. Entine, M. R. Squillante, H. B. Serreze, and E. Clarke, *IEEE Trans. Nucl. Sci.* **NS-28**(1), 558 (1981).

44. R. A. Ristinen et al., *Nucl. Instrum. Meth.* **188**, 445 (1981).

45. R. O. Bell, G. Entine, and H. B. Serreze, *Nucl. Instrum. Meth.* **117**, 267 (1974).

46. H. L. Malm and M. Martini, *IEEE Trans. Nucl. Sci.* **NS-21**(1), 322 (1974).

47. P. Siffert, J. Berger, C. Scharager, A. Cornet, R. Stuck, R. O. Bell, H. B. Serreze, and F. V. Wald, *IEEE Trans. Nucl. Sci.* **NS-23**(1), 159 (1976).

48. P. Siffert, *Nucl. Instrum. Meth.* **150**, 1 (1978).

49. R. C. Whited and M. M. Schieber, *Nucl. Instrum. Meth.* **162**, 113 (1979).

50. E. Sakai, *Nucl. Instrum. Meth.* **196**, 121 (1982).

51. J. Bojsen, *Nucl. Instrum. Meth.* **219**, 592 (1984).

52. T. Mohammed-Brahim, A. Friant, and J. Mellet, *IEEE Trans. Nucl. Sci.* **NS-32**(1), 581 (1985).

53. M. Schieber, *Nucl. Instrum. Meth.* **213**, 1 (1983).

54. H. A. Lamonds, *Nucl. Instrum. Meth.* **213**, 5 (1983).

55. A. J. Dabrowski et al., *Nucl. Instrum. Meth.* **213**, 89 (1983).

56. J. S. Iwanczyk, A. J. Dabrowski, G. C. Huth, J. G. Bradley, J. M. Conley, and A. L. Albee, *IEEE Trans. Nucl. Sci.* **NS-33**(1), 355 (1986).

57. K. Hull et al., *IEEE Trans. Nucl. Sci.* **NS-30**(1), 402 (1983).

58. J. S. Iwanczyk et al., "Continuing Development of Mercuric Iodide X-Ray Spectrometers", in *Advances in X-Ray Analysis*, Vol. 27, J. B. Cohen et al., eds., pp. 405–414, Plenum Press, New York, 1984.

59. A. Beyerle, V. Gerrish, and K. Hull, *Nucl. Instrum. Meth.* **A242**, 443 (1986).

60. V. M. Gerrish, D. J. Williams, and A. G. Beyerle, *IEEE Trans. Nucl. Sci.* **NS-34**(1), 85 (1987).

61. W. K. Warburton and J. S. Iwanczyk, *Nucl. Instrum. Meth.* **A254**, 123 (1987).

62. F. D. Becchetti et al., *Nucl. Instrum. Meth.* **213**, 127 (1983).

63. T. Kobayashi and T. Sugita, *Nucl. Instrum. Meth.* **98**, 179 (1972).

64. T. Kobayashi, I. Kuru, A. Hojo, and T. Sugita, *IEEE Trans. Nucl. Sci.* **NS-23**(1), 97 (1976).

65. F. V. Wald, J. Bullitt, and R. O. Bell, *IEEE Trans. Nucl. Sci.* **NS-22**(1), 246 (1975).

66. C. Manfredotti, R. Murri, A. Quirini, and L. Vasanelli, *IEEE Trans. Nucl. Sci.* **NS-24**(1), 126 (1977).

67. C. Manfredotti, R. Murri, and L. Vasanelli, *Nucl. Instrum. Meth.* **115**, 349 (1974).

68. E. Sakai et al., *IEEE Trans. Nucl. Sci.* **NS-35**(1), 85 (1988).

69. J. H. Yee, S. P. Swierkowski, and J. W. Sherohman, *IEEE Trans. Nucl. Sci.* **NS-24**(4), 1962 (1977).

70. A. Burger, I. Shilo, and M. Schieber, *IEEE Trans. Nucl. Sci.* **NS-30**(1), 368 (1983).

71. M. Roth and A. Burger, *IEEE Trans. Nucl. Sci.* **NS-33**(1), 407 (1986).

72. G. A. Armantrout, S. P. Swierkowski, J. W. Sherohman, and J. H. Yee, *IEEE Trans. Nucl. Sci.* **NS-24**(1), 121 (1977).

73. V. Perez-Mendez, J. Morel, S. N. Kaplan, and R. A. Street, *Nucl. Instrum. Meth.* **A252**, 478 (1986).

74. S. F. Kozlov, E. A. Konorova, Y. A. Kuznetsov, Y. A. Salikov, V. I. Redko, V. R. Grinberg, and M. L. Meilman, *IEEE Trans. Nucl. Sci.* **NS-24**(1), 235 (1977).

75. C. Canali et al., *Nucl. Instrum. Meth.* **160**, 73 (1979).

76. E. K. Konorova and S. F. Kozlov, *Sov. Phys. Semicond.* **4**, 1600 (1971).

77. F. Nava et al., *IEEE Trans. Nucl. Sci.* **NS-26**(1), 308 (1979).

78. G. C. Huth, *IEEE Trans. Nucl. Sci.* **NS-13**(1), 36 (1966).

79. P. P. Webb and A. R. Jones, *IEEE Trans. Nucl. Sci.* **NS-21**(1), 151 (1974).

80. P. P. Webb and R. J. McIntyre, *IEEE Trans. Nucl. Sci.* **NS-23**(1), 138 (1976).

81. G. C. Huth, "Avalanche Multiplying Diode," in *Semiconductor Nuclear-Particle Detectors and Circuits*, Publication No. 1593, National Academy of Sciences, Washington, DC, 1969, p. 323.

82. P. V. Hewka, G. C. Huth, and K. L. Swinth, *IEEE Trans. Nucl. Sci.* **NS-17**(3), 265 (1970).

83. P. J. Moldofsky and K. L. Swinth, *IEEE Trans. Nucl. Sci.* **NS-19**(1), 55 (1972).

84. A. R. Jones, *IEEE Trans. Nucl. Sci.* **NS-20**(1), 528 (1973).

85. S. B. Kaufman, B. D. Wilkins, M. J. Fluss, and E. P. Steinberg, *Nucl. Instrum. Meth.* **82**, 117 (1970).

86. J. L. Alberi and V. Radeka, *IEEE Trans. Nucl. Sci.* **NS-23**(1), 251 (1976).

87. E. Elad and R. Sareen, *IEEE Trans. Nucl. Sci.* **NS-21**(1), 75 (1974).

88. E. Laegsgaard, F. W. Martin, and W. M. Gibson, *IEEE Trans. Nucl. Sci.* **NS-15**(3), 239 (1968).

89. J. E. Lamport, G. M. Mason, M. A. Perkins and A. J. Tuzzolino, *Nucl. Instrum. Meth.* **134**, 71 (1976).

90. M. S. Gerber, D. W. Miller, P. A. Schlosser, J. W. Steidley and A. H. Deutchman, *IEEE Trans. Nucl. Sci.* **NS-24**(1), 182 (1977).

91. E. H. M. Heijne et al., *Nucl. Instrum. Meth.* **178**, 331 (1980).

92. J. Yorkston, A. C. Shotter, D. B. Syme, and G. Huxtable, *Nucl. Instrum. Meth. Phys. Res.* **A262**, 353 (1987).

93. E. Gatti, P. Rehak, and J. T. Walton, *Nucl. Instrum. Meth.* **226**, 129 (1984).

94. P. Rehak et al., *Nucl. Instrum. Meth.* **A248**, 367 (1986).

95. P. N. Luke, N. W. Madden, and F. S. Goulding, *IEEE Trans. Nucl. Sci.* **NS-32**(1), 457 (1985).

96. K. J. Rawlings, *Nucl. Instrum. Meth.* **A253**, 85 (1986).

97. J. Kemmer et al., *Nucl. Instrum. Meth.* **A253**, 378 (1987).

CHAPTER 14

Slow Neutron Detection Methods

Neutrons are detected through nuclear reactions which result in energetic charged particles such as protons, alpha particles, and so on. Virtually every type of neutron detector involves the combination of a target material designed to carry out this conversion together with one of the conventional radiation detectors discussed in earlier chapters. Because the cross section for neutron interactions in most materials is a strong function of neutron energy, rather different techniques have been developed for neutron detection in different energy regions. In this chapter, we discuss those methods that are of primary importance for the detection of neutrons whose energy is below the *cadmium cutoff* of about 0.5 eV. This is conventionally called the slow neutron region and is distinguished from intermediate and fast neutrons with energies above this value. Slow neutrons are of particular significance in present-day nuclear reactors and much of the instrumentation that has been developed for this energy region is aimed at the measurement of reactor neutron flux. Specific detector types that have evolved for this purpose are discussed at the end of this chapter.

We limit our discussion in this chapter to those methods that are intended to indicate only the detection of a neutron, with no attempt made to measure its kinetic energy. Devices that can measure slow neutron energies such as crystal spectrometers or mechanical monochromators are generally complex research-oriented instrumentation systems and are not covered here. In contrast, rather simple detectors can be used to measure the energy of neutrons of higher energy and these are discussed in the next chapter.

We also postpone discussion of *passive* neutron detectors, including activation foils, until Chapter 19. In this chapter we discuss only *active* detectors in which a pulse or current signal is produced by each neutron as it interacts in the device. General reviews of slow neutron detection devices and techniques may be found in Refs. 1–5. More detailed descriptions of those detectors developed specifically for reactor applications are given in Refs. 6 and 7.

I. NUCLEAR REACTIONS OF INTEREST IN NEUTRON DETECTION

In searching for nuclear reactions that might be useful in neutron detection, several factors must be considered. First, the cross section for the reaction must be as large as possible so that efficient detectors can be built with small dimensions. This is particularly important for detectors in which the target material is incorporated as a gas, of which we shall see several examples. For the same reason, the target nuclide should either be of high isotopic abundance in the natural element, or alternatively, an economic source of artificially enriched samples should be available for detector fabrication. In many applications, intense fields of gamma rays are also found with neutrons and the choice of reaction bears on the ability to discriminate against these gamma rays in the detection process. Of principal importance here is the Q-value of the reaction which determines the energy liberated in the reaction following neutron capture. The higher the Q-value, the greater is the energy given to the reaction products, and the easier is the task of discriminating against gamma-ray events using simple amplitude discrimination.

It is important to point out that all the common reactions used to detect slow neutrons result in heavy charged particles.[†] Possible reaction products are listed below:

$$\text{target nucleus + neutron} \rightarrow \begin{cases} \text{recoil nucleus} \\ \text{proton} \\ \text{alpha particle} \\ \text{fission fragments} \end{cases}$$

All the conversion reactions are sufficiently exothermic so that the kinetic energy of the reaction products is determined solely by the Q-value of the reaction and does not reflect the very small incoming energy of the slow neutron.

The distance traveled by the reaction products following their formation also has important consequences in detector design. If we are to capture the full kinetic energy of these products, the detector must be designed with an active volume that is large enough to fully stop the particles. If the detection medium is a solid, this requirement is easily achieved because the range of any of the reaction products shown does not exceed a few tenths of a millimeter in any solid material. If the detection medium is a gas, however, ranges of the reaction products (typically several centimeters) can be significant compared with detector dimensions and some may not deposit all their energy. If the detector is large enough so that these losses can be neglected, the response function will be very simple, consisting only of a single full-energy peak as shown in the sketch.

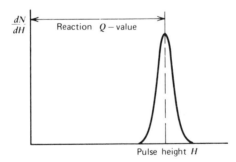

Under these circumstances the detector would exhibit a very flat counting plateau, and the ability to discriminate against low-amplitude events (such as gamma-ray-induced processes) would be maximized. If, on the other hand, a significant number of neutron-induced events do not deposit the full energy, a low-energy continuum is added to the

---

[†]Gamma rays produced by neutron capture are used in some specialized detectors, but these applications are relatively rare.

pulse height distribution and the detector performance with respect to these criteria will suffer.

A. The ^{10}B(n, α) Reaction

Probably the most popular reaction for the conversion of slow neutrons into directly detectable particles is the ^{10}B(n, α) reaction. The reaction may be written

$$Q\text{-value}$$

$$
{}^{10}_{5}\text{B} + {}^{1}_{0}\text{n} \rightarrow
\begin{cases}
{}^{7}_{3}\text{Li} + {}^{4}_{2}\alpha & 2.792 \text{ MeV (ground state)} \\
{}^{7}_{3}\text{Li*} + {}^{4}_{2}\alpha & 2.310 \text{ MeV (excited state)}
\end{cases}
\tag{14-1}
$$

where the branching indicates that the reaction product ^{7}Li may be left either in its ground state or in its first excited state.[†] When thermal neutrons (0.025 eV) are used to induce the reaction, about 94% of all reactions lead to the excited state and only 6% directly to the ground state. In either case, the Q-value of the reaction is very large (2.310 or 2.792 MeV) compared with the incoming energy of the slow neutron, so that the energy imparted to the reaction products (^{7}Li and α) is essentially just the Q-value itself. Thus, the incoming kinetic energy of the neutron is submerged in the much larger reaction energy, and it is impossible to extract any information about its original value. Also, because the incoming linear momentum is very small, the reaction products must also show a net momentum of essentially zero. Consequently, the two reaction products must be emitted in exactly opposite directions, and the energy of the reaction will always be shared in the same manner between them. Individual energies of the alpha particle and lithium nucleus can be calculated simply by conservation of energy and momentum as follows:

$$E_{\text{Li}} + E_{\alpha} = Q = 2.31 \text{ MeV} \tag{14-2}$$

$$m_{\text{Li}} v_{\text{Li}} = m_{\alpha} v_{\alpha}$$

$$\sqrt{2 m_{\text{Li}} E_{\text{Li}}} = \sqrt{2 m_{\alpha} E_{\alpha}} \tag{14-3}$$

Solving Eqs. (14-2) and (14-3) simultaneously:

$$E_{\text{Li}} = 0.84 \text{ MeV} \quad \text{and} \quad E_{\alpha} = 1.47 \text{ MeV}$$

where the calculation has been carried out for the case of populating the excited state of ^{7}Li. A similar calculation would yield larger values by 21% for reactions leading to the ground state.

Figure 14-1 is a plot of cross sections versus neutron energy for a number of nuclear reactions of interest in neutron detection. The thermal neutron cross section for the ^{10}B(n, α) reaction is 3840 barns. The cross-section value drops rapidly with increasing neutron energy and is proportional to $1/v$ (the reciprocal of the neutron velocity) over much of the range. The utility of this reaction stems from its rather large and structureless cross section and from the fact that boron, highly enriched in its ^{10}B concentration, is readily available. The natural isotopic abundance of ^{10}B is 19.8%.

[†] The excited lithium nucleus quickly returns (half-life of $\sim 10^{-13}$ s) to its ground state with the emission of a 0.48 MeV gamma ray. We assume that this photon always escapes and does not contribute to the response of the detector.

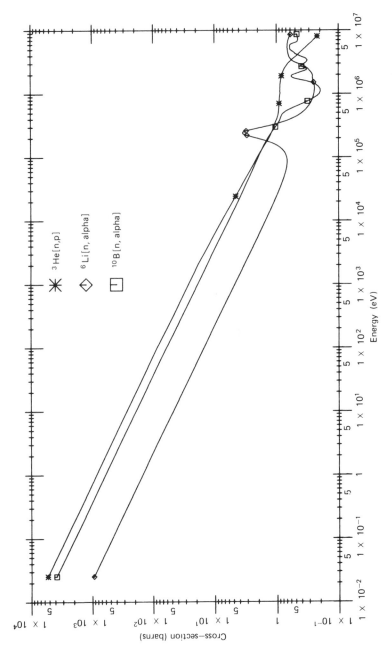

Figure 14-1 Cross section versus neutron energy for some reactions of interest in neutron detection.

B. The ^{6}Li(n, α) Reaction

The next most popular reaction for the detection of slow neutrons is the (n, α) reaction in ^{6}Li. Here the reaction proceeds only to the ground state of the product and is written simply as

$$Q\text{-value}$$

$$^6_3\text{Li} + ^1_0\text{n} \rightarrow ^3_1\text{H} + ^4_2\alpha \qquad 4.78 \text{ MeV}$$

Calculation of the reaction product energies for negligible incoming neutron energy yields the following:

$$E_{^3\text{H}} = 2.73 \text{ MeV} \quad \text{and} \quad E_\alpha = 2.05 \text{ MeV}$$

The alpha particle and triton produced in the reaction must be oppositely directed when the incoming neutron energy is low.

The thermal neutron cross section for this reaction is 940 barns. Figure 14-1 shows that the cross section remains below that for the ^{10}B reaction until the resonance region (> 100 keV). The lower cross section is generally a disadvantage but is partially offset by the higher Q-value and resulting greater energy given to the reaction products. ^{6}Li occurs with a natural isotopic abundance of 7.40% and is also widely available in separated form.

C. The ^{3}He(n, p) Reaction

The gas ^{3}He is also widely used as a detection medium for neutrons through the reaction

$$Q\text{-value}$$

$$^3_2\text{He} + ^1_0\text{n} \rightarrow ^3_1\text{H} + ^1_1\text{p} \qquad 0.764 \text{ MeV}$$

For reactions induced by slow neutrons, the Q-value of 764 keV leads to oppositely directed reaction products with energies

$$E_\text{p} = 0.573 \text{ MeV} \quad \text{and} \quad E_{^3\text{H}} = 0.191 \text{ MeV}$$

The thermal neutron cross section for this reaction is 5330 barns, significantly higher than that for the boron reaction, and its value also falls off with a $1/v$ energy dependence (see Fig. 14-1). Although ^{3}He is commercially available, its relatively high cost is a factor in some applications.

D. Neutron-Induced Fission Reactions

The fission cross sections of ^{233}U, ^{235}U, and ^{239}Pu are relatively large at low neutron energies and thus these materials can be used as the basis of slow neutron detectors. One characteristic of the fission reaction is its extremely large Q-value (approximately 200 MeV) compared with the reactions discussed previously. As a result, detectors based on the fission reaction can often give output pulses that are orders of magnitude larger than those induced from competing reactions or incident gamma rays, and very clean discrimination can be accomplished. Figure 14-2 shows a plot of fission cross sections of a variety of fissile nuclides, including some that are of primary use as fast neutron detectors. Almost all fissile nuclides are naturally alpha radioactive and consequently any detector that incorporates these materials will also show a spontaneous output signal due to decay alpha particles. The energy of the decay alpha particles, however, is always many times

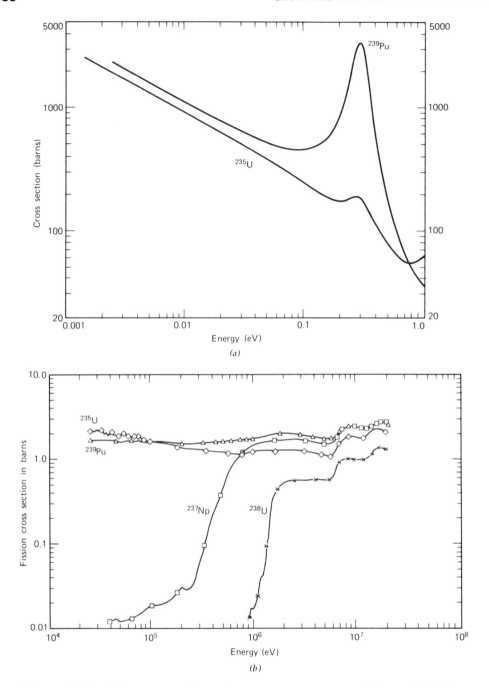

Figure 14-2 Fission cross sections of some common target nuclides used in fission chambers. (a) Slow neutron region where the cross sections shown are relatively large. (b) Fast neutron region. Chambers with ^{237}Np or ^{238}U are used as *threshold detectors* sensitive only to fast neutrons.

less than the energy given off in a fission reaction, and again these events can usually be discriminated easily on a pulse amplitude basis.

II. DETECTORS BASED ON THE BORON REACTION

A widely used detector for slow neutrons is the BF_3 proportional tube. In this device, boron trifluoride serves both as the target for slow neutron conversion into secondary particles as well as a proportional gas. A number of other boron-containing gases have been evaluated, but BF_3 is the near-universal choice because of its superior properties as a proportional gas, as well as its high concentration of boron. In nearly all commercial detectors, the gas is highly enriched in ^{10}B, resulting in an efficiency some five times greater than if the gas contained naturally occurring boron. Because the performance of BF_3 as a proportional gas is poor when operated at higher pressures, its absolute pressure in typical tubes is limited to about 0.5–1.0 atm.

A. BF₃ Tube Pulse Height Spectra — The Wall Effect

Figure 14-3a shows the ideal pulse height spectrum expected from a BF_3 tube of very large dimensions. For a large tube, nearly all the reactions occur sufficiently far from the walls of the detector to deposit the full energy of the products within the proportional gas. In that event, all the energy of the reaction is deposited in the detector and the only variation is a result of the branching of the reaction between the excited state and ground state of the ^7Li product nucleus. The branching ratio for thermal neutrons is such that about 6% of the reactions lead to the ground state and 94% to the first excited state. Therefore, the areas under the peaks shown in Fig. 14-3a should be in the ratio 94 : 6 as illustrated.

Once the size of the tube is no longer large compared with the range of the alpha particle and recoil lithium nucleus produced in the reaction, some events no longer deposit the full reaction energy in the gas. If either particle strikes the chamber wall, a smaller pulse is produced. The cumulative effect of this type of process is known as the *wall effect* in gas counters. Because the range of the alpha particle produced in the reaction is on the order of 1 cm for typical BF_3 gas pressures, almost all practical tubes are small enough in diameter so that the wall effect is significant.

Figure 14-3b shows the differential pulse height spectrum expected from a tube in which the wall effect is important. The primary change from the spectrum shown in Fig. 14-3a is the addition of a continuum to the left of the peaks corresponding to partial energy deposition in the gas of the tube. The two steps or discontinuities in the continuum are an interesting feature of the spectrum and can be explained through the following argument.

Because the incoming neutron carries no appreciable momentum, the two reaction products must be oppositely directed. If the alpha particle strikes the wall, the ^7Li recoil is therefore directed away from the wall and is very likely to deposit its full energy within the gas. Conversely, if the ^7Li recoil strikes a wall, the entire energy of the alpha particle from that same reaction is usually fully absorbed.

Thus, we expect to see wall losses for only one reaction product at a time. There are two possibilities: (1) the alpha particle hits a wall after depositing some fraction of its energy in the fill gas, whereas the ^7Li recoil is fully absorbed in the gas, or (2) the ^7Li recoil hits a wall after depositing part of its energy and the alpha particle is fully absorbed. Under case 1 above, the reaction could occur at a distance from the wall which

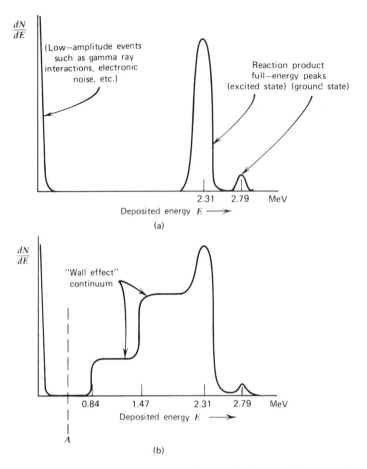

Figure 14-3 Expected pulse height spectra from BF_3 tubes. (a) Spectrum from a large tube in which all reaction products are fully absorbed. (b) Additional continuum due to the wall effect.

might be anywhere between zero and the full alpha particle range. The amount of energy deposited in the gas can correspondingly vary from $(E_{Li} + 0)$ to $(E_{Li} + E_\alpha)$, as illustrated below.

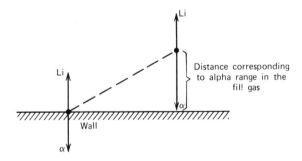

Because all locations of the reaction are more or less equally probable, the distribution of deposited energy will be approximately uniform between these two extremes.

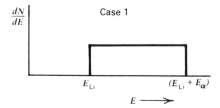

Parallel arguments can be made for case 2 to show that the energy deposited in the gas will vary from $(E_\alpha + 0)$ to $(E_\alpha + E_{Li})$.

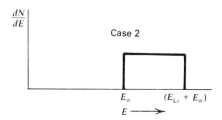

The combined energy deposition distribution of all events in which either reaction product strikes a wall will simply be the sum of the two cases.

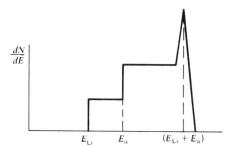

In addition to the wall effect events, the sketch above also shows the location of the full-energy peak that results from all those reactions from which both products are fully absorbed in the gas. The wall effect continuum extends from E_{Li} (0.84 MeV) up to the full-energy peak at $(E_{Li} + E_\alpha)$ (2.31 MeV). We have considered only those reactions leading to the ^{7}Li excited state because the wall effect continuum associated with the much less probable ground state is normally so small as to be submerged by the remainder of the spectrum.

The BF$_3$ tube is an example of a detector from which the differential pulse height spectrum tells us nothing about the energy spectrum of the incident radiation but is a function only of the size and geometry of the detector itself. In routine applications, there is consequently no motivation to record the pulse height spectrum from a BF$_3$ tube other than in an indirect manner. Instead, we are likely to seek a stable operating point or a counting plateau for which small drifts in operating parameters do not significantly affect the neutron sensitivity of the counter. That objective would be met by setting a fixed discriminator level at the point labeled A in Fig. 14-3b. From the arguments given in Chapter 4, we would expect a counting plateau to appear as the high voltage to the tube is varied, changing the internal gain of the proportional gas multiplication process. The flattest portion of that plateau should occur when the effective discrimination point is at the minimum in the differential pulse height spectrum, or point A. Under these conditions, all the neutrons will be counted, whereas low-amplitude events will be rejected. If the wall effect is eliminated or greatly suppressed by making the tube very large, a pulse height spectrum similar to that shown in Fig. 14-3a results. The counting plateau will then extend over a much greater range of applied voltage and thereby extend the useful operating range over which all neutron interactions are counted. A rather complete sample of spectra from a variety of BF_3 tubes is presented by Anderson and Malmskog.[8] A theoretical model has been developed by Cervellati and Kazimierski for the expected pulse height distribution from a BF_3 tube and is compared with experimentally measured distributions.[9]

B. BF₃ Tube Construction

The neutron detection efficiency can be increased and the wall effect suppressed by making the tube larger in dimension. Similar improvements can be achieved by raising the pressure of the BF_3 fill gas. Fowler[10] has reported the successful construction and operation of BF_3 tubes with diameter up to 15 cm and 180 cm long. Filling pressure ranged from 100 to 600 torr (approximately 13–80 kPa). Pressures in the range 200–300 torr (approximately 27–40 kPa) gave the best resolution in this work, whereas the full-energy peaks in the spectrum broadened considerably at higher pressure due to recombination and negative ion formation. In many counting situations, the poorer resolution is of no real consequence, and tubes with the higher gas pressure would be quite acceptable. Small-diameter tubes filled to several atmospheres pressure are commercially available, although pressures in the range 500–600 torr (approximately 67–80 kPa) are much more common.

In common with most proportional counters, BF_3 tubes are universally constructed using cylindrical outer cathodes and small-diameter central wire anodes. Aluminum is often the material of choice for the cathodes because of its low neutron interaction cross section. For low background application, other materials such as stainless steel are preferred because aluminum normally shows a small amount of low-level alpha activity. With typical anode diameters of 0.1 mm or less, operating voltages tend to be about 2000–3000 V. Larger-diameter anode wires and/or higher fill gas pressures require higher applied voltages. Typical gas multiplication at operating voltage is on the order of 100–500.

BF_3 tubes of typical construction are normally limited to operating temperature up to about 100°C, but tubes of special design can extend the operating range to as high as 150°C. However, the pulse amplitude decreases and the pulse height resolution decreases sharply[11] when operated well above room temperature. These changes may be related to the possible desorption of impurities from the counter wall or other components at elevated temperatures.

Because of the relatively high operating voltages, BF_3 tubes share some temperamental qualities with other proportional counters. Spurious pulses of about the same size as signal pulses can sometimes arise from fluctuations in leakage currents through insulators, especially under conditions of high humidity. Spurious counts can also arise in applications in which the counter is subject to vibration or shock.[12] These effects are attributed to detector microphonics and the influence of small particles of lint or dirt within the counter.

C. Gamma-Ray Discrimination

A very important consideration in many applications of BF_3 tubes is their ability to discriminate against gamma rays, which often are found together with the neutron flux to be measured. Gamma rays interact primarily in the wall of the counter and create secondary electrons that may produce ionization in the gas. Because the stopping power for electrons in gases is quite low, a typical electron will deposit only a small fraction of its initial energy within the gas before reaching the opposite wall of the counter. Thus, we should expect that most gamma-ray interactions will result in low-amplitude pulses that will lie in the tail to the left of point A in Fig. 14-3b. Simple amplitude discrimination can then easily eliminate these gamma rays without sacrificing neutron detection efficiency.

If the gamma-ray flux is sufficiently high, however, several complications can reduce the effectiveness of this amplitude discrimination. At high rates, pulse pile-up can result in apparent peak amplitudes for gamma rays which are considerably larger than any individual pulse. Brown[13] discusses the compromise that must then be struck in choosing the pulse-shaping time constant in the detector electronics. Short time constants are desirable to reduce the gamma-ray pile-up but may lead to reduction in the neutron-induced pulse amplitude due to incomplete charge integration. At very high gamma rates, there is evidence that chemical changes occur in the BF_3 gas due to molecular disassociation, leading to degraded pulse height spectra from neutron-induced events.[14] If this degradation is sufficiently severe, it may no longer be possible to separate gamma- and neutron-induced events.[15] In extreme cases, the radiation-induced chemical changes can result in permanent damage to the tube. Verghese et al.[16] report successful discrimination against gamma rays at exposure rates as high as 12 R/h using a conventional BF_3 tube. Developmental tubes that employ activated charcoal within the tube to act as an absorbing agent for contaminants have been reported.[17] These tubes exhibit good operating characteristics in gamma-ray fluxes up to 1000 R/h.

D. Detection Efficiency of a BF₃ Tube

The detection efficiency for neutrons incident along the axis of a BF_3 tube is given approximately by

$$\epsilon(E) = 1 - \exp\left[-\Sigma_a(E)L\right] \qquad (14\text{-}4)$$

where
$$\Sigma_a(E) = \text{macroscopic absorption cross}$$
$$\text{section of } {}^{10}B \text{ at energy } E$$
$$L = \text{active length of the tube}$$

Using Eq. (14-4), we find that the calculated efficiency for a 30 cm long BF_3 tube (96% enriched in ^{10}B) filled to 600 torr (80 kPa) is 91.5% at thermal neutron energies (0.025 eV) but drops to 3.8% at 100 eV. Thus, a BF_3 tube exposed to neutrons with mixed energies will respond principally to the slow neutron component. Equation (14-4) slightly over-estimates the neutron counting efficiency because there usually are regions near the end of the tube in which charge collection is inefficient, resulting in reduced neutron response. The influence of these *dead spaces* is most severe for detectors whose length is small and has been the subject of experimental investigations that lead to a more precise prediction of detector efficiency.[18,19] *End window* designs are common in which the dead space and structural materials at one end of the tube are minimized.

Most practical BF_3 counters are filled with pure boron trifluoride enriched to about 96% in ^{10}B. However, because BF_3 is not ideal as a proportional counter gas, counters are sometimes manufactured using an admixture of BF_3 with a more suitable gas such as argon. This dilution causes a decrease in detection efficiency, but the pulse height spectrum from the tube generally shows sharper peaks and consequently a more stable counting plateau than tubes filled with pure BF_3.

E. Boron-Lined Proportional Counters

An alternate approach is to introduce the boron in the form of a solid coating on the interior walls of an otherwise conventional proportional tube. This configuration has the advantage that a more suitable proportional gas than BF_3 can now be used. Some applications, particularly those in which fast timing is important, are better served by introducing one of the common proportional gases discussed in Chapter 6. Also, the chemical degradation problems in BF_3 when exposed to high gamma ray fluxes can be greatly reduced by using alternative fill gases.

Because the maximum range of the alpha particles from the boron reaction is on the order of 1 mg/cm^2, the efficiency of boron-lined counters will improve only as the coating thickness is increased to about this value. Making the deposit any thicker will simply create layers in the coating which are too far from the filling gas to permit any reaction products to reach the gas, and the efficiency will actually begin to decrease slightly due to the added attenuation of the incident neutrons. Efforts have been made to increase the surface area available for coating by introducing boron-coated plates or baffles within cylindrical tubes, but these configurations have not achieved widespread popularity.

The pulse height spectrum to be expected from a boron-lined proportional chamber with a thick boron layer is sketched in Fig. 14-4. Because the interactions are now taking place in the wall of the chamber and the reaction products are oppositely directed, only

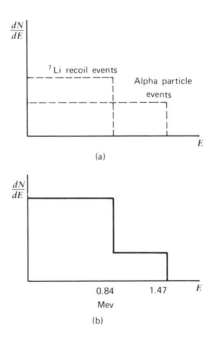

Figure 14-4 Idealized pulse height spectra from a boron-lined proportional tube. (a) Separate contributions of alpha particles and lithium recoil nuclei, which add to give the spectrum shown in plot (b).

one reaction product can be expected per interaction. If the alpha particle is directed toward the interior of the tube, the maximum energy it can deposit is its initial kinetic energy of 1.47 MeV. The actual alpha particle energy deposited in the gas will vary from this value down to zero as the possible location of the neutron interaction varies from the surface of the boron coating through those locations that are more than an alpha range away from the counter gas. Because all these locations are almost equally probable, the expected energy deposition distribution for alpha particles will be approximately rectangular in shape with a maximum at 1.47 MeV. This distribution is sketched in Fig. 14-4a as a dashed rectangle. A parallel argument can be made for the lithium recoil nucleus, with its maximum possible deposited energy equal to 0.84 MeV. The sum of these two individual rectangular distributions is shown as the solid line in Fig. 14-4b and is a somewhat idealized indication of the expected pulse height spectrum from a boron-lined chamber with a thick (greater than 1 mg/cm^2) boron lining. From the discussion given in Chapter 4, a differential pulse height spectrum without a "valley" structure does not lead to a counting plateau. Thus, it would be expected that boron-lined chambers would be less satisfactory than BF$_3$ tubes in terms of long-term counting stability. Because the average energy deposited for neutron interaction is also considerably less than for BF$_3$ tubes, the gamma-ray discrimination ability of boron-lined chambers will be inferior to that of BF$_3$ tubes.

F. Boron-Loaded Scintillators

Because the output pulse from a BF$_3$ tube originates with reaction products created with a more-or-less random location and direction, typical pulses will have rise times that vary by as much as 3–5 μs for tubes of average size. A further disadvantage for neutron time-of-flight applications is that the point of interaction of the neutron cannot be defined more precisely than somewhere within the volume occupied by the BF$_3$ fill gas. Because typical tubes are as much as 10–20 cm long to provide reasonable interaction efficiency, pathlength uncertainties can be large.

In order to circumvent both of the above limitations, other types of boron-loaded detectors have been investigated. Scintillators made by fusing B$_2$O$_3$ and ZnS have found wide application in neutron time-of-flight measurements. These scintillators are usually kept quite thin (1–2 mm) because of the relative opaqueness of this material to its own scintillation light and also to minimize pathlength uncertainty. Other work has been reported[20] on combinations of boron-containing polyester plastic and zinc sulfide. Boron-based scintillators are typically less effective in discrimination against gamma-ray backgrounds compared with BF$_3$ tubes because most scintillators have a lower scintillation efficiency for charged particles compared with that for fast electrons.

III. DETECTORS BASED ON OTHER CONVERSION REACTIONS

A. Lithium-Containing Slow Neutron Detectors

Because a stable lithium-containing proportional gas does not exist, a lithium equivalent of the BF$_3$ tube is not available. Nonetheless, the larger Q-value of the lithium reaction offers some real advantage whenever discrimination against gamma-ray pile-up and other low-amplitude events is at a premium. Also, because the lithium reaction goes exclusively to the ground state of the product nucleus, the same energy is always imparted to the reaction products for all slow neutron interactions. The resulting pulse height distribution in detectors that absorb all this energy is therefore a simple single peak. Although examples can be found[21] of the application of gas-filled detectors with solid lithium-based converters, the more common applications of this reaction employ the scintillation process to detect the products of the neutron-induced reaction.

Lithium-containing scintillators are quite common as slow neutron detectors. A logical choice, because of its chemical similarity to sodium iodide, is crystalline lithium iodide. If a small amount (less than 0.1 at. %) of europium is incorporated as an activator, light outputs of about 35% of the equivalent NaI(Tl) yield can be achieved. The scintillation mechanism is similar to that discussed earlier for sodium iodide. The scintillation decay time is approximately 0.3 μs.

Crystals of lithium iodide are generally large compared with the ranges of either of the reaction products from a neutron interaction. Therefore, the pulse height response will be free of wall effects and should be a single peak for all slow neutron interactions. The range of secondary electrons produced by gamma rays will not be large compared with typical crystal dimensions. The scintillation efficiency for lithium iodide is nearly the same for both electrons and heavy charged particles. (A 4.1 MeV electron will yield about the same light as the 4.78 MeV reaction products.) Therefore, a single gamma-ray interaction in lithium iodide is capable of producing a maximum pulse height approximately proportional to the energy of the gamma ray, whereas each neutron interaction will produce a pulse equivalent to 4.1 MeV on the same scale. The gamma-ray rejection characteristics will therefore be inferior to that of typical gas-filled neutron detectors, in which a gamma ray can deposit only a small fraction of its energy.

Similar to sodium iodide, lithium iodide is highly hygroscopic and cannot be exposed to water vapor. Commercially available crystals are hermetically sealed in a thin canning material with an optical window provided on one face. Because of the high density of the material, crystal sizes need not be large for very efficient slow neutron detection. For example, a 10 mm thick crystal prepared from highly enriched ^{6}LiI remains nearly 100% efficient for neutrons with energy from thermal through the cadmium cutoff of 0.5 eV.

Other recipes for lithium-containing scintillators have achieved some popularity. One of these reported by Stedman[22] consists of a lithium compound dispersed in a matrix of ZnS(Ag) with thickness of about 0.6 mm. These scintillators are commercially available as NE 421 (Nuclear Enterprises Limited). Their efficiency is quoted as 25–30% for 0.1 eV neutrons. Thin layer combinations of LiF and ZnS are described in Refs. 23 and 24. Because of their small thickness, gamma discrimination is very effective since a large fraction of all secondary electrons created by gamma-ray interactions will escape without depositing their full energy.

Finally, lithium-containing glass scintillators have become quite popular in neutron physics research. They are less common as slow neutron detectors, and therefore a discussion of this type of detector is postponed until Chapter 15. The same is true for combinations of lithium deposits with semiconductor detectors, and the so-called lithium sandwich spectrometer is also discussed in Chapter 15.

B. The ^{3}He Proportional Counter

With a cross section even higher than that of the boron reaction, the ^{3}He(n, p) reaction is an attractive alternative for slow neutron detection. Unfortunately, because ^{3}He is a noble gas, no solid compounds can be fabricated and the material must be used in gaseous form.

^{3}He of sufficient purity will act as an acceptable proportional gas, and detectors based on this approach have come into common use. General properties of ^{3}He propor-

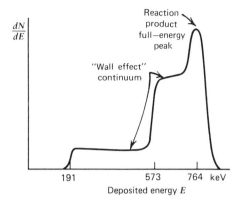

Figure 14-5 Expected pulse height spectrum from a ^{3}He tube in which the wall effect is significant.

tional tubes are surveyed in Ref. 25. In a large detector, one would expect each thermal neutron reaction to deposit 764 keV in the form of kinetic energy of the triton and proton reaction products. Because the range of these reaction products is not always small compared with the dimensions of the proportional tube, however, the wall effect discussed earlier for a BF_3 tube can also be important for ^{3}He proportional counters. The expected pulse height spectrum for a tube of typical size is illustrated in Fig. 14-5. Only a single full-energy peak should be expected for neutron energies that are small compared with 764 keV. The step structure to the left of the peak is similar to that shown in Fig. 14-3b for a BF_3 tube, except that the discontinuities will occur at energies corresponding to that of the proton (573 keV) and triton (191 keV).

The continuum in the pulse height spectrum due to the wall effect is detrimental from several standpoints. The voltage range over which an acceptable counting plateau will be observed is reduced, and the smaller pulse height for some neutron events will reduce the separation expected from low-amplitude, gamma-induced pulses. Consequently, consideration is often given in the design of ^{3}He tubes to minimize the wall effect. One obvious step is to build the counter with a diameter as large as possible so that most neutron interactions occur far away from the wall. Another is to increase the pressure of the ^{3}He gas to reduce the range of the charged particle reaction products. Because of the low atomic mass of ^{3}He, the ranges of the reaction products are unusually long and the wall effect is considerably more significant than for a BF_3 tube of the same size and fill gas pressure. One method of reducing the charged particle ranges is to add a small amount of a heavier gas to the ^{3}He to provide an enhanced stopping power. A detailed analysis of the wall effect in ^{3}He counters can be found in Ref. 26.

Compared with BF_3 tubes, ^{3}He counters can be operated at much higher pressures with acceptable gas multiplication behavior and are therefore preferred for those applications in which maximum detection efficiency is important. The lower Q-value of the ^{3}He reaction, however, makes gamma-ray discrimination more difficult than for an equivalent BF_3 tube.

The acceptable operating temperature for ^{3}He tubes has been shown[27,28] to extend as high as 200–250°C. In general, the pulse amplitude increases and the pulse height resolution decreases with increasing temperature, while the pulse rise time shows little temperature dependence.

As with all proportional counters, purity of the gas is critical, and the most typical cause of failure is leakage of air into the tube over long periods of time. Another factor is the buildup of electronegative poisons in the gas with use. As in BF_3 tubes, a layer of activated charcoal within the tube has been shown to be effective in removing these poisons and can extend the useful lifetime of a ^3He detector.[29]

C. Fission Counters

The fission reaction can also serve as a means of converting a slow neutron into ionizing reaction products that can then be detected by conventional means. One outstanding characteristic of the fission reaction is the large amount of energy (200 MeV) liberated in the reaction, about 160 MeV of which appears as kinetic energy of the fission fragments. Therefore, the neutron-induced fission reactions can be expected to be of much larger magnitude in most detectors than any other competing reaction or other event due to background or counter contamination. Under these circumstances, extremely low background rates can be achieved and neutron counting can be practically carried out at very low counting rates. In some detector types (scintillators are often an extreme example) the output pulse does not always linearly reflect the energy of the exciting particle. Nonlinearities are particularly important for densely ionizing particles such as fission fragments, and therefore the size of the neutron-induced detector pulses may not be as large as would be calculated based on simple linearity with energy. In that event, the fission fragments still may be considerably larger than competing reactions, but not by the several orders of magnitude one might ordinarily expect.

The most popular form of fission detector is an ionization chamber that has its inner surfaces coated with a fissile deposit. Because of adverse chemical and physical properties, little success has been realized in trying to incorporate the fissile material in gaseous form as part of the counter gas. In this section we stress properties of fission chambers operated in pulse mode, which is typical of nonreactor applications. Fission chambers can also function in current mode, most often as reactor flux monitors discussed later in this chapter.

The pulse height spectrum to be expected from a fission chamber depends primarily on the fissile deposit thickness and the geometric conditions under which the fission fragments are collected. For deposits that are very thin compared with fragment ranges, the familiar "double humped" fission fragment energy spectrum is observed, with the light and heavy fragment distributions peaking at about 100 and 70 MeV, respectively. If the deposit is made thicker to enhance detection efficiency, the energy loss of fragments within the deposit will reduce the average fragment energy and distort the shape of the measured distribution. Figure 14-6 illustrates the changes in spectrum shape to be expected as the thickness of a deposit of UO_2 is changed. These energy loss effects limit the practical deposit thickness to about 2–3 mg/cm². For a layer of highly enriched ^{235}U of this thickness, the corresponding detection efficiency in 2π counting geometry is about 0.5% at thermal energy, dropping to about 0.1% at 0.5 eV. Typical fission chambers employ a single layer, and are thus limited to an equivalent neutron detection efficiency. More complex fission chambers with higher detection efficiency can be designed[31-33] by providing multiple layers of fissile deposits, and detecting the fragments in each segment of the chamber between the layers.

The dimensions of the counter tend to be similar to that of alpha particle detectors, because the average range of a fission fragment is approximately half the range of 5 MeV alpha particles. Furthermore, because the fission fragment starts out with a very large

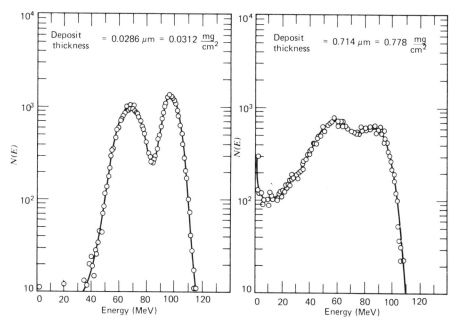

Figure 14-6 Energy spectra of fission fragments emerging from flat UO_2 deposits of two different thicknesses. A 2π detector is assumed which responds to fragments emitted in all directions from one surface of the deposit. (From Kahn et al.[30])

positive charge (on the order of 15 or 20 electronic charges) the energy loss at the beginning of the track is at its maximum, and the rate of energy loss continues to decrease as the fragment slows and additional charges are picked up. Just the reverse is true for most light charged particles, for which the rate of energy loss peaks near the end of the track. Therefore, in those detectors that do not fully stop the particle, fission fragments will yield a larger fraction of their total energy than would be expected for alpha particles or protons of the same range. For typical counter fill gases, the mean range is a few centimeters so that chambers of this dimension or larger can function reasonably as fission chambers.

The two fission fragments are always oppositely directed for slow-neutron-induced fission, and therefore detectors with a solid coating of fissionable material will respond only to the single fragment that is directed toward the active volume of the chamber. Some fission counters have been built with extremely thin backing material underneath a thin fissile deposit, so that both fragments can escape into opposite halves of a dual chamber, thereby permitting simultaneous detection of both fragments. Coincidence techniques can then discriminate against alpha particles and other background events. The very thin supports required for the fissile deposit are quite fragile, and consequently this type of fission chamber is not widely used in routine neutron detection applications.

The fissionable materials used in the construction of fission chambers are almost always alpha radioactive to some degree. As a result, typical fission chambers have an irreducible alpha-induced pulse rate that is an undesirable background. Because the average alpha particle energy is typically 5 MeV, whereas an average fission fragment energy is 10 times larger, these alpha pulses can often be simply discriminated on an amplitude basis. In those fissionable materials in which the alpha activity is relatively high (such as ^{239}Pu), problems often arise due to pile-up of alpha pulses to an amplitude that can overlap the lowest energy fission fragments. Because the maximum amplitude expected from pulse pile-up is an inverse function of detector resolving time, it can be suppressed through the use of fast fission detectors. Small methane-filled fission chambers have been reported[34] which show pulse widths as small as 10 ns. Extremely fast fission detectors can also be built by detecting fission fragments in a gaseous scintillator or by incorporating the fissionable material as part of fast solid scintillators.[35] An extensive review of various fission chamber designs may be found in Ref. 36.

IV. REACTOR INSTRUMENTATION

A. General Considerations

In thermal nuclear reactors, most of the power is generated through fission induced by slow neutrons. Therefore, nuclear sensors that are to be part of reactor control or safety systems are generally based on detectors that respond primarily to slow neutrons. In principle, many of the detector types discussed earlier in this chapter can be adapted for application to reactor measurements. However, the extreme conditions associated with reactor operation often lead to substantial design changes, and a category of slow neutron detectors designed specifically for this application has gradually evolved. Detailed reviews of reactor instrumentation are given in Refs. 6 and 7.

It is conventional to subdivide reactor instruments into two categories: *in-core* and *out-of-core*. In-core sensors are those that are located within narrow coolant channels in the reactor core and are used to provide detailed knowledge of the flux shape within the core. These sensors can be either fixed in one location or provided with a movable drive and must obviously be of rather small size (typically on the order of 10 mm diameter). Some examples are given in Ref. 37. Out-of-core detectors are located some distance from the core and thus respond to properties of the neutron flux integrated over the entire core. The detectors may be placed either inside or outside the pressure vessel, and normally will be located in a much less severe environment compared with in-core detectors. Size restrictions are also less of a factor in their design.

The majority of neutron sensors for reactor use are of the gas-filled type. Their advantages in this application include the inherent gamma-ray discrimination properties found in any gas detector, their wide dynamic range and long-term stability, and their resistance to radiation damage. Detectors based on scintillation processes are less suitable because of the enhanced gamma-ray sensitivity of solid scintillators and the radiation-induced spurious events that occur in photomultiplier tubes. Semiconductor detectors are very sensitive to radiation damage and are never used in high radiation environments.

Nearly all the gas-filled chambers, whether they are ionization or proportional counters, can be operated in a variety of modes. In pulse mode, each neutron interaction must be separated by sufficient time so that it may be resolved as an individual pulse. This mode is therefore limited to the lower ranges of neutron flux measurement but offers the benefit of gamma discrimination through simple amplitude selection of the output pulses. Pulse mode operation of most detectors is conventionally limited to rates below about 10^5 per second, although state-of-the-art techniques in chamber design and pulse-processing electronics can raise this limit to as high as 10^7 per second (see references cited in Ref. 38).

When flux levels become high enough so that pulse mode operation is no longer possible, neutron detectors are often operated in current mode. With proper ion chamber design, the range of operation can be extended to the maximum flux level of interest in reactors before serious nonlinear effects due to ion–electron recombination set in. Lower limits of current mode operation are usually determined by leakage currents that inevitably arise in the detector insulation and cable dielectric material. By operating the chamber in current mode, one sacrifices any chance of inherent gamma-ray discrimination because all pulses, whether large or small, add some contribution to the measured current.

One method of reducing gamma-ray sensitivity is to use the MSV mode of operation described in Chapter 4. This operational mode, commonly called the Campbelling technique, consists of deriving a signal that is proportional to the mean square of the current fluctuations from an ion chamber.[39-41] The mean square signal is seen from Eq. (4-8) to be proportional to the average pulse rate and the square of the ionization charge generated in each pulse. Because neutron-induced pulses result in much greater charge than pulses from gamma rays, the mean square signal will weight the neutron component by the square of the ratio of neutron- to gamma-ray-induced charge. This increase in neutron sensitivity is of greatest advantage in fission chambers. Although the MSV mode enhances the neutron sensitivity, it will not completely eliminate the gamma contribution. MSV mode operation has proved to be most useful in the intermediate reactor power ranges and in wide-range reactor control channels,[42-44] where one detector provides input to instrumentation that can operate in pulse mode, MSV mode, or current mode depending on the neutron flux level being measured.

A second approach for reducing the importance of the gamma-ray signal is to employ direct gamma-ray compensation in a specialized detector known as a compensated ion chamber (CIC). The CIC typically uses boron-lined ion chambers operating in current mode. Because of the much smaller Q-value of the neutron-induced reaction, neutron interactions in a boron-lined chamber result in an order of magnitude less charge than do neutron-induced events in a fission (uranium-lined) chamber. The effectiveness of the MSV mode of operation in discriminating against gamma rays is therefore reduced. The alternative approach of using a CIC is found to be more effective in reducing the gamma-ray contribution in boron-lined chambers than is the MSV mode of operation. The CIC, illustrated functionally in Fig. 14-7, employs a dual ion chamber from which two independent ion currents can be extracted separately. One chamber is boron-lined, whereas the construction of the second is nearly identical in terms of active volume and structural material, but without the boron lining. The current I_1 from the boron-lined chamber will consist of the sum of the current due to neutron interations in the boron and the gamma-ray interactions in the chamber walls and gas-filled region. The current I_2

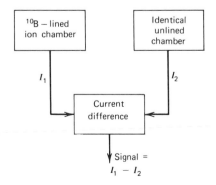

Figure 14-7 Operating principle of a compensated ion chamber (CIC).

from the unlined chamber will reflect only the gamma-ray contribution. By taking the difference between these two currents, a signal current is derived which in principle will be proportional to only the neutron contribution. The two chambers are often constructed as concentric cylinders and consequently are not identical in their response to gamma rays. Therefore, the compensation is not exact and may vary slightly with gamma-ray flux. This variation may require adjustments in the balance between the two chambers to restore exact compensation for different gamma-ray flux levels. (See Refs. 6 and 7 for further discussion of CIC design and operation.)

B. Reactor Nuclear Instrument (NI) System Overview

Before proceeding with further discussion of the specialized detectors used in reactor safety and control systems, a general discussion of the distinctly different system approaches used in pressurized water reactors (PWR) and boiling water reactors (BWR) is instructive. The reader is referred to Ref. 44 for a more detailed description of reactor NI systems.

1. PRESSURIZED WATER REACTOR NUCLEAR INSTRUMENTATION

Detectors for the routine monitoring of reactor power in a PWR are located outside the reactor pressure vessel and are characterized by the following typical environmental conditions: neutron flux up to 10^{11} n/cm^2 · s, gamma irradiation rates up to 10^6 R/h, and temperatures of approximately 100°C. Out-of-core sensors are the usual basis of reactor control and safety channels in a PWR. In choosing specific detector types, consideration must be given to the expected neutron signal level compared with noise sources, the speed of response of the detector, and the ability to discriminate against gamma-induced signals. Each of these criteria assumes different importance over various ranges of reactor power, and as a result multiple detector systems are usually provided, each designed to cover a specific subset of the power range.

Figure 14-8 illustrates a typical scheme for a PWR in which three sets of sensors with overlapping operating ranges are used to cover the entire power range of the reactor. The lowest range, usually called the source start-up range, is encountered first when bringing up reactor power from shut-down conditions. This range is characterized by conditions in which the gamma flux from the fission product inventory in the core may be large compared with the small neutron flux at these low power levels. Under these conditions, good discrimination against gamma rays is at a premium. Also, the expected neutron interaction rates will be relatively low in this range. Pulse mode operation of either fission chambers or BF$_3$ proportional counters is therefore possible, and the required gamma-ray

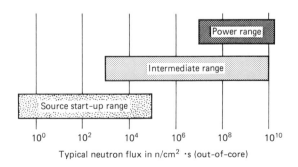

Figure 14-8   Typical ranges covered by out-of-core neutron detectors in a PWR.

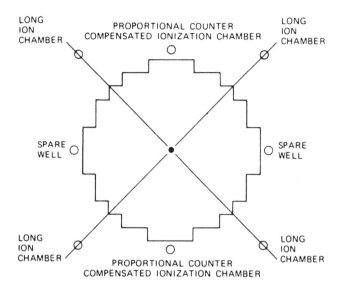

Figure 14-9 Geometric arrangement of out-of-core neutron detectors relative to the core of a PWR. (From Harrer and Beckerley.[44])

discrimination can be accomplished by accepting only the much larger amplitude neutron pulses.

As the power level is increased, an intermediate range is encountered in which pulse mode operation is no longer possible because of the excessive neutron interaction rate. In this region the gamma-ray-induced events are still significant compared with the neutron flux, and therefore simple current mode opertion is not suitable. The MSV mode of operation can reduce the importance of the gamma-ray signal in this range, but a more common method used in PWRs is to employ direct gamma-ray compensation using a CIC.

A third range of operation corresponds to the region near the full operating power of the reactor. The neutron flux here is usually so large that gamma-ray-induced currents in ion chambers are no longer significant, and simple uncompensated ion chambers are commonly used as the principal neutron sensor. Because these instruments are often part of the reactor safety system, there is a premium on simplicity which also favors uncompensated ion chamber construction.

In most PWR NI systems the geometric arrangement shown in Fig. 14-9 is employed. Two BF_3 proportional counters used in the source start-up range are placed on opposite sides of the core and two CIC detectors used in the intermediate range are placed in the same location or on the two opposing sides. Four power range monitors are then located at 90° intervals at positions between the BF_3 and CIC detectors. Each of the four power range monitors consists of two uncompensated ion chambers arranged end-to-end, resulting in a total detector length of 3–4 m. This arrangement provides both radial and axial neutron flux data for control and safety at full power as well as axial flux offset information needed for control of xenon oscillations.

Although PWR nuclear instrumentation employs out-of-core gas-filled detectors (primarily ion chambers) for control and safety channels, there remains a need for information on in-core spatial variations of the neutron flux. This information is necessary for fuel management and is provided by various types of detectors placed within the reactor core (see following descriptions).

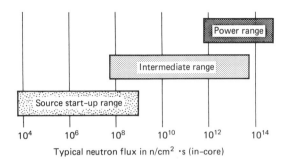

Figure 14-10   Typical ranges covered by in-core neutron detectors in a BWR.

2. BOILING WATER REACTOR NUCLEAR INSTRUMENTATION

The BWR NI system, like the PWR system, has three overlapping ranges as illustrated in Fig. 14-10. The three systems are called source, intermediate, and power range monitors. Unlike the PWR, which uses out-of-core neutron detectors, the neutron detectors are all located in-core. There are also many more detectors used in the BWR NI system than in the PWR system.

The source range monitoring system typically consists of four in-core fission chambers operating in pulse mode. Pulse mode operation provides good discrimination against gamma rays, which is necessary when measuring a relatively low neutron flux in the presence of a high gamma flux. A typical intermediate range monitoring system has eight in-core fission chambers operating in the MSV mode. The MSV mode promotes the enhanced neutron to gamma response required to provide a proper measure of neutron flux in the presence of gamma rays for both control and safety requirements. The power range monitoring system typically consists of 144–164 fission ion chambers distributed throughout the core. The fission chambers operate in current mode and are called local power range monitors (LPRM). Current mode operation provides satisfactory neutron response at the high flux levels encountered between 2 and 150% full power: In a typical system, approximately 20 LPRMs are summed to provide input to one of the seven or eight average power range monitoring (APRM) systems. The APRM system provides input for both control and reactor protection systems. Details of the neutron monitoring systems used in the BWR are described in Ref. 44.

A diagram of a typical fission chamber used in BWR NI is shown in Fig. 14-11. The physical size of the detectors used in each range are similar. The fill-gas pressures and uranium loadings are different, depending on whether the detector is designed to operate in pulse, MSV, or current mode. The LPRM system typically employs a fission chamber with a mixture of both ^{235}U and ^{234}U. The advantage of this design for extended operation in a high neutron flux is described in the next section.

C. In-Core Detectors

There is often a need to place neutron sensors within the core of a nuclear reactor to provide information on the spatial variation of the neutron flux. Because of the small size (1–7 cm) of the channel in which these instruments must be located, emphasis is placed on compactness and miniaturization in their design. They may either be left in a fixed position or provided with a motorized drive to allow traverses through the reactor core. Some may provide a continuous readout, whereas others are interrogated only at periodic intervals. Typical operating conditions are: neutron flux at full power of 5×10^{13}

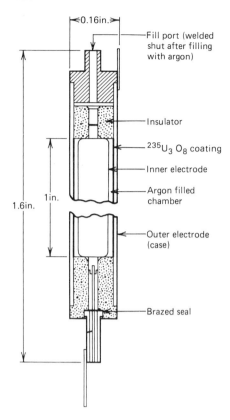

Figure 14-11 A typical in-core fission chamber used in BWR neutron monitoring systems.

n/cm^2 · s, gamma flux up to 10^8 R/h, operating temperature up to 300°C, and operating pressure up to 2500 psi (17 MPa). In-core reactor measurements are the subject of a number of papers found in Ref. 37.

1. FISSION CHAMBERS

Miniaturized fission chambers can be tailored for in-core use over any of the power ranges likely to be encountered in reactor operation. Walls of the chamber are usually lined with highly enriched uranium to enhance the ionization current. These small ion chambers are typically made using stainless steel walls and electrodes, and operating voltage varies from about 50 to 300 V. Argon is a common choice for the chamber fill gas and is used at a pressure of several atmospheres. The elevated pressure ensures that the range of fission fragments within the gas does not exceed the small dimensions of the detector.

The gradual burn up of neutron-sensitive material is a serious problem for the long-term operation of in-core detectors. For example, a fission chamber using ^{235}U will show a sensitivity decrease of about 50% after exposure to an integrated neutron fluence of about 1.7×10^{21} n/cm^2 (Ref. 45). One method of reducing the effects of burnup in fission chambers is to combine fertile and fissile material in the neutron-sensitive lining of the chamber. Use of these *regenerative* chambers will gradually convert the fertile isotopes to fissile nuclei to help compensate for the burnup of the original fissile material

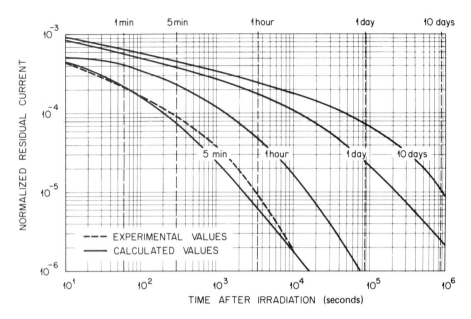

Figure 14-12 Memory effect in fission chambers corresponding to various steady-state irradiation times. (From Roux.[46])

present in the lining. The long-term response of fission chambers can greatly be improved with this method. For example, it is reported that fission chambers based on a mixture of ^{238}U and ^{239}Pu can maintain a sensitivity that does not change more than $\pm 5\%$ over an integrated neutron fluence of 4.8×10^{21} n/cm^2 (Ref. 45). Similar results have also been obtained with fission chambers based on a mixture of ^{234}U and ^{235}U.

Fission ion chambers that have been operated for long periods in high neutron fluxes will show a residual current or *memory* effect due to the buildup of fission products within the chamber. These fission products emit beta and gamma rays, which ionize the fill gas of the chamber and result in a significant ion current. Figure 14-12 shows the results to be expected if the current from a fission chamber is monitored following its removal from long-term exposure in a steady-state neutron flux. The residual current I is plotted as a fraction of the steady-state current I_0 observed during the neutron irradiation. One minute after removal about 0.1% of the signal current persists, whereas after 10 days the fission product activity has decayed sufficiently so that the residual current has dropped to about 10^{-5} of the steady-state signal.

One effect that can be important in ion chambers which must cover a wide range of irradiation rates is illustrated in Fig. 14-13. At lower rates the region of ion saturation is reached at a lower voltage than at higher rates. When the current is high, the density of ionization is correspondingly high and recombination will occur more readily than at lower currents. The electric field required to prevent recombination is therefore higher at high rates, evidenced by the increased voltage required to achieve ion saturation. It is important to select an operating voltage for these chambers at the highest irradiation rate or largest current that will be encountered. Although the change in current–voltage characteristics with increased neutron flux may be greater for in-core detectors than out of core detectors, a similar effect is observed in both the compensated and uncompensated ion chambers used in pressurized water reactors.

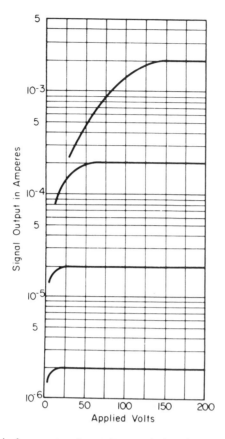

Figure 14-13 Typical current–voltage characteristics of an in-core fission chamber at different neutron irradiation levels. (Courtesy of Westinghouse Electric Corporation, Electronic Tube Division, Horseheads, NY).

2. SELF-POWERED DETECTORS

A unique type of neutron detector that is widely applied for in-core use is the *self-powered detector*. These devices incorporate a material chosen for its relatively high cross section for neutron capture leading to subsequent beta or gamma decay. In its simplest form, the detector operates on the basis of directly measuring the beta decay current following capture of the neutrons. This current should then be proportional to the rate at which neutrons are captured in the detector. Because the beta decay current is measured directly, no external bias voltage need be applied to the detector, hence the name self-powered. Another form of the self-powered detector makes use of the gamma rays emitted following neutron capture. Some fraction of these gamma rays will interact to form secondary electrons through the Compton, photoelectric, and pair production mechanisms. The current of the secondary electrons can then be used as the basic detector signal.

The self-powered detector is also known by a variety of other names. In recognition of some of the early pioneering work done by J. W. Hilborn,[47,48] they are sometimes called Hilborn detectors. Other names that can be found in the literature include beta emission detectors, collectrons, electron emission detectors, and PENA (primary emission,

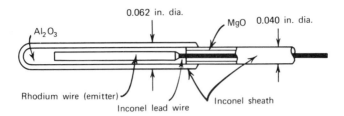

Figure 14-14 Cross-sectional view of a specific self-powered detector design. (From Stevens.[51])

neutron activation) detectors. Nonetheless, the *self-powered neutron* (SPN) detector remains the most common term applied to this family of devices.

Compared with other neutron sensors, self-powered detectors show advantages due to their small size, low cost, and the relatively simple electronics required in conjunction with their use. Disadvantages stem from the low level of output current produced by the devices, a relatively severe sensitivity of the output current to changes in the neutron energy spectrum, and, for many types, a rather slow response time. Because the signal from a single neutron interaction is at best a single electron, pulse mode operation is impractical and self-powered detectors are always operated in current mode. Various types of SPN detectors are surveyed in Ref. 49, and considerations in their application to in-core measurements in a PWR are reviewed in Ref. 50.

a. Detectors Based on Beta Decay

Figure 14-14 shows a sketch of a typical SPN detector based on beta decay. The heart of the device is the emitter, which is made from a material chosen for its relatively high cross section for neutron capture leading to a beta-active radioisotope. Ideally, the remainder of the detector does not interact strongly with the neutrons, and construction materials are chosen from those with relatively low neutron cross sections. Figure 14-15 illustrates some possible sequences which can contribute to the measured current. The principles of operation are very simple: The current corresponding primarily to the beta rays given off by the emitter is measured between the emitter and an outer shell, called the collector. The intervening space is filled with an insulator, which must be chosen to withstand the extreme temperature and radiation environment typically found in a reactor core. Various metallic oxides are often used as the insulator, with magnesium or aluminum oxide[52] most commonly used. The collector is typically high-purity stainless steel or Inconel. Great care must be taken in the fabrication of these detectors to keep the materials as clean as possible to avoid contamination by substances that might also become radioactive and add an interfering current to the measured signal. The small dimensions shown in Fig. 14-14 are necessitated by the small clearances available for instrument channels in typical reactor cores.

The key to the detector performance lies in the choice of emitter material. Factors to be considered in the selection of the emitter include its neutron capture cross section together with the energy and half-life of the resulting beta activity. The capture cross section must be neither too high nor too low, because very low cross sections will lead to detectors with low sensitivity, whereas an excessively high cross section will result in rapid burnup of the emitter material[53] in the high neutron fluxes associated with reactor cores. The beta rays produced should be of sufficiently high energy so that excessive self-absorption in the emitter or insulator is avoided, and the half-life of the induced activity should

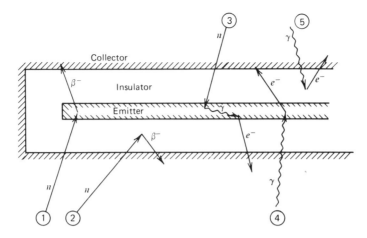

Figure 14-15 Representative events that can take place in a SPN detector. Events ① and ② are neutron capture followed by β^- decay. Event ③ shows the interaction of a prompt gamma ray emitted upon neutron capture, giving rise to a fast secondary electron. Events ④ and ⑤ show interfering fast electrons arising from interactions of external gamma rays. In standard SPN detectors, event ① is the basis of its neutron response. In those with fast response, event ③ is the preferred mode of interaction.

be as short as possible to permit the detector to respond quickly to rapid changes in neutron flux.

Based on these criteria, the two most popular choices for emitter material have been rhodium and vanadium. Table 14-1 summarizes some of the important properties of these materials when used as emitters in self-powered detectors. Vanadium produces a fairly simple beta decay with a half-life of 225 s, whereas rhodium gives rise to a more complex beta decay with an admixture of half-lives of 44 and 265 s. Figure 14-16 shows the resulting response of these materials to a step change in neutron flux level. Despite the fact that vanadium has a lower sensitivity and somewhat slower response than rhodium, vanadium emitters have become more common in reactor applications because the rate of burnup is significantly less, permitting use over periods of years in typical reactor fluxes.

In its simplest form, a self-powered detector with a single mode of induced activity and negligible burnup would behave as follows when exposed to a neutron flux for a

TABLE 14-1 Properties of Emitter Materials for SPN Detectors Based on Beta Decay

Emitter Material	Nuclide of Interest and Percent Abundance[a]	Activation Cross Section at Thermal Energy[b]	Half-Life of Induced Beta Activity[a]	Beta Endpoint Energy[a]	Typical Neutron Sensitivity[c]
Vanadium	$^{51}_{23}V$ (99.750%)	4.9 barns	225 s	2.47 MeV	$5 \times 10^{-23} \dfrac{A}{n/cm^2 \cdot s}$
Rhodium	$^{103}_{45}Rh$ (100%)	$\begin{cases} 139 \text{ barns} \\ 11 \text{ barns} \end{cases}$	$\begin{cases} 44 \text{ s} \\ 265 \text{ s} \end{cases}$	2.44 MeV	$1 \times 10^{-21} \dfrac{A}{n/cm^2 \cdot s}$

[a] Data from Lederer and Shirley, *Table of Isotopes*, 7th ed., Wiley & Sons, New York, 1978.
[b] Data from BNL-325, 3rd ed., Vol. 1 (1973).
[c] Sensitivity quoted for emitter of 1 cm length and typical diameter.

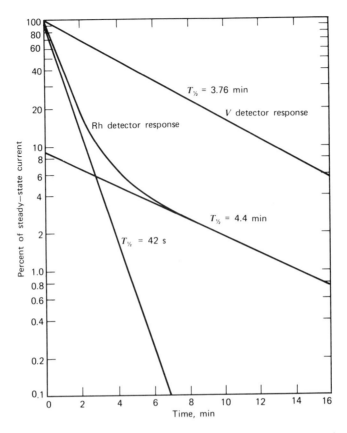

Figure 14-16 Response of rhodium and vanadium SPN detectors to an abrupt drop to zero of a steady-state neutron flux. (From Stevens.[51])

period of time t:

$$I(t) = Cq\sigma N\varphi(1 - e^{-\lambda t}) \qquad (14\text{-}5)$$

where C = dimensionless constant reflecting the specific geometry
and collection efficiency of the detector

q = charge liberated (number of beta particles $\times e$) per neutron absorbed

σ = activation cross section of the emitter material

N = number of emitter atoms

φ = neutron flux

λ = decay constant of activity produced in the emitter.

At saturation, or when the detector has been exposed to the flux for a period of time that is long compared with the half-life of the induced activity, the steady-state current is given simply by

$$I_{sat} = Cq\sigma N\varphi \qquad (14\text{-}6)$$

The saturated current is proportional to the neutron flux and consequently can serve as a corresponding monitor of the neutron flux level.

A more complex analysis of the output of these detectors must take into account a number of other factors, including neutron flux depression due to emitter self-shielding, the Compton and photoelectrons produced from gamma rays which may accompany the beta decay process, and the self-absorption probability of the beta particles within the emitter. Some electrons are stopped in the insulator before reaching the collector, whereas others can be produced within the insulator material and travel to either the emitter or the collector. After some period of operation, an equilibrium is established in which the net amount of charge flowing into the insulator is exactly balanced by the charge flowing out. Monte Carlo calculations have shown that the current due to processes taking place within the insulator is typically less than 15%[54] and consequently is not a dominant effect in determining detector response. Fairly detailed physical models have been described[54-57] which appear to give adequate predictions[58] of the response of SPN detectors, taking into account most of these effects.

b. Self-Powered Detectors Based on Secondary Electrons from Gamma Decay

One of the primary disadvantages of self-powered detectors based on beta decay is their relatively slow response time. Some efforts have been made to remedy this situation by electronic or digital processing of the signals,[59-61] but it would be preferable to speed up the inherent response of the detector itself. One method of accomplishing this objective is to rely on the secondary electrons produced by prompt capture gamma rays which can follow neutron capture in the emitter (see Fig. 14-15). These capture gamma rays are typically emitted within a small fraction of a second, as opposed to the much slower decay of typical neutron-induced beta activities. Even in vanadium and rhodium detectors, there is a prompt component of the signal that, although much smaller than the signal due to the beta current, corresponds to the prompt capture gamma rays emitted upon neutron capture in these materials. The ratio of prompt to delayed signal in commercial vanadium detectors is reported to be about 6.5%.[59]

For fast self-powered detectors, it is more common to choose a specific emitter material that will optimize the signal arising from prompt capture gamma rays. Most experience has been gained using cobalt as a prompt emitter,[62-64] with cadmium also used in commercially available detectors. The behavior and sensitivity of other prompt emitter materials have been reported by Gebureck et al.[65] With some exceptions, the neutron sensitivity is substantially less for prompt detectors compared with those based on beta decay, but the much faster response can make them the detector of choice for certain applications.

The response to external gamma rays also can be quite significant for some emitter materials. Gamma rays incident on the detector give rise to secondary electrons, which can yield a discernible signal; the process is illustrated in Fig. 14-15. The gamma-ray-induced signal can be either positive or negative, in that the net flow of current may be either in the same or the opposite direction to the neutron-induced current. Which polarity prevails depends on whether more gamma-ray-induced electrons flow from the emitter to collector or vice versa. Either situation may exist, depending on the specific construction of the detector. The commonly used emitter materials for neutron-sensitive detectors (rhodium, vanadium, or cobalt) all have gamma-ray responses that typically are less than a few percent of the neutron response.[66] Detectors made with zirconium emitters have an almost pure gamma-ray response, whereas other materials such as platinum, osmium, or cerium will give a mixed response. Shields[66] first described the properties and application of a platinum detector for in-core flux measurements in power reactors, in which the combined gamma/neutron sensitivity provides a mix of prompt and delayed

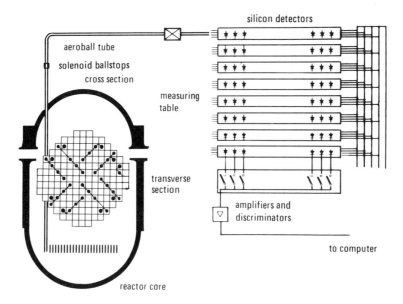

Figure 14-17 The aeroball system for continuous activation measurements of the neutron distributions in a reactor core. (From Glasow.[71])

response. Platinum detectors have gained considerable popularity, and analyses of their response have been published in Refs. 67–70.

In all types of self-powered detector, the effects of neutron and gamma-ray interactions in the connecting cable can be quite significant. As a result, considerable care must be taken in choosing the construction materials for the signal cable that lead to the detector, especially in those regions in which these cables are subjected to high radiation fields. In order to aid in the suppression of false signals arising in the connection cables, twin signal leads are often used. One lead is connected through the cable to the emitter, whereas the other is included within the same cable but is terminated without electrical contact physically near the emitter. By electronically subtracting the unconnected lead signal from the current detected from the lead connected to the emitter, effects due to cable interactions are approximately canceled out.

3. THE AEROBALL SYSTEM FOR MONITORING POWER DISTRIBUTION

A unique in-core monitoring system referred to as the computerized aeroball system[71] is used in some pressurized water reactor systems in the Federal Republic of Germany. In the aeroball system, probes containing neutron-sensitive isotopes are introduced into the reactor core and are subsequently activated. In the system shown in Fig. 14-17, the probes are columns of 1.7 mm diameter steel balls. The ball columns are piped into stainless steel tubes incorporated in selected fuel elements throughout the reactor core. The length of the columns corresponds to the height of the core. Vanadium is added to the balls as a neutron flux indicator through the 225 s half-life ^{52}V activity induced by neutron capture in ^{51}V. The activation of ^{51}V and its characteristics are summarized in Table 14-1 and the time response is shown in Fig. 14-16.

The activity of the probes is measured, following removal from the core, by using a set of silicon detectors. The count rate from each silicon detector is proportional to the relative integrated thermal neutron flux at the point where the corresponding aeroball was located during the activation process. From many such measurements, the power density distribution throughout the reactor core can be determined.

PROBLEMS

14-1. Discuss the feasibility of operating a BF_3 or ^3He tube in the ionization or Geiger regions rather than as a proportional tube. What practical considerations dictate the latter choice?

14-2. When operated at a gas multiplication factor of 1000, estimate the pulse amplitude produced by the interaction of a thermal neutron in a ^3He tube of 100 pF capacitance.

14-3. A BF_3 tube using natural boron shows a counting efficiency of 1% for 10 eV neutrons in a given application. By what factor can the efficiency be increased by using boron enriched to 96% ^{10}B?

14-4. Calculate the detection efficiency of a BF_3 tube (96% enriched in ^{10}B) filled to 600 torr (80 kPa) for incident thermal neutrons if their pathlength through the gas is 10 cm.

14-5. In which applications might one prefer to use a ^3He tube rather than a BF_3 tube for slow neutron counting?

14-6. In BF_3 tubes of small diameter, the "step" in the pulse height spectrum of Fig. 14-3b at 1.47 MeV has a finite positive slope that is much more noticeable than that for the step at 0.84 MeV. Give a physical explanation for this observation.

14-7. Sketch the pulse height spectrum expected from a boron-lined proportional tube for thermal neutrons, if the boron thickness is small enough so that energy loss of the reaction products in the layer can be neglected.

14-8. Using the data given in Table 8-3, estimate the number of scintillation photons liberated by the interaction of a thermal neutron in $^6LiI(Eu)$ and in a typical Li glass scintillator.

14-9. Why is it not possible to increase the counting efficiency of a fission chamber indefinitely simply by increasing the thickness of the fissionable deposit?

14-10. The signal current from a typical application of a self-powered neutron detector is seldom more than a nanoampere. Find the equivalent number of beta particles transferred per second between the emitter and collector.

14-11. Find the fractional decrease in sensitivity (the "burnup") of a rhodium self-powered detector if used in a neutron flux of $3 \times 10^{13}/cm^2 \cdot s$ over a period of 6 months.

REFERENCES

1. W. D. Allen, *Neutron Detection*, George Newnes, Ltd., London, 1960.
2. W. J. Price, *Nuclear Radiation Detection*, 2nd ed., Chap. 10, McGraw-Hill, New York, 1964.
3. H. Neuert, *Kernphysikalische Messverfahren*, Chap. 9, Verlag G. Braun, Karlsruhe, 1966.
4. A. Lorenz, *A Review of Neutron Detection Methods and Instruments*, UCID-16325 (1973).
5. G. Grosshoeg, *Neutron Ionization Chambers*, North-Holland, Amsterdam, 1979.
6. J. F. Boland, *Nuclear Reactor Instrumentation (In-Core)*, Gordon & Breach, New York, 1970.
7. J. M. Harrer and J. G. Beckerley, *Nuclear Power Reactor Instrumentation Systems Handbook*, Vol. 1, Chaps. 2 and 3, TID-25952-P1 (1973).

8. I. O. Anderson and S. Malmskog, "Investigation of the Pulse Height Distribution of Boron Trifluoride Proportional Counters," AE-84 (1962).

9. R. Cervellati and A. Kazimierski, *Nucl. Instrum. Meth.* **60**, 173 (1968).

10. I. L. Fowler, *Rev. Sci. Instrum.* **34**, 731 (1973).

11. E. Sakai, S. Usui, H. Ohkado, Y. Hayashi, and H. Nakatani, *IEEE Trans. Nucl. Sci.* **NS-30** (1), 802 (1983).

12. C. W. Peters, A. L. Snyder, and A. S. Gallia, Jr., *IEEE Trans. Nucl. Sci.* **NS-13**(1), 636 (1966).

13. D. P. Brown, *IEEE Trans. Nucl. Sci.* **NS-21** (1), 763 (1974).

14. T. Tomoda and S. Fukakusa, *Nucl. Instrum. Meth.* **224**(3), 557 (1984).

15. J. Csikai and M. Buczko, *Nucl. Instrum. Meth.* **8**, 73 (1960).

16. K. Verghese, J. R. Bohannon, and A. D. Kowalczuk, *Nucl. Instrum. Meth.* **74**, 355 (1969).

17. A. J. Stokes, T. J. Meal, and J. E. Myers, Jr., *IEEE Trans. Nucl. Sci.* **NS-13**(1), 630 (1966).

18. I. L. Fowler and P. R. Tunnicliffe, *Rev. Sci. Instrum.* **21**, 734 (1950).

19. T. E. Sampson and D. H. Vincent, *Nucl. Instrum. Meth.* **95**, 563 (1971).

20. K. H. Sun, P. R. Malmberg, and F. A. Pecjak, *Nucleonics* **14**, 7, 46 (1956).

21. O. P. Joneja, R. Hecker, and A. Mohsin, *Nucl. Instrum. Meth.* **193**, 563 (1982).

22. R. Stedman, *Rev. Sci. Instrum.* **31**, 1156 (1960).

23. R. H. Bossi and A. H. Robinson, *Trans. Am. Nuc. Soc.* **22**, 153 (1975).

24. F. Mantler-Niederstatter, F. Bensch, and F. Grass, *Nucl. Instrum. Meth.* **142**, 463 (1977).

25. W. R. Mills, Jr., R. I. Caldwell, and I. L. Morgan, *Rev. Sci. Instrum.* **33**, 866 (1962).

26. S. Shalev, Z. Fishelson, and J. M. Cuttler, *Nucl. Instrum. Meth.* **71**, 292 (1969).

27. E. Sakai, K. Kubo, and H. Yoshida, *IEEE Trans. Nucl. Sci.* **NS-27**(1), 776 (1980).

28. F. L. Glesius and T. A. Kniss, *IEEE Trans. Nucl. Sci.* **NS-35**(1), 867 (1988).

29. A. E. Evans, H. O. Menlove, R. B. Walton, and D. B. Smith, *Nucl. Instrum. Meth.* **133**, 577 (1976).

30. S. Kahn, R. Harman, and V. Forgue, *Nucl. Sci. Eng.* **23**, 8 (1965).

31. H. Ries, J. Drexler, R. Fischer, W. Günther, K. Huber, U. Kneissl, H. Ströher, and W. Wilke, *Nucl. Instrum. Meth.* **185**, 373 (1981).

32. A. A. Bogdzel, A. Duka-Zólyomi, J. Kliman, V. Presperin, S. P. Avdeev, V. D. Kuznetsov, Z. Dlouhý, *Nucl. Instrum. Meth.* **200**, 407 (1982).

33. K. H. Valentine, M. K. Kopp, G. W. Allin, W. T. Clay, and V. C. Miller, *IEEE Trans. Nucl. Sci.* **NS-32** (1), 384 (1985).

34. N. W. Hill, J. T. Mihalczo, J. W. Allen, and M. M. Chiles, *IEEE Trans. Nucl. Sci.* **NS-22** (1), 686 (1975).

35. E. Catalano and J. B. Czirr, *Nucl. Instrum. Meth.* **143**, 61 (1977).

36. R. W. Lamphere, "Fission Detectors," in *Fast Neutron Physics*, Part I, p. 449 (J. B. Marion and J. L. Fowler, eds.), Interscience Publishers, New York, 1960.

37. *Proceedings of the Specialists' Meeting on In-Core Instrumentation and Reactor Assessment*, Nuclear Energy Agency (NEA), Organization for Economic Co-Operation and Development (OECD), 1984.

38. W. H. Ellis, J. L. Cooper, Jr., and G. H. Sanders, *IEEE Trans. Nucl. Sci.* **NS-20**(1), 639 (1973).

39. R. A. DuBridge, *IEEE Trans. Nucl. Sci.* **NS-14**(1), 241 (1967).

40. H. A. Thomas and A. C. McBride, *IEEE Trans. Nucl. Sci.* **NS-15**(1), 15 (1968).

41. N. R. Campbell and V. J. Francis, *JIEE* **93**, Part III (1946).

42. M. Oda, M. Wada, and S. Badono, *IEEE Trans. Nucl. Sci.* **NS-23**(1), 304 (1976).

43. S. Shirayama, T. Itok, C. C. Wimpee, and J. Sturz, Proceedings of the Specialists' Meeting on In-Core Instrumentation and Reactor Assessment, NEA/OECD, pp. 66–77 (1984).

44. J. M. Harrer and J. G. Beckerley, *Nuclear Power Reactor Instrumentation Handbook*, Vol. 2, Chaps. 15 and 16, TID-25952-P2 (1974).

45. H. Böck and E. Balcar, *Nucl. Instrum. Meth.* **124**, 563 (1975).

46. D. P. Roux, ORNL-3929 (1966).

47. J. W. Hilborn, *Nucleonics* **22**, 2, 69 (1964).

48. J. W. Hilborn, "Self-Powered Neutron Detector," U.S. Patent 3,375,370 (March 26, 1968).

49. N. P. Goldstein and W. H. Todt, *IEEE Trans. Nucl. Sci.* **NS-26**(1), 916 (1979).

50. D. P. Bozarth and H. D. Warren, *IEEE Trans. Nucl. Sci.* **NS-26**(1), 924 (1979).

51. H. H. Stevens, "Neutron Sensors—In-Core," Chap. 3 in *Nuclear Power Reactor Instrumentation Systems Handbook*, TID-25952-P1 (1973).

52. D. P. Bozarth and H. D. Warren, *Trans. Am. Nucl. Soc.* **23**, 517 (1976).

53. H. D. Warren, *Trans. Am. Nucl. Soc.* **23**, 460 (1976).

54. N. P. Goldstein, *IEEE Trans. Nucl. Sci.* **NS-20**(1), 549 (1973).

55. W. Jaschik and W. Seifritz, *Nucl. Sci. Eng.* **53**, 61 (1974).

56. H. D. Warren and N. H. Shah, *Nucl. Sci. Eng.* **54**, 395 (1974).

57. H. D. Warren, *Nucl. Sci. Eng.* **48**, 331 (1972).

58. P. S. Rao and S. C. Misra, *Nucl. Instrum. Meth. Phys. Res.* **A253**, 57 (1986).

59. W. Seifritz, *Nucl. Sci. Eng.* **49**, 358 (1972).

60. J. M. Carpenter, R. F. Fleming, and H. Bozorgmanesh, *Trans. Am. Nucl. Soc.* **22**, 606 (1975).

61. L. A. Banda and B. I. Nappi, *IEEE Trans. Nucl. Sci.* **NS-23**(1), 311 (1976).

62. H. Böck and M. Stimler, *Nucl. Instrum. Meth.* **87**, 299 (1970).

63. H. Böck, *Nucl. Instrum. Meth.* **125**, 327 (1975).

64. J. C. Kroon, F. M. Smith, and R. I. Taylor, *Trans. Am. Nucl. Soc.* **23**, 459 (1976).

65. P. Gebureck, W. Hofmann, W. Jaschik, W. Seifritz, and D. Stegemann, IAEA-SM-168/G-8, p. 783 (1973).

66. R. B. Shields, *IEEE Trans. Nucl. Sci.* **NS-20**(1), 603 (1973).

67. G. F. Lynch, R. B. Shields, and P. G. Coulter, *IEEE Trans. Nucl. Sci.* **NS-24**(1), 692 (1977).

68. N. P. Goldstein, *IEEE Trans. Nucl. Sci.* **NS-25**(1), 292 (1978).

69. N. P. Goldstein, C. L. Chen, and W. H. Todt, *IEEE Trans. Nucl. Sci.* **NS-28**(1), 752 (1981).

70. D. S. Hall, *IEEE Trans. Nucl. Sci.* **NS-29**(1), 646 (1982).

71. P. A. Glasow, *Nucl. Instrum. Meth.* **226**, 17 (1984).

CHAPTER · 15

Fast Neutron Detection and Spectroscopy

In Chapter 14 on slow neutron detection, a number of neutron-induced reactions were discussed which can serve as the basis for the conversion of neutrons to directly detectable charged particles. In principle, all these reactions could be applied to detect fast neutrons as well. As shown in the cross-section plot of Fig. 14-1, however, the probability that a neutron will interact by one of these reactions decreases rapidly with increasing neutron energy. As a result, conventional BF_3 tubes have an extremely low detection efficiency for fast neutrons and consequently are almost never used for this purpose. For reasons to be discussed later in this chapter, the ^{3}He proportional counter is useful both for thermal neutron detection and for fast neutron spectroscopy. As a rule, however, fast neutron devices must employ a modified or completely different detection scheme to yield an instrument with acceptable detection efficiency.

The most important additional conversion process useful for fast neutrons is elastic neutron scattering. In this interaction an incident neutron transfers a portion of its kinetic energy to the scattering nucleus, giving rise to a *recoil nucleus*. The energy that can be transferred from a slow neutron is therefore very small, and the resulting recoil nuclei are too low in energy to generate a usable detector signal. Once the neutron energy reaches the keV range, however, recoil nuclei can be detected directly and assume a large importance for fast neutron detection. By far the most popular target nucleus is hydrogen. The cross section for neutron elastic scattering from hydrogen is quite large and its energy dependence is accurately known. More important, however, is the fact that an incident neutron can transfer up to its entire energy in a single collision with a hydrogen nucleus, whereas only a small fraction can be transferred in collisions with heavy nuclei. Therefore, the resulting *recoil protons* are relatively easy to detect and serve as the basis for a wide variety of fast neutron detectors.

An important distinction in the application of fast neutron detectors is whether or not an attempt is made to measure the energy of the incoming neutron. For all the slow neutron detection methods discussed in Chapter 14, the information on initial neutron kinetic energy is hopelessly lost in the conversion process, because the neutron energy is extremely small compared with the energy liberated in the conversion reaction itself (the Q-value). Once the incoming neutron energy is no longer negligible compared with the reaction Q-value (that means at least 10–100 keV for most of the reactions discussed in

Chapter 14), the energy of the reaction products begins to change appreciably with changes in the neutron energy. An accurate measure of the reaction product energies can then, in principle, be used to deduce the incoming neutron energy. In elastic scattering the reaction Q-value is zero, so that neutron energies can begin to be measured by this technique at the point at which the resulting recoils have measurable kinetic energy. The collection of instruments and techniques applied to the measurement of fast neutron energy is conventionally included in the category of *fast neutron spectroscopy*.

In some instances, however, the purpose of the measurement is simply to record the presence of fast neutrons without a corresponding measurement of their energy. Such fast neutron counters can employ any of the methods discussed to convert neutrons to charged particles, and then simply record all pulses from the detector. Fast neutron counters of this type will have a severe variation in efficiency with neutron energy, but if the incident neutron energy is not likely to change greatly between measurements, they can provide useful information on the relative intensity of a fast neutron flux. Other applications in which the fast neutron spectrum may change considerably between measurements benefit from counters tailored to the application. We begin our discussion of fast neutron devices with counters of this type.

I. COUNTERS BASED ON NEUTRON MODERATION

A. General Considerations

The inherently low detection efficiency for fast neutrons of any slow neutron detector can be somewhat improved by surrounding the detector with a few centimeters of hydrogen-containing moderating material. The incident fast neutron can then lose a fraction of its initial kinetic energy in the moderator before reaching the detector as a lower-energy neutron, for which the detector efficiency is generally higher. By making the moderator thickness greater, the number of collisions in the moderator will tend to increase, leading to a lower value of the most probable energy when the neutron reaches the detector. One would therefore expect the detection efficiency to increase with moderator thickness if that were the only factor under consideration. A second factor, however, tends to decrease the efficiency with increasing moderator thickness: The probability that an incident fast neutron ever reaches the detector will inevitably decrease as the moderator is made thicker. Several effects are at work here, as illustrated in Fig. 15-1. As the detector becomes a smaller and smaller fraction of the total volume of the system, there will be a lower probability that a typical neutron path will intersect the detector before escaping from the surface of the moderator. Furthermore, a neutron may be absorbed within the moderator before it has a chance of reaching the detector. The absorption probability will increase rapidly with increasing moderator thickness because absorption cross sections generally are larger at lower neutron energies.

As a result of all these factors, the efficiency of a moderated slow neutron detector when used with a monoenergetic fast neutron source will show a maximum at a specific moderator thickness. Assuming that the moderator is the usual choice of a hydrogenous material such as polyethylene or paraffin, we find that the optimum thickness will range from a few centimeters for keV neutrons up to several tens of centimeters for neutrons in the MeV energy range.

If the thickness of the moderator is fixed at a fairly large value, the overall counting efficiency of the system versus incident neutron energy will also tend to show a maximum. Low-energy neutrons will not penetrate far enough into the moderator before they are

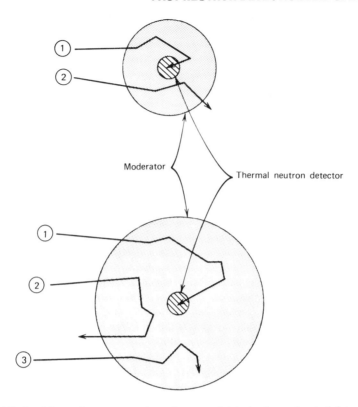

Figure 15-1 Schematic representation of neutron histories in moderated detectors. The small thermal neutron detector at the center is shown surrounded by two different thicknesses of moderator material. Histories labeled ① represent incident fast neutrons that are successfully moderated and detected. Those labeled ② are partially or fully moderated but escape without reaching the detector. History ③ represents those neutrons that are parasitically captured by the moderator. Larger moderators will tend to enhance process ③ while reducing process ②. See text for a discussion of the variation of process ①.

likely to be captured in the moderator itself, whereas high-energy neutrons will not be adequately moderated for efficient detection. By careful choice of the diameter and composition of the moderator–detector system, its overall efficiency versus energy curve can often be shaped and tailored to suit a specific application.

B. The Spherical Dosimeter

In an effort to develop a useful neutron spectrometer, Bramblett, Ewing, and Bonner[1] first investigated the properties of a small lithium iodide scintillator placed at the center of polyethylene moderating spheres of different diameters. The general behavior seen in this type of study is shown in Fig. 15-2. The difference in the shapes and position of the maxima in these response curves serves as the basis for using the set of spheres as a simple neutron spectrometer. By measuring the count rate with each sphere individually, an unfolding process can, in principle, provide some information about the energy distribution of the incident neutrons.[2-5] Because the response functions are rather broad, however, this approach to neutron spectroscopy has received only limited application.

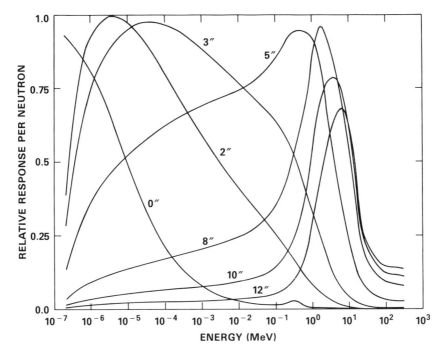

Figure 15-2 The energy dependence of the relative detection efficiencies of Bonner sphere neutron detectors of various diameters up to 12 inches. (From Johnson et al.[5])

A very useful type of counter has emerged from these studies, however. The response curve shown in Fig. 15-2 for the 12 in. sphere turns out to have a similar shape compared with the dose equivalent delivered per neutron as a function of energy. This resemblance is entirely coincidental, in that there is no fundamental relation between the probability of neutron detection at the center of the sphere and the magnitude of the dose delivered by that neutron in a biological medium. Nonetheless, this happy circumstance can be used to good advantage when it is desired to measure the dose equivalent due to neutrons with an unknown or variable energy spectrum. Because of the similarity of the two curves, the efficiency of the detector is high for those neutrons whose biological importance is high and is low for neutrons that deliver less dose. Therefore, the overall count from the detector in a polyenergetic spectrum will automatically include proper weighting factors for all energies and give a meaningful measure of the combined dose due to all the neutrons. The small size (typically 4 mm × 4 mm) of the LiI scintillator and the relatively high Q-value of the lithium capture reaction for neutrons allow for very effective discrimination against gamma rays, even in relatively large gamma-ray fields. The spherical geometry provides for a reasonably nondirectional detector response. Typical intrinsic efficiency for a 12 in. sphere, defined as the fraction of neutrons that strike the surface of the sphere which ultimately result in a count, is 2.5×10^{-4} at the peak of the 12 in. detector response. Translated into dose, the average response corresponds to about 3×10^3 counts/mrem. It is virtually the only monitoring instrument that can provide realistic neutron dose estimates over the many decades of neutron energy ranging from thermal energies to the MeV range. However, because the detection mechanism is not fundamentally related to radiation dose, substantial errors can arise when applied to widely different source conditions.[6]

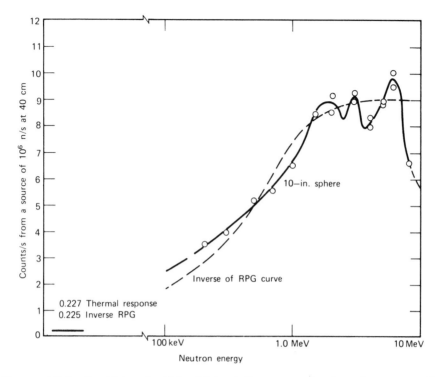

Figure 15-3 Sensitivity of a 10 in. (25.4 cm) diameter moderating sphere surrounding a 4 mm·× 4 mm LiI scintillator. Also shown is the relative dose per neutron labeled "inverse of RPG curve." (From Hankins.[11])

These spherical moderator systems are generally known as Bonner spheres after one of the authors (T. W. Bonner) of the original paper describing its experimental investigation.[1] More recent work[7-10] has reexamined the response of spherical moderators used in combination with several different thermal neutron detectors, both experimentally and by neutron transport calculational techniques.

Hankins[11,12] chose to study the response of a 10 in. diameter (25.40 cm) sphere used with a 4 mm × 4 mm lithium iodide scintillator. The moderator size was selected as that likely to give the closest fit between the efficiency and the dose equivalent per incident neutron over a wide range of neutron energy. Figure 15-3 shows the estimated response of the 10 in. sphere detector together with the dose per neutron curve. Obviously, the match is quite good over several decades of neutron energy. Below 100 keV, direct measurements of the detector efficiency are difficult. Hankins[11] has applied a multigroup neutron transport code to calculate the detector response and these results are shown in Fig. 15-4. The calculations confirm the good match between the two curves above 100 keV and at thermal energy but show a sizable deviation in the intermediate energy range between 0.1 eV and 100 keV. The deviation is such as to lead to an overestimate of the neutron dose if the spectrum contains a significant component over this energy range. Although the deviation at 10 keV is as much as a factor of 5, measurements made when the neutron spectrum covers a broad range in energies will show a considerably lower average deviation. Hankins reports that, in a variety of applications, use of the 10 in. instrument leads to a maximum overestimate of the dose of 65%.

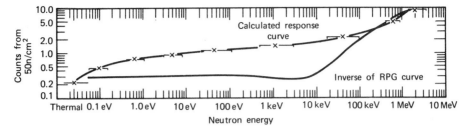

Figure 15-4 Calculated sensitivity of a 10 in. (25.4 cm) diameter moderating sphere, together with the relative dose per neutron (inverse of RPG curve). (From Hankins.[11])

An alternate version of the spherical neutron dosimeter developed by Leake[13] is shown in Fig. 15-5. This design has come to be widely implemented as the Harwell type 95/0075 neutron survey meter. A spherical ^{3}He proportional counter is substituted for the lithium iodide scintillator as the slow neutron detector. This substitution is made to minimize the response of the detector to gamma rays, and Leake reports application of the dosimeter in gamma-ray fields as high as 20 R/h. Used with a simple 20.8 cm diameter polyethylene moderator, the energy response of the system to thermal and epithermal neutrons is higher than ideal. Therefore, a spherical cadmium absorber, perforated with holes, is placed around the ^{3}He detector to shape the response curve. Although the instrument still overresponds to neutrons in the keV energy range (by a factor of 4.9 at 10 keV), the response to broad neutron spectra typical of shielded fission sources does not deviate by more than ±40% for a very wide range of experimental and calculated spectra.[14,15] The response at high neutron energies drops off somewhat; the instrument underestimates the dose from 14 MeV neutrons shielded by concrete by about a factor of 2 (Ref. 14).

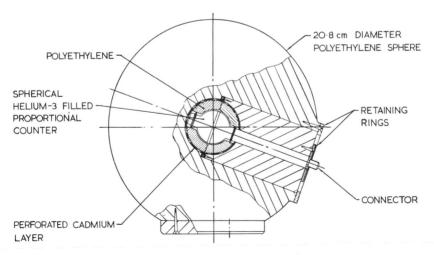

Figure 15-5 A spherical neutron dosimeter based on a ^{3}He neutron detector. (From Leake.[13])

C. The Long Counter

A detector whose counting efficiency does not depend on the neutron energy can be a very useful device in many areas of neutron physics. For an ideal detector of this type, a graph of the detection efficiency versus neutron energy is a horizontal line, which has led to the name *flat response* detectors. Although no real detector exists with a perfectly flat response over the entire range of possible neutron energies, several designs have evolved that come close to this ideal.

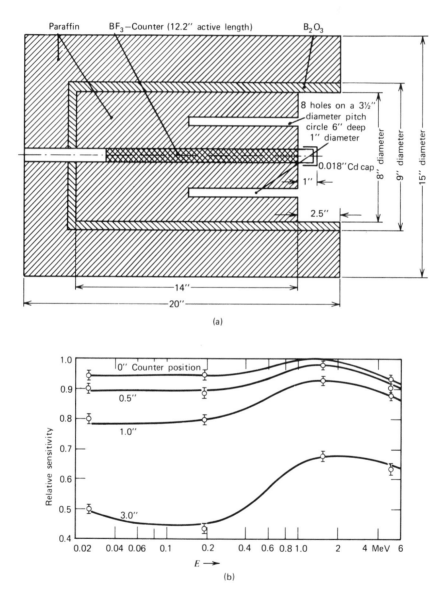

(a)

(b)

Figure 15-6 (a) Cross-section of the long counter developed by McTaggart. (b) Relative sensitivity of McTaggart long counter versus neutron energy. The parameter varied for the different curves is the distance the end of the BF_3 tube is shifted in from the front of the moderator face. The flattest response occurs when the tube is flush with the front face. (From K. H. Beckurts and K. Wirtz, *Neutron Physics*. Copyright 1964 by Springer-Verlag, Inc. Used with permission.)

Over the years, the most popular flat response neutron detector has been the *long counter*. Like the spherical neutron dosimeter, it is based on the principle of placing a slow neutron detector at the center of a moderating medium. For the long counter, however, the slow neutron detector is a BF_3 tube, and the system is designed to respond properly to neutrons only when they are incident from a specific direction.

The combination of a BF_3 tube and cylindrical moderator was first suggested as a flat response neutron detector by Hanson and McKibben.[16] A later design by McTaggart[17] is shown in Fig. 15-6a and has achieved fairly widespread acceptance as the standard long counter. The counter is designed to be sensitive only to neutrons incident on the right-hand face of the counter within the boron oxide shell. Those incident from other directions tend to be moderated by the outer annulus of paraffin and are subsequently captured in the boron layer without giving rise to a count. Neutrons incident on the front face parallel to the cylindrical axis will penetrate some distance before undergoing moderation. The average distance of penetration will increase as the neutron energy increases. If the BF_3 tube and cylindrical moderator are sufficiently long, then a typical cross section through the cylinder at the point of moderation will not be different for various energy neutrons. Therefore, the probability that the moderated neutron will find its way to the BF_3 tube and produce a count should not depend strongly on neutron energy. It is this property that leads to the flat energy response of the detector. The holes provided in the front surface prevent a fall-off in the efficiency at neutron energies below 1 MeV by allowing lower-energy neutrons to penetrate farther into the moderator, away from the front surface from which they might otherwise escape. Figure 15-6b shows a plot of the efficiency of a McTaggart long counter versus neutron energy for various axial positions of the BF_3 tube. With the BF_3 tube flush with the front surface, the efficiency does not change by more than 10% over the neutron energy range shown. A long counter of similar design by DePangher and Nichols[18] has also achieved some recognition as a standard in health physics measurements, and documentation of its flat response between about 2 keV and 6 MeV is given in Refs. 19–21.

Long counters derive many of their operational characteristics from the BF_3 tube on which their design is based. Sensitivity to relatively high levels of gamma rays can be

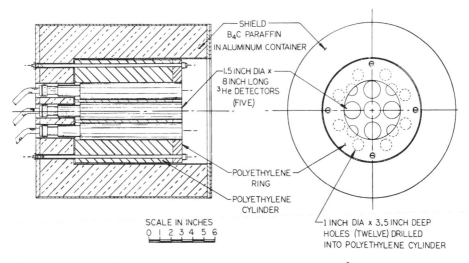

SHIELD
B_4C PARAFFIN
IN ALUMINUM CONTAINER

1.5 INCH DIA x
8 INCH LONG
^{3}He DETECTORS
(FIVE)

POLYETHYLENE
RING

POLYETHYLENE
CYLINDER

SCALE IN INCHES
0 1 2 3 4 5 6

1 INCH DIA x 3.5 INCH DEEP
HOLES (TWELVE) DRILLED
INTO POLYETHYLENE CYLINDER

Figure 15-7 A high-efficiency long counter utilizing multiple ^{3}He tubes. (From East and Walton.[22])

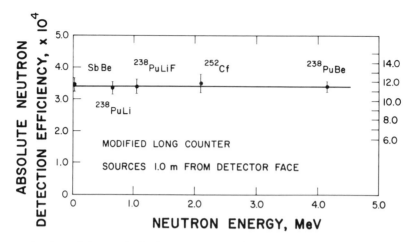

Figure 15-8 Efficiency of the long counter shown in Fig. 15-7 versus the average energy of some neutron sources. The efficiency figures are for a point source located 1 m from the detector face. (From East and Walton.[22])

eliminated by simple amplitude discrimination, while continuing to count all the neutron interactions in the tube. The long counter normally displays good long-term stability and traditionally has achieved widespread application as a neutron flux monitor in a wide variety of neutron physics experiments.

A modified long counter with some improved characteristics has been developed by East and Walton[22] and is shown in Fig. 15-7. It substitutes ^{3}He detectors for the BF_3 tube used in McTaggart design and provides five separate detectors near the center of the cylindrical moderator. By using high-pressure ^{3}He tubes, the multiple detector arrangement leads to a rather high overall neutron detection efficiency of 11.5%, compared with a standard long counter efficiency of about 0.25%. The 12 holes that penetrate the inner polyethylene cylinder are covered on the front face by a 19 mm thick ring of polyethylene to provide a geometry of moderation which best favors an overall flat response. The measured efficiency of this detector is plotted in Fig. 15-8 for a number of neutron sources with broad energy distributions. While the measured efficiency is essentially constant for these sources, greater fluctuation has been observed[23] for more nearly monoenergetic neutrons. These variations can arise because of narrow resonances in the cross sections of some of the constituent materials.

D. Other Detectors Based on Moderation

A number of detection systems have evolved which consist of a spherical assembly into which the neutron source is inserted. This approach can obviously be used for small portable neutron sources and can also be adapted for a highly collimated parallel beam of neutrons by providing a small-diameter entrance channel through which the beam can pass to reach the center of the assembly. A typical design consists of placing multiple ^{3}He or BF_3 counters in a hydrogenous moderator,[24-29] and in some sense is a spherical version of the long counter. Overall counting efficiencies on the order of 1% can be achieved with a response that is flat to within ±1% over a neutron energy range from 30 keV to 5 MeV.[25]

Other flat response detectors that also rely on neutron moderation have been developed. The *grey neutron detector* of Poenitz[30,31] uses a NaI(Tl) scintillator to detect the 2.2 MeV capture gamma rays produced when neutrons are thermalized in a hydrogenous sphere. To provide a faster response, the *black neutron detector* was also introduced,[32] which is based on the light produced in a hydrogenous spherical scintillator as the neutron is moderated. Both types of detector can provide a very flat efficiency curve over several decades of neutron energy.

II. DETECTORS BASED ON FAST NEUTRON-INDUCED REACTIONS

The detectors described in the previous section rely on the slowing down of a fast neutron in a moderating material before its detection as a thermal neutron. The moderating process eliminates all information on the original energy of the fast neutron and normally cannot be used if an attempt is made to extract energy information. Furthermore, the detection process is relatively slow. In most designs, the neutron must undergo multiple collisions with moderator nuclei followed by diffusion as a thermal neutron before the detection signal is generated. As a result, such detectors cannot provide a fast detection signal required in many neutron detection applications.

Both these limitations may be overcome if the fast neutron can be made to induce directly a suitable nuclear reaction without the moderation step. The reaction products will then have a total kinetic energy given by the sum of the incoming neutron kinetic energy and the Q-value of the reaction. Provided the neutron energy is not a hopelessly small fraction of the Q-value, a measurement of the reaction product energies will give the neutron energy by simple subtraction of the Q-value. Additionally, the detection process can potentially be fast because the incoming fast neutron will typically spend no more than a few nanoseconds in the active volume of the detector, and only a single reaction need occur to provide a detector signal. However, the cross sections for typical fast-neutron-induced reactions are orders of magnitude lower than the corresponding thermal neutron cross sections, and such detectors will inevitably show a much lower detection efficiency than their thermal neutron counterparts.

Excluding elastic scattering, which will be the topic of the next section, there are only two reactions of major importance in fast neutron spectroscopy: ^{3}He(n, p), and ^{6}Li(n, α). Both of these reactions were discussed at the beginning of Chapter 14. A plot of the cross-section variation with neutron energy for the fast region is shown in Fig. 15-9. We now emphasize the application of these reactions in neutron spectroscopy, where the neutron energy is inferred by measuring the energy of the reaction products. It should be clear that the same detectors can be used simply to detect the presence of fast neutrons by arranging to count all (or some fixed fraction) of the neutron-induced reactions in the detector. A third reaction, neutron-induced fission, is not of interest in spectroscopy because of the very high Q-value associated with the reaction. The fission process, however, can serve as the basis of a fast neutron counter if energy information is not required.

Finally, a class of fast neutron detectors (called *activation counters*) is based on detecting the radioactivity induced in certain materials. These detectors have proved to be useful when applied to sources producing short pulses of fast neutrons. Because these devices do not produce prompt signals, a discussion of their properties is postponed until Chapter 19.

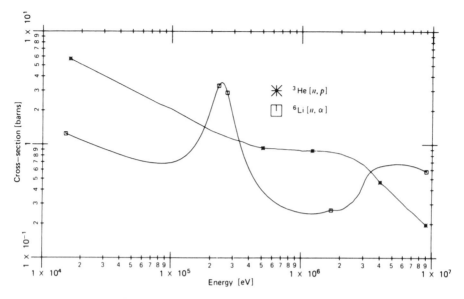

Figure 15-9 The ^{3}He(n, p) and ^{6}Li(n, α) cross sections for the fast neutron region.

A. Methods Based on the ^{6}Li(n, α) Reaction

As seen from Fig. 15-9, the cross section for the ^{6}Li(n, α) reaction drops off rather smoothly with increasing neutron energy, except for the pronounced resonance at a neutron energy of about 250 keV. The relatively large Q-value of 4.78 MeV is an advantage in thermal neutron detection but limits the application in fast neutron spectroscopy to neutrons with energy of at least several hundred keV. A competing reaction, ^{6}Li(n, n'd)^{4}He, has a Q-value of −1.47 MeV and becomes the dominant neutron-induced reaction at energies above about 2.5 MeV. Because this reaction leads to three products, one of which is a neutron that normally escapes, a continuum of deposited energy should be expected even for monoenergetic incident neutrons. Therefore, although it will contribute some neutron pulses, this additional reaction is generally an undesirable part of the response of any detector that attempts to measure the incident neutron energy.

If we neglect the continuum that may be introduced by this latter reaction, the response function of fast neutron detectors based on the lithium reaction should be a single peak located at an energy equal to the neutron energy plus the Q-value of the (n, α) reaction (4.78 MeV). In practical situations, an additional peak is often observed at 4.78 MeV from reactions induced by neutrons whose energy has been reduced to the thermal range by moderation in the laboratory walls, shielding, and any other material in the vicinity of the detector. Unless special care is taken to eliminate those low-energy neutrons, their large interaction cross section will result in many events, all of which deposit the same energy (the reaction Q-value) in the detector. The resulting peak is usually called the *epithermal peak* and can provide a convenient energy calibration point for the detector output.

1. THE LITHIUM IODIDE SCINTILLATOR

The lithium iodide (europium-activated) scintillation crystals discussed in Chapter 14 in connection with thermal neutron detection can also be applied to fast neutron spectroscopy.[33,34] However, the use of this material at room temperature is severely hampered by its nonlinear response to the tritons and alpha particles produced from the lithium reaction. This nonlinearity results in a resolution of about 40% for the full-energy peak due to incident thermal neutrons. This very broad response function seriously limits the application of LiI(Eu) in fast neutron spectroscopy. It has been shown[35] that the nonlinearities in the crystal response can be substantially reduced by cooling the crystal to liquid nitrogen temperature, which improves the resolution of the full-energy peak to about 20%. The practical problems involved in cooling the crystal are significant, but fast neutron spectra measurements have been reported using this technique.[34]

2. LITHIUM GLASS SCINTILLATORS

Lithium can also be incorporated in other scintillation matrices and used as a fast neutron detector. Because of their relatively poor light output and nonlinearity of response, these scintillators have not been used directly for neutron spectroscopy in the manner described above for lithium iodide scintillators. In neutron time-of-flight spectroscopy, however, the detector need register only the arrival time of a neutron, and various lithium scintillators have evolved for this application. Granular scintillators consisting of mixtures of lithium fluoride and zinc sulfide have been developed[36, 37] for time-of-flight use. Lithium-containing glass scintillators have become much more popular in these applications, however, because of the relatively fast response time and large areas that can easily be fabricated. Silicate glasses of various compositions, generally with cerium activation, are used as the scintillation medium.

General characteristics of glass scintillators are described in Refs. 38–42. Table 15-1 lists some properties of commercially available lithium glass scintillators and shows that lithium concentrations of up to 7.7% can be obtained. The low-background properties of NE912 are important for low-level neutron counting and are achieved through the use of materials that are low in natural thorium. In addition to the highly enriched ^{6}Li formulations shown in Table 15-1, equivalent scintillators are also available in which the lithium is present as natural lithium (7.5% ^{6}Li, 92.5% ^{7}Li) or depleted lithium (> 99.9% ^{7}Li). The latter are neutron insensitive and can be used to measure separately the gamma contribution in a mixed neutron–gamma irradiation.

TABLE 15-1 Properties of Some Commercially Available Lithium Glass Scintillators[a]

Manufacturer's Identification[b]	Type NE902	NE905	NE908	NE912
Density (g/cm³)	2.6	2.48	2.674	2.55
Refractive index	1.58	1.55	1.57	1.55
Melting point (°C)	1200	1200	1200	1200
λ of emission maximum (nm)	395	395	395	397
Light output relative to anthracene	22–34%	20–30%	20%	25%
Decay constant (ns)	75	100	5 + 75	75
Content of Li	2.2 wt %	6.6 wt %	7.5 wt %	7.7 wt %
^{6}Li enrichment	95%	95%	95%	95%
Background α activity per 100 g of glass (/min)	100–200	100–200	100–200	20
Resolution expected for thermal neutrons (depends on glass thickness)	13–22%	15–28%	20–30%	20–30%

[a] Data from McMurray et al.[39]

[b] Commercial identification numbers are those used by Nuclear Enterprises, Ltd. Lithium glass scintillators of similar properties are also available through Koch-Light Laboratories, Ltd.

The time resolution that can be achieved from these detectors depends somewhat on the pulse amplitude distributions produced by the incident neutrons but can be as low as a few nanoseconds. Unfortunately, glass scintillators show a much reduced light output per unit energy for charged particles compared with electrons, and reaction products with 4.78 MeV energy will yield about the same light output as a 1.2 MeV gamma ray. The gamma-ray discrimination ability of these scintillators[43] is therefore not as good as other detectors in which the response is more uniform for all particles. The detection efficiency for thick lithium glasses is of considerable interest in many applications and is difficult to calculate accurately due to the influence of multiple scattering within the glass. Resonances in the scattering cross sections of various materials in the glass lead to sharp peaks in the detection efficiency at neutron energies above about 100 keV.

3. LITHIUM SANDWICH SPECTROMETER

Another way in which the lithium reaction has been widely used to measure fast neutron energies is outlined in Fig. 15-10. A thin layer of lithium fluoride or other lithium-containing material is prepared on a very thin backing and placed between two semiconductor diode detectors. When the neutron energy is low, the two reaction products are oppositely directed, and coincident pulses should be observed from the two semiconductor detectors. Neglecting energy loss of the charged particles before they reach the active volume, the sum of the energy deposited in the two detectors should be equal to the incoming neutron energy plus the Q-value of the lithium reaction.

In practice, complications arise because of the necessity to use finite thicknesses for both the lithium deposit and the backing on which it is supported. Figure 15-10 illustrates that the energy loss in the target materials, which does not contribute to the detected

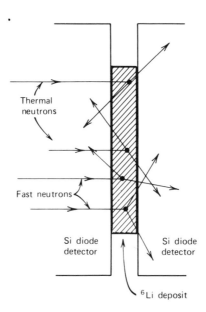

Figure 15-10 Elements of a lithium sandwich spectrometer. Reaction products from thermal neutrons are always oppositely directed, whereas fast neutron interactions will lead to some net forward momentum for the triton and alpha particle. If both are emitted in the forward direction, no coincidence will exist.

signal, will vary as the angle of the emitted reaction products changes through all possible values. Particularly troublesome are those reaction products that are emitted near the plane of the target foil, which will have long pathlengths through the deposit and backing before escaping into the detectors. These energy loss effects can be minimized by making the deposit very thin, but only at the expense of a reduced counting efficiency. Alternatively, the paths near the plane of the foil can be eliminated by geometric collimation between the deposit and detectors, but again, the counting efficiency will be reduced.

The employment of coincidence detection of the two reaction products greatly reduces the background in semiconductor sandwich spectrometers. Any background events that occur only in one detector will automatically be eliminated. If the neutron energy is significant compared with the Q-value, the reaction products must have some momentum in the direction of the incoming neutron and will not be exactly oppositely directed. Then, some fraction of all reactions will lead to two products, both of which strike the same detector and do not give rise to coincidences (see Fig. 15-10). The fraction of neutron events lost to this effect becomes more significant as the neutron energy increases.

Several methods for processing the data from semiconductor sandwich spectrometers have been described. The simplest is to record the sum signal from coincident pulses and deduce the neutron energy by subtracting the Q-value of the reaction. At neutron energies below several hundred keV, this method becomes quite sensitive to small errors and uncertainties. An alternative scheme first proposed by Maroni[44] is based on measuring the difference in energy between the triton and alpha particle. For reaction products that are collinear with the incoming neutron direction, this difference is very sensitive to small changes in neutron energy below about 100 keV. Rickard[45] discusses several other approaches to the analysis of the detector signals which can be advantageous in some applications. General discussions of the application of lithium sandwich detectors in fast neutron spectroscopy can be found in Refs. 46–52.

B. Detectors Based on the ^{3}He(n, p) Reaction

The ^{3}He(n, p) reaction discussed in Chapter 14 has also been widely applied to fast neutron detection and spectroscopy. The fast neutron cross section plotted in Fig. 15-9 falls off continuously with increasing neutron energy. Several competing reactions must be considered in any detector based on this reaction. The most significant of these is simple elastic scattering of the neutrons from helium nuclei. The cross section for elastic scattering is always larger than that for the (n, p) reaction, and this predominance becomes more pronounced as the neutron energy becomes larger. For example, the cross sections are about equal at a neutron energy of 150 keV, but elastic scattering is about three times more probable at 2 MeV. In addition, a competing (n, d) reaction on ^{3}He is possible at neutron energies exceeding 4.3 MeV, but the cross section is low for energies below about 10 MeV. The (n, p) reaction and elastic scattering therefore account for all the important features of ^{3}He detector response for all but the highest neutron energies.

The pulse height spectrum from a detector based on the ^{3}He reaction should show three distinct features. Neglecting the wall effect, the full energy of the reaction products is always totally absorbed within the detector and a spectrum similar to that shown in Fig. 15-11 should be expected. The first feature is a full-energy peak corresponding to all the (n, p) reactions induced directly by the incident neutrons. This peak occurs at an energy equal to the neutron energy plus the Q-value of the reaction. Second, a pulse height continuum results from elastic scattering of the neutron and a partial transfer of its

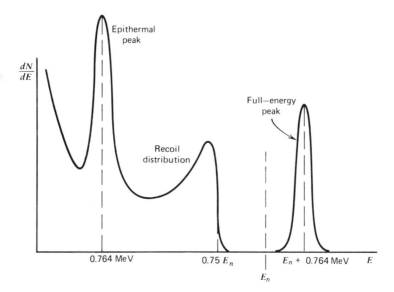

Figure 15-11 Differential energy spectrum of charged particles expected from fast neutrons incident on a ^{3}He detector.

energy to a recoiling helium nucleus. The maximum energy of the continuum can be calculated from Eq. (15-4) (given later in this chapter) and is 75% of the incoming neutron energy for ^{3}He. Third, an epithermal peak appears in virtually every spectrum taken with a ^{3}He detector (as in detectors based on the lithium reaction) and corresponds to the detection of incident neutrons which have been reduced to the thermal range by moderation in external materials. All such neutron interactions deposit an energy equal to the Q-value, or 764 keV.

Wall effects arise whenever the dimensions of the detector are not large compared with the ranges of the secondary particles produced in these reactions. As described in Chapter 14 in connection with BF_3 tubes, the effect on the pulse height spectrum is to fill in the region to the left of the peaks shown in Fig. 15-11.

1. THE ^{3}He PROPORTIONAL COUNTER

If a large ^{3}He-filled proportional counter is irradiated by fast neutrons, a spectrum similar to that shown in Fig. 15-11 will be observed. To minimize the wall effect in smaller counters, several atmospheres of pressure are usually used, and a second component consisting of a heavier gas such as krypton is often added to the ^{3}He to reduce the reaction product ranges. Specific designs and applications of ^{3}He proportional tubes are described in Refs. 53–56.

The undesirable influences of the wall effect and elastic scattering in ^{3}He proportional counters can be ameliorated considerably at the price of added complexity. The additional information carried by the rise time of the output pulses can be used to eliminate a large number of these unwanted events, while retaining virtually all the direct ^{3}He neutron capture events. In any proportional counter, the charge collection time depends on the radial distance at which the ions are formed by the original charged particle. Those tracks that are either very short or parallel to the axis of the proportional tube will generate ions, all of which are collected in about the same time. The corresponding pulse rise time will therefore be small. Tracks that cover a wide range of radii will generate ions with widely different collection times and pulses with longer rise times. Rejection of the elastically scattered ^{3}He recoils is possible because their specific ionization is greater than protons of the same energy. As a result, recoils have a shorter range and are likely to be limited to a smaller variation in radii per track. They will therefore tend to produce pulses with a shorter rise time than the preferred proton events and can be discriminated on that basis. Because there is some overlap in these rise times, the rejection will also eliminate some true signals and will reduce the overall counting efficiency. With proper selection criteria, it has been possible to reject virtually all recoil events with no more than about a factor of 2 reduction in the (n, p) detection efficiency.[53]

There is also evidence[57] that wall effect pulses have an average rise time considerably slower than that of the full-energy pulses and can also be suppressed by eliminating long rise time events. As a bonus, long rise time rejection will also effectively eliminate gamma-ray background because fast electrons will generally travel completely across the active volume of the tube. More elaborate data recording involving two-dimensional storage of both amplitude and rise time information for each pulse[58] allows for a more selective choice of acceptance parameters but represents a considerable added complexity.

2. THE ^{3}He IONIZATION CHAMBER

Although ^{3}He filled chambers were first developed as proportional counters, there is some advantage in designing such chambers to operate instead as gridded ionization chambers of the type described beginning on p. 154. By avoiding the added fluctuations introduced by avalanche formation, gridded ion chambers can display significantly better pulse height resolution than the equivalent proportional counter. When applied to fast neutron spectroscopy, this advantage translates into superior energy resolution. Some examples of the design and application of ^{3}He ionization chambers are given in Refs. 59–65.

Based on an original design by Shalev and Cuttler,[66] a widely used version of this type of chamber uses a mixture of ^{3}He, argon, and methane at partial pressures of 3, 6, and 0.5 atm, respectively. The predominance of the heavy gas argon reduces the ranges of the reaction products and minimizes the complications of wall and end effects or partial energy loss in the gas. The FWHM of the full-energy peak identified in Fig. 15-11 ranges from about 12 keV for thermal neutrons to 20 keV for incident 1 MeV neutrons.[59] The slow charge collection time and small pulse amplitude from this type of detector create more severe problems compared with proportional tubes. Relatively long shaping times (5–10 μs) are necessary to fully develop the pulse, making such chambers susceptible to microphonic noise and pulse pileup. Sensitive low-noise preamplifiers are also needed to preserve the good energy resolution of the chamber.

3. THE ^{3}He SCINTILLATOR

As described previously in Chapter 8, the noble gases including helium can be used as scintillators. Pure helium has a rather poor light yield, but the addition of xenon with as little as a few percent concentration can enhance the light yield by as much as a factor of 5 (Ref. 67). The emitted light is relatively low in intensity compared with conventional scintillation materials, and appears mostly in the ultraviolet region of the spectrum. It is common to use additives in the gas (such as nitrogen) or wavelength shifting materials (such as p-terphenyl) as reflecting layers on the inner surfaces of the scintillation chamber to convert much of the ultraviolet to the visible band. Purity of the gas is very important, since trace amounts of oxygen or organic vapors are known to reduce the light yield significantly.

The decay time of the scintillation is only several nanoseconds, leading to very fast risetime of the output pulse. This advantage in timing compared with ^{3}He proportional or ion chambers is offset by a poorer pulse height resolution. One design[68] has shown a FWHM of 121 keV for the full energy peak from 2.5 MeV neutrons, limited largely by the light collection efficiency variations throughout the volume of the gas. In order to increase the neutron detection efficiency, ^{3}He scintillation chambers have been designed[67] to withstand up to 150 atm pressure.

4. THE ^{3}He SEMICONDUCTOR SANDWICH SPECTROMETER

The configuration of a neutron-sensitive target surrounded on both sides by semiconductor detectors is most commonly used with ^{6}Li as the target. This type of neutron detector was discussed earlier as the lithium semiconductor sandwich spectrometer. Less attention has been given to the use of ^{3}He as the target material, but some potential advantages have spurred efforts in this direction. These advantages can include a considerably higher detection efficiency for equivalent neutron energy resolution and a cross section that is smooth and well known. The gain in efficiency, which can be as much as a factor of 20–50 (Ref. 69) is due mostly to the lower specific energy loss of the proton and triton reaction products in helium compared with that for the alpha particle and recoil triton in lithium targets. Consequently, thicker targets can be used which also can consist of pure elemental ^{3}He. Also from Fig. 15-9, the fast neutron cross section for the ^{3}He reaction is larger than that for the ^{6}Li reaction. Disadvantages include the lower Q-value of the ^{3}He reaction which makes discrimination against gamma rays much more difficult. Furthermore, the larger volume of pressurized helium gas which must be substituted for the solid lithium target makes efficiency calculations more complicated and subject to uncertainties. Descriptions of the design and application of ^{3}He semiconductor sandwich spectrometers can be found in Refs. 69–71.

III. DETECTORS THAT UTILIZE FAST NEUTRON SCATTERING

A. General Properties

The most common method of fast neutron detection is based on elastic scattering of neutrons by light nuclei. The scattering interaction transfers some portion of the neutron kinetic energy to the target nucleus, resulting in a *recoil nucleus*. Because the targets are always light nuclei, this recoil nucleus behaves much like a proton or alpha particle as it loses its energy in the detector medium. Hydrogen, deuterium, and helium are all of interest as target nuclei, but hydrogen is by far the most popular. The recoil nuclei that result from neutron elastic scattering from ordinary hydrogen are called *recoil protons*, and devices based on this neutron interaction are often referred to as *proton recoil detectors*.

The Q-value of elastic scattering is zero because the total kinetic energy after the reaction by definition is the same as the kinetic energy before. For all practical purposes the target nuclei are at rest, and therefore the sum of the kinetic energies of the reaction products (recoil nucleus and scattered neutron) must equal that carried in by the incident neutron. For single scattering in hydrogen, the fraction of the incoming neutron energy that is transferred to the recoil proton can range anywhere between zero and the full neutron energy, so that the average recoil proton has an energy about half that of the original neutron. Therefore, it is usually possible to detect preferentially fast neutrons in the presence of gamma rays or other low-energy background, but the discrimination becomes more difficult as the incoming neutron energy drops below a few hundred keV. By employing techniques such as pulse shape or rise time discrimination to eliminate gamma-ray-induced events, specialized proton recoil detectors can be used to a neutron energy as low as 1 keV. Recoil methods are insensitive to thermal neutrons except through any competing reactions that might be induced in the target material or other parts of the detector.

1. KINEMATICS OF NEUTRON ELASTIC SCATTERING

We first define some symbols to be used in the equations that follow:

A = mass of target nucleus/neutron mass

E_n = incoming neutron kinetic energy (laboratory system)

E_R = recoil nucleus kinetic energy (laboratory system)

Θ = scattering angle of the neutron in the center-of-mass coordinate system

θ = scattering angle of the recoil nucleus in the lab coordinate system

These definitions are illustrated graphically in Fig. 15-12.

For incoming neutrons with nonrelativistic kinetic energy ($E_n \ll 939$ MeV), conservation of momentum and energy in the center-of-mass coordinate system gives the following relation for the energy of the recoil nucleus:

$$E_R = \frac{2A}{(1 + A)^2}(1 - \cos \Theta)E_n \tag{15-1}$$

To convert to the more familiar laboratory coordinate system in which the original target nucleus is at rest, we use the following transformation:

$$\cos \theta = \sqrt{\frac{1 - \cos \Theta}{2}} \tag{15-2}$$

which, when combined with Eq. (15-1), gives the following relation for the recoil nucleus

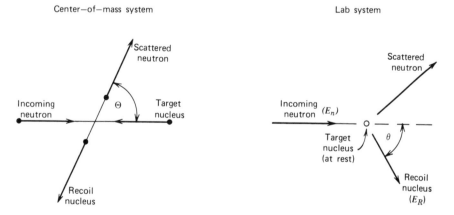

Figure 15-12 Neutron elastic scattering diagrams for the center-of-mass and laboratory coordinate systems.

TABLE 15-2   Maximum Fractional Energy Transfer in Neutron Elastic Scattering

Target Nucleus	A	$\left.\dfrac{E_R}{E_n}\right\vert_{max} = \dfrac{4A}{(1+A)^2}$
$_{1}^{1}\text{H}$	1	1
$_{1}^{2}\text{H}$	2	$8/9 = 0.889$
$_{2}^{3}\text{He}$	3	$3/4 = 0.750$
$_{2}^{4}\text{He}$	4	$16/25 = 0.640$
$_{6}^{12}\text{C}$	12	$48/169 = 0.284$
$_{8}^{16}\text{O}$	16	$64/289 = 0.221$

energy in terms of its own angle of recoil:

$$E_R = \frac{4A}{(1+A)^2}(\cos^2\theta)\,E_n \tag{15-3}$$

From Eq. (15-3) we can see that the energy given to the recoil nucleus is uniquely determined by the scattering angle. For a grazing angle encounter in which the neutron is deflected only slightly, the recoil is emitted almost perpendicular to the incoming neutron direction ($\theta \cong 90°$), and Eq. (15-3) predicts that the recoil energy is near zero. At the other extreme, a head-on collision of the incoming neutron with the target nucleus will lead to a recoil in the same direction ($\theta \cong 0$), resulting in the maximum possible recoil energy,

$$E_R\vert_{max} = \frac{4A}{(1+A)^2}E_n \tag{15-4}$$

Table 15-2 lists the maximum fraction of the incoming neutron energy that can be transferred to a recoil nucleus in a single collision for a variety of target nuclei. As the target nucleus mass increases, the maximum fractional energy transfer decreases. Only in collisions with ordinary hydrogen can the neutron transfer all its energy in a single encounter. The trend shown in the table explains why only light nuclei are of primary interest in recoil detectors, with hydrogen assuming the predominant role.

2. ENERGY DISTRIBUTION OF RECOIL NUCLEI

We must also be concerned with the way in which the recoil energies are distributed between a minimum of zero and the maximum given in Table 15-2. Because all scattering angles are allowed, in principle, a continuum of possible recoil energies between these extremes should be expected. If we define $\sigma(\Theta)$ as the differential scattering cross section in the center-of-mass system, then, by definition, the probability that the neutron will be scattered into $d\Theta$ about Θ is

$$P(\Theta)\,d\Theta = 2\pi \sin\Theta\,d\Theta\,\frac{\sigma(\Theta)}{\sigma_s} \tag{15-5}$$

where σ_s is the total scattering cross section integrated over all angles. We are more interested in the distribution in recoil nucleus energy and will let $P(E_R)\,dE_R$ represent the probability of creating a recoil with energy in dE_R about E_R. Now, because $P(E_R)\,dE_R =$

$P(\Theta)d\Theta$, it follows that

$$P(E_R) = 2\pi \sin\Theta \frac{\sigma(\Theta)}{\sigma_s} \cdot \frac{d\Theta}{dE_R} \tag{15-6}$$

Now, evaluating $d\Theta/dE_R$ from Eq. (15-1) and substituting, we obtain

$$P(E_R) = \frac{(1+A)^2}{A} \frac{\sigma(\Theta)}{\sigma_s} \cdot \frac{\pi}{E_n} \tag{15-7}$$

Equation (15-7) shows that the shape expected for the recoil energy continuum is just the same as the shape of the differential scattering cross section $\sigma(\Theta)$ as a function of the center-of-mass scattering angle of the neutron. For most target nuclei, the shape of $\sigma(\Theta)$ will tend to be somewhat peaked to favor forward and backward scattering as shown in Fig. 15-13.

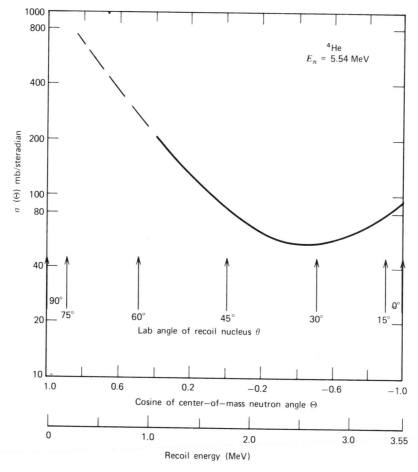

Figure 15-13 The differential scattering cross section of ^{4}He at a neutron energy of 5.54 MeV. Also indicated are the corresponding angle and energy of the helium recoil nucleus in the laboratory system.

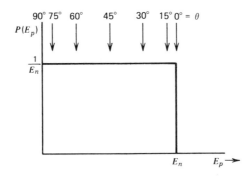

Figure 15-14 Energy distribution of recoil protons produced by monoenergetic neutrons. Recoil energies are indicated for various values of the recoil emission angle θ as given by Eq. (15-3).

A very important simplification holds if the scattering process is *isotropic in the center-of-mass-coordinate system*. Then, $\sigma(\Theta)$ does not change with Θ and is equal to a constant $\sigma_s/4\pi$. This fortunate circumstance is not generally the case but does hold for scattering from hydrogen over most of the energy range of interest ($E_n < 10$ MeV). Because hydrogen is by far the most important target nucleus, the simplifications that result are of real significance. *The expected proton recoil energy distribution is therefore a simple rectangle*, extending from zero to the full incident neutron energy, as sketched in Fig. 15-14. The response function of a detector based on simple hydrogen scattering should therefore have a correspondingly simple rectangular shape. As discussed in the following sections, however, there are a number of complicating factors that can distort this simple rectangular response.

3. DETECTION EFFICIENCY

The detection efficiency of a device based on recoil protons or other recoil nuclei can be calculated from the scattering cross section σ_s. If nuclei of only one species are present in the detector, the intrinsic efficiency is given simply by

$$\epsilon = 1 - \exp(-N\sigma_s d) \qquad (15\text{-}8a)$$

where N is the number density of target nuclei, σ_s is the scattering cross section for these nuclei, and d is the pathlength through the detector for incident neutrons. Carbon often appears in combination with hydrogen in proton recoil detectors, and competing effects due to carbon scattering must then be taken into account. The counting efficiency, neglecting multiple scattering, is then given by

$$\epsilon = \frac{N_H\sigma_H}{N_H\sigma_H + N_C\sigma_C}\left\{1 - \exp\left[-(N_H\sigma_H + N_C\sigma_C)d\right]\right\} \qquad (15\text{-}8b)$$

where the subscripts H and C refer to the separate hydrogen and carbon values for the quantities defined above.

Plots of the scattering cross section for several materials of interest in fast neutron detectors are given in Fig. 15-15. An empirical fit to the hydrogen scattering cross section suggested by Marion and Young[72] is

$$\sigma_s(E_n) = \frac{4.83}{\sqrt{E_n}} - 0.578 \text{ barns} \qquad (15\text{-}9)$$

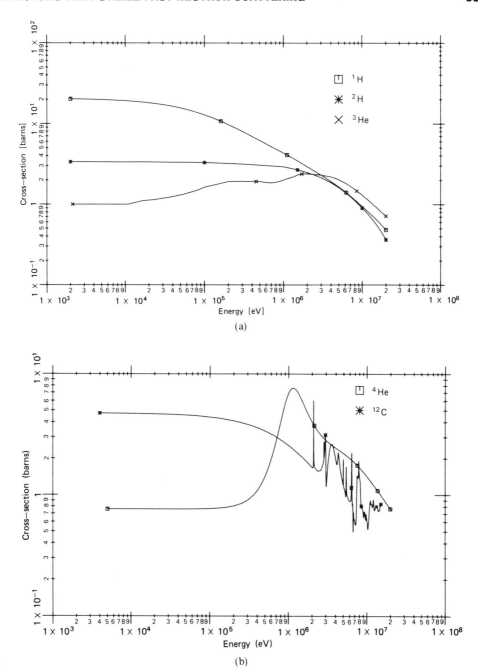

Figure 15-15 (a) Elastic scattering cross sections for ^{1}H, ^{2}H, and ^{3}He. (b) Elastic scattering cross sections for ^{4}He and ^{12}C.

where E_n is in MeV. This approximation is accurate to within about 3% over the neutron energy range of $0.3 < E_n < 30$ MeV.

B. Proton Recoil Scintillators

One of the easiest ways to use proton recoil in the detection of fast neutrons is through the application of hydrogen-containing scintillators. Fast neutrons incident on the scintillator give rise to recoil protons whose energy distribution should be approximately rectangular, ranging from zero to the full neutron energy. Because the range of the recoil protons is usually small compared with the dimensions of the scintillator, their full energy is deposited in the scintillator and the expected pulse height distribution is also approximately rectangular. Hydrogen-containing scintillation detectors are widely applied in many areas of fast neutron physics research, as indicated in the extensive review by Harvey.[73]

1. SCINTILLATION MATERIALS

Because scintillation materials that contain hydrogen are quite common, there is no shortage of candidates for use as fast neutron scintillators. Successful applications have been reported using organic crystals such as anthracene or stilbene, liquid scintillators that combine an organic scintillant dissolved in a hydrogen-containing organic solvent, and plastic scintillators in which an organic scintillant is incorporated in a bulk matrix of polymerized hydrocarbon. A general discussion of the basic properties of these scintillators can be found in Chapter 8.

Much of the early work in developing proton recoil scintillators was done using crystals of anthracene or stilbene. Anthracene has the largest light output of any organic scintillator, but attention gradually shifted to stilbene because of its superior gamma-ray rejection characteristics. However, both crystals are difficult and expensive to obtain in large sizes (greater than a few centimeters in dimension) and are also subject to damage from thermal and mechanical shock. A further disadvantage stems from the directional variation of the light output from such crystals, which depends on the orientation of the path of the charged particles with respect to the crystal axis. It is not unusual to observe variations as large as 25% as the charged particle orientation is varied. This effect, although not serious if the detector is used only to count neutrons, greatly complicates the job of unfolding an observed pulse height spectrum to derive the incident fast neutron energy spectrum.

Consequently, emphasis has shifted toward the use of liquid and plastic organic scintillators for fast neutron spectroscopy. These materials are relatively inexpensive, can be tailored to a wide variety of sizes and shapes, and are totally nondirectional in their scintillation response. Table 15-3 lists some properties of representative liquid and plastic scintillators that are commercially available and are formulated specifically for use in fast neutron spectroscopy.

For specialized purposes, liquid scintillators are also available with deuterium substituted for the normal hydrogen content. Other formulations based on hexafluorobenzene are totally free of hydrogen and can be used to measure separately the gamma-ray contribution in a mixed fast neutron–gamma-ray field.

2. SCINTILLATOR SIZE

In choosing a size for a recoil proton scintillator, several compromises must be struck. The first involves a trade-off of counting efficiency versus energy resolution. By making the scintillator thick, the efficiency calculated from Eq. (15-8) is obviously enhanced. For

TABLE 15-3 Some Representative Plastic and Liquid Scintillators of Interest as Fast Neutron Detectors[a]

Manufacturer	Identification	Light Output	Decay Constant (ns)	Wavelength of Maximum Emission (nm)	H/C Atomic Ratio	Comments
PLASTIC						
Nuclear Enterprises, Ltd.	NE 102 A	65% of anthracene	2.4	425	1.104	General applications
Nuclear Enterprises, Ltd.	NE 104	68%	1.8	405	1.099	Fast timing
Nuclear Enterprises, Ltd.	NE 111	55%	1.7	375	1.093	Ultrafast timing (cannot be used in large sizes due to light self-absorption)
Nuclear Enterprises, Ltd.	Pilot B	68%	1.8	408	1.100	General applications
Nuclear Enterprises, Ltd.	Pilot U	67%	1.36	391	1.100	Ultrafast timing
Amperex Electronic Corp.	SPF	55–65%	4	430	1.0	
Koch–Light Laboratories, Ltd.	KL 211	60%	2.2	420	0.992	General applications
Koch–Light Laboratories, Ltd.	KL 236	65%	1.87	410	0.992	Ultrafast timing
LIQUID						
Nuclear Enterprises, Ltd.	NE 211	78% of anthracene	2.6	425	1.248	General use
Nuclear Enterprises, Ltd.	NE 213	78%	3.7	425	1.213	For pulse shape discrimination against gamma rays
Nuclear Enterprises, Ltd.	NE 228	45%	—	385	2.11	High H/C ratio

[a]For all the plastic scintillators shown, density = 1.03–1.06 g/cm^3; refractive index = 1.58–1.59; and softening temperature = 75–85°C.

example, if the scintillator is made about 5 cm thick in the direction of the neutron path, interaction probabilities of at least 40% will hold for neutrons whose energy is less than 2 or 3 MeV. As the neutron energy increases, the detection efficiency will decrease, and consequently there will be strong motivation to make the scintillator larger. However, it is more difficult to achieve uniform light collection from a large-volume scintillator, and the energy resolution will worsen. Another factor that often limits scintillator size is the pulse rate due to gamma rays interacting within the detector. In many applications, this rate exceeds that from fast neutrons, and the scintillator must be kept sufficiently small so that the pileup of gamma-ray events is not a problem.

If the scintillator is used as a fast neutron spectrometer, other factors enter into the choice of size. In small crystals, a typical neutron is likely to scatter only once, and the energy spectrum of proton recoils will closely approximate the rectangular distribution discussed earlier. As long as the scintillator dimensions are larger than a few millimeters, escape of protons from the surface is unlikely, and the response function of the detector is simple and easily calculated. As the detector dimensions are increased, multiple scattering of the neutrons becomes more likely and the response function is more complicated and harder to predict. Because an accurate knowledge of the response function is critical for the unfolding process, one would like to keep the scintillator small so that these complicating effects do not introduce large uncertainties.

3. RESPONSE FUNCTION

For that subset of detectors for which the rectangular response function is a reasonable approximation, the task of deriving the incident neutron energy spectrum is particularly simple. Because the derivative with respect to energy of a rectangular distribution is zero everywhere except at the maximum, the derivative of the recoil proton spectrum will give a narrow peak located at the incident neutron energy. The derivative of the spectrum recorded from a complex source therefore gives an easily calculated representation of the incident neutron spectrum.

For response functions that are more complex, the general techniques of deconvolution or unfolding, discussed in Chapter 18, must be applied. Representative examples of the application of unfolding methods to the neutron response of organic scintillators are given in Refs. 74–77. Because of their complexity, unfolding calculations of this type are inevitably cast in the form of large computer codes.[78,79] An accurate knowledge of the detector response function is a necessary input to all these unfolding methods. We therefore list some of the factors that distort the simple rectangular distribution in organic scintillators and show the qualitative effect on the response function.

a. Nonlinear Light Output with Energy

As discussed in Chapter 8, the light output from most organic scintillators does not increase linearly as the deposited energy increases. This nonlinear behavior distorts the expected rectangular proton energy distribution of Fig. 15-16c into the pulse height distribution shape sketched in Fig. 15-16d. For many organic scintillators, the light output H is approximately proportional to $E^{3/2}$, in which case

$$H = kE^{3/2} \tag{15-10}$$

and the pulse height distribution shape is given by

$$\frac{dN}{dH} = \frac{dN/dE}{dH/dE} = \frac{\text{constant}}{\frac{3}{2}kE^{1/2}} = k'H^{-1/3} \tag{15-11}$$

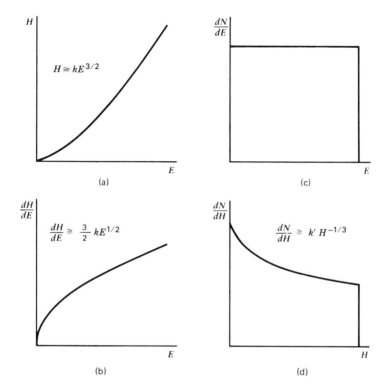

Figure 15-16 Part (a) shows a plot of pulse height versus energy for a typical organic scintillator. This nonlinear response leads to a distortion of the rectangular proton recoil spectrum of part (c) into the spectrum shown in part (d).

where k and k' are proportionality constants. This relation is only an approximation to the actual distortion caused by the nonlinearity, and a more detailed description of the light output versus energy is required as a basis for accurate spectrum unfolding.[80]

b. Edge Effect

If the scintillator is sufficiently small or the neutron energy very high so that the range of the recoil protons is not small compared with detector dimensions, some effects can be expected due to escape of protons from the surface of the scintillator. The event is not lost but simply shifted to an energy lower than would normally have been observed. The effect on the response function will be to shift events from high pulse height to low pulse height, further increasing the slope of Fig. 15-16d.

c. Multiple Scattering from Hydrogen

For detectors that are not small, it is possible for an incident neutron to scatter more than once from hydrogen nuclei before escaping from the scintillator. Because all such events normally occur within a very short period of time compared with the scintillation decay time, the light from all recoil protons is summed and a pulse produced whose amplitude is proportional to the total light output. Multiple scattering will therefore increase the average pulse height and change the expected response function by adding events at large pulse heights at the expense of those at lower amplitudes.

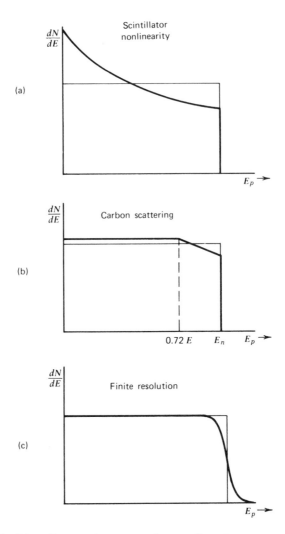

Figure 15-17 Distortions to the rectangular recoil proton energy spectrum due to three separate factors.

d. Scattering from Carbon

All organic scintillators contain carbon as well as hydrogen. Because of the decreased scintillation efficiency for high dE/dx particles, carbon recoils produced by neutron elastic scattering do not contribute much to the detector output. However, scattering from carbon does affect the detector response function indirectly in that the scattered neutrons may still undergo a hydrogen scattering before escaping from the scintillator. Because the neutron energy has been decreased in the initial carbon scatter, the proton recoil spectrum produced from carbon-scattered neutrons will not extend to as high an energy as that from unscattered neutrons. Because the incident neutron can lose between 0 and 28% of its initial energy in a carbon scattering, the maximum energy of a subsequent recoil proton will vary between 100 and 72% of the original energy. The corresponding effect on the detector response function is shown in Fig. 15-17b.

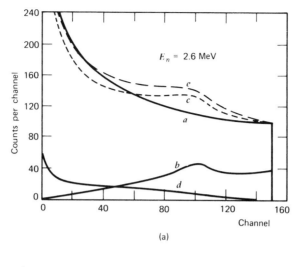

(a)

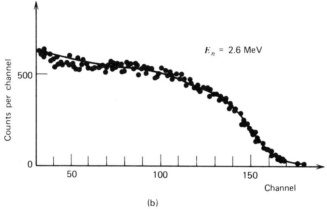

(b)

Figure 15-18 (a) Calculated response function at 2.6 MeV (for perfect detector resolution) for a 2.54 cm × 2.54 cm cylindrical stilbene crystal. Separate components shown are: *a*—single scattering with detector nonlinearity and edge effects; *b*—double scattering from hydrogen; *c*—single plus double scattering from hydrogen; *d*—proton recoils from carbon-scattered neutrons; *e*—composite spectrum. (b) Measured pulse height spectrum at 2.6 MeV for the same crystal included in the calculation of part (a). The added influence of imperfect detector resolution is evident. (From Bormann et al.[81])

e. Detector Resolution

The discussion to this point has dealt with ideal detector response without considering the spread introduced by nonuniform light collection, photoelectron statistics, and other sources of noise. These sources of dispersion will tend to wash out some of the distinct structure expected in the response function, as illustrated in Fig. 15-17c.

f. Overall Response Function

The manner in which all these distorting effects combine to give the overall detector response function is illustrated in Fig. 15-18, for a 2.54 × 2.54 cm cylindrical stilbene crystal at a neutron energy of 2.6 MeV. This example is fairly typical of any organic or

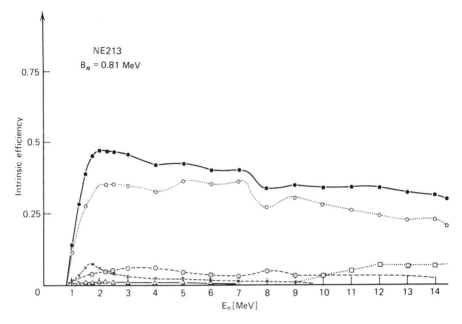

Figure 15-19   The calculated efficiency of a NE213 liquid scintillator (1.9 cm radius, 10 cm length) for a discrimination level of 0.81 MeV. Identification of the symbols is as follows: ● combined efficiency from all processes; ○ single hydrogen scattering; × n-H, n-H double scattering; ⊙ n-C, n-H double scattering; △ n-C, n-H, n-H triple scattering; □ (n, α) and (n, n′)3α reactions. (From Hermsdorf et al.[82])

plastic scintillator response to neutrons whose energy lies below the threshold of competing reactions discussed in the following section.

g. Competing Reactions at High Neutron Energies

Once the neutron energy exceeds 8 or 9 MeV, two competing reactions must be considered in the overall response of organic scintillators:

	Neutron Energy at Threshold
$^{12}C(n, \alpha)^9Be$	6.17 MeV
$^{12}C(n, n')3\alpha$	7.98 MeV

The combined effects of multiple scattering and the competing reactions can be seen in Fig. 15-19, which plots the counting efficiency of a liquid organic scintillator versus neutron energy. In the figure, the competing reactions $^{12}C(n, \alpha)$ and $^{12}C(n, n')3\alpha$ are seen to contribute to the detection efficiency at high neutron energies. The threshold energies for these reactions are 6.17 and 7.98 MeV, but they become significant only above about 9 MeV. An inhibiting factor is the lower light output per unit energy that most organic scintillators exhibit for alpha particles compared to recoil protons.

4. DETECTOR COUNTING EFFICIENCY VERSUS DISCRIMINATOR BIAS LEVEL

A common mode of application of proton recoil detectors is to set a fixed discrimination level to eliminate all pulse amplitudes below a given size. Some finite discrimination level is always required to eliminate inevitable noise pulses that arise spontaneously in the

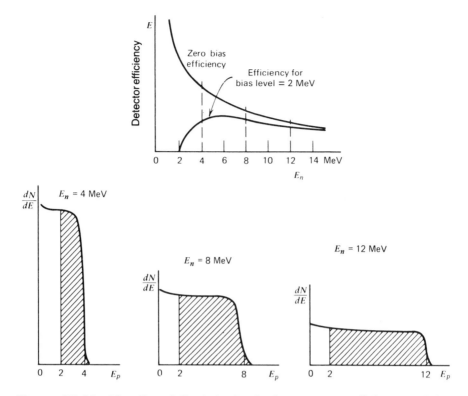

Figure 15-20 The effect of discrimination level on proton recoil detector efficiency. The upper graph gives both the zero bias efficiency and the reduced efficiency when a discrimination level corresponding to a proton energy of 2 MeV is used. See text for further elaboration.

counting system, but gamma rays and other background events may dictate discrimination at a substantially higher level. Because the proton recoil energy distribution extends all the way to zero, some recoil events will inevitably be eliminated in the discrimination process. Therefore, the efficiency assuming that all pulses are counted is sometimes called the *zero bias* efficiency, and a real detector with a finite discrimination level will always have a somewhat lower efficiency, particularly at lower neutron energies.

The effect of discrimination level on counting efficiency is shown graphically in Fig. 15-20. Sketched along the bottom of the figure are three differential pulse height spectra representing the response of a scintillator to three different neutron energies. Neglecting more complex interactions, the total area under each of these curves is proportional to the number of neutron–hydrogen scattering events in the detector at that energy. This number is proportional to the zero bias efficiency, shown plotted versus neutron energy at the top of the figure.

When a discriminator is used, that area under the differential spectrum which lies to the right of the discrimination level is proportional to the number of recorded counts. For the example shown, this reduced efficiency is also plotted versus neutron energy on the upper graph. This efficiency obviously drops to zero for neutron energies below the discrimination level and approaches the zero bias efficiency at neutron energies that are large by comparison. More detailed analyses of detection efficiency of organic scintillators can be found in Refs. 83–91.

5. ENERGY CALIBRATION WITH GAMMA RAYS

The light output of organic scintillators for electrons is always higher per unit energy than for heavy charged particles. Therefore, the problem of gamma discrimination is made more severe because it may require a neutron with energy of 2 or 3 MeV to give the same light output as a 1 MeV gamma ray. The low Z value of the constituents of organic scintillators (hydrogen, carbon, and perhaps oxygen) results in a very low photoelectric cross section, so that virtually all gamma-ray interactions are Compton scatterings. A gamma-ray spectrum taken with an organic scintillator will therefore show no photo-peaks, and Compton edges are the only distinguishable features. Because the scintillation response to electrons is fairly linear, a gamma-ray source is often used to calibrate the energy scale of the detector output. Because there are no photopeaks, some point on the Compton edge must be selected and associated with the maximum energy of a Compton recoil electron. Details of the shape of the Compton continuum and the effects of detector resolution have led Flynn[92] to suggest a standard procedure in which the channel number at which the Compton continuum has fallen to half its plateau value is associated with an energy equal to 104% of the maximum Compton electron energy. A more detailed procedure that takes into account variability of the detector resolution is described in Ref. 93.

6. PULSE SHAPE DISCRIMINATION AGAINST GAMMA RAYS

As discussed in Chapter 8, the relative intensity of the fast and slow components of the light yield of some organic scintillators depends on the specific ionization of the ionizing particle (see Fig. 8-5). Therefore, particles of different mass or charge will produce signal pulses with different time characteristics. Specifically, gamma-ray-induced fast electrons generate a larger fraction of their scintillation light in the prompt component as compared with recoil protons. The methods of pulse shape discrimination discussed in Chapter 17 can be very effective in rejecting gamma-ray pulses from organic scintillators, while retaining reasonable fast neutron efficiency.[94-100]

If the gamma-ray pulses are not simply rejected but instead are analyzed in a separate spectrum, simultaneous spectroscopy on both neutrons and gamma rays can be carried out. Because of the lack of full-energy gamma-ray peaks, the gamma-ray spectra must be unfolded using the same type of computation as is applied to the neutron spectrum. Both calculated[101,102] and measured[103] response functions to gamma rays for NE213 liquid scintillator have been published and can be used as input to the unfolding codes.

C. Gas Recoil Proportional Counters

1. GENERAL

As an alternative to the organic scintillator, gas proportional counters can also be used to measure fast neutrons through the recoil process. In these applications, the fill gas is usually hydrogen, a hydrogen-containing gas such as methane, or some other low-Z gas such as helium. In the case of hydrogen, the expected proton recoil spectrum should again be the simple rectangular shape described earlier. As in the case of organic scintillators, however, complicating effects often arise which distort this simple response function and make the task of unfolding observed pulse height spectra a great deal more complicated.

Because the detection medium is a gas, recoil proportional counters inevitably have a lower counting efficiency than typical organic scintillators. Equation (15-8) can again be used to estimate the neutron detection efficiency for simple hydrogen-filled counters, and typical values for neutrons in the MeV energy range will be less than 1%. This reduced interaction probability is not entirely lacking some compensating advantages. For example, the interaction probability for scattered neutrons will also be quite low, and thus the response function of gas detectors will be largely free of the multiple scattering complications discussed earlier for scintillators.

Recoil proportional counters are undoubtedly more sensitive detectors than organic scintillators. As discussed in Chapter 6, purity of the fill gas is of utmost importance and microscopic air leaks will ultimately lead to detector failure. Although recoil proportional counters are commercially available, there is a common tendency for experimenters to design their own counters specifically tailored to the application required. The reader should note that considerable attention to construction details such as surface preparation, high vacuum pumping before gas filling, and high voltage insulator design are necessary to have a successful detector. By comparison, scintillators are relatively easy to assemble and operate. The fact that recoil proportional counters are not large compared with the range of recoil nuclei in a gas means that the correction for wall and end effects, usually quite small in scintillators, becomes an important consideration in determining the response of these detectors. An extensive review of response functions for recoil proportional counters of different designs is given in Ref. 104.

2. GAMMA-RAY SENSITIVITY AND PULSE SHAPE DISCRIMINATION

Another consequence of the low-density gas detection medium is the behavior of the detector in the presence of gamma rays. Typical recoil scintillators will have dimensions that are large with respect to both the recoil nuclei produced by neutron interactions as well as to fast electrons created in gamma-ray interactions. Therefore, the full energy of each is almost always deposited within the detector active volume. Gas proportional counters, on the other hand, are often of a size that is not extremely large compared with recoil nuclei ranges, and usually they are much smaller than the range of gamma-ray-produced electrons. Therefore, it is likely that neutron-induced events deposit all their energy in the gas, whereas gamma rays will deposit only a small part of their energy. Furthermore, the inherent response of organic scintillators is such that the light output may be as much as a factor of 2 or 3 times greater for electrons than for charged particles of the same energy. This extreme variation in response does not exist in proportional counters, in which the energy expended to create one ion pair changes only slightly with the nature of the ionizing particle.

A difference that works to the disfavor of the proportional counter is the nature of gamma-ray interactions in the detector.[105] For the organic scintillator, both neutrons and gamma rays must interact within the scintillator volume in order to give rise to detected pulses. In the gas counter, neutrons must interact within the fill gas, but gamma rays may interact either in the gas or, more likely, in the walls and other construction materials of the counter, leading to secondary electrons that can escape into the gas volume. Therefore, if the fill gas of a recoil proportional counter with the same elemental composition as that for an organic scintillator is chosen, one would expect a considerably larger ratio in the number of gamma rays to neutron pulses in the proportional counter compared with the organic scintillator. In the proportional counter, however, the average gamma-ray pulse amplitude would be considerably smaller.

As in all proportional counters, the rise time of observed signal pulses depends primarily on the radial distribution of the ionization track that gives rise to that signal pulse. As discussed for ^{3}He tubes, this property can be used to count preferentially neutron-induced events in the presence of potentially interfering gamma radiation. Provided the neutron energy is fairly low, proton recoil tracks will tend to be rather short and consequently will be confined to a limited range in radius. Gamma-ray-induced fast electrons will almost always pass completely through the gas and, on the average, involve a much greater range of radii in the tube. As a result, neutron-induced pulses will tend to have shorter rise times than those induced by gamma rays, and pulse shape discrimination methods will serve to differentiate between the two. Extension of the proton recoil technique in proportional counters down to a neutron energy as low as 1 keV, as pioneered by Bennett,[106] is critically dependent on elimination of the majority of the gamma-ray pulses by this method.

3. COUNTER DESIGN AND CHOICE OF FILL GAS

Although some results have been reported for spherically shaped chambers,[107,108] the majority of recoil proportional counter designs incorporate cylindrical geometry similar to that discussed in Chapter 6 and illustrated in Fig. 6-5. As in all proportional counters, the axial anode wire must be of very uniform and small diameter to ensure uniform gas multiplication along its length. The field tubes at either end of the anode wire are common devices used to abruptly terminate the region in which gas multiplication can occur, so that the active volume of the counter is well defined. Nonetheless, there remains some nonuniformity (sometimes called the *tip effect*) due to the curvature of the electric field lines at the point of discontinuity between the field tube and anode wire. Bennett and Yule[109,110] report that the electric field distortions can be minimized by reducing the outer diameter of the cathode in these regions. Nonetheless, the distortion of the electric field in the end regions of the tube remains one of the important complicating factors that must be considered in analyzing the efficiency and response function of the detector.

By far the majority of all recoil detectors utilize hydrogen or some compound of hydrogen to make use of recoil protons produced by neutron elastic scattering. The simple rectangular energy distribution for proton recoils illustrated in Fig. 15-14 is distorted by several important effects in proportional counters. The most important of these is the effect of the finite size of the active volume of the chamber which leads to proton tracks that are truncated either in the walls or end of the tube. There are several approaches to this problem. One is simply to live with these losses and either attempt to calculate or measure their effects on the overall detector response function. Reasonably good success has been achieved with this approach under some circumstances,[111-113] but it would clearly be preferable to eliminate these events as much as possible. If those proton tracks that leave the active volume can be identified and discarded, a much simpler response function will result. A track that leaves the active volume at the cylindrical cathode can be recognized if a somewhat more complex chamber is constructed, in which the cathode consists of a cage of fine wire so that the proton can pass through and continue to produce ionization outside the cathode. A ring of counters around the central region can be used to detect this leaked ionization and thus to reject the original event by anticoincidence. Such a chamber has been described by Heiberg,[114] in which losses to the ends of the tubes were also eliminated through the use of signals derived from the field tubes. Tracks that overlap the active volume and end region of the tube will give coincident pulses on both the anode and field tubes and therefore can also be rejected through anticoincidence.

Although hydrogen is in many ways the ideal target for neutron scattering, its use as a fill gas in proportional counters is limited by its low density and relatively low stopping power for recoil protons. Methane is a more common choice, but complicating effects introduced by the carbon nuclei must then be taken into account. It has been estimated[109] that carbon ions will create approximately 75% as much ionization in the gas as the

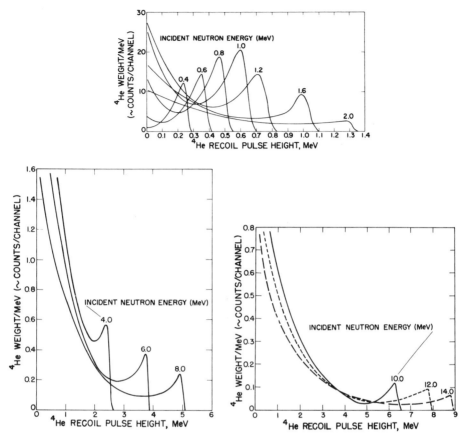

Figure 15-21 Calculated pulse height spectra for a recoil proportional counter filled with ^{4}He at 8 atm pressure. (From Atwater.[115])

equivalent energy proton. Because the maximum energy of a carbon ion is 28% of the neutron energy, we would then expect that all carbon recoil pulses should lie below about 21% of the maximum proton recoil pulse amplitude. Helium is also widely used as a fill gas, and typical recoil spectra are shown in Fig. 15-21. Because elastic scattering from helium is not isotropic in the center-of-mass system (see Fig. 15-13), these distributions will not be rectangular even in the absence of complicating effects.

4. ENERGY CALIBRATION AND LINEARITY

Because sources of monoenergetic neutrons are not commonly available, the energy calibration of a proton recoil proportional counter is not a straightforward process. Gamma rays cannot be used as previously described in the calibration of proton recoil scintillators because the secondary electrons created in gamma ray interactions generally cannot be stopped in the fill gas. One possible method is to incorporate a small amount of ^{3}He into the fill gas so that, when irradiated with thermal neutrons, proton and triton pairs sharing 764 keV kinetic energy (see p. 485) are generated internally. The corresponding peak in the recorded spectrum then provides an energy marker. Although this technique has proven very useful, errors in the calibration can result[116] because one of these particles is not a proton. Differences then can arise that are related to differences in the energy required to form an ion pair or due to disparities in the particle ranges. At low energies, the addition of a small amount of radioactive ^{37}Ar will result in a peak from the 2.82 keV X-rays that are emitted in its decay.[112] This approach has the disadvantage that the calibration peak is present in all measurements and may interfere with an interesting region of the spectrum. The calibration is also sensitive to potentially large differences in the energy loss behavior of the low energy photoelectrons generated by the X-rays compared with the recoil protons of interest.

Above about 10 keV, there seems to be little doubt that the ionization produced by recoil protons in standard proportional counters is essentially linear with the proton recoil energy. Stated another way, the *W*-value (defined as the energy expended per ion pair produced) in hydrogen is a constant for protons above 10 keV. Below this energy, however, there is evidence[117] that the value of *W* falls off and may be as low as about 70% of its high-energy value at a proton energy of 1 keV. These nonlinearities in the relation between energy and ionization must obviously be taken into account when unfolding the detector response to low-energy neutrons.

D. Proton Recoil Telescopes

In conventional organic scintillators or recoil proportional counters, one has little choice but to accept all angles of neutron scattering as they occur. Therefore, the response function of these detectors to monoenergetic neutrons incorporates all recoil proton energies up to the neutron energy and has the approximately rectangular shape discussed earlier. For spectroscopy purposes, however, it would be preferable if the response function were a simple narrow peak to avoid the problems of spectrum unfolding otherwise required. If only those proton recoils that occur at a fixed angle with respect to the neutron direction can be singled out, the recoil proton energy will be fixed for monoenergetic neutrons and the response function will approach the ideal narrow peak. Devices based on a narrow selection of recoil directions are generally known as *proton recoil telescopes* and have been applied to a wide variety of fast neutron measurements.

From Eq. (15-3), the energy of recoil protons observed at an angle θ with respect to the incoming neutron direction is given simply by

$$E_p = E_n \cos^2 \theta \qquad (15\text{-}12)$$

Recoil proton telescopes can be applied only to situations in which the incoming neutron direction has been defined by collimation or other means.

A schematic diagram of a common form of recoil telescope is shown in Fig. 15-22. Neutrons are incident on a thin film, usually made from an organic polymer, whose thickness is kept small compared with the range of the lowest energy recoil proton to be measured. The angle θ at which recoil protons are observed is defined by positioning a detector some distance from the radiator, with the intervening space evacuated to prevent proton energy loss. Because of the $\cos^2 \theta$ fall-off of recoil proton energy, the detector is usually positioned at a small angle with respect to the neutron direction. Many designs put the proton detector at $\theta = 0$, but others choose a finite observation angle to avoid neutron-induced background events in the detector from the primary beam.

Although a single detector can, in principle, be used to measure the proton energy and hence the neutron energy through Eq. (15-12), multiple detectors are often used in coincidence to reduce backgrounds from competing reactions and other unwanted events. The arrangement shown in Fig. 15-22 is a common one in which a very thin ΔE detector is placed in front of a thicker E detector, which fully stops the recoil protons. By

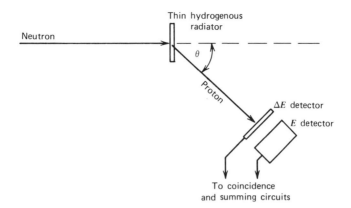

Figure 15-22 A proton recoil telescope.

operating the two detectors in coincidence, only particles incident from the direction of the radiator are recorded. If both detector responses are linear, the sum of the two signals will be proportional to the total proton energy. More elaborate particle identification schemes (see p. 380) can be applied to this arrangement by demanding that the ΔE signal be of proper size relative to the E signal to correspond to the energy loss expected from protons. This requirement will then effectively eliminate other charged particle events such as alpha particles produced in (n, α) reactions in the radiator or other structural components. Proton recoil telescopes of this design, which use semiconductor diode detectors to measure recoil proton energy, are described in Refs. 118 and 119. Other detector types, including gas proportional counters and scintillation detectors, also have been incorporated as the proton detector.

The dominant disadvantage of the proton recoil telescope is its extremely low detection efficiency (typically one count per 10^5 incident neutrons.) This low efficiency stems from two factors, neither of which can be improved without sacrificing energy resolution of the device. First, the radiator thickness must be kept small to avoid appreciable energy loss of the recoil protons before they leave the radiator. Usable radiator thicknesses lead to a probability of about 10^{-3} or 10^{-4} that the incident neutron undergoes a scattering event within the radiator. Second, the solid angle subtended by the recoil proton detectors must be kept relatively small to avoid including too large a spread in recoil angles, and consequently smearing the peak response function.

One of the attractive features of proton recoil telescopes is the fact that their detection efficiency can be calculated quite accurately. Because complications such as multiple scattering or wall effects are largely avoided, the probabilities of neutron scattering and subsequent proton recoil detection are quite easily calculated from the accurately known hydrogen scattering cross section and geometric evaluation of the detector solid angle. Useful analyses of these factors can be found in Refs. 120–123.

One way to improve the detection efficiency is to make the radiator thick and to record separately the energy loss of the protons before they escape. Then the radiator takes the form of a hydrogenous detector such as an organic scintillator[124] or methane-filled proportional counter.[125] Protons that originate at any depth up to their range can deposit part of their energy in the radiator-detector, and then leave its surface to deposit the remainder in a second detector located some distance away to define the recoil angle. By analyzing the coincident pulses observed from the two detectors, the neutron energy that gave rise to the recoil proton can be inferred.

PROBLEMS

15-1. Calculate the efficiency of a 4 mm thick ^{6}LiI scintillator for incident 1 MeV neutrons. Repeat for thermal neutrons.

15-2. A thermal neutron detector is placed at the center of a spherical moderator that is exposed to a source of 5 MeV neutrons. If the moderator diameter is varied while holding all other conditions constant, sketch the corresponding expected variation of the counting rate. Offer physical explanations for the behavior of this curve at both large and small diameters.

15-3. A lithium iodide scintillator is often used as the central detector in the neutron spherical dosimeter. Sketch the expected pulse height spectrum from the scintillator in this application.

15-4. An incident fast neutron is moderated and then diffuses a total pathlength of 10 cm before being captured in the BF_3 tube of a long counter. Estimate the time delay between the time of neutron incidence and the leading edge of the output pulse.

15-5. Incident 3 MeV neutrons interact in a lithium sandwich spectrometer. Calculate the reaction product energies for the case in which the alpha particle is emitted in the forward direction at $0°$ and the triton at $180°$.

15-6. Calculate the maximum proton energy from the ^{3}He(n, p) reaction when induced by 1.5 MeV neutrons.

15-7. Explain the physical origin of the epithermal peak observed in most pulse height spectra from ^{3}He proportional tubes when used with fast neutron sources.

15-8. Calculate the detection efficiency of a methane-filled proportional counter for incident 100 keV neutrons if the gas pressure is 1 atm and the neutron pathlength through the gas is 5 cm.

15-9. A 1 MeV neutron enters a plastic scintillator and undergoes two sequential scatterings from hydrogen nuclei before escaping. If the first scattering deflects the neutron at an angle of $40°$ with respect to its original direction and the scattering sites are 3 cm apart, calculate the time that separates the two events. If the PM anode time constant is 20 ns, will the two events be resolved?

15-10. Sketch the differential pulse height spectrum you would expect from a proton recoil detector if the incident neutron energy spectrum is known to have three very prominent and narrow peaks at 75, 150, and 300 keV.

15-11. Show that the angle (in the laboratory frame) between a recoil proton and the corresponding scattered neutron is always $90°$.

15-12. Using the parameters listed in Table 6-1, estimate the maximum pulse amplitude expected if a methane-filled proportional counter with the following properties is irradiated by 1 MeV neutrons: gas pressure, 0.75 atm; applied voltage, 2000 V; anode radius, 0.005 cm; cathode radius, 2 cm; tube capacitance, 60 pF.

15-13. A silicon detector is irradiated by 1 MeV neutrons. Find the minimum and maximum energies expected for the silicon recoil nuclei produced in elastic scattering of the incident neutrons.

15-14. What basic physical difference leads to the observation that the recoil energy distribution from 5 MeV neutron scattering from hydrogen is uniform or rectangular shaped, while it is highly nonuniform for scattering from helium?

15-15. What factor limits increasing the detection efficiency of a proton recoil telescope by simply increasing the thickness of the hydrogenous radiator?

REFERENCES

1. R. L. Bramblett, R. I. Ewing, and T. W. Bonner, *Nucl. Instrum. Meth.* **9**, 1 (1960).
2. G. J. H. Jacobs and R. L. P. van den Bosch, *Nucl. Instrum. Meth.* **175**, 483 (1980).
3. L. W. Brackenbush and R. I. Scherpelz, PNL-SA-11645, CONF-840202-13 (1983).
4. W. H. Miller and R. M. Brugger, *Nucl. Instrum. Meth.* **A236**, 333 (1985).
5. T. L. Johnson, Y. Lee, K. A. Lowry, and S. C. Gorbics, *Proceedings of the American Nuclear Society Topical Meeting on Theory and Practices in Radiation Protection and Shielding*, April 1987.
6. D. W. O. Rogers, *Health Phys.* **37**, 735 (1979).
7. M. A. Gomaa and E. Moustafa, *Nucl. Instrum. Meth.* **136**, 379 (1976).
8. M. P. Dhairyawan, P. S. Nagarajan, and G. Venkataraman, *Nucl. Instrum. Meth.* **169**, 115 (1980).
9. P. M. Thomas, K. G. Harrison, and M. C. Scott, *Nucl. Instrum. Meth.* **224**, 225 (1984).
10. N. E. Hertel and J. W. Davidson, *Nucl. Instrum. Meth.* **A238**, 509 (1985).
11. D. E. Hankins, LA-2717 (1962).
12. D. E. Hankins and R. A. Pederson, LAMS-2977 (1964).
13. J. W. Leake, *Nucl. Instrum. Meth.* **63**, 329 (1968).
14. K. G. Harrison, *Nucl. Instrum. Meth.* **166**, 197 (1979).
15. J. W. Leake, *Nucl. Instrum. Meth.* **178**, 287 (1980).
16. A. O. Hanson and M. L. McKibben, *Phys. Rev.* **72**, 673 (1947).
17. M. H. McTaggart, AWRE NR/Al/59 (1958).
18. J. De Pangher and L. L. Nichols, BNWL-260 (1966).
19. J. B. Hunt and J. C. Robertson, *Proceedings of the First Symposium on Neutron Dosimetry in Biology and Medicine*, EUR 4896 d-f-e, 935 (1972).
20. D. R. Slaughter and D. W. Rueppel, *Nucl. Instrum. Meth.* **145**, 315 (1977).
21. J. B. Hunt and R. A. Mercer, *Nucl. Instrum. Meth.* **156**, 451 (1978).
22. L. V. East and R. B. Walton, *Nucl. Instrum. Meth.* **72**, 161 (1969).
23. A. E. Evans, *Nucl. Instrum. Meth.* **199**, 643 (1982).
24. H. H. Thies and K. J. Böttcher, *Nucl. Instrum. Meth.* **75**, 231 (1969).
25. R. F. Barrett, J. R. Birkelund and H. H. Thies, *Nucl. Instrum. Meth.* **68**, 277 (1969).
26. B. K. Kamboj, M. G. Shahani, U. V. Phadnis, and D. Sharma, *Nucl. Instrum. Meth.* **148**, 57 (1978).
27. E. Hochhäuser and E. Schönfeld, *Nucl. Instrum. Meth.* **80**, 347 (1970).
28. K. K. Sekharan, H. Laumer, B. D. Kern, and F. Gabbard, *Nucl. Instrum. Meth.* **133**, 253 (1976).
29. E. A. Sokol et al., *Nucl. Instrum. Meth.* **219**, 336 (1984).
30. W. P. Poenitz, *Nucl. Instrum. Meth.* **58**, 39 (1968).
31. W. P. Poenitz, *Nucl. Instrum. Meth.* **72**, 120 (1969).
32. W. P. Poenitz, ANL-7915 (1972).
33. J. R. P. Eaton and J. Walker, *Proc. Phys. Soc.* (*London*) **83**, 301 (1964).
34. D. R. Johnson, J. H. Thorngate and P. T. Perdue, *Nucl. Instrum. Meth.* **75**, 61 (1969).
35. R. B. Murray, *Nucl. Instrum. Meth.* **2**, 237 (1958).

36. A. R. Spowart, *Nucl. Instrum. Meth.* **75**, 35 (1969).

37. A. R. Spowart, *Nucl. Instrum. Meth.* **82**, 1 (1970).

38. J. M. Neill, D. Huffman, C. A. Preskitt, and J. C. Young, *Nucl. Instrum. Meth.* **82**, 162 (1970).

39. W. R. McMurray, N. J. Pattenden, and G. S. Valail, *Nucl. Instrum. Meth.* **114**, 429 (1974).

40. A. R. Spowart, *Nucl. Instrum. Meth.* **135**, 441 (1976).

41. A. R. Spowart, *Nucl. Instrum. Meth.* **140**, 19 (1977).

42. E. J. Fairley and A. R. Spowart, *Nucl. Instrum. Meth.* **150**, 159 (1978).

43. G. L. Jensen and J. B. Czirr, *Nucl. Instrum. Meth.* **205**, 461 (1983).

44. C. Maroni, F. Russo, and E. Verondini, *Nucl. Instrum. Meth.* **74**, 256 (1969).

45. I. C. Rickard, *Nucl. Instrum. Meth.* **113**, 169 (1973).

46. M. G. Silk, *Nucl. Instrum. Meth.* **66**, 93 (1968).

47. G. B. Bishop, *Nucl. Instrum. Meth.* **62**, 247 (1968).

48. R. A. Wolfe and W. F. Stubbins, *Nucl. Instrum. Meth.* **60**, 246 (1968).

49. H. Bluhm and D. Stegemann, *Nucl. Instrum. Meth.* **70**, 141 (1969).

50. G. Koutzoukos and C. B. Besant, *J. Br. Nucl. Energy Soc.* **14**, 83 (1975).

51. P. J. Clements, *Nucl. Instrum. Meth.* **127**, 61 (1975).

52. T. Pinelli et al., *Nucl. Instrum. Meth.* **150**, 497 (1978).

53. A. Sayres and M. Coppola, *Rev. Sci. Instrum.* **35**, 431 (1964).

54. J. L. Friedes and R. E. Chrien, *Rev. Sci. Instrum.* **35**, 469 (1964).

55. T. Fuse, T. Miura, A. Yamaji, and T. Yoshimura, *Nucl. Instrum. Meth.* **74**, 322 (1969).

56. S. Nishino, T. Nakamura, and T. Hyodo, *Mem. Fac. Eng. (Kyoto Univ.)* **35**(3), 309 (1973).

57. S. Izumi and Y. Murata, *Nucl. Instrum. Meth.* **94**, 141 (1971).

58. J. M. Cuttler, S. Greenberger, and S. Shalev, *Nucl. Instrum. Meth.* **75**, 309 (1969).

59. H. Franz, W. Rudolph, H. Ohm, K.-L. Kratz, G. Herrmann, F. M. Nuh, D. R. Slaughter, and S. G. Prussin, *Nucl. Instrum. Meth.* **144**, 253 (1977).

60. J. G. Owen, D. R. Weaver, and J. Walker, *Nucl. Instrum. Meth.* **188**, 579 (1981).

61. W. A. Fisher, S. H. Chen, D. Gwinn, and R. R. Parker, *Nucl. Instrum. Meth. Phys. Res.* **219**, 179 (1984).

62. A. E. Evans, *IEEE Trans. Nucl. Sci.* **NS-32**(1), 54 (1985).

63. K.-H. Beimer, G. Nyman, and O. Tengblad, *Nucl. Instrum. Meth. Phys. Res.* **A245**, 402 (1986).

64. H. Ohm, K.-L. Kratz, and S. G. Prussin, *Nucl. Instrum. Meth. Phys. Res.* **A256**, 76 (1987).

65. F. Hoenen and W. Bieger, *Nucl. Instrum. Meth. Phys. Res.* **A259**, 529 (1987).

66. S. Shalev and J. Cuttler, *Nucl. Sci. Eng.* **51**, 52 (1973).

67. A. E. Evans, Jr., "Development of a High-Pressure ^{3}He Neutron Scintillator Spectrometer," Los Alamos National Laboratory Program Technical Note, LA-Q2TN-82-109, Apr. 29, 1982.

68. M. S. Derzon, D. R. Slaughter, S. G. Prussin, *IEEE Trans. Nucl. Sci.* **NS-33**(1), 247 (1986).

69. H. Bluhm, *Nucl. Instrum. Meth.* **115**, 325 (1974).

70. M. G. Silk, "The Determination of the Fast Neutron Spectrum in Thermal Reactors Using ^{6}Li and ^{3}He Semiconductor Spectrometers," AERE-R-5183 (1966).

71. T. R. Jeter and M. C. Kennison, *IEEE Trans. Nucl. Sci.* **NS-14**(1), 422 (1967).

72. J. B. Marion and F. C. Young, *Nuclear Reaction Analysis*, North-Holland, Amsterdam, 1968.

73. J. A. Harvey and N. W. Hill, *Nucl. Instrum. Meth.* **162**, 507 (1979).

74. N. R. Stanton, COO-1545-92 (1971).

75. J. Devos et al., *Nucl. Instrum. Meth.* **135**, 395 (1976).

76. R. H. Johnson et al., *Nucl. Instrum. Meth.* **145**, 337 (1977).

77. D. Slaughter and R. Strout II, *Nucl. Instrum. Meth.* **198**, 349 (1982).

78. M. J. Coolbaugh, R. E. Faw, and W. Meyer, "Fast Neutron Spectroscopy in Aqueous Media Using an NE213 Proton Recoil Spectrometer System," COO-2049-7 (1971).

79. M. E. Toms, *Nucl. Instrum. Meth.* **92**, 61 (1971).

80. N. Sasamoto and S. Tanaka, *Nucl. Instrum. Meth.* **148**, 395 (1978).

81. M. Bormann, R. Kühl, K. Schäfer, and U. Seebeck, *Nucl. Instrum. Meth.* **88**, 245 (1970).

82. D. Hermsdorf, K. Pasieka, and D. Seeliger, *Nucl. Instrum. Meth.* **107**, 259 (1973).

83. M. Reier, *Nucl. Instrum. Meth.* **65**, 119 (1968).

84. S. Mubarakmand and M. Anwar, *Nucl. Instrum. Meth.* **93**, 515 (1971).

85. S. T. Thornton and J. R. Smith, *Nucl. Instrum. Meth.* **96**, 551 (1971).

86. R. Plasek, D. Miljanic, V. Valkovic, R. B. Liebert, and G. C. Phillips, *Nucl. Instrum. Meth.* **111**, 251 (1973).

87. R. De. Leo, G. D'Erasmo, A. Pantaleo, and G. Russo, *Nucl. Instrum. Meth.* **119**, 559 (1974).

88. P. Leleux, P. C. Macq, J. P. Meulders, and C. Pirart, *Nucl. Instrum. Meth.* **116**, 41 (1974).

89. R. A. Cecil, B. D. Anderson, and R. Madey, *Improved Predictions of Neutron Detection Efficiency for Hydrocarbon Scintillators from 1 MeV to About 300 MeV*, North-Holland, Amsterdam, 1979.

90. J. L. Fowler, J. A. Cookson, M. Hussain, R. B. Schwartz, M. T. Swinhoe, C. Wise, and C. A. Uttley, *Nucl. Instrum. Meth.* **175**, 449 (1980).

91. M. Drosg, D. M. Drake, and P. Lisowski, *Nucl. Instrum. Meth.* **176**, 477 (1980).

92. K. F. Flynn, L. E. Glendenin, E. P. Steinberg, and P. M. Wright, *Nucl. Instrum. Meth.* **27**, 13 (1964).

93. G. Dietze and H. Klein, *Nucl. Instrum. Meth.* **193**, 549 (1982).

94. Y. Furuta, S. Kinbara and K. Kaieda, *Nucl. Instrum. Meth.* **84**, 269 (1970).

95. R. St. Onge, A. Galonsky, R. K. Jolly, and T. M. Amos, *Nucl. Instrum. Meth.* **126**, 391 (1975).

96. J. B. Dance and P. E. Francois, *IEEE Trans. Nucl. Sci.* **NS-23**(4), 1433 (1976).

97. A. Chalupka, G. Stengl, M. R. Maier, and P. Sperr, *Nucl. Instrum. Meth.* **150**, 209 (1978).

98. L. J. Perkins and M. C. Scott, *Nucl. Instrum. Meth.* **166**, 451 (1979).

99. R. E. Howe, *Nucl. Instrum. Meth.* **190**, 309 (1981).

100. A. G. Da Silva, L. T. Auler, J. C. Suita, L. J. Antunes, and A. A. Da Silva, *Nucl. Instrum. Meth. Phys. Res.* **A264**, 381 (1988).

101. N. A. Lurie, L. Harris, Jr., and J. C. Young, *Nucl. Instrum. Meth.* **129**, 543 (1975).

102. C. Chen, J. A. Lockwood, and L. Hsieh, *Nucl. Instrum. Meth.* **138**, 363 (1976).

103. D. T. Ingersoll and B. W. Wehring, *Nucl. Instrum. Meth.* **147**, 551 (1977).

104. V. V. Verbinski and R. Giovannini, *Nucl. Instrum. Meth.* **114**, 205 (1974).

105. P. K. Ray and E. S. Kenney, *Nucl. Instrum. Meth.* **144**, 579 (1977).

106. E. F. Bennett, *Nucl. Sci. Eng.* **27**, 16 (1967).

107. P. W. Benjamin, C. D. Kemshall, and J. Redfearn, *Nucl. Instrum. Meth.* **59**, 77 (1968).

108. E. Korthaus, EURFNR-1197; KFK-1994 (1974).

109. E. F. Bennett and T. J. Yule, ANL-7763 (1971).

110. E. F. Bennett and T. J. Yule, *Nucl. Instrum. Meth.* **98**, 393 (1972).

111. N. L. Snidow and H. D. Warren, *Nucl. Instrum. Meth.* **51**, 109 (1967).

112. R. Gold and E. F. Bennett, *Nucl. Instrum. Meth.* **63**, 285 (1968).

113. D. W. Vehar and F. M. Clikeman, *Nucl. Instrum. Meth.* **190**, 351 (1981).

114. S. A. Heiberg, *Nucl. Instrum. Meth.* **63**, 71 (1968).

115. H. F. Atwater, *Nucl. Instrum. Meth.* **100**, 453 (1972).

116. I. R. Brearley, A. Bore, N. Evans, and M. C. Scott, *Nucl. Instrum. Meth.* **192**, 439 (1982).

117. H. Werle, G. Fieg, H. Seufert, and D. Stegemann, *Nucl. Instrum. Meth.* **72**, 111 (1969).

118. T. B. Ryves, *Nucl. Instrum. Meth.* **135**, 455 (1976).

119. M. Cambiaghi, F. Fossati, and T. Pinelli, *Nucl. Instrum. Meth.* **82**, 106 (1970).

120. H. Gotoh and H. Yagi, *Nucl. Instrum. Meth.* **97**, 419 (1971).

121. H. Gotoh and H. Yagi, *Nucl. Instrum. Meth.* **101**, 395 (1972).

122. D. Sloan and J. C. Robertson, *Nucl. Instrum. Meth.* **198**, 365 (1982).

123. B. R. L. Siebert, H. J. Brede, and H. Lesiecki, *Nucl. Instrum. Meth. Phys. Res.* **A235**, 542 (1985).

124. K. N. Geller, D. Eccleshall, and T. T. Bardin, *Nucl. Instrum. Meth.* **69**, 141 (1969).

125. H. Borst, *Nucl. Instrum. Meth.* **169**, 69 (1980).

CHAPTER · 16

Pulse Processing and Shaping

In this chapter we begin discussion of the methods used to extract information from the pulses produced by radiation detectors. Many of the topics discussed in this and the following two chapters are more thoroughly covered in texts on nuclear electronics, of which Refs. 1 and 2 are examples. Leaving more detailed circuit descriptions to these specialized texts, we stress here the implementation of pulse-handling and processing methods from a *user's* rather than a designer's point of view. A functional or "blackbox" approach will be taken to most electronic components, with the exceptions limited only to those cases in which a more detailed description is necessary to permit an intelligent choice between alternatives, or to allow optimization of adjustments normally carried out by the user.

I. DEVICE IMPEDANCES

A basic concept in the processing of pulses from radiation detectors is the impedance of the devices that comprise the signal-processing chain. A simplified representation of the input and output configurations of a typical component is shown in Fig. 16-1. Both the input and output impedances can in general involve capacitive or inductive components, but for the sake of simplicity, a purely resistive impedance will be assumed.

The input impedance Z_i represents the extent to which a device loads a given signal source. A high input impedance will draw very little current from the source and therefore present only a very light load. For example, the input impedance of an oscilloscope is always very high to avoid perturbing the signals that are being inspected. For most applications, input impedances of devices are kept high to avoid excessive loading, but other factors may sometimes dictate situations in which the input impedance must be low enough to load the source significantly.

The output impedance can be thought of as an internal resistance in series with a voltage generator representing the output stage of a given component (see Fig. 16-1). For most applications, one would ideally want this output impedance to be as low as possible to minimize the signal loss when the output is loaded by a subsequent component. In Fig. 16-1, the voltage V_L appearing across a loading Z_L is given by the voltage-divider relation

$$V_L = V_S \frac{Z_L}{Z_0 + Z_L} \qquad (16\text{-}1)$$

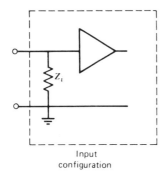 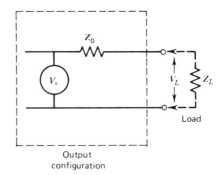

Input   Output
configuration   configuration

Figure 16-1 Idealized input and output configurations. For the input configuration, the triangle represents an ideal operational amplifier with infinite input impedance. Therefore, the effective device input impedance is Z_i. For the output configuration, Z_0 represents the output impedance in series with an ideal voltage generator V_S.

The open-circuit or unloaded ($Z_L = \infty$) voltage appearing at the output of the device is simply V_S. To preserve maximum signal level, one normally wants V_L to be as large a fraction of V_S as possible. If the output impedance is low compared with the load ($Z_0 \ll Z_L$), then $V_L \cong V_S$ and essentially all the signal voltage is transmitted to the load. If the output impedance is equal to the load ($Z_0 = Z_L$), then $V_L = V_S/2$ and only half the unloaded output voltage of the device appears across the load. Output stages with low output impedance thus are often an important design goal. As an example, an emitter–follower configuration is a common type of output circuit that can have an output impedance of less than 1 ohm. It can therefore drive loads of 50 to 100 ohms or larger without significant signal attenuation.

When devices are interconnected in a signal chain, the load Z_L presented to a given component is the impedance of the following component. If all output impedances are low compared with input impedances, the maximum signal level is preserved throughout the chain. This condition is often realized in normal nuclear pulse-processing systems. If very fast pulses are being handled, however, considerations involving reflections in coaxial cables can dictate impedance-matching conditions in which the output/input impedance ratio is not always small, and some signal attenuation will occur.

II. COAXIAL CABLES

A. Cable Construction

Virtually all interconnection of components in a signal chain for nuclear detector pulses is carried out using shielded coaxial cable. A diagram of a typical cable construction is shown in Fig. 16-2. The shielded construction is designed to minimize pickup of noise from stray electric and electromagnetic fields. To preserve the flexibility of the cable, the outer shield is usually made of braided strands of fine copper wire. The effectiveness against low-frequency electric fields is determined primarily by the tightness of the braided shield. High-frequency electromagnetic fields are shielded by virtue of the skin effect. At frequencies at which the skin depth is comparable to or smaller than the braid strand thickness (say, greater than 100 kHz), the shielding is quite effective but will become less so at lower frequencies. Under extreme conditions, it is sometimes necessary to surround the braid with a second shield to fully exclude the effects of very strong fields

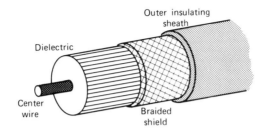

Figure 16-2 Construction of a standard coaxial cable.

through which the cable must pass. Doubly shielded coaxial cables are commercially available (see Table 16-1) in which a second braided shield is provided, but a better solution for difficult cases is to run a conventional cable inside tubing made of a solid conductor. For most routine applications, single-shielded cable provides entirely satisfactory isolation from unwanted signal pickup.

B. Cable Properties

The velocity of propagation for pulses through a coaxial cable is a function only of the dielectric materials separating the central conductor and the outer shield and is inversely proportional to the square root of the dielectric constant. Cables using air or some other gas as a dielectric have a propagation velocity very close to the velocity of light in a vacuum (3.00×10^8 m/s). Virtually all general-duty signal cables utilize a solid such as polyethylene for the dielectric, in which case the velocity of propagation is about 66% of that of light in a vacuum. Some special-duty cables with polyethylene foam dielectric have a somewhat higher velocity of propagation. At the other extreme, special delay cables with helically wound central conductors can reduce the velocity of propagation by factors of 100 or more.

Table 16-1 lists important properties of coaxial cable types commonly used for nuclear instrumentation. The "RG/U" designation arose originally as a military specification and the numbers were assigned to various cable types in order of approval. Consequently, they bear no relation with specific cable properties. The RG/U designation implies adherence to certain quality standards not usually found in lower-cost coaxial cables intended for consumer electronics applications. The historical trend has been to replace the older large-diameter signal cables (such as RG-8/U) with smaller sized cables such as RG-59/U or RG-62/U, or with miniature types such as RG-178/U. In signal cables, the important specifications are usually the characteristic impedance (see discussion later in this chapter) and the capacitance per unit length. In cables intended to carry bias voltage to detectors, the maximum voltage rating is also important.

No real cable is a perfect transmission line. There will always be dissipative losses due to imperfect dielectrics and resistance of the center conductor that will result in some attenuation and distortion of the transmitted pulse, especially its high-frequency components. For most applications, these effects are small and can often be neglected for cables shorter than a few tens of meters. However, for demanding situations involving the transmission of fast rise time pulses, some attention should be paid to the high-frequency attenuation specifications of the cable to prevent deterioration of the fast characteristics of the transmitted pulse. As an example, the distortion of a 1 ns leading edge pulse after transmission through 3 m of RG-174/U cable is easily seen on a fast oscilloscope.[3]

TABLE 16-1   Properties of Coaxial Cables[a]

	Insulating Material	Cable Diameter (cm)	Characteristic Impedance (ohms)	Signal[b] Propagation	HV Rating	Cable Capacitance (pF/m)	Signal Attenuation per Meter	
							MHz	dB
RG-8/U	Polyethylene	1.03	52	0.659	5000	96.8	100	0.066
							400	0.154
RG-11/U	Polyethylene	1.03	75	0.659	5000	67.3	100	0.066
							400	0.138
RG-58/U	Polyethylene	0.50	53.5	0.659	1900	93.5	100	0.135
							400	0.312
RG-58C/U	Polyethylene	0.50	50	0.659	1900	100.1	100	0.174
							400	0.413
RG-59/U	Polyethylene	0.61	73	0.659	2300	68.9	100	0.112
							400	0.233
RG-62/U	Semisolid polyethylene	0.61	93	0.840	750	44.3	100	0.102
							400	0.207
RG-174/U	Polyethylene	0.25	50	0.659	1500	101.0	100	0.289
							400	0.656
RG-178/U	TFE teflon	0.18	50	0.694	1500	95.1	400	0.951
DOUBLE SHIELDED COAXIAL CABLES								
RG-9/U	Polyethylene	1.07	51	0.659	5000	98.4	100	0.062
							400	0.135
RG-223/U	Polyethylene	0.52	50	0.659	1900	101.0	100	0.157
							400	0.328

[a]Data derived in part from Coaxial Cable Catalog, Belden Corporation, Richmond, IN.
[b]Fraction of speed of light in a vacuum (3.00×10^8 m/s).

Specially made rigid transmission lines using solid copper tubing for both the shield and center conductor, which display superior high-frequency characteristics compared with conventional coaxial cables, are described in Ref. 4.

C. Noise Pickup and Component Grounding

The outer shield also serves to interconnect the chassis of each component with that of the next. When all components are mounted in the same instrument bin or rack, this electrical connection is often redundant. When components are physically separated, however, the shield will tend to establish a common ground potential for all components. If all chassis are not grounded internally to the same point, some dc current may need to flow in the shield to maintain the common ground potential. In many routine applications, this ground current is small enough to be of no practical consequence. However, if components are widely separated and internally grounded under widely different conditions, the shield current can be large and its fluctuations may induce significant noise in the cable. Under these conditions, such *ground loops* must be eliminated by ensuring that all components are internally grounded to a single common point for the entire system. Undesirable transient signals can also be induced in cable shields (especially loop configurations) if nearby equipment involves the fast switching of large currents. Brookshier[5] presents a general analysis of these and other sources of noise pickup in instrument systems interconnected by coaxial cables.

A technique known as *common mode rejection* is sometimes helpful in reducing the effects of noise pickup on interconnecting cables. The receiving device (often the linear amplifier) is designed with a differential input so that the signal is measured on one input relative to a reference voltage at the second input. Connection to the two inputs is made by identical cables that are run side-by-side to the signal source (often the preamplifier). Only the first cable is connected to the signal source, whereas the cable from the reference input is left open at the sending end. In principle, much of the pickup will appear on both cables and will therefore be eliminated by the differential input.

D. Characteristic Impedance and Cable Reflections

A general discussion of pulse transmission through coaxial cables is best divided into two extremes: cases in which low-frequency or slow pulses are transmitted, or those in which high-frequency or fast pulses are involved. The distinction of whether a given application involves fast or slow pulses depends on a comparison of the fastest pulse component (usually the rise time) with the transit time of the pulse through the cable. For cables with solid polyethylene dielectric, the transit time is about 5.1 ns/m. Pulses having rise times that are large compared with the transit time are slow pulses, whereas those having a rise time comparable to or shorter than the transit time are fast pulses. For cables of a few meters in length, only the pulses from very fast detectors will qualify as fast pulses, whereas even relatively slow detector outputs must be considered as fast pulses if the cable is several hundred meters long. However, most routine situations involve pulses which by this definition are slow.

For slow pulses, the cable acts much like a simple conductor interconnecting components. Its important properties then are simply its series resistance and capacitance to ground. The resistance of the central conductor is very small for cables less than a few hundred meters in length, so the most significant parameter usually is the cable capacitance. This capacitive loading will increase linearly with cable length, but is seldom a

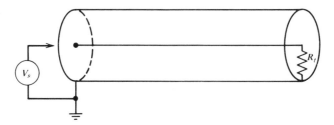

Figure 16-3   Application of a step voltage V_S to a coaxial cable terminated in R_t.

practical problem except between the detector and preamplifier. Because the noise characteristics of a typical preamplifier will deteriorate with increasing input capacitance, there is a premium on keeping the additional capacitive loading to a minimum. In most other aspects, the choice of cable for slow pulse applications is not critical, and virtually any of the cables listed in Table 16-1 will suffice for routine interconnection of components.

For fast pulses, a number of other considerations become important. Prominent among these is the *characteristic impedance* of the cable. It is a property that depends on the dielectric material and diameters of the inner conductor and outer shield of the cable but is independent of the cable length. To illustrate the physical meaning of characteristic impedance, imagine that the voltage generator shown in Fig. 16-3 is capable of generating a step voltage change from zero to some finite value V_0 at a time $t = 0$. If this voltage step is applied to the input of a coaxial cable, it will travel down the length of the cable with the velocity of propagation discussed in the previous section. During the time this voltage step is traveling down the cable, current is being drawn from the signal source because a finite charge per unit length of the cable is required to raise the central conductor voltage from zero to V_0. If the cable were infinitely long, this current drain would be continuous because the step voltage would propagate indefinitely and continue to draw current from the source. In this example, the characteristic impedance of the cable is simply given by the ratio of the step voltage V_0 divided by the current drawn by the infinitely long cable.

For cables of finite length, we must pay some attention to the conditions that exist at the far end of the cable. The cable is *terminated* by the effective resistance R_t that appears between the central conductor and outer shield of the cable at the far end. If the cable is connected to an electronic component, then the termination resistance is effectively just the input impedance of that component. If the cable is simply left unconnected, its termination resistance is infinite, or it may be terminated by connecting a resistor between the central conductor and the outer shield. When it is desired to terminate a cable that is connected to a component with high input impedance with a lower termination resistance, a resistor-to-ground (called a *shunt terminator*) can be inserted parallel to the input of the device so that the effective termination is the parallel combination of the input impedance and the shunt terminator resistance.

Let us next consider what happens when a cable is terminated in its own characteristic impedance. For the sake of example, assume that we have a long cable with 50 ohm characteristic impedance which is terminated by connecting a 50 ohm resistor across the far end. If we apply a 5 V step input to the sending end, then during the time the step is propagating along the length of the cable a current of 100 mA is drawn from the source. No power is being dissipated at this stage because the current simply goes into charging successive segments of the cable as the voltage step propagates. During this period the 50

TABLE 16-2   Reflection Conditions Created by Various Terminations at the End of a Coaxial Cable with Characteristic Impedance Z_0. Step Input Waveform with Amplitude A Is Assumed

Termination Resistance R	Reflected Step Amplitude
0	$-A$
Between 0 and Z_0	Between $-A$ and 0
Z_0	0
Between Z_0 and ∞	Between 0 and $+A$
∞	$+A$

ohm resistor does not draw current because the center conductor voltage at the far end of the cable remains at zero. When the voltage step finally reaches the far end, the situation changes. Current no longer goes into charging up the cable because it is already fully charged over its entire length. However, the 50 ohm resistor now feels a 5 V potential and will therefore begin to draw a current of 100 mA. If the 5 V step is held by the signal generator at the sending end of the cable, this current through the 50 ohm resistor will continue indefinitely. Therefore, the signal generator must continuously supply a 100 mA current to the cable from the instant the step is applied, and it must maintain this current even after the step has propagated along the entire length of the cable. Therefore, as far as the signal generator is concerned, a *coaxial cable terminated in its own characteristic impedance behaves like an infinitely long cable of the same impedance*.

Other conditions prevail if the cable is not terminated in its characteristic impedance. There will then be an abrupt change in the properties of the medium through which the step is propagating, and reflections from the end of the cable will therefore be generated. If the cable is shorted (R_t is made equal to zero), then the step is inverted and reflected back down the cable toward the sending end with an amplitude equal to the original step. If the end of the cable is simply left unterminated (R_t is infinite), then a reflection of the same polarity and amplitude will propagate back toward the sending end. Only when the cable is terminated in its own characteristic impedance are reflections completely avoided. Table 16-2 shows the reflection conditions for a step voltage of amplitude A for termination conditions between the extremes of zero and infinity.

When transmitting fast pulses through coaxial cables, these reflections can be very undesirable and may lead to distortions of the transmitted pulse form. In these applications, therefore, the experimenter must pay careful attention to the impedance of the cable and the termination conditions at each end. Because the majority of cables used for nuclear pulse applications have characteristic impedances of either 50 or 93 ohms (see Table 16-1), most of the commercial circuits intended for fast pulse applications are designed with input and/or output impedances of 50 or 93 ohms as well. It is usually sufficient to properly terminate only the receiving end of the cable because, in principle, all reflections are thereby avoided. If slight mismatches occur, however, a small reflected component will be propagated back to the sending end where the termination conditions now become important. The output impedance of the generating device determines this termination value, and if it is also equal to the characteristic impedance of the cable, further reflections are suppressed. Instrument systems in which all the input, output, and cable impedances are the same value can therefore be rather generally interconnected without fear of significant reflections.

If the device impedance does not match the cable impedance and reflections must be avoided, external termination resistors may be used at the point where the cable connects to the device in order to create proper termination conditions. A termination resistor

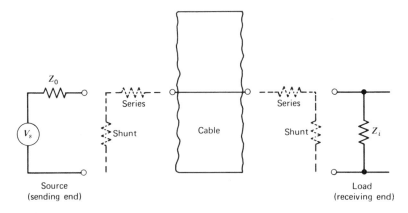

Figure 16-4 Methods of inserting terminators at either end of a coaxial cable.

added in series will add its resistance value to the device impedance at either end of the cable, while a shunt resistor added in parallel to the output or input will lower the device impedance to the equivalent resistance of the parallel combination (see Fig. 16-4). For example, a 50Ω cable connected to the high-impedance (say > 1000Ω) input of a device will be properly terminated by using a 50Ω shunt resistor, since the parallel combination of 50Ω and 1000Ω has an equivalent resistance value very close to 50Ω.

E. Useful Coaxial Cable Accessories

1. TERMINATORS
As described in the previous section, it is often necessary to convert a high impedance input at the end of a coaxial cable to an impedance that matches the characteristic impedance of the cable to prevent reflections. Because of the widespread use of 50 ohm systems for fast pulse transmission, 50 ohm termination is a common requirement. Consequently, 50 ohm terminators are widely used in fast pulse work and consist of a compact cylindrical 50 ohm resistance to ground fitted with a standard coaxial connector. These can be of the simple one-ended type with a plug connection, or the "feedthrough" type with a plug and jack on opposite ends. The latter type can be connected directly to a high-impedance input to provide a shunt termination and a substitute jack for cable attachment.

2. PULSE ATTENUATOR
In some pulse-handling situations, the need arises to reduce the amplitude of a pulse to match the input requirements of a given component. For linear signals, this practice should be avoided if possible because attenuation will inevitably deteriorate the signal-to-noise characteristics of the pulses. At times, however, an attenuation step cannot be avoided.

A simple resistive voltage divider, as shown in Fig. 16-5a, will work as an attenuator, but some attention must be paid to impedance levels and high-frequency performance. To provide an invariant multiplication factor of

$$\frac{R_2}{R_1 + R_2}$$

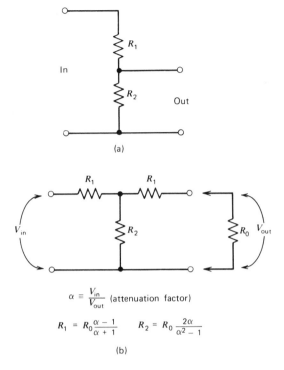

$$\alpha \equiv \frac{V_{in}}{V_{out}} \quad \text{(attenuation factor)}$$

$$R_1 = R_0\frac{\alpha - 1}{\alpha + 1} \qquad R_2 = R_0\frac{2\alpha}{\alpha^2 - 1}$$

(b)

Figure 16-5 (a) A simple voltage-divider attenuation network. (b) A T-attenuator network. The attenuation factor holds provided the network is loaded with the value of R_0 shown. The input and output impedances are then also equal to R_0.

the divider must be used with a source impedance $Z_0 \ll R_1$, and the attenuated pulse must be supplied to a following component whose input impedance $Z_i \gg R_2$. Because of resistor nonidealities and stray capacitance, a simple divider configuration is not usable for pulses with rise or fall times much less than about 100 ns, due to nonlinear attenuation of higher-frequency components and a resulting distortion of the pulse shape. Although the high-frequency performance can be improved with the addition of parallel compensating capacitors (e.g., see Ref. 6), attenuators of this type are not widely used for fast pulse work.

The configuration of Fig. 16-5b is known as a *T-section* attenuator and is more popular as a fast pulse attenuator. It has the advantage that symmetric and equal input and output impedances are realized for convenient matching to coaxial cables. Attenuators of this type, using thin-film resistors mounted in closely fitting metallic cylinders and provided with standard coaxial connections, have excellent high-frequency characteristics and can be used with nanosecond pulses.

3. PULSE SPLITTER
It is occasionally necessary to split the signal chain into two branches at a given point. For slow pulses for which impedance matching is not important, the familiar coaxial "tee" can be used without fear of pulse distortion. When impedance matching is required to prevent cable reflections, the configuration shown in Fig. 16-6 serves to distribute a pulse applied to any terminal to the other two while maintaining a constant impedance level. For example, in a 50 ohm splitter, the resistance values should each be 16.6 ohms so

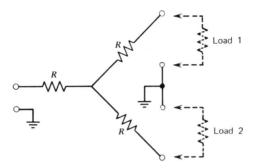

Figure 16-6 A symmetric pulse splitter that can be used to drive two loads while maintaining matched impedance levels.

that the impedance looking into any terminal will be 50 ohms if the other two terminals are connected to 50 ohm loads. However, the signal level delivered to each load will be only half that if the load were directly coupled to the source.

4. INVERTING TRANSFORMER

When dealing with signal pulses of various types, it is sometimes desirable to invert the polarity to be compatible with the input requirements of a particular component. In general, an active circuit involving an amplifier stage is required to carry out a general reversal of the polarity of a pulse. However, in the special case that fast pulses with rise and fall times of a few tens of nanoseconds are involved, the *inverting transformer* shown in Fig. 16-7 is a useful, simple, and compact device. The inversion is carried out simply by grounding opposite extremes of the primary and secondary windings on a pulse transformer and will do a creditable job of inverting a pulse as long as its duration is rather short (say, less than 100 ns). For longer pulses, some nonlinearities and distortion of the output pulse are to be expected.

III. PULSE SHAPING

In dealing with signal pulses from radiation detectors, it is often desirable to change the shape of the pulse in some predetermined fashion. By far the most common application is in processing a train of pulses produced by a preamplifier. In order to assure that complete charge collection occurs, preamplifiers are normally adjusted to provide a decay time for the pulse which is quite long (typically 50 μs). If the rate of interaction in the detector is not small, these pulses will tend to overlap one another and give rise to a pulse train that has the appearance shown in Fig. 16-8a. Because it is the amplitude that carries

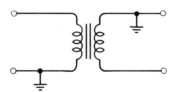

Figure 16-7 The basic configuration of an inverting pulse transformer.

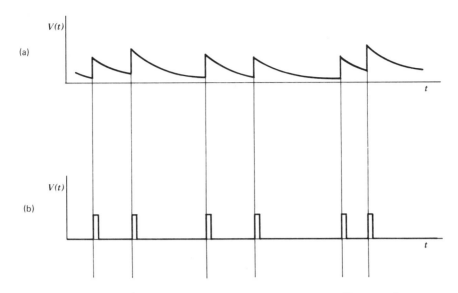

Figure 16-8 The pulses with long tails shown in part (a) illustrate the apparent variation in amplitude due to pulse pileup. These effects can greatly be reduced by shaping the pulses as in part (b).

the basic information (the charge Q deposited in the detector), the "pileup" of pulses on the tails of preceding pulses, which have not fully decayed to zero, can be a serious problem. Because the time spacing between nuclear pulses is random, each pulse can be superimposed on a different residual tail and the resulting amplitude no longer is a good measure of Q from that event.

The ideal solution is to *shape* the pulses in such a way as to produce a pulse train similar to that shown in Fig. 16-8b. Here all the long tails have been eliminated, but the information carried by the maximum amplitude of the pulse has been preserved. The pulses have been shaped in the sense that their total length has been reduced drastically but in a way that does not affect the maximum amplitude.

This type of pulse shaping is conventionally carried out in the linear amplifier element of a nuclear pulse signal chain and will be discussed further in the following chapter. At this point we discuss two categories of pulse-shaping methods: *RC* shaping networks and the use of shorted transmission lines. Both of these techniques are used in nuclear pulse amplifiers but can also be employed generally by the experimenter to carry out pulse-shaping operations at other points in the signal chain.

A. *CR* and *RC* Shaping

In general electrical circuits, the term *RC shaping* refers to the use of passive resistor–capacitor networks to carry out a desired alteration in pulse shape. When discussing nuclear pulse shaping, it is convention to make a semantic distinction between *differentiator* or *CR* networks on one hand, and *integrator* or *RC* networks on the other. Both operations can also be thought of as filtering in the frequency domain, and one purpose of pulse shaping is to improve signal-to-noise ratio by limiting the response of the instrumentation to those frequency ranges in which the signal has useful components.

In both types of network, the *time constant* given by the product of resistance and capacitance plays a critical role. In the analysis that follows, we represent this time

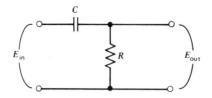

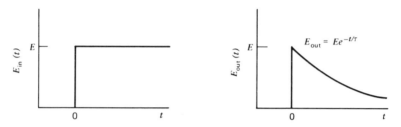

Figure 16-9 A high-pass CR filter or differentiator network. The response to a step function input is illustrated.

constant as τ, or

$$\tau \equiv RC \tag{16-2}$$

(The units of τ are seconds if R is in ohms and C is in farads.)

1. *CR* DIFFERENTIATOR OR HIGH-PASS FILTER

A basic CR differentiator network is diagrammed in Fig. 16-9. From the circuit equations, the input voltage E_{in} and output voltage E_{out} are related by

$$E_{in} = \frac{Q}{C} + E_{out} \tag{16-3}$$

where Q represents the charge stored across the capacitor. Now, differentiating with respect to time,

$$\frac{dE_{in}}{dt} = \frac{1}{C}\frac{dQ}{dt} + \frac{dE_{out}}{dt} \tag{16-4}$$

$$\frac{dE_{in}}{dt} = \frac{1}{C}i + \frac{dE_{out}}{dt} \tag{16-5}$$

Noting that $E_{out} = iR$ and setting $RC = \tau$, we obtain

$$E_{out} + \tau\frac{dE_{out}}{dt} = \tau\frac{dE_{in}}{dt} \tag{16-6}$$

Now, if we make RC sufficiently small, we can neglect the second term on the left and

$$E_{out} \cong \tau\frac{dE_{in}}{dt} \tag{16-7}$$

Thus, in the limit of small time constant τ, the network acts to produce an output E_{out} that is proportional to the time derivative of the input waveform E_{in}—hence the name *differentiator*. In order to meet these conditions, the time constant should be small compared with the duration of the pulse to be differentiated.

In the opposite extreme of large time constant, the first term on the left of Eq. (16-6) can be neglected and we have

$$\tau \frac{dE_{out}}{dt} \cong \tau \frac{dE_{in}}{dt} \tag{16-8}$$

and setting the constant of integration equal to zero

$$E_{out} \cong E_{in} \tag{16-9}$$

Therefore, if the conditions for differentiation are not met, the network will tend to pass the waveform without alteration.

We can solve Eq. (16-6) for arbitrary E_{in} waveforms. Let us state two specific results.

(a) *Sinusoidal E_{in}*
For

$$E_{in} = E_i \sin 2\pi ft \tag{16-10}$$

it can be shown that

$$\frac{E_{out}}{E_i} = |A| \sin (2\pi ft + \theta) \tag{16-11}$$

where

$$|A| = \frac{1}{\left[1 + (f_1/f)^2\right]^{1/2}} \qquad \theta = \tan^{-1}\left(\frac{f_1}{f}\right)$$

$$f_1 \equiv \frac{1}{2\pi\tau}$$

For high-frequency inputs, $f \gg f_1$, and $|A| \cong 1$. Hence, high frequencies are passed to the output with little attenuation, and we have a *high-pass filter*. Low frequencies are attenuated, because for $f \ll f_1$, $|A| \cong 0$.

(b) *Step Voltage Input*
For

$$E_{in} = \begin{cases} E & (t \geq 0) \\ 0 & (t < 0) \end{cases}$$

the output is

$$E_{out} = E e^{-t/\tau} \tag{16-12}$$

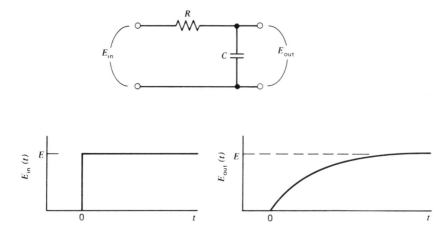

Figure 16-10 A low-pass RC filter or integrator network. The response to a step function input is illustrated.

This case approximately represents the shaping of a fast-rising signal pulse with a long tail by a single CR differentiator. Note that the fast leading edge of the step is *not* differentiated because τ is not small compared with its rise time. The leading edge is therefore simply passed through, and the shaping consists of "differentiating away" the long tail. The amplitude E is not affected provided τ does not become small compared with the finite rise time of the actual signal pulse.

2. RC INTEGRATOR OR LOW-PASS FILTER

When configured as shown in Fig. 16-10, a passive RC network can also serve as an integrator. The circuit equation is now

$$E_{in} = iR + E_{out} \tag{16-13}$$

The current i also represents the rate of charging or discharging of the capacitor.

$$i = \frac{dQ}{dt} = C\frac{dV_c}{dt} \tag{16-14}$$

or

$$i = C\frac{dE_{out}}{dt} \tag{16-15}$$

Now combining Eqs. (16-13) and (16-15) and setting $RC = \tau$, we obtain

$$E_{in} = \tau\frac{dE_{out}}{dt} + E_{out} \tag{16-16}$$

Rearranging, we have

$$\frac{dE_{out}}{dt} + \frac{1}{\tau}E_{out} = \frac{1}{\tau}E_{in} \tag{16-17}$$

Now, if RC is sufficiently large, only the first term on the left is significant, and

$$\frac{dE_{out}}{dt} \cong \frac{1}{\tau}E_{in}$$

or

$$E_{out} \cong \frac{1}{\tau}\int E_{in}\,dt \tag{16-18}$$

Hence, the name *integrator*. The network will integrate provided the time constant τ is large compared with the time duration of the input pulse.

In the opposite extreme of small time constant only the second term on the left of Eq. (16-17) is significant, and therefore

$$\frac{1}{\tau}E_{out} \cong \frac{1}{\tau}E_{in}$$

or

$$E_{out} \cong E_{in} \tag{16-19}$$

Thus, if the conditions for integration are not met, the network tends to pass the waveform without change.

Again we apply Eq. (16-17) to some specific input waveforms.

(a) *Sinusoidal E_{in}*
For

$$E_{in} = E_i \sin 2\pi ft$$

the solution is

$$\frac{E_{out}}{E_i} = |A|\sin\left(2\pi ft + \theta\right) \tag{16-20}$$

where

$$|A| = \frac{1}{\left[1 + (f/f_2)^2\right]^{1/2}} \qquad \theta = -\tan^{-1}\left(\frac{f}{f_2}\right)$$

$$f_2 \equiv \frac{1}{2\pi\tau}$$

Again, we examine the extremes of frequency response. If

$$f \gg f_2, \quad \text{then} \quad |A| \cong 0$$

and if

$$f \ll f_2, \quad \text{then} \quad |A| \cong 1$$

Therefore, the network blocks high frequencies, passes low frequencies without attenuation, and is thus a low-pass filter.

(b) *Step Voltage Input*
For

$$E_{\text{in}} = \begin{cases} E & (t \geq 0) \\ 0 & (t < 0) \end{cases}$$

the output is

$$E_{\text{out}} = E(1 - e^{-t/\tau}) \tag{16-21}$$

This response is also plotted in Fig. 16-10. Recall that the circuit performs as an integrator in the limit of large τ. The mathematical integral of the step input should be a linearly increasing ramp. The actual response starts out as a linear ramp, but over a sufficiently long time scale the circuit time constant is no longer large by comparison, and the integration begins to fail. The output voltage then approaches the input step as a limit.

3. CR-RC SHAPING

The output of a single differentiating network shown in Fig. 16-9 is not a very attractive waveform for pulse analysis systems. The sharply pointed top makes subsequent pulse height analysis difficult because the maximum pulse amplitude is maintained only for a very short time period. Furthermore, because the differentiation allows all high-frequency components of any noise mixed with the signal to be passed by the network, the signal-to-noise characteristics of the network in practical applications are usually very poor. If a stage of *RC* integration is added following the differentiation, however, both of these drawbacks are considerably improved. The combination of a single stage of differentiation followed by a single stage of integration is in fact a common method of shaping preamplifier pulses.

Figure 16-11 shows the elements of the basic *CR-RC* shaping network. An ideal unity-gain operational amplifier (with infinite input impedance and zero output imped-

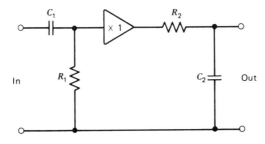

Figure 16-11 A shaping network consisting of sequential differentiating and integrating stages, sometimes denoted a *CR-RC* network.

ance) separates the two individual networks for impedance isolation so that neither network influences the operation of the other. The general solution of the response of the combined network to a step voltage of amplitude E at $t = 0$ is

$$E_{\text{out}} = \frac{E\tau_1}{\tau_1 - \tau_2}\left(e^{-t/\tau_1} - e^{-t/\tau_2}\right) \tag{16-22}$$

where τ_1 and τ_2 are time constants of the differentiating and integrating networks, respectively. Plots of this response for several different combinations of τ_1 and τ_2 are shown in Fig. 16-12.

In nuclear pulse amplifiers, *CR-RC* shaping is most often carried out using equal differentiation and integration time constants. In that event, Eq. (16-22) becomes indeterminant, and a particular solution for this case is

$$E_{\text{out}} = E\frac{t}{\tau}e^{-t/\tau} \tag{16-23}$$

Plots of this step input response for two different values of time constant τ are also shown in Fig. 16-12.

The proper choice for the time constant of the shaping circuits depends primarily on the charge collection time in the detector being used. In the interests of reducing pileup,

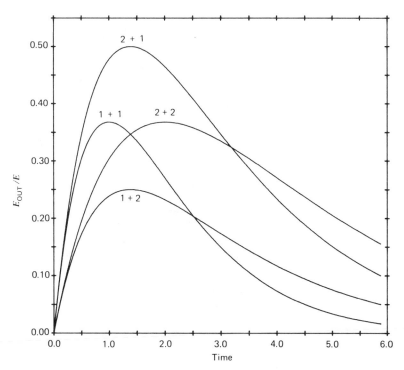

Figure 16-12 The response of a *CR-RC* network to a step voltage input of amplitude E at time zero. Curves are shown for four pairs of differentiator + integrator time constants. Units of the time constants and time scale are identical.

one would like to keep these time constants short so that the shaped waveform can return to the baseline as quickly as possible. On the other hand, once the shaping time constants become comparable with the rise time of the pulse from the preamplifier, the input to the network no longer appears as a step voltage and some of its amplitude is lost. This loss is called the *ballistic deficit* (see p. 606) and can be avoided only by keeping the time constants long compared with the charge collection time in the detector. Typical values for τ might vary from a few tenths of a microsecond, for use with very fast semiconductor diode detectors, to tens of microseconds, more suitable for some types of proportional counters in which the charge collection time is quite large.

4. GAUSSIAN OR CR-(RC)^n SHAPING

If a single CR differentiation is followed by several stages of RC integration, a pulse shape that approaches a mathematical Gaussian is realized. If the differentiation and n integration time constants are all the same value τ, the particular solution of the corresponding circuit equation is

$$E_{\text{out}} = E\left(\frac{t}{\tau}\right)^n e^{-t/\tau} \qquad (16\text{-}24)$$

In practice, four stages of integration ($n = 4$) are sufficient so that the difference between the resulting pulse shape and a true Gaussian is negligible. The time required for the shaped pulse to reach its maximum amplitude (often called the *peaking time*) is equal to $n\tau$.

For equal time constants throughout, a $CR\text{-}(RC)^4$ network results in a peaking time that is a factor of 4 longer than that for a simple $CR\text{-}RC$ network. However, if the time constants are adjusted to give equal peaking times for the two methods, the more symmetric shape of the Gaussian pulse results in a faster return to the baseline. Pulse pileup at high counting rates is thereby reduced. Gaussian shaping also has the advantage of better signal-to-noise characteristics for individual pulses compared with $CR\text{-}RC$ shaping (see Fig. 17-14). For these two reasons, Gaussian shaping has become a popular choice in gamma-ray spectroscopy systems employing high-resolution germanium detectors.

5. ACTIVE FILTER PULSE SHAPING

Earlier in this chapter, it was shown that the processes of differentiation and integration correspond to filtering steps in the frequency domain. The CR differentiator functions as a high-pass filter, while the RC integrator is a low-pass filter. Alternative methods are available to the circuit designer to carry out similar operations using active circuits that incorporate elements such as transistors and diodes. In the design of the shaping networks of modern linear amplifiers, most designers have now substituted these active filters for the passive RC networks described earlier in this chapter. Their function, however, remains similar to that of the corresponding passive network. For example, one of the most common choices is the combination of a high-pass filter (differentiator) followed by several stages of active low-pass filters (integrators) to produce the Gaussian pulse shaping discussed above.

6. TRIANGULAR SHAPING

The discussion of signal-to-noise behavior of different pulse-shaping schemes in linear amplifiers given in Chapter 17 points out the theoretical advantage that a symmetrical triangular shape has over the Gaussian shape (see Fig. 17-14). It is practically impossible

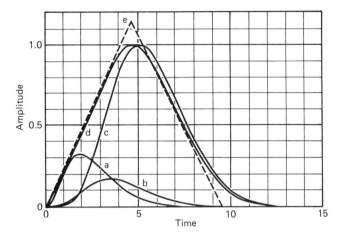

Figure 16-13 Synthesis of a near triangular waveform (curve d, normalized to unity) through the summation of active integrator output waveforms a, b, and c. Curve e is a symmetrical triangle for comparison. Time is in units of the initial differentiation time constant. (From Goulding et al.[7])

to achieve a triangular shape using passive circuit elements alone. If active filter shaping elements are provided, however, it has been shown[7] that a waveform approximating a symmetrical triangle can be synthesized. After an initial differentiation, the outputs from active integrators with different characteristics are weighted and mixed as shown in Fig. 16-13. Similar schemes for producing a triangular pulse shape are now offered as a feature of some commercially available linear amplifiers.

B. Pole-Zero Cancellation

Thus far our comments on pulse shaping have assumed that the input pulse from the preamplifier consists of a step voltage. Although the decay of the preamplifier pulse is usually long, it is not infinite and the finite decay will have a measurable effect on the response of the networks discussed above. For example, a basic CR-RC differentiator–integrator will no longer produce a strictly unipolar response if the input pulse has the finite decay shown in Fig. 16-14. Instead, there will be a slight zero crossover or *undershoot* of the pulse, which then recovers back to zero with a time characteristic of the preamplifier decay time. Because preamplifiers have long decays (of the order of 50 μs), the undershoot persists for a relatively long time. If another pulse arrives during this period of time, it will be superimposed on the undershoot and an error will be induced in its amplitude. The problem is particularly severe for very large signal pulses that overload the amplifier and consequently lead to rather large following undershoots.

Our previous analysis of the response of a CR differentiator to a step input [Eq. (16-12)] showed that the output was a simple exponential decay with no undershoot. The transfer function for this network is $\tau_1 s/(1 + s\tau_1)$, where τ_1 is the time constant and s is the Laplace variable. Assume that this output is now supplied to a second CR differentiator with time constant τ_2 and transfer function $\tau_2 s/(1 + s\tau_2)$. The overall transfer function is then $\tau_1\tau_2 s^2/(1 + s\tau_1)(1 + s\tau_2)$, and it converts a step input to an output pulse with undershoot. The poles in the denominator of the transfer function assures that a simple exponential decay will not be possible.

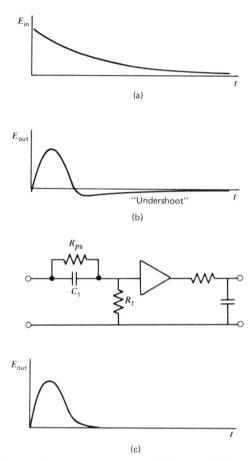

Figure 16-14 Application of pole-zero cancellation to eliminate the undershoot (b) normally generated by a *CR-RC* shaping network for an input step with finite decay time. By adding an appropriate resistance R_{pz} to the differentiator stage, a waveform without undershoot (c) can be obtained.

The above picture represents the origin of the undershoot if a preamplifier pulse with finite decay is presented to a *CR* differentiator in a shaping network. The term *pole-zero cancellation*[8] describes a technique in which the network is modified as in Fig. 16-14 to again restore the simple exponential output without undershoot. A resistance R_{pz} is added in parallel with the capacitor of the *CR* network, resulting in a modified transfer function of the form

$$\frac{\tau_1\left(1 + sR_{pz}C_1\right)}{(1 + s\tau_2)\left(R_{pz}C_1s\tau_1 + R_{pz}C_1 + \tau_1\right)}$$

If the value of R_{pz} is now chosen such that

$$R_{pz} = \frac{\tau_2}{C_1}$$

the transfer function reduces to

$$\frac{1}{\tau_2(s + k)} \quad \text{where} \quad k \equiv \frac{\tau_1 + \tau_2}{\tau_1\tau_2}$$

This result, with its single pole in the denominator, assures that the network once again produces a simple exponential decay for a step input.

A somewhat more heuristic way of describing the function of the modified network of Fig. 16-14 is to note that R_{pz} allows an attenuated replica of the input pulse to pass through to the output. If the input is a positive step with finite decay, then the slight undershoot from a normal CR differentiator can be cancelled by this slight admixture of the input pulse. A proper choice for the resistance value depends on the value of the input decay time from the particular preamplifier in use, and therefore in practice it is often chosen empirically. The pole-zero network is simply adjusted while the output pulse is inspected on an oscilloscope and adjustments are made to eliminate the long undershoot by visual observation.

C. Baseline Shift

1. THE ORIGIN OF THE PROBLEM

The discussion to this point has dealt with the shaping of single isolated pulses. In any radiation detector system, pulses will occur in time sequence with many other pulses. This pulse train leads to a potential problem not encountered with single pulses. In Fig. 16-15a, we show the response of the CR-RC network to a train of regularly spaced rectangular pulses. Because capacitors cannot conduct direct current, the average dc voltage of any point to the right of the capacitor of the differentiating network in Fig. 16-11 must be zero. If the dc voltage were nonzero, a finite dc current must flow through the resistor to ground. Any such current would have to come through the capacitor, which

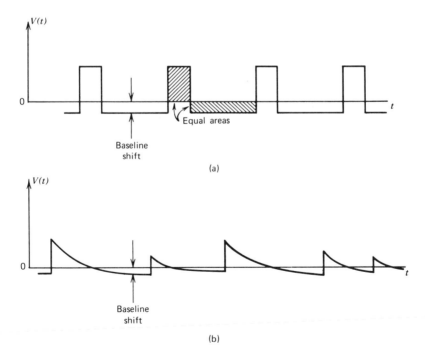

Figure 16-15 Illustrating baseline shift which must take place in ac-coupled circuits. In part (a) uniform and regular pulses lead to fixed baseline shift, but the random pulses of part (b) give rise to variable shift.

Unipolar   Bipolar

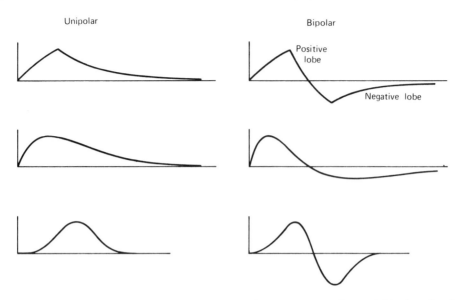

Figure 16-16 Illustration of the distinction between monopolar and bipolar signal pulses of various shapes.

is a physical impossibility. Therefore, the baseline on which these rectangular pulses are superimposed must be suppressed below the true zero level such that the areas enclosed by the output waveform above and below the zero axis are equal. The amount by which this apparent baseline is depressed below true zero is called the *baseline shift* and will obviously become more severe as the average spacing between pulses is made smaller.

In nuclear pulse analysis systems, the amplitude of the pulse carries much of the basic information. Because the amplitude is measured relative to a true zero baseline, the presence of any baseline shift will reduce the apparent pulse amplitude. For regular periodic pulses such as those shown in Fig. 16-15a, the problem would not be too severe because the baseline shift would be constant and appropriate compensation could be made. However, for pulses from radiation detectors, both the amplitude and spacing are variable and the situation will appear more like that shown in Fig. 16-15b. Now the degree of baseline shift is not constant and no adequate compensation can be carried out.

In principle, baseline shifts can be eliminated if the pulse shape is made to be *bipolar* rather than *monopolar*. Bipolar pulses contain both positive and negative lobes, as shown in Fig. 16-16, and differ from monopolar pulses, which are restricted to voltages only on one side of the zero axis. If the positive and negative lobes of a bipolar pulse are of equal area, its average dc value is zero and it can be passed by a capacitor without alteration of the baseline.

Effects of baseline shifts could also be avoided if all elements in the signal chain were dc coupled rather than interconnected by coupling capacitors. Some pulse-processing components are in fact dc coupled for this reason, but it is impractical to dc couple some elements such as a linear amplifier. One reason is that very small shifts in the input zero level to a high-gain amplifier will, if dc coupled, give rise to very large offsets at the output. Furthermore, if *CR* differentiation is used at any stage in the amplifier, it automatically becomes ac coupled. Therefore, ac coupling is the usual rule, and baseline shift must be countered at high pulse rates through the use of bipolar pulse shaping or through active baseline restoration.

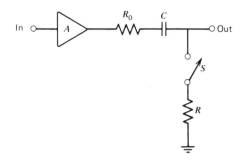

Figure 16-17 Equivalent circuit of a baseline restorer. R_0 is the output impedance of the operational amplifier, normally with gain $A = 1$. The time constant is given by the product $(R + R_0)C$.

2. BASELINE RESTORATION

One method to avoid baseline shift is to use shaping networks (described later) that produce bipolar pulses. Inevitably, the added shaping step required to produce the bipolar shape deteriorates the signal-to-noise characteristics of the system compared with simple monopolar shaping. Therefore, it is sometimes attractive to retain monopolar pulses and instead make use of an active electronic circuit to eliminate the resulting baseline shift. Such a *baseline restorer* circuit has as its primary purpose the return to true zero of the baseline between pulses in as short a time as possible. As a bonus, the use of baseline restoration also greatly reduces the deleterious effects of low-frequency disturbances, such as power-line hum and vibrational microphonics, which may be present along with the signal.

The equivalent circuit of a baseline restorer is diagrammed in Fig. 16-17. In principle, the switch is open only during the duration of each pulse, and its closing restores the output voltage to zero, with a time constant given by product of $(R + R_0)$ and the coupling capacitance C, where R is the equivalent series resistance of the switch and R_0 is the output impedance of the operational amplifier. In actual circuits, the role of the switch is carried out by diodes[9] or by more complex nonlinear circuitry.[10-12] Other restorer designs are described in Refs. 13–16. Although the use of these circuits introduces some degree of additional noise into the system, the benefits derived normally predominate, and most high-resolution spectrometry systems now employ some type of active baseline restoration.

To be effective, baseline restoration must take place near the end of the signal chain so that no further ac coupling takes place between the restorer and the point at which the pulses are analyzed or measured. If another capacitively coupled component does exist after the active restoration, the carefully restored baseline level will again shift in the manner illustrated in Fig. 16-15. Baseline restorers are therefore most commonly found at the output stage of linear amplifiers, which are intended to supply pulses directly to the analog-to-digital converter (ADC) of a multichannel analyzer.

D. Other Pulse Shaping Methods

1. DOUBLE DIFFERENTIATION, OR *CR-RC-CR* SHAPING

In order to achieve a bipolar pulse shape, a second state of differentiation is sometimes added to the *CR-RC* configuration discussed previously. Again, the most common choice

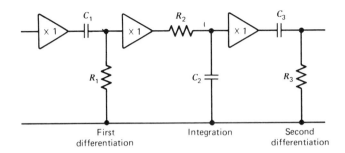

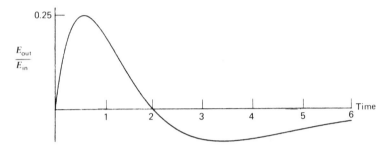

Figure 16-18 The *CR-RC-CR* network and its output waveform for a step input when all three time constants are equal. Units of the time scale are in values of this time constant.

is to make all three time constants approximately the same, which results in the pulse form sketched in Fig. 16-18. The bipolar shape makes baseline shifts much less severe, but because the two lobes of the pulse are not of exactly equal area, some baseline shift will remain. This type of shaping is most useful at high counting rates, but at lower rates the signal-to-noise characteristics are usually inferior to single-stage differentiation and integration.

2. SINGLE DELAY LINE (SDL) SHAPING

The properties of coaxial cables discussed earlier can also be applied to carry out pulse shaping. In particular, recall that a coaxial cable that is shorted at the receiving end will give rise to a reflection when a step voltage reaches that end of the cable. The reflection is a step moving back toward the sending end of the cable, with an amplitude equal to the initial step but opposite in polarity.

A configuration in which a shorted transmission line can be used to shape a step input is shown in Fig. 16-19. The transmission line or coaxial cable is assumed to be long enough so that the propagation time through its entire length is long compared with the rise time of the step voltage applied to the input of the network. The unity-gain operational amplifier simply provides impedance isolation at the input. Because the amplifier is assumed to have zero output impedance, the resistor Z_0 terminates the cable in its own characteristic impedance at the sending end. The far end of the cable is assumed to be shorted, or terminated in zero resistance.

The waveforms shown in Fig. 16-19 illustrate the sequence of events if a step voltage is applied to the input of the network. The initial positive step is assumed to be applied to the network at time $t = 0$ and to persist for a long period of time. The step propagates

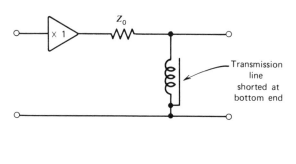

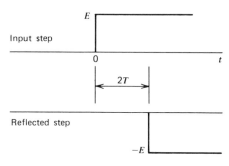

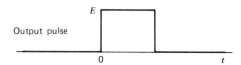

Figure 16-19 Single delay line (SDL) network applied to shaping an input step waveform.

down the cable, is reflected from the shorted end, and travels back to the sending end as an inverted or negative step. The voltage observed at the output of the network is simply the algebraic sum of the original and reflected waveforms, or the rectangular pulse shown in the figure. The time width of this shaped pulse is the down-and-back propagation time through the length of cable used for the shaping. For many applications, this propagation time should be of the order of a microsecond. In order to avoid excessively long cables, special delay lines with much reduced propagation velocity are often used for this purpose. Hence, the process just described is often known as *single delay line* (SDL) shaping.

The propagation time of the delay line must always be larger than the rise time of the pulse from the preamplifier to avoid the ballistic deficit discussed under *RC* shaping. Also, if the decay time of the preamplifier pulse is not many times larger than the propagation time, a situation will result as illustrated in Fig. 16-20, in which there is an undesirable undershoot following the shaped pulse. If the preamplifier decay time is always a fixed value, this undershoot can virtually be eliminated by slightly attenuating the reflected pulse as shown in the figure.

Similar shaping is also commonly used to reduce the length of pulses with very fast (of the order of nanoseconds) leading edges. In this case the role of the delay line is carried out by a normal coaxial cable of a few meters length, commonly referred to as a

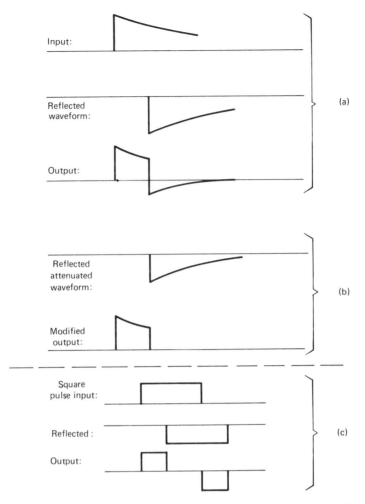

Figure 16-20 (a) Effect of applying simple SDL shaping to an input tail pulse with decay time comparable with the propagation time of a delay line. (b) Remedy to the undershoot of part (a) accomplished by partially attenuating the reflected waveform. (c) Effect of SDL shaping on a rectangular shaped input pulse whose length exceeds the delay line down-and-back time.

shorted stub. As is also shown in Fig. 16-20, this type of shaping applied to a rectangular input pulse gives rise to two shaped pulses of opposite polarity separated in time by the length of the original rectangular pulse.

3. DOUBLE DELAY LINE (DDL) SHAPING

Because the output of a single delay line shaping network is a unipolar pulse for typical preamplifier pulses, the same comments apply as were made earlier for single differentiation networks. At high counting rates, it may be desirable to substitute a shaping network that results in bipolar pulses to eliminate baseline shift. Bipolar pulses can be produced if two delay lines are used in the configuration shown in Fig. 16-21, known as *double delay line* (DDL) shaping. Both delay lines should have equal propagation time, and the resulting pulse will have positive and negative lobes of equal amplitude and duration.

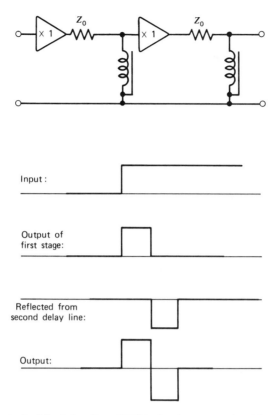

Figure 16-21 A double delay line (DDL) shaping network with equal delay times applied to a step input waveform.

Therefore, an average dc level of zero can accurately be maintained, and baseline shift in subsequent ac-coupled circuits will virtually be eliminated. Although DDL shaping therefore has excellent high counting rate characteristics, it applies no high-frequency filtering to the signal and consequently is usually inferior from a signal-to-noise standpoint compared with methods based on RC networks. Therefore, one seldom sees delay line shaping used in conjunction with detectors with inherently high energy resolution (such as germanium detectors), but with poorer resolution detectors (such as scintillators) the difference in overall system resolution can be negligible. In these circumstances the better high counting rate characteristics of DDL shaping may dictate its choice.

4. TRAPEZOIDAL (DDL-RC) SHAPING

At very high counting rates, the somewhat gradual return of the common Gaussian pulse shape to the baseline can lead to excessive pileup. One remedy would be to shorten the time constants, but the variable ballistic deficit that is encountered when processing pulses from detectors (such as germanium gamma-ray spectrometers) with variable charge collection time will set a minimum shaping time. An alternative that has proved useful[17] in this application is a modified combination of a double delay line stage followed by a single RC integrator. The delay line output is modified by attenuating the second lobe by a factor of e before sending it to an integrator stage with a time constant equal to the propagation time of the prior delay lines.[18] The result is a unipolar pulse shape that closely resembles a triangle. A duplicate of this pulse is sent through a branch with a fixed delay and added to the undelayed version to produce a symmetric waveform of trapezoidal shape. The flat top helps accommodate different input pulse rise times without excessive ballistic deficit. Although this method of shaping is inferior to others discussed earlier from a signal/noise standpoint at modest counting rates, its advantages at high rates can be significant.[18]

PROBLEMS

16-1. Find the transit time of a pulse through 15 m of RG-59/U coaxial cable.

16-2. Describe a convenient method of measuring the characteristic impedance of a coaxial cable by varying its termination conditions.

16-3. For each instance cited below, determine whether termination to prevent cable reflections is potentially needed.

   (a) Transmission of 0.5 μs rise time pulses through 20 m of RG-59/U cable.
   (b) Transmission of a 10 ns rise time pulse through 10 m of RG-62/U cable.

16-4. **(a)** Fast pulses are to be transmitted from a source with output impedance of 50 Ω to another component with input impedance of 1 kΩ. Choose one of the coaxial cable types listed in Table 16-1, and show the termination conditions required to avoid cable reflections at either end.
   (b) If the output pulse amplitude from the source without any load attached is 5 V, find the amplitude that appears across the input of the second component after the termination required in part (a).

16-5. Find the resistance values R_1 and R_2 in the T-attenuator network of Fig. 16-5 if an attenuation factor of 10 is needed while preserving a 50 Ω impedance level.

16-6. Prove that the resistance values in the pulse splitter of Fig. 16-6 should be 16.6 Ω in order to distribute a pulse to two 50 Ω loads while maintaining a 50 Ω impedance level.

16-7. Pulses from a preamplifier are produced with exponential tails with 50 μs time constant. What is the minimum spacing between adjacent pulses so that amplitude change due to pileup is less than 1%?

16-8. An input voltage of the form $V(t) = E[1 - \exp(-t/k)]$, where E and k are constants, is supplied to a simple RC integrator circuit. Derive the form of the output voltage.

16-9. A differentiator circuit has component values of $C = 500$ pF and $R = 500$ Ω. Find the frequency of a sinusoidal input voltage that will be attenuated by a factor of 2 by this circuit.

16-10. A step voltage of 1 V amplitude is applied to the input of a CR-RC differentiator–integrator network with equal time constants. What is the amplitude of the shaped pulse?

16-11. Derive Eq. (16-22) for the response of a CR-RC network to a step voltage input by using Eq. (16-12) as the input waveform to the RC stage.

16-12. What is the principal advantage of bipolar shaped pulses over monopolar shaped pulses?

16-13. What property of the detector is reflected in the rise time of the tail pulse obtained by a large time-constant collection circuit?

16-14. Pulses from a detector are shaped to produce monopolar pulses of approximately triangular shape with 10 V amplitude and width of 5 μs. Find the *average* value of the baseline shift after these pulses are passed through a capacitor if the pulse rate is 100/s. Repeat for a rate of 50,000/s.

16-15. Sketch the result of shaping a rectangular monopolar pulse of 200 ns width by using a shorted stub of RG-59/U of 10 m length.

REFERENCES

1. E. Kowalski, *Nuclear Electronics*, Springer-Verlag, New York, 1970.

2. P. W. Nicholson, *Nuclear Electronics*, Wiley, New York, 1974.

3. F. A. Kirsten, *IEEE Trans. Nucl. Sci.* **NS-20** (5), 22 (1973).

4. K. Porges, W. Corwin, L. Burkel and E. Lewandowski, *Rev. Sci. Instrum* **41**, 138 (1970).

5. W. K. Brookshier, *Nucl. Instrum. Meth.* **70**, 1 (1969).

6. J. Millman and H. Taub, *Pulse, Digital, and Switching Waveforms*, McGraw-Hill, New York, 1965.

7. F. S. Goulding, D. A. Landis, and N. W. Madden, *IEEE Trans. Nucl. Sci.* **NS-30** (1), 301 (1983).

8. C. H. Nowlin and J. L. Blankenship, *Rev. Sci. Instrum.* **36**, 1830 (1965).

9. L. B. Robinson, *Rev. Sci. Instrum.* **32**, 1057 (1961).

10. R. L. Chase and L. R. Poulo, *IEEE Trans. Nucl. Sci.* **NS-14** (1), 83 (1967).

11. E. A. Gere and G. L. Miller, *IEEE Trans. Nucl. Sci.* **NS-14** (1), 89 (1967).

12. F. S. Goulding, D. A. Landis, and R. H. Pehl, UCRL-17560 (1967).

13. M. Bertolaccini and C. Bussolati, *Nucl. Instrum. Meth.* **100**, 349 (1972).

14. E. R. Semple, *IEEE Trans. Nucl. Sci.* **NS-19** (1), 445 (1972).

15. N. Karlovac and T. V. Blalock, *IEEE Trans. Nucl. Sci.* **NS-22** (1), 457 (1975).

16. E. Fairstein, *IEEE Trans. Nucl. Sci.* **NS-22** (1), 463 (1975).

17. N. Taccetti, *Nucl. Instrum. Meth. Phys. Res.* **225**, 118 (1984).

18. M. Bruschi, P. Calonaci, G. Poggi, and N. Taccetti, *Nucl. Instrum. Meth. Phys. Res.* **A267**, 171 (1988).

CHAPTER · 17

Linear and Logic Pulse Functions

I. LINEAR AND LOGIC PULSES

In any pulse-processing system, it is important to distinguish between two types of signal pulse. A *linear* pulse is defined as a signal pulse that carries information through its amplitude, and sometimes by its shape as well. A sequence of linear pulses may therefore differ widely in size and shape characteristics. On the other hand, a *logic* pulse is a signal pulse of standard size and shape which carries information only by its presence or absence, or by the precise time of its appearance. Virtually all radiation detector signal chains start out with linear pulses and, at some point, a conversion is made to logic pulses based on some predetermined criteria.

A. Linear Pulses

Although there are a number of exceptions, linear pulses in radiation detector signal chains are usually one of three general types. *Fast linear pulses* are those that correspond to collecting the output current of a radiation detector using a collection circuit whose time constant is small (see discussion in Chapter 4). The rise and fall characteristics of these pulses are therefore determined primarily by the time history of the current generated directly by the interaction of the radiation with the specific detector.[†] Because these times are typically a few microseconds or less in most radiation detectors, fast linear pulses are seldom longer than several microseconds and may be much narrower for very fast detectors. For reasons outlined in Chapter 4, the signal-to-noise properties of fast linear pulses are always far inferior to those of the corresponding tail pulse derived by integrating the charge output of the detector across a large time constant collection circuit (see the following discussion). However, the rapid rise and fall can often offset this disadvantage when timing information and high counting rates are more important than amplitude resolution. The polarity of fast linear pulses often depends on the polarity of the bias voltage applied to the detector and can, in general, be either positive or negative.

A linear *tail pulse* is the type of linear pulse generated when the output of a radiation detector is collected across an equivalent circuit of large time constant. As discussed in Chapter 4, the leading edge of such a pulse corresponds to the time over which the charge produced by the detector is integrated across the capacitance of the collection circuit. Therefore, the time characteristics of the leading edge are determined exclusively by the

---

[†] For this reason, such pulses are sometimes also called *current pulses*.

584

charge collection time within the detector. The decay or return to zero of the pulse is determined by the time constant of the collection circuit. In order to assure complete charge collection or to avoid ballistic deficit, this time constant must be large compared with the detector charge collection time. Therefore, such pulses have a long tail compared with their leading edge, giving rise to the name tail pulse. From the arguments given in Chapter 4, the amplitude reached by a tail pulse in a given situation will always be many times larger than that of the equivalent fast linear pulse. Polarity can again be either positive or negative, but negative tail pulses are generally encountered more frequently in common applications.

For linear tail pulses, the most significant parameter besides amplitude is the *rise time* of the leading edge of the pulse. It should be pointed out that the term *rise time* always refers to the leading edge of pulses whether they are positive or negative polarity, even though the leading edge may have a negative slope for negative polarity pulses. The most widely used definition of rise time is the interval between the times at which the pulse reaches 10% and 90% of its final amplitude. For fast linear pulses, the full width at half maximum amplitude (FWHM) is often quoted as a measure of the overall pulse width.

A *shaped* linear pulse is a tail pulse whose time width has drastically been reduced by one of the shaping methods discussed in Chapter 16. Rather than the 50 or 100 μs decay times of tail pulses, the time width of a shaped linear pulse is typically only several microseconds. Shaped linear pulses are usually produced by a linear amplifier and consequently, within limits, can have arbitrary amplitude and polarity. Their dynamic range or "span" can often be chosen to be consistent with the dynamic range of circuits to which the linear pulse will be supplied. The nuclear instrument module (NIM) standard (outlined in Appendix A) recommends that positive polarity linear pulses be used with a span of either 0–1 or 0–10 V, with the latter choice by far the more popular.

B. Logic Pulses

There are two main classes of logic pulses specified in the NIM standard. *Standard logic pulses* are used in normal applications when the potential counting rate does not exceed 20 MHz. They are to be positive in polarity and within the amplitude limits specified in Appendix A. Their width is not standardized but is usually of the order of 1 μs. Pulse shape tends to be a unipolar square pulse to allow the most abrupt switching between logic states.

Fast logic pulses arise primarily in fast timing situations or when the counting rate in a particular system may potentially be greater than 20 MHz. The rise time of these logic pulses should be as short as possible and is usually of the order of a few nanoseconds. Because most timing is carried out from the leading edge of the pulse, their width is not of primary importance and is a factor only in determining the maximum practical counting rate. Because of the importance of cable reflection when dealing with short rise time pulses, fast logic pulses are usually handled in systems in which the input, output, and cable impedances are all set to be 50 ohms. In this way, simple interconnection of the components assures undistorted transmission of fast logic pulses from one unit to the next. The NIM standard calls for negative polarity for fast pulses, and more detailed specifications are given in Appendix A.

Although they are technically not logic pulses, *gate pulses* share some common properties. A gate pulse is usually rectangular in shape and, when applied to the gate pulse input of a linear gate, will cause the gate to open for a period of time equal to the width of the gate pulse. Its amplitude and polarity must correspond to the transition between closed and open conditions of the linear gate and are usually positive and of the order of several volts. Its width, however, may be variable and is usually adjusted to the requirements of the particular application.

II. INSTRUMENT STANDARDS

Before the widespread use of solid-state circuitry, most electronic instrumentation used for nuclear measurements consisted of vacuum tube circuits housed in a self-contained chassis and required only connection to a power line. To facilitate mounting in a vertical stack, much of this equipment was built to fit a frame commonly known as the standard *19 in. relay rack*. Individual chassis are screw mounted one above the other in a relay rack, and widespread use persists to this day for the convenient mounting of nuclear electronic components. Some bulky units such as high-voltage power supplies continue to be manufactured in the full 19 in. width, but the advent of solid-state circuitry has led to more compact packaging in which only a fraction of the full width need be occupied by a single electronic unit.

It has become common practice to manufacture most nuclear electronics in standard modules, a number of which can fit into a housing called a *bin* or *crate*, which occupies the full 19 in. width. Two important international standards, known as NIM (Nuclear Instrument Module) and CAMAC (Computer Automated Measurement and Control), have come to be widely accepted, and now virtually all modular nuclear electronics are manufactured according to one of these standards. These systems have resulted in a great simplification for the user, and one can quite generally intermix components and bins from different manufacturers with assurance that those adhering to a common standard will be mutually compatible. The NIM system is usually better suited for small-scale linear pulse processing normally encountered in the routine application of radiation detectors. The more costly CAMAC system, which is strongly oriented toward large digital systems and computer interfacing, is usually necessary only when large-scale signal processing involving many detectors or logical operations is required.

As a self-contained system, each of these standards shares the following attributes:

1. The basic dimensions for the bin and modules are specified so that all modules and bins within a given system are mechanically interchangeable.

2. The philosophy generally adopted is that only the bin (or crate) will be connected to the laboratory ac power, and it will contain the necessary power supplies to generate all the dc supply voltages required by all modules contained within that bin. Individual modules therefore do not contain their own power supplies, but draw power from the bin in which they are located.

3. The connector interface between the module and bin must therefore be standardized both electrically and mechanically, so that any standard module can draw its required power when plugged into any one of the available bin locations.

4. Specifications are included for the polarity and span of both logic and linear pulses of various types.

Details of both the NIM and CAMAC standards are included in Appendix A.

III. SUMMARY OF PULSE-PROCESSING UNITS

In almost all applications of radiation detectors operated in pulse mode, the output of the detector is converted to a linear pulse whose amplitude and shape may carry information of interest to the user. This linear pulse may be processed in the signal chain before its

TABLE 17-1 Summary of Common Pulse-Processing Functions

(A) Linear–Linear	In	Out
PREAMPLIFIER	Linear charge pulse from the detector	Linear tail pulse
LINEAR AMPLIFIER	Linear tail pulse	Amplified and shaped linear pulse
BIASED AMPLIFIER	Shaped linear pulse	Linear pulse proportional to amplitude of input pulse that lies above input bias level
PULSE STRETCHER	Fast linear pulse	Conventional shaped linear pulse of amplitude equal to input pulse
SUM AMPLIFIER	Two or more shaped linear pulses	Shaped linear pulse with amplitude equal to the sum of coincident input pulses
DELAY	Fast linear or shaped linear pulse	Identical pulse after a fixed time delay
Linear input LINEAR GATE Gate input	(1) Shaped linear pulse (2) Gate pulse	Linear pulse identical to linear input if gate pulse is supplied in time overlap

(B) Linear–Logic	In	Out
INTEGRAL DISCRIMINATOR	Shaped linear pulse	Logic pulse if input amplitude exceeds discrimination level
DIFFERENTIAL DISCRIMINATOR (SINGLE-CHANNEL ANALYZER)	Shaped linear pulse	Logic pulse if input amplitude lies within acceptance window
TIME PICK-OFF (TRIGGER)	Fast linear or shaped linear pulse	Logic pulse synchronized with some feature of input pulse

(C) Logic–Linear	In	Out
START STOP TIME–AMPLITUDE CONVERTER	Logic start and stop pulses separated by time Δt	Shaped linear pulse with amplitude proportional to Δt

TABLE 17-1 (Continued)

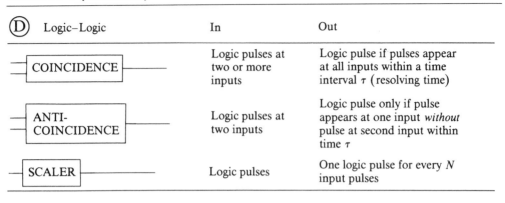

⒟ Logic–Logic	In	Out
COINCIDENCE	Logic pulses at two or more inputs	Logic pulse if pulses appear at all inputs within a time interval τ (resolving time)
ANTI-COINCIDENCE	Logic pulses at two inputs	Logic pulse only if pulse appears at one input *without* pulse at second input within time τ
SCALER	Logic pulses	One logic pulse for every N input pulses

properties are recorded in some way, or logic pulses may be derived to convey other information. Of the various electronic units in the signal chain, some are linear devices intended to perform operations on a linear pulse input to provide a linear pulse output. Other modules are strictly logic units which examine logic pulse inputs using set selection criteria to determine whether a logic pulse will be produced at the output. A third general category are linear-to-logic converters, in which logic output pulses are produced only when the input linear pulse meets predetermined conditions. In some circumstances, supplying a linear pulse to a device intended for logic pulse inputs may cause the device to function in some manner, but the variable size and shape of the linear pulse will usually result in unreliable operation. It is therefore important to maintain a clear distinction between linear and logic pulse functions, and to carry out linear-to-logic conversion only at a definite point in the signal chain.

To provide a general overview, Table 17-1 lists the summarized properties of some of the more popular types of nuclear instrumentation modules. Each of these is discussed in greater detail in the applications sections that follow. These modules represent the basic building blocks from which common pulse-processing chains are assembled. Although we treat each as a separate module, some of the functions shown are often combined into a single NIM module for convenience and versatility (e.g., *timing single-channel analyzers* are widely available which combine the time pick-off and SCA functions). All the functions shown are commonly available in NIM format, and most of the logic functions are available in CAMAC as well.

The largely digital operations that are the speciality of CAMAC require modules designated as scalers, coincidence units, and analog-to-digital converters (see Chapter 18). The common digital manipulations involving registers, buffers, encoders, and so on are explored in texts on digital methods (e.g., Ref. 1) and are not discussed here.

The remainder of this chapter is divided into sections centering on application of various modules in different signal chains. Each is intended to typify a commonly encountered situation in radiation measurements, but no attempt is made to provide an exhaustive review of the wide assortment of applications to which the modules may be put. Most basic module types are detailed under the application in which their use is most critical. However, the discussion begins with one element common to nearly all applications (the preamplifier) and two ancillary elements (the detector bias supply and the test pulse generator) that are also important in virtually any instrumentation system.

IV. COMPONENTS COMMON TO MANY APPLICATIONS

A. Preamplifiers

1. GENERAL

As outlined in Chapter 4, the fundamental output of all pulse type radiation detectors is a burst of charge Q that is liberated by the incident radiation. For G-M tubes and many scintillation counter applications, Q is sufficiently large so that a fairly large voltage pulse is produced by integrating this charge pulse across the summed capacitance represented by the detector, connecting cable, and input of the recording circuitry. For most other detectors, however, the charge is so small that it is impractical to deal with the signal pulses without an intermediate amplification step. The first element in a signal-processing chain is therefore often a *preamplifier* provided as an interface between the detector and the pulse-processing and analysis electronics that follow.

The preamplifier is usually located as close as possible to the detector. From a signal-to-noise standpoint, it is always preferable to minimize the capacitive loading on the detector, and therefore long interconnecting cables between the detector and pre-amplifier should be avoided if possible. One function of the preamplifier is to terminate the capacitance quickly and therefore to maximize the signal-to-noise ratio. It also serves as an impedance matcher, presenting a high impedance to the detector to minimize loading, while providing a low impedance output to drive succeeding components.

The preamplifier conventionally provides no pulse shaping, and its output is a linear tail pulse. The rise time of the output pulse is kept as short as possible, consistent with the charge collection time in the detector itself. The decay time of the pulse is made quite large (typically 50 or 100 μs) so that full collection of the charge from detectors with widely differing collection times can occur before significant decay of the pulse sets in.

2. VOLTAGE- AND CHARGE-SENSITIVE CONFIGURATIONS

Preamplifiers can be of either the *voltage-sensitive* or *charge-sensitive* type. Historically, the voltage-sensitive type is the more conventional in many electronic applications and consists simply of a configuration that provides an output pulse whose amplitude is proportional to the amplitude of the voltage pulse supplied to its input terminals. A schematic diagram of a voltage-sensitive configuration is shown in Fig. 17-1. If the time constant of the input circuit (the parallel combination of the input capacitance and resistance) is large compared with the charge collection time, then the input pulse will have an amplitude equal to

$$V_{\text{max}} = \frac{Q}{C} \tag{17-1}$$

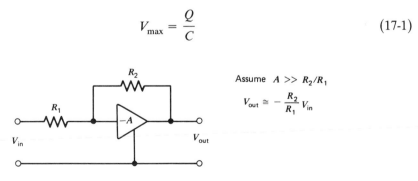

Figure 17-1 Schematic diagram of a simplified voltage-sensitive preamplifier config-uration. R_2 is the feedback resistance.

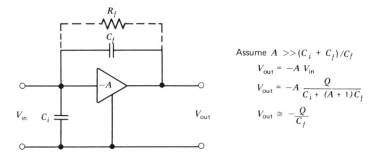

Assume $A \gg (C_i + C_f)/C_f$

$$V_{out} = -A \, V_{in}$$

$$V_{out} = -A \, \frac{Q}{C_i + (A + 1)C_f}$$

$$V_{out} \cong -\frac{Q}{C_f}$$

Figure 17-2 Simplified diagram of a charge-sensitive preamplifier configuration. If the conditions indicated are met, the output pulse amplitude becomes independent of the input capacitance C_i. The time constant given by the product $R_f C_f$ determines the decay rate of the tail of the output pulse.

where C is the input capacitance. For most detectors, the input capacitance is fixed so that the output pulse produced by a voltage-sensitive preamplifier is proportional to the charge Q liberated by the incident radiation. If the input capacitance were to change, however, this desirable proportionality would no longer hold. In semiconductor diode detectors, for example, the detector capacitance may change with operating parameters. In these situations, a voltage-sensitive preamplifier is undesirable because the proportionality between V_{max} and Q is lost.

The elements of a charge-sensitive configuration that can remedy this situation are shown in Fig. 17-2. For this circuit, the output voltage is proportional to the total integrated charge in the pulse provided to the input terminals, as long as the duration of the input pulse is short compared with the time constant $R_f C_f$. Changes in the input capacitance no longer have an appreciable effect on the output voltage. Although originally developed for use with semiconductor diode detectors, this charge-sensitive configuration has proved its superiority in a number of other applications, so that preamplifiers used with other detectors in which the capacitance does not necessarily change are also often of the charge-sensitive design.

3. NOISE CHARACTERISTICS

Probably the most important specification for a preamplifier is its noise figure. This specification is normally quoted as the FWHM of the response function of the system due only to the preamplifier noise. The figure is normally given as the equivalent energy spread in the type of detector for which the preamplifier is designed. The noise figure is a strong function of the capacitance with which the preamp input is loaded. For example, a good quality preamplifier used with silicon diode detectors may have a noise figure of 2 keV with zero input capacitance, but this figure may double if the input is loaded with 100 pF. The input capacitance arises from both the inherent detector capacitance and from the connecting cable between the detector and preamplifier. It is therefore important to keep the interconnecting cable as short as possible and to choose a detector whose inherent capacitance is no larger than necessary. The rise time for charge-sensitive preamplifiers also normally increases with input capacitance.

For a wide assortment of applications, the noise level of commercially available preamplifiers is sufficiently low so that their contribution to the FWHM of the system response is small compared with the inherent contributions of the detector itself. This is not the case for very-low-energy measurements, however, and, as an example, the preamplifier noise contribution is very significant in the measurement of energy deposition in a semiconductor diode detector below a few keV. Therefore, a strong motivation remains to reduce the inherent preamplifier noise for these low-energy measurements.

One of the significant contributors to preamplifier noise is the Johnson noise associated with the feedback resistor (R_f in Fig. 17-2). The noise is made smaller by increasing the resistance value, but the longer time constant leads to very long tails on the output pulses. Problems can then arise with respect to overload recovery and pileup in the preamplifier (see below). An alternative approach is to eliminate the feedback resistor altogether. Two preamplifier designs that do not require resistive feedback are called *pulsed optical feedback* and *transistor reset* preamplifiers. Their operation is described beginning on p. 592. In both cases, the elimination of the feedback resistor and its Johnson noise permits a lower preamplifier noise contribution compared with conventional resistive feedback preamplifiers.

In demanding situations, the noise generated in the preamplifier input stage can also be reduced by cooling. The practical problems involved generally make this approach unattractive except in the case of applications in which the detector itself is operated at reduced temperature. Most cooled preamplifier applications are therefore in conjunction with germanium or Si(Li) detectors, which are normally held at liquid nitrogen temperature.

4. OVERLOAD RECOVERY AND PILEUP

The charge-sensitive preamplifier is relatively susceptible to saturation when very large pulses are supplied to its input. For example, large-amplitude cosmic-ray events in a scintillation counter can interfere with the processing of smaller pulses that are of real interest. The overload recovery properties of the preamplifier are therefore an important specification for those applications in which frequent large pulses might obscure the signal to be measured. Methods to reduce the sensitivity of charge-sensitive preamplifiers to overloading pulses are described in Refs. 2 and 3.

Because the output of a preamplifier is a tail pulse with a rather long decay time, the pileup of pulses is inevitable, except at very low signal rates. Although the effects of this pileup are largely removed by subsequent shaping (see Chapter 16), one potential effect cannot be dealt with so easily. In Fig. 17-3, the piled-up output of a preamplifier at high rates is sketched. The average level of this waveform will increase with rate and may approach the limit of linear operation of the preamplifier. In that event, some piled-up pulses may drive the preamplifier into saturation and thus will be seriously distorted.

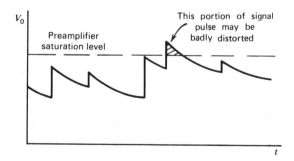

Figure 17-3 The pileup of pulses within the preamplifier at high rates. If the saturation level of the preamplifier is exceeded, some pulses can be seriously distorted.

Choosing small values for the feedback resistor R_f will minimize this effect by ensuring rapid decay of the pulse, but at the expense of an increased noise contribution.

5. ACTIVE RESET TECHNIQUES

There are two potential advantages to be gained if the feedback resistance R_f is eliminated from the charge-sensitive preamplifier configuration shown in Fig. 17-2. One is the improvement in the noise level due to the removal of the Johnson noise associated with the resistor. The second relates to the performance at very high pulse rates, where the pileup and overload problems discussed in the preceding section limit the maximum rate at which a preamplifier with resistive feedback can be used.

Without the feedback resistor, the current pulses from the detector are simply accumulated on the feedback capacitance (C_f in Fig. 17-2). The output voltage then grows in the random staircase fashion shown in Fig. 17-4, with each upward step corresponding to a separate pulse. This staircase voltage corresponds to the sequence of pulses illustrated in Fig. 17-3, where the tails now have infinite length. (As before, the tails are later removed by the shaping process in a following linear amplifier, producing a series of shaped pulses with amplitudes proportional to each individual step.)

Some method must now be provided to reset the preamplifier voltage to zero when the staircase approaches some maximum voltage short of the saturation level. One method[4-6] is to discharge the feedback capacitor by momentarily illuminating the drain–gate junction of the input stage FET with light. A light pulse of a few microsecond duration is produced by a light-emitting diode connected to the output stage. This *pulsed optical feedback* avoids any physical connection to the input stage and does not add to the input capacitance, keeping noise levels very low. It is therefore widely used in demanding situations where minimum noise from the preamplifier is critical. One common application is in X-ray spectroscopy with Si(Li) detectors, where the preamplifier contribution to the system noise can be as low as 100 eV equivalent.[5]

An alternative reset method generally called *transistor reset* is often applied[7,8] in preamplifiers intended for high-rate applications. Here the voltage is reset using an active circuit with a transistor connected to the input stage. While this approach cannot match the low capacitance and noise figures of the pulsed light feedback technique, it can

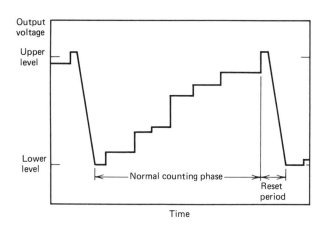

Figure 17-4 The output voltage of an active reset preamplifier. The reset phase is triggered when the voltage crosses the upper-level limit during the normal counting phase. Each upward step is a separate signal pulse.

provide adequately low noise performance when used with detectors whose inherent capacitance is not extremely small.[7]

In both the pulsed optical and transistor reset techniques, the sudden large drop in the output voltage during the reset can cause overloading problems if processed by a following linear amplifier. Therefore, such preamplifiers provide an "inhibit" output signal that is turned on just before the reset occurs and is maintained until the reset is completed. The inhibit signal is then supplied to a gate input on the amplifier to block signals during its duration. This period represents a dead time that can become significant at high rates and for large amplitude pulses when frequent resets are required. One advantage of transistor reset compared with pulsed optical techniques of present design is that this dead time can be made shorter, allowing higher-rate operation. Reset dead times of a few microseconds are typical.

Another advantage inherent to both pulsed optical and transistor reset is that the pole-zero cancellation adjustment (see p. 573) in the following amplifier is no longer required. The need for the adjustment is normally brought about by the finite decay time of the tails of pulses produced by resistive feedback preamplifiers. This adjustment is

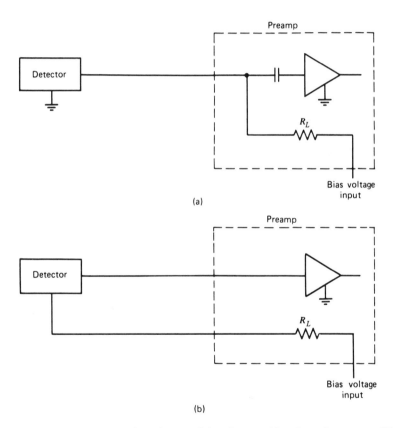

Figure 17-5 Two configurations for supplying detector bias through a preamplifier. (a) An ac-coupled arrangement, in which a coupling capacitor is provided between the detector and preamplifier circuits. This allows interchanging different values of R_L without affecting the preamplifier input. (b) A dc-coupled configuration that eliminates the coupling capacitor and generally leads to better noise performance for critical applications. Now the detector must be isolated from ground, and changing the bias resistor R_L may affect the input stage characteristics.

critical to achieving good resolution performance, especially at high rates. The active reset methods produce pulses that have infinite decay time, removing the pole-zero cancellation requirement.

6. DETECTOR BIAS VOLTAGE

Another function normally carried out by the preamplifier is to provide a means for supplying bias voltage to the detector. Arrangements can always be made to supply voltage to the detector independent of the preamplifier, but it is usually convenient to do so through the preamplifier. In Fig. 17-5, two configurations are shown in which the bias voltage is supplied through a load resistance R_L. A single cable between the preamplifier and detector provides both the voltage to the detector and the signal pulse to the input of the preamplifier. The load resistance, together with the input capacitance, determines the time constant across which the detector current is collected. From the standpoint of minimum noise, R_L should be as large as possible. Practical limitations always dictate that its value be no more than a few thousand megohms due to the fact that any dc signal or leakage current drawn by the detector must also flow through this resistor. Leakage currents are especially troublesome for semiconductor diode detectors and, in those cases in which R_L is large, can lead to a substantial dc voltage drop across R_L. In that event, the voltage actually applied to the detector is less than the supply voltage, and experimenters must be aware of the magnitude of the leakage current so that they can compensate for the voltage drop across R_L by raising the supply voltage. In some preamplifiers, it is recommended that R_L be changed by the user to suit the specific application to which the preamplifier is put.

The best noise performance in present designs results from the use of a field effect transistor (FET) as the input stage. FETs are notoriously sensitive to overvoltage transients and can easily be damaged by the transients that can be generated by switching a detector bias supply in coarse steps or abruptly disconnecting or turning off the voltage. As a result, many commercial preamplifiers are provided with overvoltage protection circuits. For the ultimate in low noise performance, however, it is often necessary to switch out the protection circuits, and in such circumstances the bias supply must be changed only gradually and in a continuous fashion.

7. TEST PULSE INPUT

Most preamplifiers are also provided with an input labeled *test pulse*, which is intended to receive the output of a pulse generator for system test purposes. Figure 17-6 shows one means of applying this test pulse to the input stage of a charge-sensitive configuration. If the pulser is sufficiently stable, its output should be resolved into a single channel by a pulse height analyzer connected to the system output. Any broadening of this response into more than one channel can be attributed to the inherent electronic noise of the measurement system. It is often interesting to measure this inherent electronic noise width because any measured detector response can never be better resolved than this electronic limit. It is important to make this measurement with the detector attached to the input of the preamplifier because its capacitance is often a significant factor in the overall noise characteristics of the preamplifier. The test pulse input provides a convenient means of carrying out this noise determination as well as simply checking the overall function of the signal-processing system prior to a measurement. In more sophisticated systems, a test pulse input may also be required during the course of a measurement for purposes of dead time determination or gain stabilization (see Chapter 18).

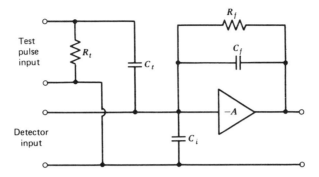

Figure 17-6 A charge-sensitive preamplifier that has been provided with a test pulse input. If a step voltage pulse of amplitude V_t is applied to this input, a charge equal to $V_t C_t$ is supplied to the preamplifier input stage. R_t is a small-value termination resistance.

8. CONSIDERATION IN THE USE OF PREAMPLIFIERS

The signal-processing section of preamplifiers intended for either solid-state diode detectors or gas-filled ionization or proportional counters is quite similar. The primary differences are in the value chosen for the load resistor R_L in Fig. 17-5 and in the degree of isolation provided along the high-voltage path through the preamplifier. Preamplifiers intended for proportional counter applications must be designed to withstand several thousand volts, whereas silicon junction detectors seldom require bias supplies of more than a few hundred volts. The value of the load resistance R_L can be higher (to minimize its noise contribution) for gas-filled detectors because their leakage current is normally lower than that of solid-state detectors.

Preamplifiers intended for scintillation counters are usually quite different. Because the signal level from a photomultiplier (PM) tube is rather high, the gain and noise specifications required of a scintillation preamplifier are relatively undemanding. Although it is quite possible to use a scintillation counter without a preamplifier in many applications, it is convenient to include a preamplifier simply to avoid the changes that can occur in the time constant of the equivalent anode circuit of the PM tube if it is connected directly to a linear amplifier or other measuring circuits. Without the preamplifier, the length of the cable used and the input impedance of the measuring circuits both affect the time constant and can lead to situations in which nonoptimum time constants result from equipment rearrangement. The high voltage to scintillation counters is also supplied somewhat differently as compared with semiconductor diode or gas-filled detectors. The signal from a PM tube is derived from the anode and is independent of the high voltage, which is supplied directly to the resistor divider string in the tube base. Therefore, scintillation counter preamplifiers do not have a bias voltage supply input; instead, the high voltage is supplied directly to the scintillation counter base.

Reflecting the usual interests of the user, most commercial preamplifiers are optimized with respect to the pulse height resolution for pulses derived from a specific detector. In some applications, however, information on the timing of the pulse is more important than an accurate measure of its amplitude, and modifications must be made in the way the basic detector pulse is handled. Sherman and Roddick[9] discuss the methods that can be used to derive a fast timing signal from preamplifiers, while at the same time making minimum compromises in energy resolution.

It is conventional to supply the power required for the operation of the preamplifier through a multiconductor cable connected to the chassis of a subsequent electronic component, most often the linear amplifier. The type of connector and specific pin assignments for this power cable are covered in the NIM standard; see Appendix B. When there is a large separation between the preamplifier and subsequent electronics, this long interconnecting power cable is occasionally the source of noise problems due to ground loops. In that event, it is better policy to provide a separate local preamplifier power supply.

B. Detector Bias and High-Voltage Supplies

With few exceptions (such as the self-powered neutron detector), virtually all radiation detectors require the application of an external high voltage for their proper operation. This voltage is conventionally called *detector bias*, and high-voltage supplies used for this purpose are often called detector bias supplies.

Some characteristics of detector bias supplies which can be important in specific applications are the following:

1. The maximum (and minimum) voltage level and its polarity.

2. The maximum current available from the supply.

3. The degree of regulation against long-term drifts due to changes in temperature or power line voltage.

4. The degree of filtering provided to eliminate ripple at power line frequency or other low-frequency noise.

The sophistication required of the bias supply varies greatly with the detector type. For detectors that draw very little current (such as an ion chamber) the bias supply can be as simple as a dry cell battery. On the other hand, supplies that must simultaneously provide high voltage and relatively high current involve a substantial amount of design engineering and can be among the heaviest and bulkiest of the equipment normally found in a nuclear instrumentation system.

The most demanding common application is to provide high voltage for the operation of PM tubes in connection with scintillation counters (see Chapter 9 for design details). Typical scintillation high-voltage supplies must be capable of providing up to 3000 V with a current of a few milliamperes. The output must also be well regulated to prevent gain shifts in the PM tube due to drifts in the high-voltage level. Bias supplies for proportional counters must also supply relatively high voltages, but the current demands are considerably less. However, the degree of regulation and filtering is again important because any high-voltage fluctuations appear superimposed on the signal. Semiconductor diode detectors draw relatively little current and the voltage demands seldom exceed 1000 V. The high-voltage supply for G-M tubes is the least demanding of all in several respects. Although the voltage may be as high as several thousand volts, the current requirements are very small. Furthermore, stability can be very poor because the plateau characteristics of the tube assure that the counting rate will be relatively independent of bias voltage perturbations.

The voltage level on most bias supplies is adjustable either through switching in steps or by means of a continuously adjustable helipot. When used with preamplifiers with FET input stages, continuous adjustment avoids switching transients that may be potentially damaging to the FET. Some designs provide alternate protection by limiting the rate of change of the voltage between steps.

C. Pulse Generators

An electronic pulse generator is indispensable in the initial setup and calibration of virtually any nuclear instrumentation system. Furthermore, some methods of gain stabilization and dead time determination require the output of a pulse generator to be mixed with signal pulses during the course of a measurement. Pulsers are therefore a very common element in most radiation instrumentation systems.

A tail pulse generator with adjustable rise and decay times is probably the most useful of all pulser types. Its output is conveniently fed to the test pulse input on preamplifiers or used directly in place of the preamplifier output. In its simplest form, the pulser output provides for convenient adjustment of system parameters such as shaping time constants, pole-zero parameters, and various timing and delay adjustments. If the output amplitude is truly constant, a measurement of the amplitude distribution recorded by the pulse analysis system determines the electronic noise level present in the system.

Most pulse generators also provide a front panel adjustment for the amplitude of the pulse. In some designs, often called *precision pulsers*, this adjustment is accurately controlled by a front panel dial. Such pulsers can then be used to check the integral linearity of a pulse-handling system simply by recording the output amplitude for several different settings of the input pulse amplitude. The output of most precision pulsers is sufficiently stable so that, in the absence of electronic noise, all pulses of constant amplitude would be resolved into a single channel in a multichannel analysis system.

Some designs can also provide a pulse source of other than constant amplitude. One of the more useful is called a *sliding pulse generator*, which produces pulses whose amplitude is uniformly distributed between zero and some set maximum. Recording these pulses in a multichannel pulse analysis system gives a direct measure of the differential linearity of the multichannel analyzer (see Chapter 18).

For purposes of determining the pulse resolving time of a given measurement system, a *double pulse generator* can be very useful. Here, two pulses of fixed amplitude are produced periodically with a variable time spacing between the two pulse leading edges. The resolving time can be determined simply by varying the time spacing between the double pulses and observing the transition point at which the system begins to recognize two distinct pulses rather than only one.

For normal pulse generators, the interval between pulses is uniform and periodic. However, for determining some system parameters such as pileup behavior and other time-dependent phenomena, a periodic source does not adequately represent the random time spacing encountered from actual radiation detector pulses. Therefore, designs have evolved which provide a source of randomly spaced pulses of constant amplitude. In such *random pulsers*, the noise signal from an internal component is often used to trigger randomly the time at which a pulse is produced at the output. Another method of producing randomly spaced pulses is to use the triggering feature provided on many pulse generators. In this mode, the pulse generator will produce an output pulse only when provided with an external trigger pulse. If this trigger pulse is derived from a random source such as a separate radiation detector, the output of the pulser can again be an accurate simulation of the random pulse spacing normally encountered in radiation detector signal chains.

Pulsers other than tail pulse generators can also play a useful role in detector pulse processing. A *gate generator* produces a square pulse of variable amplitude and width when triggered by an external source, usually a standard logic pulse. As its name implies, it is most useful in providing the gate pulse necessary to open or close a linear gate based

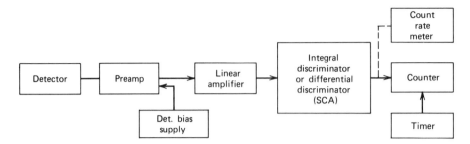

Figure 17-7 Elements of a typical signal chain for pulse counting.

on external logic criteria. Other square pulse generators, especially those that can be triggered and that provide a variable delay between the trigger pulse and output pulse, can be very useful in providing logic pulses to various points in the counting system under variable conditions.

V. PULSE COUNTING SYSTEMS

The signal chain shown in Fig. 17-7 represents a basic measurement scheme in which only the number or rate of pulses from a radiation detector are to be recorded. The tail pulse output of the preamplifier typically has an amplitude of a few tens or hundreds of millivolts and is too small to be counted directly. Furthermore, the pileup of these long pulses at high rates could cause stability problems. Therefore, the next step is normally to process the pulses through a linear amplifier. Here a voltage gain of 1000 or more can be provided so that the shaped linear pulse at its output can easily cover a span of 0–10 V. The shaping requirements in a simple counting system are usually not severe, and only at relatively high counting rates must one pay close attention to the specific method of shaping chosen. On the other hand, spectroscopy systems described later place much more emphasis on shaping, and a detailed description of this and other functions of linear amplifiers is therefore postponed until the following section on pulse spectroscopy.

A. Integral Discriminator

In order to count the pulses properly, the shaped linear pulses must be converted into logic pulses. The *integral discriminator* is the simplest unit that can be used for this conversion and consists of a device that produces a logic output pulse only if the linear input pulse amplitude exceeds a set discrimination level.[†] If the input pulse amplitude is below the discrimination level, no output appears. This selection process is illustrated in Fig. 17-8a. Unless specifically designed otherwise, the logic pulse is normally produced shortly after the leading edge of the linear pulse crosses the discrimination level. This *leading edge timing* is compared with other schemes of generating the logic pulse in the later section on time pick-off methods.

The discrimination level is normally adjustable by a front-panel control. In many counting situations, the level is set just above the system noise so that the maximum sensitivity for counting detector pulses of all sizes is realized. Other situations may call for

---

[†]The discrimination level is sometimes referred to as the *pulse height bias* level of the counting system. This terminology is somewhat unfortunate because of the potential confusion with the phrase *detector bias*, commonly used to describe the external dc voltage supplied to the detector, and is not used here.

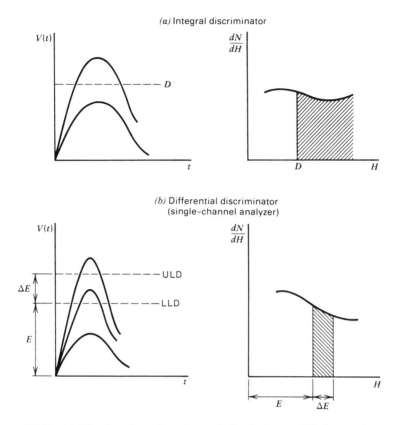

Figure 17-8 (a) The function of an integral discriminator. Of the two input pulses shown, only the larger clears the discrimination level D and produces a logic pulse output. The effect is to select only the area to the right of the amplitude D in the differential pulse height spectrum. (b) The function of a differential discriminator or single-channel analyzer. For the three input pulses shown, only the intermediate amplitude lies within the acceptance window and leads to an output logic pulse. In the pulse height spectrum, only the pulses falling within the cross-hatched area are accepted.

a higher discrimination level to count selectively only events above a set minimum size. For example, much of the background may be limited to relatively low pulse amplitudes so that some finite discrimination level may greatly enhance the signal-to-background counting ratio.

The integral discriminators must be designed to accept shaped linear input pulses of a specific span (usually 0–10 V positive). The stability and linearity of the discriminator adjustment are usually adequate for routine applications but may become important specifications for demanding situations.

B. Differential Discriminator (Single-Channel Analyzer)

Another linear-to-logic converter in widespread use involves two independent discrimination levels. As illustrated in Fig. 17-8b, a *differential discriminator* or *single-channel analyzer* (SCA) produces a logic output pulse only if the input linear pulse amplitude lies between the two levels. The action of the unit is therefore to select a band of amplitudes or window in which the input amplitude must fall in order to produce an output pulse.

Several systems of nomenclature and adjustment persist for SCAs. In some units, the *lower-level discriminator* (LLD) and *upper-level discriminator* (ULD) are independently adjustable from front-panel controls. In others, the lower level is labeled the E level, and the window width or difference between levels is labeled ΔE and can be varied separately without affecting the E level.

In counting systems, the SCA can serve to select only a limited range of amplitudes from all those generated by the detector. A common example is one in which the window is set to correspond only to those events in the detector which deposit the full energy of an incident radiation. In this way, one type or energy of radiation often can be measured selectively in the presence of other radiations.

In normal SCAs, the time of appearance of the logic pulse is not closely coupled to the actual event timing, and use of these logic pulses in timing measurements will often give imprecise results. If one of the time pick-off methods discussed later in this chapter is incorporated into the SCA design, the logic pulse can be much more closely correlated with the actual event time. Modules with this feature are often called *timing SCAs* and are widely used in coincidence applications or other timing measurements.

Most SCAs provide the option of switching out the upper-level discriminator and using the unit as a simple integral discriminator controlled by the lower level. The input linear pulses are intended to be shaped with typical 0.5–10 μs widths and a span that most commonly is 0–10 V positive. Bipolar pulses with positive leading edges are also normally acceptable. Other specifications that can be important in some applications include the linearity of the discriminator level adjustments, the stability of the levels with respect to temperature changes, and the SCA dead time (usually 1 or 2 μs larger than the input pulse width).

C. Scalers or Counters

As the final step in a counting system, the logic pulses must be accumulated and their number recorded over a fixed period of time. The device used for this purpose may be a simple digital register that is incremented by one count each time a logic pulse is presented to its input. In nuclear pulse counting applications, such devices are often called *scalers* as a historic anomaly that dates from the time when digital registers of reasonable size were not widely available. Then it was common to use a scaling circuit to divide the input pulse repetition rate by a fixed factor such as 100 or 1000 so that the rate would be low enough to be directly recorded by an electromechanical register. These systems have all but disappeared and have been replaced by all-electronic digital registers. The scaling function persists only in the sense that an overflow output pulse is often provided when the maximum content of the register is exceeded (usually no less than 10^5 or 10^6 counts). We henceforth refer to such units as counters because the term more adequately describes the actual function.

Counters are commonly operated in one of two modes: preset time and preset count. In the preset time mode, the counting period is controlled by an external timer. This timer may be built as part of a common chassis with the counter, or separate timers can be obtained as individual modules. In the preset count mode, the counter will accumulate pulses until a specified total has been achieved, at which point the counting period is terminated. If the period of time over which these counts have been accumulated can be recorded independently, the counting rate can be determined. The preset count mode has the advantage that a given statistical precision can be specified before the start of the measurement, and the duration of counting will be prolonged until enough counts have been accumulated to guarantee the desired statistical precision.

Counters can also be of the *blind* or *display* type. In a display counter, the contents of the register are continuously displayed on a front-panel numeric indicator. Display counters provide the advantage that the progress of a measurement can be monitored visually and malfunctions can often be detected quickly by aberrant behavior of the digital display. Blind counters do not provide a visual display but, instead, can generate a coded logic readout of the register content when triggered by an external command. Because blind counters are less expensive than the display type, they have found favor in large-scale systems in which many independent counts must simultaneously be accumulated. In that event, the blind counters are often part of a CAMAC system in which the interrogation and readout take place over the dataway.

A *printing counter* is one in which an interface has been provided to generate the proper readout signals to drive a conventional line printer or other device. Other features found in some counters include an internal input gate that can be controlled by a gate pulse supplied to the unit, or a built-in integral discriminator to eliminate any noise that may appear along with the input pulses. Other specifications of importance are the minimum time separation between the leading edge of two logic pulses in order that they be counted as separate events (the *pulse pair resolving time*) and the maximum counting rate at which the counter may be driven.

D. Timers

The function of a timer is simply to start and stop the accumulation period for an electronic counter or other recording device. Obviously its most important property is the precision to which the time interval is controlled. Two general methods of control are commonly encountered. The simplest and least expensive method is to base the timing interval on the frequency of the alternating current of the power line to which the unit is connected. The precision of the timing is therefore determined solely by the accuracy and stability of the power line frequency. Utility companies usually do a good job of controlling the accuracy of the power line frequency when integrated over a day or more in order to maintain the accuracy of clocks also synchronized to the line frequency. On the other hand, the frequency may wander substantially over short periods of time, and timers based on power line synchronization thus may give rise to substantial timing interval errors if the interval is less than a few hours. In order to guarantee better accuracy, timers based on internal crystal-controlled circuits are preferred for more exacting measurements. The most important specification in this case then becomes the stability of the timing frequency to changes in temperature.

E. Counting Rate Meter

In some situations it is advantageous to have a visual indication of the rate at which pulses are being counted in the system. This function can sometimes be provided by a display counter in which the experimenter visually observes the rate at which counts are accumulated. Because of the random spacing between nuclear events, small changes in counting rate are difficult to observe in this way, particularly at low counting rates.

A *counting rate meter* provides a more direct means of indicating the rate at which pulses are being accumulated. In its most common form, a rate meter can be represented by the diode pump circuit outlined in Fig. 17-9. The output stage of the logic device driving the rate meter is represented by the voltage generator and series output impedance R_f. Each logic pulse, as it enters the circuit, deposits a small fixed amount of charge on the storage capacitor C_t. This capacitance is also continuously discharged by a current that flows through the resistance R. If the rate of arrival of logic pulses is constant, an

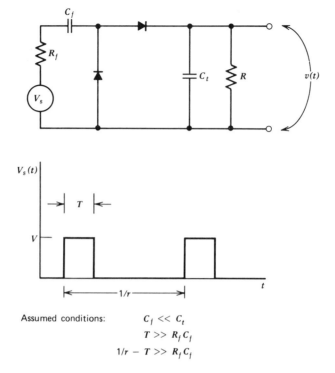

Assumed conditions:

$$C_f \ll C_t$$
$$T \gg R_f C_f$$
$$1/r - T \gg R_f C_f$$

Figure 17-9 The diode pump rate meter circuit. The assumed input voltage waveform V_s is also shown.

equilibrium will eventually be established in which the rate of charge deposition on the capacitor is just equal to the rate of its discharge through the resistance. Equilibrium is reached after several values of the time constant of the circuit have elapsed following an increase or decrease in the rate. This time constant is given simply by the product of the capacitance C_t and the parallel resistance R.

If the conditions shown in the figure are met, the average voltage appearing at the output of the circuit is

$$\bar{v} = iR = QrR = C_f VrR \tag{17-2}$$

where r is the average rate at which pulses are supplied to the circuit, and Q is the charge deposited per pulse given by the product of the coupling capacitor C_f and the pulse amplitude V. This output voltage is therefore proportional to the rate of arrival of the input logic pulses. If the input pulses were regularly spaced in time, the output voltage would have the appearance sketched in Fig. 17-10a. Longer time constants result in a more nearly constant signal, but the response to abrupt changes in rate will be slower. The full-scale range of the meter is normally varied by selecting the value of R with a front-panel control. Other rate meter circuits have been developed[10] which provide an output proportional to the logarithm of the count rate. These meters allow compression of the counting rate scale so that several decades may be monitored without the inconvenience of switching between scales.

When dealing with events from a radiation detector, the spacing between pulses is irregular and fluctuations in the output voltage arise due to the random variation in

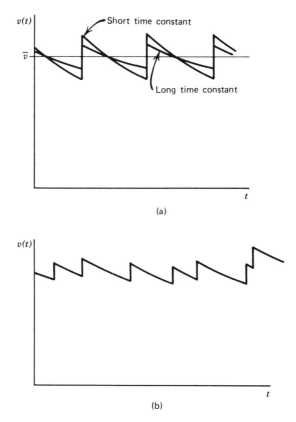

v(t)

Short time constant

v̄

Long time constant

t

(a)

v(t)

t

(b)

Figure 17-10 Output voltage from the diode-pump rate meter for (a) periodic pulses and (b) random pulses.

spacing. The rate meter signal then has an appearance typical of that sketched in Fig. 17-10b. The standard deviation σ of this signal can be defined as the square root of the variance of the values derived by sampling the signal many times at random and may be derived as follows.

The differential contribution to the output voltage produced by a rate r during the time between t and $t + dt$ is $(Qr/C_t)\,dt$. Because C_t is continuously discharged through R, the importance of this contribution decreases exponentially with time and at a later time t_0 is multiplied by a factor $\exp[-(t_0 - t)/RC_t]$. Similarly, the differential contribution to the standard deviation is $(Q/C_t)\sqrt{r}\,dt$ with an importance that also decreases exponentially with time. If the mean rate r does not change, the total variance of the output voltage can be obtained by integrating the independent weighted contributions of all prior time intervals:

$$\sigma_v^2 = \left(\frac{Q}{C_t}\right)^2 r \int_{-\infty}^{t_0} \exp\left(-\frac{2(t_0 - t)}{RC_t}\right) dt \qquad (17\text{-}3)$$

$$= \frac{Q^2 R}{2C_t} r \qquad (17\text{-}4)$$

The standard deviation is the square root of the variance:

$$\sigma_v = Q\sqrt{\frac{Rr}{2C_t}} \qquad (17\text{-}5)$$

The fractional standard deviation is usually of more interest and can be derived by combining Eqs. (17-2) and (17-5),

$$\frac{\sigma_v}{\bar{v}} = \frac{1}{\sqrt{2rRC_t}} \qquad (17\text{-}6)$$

which is often expressed as a percentage. The fractional standard deviation is usually selectable through a front-panel switch that varies the value of C_t. To achieve small fluctuations, C_t must be large, but the resulting long time constant limits the rate meter speed of response to rate changes.

An interesting digital equivalent of the rate meter circuit has been described by White.[11] It is based on the observation that in the diode pump circuit (Fig. 17-9) the amount of charge removed from the capacitor per unit time is always a fixed fraction of the total charge on the capacitor. By analogy, the digital rate meter consists of a register into which input pulses are gated for a fixed period of time. At the end of this time, a fixed fraction of the register content is subtracted from the accumulated content. The cycle of accumulation and fixed fraction subtraction is then repeated continuously. As in the analog rate meter, an equilibrium is exponentially approached in which the rate at which pulses are added to the register becomes equal to the rate at which they are subtracted. The equivalent time constant is given by

$$\tau = TF$$

where T is the accumulation gating time and F is the fraction of pulses subtracted at each step. With this substitution for the time constant $\tau = RC_t$, the standard deviation predicted by Eq. (17-6) for the analog rate meter also applies to the digital equivalent.

F. Dead Time in Counting Systems

An important consideration in many counting applications is the loss of events due to the dead time of the system. For some detectors (notably the G-M tube) the detector mechanism itself limits the minimum interval between events for which two distinct pulses can be counted. More often, however, the detector will be capable of producing pulses that are separated by a time that is less than the dead time inherent in the operation of an electronic component in the signal chain, and therefore it will be this component that determines the system dead time. In the simple counting systems shown in Fig. 17-7, this limiting component is usually the integral discriminator or the SCA. Although there are many exceptions, the dead time of a discriminator or SCA is typically related to the width of the linear pulse presented to its input and is characteristically a microsecond or two larger than this width.

The validity of the corrections for dead time losses discussed in Chapter 4 depends on the assumption that the dead time is constant for all events. The inherent dead time of electronic units can sometimes vary with the amplitude or shape of the input pulse, so that steps to set the dead time artificially are warranted in some critical applications. In this approach, the dead time is standardized by an element such as a linear gate, which is held closed for a fixed period of time following each pulse. This time is chosen to be larger than the dead time of any component in the system, so that accurate corrections can be made, even under conditions in which wide variations in pulse amplitude are encountered. The dead time of this element may also be measured conveniently by direct observation on an oscilloscope.

VI. PULSE HEIGHT ANALYSIS SYSTEMS

A. General Considerations

Next to the simple counting of pulses, the most common procedure in nuclear measurements involves recording the amplitude distribution of pulses produced by a radiation detector. Most often the object is to deduce properties of the incident radiation from the position of peaks in the recorded spectrum, although other aspects of the spectrum may be of interest in different situations. The performance required of the instrument system used to record the pulse height spectrum is largely dependent on the inherent energy resolution of the detector. If the detector energy resolution is relatively poor, the requirements of the recording system are undemanding and easy to meet. On the other hand, detectors with good energy resolution require careful attention to the pulse-processing system to assure that additional degradation of the resolution is minimized.

A simple pulse height analysis system is shown in Fig. 17-11a. The key element in this signal chain is the linear amplifier, which shapes the pulses from the preamplifier and provides enough amplification to match the input span for which the multichannel

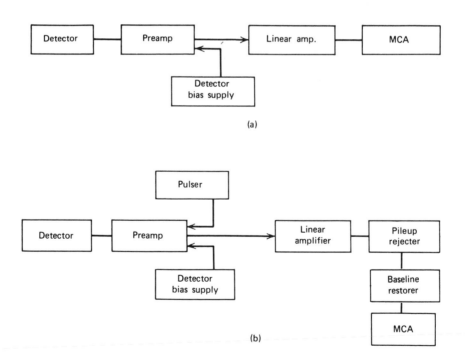

(a)

(b)

Figure 17-11 Signal chains for pulse height spectrometry. Part (a) shows a rudimentary system for noncritical applications, whereas system (b) includes some added functions that can improve performance for high-resolution detectors when operated at high pulse rates.

analyzer has been designed. It is the shaping function of the linear amplifier that often dominates the performance of the pulse-processing system. For low-resolution detectors, virtually any of the shaping techniques discussed in Chapter 16 can be used to reduce the long tails of the preamplifier output pulses. For high-resolution detectors, however, consideration must be given to the effect of various shaping methods on the pulse properties with regard to signal-to-noise and pileup. Also shown in Fig. 17-11b is a pulse spectrometry system to which several components have been added to mitigate the effects of pulse pileup and baseline shift.

The strategies employed in choosing parameters for the pulse-processing system change drastically with the expected counting rate. If the counting rate is low (say 100 counts/s or less), the system can be optimized with regard to processing each individual pulse without much concern for interfering effects between pulses.[†] The problem becomes much more complicated at high rates, where instrument parameters chosen to minimize pulse pileup and to assure rapid return of the pulse waveform to the baseline often conflict with the choices that would be made on a signal-to-noise basis alone. Thus, it is in the high-resolution–high-rate situation that the demands are greatest, and the proper choices are often difficult to predict in advance. Decisions are then frequently made on an empirical basis by varying the system parameters to find the best combination for each individual situation.

The most critical choices involve the method chosen for pulse shaping within the linear amplifier. The following section on linear amplifier operation therefore stresses the various compromises that must be struck in choosing the type and time characteristics of the pulse shaping. These choices necessarily involve the general concepts of ballistic deficit, signal-to-noise, and pileup of signal pulses.

B. Ballistic Deficit

The rise time of the pulse from the preamplifier normally corresponds to the charge collection time in the detector itself. If the full amplitude of the preamplifier pulse is to be preserved through the shaping process, the shaping time constants must be large compared with the preamplifier pulse rise time. Because the shaping time constants cannot always be chosen as arbitrarily large, the amplitude of the shaped pulse can sometimes be slightly less than that attainable with very long time constants. The degree to which the infinite time constant amplitude has been decreased by the shaping process is called the *ballistic deficit* (see Fig. 17-12).

In detectors with a constant charge collection time, relatively large ballistic deficits can often be tolerated because a constant fraction of the amplitude for each pulse is lost. Should the charge collection time vary, however, a variable amount of each pulse will be lost, which can lead to resolution degradation. In these cases, one must then choose longer time constants than might be optimum on the basis of signal-to-noise or pileup considerations.

This problem is obviously most severe for those detectors with the largest variation in charge collection time. Rise time variations are often important in proportional counters, where the radial variation of the primary ionization determines the spread in the charge

---

[†]A quick estimate of the importance of rate-related effects can be made by calculating the *duty cycle* obtained by multiplying the effective width of the shaped pulse by the rate. If this product is less than about 10^{-3}, these effects should be minimal and can often be neglected. A duty cycle of 10^{-2} is a moderate rate, whereas in high-rate situations it may approach 10^{-1}. For a typical pulse width of 5 μs, the corresponding rates are 200, 2000, and 20,000 per second.

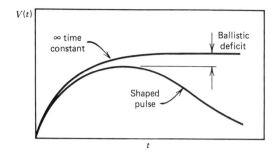

Figure 17-12 Definition of the ballistic deficit. If the shaping times are fixed, the ratio of the deficit to the amplitude of the shaped pulse will be constant for pulses of the same leading edge shape but will vary if this shape changes.

collection time. Perhaps the biggest variations occur in coaxial germanium detectors, where large differences in collection time correspond to different radial locations of the radiation interaction in the detector. Optimum shaping times then tend to be many times greater than the average collection time.

In the common pulse-shaping methods, events leading to the largest ballistic deficit will also produce the longest peaking time of the shaped pulse. Because of this correlation, it is theoretically possible to correct for ballistic deficit on a pulse-to-pulse basis by estimating its magnitude by sensing the peaking time. A corresponding voltage can then be added to the shaped pulse to compensate for the amplitude loss. Ballistic deficit variations can be a dominant contributor to the energy resolution of gamma-ray spectra taken under high rate conditions where short shaping times are needed to minimize pulse pileup. Goulding and Landis[12] have shown that use of a correction circuit based on this principle can significantly improve energy resolution under these conditions.

C. Signal / Noise Considerations

By definition, noise is any undesired fluctuation that appears superimposed on a signal source. Fig. 17-13 gives a graphical representation of the way in which a random noise component can degrade the amplitude information carried by pulses from a radiation detector. A general analysis of noise sources in radiation systems is beyond the scope of this discussion but may be found in texts such as Ref. 13. At this point we outline only some of the more important considerations regarding noise sources and the practical steps that can be taken to minimize their influence on measured pulse height spectra.

The important sources of noise occur near the beginning of the signal chain where the signal level is at a minimum. Noise generated at this point undergoes the same amplification as the signal, whereas noise generated further along the signal chain is usually much smaller than the signal. Therefore, discussions of electronic noise sources generally center on the preamplifier and, most importantly, its input stage. As outlined by Radeka,[14] the noise sources can conveniently be categorized into those that are effectively in parallel with the input and those that are in series with the signal source. For example, sources of parallel noise include fluctuations in the leakage current of the detector and in the current drawn by the input stage of the preamplifier, whereas series noise sources include the thermal noise in the FET of the preamplifier input. The frequency spectrum of many of these noise sources is very broad, and for some analytic purposes it is often assumed that a "white" or uniform distribution in frequency can represent their cumulative effect.

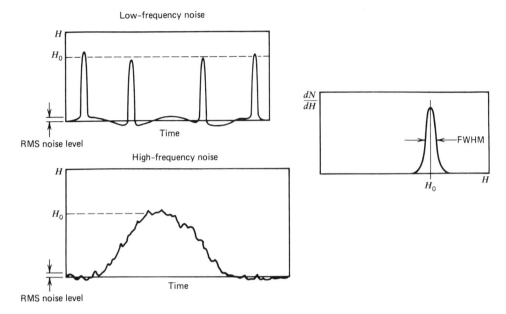

Figure 17-13 Sketches of the effect of low-frequency and high-frequency noise on signal pulses of constant amplitude. The effect is to broaden the peak recorded in the differential pulse height spectrum for these pulses. The FWHM of the peak of Gaussian-distributed noise will be equal to the RMS noise level multiplied by a shape factor of 2.35, provided noise is the only factor in broadening the peak.

In contrast, the frequency spectrum of the signal is confined to a much narrower band. For example, the signal from a detector whose charge collection time is about a microsecond can never contain useful information on a time scale of nanoseconds. Therefore, low-pass filtering of the output of such a detector will eliminate the contribution of high-frequency noise but will not affect the information-carrying components of the signal. Similarly, low-frequency power line pickup is a noise source that can potentially degrade the signal, but high-pass filtering can remove this component without substantially affecting the signal pulse shape from most detectors. Therefore, the pulse-shaping role traditionally carried out by the linear amplifier actually amounts to selective filtering to remove as much broad spectrum noise as possible without severely attenuating the useful signal components. The pulse shaping is normally carried out through a combination of differentiating and integrating circuits, but from the discussion of Chapter 16, it is equally valid to regard these operations as high-pass and low-pass filtering, respectively.

The amount of noise added by the preamplifier–amplifier combination is often expressed in terms of the *equivalent noise charge* (ENC). This is defined as the amount of charge that, if applied suddenly to the input terminals of the system, would give rise to an output voltage equal to the RMS level of the output due only to noise. For specific types of detector, the ENC is sometimes translated into the equivalent energy deposition by a charged particle in the detector.

The effect of pulse shaping on signal-to-noise ratios in nuclear pulse systems has been the subject of extensive theoretical studies. The reader is referred to detailed discussions of this topic in Refs. 15 and 16, and only a summarized statement of the results is given here. These analyses are generally based only on the assumption that the noise is broadly

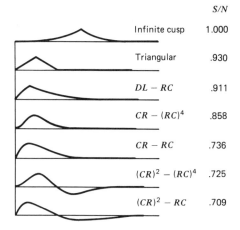

	S/N
Infinite cusp	1.000
Triangular	.930
$DL - RC$	.911
$CR - (RC)^4$	.858
$CR - RC$	.736
$(CR)^2 - (RC)^4$	.725
$(CR)^2 - RC$	.709

Figure 17-14 Various pulse shapes and their signal-to-noise ratio (S/N) relative to the infinite cusp. Various time constants are chosen to yield minimum noise, and pulse shapes are normalized to constant height. (Reproduced with permission from *Nuclear Electronics*, by P. W. Nicholson. Copyright 1974, by John Wiley & Sons Ltd.)

distributed in frequency and deal with the problem from the standpoint of signal pulses that are widely separated in time. As we shall see, the interfering effects of signal pulses at high rates can sometimes override the considerations of signal-to-noise ratios for single isolated pulses given below.

It can be shown[16] that the best possible signal-to-noise ratio is achieved if the signal pulses are shaped to the form of an infinite cusp (see Fig. 17-14). As a practical matter, the pulse shape must have finite width, and the finite cusp has the best signal-to-noise properties if pulse shapes of finite width are compared. Because it has been shown[17] that no other type of pulse shaping (including time-dependent or nonlinear processes) can yield a pulse shape superior to the cusp in its signal-to-noise properties, this pulse shape has become the standard by which the performance of other methods of pulse shaping are compared.

The cusp is not particularly practical for several reasons. The top is sharply pointed, which makes pulse amplitude measurement difficult unless pulse-stretching methods are employed. Furthermore, the return to the baseline is rather slow, which is undesirable from a pileup standpoint. The cusp shape is also difficult to achieve with practical shaping circuits. For all these reasons, other methods of pulse shaping are more commonly used which are inferior to the cusp in signal-to-noise properties but are preferable because some of the above drawbacks are avoided.

Figure 17-14 lists some of the common methods of pulse shaping generally found in linear pulse amplifiers, together with the signal-to-noise performance compared with the infinite cusp. All those shown in the figure can be achieved using the methods of passive shaping discussed in Chapter 16. More complicated methods of pulse shaping involving active circuits are useful in systems in which the ultimate in signal-to-noise must be attained. Of these, active filters that approximate a Gaussian or triangular shape (see p. 572) have been incorporated into commercially available linear amplifiers.

A specific type of noise, called *microphonics*, can sometimes be a problem in practical detector systems. Mechanical vibrations transmitted to the detector–preamplifier input stage can produce small fluctuations in capacitance that cause a modulation of the output signal. The effects are most pronounced for systems such as ionization chambers or semiconductor spectrometers for which the small charge per pulse requires keeping the input stage capacitance at a minimum. For example, it has been demonstrated[18] that a capacitance change of only 5×10^{-7} pF between the FET gate and the high-voltage bias of a Si(Li) system corresponds to a microphonic signal equivalent to the deposition of 10 eV in the detector. For stray capacitance values on the order of 1 pF, such a change corresponds to a mechanical motion of only one part in 10^7. Fortunately, much of this noise will occur at low frequencies and will lie outside the frequency band passed by the shaping networks in the linear amplifier. In those cases in which microphonic noise is a problem, choosing short shaping times (corresponding to raising the low-frequency cutoff) will help minimize this contribution to the system noise level.

D. Pileup

1. THE PROBLEM

The fact that pulses from a radiation detector are randomly spaced in time can lead to interfering effects between pulses when counting rates are not low. These effects are generally called *pileup* and can be minimized by making the total width of the pulses as small as possible. Other considerations of ballistic deficit and signal-to-noise prevent reduction of the pulse width beyond a certain point, and therefore the effects of pulse pileup at high rates are often very significant.

Pileup phenomena are of two general types, which have somewhat different effects on pulse height measurements. The first type is known as *tail pileup* and involves the

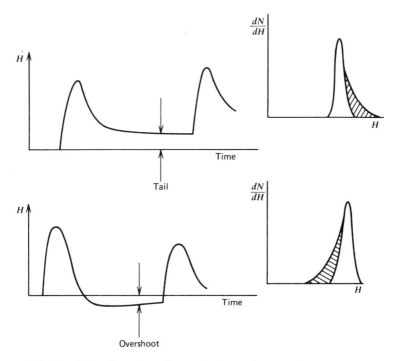

Figure 17-15 Pileup from the tail or undershoot of a preceding pulse. Both are usually categorized as tail pileup. The effect on the differential pulse height spectrum for constant-amplitude pulses is shown as the cross-hatched area at the right.

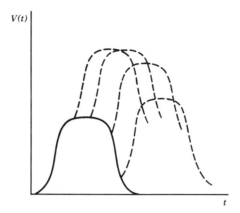

Figure 17-16 Peak pileup, in which two closely spaced signal pulses combine to form one distorted pulse. Several different cases are sketched with increasing overlap between the first and second pulse.

superposition of pulses on the long-duration tail or undershoot from a preceding pulse (see Fig. 17-15). Under conditions described earlier, tails or undershoots can persist for relatively long periods of time so that tail pileup can be significant even at relatively low counting rates. The effect on the measurement is to worsen the resolution by adding wings to the recorded peaks in the pulse height spectra as shown in the figure. The remedy for tail pileup is to eliminate residual tails or undershoots through the use of pole-zero cancellation or active baseline restoration techniques discussed in Chapter 16. Pulse shapes that return quickly to the baseline will also help eliminate the irreducible tail pileup due to the normal decay of the shaped pulse.

A second type of pileup, generally called *peak pileup*, occurs when two pulses are sufficiently close together so that they are treated as a single pulse by the analysis system. As shown in Fig. 17-16, the superposition of pulses with relatively flat tops will lead to a combined pulse with an apparent amplitude equal to the sum of the two individual amplitudes. Lesser degrees of overlap will give a combined pulse with an amplitude somewhat less than the sum. Not only does this type of pileup lead to distortions in the recorded spectrum, including an occasional sum peak, but it also interferes with quantitative measurements based on measuring the area under full-energy peaks. The pileup of two full-energy pulses effectively removes both from the proper position in the pulse height spectrum, and the area under the full-energy peak in the spectrum will no longer be a true measure of the total number of full-energy events. Because peak pileup leads to the recording of one pulse in place of two, the total area under the recorded spectrum is also smaller than the total number of pulses presented to the system during its live time.

The seriousness of peak pileup for a given situation can be estimated from the counting rate and the effective width of the signal pulses. The effective width is difficult to define, except for rectangular pulses, but can be approximated by the FWHM of the first lobe of the shaped pulse.[19] If we neglect second-order effects, the interval distribution [Eq. (3-60)] can be applied to estimate the degree of pileup to be expected. The probability of observing an interval greater than τ is given by

$$P(>\tau) = \exp(-n\tau) \qquad (17\text{-}7)$$

where n is the true rate of signal pulses. For pileup to be avoided, the interval following

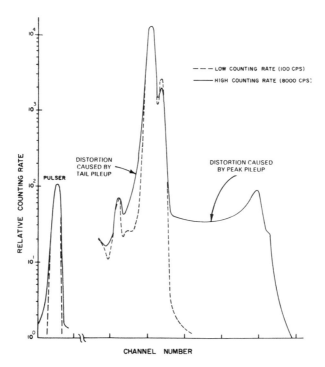

Figure 17-17 Spectral effects of peak and tail pileup. The dashed curve shows a ^{55}Fe spectrum taken at a low counting rate at which pileup is negligible. The solid curve shows a high-rate spectrum and illustrates the sum peak and continuum caused by peak pileup. The low-energy tail added to the primary peak by overshoot or tail pileup is also observed. (From Wielopolski and Gardner.[21])

each pulse must be greater than the effective pulse width. For $n = 20{,}000$ s^{-1} and an effective width of 5 μs, Eq. (17-7) predicts that 90.5% of all intervals will be greater than this width. However, two pulses are affected if a pileup event takes place, so that only about 81% of all true events escape pileup and are properly recorded. At high rates, higher-order pileup events must be considered in which more than two pulses may overlap simultaneously.

If all signal pulses were rectangular in shape with constant amplitudes, the spectral effects of pileup would be relatively simple. At very low rates, only a single peak should appear in the recorded pulse height spectrum because pileup is then negligible. At higher rates, any pileup that occurs can result only in a total amplitude that is an integral multiple of the single pulse amplitude. Therefore, the effect of pileup would be to introduce sum peaks that would appear only at uniformly spaced positions in the recorded spectrum. Of these, the *double sum peak* corresponding to the simple pileup of two pulses will be the most intense and will appear at a position corresponding to twice the basic amplitude.

A more realistic representation must account for the variable amplitude and nonrectangular shape of most signal pulses.[20] The pileup spectrum will then have the continua shown in Fig. 17-17, which fill in the space between expected sum peaks and the single amplitude peak. These events correspond to the partial pileup of pulses with leading and trailing edges that are not vertical, and to pileup of individual pulses with different amplitudes.

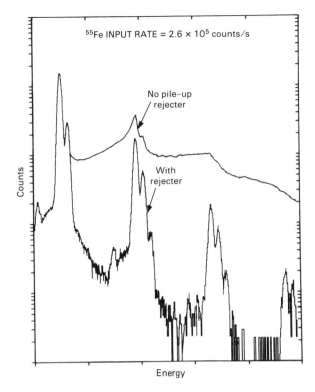

Figure 17-18 Pulse height spectra recorded with and without a pileup rejecter. The true event rate n was $2.6 \times 10^5/s$ and the rejecter resolution time τ was 300 ns. Note that the contributions to the spectrum from pileup are greatly reduced by the rejecter, while the numbers of counts in the primary peaks (free of pileup) are unaffected. (From Goulding and Landis.[28])

Pileup can also influence measurements other than the pulse height distribution. In particular, timing measurements can be affected by the resulting changes in pulse shape that accompany pileup. For example, Refs. 22 and 23 describe the quantitative losses that can occur in coincidence experiments based on crossover timing when pileup distorts the pulse shape. Wilkinson[24] also points out the errors that can occur in time interval measurements when variable counting rates are encountered in the course of the measurement.

2. PILEUP REJECTION
One method of avoiding the pulse height spectrum effects arising from a peak pileup is to use some means of *pileup rejection*. Most commercial linear amplifiers are provided with active circuitry that can discard pulses that are expected to be affected by pileup. Some practical methods of pileup rejection are reviewed in Refs. 25–27. Circuits of this type can dramatically reduce the spectral distortions due to pileup,[21] but the quantitative losses must be dealt with in the same manner as dead time losses. (See analysis below.) An example of the degree of pileup rejection that can be achieved is shown in Fig. 17-18.

A common technique employed in many pileup rejecters is to pass the signal through two parallel branches—a *fast* and a *slow* branch. In the fast branch, one of the time pick-off techniques described later in this chapter is used to generate a very short logic pulse as quickly as possible. In the slow branch, the event is processed using conventional (and somewhat slower) shaping techniques designed to best measure its amplitude. This shaped pulse is then passed through a linear gate only if a second pulse is *not* triggered in the fast branch during the processing time in the slow branch. In this way, the output pulses should not be contaminated by events that occur during the processing time, and the corresponding spectrum is, in principle, free of pileup distortions.

In reality, however, the fast–slow pileup rejection technique only reduces the problem but cannot totally eliminate it. There is still some possibility that two or more events will occur within the pulse resolution time of the fast branch, and only one fast pulse will be produced. The slow branch will then process an amplitude that is some combination of the multiple events, and a pulse resulting from pileup is again recorded. The likelihood of this occurrence depends on the time behavior of the fast branch, which can be approximated using one of the two models for dead time behavior introduced in Chapter 4. It is therefore instructive to look in somewhat greater detail at the nature of the events lost in dead time processes, because these same combinations of closely spaced events will also contribute to pileup.

3. STATISTICAL ANALYSIS OF PILEUP EVENTS

We begin by defining the terms that will be used throughout this analysis. An *event* is the consequence of a radiation interaction in the detector that should lead to a recorded pulse, in the absence of dead time or pileup. A determination of the amplitude spectrum of true events is often the object of our measurement. A *count* is a pulse as actually registered by the recording system. Due to dead time and/or pileup, fewer counts are recorded than the number of true events. Also, some recorded counts correspond to the pileup of more than one event, and the recorded amplitude distribution is distorted compared with the true event spectrum.

The analysis also assumes that any inherent dead time of the detector/preamplifier is small compared with the pileup resolution time τ of the pulse-processing system. This time τ is now defined as the minimum time that must separate two events so that they do not pile up. Thus, events arrive at the amplifier as a Poisson distributed random process and are assumed to pile up if they occur with a time spacing less than τ following a previous event. The discussion of dead time losses given previously in Chapter 4 assumed that events were simply lost under these conditions, but we now extend the analysis to examine the categories of counts that may result. True events are again assumed to occur at a rate n. Due to pileup, the recording system will perceive counts at a lower rate m. We now seek to classify these counts according to the number of true events that contribute to each count. The results will depend on whether the system behaves in a paralyzable or nonparalyzable manner, as defined in Chapter 4.

a. Nonparalyzable System

In Chapter 4 [see Eq. (4-24)], it was shown that the event rate n and counting rate m are related by

$$n = \frac{m}{1 - m\tau} \quad \text{or} \quad m = \frac{n}{1 + n\tau} \tag{17-8}$$

We now derive the probability that a typical count corresponds to the pileup of two or more true events. Over the time τ, the average number of true events is simply $n\tau$. From the Poisson distribution [Eq. (3-24)] we can write the probability that exactly x events occur over this time as

$$P(x) = \frac{\bar{x}^x e^{-\bar{x}}}{x!} = \frac{(n\tau)^x e^{-n\tau}}{x!} \tag{17-9}$$

Each of these time intervals is started by a true event. If *no* additional events occur over the following time τ, a count is recorded that is free of pileup. The probability that a given *count* falls in this category is

$$P(0) = e^{-n\tau} \tag{17-10}$$

Note that n is the true event rate, not the observed counting rate. Similarly, the probability that a given count represents the pileup of exactly two events is

$$P(1) = n\tau e^{-n\tau} \tag{17-11}$$

In general, the probability that $(x + 1)$ true events pile up to produce a given count is

$$P(x) = \frac{(n\tau)^x e^{-n\tau}}{x!} \tag{17-12}$$

Since the Poisson distribution is normalized,

$$\sum_{x=0}^{\infty} P(x) = 1$$

and we have accounted for all counts.

As a cross check, we note that a count corresponds to $(x + 1)$ events with a probability of $P(x)$. Therefore, the average number of true events per count is

$$\langle x \rangle = \sum_{x=0}^{\infty} (x + 1) P(x) = \sum_{x=0}^{\infty} (x + 1) \frac{(n\tau)^x e^{-n\tau}}{x!}$$

$$= e^{-n\tau} \sum_{x=0}^{\infty} (x + 1) \frac{(n\tau)^x}{x!} = e^{-n\tau} e^{n\tau} (n\tau + 1)$$

$$= n\tau + 1 \tag{17-13}$$

But, by definition, $\langle x \rangle = n/m$, so

$$\frac{n}{m} = n\tau + 1 \quad \text{or} \quad m = \frac{n}{1 + n\tau}$$

Since this expression is identical to Eq. (17-8) above, we have correctly accounted for all true events as well.

b. Paralyzable System

From Eq. (4-27), we have previously shown that for a paralyzable system

$$m = ne^{-n\tau} \tag{17-14}$$

The arguments leading to Eq. (17-10) are still valid for a paralyzable system, so the probability that a given *count* is free of pileup remains

$$P(0) = e^{-n\tau}$$

Two (and only two) events will pile up under paralyzable conditions if the following sequence occurs: a true event at $t = 0$, with a second event sometime in the interval $0 < t < \tau$ that is followed by an event-free interval of length τ. The probability that a given count originates in this way can be written as

$$P(1) = \int_0^\tau \left(\begin{array}{c} \text{probability of} \\ \text{no event } 0 \to t \end{array} \right) \left(\begin{array}{c} \text{probability of} \\ \text{event in } dt \end{array} \right) \left(\begin{array}{c} \text{probability of} \\ \text{no event } t \to (t + \tau) \end{array} \right)$$

$$= \int_0^\tau e^{-nt} n \, dt \cdot e^{-n\tau}$$

$$= e^{-n\tau}(1 - e^{-n\tau})$$

We can extend this analysis to predict the probability that a count results from the pileup of *three* true events. In this case, following an initial event at $t = 0$, one additional event must occur with time spacing less than τ, followed by the sequence described above for two-event pileup. Thus, we can write

$$P(2) = \int_0^\tau \left(\begin{array}{c} \text{probability of} \\ \text{no event } 0 \to t \end{array} \right) \left(\begin{array}{c} \text{probability of} \\ \text{event in } dt \end{array} \right) \cdot P(1)$$

$$= \int_0^\tau e^{-nt} n \, dt \cdot e^{-n\tau}(1 - e^{-n\tau})$$

$$= e^{-n\tau}(1 - e^{-n\tau})^2$$

To generalize, the probability that a given count is formed from the pileup of $(x + 1)$ events is

$$P(x) = e^{-n\tau}(1 - e^{-n\tau})^x \tag{17-15}$$

Note that

$$\sum_{x=0}^{\infty} P(x) = e^{-n\tau} \sum_{x=0}^{\infty} (1 - e^{-n\tau})^x = e^{-n\tau} e^{+n\tau} = 1$$

so that we have accounted for all counts.

To check the accounting of true events, we first note that each recorded count results from $(x + 1)$ true events with a probability $P(x)$. The average number of events per count is therefore

$$\langle x \rangle = \sum_{x=0}^{\infty} (x + 1) P(x) = e^{-n\tau} \sum_{x=0}^{\infty} (x + 1)(1 - e^{-n\tau})^x$$

$$= e^{-n\tau} e^{+2n\tau} = e^{+n\tau} \tag{17-16}$$

This expression is consistent with the result obtained directly from Eq. (17-14):

$$\langle x \rangle = \frac{n}{m} = \frac{n}{ne^{-n\tau}} = e^{+n\tau}$$

so the check on true events is verified.

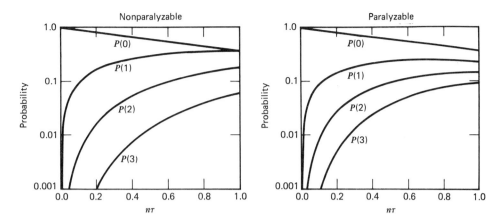

Figure 17-19 Plots of the probabilities that a recorded count is free of pileup [$P(0)$] or due to ith-order pileup [$P(i)$]. The abscissa is the product of true event rate n and the pileup resolution time τ. The left plot is from Eq. (17-12) for nonparalyzable systems, and the right plot is from Eq. (17-15) for paralyzable systems.

The pulse height spectrum measured in the presence of pileup can be thought of as the sum of a number of individual spectra due to different categories of counts. One of these is the original event spectrum that is free of pileup, and its relative contribution to the measured spectrum is given by $P(0)$ in Eq. (17-10). In addition, the spectrum created by the pileup of two events will be added with a relative contribution equal to $P(1)$. Spectra due to higher-order pileup will also become significant as the event rate increases and will add individual spectra due to the pileup of $(x + 1)$ events with a relative contribution given by the expressions for $P(x)$ in Eqs. (17-12) and (17-15). Each of these contributions is plotted as a function of true event rate in Fig. 17-19, both for paralyzable and nonparalyzable systems. In the pulse height spectrum previously shown in Fig. 17-18, the individual contributions are evident for the pileup of up to four true events [corresponding to $P(3)$ above].

If no pileup rejection is used, then τ in the previous expressions represents the effective width of the shaped pulse at the amplifier output (see Fig. 17-16). Typical values for germanium detectors are in the 3–10 μs range. The system is most likely to behave as a paralyzable one, since the arrival of a true event during a previous pulse will extend the period for pileup by another time τ. When pileup rejection is applied, the effect is to greatly suppress the pileup contributions in the recorded spectrum by imposing a much smaller value for τ. This time now becomes the pulse resolution time in the fast branch of the rejecter, and typical values are fractions of a microsecond. The system behavior (whether paralyzable or nonparalyzable) will depend on the specific design of the rejection circuits.

One of the effects of pileup rejection is to assure that partially overlapped pulses are efficiently discarded, leaving only pileup corresponding to the nearly complete overlap of two pulses. Thus, the bottom spectrum in Fig. 17-18 more cleanly separates the contributions of each order of pileup that remains. Tenney[29] has demonstrated how individual pileup spectra can be calculated under certain assumed conditions and gives examples for some simple true event pulse height spectra. Other discussions of spectral distortions caused by pileup are given in some detail in Refs. 21 and 30–33.

It should be emphasized that the results derived above are based on recorded counts. The throughput of the system is better gauged by considering the effect of pileup on the fraction of true events that are properly recorded free of pileup. For the nonparalyzable case, recall that the average number of true events per count is, from Eq. (17-13),

$$\langle x \rangle = n\tau + 1$$

The probability (per count) of recording a pileup-free event is $P(0) = e^{-n\tau}$. Therefore, the fraction of true events that escape pileup is

$$f_e|_{\text{nonpara}} = \frac{P(0)}{\langle x \rangle} = \frac{e^{-n\tau}}{n\tau + 1} \tag{17-17}$$

For $n\tau \ll 1$, the first-order expression is

$$f_e|_{\text{nonpara}} \cong \frac{1 - n\tau}{1 + n\tau} \cong 1 - 2n\tau$$

This result can be compared with the first-order expression for the fraction of *counts* that are free of pileup:

$$f_c = P(0) = e^{-n\tau} \cong 1 - n\tau$$

Thus, the first-order losses to pileup are twice as great for true events compared with those for recorded counts. This comparison is in agreement with the heuristic observation made earlier that, at low count rates, two events are lost for every pileup count.

In the case of a paralyzable system, Eq. (17-16) gives the average number of events per count as

$$\langle x \rangle = e^{n\tau}$$

so the fraction of true events that escape pileup is now

$$f_e|_{\text{para}} = \frac{P(0)}{\langle x \rangle} = \frac{e^{-n\tau}}{e^{n\tau}} = e^{-2n\tau} \tag{17-18}$$

Again, the first-order expression for $n\tau \ll 1$ is

$$f_e|_{\text{para}} \cong 1 - 2n\tau$$

which is identical to the nonparalyzable case. Plots of $f_e|_{\text{nonpara}}$, $f_e|_{\text{para}}$ and f_c are shown in Fig. 17-20.

Finally, the rate at which true events are recorded free of pileup (r_{pf}) is of primary interest to the user. In systems with an effective pileup rejecter, r_{pf} is approximately the observed output rate, where τ represents the processing time in the slow branch. This rate is given by the product of the true rate n and the fraction of true events that escape pileup (f_e). For the nonparalyzable case, we use Eq. (17-17) for this fraction and find

$$r_{pf}|_{\text{nonpara}} = nf_e|_{\text{nonpara}} = \frac{ne^{-n\tau}}{1 + n\tau} \tag{17-19}$$

As a check, this rate should also be given by the product of the counting rate m and the probability (per count) of escaping pileup $P(0)$:

$$r_{pf}|_{\text{nonpara}} = mP(0) = \frac{n}{1 + n\tau}e^{-n\tau}$$

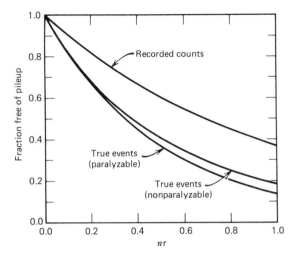

Figure 17-20 The fraction of recorded counts (upper curve) and true events (lower two curves) that escape pileup as a function of true event rate n.

and we obtain the same result. As the true rate is increased, r_{pf} goes through a maximum. Increasing the true rate further will continue to increase the total counting rate, but the growing importance of pileup actually decreases r_{pf}. This behavior is illustrated in Fig. 17-21. The maximum value of r_{pf} is given by $0.206/\tau$, and it is reached at a true event rate of $0.618/\tau$.

For the paralyzable case, the same type of analysis leads to

$$r_{pf}|_{\text{para}} = nf_e|_{\text{para}} = ne^{-2n\tau} \qquad (17\text{-}20)$$

Now r_{pf} reaches a maximum value of $0.184/\tau$ at a true event rate of $0.500/\tau$. Figure 17-21 also plots this case, together with the total counting rate (without pileup rejection) for both models. Recall (see Fig. 4-8) that the counting rate for the paralyzable model also goes through a maximum given by $0.368/\tau$ at a true rate of $1/\tau$. At that point, Eq. (17-18) predicts that only 13.5% of the true events are recorded free of pileup.

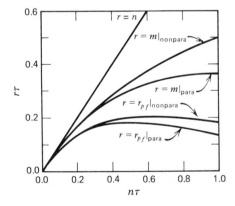

Figure 17-21 Variation of the recorded count rate (m) and the rate at which true events are recorded free of pileup (r_{pf}) for both paralyzable and nonparalyzable systems. The true event rate is n and the pileup resolution time is τ.

4. QUANTITATIVE CORRECTION FOR PILEUP EFFECTS

The most effective technique of experimentally correcting for the quantitative errors in peak areas introduced by pileup is to employ the *pulser method*. The output of a fixed amplitude pulse generator is mixed with the pulses from the detector, normally at the preamplifier stage. The amplitude of the pulses from the generator is adjusted to create an artificial peak in the recorded pulse height spectrum at a location that does not interfere with peaks or other spectral features from the signal pulses. Since the number of pulses injected from the generator can be accurately determined, a measurement of the area under the artificial peak in the spectrum yields the fraction of those pulses that have escaped pileup. The assumption is then made that this same fraction also applies to the signal pulses to allow correction of their measured peak areas for pileup effects. This assumption is rigorously valid only for a pulse generator that produces pulses that are randomly distributed in time, and *random pulse generators* are commercially available for such applications. Periodic pulse generators may also be used provided that the pulse rate is kept small compared with the signal pulse rate. (Differences arise because random pulses can pile up with each other, but periodic pulses cannot. Monte Carlo simulations[34] show that a pulser rate of less than 10% of the signal rate generally results in a negligible difference.) An extensive analysis of the correction of pulse height spectra for pileup effects for both periodic and random pulse generators can be found in Ref. 33.

E. Linear Amplifier

Two primary functions are conventionally provided by the linear amplifier element in the pulse-processing chain: pulse shaping and amplitude gain. The unit accepts tail pulses as an input, often of either polarity, and produces a shaped linear pulse with standard polarity and span (the NIM standard is positive polarity, 0–10 V amplitude). The amplification factor or gain required varies greatly with application but is typically a factor between 100 and 5000. The gain is normally adjustable over a wide range through a combination of coarse and fine controls. If the product of input amplitude and gain exceeds the maximum design output amplitude, the amplifier will *saturate* or *limit* and produce a distorted output pulse with a flat top at the amplitude at which saturation occurs (typically 10 V). Linear amplification will be realized only for those pulses that are short of this saturation level. Amplifier specifications normally include a figure for maximum nonlinearity over the design span, which is usually small enough to have little practical consequence.

Of all the elements in a pulse height analysis system, the linear amplifier presents the user with the greatest variety of operational characteristics and, as a result, the most difficult choices. The conflicting requirements placed on pulse shaping from consideration of signal-to-noise, ballistic deficit, and pulse pileup must be weighed for each potential application. Fundamental to these considerations is the inherent trade-off between counting rate and resolution. High rates favor bipolar pulses with small width, whereas optimum shaping for best resolution points toward unipolar pulses with relatively large widths. On many commercial linear amplifiers, the shaping times are adjustable. Then the choice may be optimized for a given application by measuring the pulse height resolution of the system as the shaping times are varied. Of course, it is important to carry out these measurements at a counting rate equivalent to that expected in the application itself.

Examples of this optimization are given in Fig. 17-22, where the effects of varying the shaping time in the amplifier are shown for two different pulse rates. In both cases, there is an optimum shaping time that minimizes the FWHM of the measured gamma-ray peak from a germanium detector. For the low-rate case, pileup and baseline shift effects are negligible. As the shaping time is increased from small values, the FWHM decreases

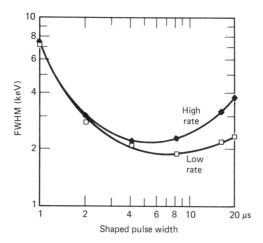

Figure 17-22 Variation of the energy resolution of a typical germanium detector system as the amplifier shaping time is changed. The horizontal axis is the width of the shaped pulse at half maximum. The upper curve is for a relatively high rate (62,000/s) and the lower curve is for a low rate (2000/s). (Data from Fairstein.[35])

(resolution improves) for two reasons: The ballistic deficit decreases (see p. 606) and the contribution of series noise (due primarily to fluctuations in the channel current of the input stage FET) is reduced as the high-pass filter cutoff frequency is lowered. At the other extreme, increasing the shaping time causes a gradual increase in the FWHM due to a greater sensitivity to parallel noise (originating primarily from fluctuations in detector leakage current). At an optimum shaping time between these extremes, the observed FWHM goes through a minimum value. For the high-rate case, the minimum occurs at a smaller value of the shaping time, and the corresponding best FWHM is somewhat larger than for the low rate. Now the effects of pileup and baseline shift are important for all but the smallest values of shaping time near the left of the plot. For larger values of the shaping time, these effects increasingly contribute to the measured FWHM and are reflected in the steeper rise to the right of the minimum. In general, one would expect the optimum shaping time to move to smaller values as the counting rate is increased and the effects of pileup and baseline shift become more severe.

Most amplifiers in use today employ one of the pulse-shaping methods discussed in Chapter 16. For low-resolution systems (such as scintillation detectors), virtually any method of pulse shaping will be acceptable at low rates, since the detector itself will dominate the observed resolution. For high rates, double delay line shaping can be useful because of its relatively narrow pulse width and almost complete freedom from baseline shift due to the symmetrical bipolar pulse shape. For high-resolution systems (such as germanium gamma-ray spectrometers), shaping choices are more critical. Delay line shaping is not used because of its relatively inferior signal-to-noise characteristics. At low counting rates, the unipolar near-Gaussian pulse shape produced by one differentiation followed by several integration networks is a common choice. Alternatively, triangular shaping (see p. 572) can offer a slightly better noise suppression. At high rates, the baseline shift that accompanies unipolar pulses requires either the substitution of a network that results in bipolar shape (provided as an alternative output on many units), or the use of some type of baseline restoration. Increasingly, active baseline restoration is included as a standard feature in the design of modern linear amplifiers.

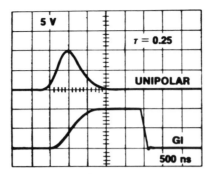

Figure 17-23 Waveforms in the operation of a gated integrator. The unipolar output of a standard Gaussian shaping network is integrated in an active circuit to give the waveform shown in the lower plot. (From Britton et al.[8])

Performance at high rates can also benefit from use of a *gated integrator* at the output stage of the amplifier shaping networks. Its function is outlined in Fig. 17-23. The unipolar pulse from a Gaussian shaping network is integrated on a capacitance that is part of an active circuit. The integration is continued for a time interval that is typically 8–10 times the shaping time used in the Gaussian network (2–3 times the peaking time). If an additional event occurs during this period, the process is aborted by the pileup rejection circuits in the amplifier. At the end of this interval, the capacitance is abruptly discharged by closing a switch. The resulting pulse amplitude is now proportional to the area rather than the peak value of the shaped pulse and is therefore much less sensitive to changes in ballistic deficit caused by variable charge collection times in the detector. Consequently, the shaping times can be made much shorter than could be tolerated without the gated integrator to minimize the deleterious effects of pileup on energy

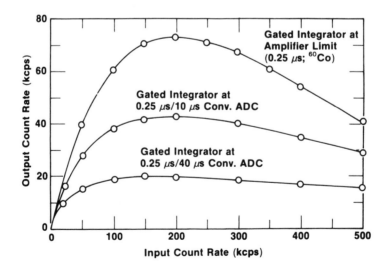

Figure 17-24 An example of the throughput of a pulse-processing system at high input rates. In this case, a gated integrator allows the use of a short (0.25 μs) Gaussian shaping time. The top curve is for the amplifier alone. The remaining plots show the additional losses due to the conversion time of the ADC of the multichannel analyzer used to record the pulses. (From Britton et al.[8])

resolution. An example of the high-rate performance of a specific system is shown in Fig. 17-24. Here the pileup rejection causes the output rate to pass through a maximum as a function of input rate (see Fig. 17-21). Pulse-processing systems employing fast pileup rejection and gated integrators to minimize shaping times have proved useful in limiting the deterioration in energy resolution from germanium gamma-ray detectors for input rates as high as 10^6 pulses/s.

The response of a linear amplifier to very large overloading pulses can also be an important factor in determining its overall performance, especially at high rates. The overload recovery time is usually included in a list of amplifier specifications. The output pulse shape may return quickly to zero for normal pulses, but strongly saturating pulses can substantially extend the time required to reestablish true zero baseline. Even for bipolar shaped pulses, the two lobes may not saturate in a symmetric way. Then equal areas above and below the zero level are no longer maintained, and some baseline shift must occur. Thus, active baseline restoration can be beneficial for high-rate operation with bipolar as well as unipolar pulse shapes.

In summary, a linear amplifier is subject to performance criteria that often require trade-offs depending on the particular detector in use and the details of its application. Important amplifier functions include:

1. Signal amplification.

2. Pulse shaping that promotes accurate measurement of collected charge with maximum immunity from ballistic deficit and variations in shape of the current–time profile. (This consideration favors long shaping times.)

3. Pulse shaping to minimize pileup and overload and therefore maximize performance at high counting rates. (This factor argues for short shaping times.)

4. Pulse shaping to provide for best signal-to-noise ratio in the analysis of individual pulse amplitude. (This criterion favors the shaping methods listed in Fig. 17-14 that show the best noise immunity.)

5. For demanding applications, the provision of active circuits for pileup rejection and baseline restoration to help preserve performance at high rates.

F. Biased Amplifier

The function of a biased amplifier is outlined in Fig. 17-25. The input is usually a shaped linear pulse as produced at the output of a normal linear amplifier. A biased amplifier, however, produces no output pulse for any input pulse that lies below a given *bias level*, which normally can be adjusted by a front panel control. The entire span of the biased amplifier output is assigned to the range of input pulses between the bias level and the specified maximum input level. The effect is to magnify this limited range of input amplitudes by spreading it out over the entire output range. This function is especially useful if a small region of a pulse height spectrum must be examined in great detail. The small region can then be spread over the total number of amplitude channels available in the multichannel analyzer.

Because the biased amplifier responds only to that portion of the input pulse which lies above the bias level, the output pulse is usually significantly narrower than the input pulse width. Consequently, some cases may require the use of a pulse stretcher (see below) to restore the pulse width necessary for subsequent amplitude measurement.

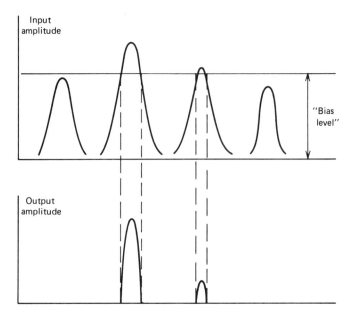

Figure 17-25   The function of a biased amplifier is to provide linear amplification for only that portion of each input pulse which lies above the bias level.

G. Sum or Difference Amplifier

In some measurements it may be advantageous to sum the outputs of several detectors to provide a single pulse for later analysis. Sum amplifiers, which can accept multiple inputs (usually the shaped outputs of linear amplifiers) and provide an output pulse that is the algebraic sum of all coincident inputs, are commercially available. A sum amplifier can be converted into a difference amplifier by inverting one of the pulses before the summation.

H. Linear Gate

In principle, a linear gate functions as a switch for linear pulses. When the gate is open, linear pulses pass through the gate without change or distortion. While the gate is closed, all pulses are blocked. The opening of the gate is controlled by the presence or absence of a gate pulse at a second input at the time the linear pulse is presented to the linear input. The gate pulse is a standard logic pulse whose width is usually adjustable and chosen to be somewhat longer than the linear pulse. Proper time overlap may require delay of the linear pulse to allow the gate pulse to arrive first. The *pedestal* of a gate is an undesirable dc level which can be added to the linear pulse in the process of passing though the gate. Ideal linear gates should have zero pedestal, or a means should be provided for the adjustment of the pedestal to zero under experimental conditions. Other desirable properties are low feedthrough under closed gate conditions, low nonlinearity, and fast switching between open and closed states.

I. Pulse Stretcher

A stretcher unit produces a linear pulse output whose shape is set to a standard compatible with following components, and whose amplitude is equal to the maximum amplitude of the input pulse. Its most common application is to accept fast linear pulses or those that have been shaped with a sharply peaked waveform and convert them into a nearly flat-topped pulse of equal amplitude, which suits the ADC requirements of most multichannel analyzers.

VII. SYSTEMS INVOLVING PULSE TIMING

In a large number of applications, information on the precise arrival time of a quantum of radiation in the detector is of particular interest. When timing information is the major purpose, detector pulses are often handled quite differently than when accurate pulse height measurement is the object. The accuracy with which timing can be performed depends both on the properties of the specific detector and the type of electronics used to process the signal. The best timing performance is obtained for the fastest detectors, that is, those in which the signal charge is collected most rapidly. For detectors with equal charge collection time, those that generate the greatest number of information carriers (ion pairs, electron–hole pairs, etc.) per pulse will be influenced least by the "graininess" of the signal and therefore will demonstrate superior timing properties. The timing characteristics of a given system depend greatly on the dynamic range (ratio of maximum to minimum pulse height) of the signal pulses. If all the signal pulses are confined to a narrow range of amplitudes, many different timing schemes can give good results. If the pulse amplitude must cover a wide range, however, some sacrifice in timing accuracy nearly always results.

A. Time Pick-Off Methods

The most fundamental operation in timing measurements is the generation of a logic pulse whose leading edge indicates the time of occurrence of an input linear pulse. Electronic devices that carry out this function are called *time pick-off* units or *triggers*.

Factors that lead to some degree of uncertainty in deriving the timing signal are always present. Sources of timing inaccuracy are conveniently divided into two categories. Those that apply when the input pulse amplitude is constant are usually called sources of *time jitter*, whereas those effects that derive primarily from the variable amplitudes of input pulses are grouped together in a category called *amplitude walk* or *time slewing*. Relative importance of these two categories depends on the dynamic range expected in the input pulse amplitude. The best timing performance will be achieved if the input pulses are confined to a very narrow range in amplitude, because then only sources of time jitter contribute to uncertainty. More often, however, practical applications require that pulses of different amplitudes be processed, and the additional contribution of walk will worsen the overall time resolution of the system.

An important source of time jitter is the random fluctuations in the signal pulse size and shape. These fluctuations can arise from several sources. One is the electronic noise added by components that process the linear pulse prior to the time pick-off. Another is the discrete nature of the electronic signal as generated in the detector. When the number of information carriers that make up the signal is low, statistical fluctuations in their number and time of occurrence will also be reflected in the size and shape fluctuations of the pulse. This effect will obviously be greatest for small-amplitude pulses and for detectors that generate relatively few information carriers such as scintillation counters.

1. LEADING EDGE TRIGGERING

The easiest and most direct time pick-off method is to sense the time that the pulse crosses a fixed discrimination level. Such *leading edge* timing methods are in common use

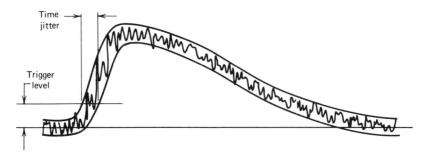

Figure 17-26   The time jitter in leading edge triggering arising from random noise. An envelope is shown of many repeated signal pulses of the same amplitude and shape, but with a random contribution of noise.

and can be quite effective, especially in situations in which the dynamic range of the input pulses is not large.

The effect of time jitter on leading edge timing is shown in Fig. 17-26. The random fluctuations superimposed on signal pulses of identical size and shape may cause the generation of an output logic pulse at somewhat different times with respect to the centroid of the pulse. The timing errors will be approximately symmetrical and will increase if the slope of the leading edge of the pulse is decreased.

The amplitude walk associated with a leading edge trigger is graphically demonstrated in Fig. 17-27. The two pulses shown have identical true time of origin but give rise to output logic pulses that differ substantially in their timing. Under extreme conditions, the amplitude walk can amount to the full rise time of the input pulse and is often unacceptable in situations where accurate timing is needed over a wide amplitude range.

Even if the input amplitude is constant, walk can still take place if changes occur in the shape of the pulse. Detectors with variable charge collection time, such as germanium detectors, produce output pulses with variable rise time as illustrated in Fig. 17-28. Changes that occur in the pulse shape before the discrimination point will affect the timing and can constitute another source of timing uncertainty. The sensitivity of leading edge triggering to timing walk due to amplitude and shape variations is minimized by setting the discrimination level as low as possible. However, the discrimination point should be in a region of steep slope on the pulse leading edge to minimize uncertainties due to jitter. Practical compromises in these somewhat conflicting requirements often lead

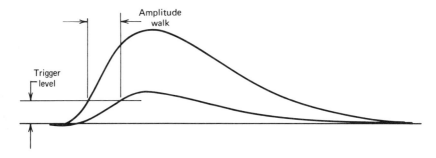

Figure 17-27   Amplitude walk in leading edge triggering. Two pulses with identical shape and time of occurrence but different amplitude are seen to cross the trigger level at different times.

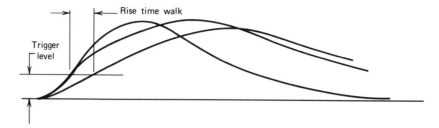

Figure 17-28 Time walk in leading edge triggering due to variations in pulse shape or rise time. Changes in either the overall pulse rise time or the shape of the leading edge for pulses with the same rise time can be responsible for time walk.

to optimum time resolution for levels that are set at about 10–20% of the average pulse amplitude.

2. CROSSOVER TIMING

When leading edge triggering is applied to pulses with a wide amplitude range, amplitude walk often results in large timing uncertainties. Another time pick-off method, known as *crossover timing*, can greatly reduce the magnitude of amplitude walk but requires the input pulse to have bipolar shape. Figure 17-29 shows two pulses with different amplitudes as shaped by a *CR-RC-CR* double differentiating network. Although the pulse amplitudes differ widely, the time at which the waveform crosses from the positive to the negative side of the axis is theoretically independent of the amplitude and depends only on the shaping time constant chosen for the network. Double delay line shaping produces the same effect and is usually preferred in crossover timing. Compared with leading edge timing, crossover methods greatly reduce amplitude walk, but only at the expense of increased jitter. The noise introduced by the required shaping stage and the increased susceptibility of the crossover point to statistical fluctuations in the signal contribute to the increased jitter.

A related method known as *fast crossover* can be implemented for the specific case of fast scintillator–photomultiplier tube signals. A fast bipolar pulse is produced directly at the PM tube anode using a "tee" connection and a shorted length of coaxial cable. This *shorted stub* technique is described in Chapter 16, and the resulting waveform is shown in

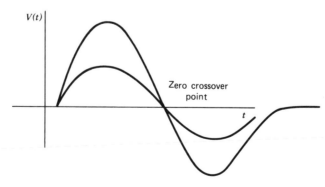

Figure 17-29 Bipolar pulses of different amplitude showing the same time of zero crossover.

Fig. 16-20a. Because the time to the zero crossover point is determined by the length of the shorted cable, it is also independent of the signal pulse amplitude. However, the pulse shape must not vary if this freedom from walk is to be preserved.

3. CONSTANT FRACTION TIMING

If the input dynamic range is small, the lower jitter of leading edge timing results in superior timing performance compared with crossover timing. Furthermore, it is empirically found that the best leading edge timing characteristics are obtained when the timing discriminator is set at about 10–20% of the pulse amplitude. These observations have led to the development[36] of a time pick-off method that produces an output signal a fixed time after the leading edge of the pulse has reached a *constant fraction* of the peak pulse amplitude. This point is then independent of pulse amplitude for all pulses of constant shape. Therefore, pulses over a wide dynamic range can be accepted with much the same freedom from amplitude walk as exhibited by crossover timing, but with lower jitter. The electronic shaping steps required to carry out constant fraction timing are diagrammed in Fig. 17-30. The process involves taking the preamplifier output [part (a)]

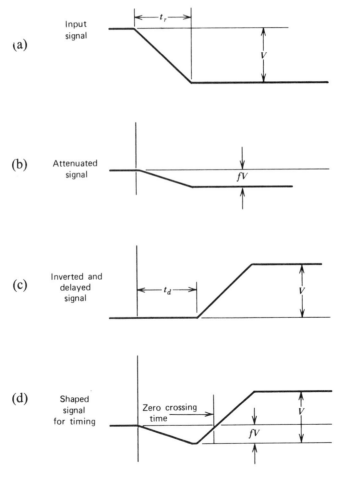

Figure 17-30 Waveforms in the constant fraction time pick-off method. For clarity, only the leading edge of the pulse is shown.

and multiplying it by the fraction f which is to correspond to the desired timing fraction of full amplitude. The input waveform is also inverted and delayed for a time greater than the pulse rise time to give the waveform shown in part (c). The sum of waveforms (b) and (c) is then taken to give the pulse that is sketched in part (d). The time that this pulse crosses the zero axis is independent of the pulse amplitude and corresponds to the time at which the pulse reaches the fraction f of its final amplitude. Detailed descriptions of specific circuits and their use in several timing applications can be found in Refs. 37–40.

4. AMPLITUDE AND RISE TIME COMPENSATED (ARC) TIMING

In situations where the shape or rise time of the pulse can change, even constant fraction timing methods cannot eliminate walk. To meet the need for accurate timing for germanium detectors, in which the rise time variations are large, a system known as ARC timing, first described by Chase,[41] has gained considerable popularity.[42] The scheme amounts to basing the timing signal on a fixed fraction of only the early portion of the

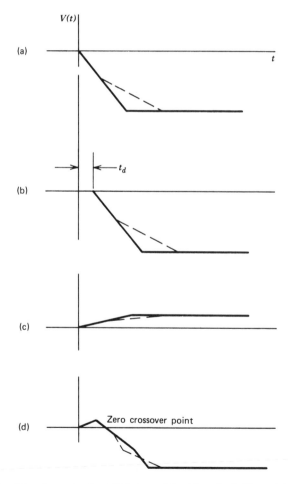

Figure 17-31 Waveforms in the ARC time pick-off method. The original waveform (a) is delayed (b), inverted, and attenuated (c). The sum of (b) and (c) give waveform (d) with a zero crossover point used to generate the timing signal. The dashed waveform has a slower rise time than the solid, but both give the same timing signal.

pulses and is therefore unaffected by shape changes that may occur at a late point in the waveform.

The pulse-processing methods are similar to those used in constant fraction timing and are shown in Fig. 17-31. The only difference is that the delay time illustrated in the figure is chosen to be as small as practical (of the order 5–10 ns). The delay must be sufficiently long so that the initial portion of the output pulse is allowed to significantly exceed the noise level, so that a simple discriminator signal can be generated which will actuate the circuits that sense the crossing of the waveform back across zero. If we use this method, the time of zero crossing is given by $t_d/(1 - f)$, provided that the input pulse rises linearly to that point. Therefore, this crossover time again occurs at a constant fraction of the amplitude of all input pulses and will furthermore be insensitive to any slope change in the rise of the input pulse which occurs after the zero crossing point.

5. EXTRAPOLATED LEADING EDGE (ELET) TIMING

As an alternative to ARC timing, an extrapolation principle has also been applied successfully to the almost walk-free derivation of timing signals from germanium detectors with variable rise time.[43] The method is similar to ARC timing in that it is based only on the initial portion of the pulse rise and is not affected by shape variations that may follow that initial portion. ELET timing uses two independent discriminators set at different discrimination levels to carry out an extrapolation of the pulse waveform back to its time of origin. For purposes of illustration, assume that the second discrimination level is set at twice the level of the first. For pulses with a linear rise, the time difference between the discrimination points should be equal to the time delay between the true start of the pulse and the time at which the first discriminator is crossed. The technique employs time-to-amplitude converters to carry out an effective extrapolation back to zero, so that a timing pulse corresponding to the true time of origin of the input pulse is generated. The two discrimination levels are ideally set as low as possible, so that the assumption of linearity need be met for only a short portion of the leading edge of the pulse.

6. FIRST PHOTOELECTRON (FPET) TIMING

A unique method for providing timing information from scintillation detectors can be applied when the conditions of the experiment limit the possible triggering time to a relatively small time interval. If this interval is small compared with the average time spacing between photomultiplier tube noise pulses, excellent timing can be carried out simply by sensing the arrival of the photomultiplier signal corresponding to a single photoelectron. This is, in effect, leading edge timing with a trigger level that is as low as physically possible. Because photomultiplier tube noise would also trigger at this level, the probability of a noise pulse within the time range of interest must remain small. This condition has been made practical by low-noise bialkali photomultiplier tubes in which noise rates may be as low as tens or hundreds per second. The use of FPET was pioneered by Lynch[44] and is particularly advantageous for timing from scintillators when low-energy gamma rays are involved. Further studies of FPET with sodium iodide scintillators are given in Ref. 45.

7. COMPARISON OF TIME PICK-OFF METHODS

As a general rule, leading edge triggering gives the best time resolution for signal pulses whose amplitudes are restricted to a narrow range and whose shape characteristics do not vary. When pulses with a wide range of amplitude are processed, leading edge methods show large amplitude walk. Constant fraction timing methods are very effective at reducing amplitude walk when the pulse shape does not change, and this technique has largely replaced crossover timing originally introduced for the same purpose. For example, scintillation detectors produce pulses of fixed shape when a given type of radiation is involved. A useful comparison of leading edge and constant fraction pick-off techniques applied in timing measurements using scintillators can be found in Ref. 46. For detectors that produce pulses with variable rise time (such as germanium detectors), both ARC and ELET timing methods can be used to good advantage.[47-49] Alternatively, more elaborate methods that sense the rise time (often using two independent discriminators set at different levels) can be employed to make corrections in the timing signal[50,51] to reduce the influence of variable rise time.

The time resolution achievable with any of these pick-off methods varies greatly with detector type and is determined primarily by the characteristics of the charge collection process that are reflected in the output pulse rise time. A discussion of the time behavior of each major detector type is included with its description given in other chapters. The best timing is attainable from detectors with fast and nonvariable rise time. Among detectors with equal rise time, those producing the largest signal will generally result in better timing performance. For example, plastic scintillators can produce a best time resolution of around 100 ps (defined as the FWHM of the prompt coincidence peak, as described in the following section) for large amplitude pulses, but the resolution will worsen for smaller pulses. Slower scintillators such as NaI(T1) and BGO can show time resolution values of 1–2 ns if pulses of nearly constant amplitude are involved but will have poor timing performance if pulses covering a broad range of amplitude are processed. One study[52] of 14 different germanium detector systems measured timing resolutions ranging from about 2 to 10 ns using a version of the constant fraction triggering technique. Gas-filled detectors generally lead to poorer timing resolution because of the slower charge collection and variability of the pulse rise time.

B. Measurement of Timing Properties

1. MULTICHANNEL TIME SPECTROSCOPY

We begin our discussion of pulse timing systems by considering some basic concepts of time interval measurement. The time-to-amplitude converter (TAC) is a device (discussed in greater detail later in this chapter) that produces an output pulse with an amplitude proportional to the time interval between input "start" and "stop" pulses. The differential amplitude distribution of the output pulses as recorded by a multichannel analyzer is thus a measure of the distribution of time intervals between start and stop pulses and is often called a *time spectrum.*

The time spectrum bears a close relation to the pulse height spectrum (or differential pulse height distribution) discussed in Chapter 4. The abscissa, rather than pulse height is the time interval length T. The ordinate is dN/dT, the differential number of intervals whose length lies within dT about the value T. The area under the time spectrum between any two limits is the number of intervals of length between the same limits. The discrete approximation to the ordinate, $\Delta N/\Delta T$, is the number of intervals of length within a finite increment ΔT, normalized to the length of that increment. During the discussion that follows, it is also convenient to divide the number of recorded intervals by the measurement time to obtain the corresponding rate. The ordinate then becomes dr/dT for the time spectrum, to be interpreted as the differential rate of occurrence of intervals whose length lies within dT about T.

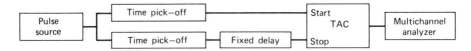

Figure 17-32 A simplified system for recording multichannel time spectra from a split common source.

A simple system for recording a time spectrum is shown in Fig. 17-32. For the sake of illustration we assume that a single source of pulses such as a radiation detector provides a split output that is simultaneously sent down two signal branches. A time pick-off unit in each branch provides timing logic pulses that are supplied to the start and stop inputs of a TAC. A fixed time delay is also present in the stop branch. The multichannel analyzer (MCA) records the number of time intervals that lie within many contiguous increments or channels of width ΔT, ranging from zero to some maximum interval length set by the range of the TAC. The maximum time range of the TAC is assumed to be small compared with the average spacing between signal pulses, so that the probability of more than one signal pulse per TAC interval is small. Alternatively, the output of a pulse generator may be substituted for the random pulse source and adjusted to have a period between pulses which is larger than the TAC maximum range.

Under these conditions, the time spectrum recorded by the MCA is exceedingly simple. In the absence of time jitter or walk, each timing pulse is produced at precisely the same time in each branch, and therefore the start and stop pulses are always separated by the fixed delay value. The TAC always produces a constant-amplitude output that is therefore stored in a single channel of the MCA. If the delay is made smaller, the interval between start and stop pulses also decreases, producing a TAC pulse of smaller amplitude which will be stored in a lower-numbered channel.

A more realistic time spectrum shown in Fig. 17-33, takes into account the fact that any time pick-off method will always involve some degree of jitter and time walk. If these timing uncertainties in each branch are independent of each other, the time spectrum will display a distribution as shown in the sketch. Because the time pick-off in each branch is likely to behave similarly with regard to time walk due to amplitude or shape variations, most of the distribution spread measured in this conceptual experiment would be caused

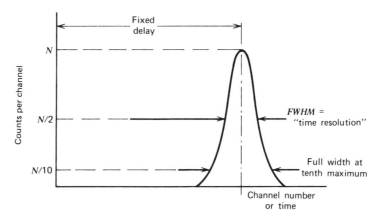

Figure 17-33 The multichannel time spectrum for the system of Fig. 17-32.

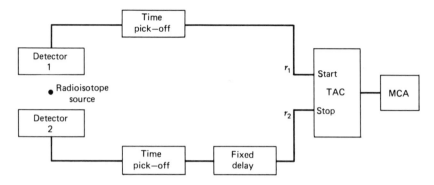

Figure 17-34 A simplified system to record multichannel time spectra from a radio-isotope source emitting coincident radiation.

by those random sources of uncertainty or jitter which are independently generated in each branch.

The full width at half maximum of the time distribution is often used as a measure of the overall timing uncertainty in the measurement system and is called the *time resolution*. Another widely quoted specification is the full width at $\frac{1}{10}$ maximum, which more fairly accounts for tails sometimes observed at either extreme of the distribution. The recording of the time interval distribution in the manner just described is often called *multichannel time spectroscopy*.

2. PROMPT AND CHANCE COINCIDENCE SPECTRA

We now imagine the system to be reconfigured as shown in Fig. 17-34. Two independent detectors are irradiated by a common radioisotope source which is assumed to emit at least two detectable quanta in *true coincidence*; that is, both radiations arise from the same nuclear event within the source. We further assume that for all true coincidences, the nuclear decay scheme is such that there is no appreciable time delay between the emission of both radiations.

The time spectrum taken under these conditions will have the general appearance shown in Fig. 17-35. Some fraction of all the true coincidence events will give rise to radiations that are detected simultaneously in both detectors. These true coincidence

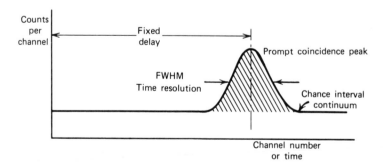

Figure 17-35 The multichannel time spectrum for a radioisotope source emitting some radiation in prompt coincidence. The cross-hatched area gives the total number of recorded coincident events. The time resolution of the system is conventionally defined as the FWHM of the prompt coincidence peak.

events will appear in the same region of the time spectrum as did the split pulse output in the previous example, producing a *prompt coincidence* peak. If there were no delay difference between the two branches, this peak would be centered about ·time zero, and therefore only about half of its shape would be measured. Introducing the fixed delay into the stop channel moves the entire time spectrum to the right by an amount equal to the delay and allows both sides of the prompt coincidence peak to be recorded. The area under the peak, after subtraction of the continuum discussed later in this section, gives the total number of detected coincidences.

The width of the prompt distribution is likely to be somewhat greater than in the previous example because separate detectors are now involved. Amplitude walk, for example, will no longer be identical in both branches and therefore will also contribute to the width of the distribution together with the time jitter. If detectors, timing electronics, and triggering conditions are nearly identical in both branches, then all sources of time jitter and walk should be symmetric. Under these conditions the prompt coincidence peak should also be symmetric with a width that indicates the total contribution of all sources of time uncertainty. If one branch differs substantially from the other, then asymmetric prompt coincidence peaks will often result. For example, the effect of amplitude walk in leading edge triggering is to produce a small number of timing pulses that occur substantially later than the majority. If a system subject to amplitude walk is used in the stop branch in conjuction with a system with little time uncertainty in the start branch, the measured prompt coincidence distribution will have an asymmetric tail in the direction of longer intervals.

In addition to the coincident events, each detector will typically produce a much larger number of pulses that correspond to the detection of one quanta for which there may not be a corresponding coincident emission, or for which the coincident radiation escapes detection in the opposite detector. For these events there can be no true coincidence. Because of their random distribution in time, however, some sequences will occur by chance in which a stop pulse will be generated within the TAC time range following an unrelated start pulse. These events are called *chance intervals*, and their intensity depends on the rates at which pulses are generated in either branch leading to the TAC. If these rates, often called *singles rates*, are not high compared with the reciprocal of the time range of the TAC, the chance interval distribution will be uniform over the entire time range, as shown in Fig. 17-35.

The amplitude of the chance distribution can be derived as follows: Let r_1 and r_2 represent the rates of arrival of uncorrelated start and stop pulses, respectively, at the TAC inputs. For typical applications, r_1 and r_2 will be much larger than the true coincidence rate and thus are essentially equal to the singles rate in either branch. (If r_1 is not small compared with the inverse of the TAC time range, account must be taken of those start pulses that are lost because the TAC is busy when the start pulse arrives. We ignore such losses here.) After each start, the probability that an interval of length T will elapse without a stop pulse is simply e^{-Tr_2}. The differential probability of a stop pulse arriving within the next differential time dT is just $r_2\,dT$. Because both independent events must occur, the overall differential probability of generating an interval within dT about T is simply $r_2 e^{-Tr_2}\,dT$. The differential rate of these intervals is then simply the product of the rate of arrival of start pulses multiplied by this probability, or $r_1 r_2 e^{-Tr_2}\,dT$. Now as long as r_2 is not large compared with the reciprocal of the TAC time range, $r_2 T$ will be small and the exponential can be approximated by unity. The differential distribution dr/dT is then constant and equal to $r_1 r_2$. If the output of the TAC is recorded by an MCA with a time width of ΔT per channel, then the chance interval rate per channel is simply $r_1 r_2 \Delta T$.

In a multichannel time spectroscopy measurement, one would normally like to maximize the true coincidence peak compared with the chance interval background. For equal areas, peaks with the narrowest width will be most prominent, so that improvements in the timing accuracy of either branch, which diminish the overall timing uncertainty, are always beneficial. Other experimental factors can also help improve the ratio of the prompt peak to the chance continuum. For example, use of energy selection criteria in each branch can limit r_1 and r_2 by discarding any events that cannot correspond to true coincidences. The chance continuum will therefore be reduced without affecting the area under the true coincidence distribution. The true rate also scales linearly with source activity, whereas the chance rate is proportional to the product of r_1 and r_2 or the square of the source activity. Consequently, using as low a source activity as permitted by reasonable counting statistics will also enhance the prominence of the true coincidence peak. Varying the rates by changing the counting geometry, however, affects both the true and chance rates by an equal factor.

3. MEASUREMENTS USING A COINCIDENCE UNIT

An equivalent single-channel method is available to carry out the type of time spectroscopy just described, but at the price of a somewhat increased total measurement time. The single-channel method, in effect, consists of setting a time window to accept only those sequences in which the interval between start and stop pulses lies within a narrow band. The situation is somewhat analogous to pulse height analysis, where both single-channel and multichannel methods are used. As discussed earlier in this chapter, the unit used to perform narrow band pulse height selection is called a single-channel analyzer.

For time spectroscopy, the *coincidence unit* performs the equivalent function and selects from all intervals only those for which the time difference between inputs is less than a circuit parameter, known as the *resolving time*. A coincidence unit, in its simplest form, has two identical logic inputs. Whereas the TAC must receive a start pulse and stop pulse within the time range in that specific order, a coincidence unit will produce a logic output if pulses arrive at either input within the resolving time of a second pulse at the opposite input. The order of the arrival is not significant.

The system shown in Fig. 17-36 illustrates the application of a coincidence unit to time spectroscopy. The fixed delay is assumed to be the same as in the previous example, and a variable delay has been inserted into the opposite branch. We initially assume that the coincidence unit is set with a resolving time equal to one-half the time width of one channel in the multichannel time spectroscopy example. We now record the rate from the coincidence unit as a function of the variable delay value. If the delay is set to a value that corresponds to a time interval at the midpoint of one of the channels in the multichannel

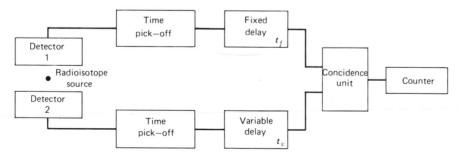

Figure 17-36 A simplified system to record coincidence-delay curves from a radioisotope source emitting coincident radiation.

spectrum, then the rate measured by the coincidence unit will be exactly the same as the rate recorded by that specific channel. Under these conditions, the coincidence unit produces an output for predelay intervals that range from $(t_v - t_f - \tau)$ to $(t_v - t_f + \tau)$ where t_f and t_v are the fixed and variable delay times, and τ is the coincidence resolving time. If a series of measurements is now made in which the coincident rate is measured as t_v is varied in increments of 2τ, then a curve that is exactly equivalent to the multichannel time spectrum of Fig. 17-35 will be generated by plotting this rate versus the delay setting. This type of plot is often called a *coincidence-delay curve* and can be used to reproduce the multichannel time spectrum as just described, provided the resolving time τ is small compared with the overall time resolution of the system.

In common coincidence measurements, however, the object is not to map out fully the time interval spectrum but rather simply to record the number of true coincidence events. The coincidence resolving time is therefore chosen to be larger than the system time resolution, so that the acceptance time window can fully encompass all true coincidences.

Figure 17-37 graphically illustrates the interrelation between the measured coincidence rate and the differential distribution dr/dT versus T for the case of a relatively large resolving time τ. This differential distribution is the same as the multichannel spectrum of Fig. 17-35, normalized by the measurement time so that the distribution now is in units of rate. At a given delay setting t_v the measured coincidence rate corresponds to the area under the differential spectrum between the time window limits of $t_v - \tau$ and $t_v + \tau$.

Ideally, coincidence measurements are carried out with delay t_v adjusted to the point indicated as ① on the figure. Here, the acceptance time window is centered around the prompt coincidence peak in the spectrum. The measured coincidence rate corresponds to the cross-hatched area and consists of two additive terms. The *true coincidence rate* corresponds to the net area under the prompt coincidence peak, whereas the *chance coincidence rate* corresponds to the area of the flat chance continuum on which the peak is superimposed.

If the delay is changed to a point well away from the prompt peak (such as point ② on Fig. 17-37), only chance coincidences are measured. The rate can be deduced from the results of the previous section, where it was shown that the amplitude of the differential distribution dr/dT for chance intervals is the product of the two singles rates $r_1 r_2$. Therefore, the area under the differential spectrum in the chance continuum region is this amplitude multiplied by the width of the time window 2τ. The general result for any twofold coincidence unit is therefore that the chance coincidence rate from uncorrelated inputs at rates r_1 and r_2 is given by

$$r_{ch} = 2\tau r_1 r_2 \qquad (17\text{-}21)$$

This chance contribution must be subtracted from the rate measured at point ① to derive the net true coincidence rate.

The remainder of the coincidence-delay curve shown at the bottom of Fig. 17-37 can be traced out by measuring the total coincidence counting rate as t_v is varied. This is a very useful calibration procedure during the initial setup of a standard coincidence measurement. Starting at ①, assume that the delay is changed very gradually. For a while nothing changes because the amount of area under the chance distribution that is lost by moving one limit of the acceptance window is made up by that area gained at the opposite limit. However, when the delay is changed sufficiently so that some portion of

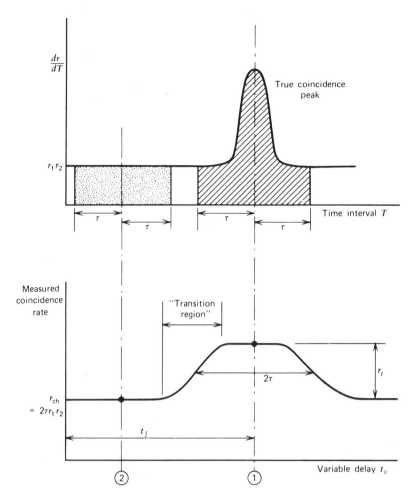

Figure 17-37 Relation between the differential time spectrum (upper plot) and the coincidence-delay curve recorded with the system of Fig. 17-36 (lower plot). The recorded coincidence rate at any specific delay value t_v is equal to the area under the differential spectrum between the limits $t_v - \tau$ and $t_v + \tau$.

the true coincidence peak begins to be lost, the measured coincidence rate starts to drop off. When the delay has been changed sufficiently, the entire peak is excluded and only the chance rate remains. The transition region at either side extends over a range of delay equal to the full width (at its base) of the true coincidence peak. The midpoint of a transition region corresponds to the delay setting at which one edge of the time window exactly bisects the prompt coincidence peak. The time difference between these two midpoints (as illustrated on the figure) is therefore a measure of the time window width and is equal to 2τ, twice the coincidence unit resolving time.

Coincidence-delay curves generated by assuming different values for τ are shown in Fig. 17-38. The minimum value of τ for which all true coincidences can be recorded is half the total width (at its base) of the true coincidence peak. The curve labeled C in the figure corresponds to this case. With the resolving time set to this minimum, only one specific value of the delay will lead to counting all true coincidences. Should any time

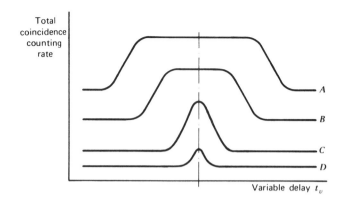

Figure 17-38 Coincidence-delay curves for different values of the coincidence resolving time τ. Curve A corresponds to the largest value of τ, curve D to the smallest. Values of τ larger than half the width (at its base) of the prompt coincidence peak in the differential time spectrum of Fig. 17-37 lead to the flat-topped plateau shape of curves A and B. In curve C, τ is just equal to this value and the full true coincidence rate is obtained only at one specific delay value. In curve D, τ is too small to obtain all the true coincidences at any delay value.

drift occur in either branch, true coincidence events would begin to be lost. Therefore, in most coincidence measurements one would like to have the resolving time τ somewhat larger than this minimum to allow some leeway for such drifts or other timing changes. On the other hand, the chance coincidence rate increases linearly with τ, and the resolving time should be kept as small as possible to maximize the true-to-chance coincidence ratio. The usual compromise between these conflicting considerations is one in which τ is chosen to be several times the system time resolution as in curve B. The coincidence-delay curve then has a flat top or plateau, which represents the range in delay settings that can be tolerated without losing true coincidences. In the initial calibration of the coincidence system, delays are then adjusted to choose an operating point near the middle of this plateau. Because of slight differences in the inherent delay of pulse-processing components, this point may correspond to some apparent delay difference between the two branches.

In practice, the fixed delay t_f in Fig. 17-36 is usually omitted and the coincidence-delay curve is obtained by using only the variable delay t_v. The curve is then approximately centered about zero, and only the right half is recorded with t_v in the lower signal branch. The remainder of the curve is then obtained by shifting t_v to the upper branch, where its value represents "negative delay."

Coincidence measurements are not necessarily confined to two signal branches but can, in general, involve inspecting any number of signals for true coincidence. For such multiple inputs, all signals must arrive within a total time interval corresponding to the resolving time of the unit for an output logic pulse to be produced. The measured rate will again be a mixture of true and chance coincidences, but correction for the latter is more difficult, as discussed in the following section.

4. CHANCE COINCIDENCE CORRECTIONS

In standard coincidence measurements it is essential to correct the recorded coincidence rate for the contribution due to chance events to derive the rate due to true coincidences alone. For twofold coincidence systems, Eq. (17-21) allows calculation of the chance coincidence contribution if each singles rate is measured and the resolving time of the coincidence unit is known. Alternatively, the chance rates can be measured directly by temporarily inserting a large delay in either branch of the system so that the true coincidence peak occurs well away from the acceptance time window of the coincidence unit. The latter approach is usually preferable because it can be applied more easily to situations in which the singles counting rate may not be constant over the period of the measurement. Several electronic schemes have been suggested[53-55] for simultaneously measuring the total coincidence and chance coincidence rates throughout the course of the measurement by periodically spoiling the true coincidences in a similar manner.

In multiple coincidence systems, it is usually not possible to make an analytic correction for chance contributions. Instead, supplemental measurements must be carried out in which the separate twofold coincidences between various inputs are individually determined and applied in a more complex analysis, such as that given in Refs. 56 and 57. For the example of a triple coincidence unit, the chance rate from totally uncorrelated inputs can be calculated from the resolving time and individual singles rates as $3\tau^2 r_1 r_2 r_3$. If this were the only source of chance events, the correction could be made analytically as in the case of twofold coincidences. However, the added complexity arises due to partially correlated events in which a true coincidence in two branches is accompanied by a chance event in the third branch within the system resolving time. Because the probability of these events depends on the specifics of each experiment, they must, in general, be experimentally evaluated for each case.

5. DETERMINATION OF COINCIDENCE RESOLVING TIME

Several methods are available to the experimenter for the determination of the resolving time of a coincidence circuit. One is to provide totally uncorrelated inputs and to measure simultaneously both the singles rates and the resulting chance coincidence rate. The resolving time can then be calculated from Eq. (17-21). In setting up such a measurement, care must be taken to exclude any possibility of true coincidences between the two branches. The source must be incapable of generating coincident radiations that can interact in both detectors, or alternatively, separated sources with adequate shielding should be used to prevent true coincidences. Consideration should also be given to the possibility that true coincidences may arise from the scattering of radiation from one detector to the other.

The resolving time can also be measured by recording the coincidence-delay curve. As shown in Fig. 17-37, the FWHM of the portion of the curve that corresponds only to true coincidences is equal to twice the resolving time. For this method to be practical, a source with a sufficiently high probability of true coincidence emission must be used to ensure that the true coincidence rate stands out well above the chance background.

6. DELAYED COINCIDENCE AND OTHER INTERVAL MEASUREMENTS

In the previous example, the time spectrum was illustrated for a source that emitted prompt coincidence radiation. One way of defining prompt coincidence is to include any events that are separated by a delay time that is small compared with the instrumental time resolution. There are many occasions in which radiations are emitted in the same nuclear decay but are separated in time due to an intermediate nuclear state of finite lifetime. The time distribution should then show an exponential tail to the right of the prompt peak shown previously in Fig. 17-35. By measuring the time constant of this tail, the decay constant of the intermediate state can be deduced. These measurements can be carried out either with the multichannel technique using a TAC or as a series of single-channel measurements using a coincidence unit and variable delay. The latter method has historically been known as the *delayed coincidence* technique, and before the days of multichannel time spectroscopy was widely used to measure time interval distributions.

Multichannel time spectroscopy can be applied quite generally to any situation in which a time interval is to be measured. For example, in neutron time-of-flight spectroscopy, the start pulse is supplied from a detector that senses the time at which the neutron is generated, whereas the stop pulse is taken from a detector in which the neutron interacts after traveling some distance. The time interval between these two events is then a measure of the flight time of the neutron from which its energy can be calculated. There are many other examples of physical measurements in which the time interval distribution is important, and methods originally developed for nuclear measurements have been applied to a large assortment of determinations in other scientific fields.

7. MEASUREMENT OF ABSOLUTE SOURCE ACTIVITY USING COINCIDENCE TECHNIQUES

As an illustration of one of the common applications of coincidence circuits, we consider the problem of measuring the absolute activity of a given radioisotope source. If the source emits two coincident radiations that are distinguishable, methods can be applied which eliminate the need to know absolute detector efficiencies in order to calculate the source activity. Because detector efficiencies are often uncertain and hard to determine, use of these methods can provide a more accurate means of source activity determination than otherwise available.

Assume that the source activity (disintegrations per second) is given by S. Further assume that each such disintegration gives rise to two coincident radiations with no correlation of the angle of emission of one with respect to the other. We now arrange two detectors, such that detector 1 records pulses only from radiation 1, whereas detector 2 records pulses from radiation 2 only. The output of these detectors after appropriate pulse processing is fed to a coincidence unit with a resolving time τ. By separately measuring the two singles rates and the coincidence counting rate, the source activity can be determined as follows.

The singles counting rate in branch 1 (r_1), corrected for background and dead time losses, can be written as the product of the source activity S multiplied by some overall efficiency ϵ_1. This efficiency includes the solid angle subtended by the detector, the interaction probability within the detector, and the fraction of detector pulses accepted by the subsequent circuitry. A similar relation can be written for branch 2. Therefore, we have

$$r_1 = \epsilon_1 S \tag{17-22}$$

$$r_2 = \epsilon_2 S \tag{17-23}$$

The true coincidence rate r_t can be predicted by noting that two independent events must occur: Radiation 1 must be detected in branch 1, whereas at the same time radiation 2 must be detected in branch 2. The independent probabilities (they are independent if there is no angular correlation) of these two events are ϵ_1 and ϵ_2. Therefore, the probability that both occur is simply their product $\epsilon_1\epsilon_2$. The true coincidence rate is thus the product of this combined probability and the source activity S:

$$r_t = \epsilon_1\epsilon_2 S \tag{17-24}$$

The measured coincidence rate r_{12} is the sum of the true coincidence rate and the chance coincidence rate. Therefore,

$$r_{12} = r_t + r_{ch}$$

$$r_{12} = \epsilon_1 \epsilon_2 S + r_{ch} \qquad (17\text{-}25)$$

Now we can solve the three equations (17-22), (17-23), and (17-25) simultaneously and thereby eliminate any two variables. By eliminating the efficiencies ϵ_1 and ϵ_2, we can write

$$S = \frac{r_1 r_2}{r_{12} - r_{ch}} \qquad (17\text{-}26)$$

This expression gives the source activity in terms of directly measured rates and the chance coincidence rate which can be measured using the methods described earlier. In many applications, the two coincident radiations selected are beta (β) and gamma (γ) rays emitted in the decay of a given isotope, and the method is often called β-γ *coincidence*. The requirement of no angular correlation can be relaxed if one of the radiations is detected over a 4π solid angle. Therefore, a common implementation of the method is to use a 4π proportional counter as the beta detector in connection with a gamma-ray detector subtending a smaller solid angle. Accuracies approaching 1% can be obtained with this method, whereas activity measurements that rely on a prior knowledge of detector efficiency are nearly always less precise.

The analysis above is somewhat oversimplified, because it is often difficult to meet the condition that each detector responds only to one of the two radiations. In β-γ coincidence measurements, it is normally quite easy to eliminate the β sensitivity of the γ detector by interposing an absorber that is thicker than the maximum distance of penetration of the beta particles involved. However, most beta particle detectors show some response to gamma rays. Two gamma rays of distinguishable energies are also sometimes used as the two radiations (γ-γ coincidence) by selecting each photopeak separately in two gamma-ray detectors. It is then difficult to avoid including some of the higher-energy gamma rays in the lower-energy branch because of the contribution of the Compton continuum from the higher-energy gamma rays. Also, the complications of true and chance summing (see p. 302) that may occur in either detector must be considered. For a more detailed discussion of these and other complications in absolute activity measurements using coincidence methods, see Ref. 58.

C. Modular Instruments for Timing Measurements

Pulse-processing systems composed of readily available nuclear instrumentation modules are shown in Fig. 17-39 for simple coincident measurements and time spectroscopy. If there is no interest in pulse height information, these systems can be designed exclusively for their timing properties, using procedures that often compromise the pulse height information. As an example, the time constant for the output pulses from a scintillation counter can be chosen to be very small so that fast linear pulses are produced. These pulses follow the time profile of the collected electrons from the photomultiplier tube and provide the best performance in leading edge time pick-offs. However, as discussed in Chapter 4, fast linear pulses have inferior pulse height information when compared with the integrated tail pulse, which is usually provided when pulse height information is desired.

STANDARD TWOFOLD COINCIDENCE:

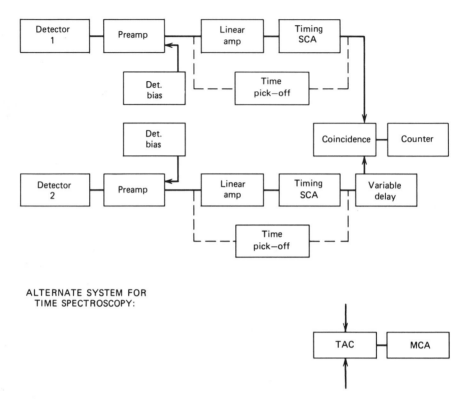

ALTERNATE SYSTEM FOR
TIME SPECTROSCOPY:

Figure 17-39 Basic modular system for twofold coincidence measurements. The use of a time pick-off in place of the linear amplifier and timing SCA, as indicated by the dashed lines, is warranted if no pulse amplitude selection is desired. An alternative arrangement for multichannel time spectroscopy is indicated at the bottom, in which the coincidence unit and counter are replaced by a TAC and MCA.

In applications in which both timing and pulse height information must be extracted, it has become common to arrange a *fast–slow* instrumentation scheme as outlined in Fig. 17-40. In the example shown, it is desired to accept timing pulses only for those detector pulses of a specific amplitude. In this case, separate signals for timing and amplitude are processed through the fast and slow branches so that optimum choices on pulse shaping, and so on, may be made independently to optimize the performance with regard to timing or amplitude information. The output from the amplitude branch is then used to operate a gate that can accept or reject the corresponding timing pulse. The time delays through the slower amplitude branch are such that it is normally required that a fixed delay be added to the timing branch before the two are combined.

1. TIME PICK-OFF MODULES

For optimum performance, the time pick-off is often best located directly at the detector. This is seldom done in practice, however, because of the practical problems in preserving good amplitude resolution at the same time. One exception is the scintillation counter, in which both the anode and a preceding dynode of the photomultiplier tube can be tapped

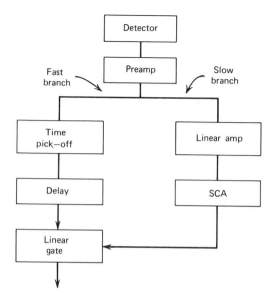

Figure 17-40 A simple fast–slow pulse-processing system in which a slow amplitude branch is used to select only those fast timing pulses that correspond to events of a predetermined amplitude set by the SCA window.

to provide output signals. By choosing the anode load resistance to be small (say, 50 ohms), the anode circuit time constant will normally be smaller than the time over which electrons are collected and a fast linear pulse will therefore result. A preceding dynode can also be used to generate a signal by inserting a load resistance between that dynode and the voltage divider or other source of dynode dc voltage. If the value of this load resistor is high, the resulting time constant can be large and tail pulses will be generated. These tail pulses then serve as the primary signal for amplitude processing or the slow branch, whereas the fast linear pulses are used for timing purposes. (Note that dynode pulses are of opposite polarity from the anode pulse because more electrons leave each dynode than are collected from the preceding stage.)

It is more likely that the time pick-off follows the preamplifier and operates on the leading edge of the preamplifier output pulse. Commercial preamplifiers are available in which a premium is placed on fast rise time performance, so that the output leading edge reflects the charge collection time from the detector as closely as possible.

Conventional crossover methods of time pick-off require a shaped linear pulse to provide the bipolar shape on which the crossover method is based. Such pick-offs must then obviously follow the shaping networks in the linear amplifier. Other time pick-off methods are also occasionally built into single-channel analyzers or integral discriminators intended to accept a shaped linear pulse. These units are a convenience in timing situations in which the ultimate in performance is not warranted and can be used with virtually no sacrifice in the separate amplitude resolution of the pulses.

2. COINCIDENCE UNITS

Many commercal coincidence units are based on an overlap principle in which the width of the input pulses directly determines the resolving time of the coincidence circuit. Other types of circuits are sensitive only to the leading edge of the input pulse, and the resolving times can be chosen independently. The latter type is more flexible for situations in which detectors with different time properties may be interchanged.

Most coincidence units provide multiple inputs (often up to four) that can independently be switched in or out of the circuit. When only one input is switched in, every logic pulse presented to that input is simply routed through to the output of the coincidence unit. This mode of operation provides a convenient way of recording the singles rate in any input branch. When two inputs are switched in, the unit functions as a simple twofold coincidence, three switched inputs lead to threefold coincidence, and so on.

Coincidence units often provide for at least one *anticoincidence* input. Pulses provided to this input within a predetermined time act to cancel the normal output of the unit. For example, if one normal input is selected, an output pulse will appear for each input pulse that is not accompanied by an anticoincidence pulse within the resolving time of the circuit.

3. TIME-TO-AMPLITUDE CONVERTER (TAC)

The great popularity of TACs for the measurement of time intervals stems from the wide availability of multichannel analyzers in most measurement laboratories. By converting the time interval to a proportional pulse amplitude, the TAC allows the use of well-developed methods (discussed in Chapter 18) for the analysis and storage of pulse amplitudes as a substitute for the direct measurement of the time interval.

One of the more important properties of the TAC is the linearity of its time interval to amplitude conversion. In order to test the linearity, means must be provided for introducing fixed delays of known magnitude between start and stop pulses. For timing periods up to a few hundred nanoseconds, this can be accomplished by using variable lengths of coaxial cables. Other methods applicable for testing TAC linearity over wider time ranges are given in Ref. 59.

TACs are generally of two distinct types. The *overlap* type is based on supplying start and stop pulses of standard rectangular shape to the converter and measuring the area of overlap between the two. If the two pulses are coincident, the overlap will be complete, whereas if they are separated by a pulse width, there will be no overlap. Therefore, if an output pulse is generated whose amplitude integrates the area of overlap, the time to amplitude conversion is carried out. The principal merit of the overlap scheme is that it is very fast compared with other methods. Unfortunately, it tends to have poor linearity and accuracy specifications, and therefore it is used mainly in those applications in which maximum counting rates are of primary interest.

In the *start–stop* type of TAC, the start pulse initiates some circuit action, such as the charging of a capacitor by a constant-current source. This action continues until terminated by the appearance of the stop pulse. The constant current generates a linear ramp voltage, which is stopped at an amplitude proportional to the interval between the start and stop pulses. Designs of this type tend to have better linearity characteristics than overlap types and are more commonly encountered in routine time spectroscopy measurements.

4. DIRECT TIME TO DIGITAL CONVERTER

In the time spectroscopy system of Fig. 17-34 the time interval between start and stop pulses is first converted to a pulse amplitude. In the multichannel analyzer, this pulse amplitude is digitized and recorded. It is ironic that the digitization is often carried out by converting the amplitude back into a time interval over which clock oscillator pulses are accumulated (see Chapter 18). Therefore, it is only logical to consider eliminating the step of converting the time interval to an amplitude and instead to digitize the interval directly by using it to gate the output of a constant-frequency clock. Utilization of this method, however, is limited by the highest frequency at which clock pulses can be reliably accumulated and counted. Without special techniques, the upper limit[60, 61] is about 500 MHz, corresponding to a period of 2 ns. Therefore, intervals of the order of a microsecond or more can be measured accurately in this manner, but the digitization of nanosecondintervals suffers from the coarseness of the time scale imposed by the period of the clock frequency. Instead, more complex interpolation techniques have been developed[61-63] to stretch the basic interval by typical factors of 256 or more. General methods for the measurement of very small time intervals are elaborated in the review by Porat.[64]

5. TIME DELAYS

In many timing applications it is necessary to introduce delays at one or more points in the signal chain for timing adjustment or calibration purposes. On a nanosecond scale, the length of coaxial cables used to interconnect the various modules can be varied to provide differences in transit time. Beyond about 100 ns, cable lengths become excessively long (> 30 m) for convenient use. Use of delay lines with helically wound center conductors can greatly reduce the physical length of the cable for a given transit time, but they have relatively poor high-frequency transmission characteristics and will badly distort fast pulses. When used with shaped linear pulses that have already been high-frequency filtered, the distortions are less serious, and passive delay lines of various designs can provide adjustable delays up to several microseconds. Linear amplifiers often incorporate a delay line to provide a supplemental delayed output for convenience in setting up pulse timing systems.

Delay elements for logic pulses are less demanding because the shape of the pulse carries no information. One type of logic pulse delay unit is based on generating a ramp voltage and sensing the point at which the ramp exceeds a discrimination level. The ramp is initiated by the input logic pulse, and its termination produces a similar logic pulse after a delay time that can be continuously adjusted by varying the discrimination level.

6. FAST AMPLIFIERS

When the need for timing information is paramount, the detector output is often chosen to be a fast linear pulse of short duration. This type of pulse is commonly supplied to a leading edge time pick-off unit to provide a prompt timing signal. If the amplitude of the fast linear pulse is insufficient to reach a convenient trigger level in the time pick-off, it is necessary to provide some prior amplification. The linear amplifiers discussed previously will not suffice for this application because their shaping circuits are designed to eliminate the high-frequency components in the signal for optimum amplitude resolution. In order to preserve the fast linear pulse shape, an amplifier with linear response to as high a frequency as possible is desired. From the standpoint of baseline stability, it can also be desirable to dc couple all stages of the amplifier and thereby extend the frequency response to zero. Such amplifiers are called *wide band amplifiers*, and because they filter neither high nor low frequencies from the signal, they provide no shaping to the input pulse. When applied to nuclear pulse processing, they are more conventionally called *fast amplifiers* and ideally produce an output pulse that is a faithful replica of the input pulse, but with amplitude gain. Because a single stage of amplification usually reverses the polarity of the input pulse, fast amplifiers can also be a convenient means of inverting a fast linear pulse.

VIII. PULSE SHAPE DISCRIMINATION

The usual information carried by a linear pulse is its amplitude (and time of occurrence). There are occasions in nuclear measurements, however, when the shape of the pulse also assumes some importance. Most shape differences arise because of differences in the time profile of the current produced at the detector by a radiation interaction. If a fast linear pulse is generated with a short collection time constant at the detector, the shape of the pulse is a reproduction of the time history of this current and will therefore directly reproduce these differences. In the more common application, the linear tail pulse obtained by collecting this current across a large time constant will show changes only in its leading edge characteristics. Therefore, *pulse shape discrimination* (PSD) methods designed to sense the difference between such events are sometimes also called *rise time discrimination* methods.

Sensing differences in pulse shape can serve a useful purpose when applied to the output of a number of different detector types. Some applications have been outlined in previous chapters and include (1) discrimination against gamma-ray background in organic scintillators used as fast neutron detectors; (2) separation of various particle types in scintillators such as CsI(Tl); (3) discrimination between short-range and long-range particles in proportional counters; (4) discarding defective pulses from silicon or germanium detectors to improve resolution; (5) separation of radiations of different range in the phoswich detector; (6) rejection of piled-up pulses, which can occur in any signal chain at high pulse rates.

The most common PSD method is based on passing the pulse through a shaping network to produce a bipolar shape. This can be either a *CR-RC-CR* network or double delay line shaper, as discussed in Chapter 16. In either case, the time at which the bipolar pulse crosses zero does not depend on pulse amplitude, but instead is a function of the pulse shape and rise time. The time interval between the beginning of the pulse and the zero crossover point will therefore be an indication of the differences in pulse shape prior to the shaping network. Conventionally, the time difference is measured between a leading edge trigger set as low as possible at the beginning of the pulse and a second

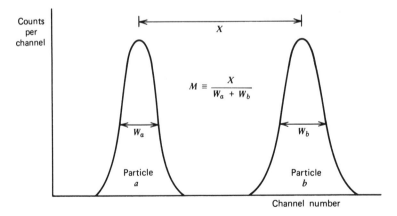

Figure 17-41 Definition of the figure of merit M for pulse shape discrimination applications. In the most common PSD method, the abscissa corresponds to the crossover time of a doubly differentiated input pulse.

trigger that senses the crossover point. This time difference is then converted by a TAC into a pulse amplitude. Systems of this type have been described in Refs. 65-68. A representation of a typical multichannel analysis of the TAC output is shown in Fig. 17-41. The TAC output is normally fed to a single-channel analyzer to select only those time intervals that correspond to the radiation of interest. This step is equivalent to selecting a band of channels which includes only one of the peaks in Fig. 17-41. The SCA output can then be used to gate the pulse-recording system only when the acceptance criteria are met. More elaborate implementations involve the simultaneous recording of the pulse shape and pulse height information in a two-parameter multichannel analyzer, as described in Chapter 18.

Several specifications are of importance in discussing PSD circuits. One is the figure of merit M defined in Fig. 17-41. This factor, as introduced in Ref. 69, is a measure of the separation that can be achieved between different types of events in a given application. The figure of merit is likely to depend on the dynamic range of the input pulses, defined as the ratio between the maximum and minimum amplitude pulses acceptable by the system. Effective pulse shape discrimination systems will operate over a dynamic range of 100 or more.

A second major approach to the design of pulse shape discriminators is based on independent measurements of the integrated charge over two different time regions of the pulse. The ratio of these signals will be approximately constant for pulses of common shape, independent of pulse amplitude. The distribution of ratio values will have the appearance shown in Fig. 17-41 if pulses of two specific shapes are involved. Such measurements can either be carried out with analog circuitry (for example, as in Refs. 70 and 71) or using digital techniques.[72–74] The performance of pulse shape discriminators deteriorates at high pulse rates due to pile-up effects, but can be improved through the rejection of closely spaced pulsed pairs at the PSD input.[75]

A rather different method for pulse shape discrimination in scintillators was first developed by Owen.[76] It is based on the operation of a photomultiplier tube with lower than normal voltage between the last dynode and anode. During the fast portion of the current pulse, space charge effects strongly suppress the secondary electron current from the last dynode. Once the space charge has dissipated, however, the longer components of the pulse are less affected and are collected normally at the anode. The voltage pulse from the last dynode therefore has a strong negative component followed by a broad positive peak (sometimes called the Owen pulse). By measuring the amplitude and zero crossing time of this pulse, effective discrimination between scintillations of different decay times has been demonstrated. Although the Owen method has largely been supplanted by newer methods, it is still suitable for some applications.[77,78]

PROBLEMS

17-1. Derive the expression $V_{out} \cong -(R_2/R_1)V_{in}$ for the voltage-sensitive preamplifier configuration of Fig. 17-1. (Note that the input current must flow through both R_1 and R_2.)

17-2. Derive the expression $V_{out} \cong -Q/C_f$ for the charge-sensitive preamplifier configuration of Fig. 17-2. (Note that the input charge is divided between C_i and C_f.)

17-3. Why is it preferable to locate a preamplifier as close as possible to the detector with which it is used?

17-4. Sketch a configuration consisting of two integral discriminators and an anticoincidence unit that will perform the function of a single-channel analyzer.

17-5. A rate meter circuit can be made to respond to rapid changes more quickly by reducing its time constant. What penalty is normally also associated with this same change?

17-6. A given application involves processing pulses at a low average rate from a high-resolution detector. If the objective is to preserve the pulse height resolution as much as possible, indicate which of the options given below is the better choice:

(a) Short or long shaping times.
(b) Monopolar or bipolar shaping.
(c) With or without active baseline restoration.

17-7. Pulses corresponding to a particular full-energy peak occur at a rate of 8000/s in a system in which the total pulse rate is 25,000/s. Estimate the fraction of the full-energy events that are lost due to pileup, if the effective pulse width is 4 μs.

17-8. A given application involves signal pulses with constant shape and only a small variation in amplitude. Which of the time pick-off methods discussed in the text is likely to give the best timing performance?

17-9. List two independent methods of pulse shaping that can yield the bipolar shape needed for crossover timing.

17-10. A single-channel analyzer produces an output logic pulse for any input pulse that lies within a narrow *acceptance window* in the full pulse height spectrum. What electronic unit carries out the analogous function within the time spectrum?

17-11. Random and uncorrelated pulses are supplied at rates r_1 and r_2 to the inputs of an anticoincidence unit with resolving time τ. Taking into account the effects of chance coincidences, what should be the observed output rate?

17-12. Give two independent methods of determining the chance coincidence contribution to a standard twofold coincidence measurement.

17-13. What length of conventional coaxial cable is necessary to shift the position of the prompt coincidence peak in a time spectrum by 100 ns?

17-14. A sample of ^{60}Co will show a gamma-ray spectrum with peaks at 1.17 and 1.33 MeV, plus a β^- spectrum with endpoint energy of 0.31 MeV. Draw two independent decay schemes that are consistent with the above observations, and outline an experiment that could help decide among the alternatives.

17-15. Two sodium iodide scintillators are arranged to count one each of two gamma rays emitted in coincidence with no angular correlation by a radioactive source. In each of the cases listed below, describe the effect of the indicated change on the ratio of true coincidence to chance coincidence counting rates, and indicate the practical limits to which each of the changes may be carried in order to improve this ratio.

(a) Moving the detector positions to change the solid angle subtended by each.
(b) Varying the amount of source material.
(c) Changing the coincidence circuit resolving time.
(d) Varying the pulse height window width used to select a given gamma energy.

17-16. The following data were obtained for the coincidence-delay curve in a coincidence experiment:

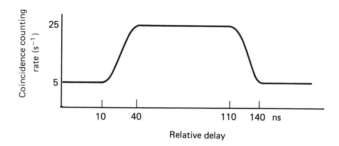

(a) What is the resolving time of the coincidence unit?
(b) What is the width of the prompt coincidence peak in the time spectrum?
(c) What is the singles rate, assuming that it is about the same in both branches supplied to the coincidence unit?

REFERENCES

1. J. Millman, *Microelectronics*, McGraw-Hill Book Co., New York (1979).
2. M. M. Satterfield, G. R. Dyer, and W. J. McClain, *Nucl. Instrum. Meth.* **75**, 312 (1969).
3. V. Radeka, *IEEE Trans. Nucl. Sci* **NS-17** (1), 269 (1970).
4. D. A. Landis, F. S. Goulding, R. H. Pehl, and J. T. Walton, *IEEE Trans. Nucl. Sci.* **NS-18** (1), 115 (1971).
5. K. Kandiah, A. J. Smith, and G. White, *IEEE Trans. Nucl. Sci.* **NS-22** (5), 2058 (1975).
6. D. A. Landis, N. W. Madden, and F. S. Goulding, *IEEE Trans. Nucl. Sci.* **NS-26** (1), 428 (1979).
7. D. A. Landis, C. P. Cork, N. W. Madden, and F. S. Goulding, *IEEE Trans. Nucl. Sci.* **NS-29** (1), 619 (1982).
8. C. L. Britton, T. H. Becker, T. J. Paulus, and R. C. Trammell, *IEEE Trans. Nucl. Sci.* **NS-31** (1), 455 (1984).
9. I. S. Sherman and R. G. Roddick, *IEEE Trans. Nucl. Sci.* **NS-17** (1), 252 (1970).
10. J. J. Eichholz, ANL-6968, (1966).
11. G. White, *Nucl. Instrum. Meth.* **125**, 313 (1975).
12. F. S. Goulding and D. A. Landis, *IEEE Trans. Nucl. Sci.* **NS-35**(1), 119 (1988).
13. A. B. Gillespie, *Signal, Noise and Resolution in Nuclear Counter Amplifiers*, Pergamon Press, London, 1953.
14. V. Radeka, *Nucleonics* **23**(7), 53 (1965).
15. E. Kowalski, *Nuclear Electronics*, Springer-Verlag, New York, 1970.
16. P. W. Nicholson, *Nuclear Electronics*, Wiley, London, 1974.
17. M. Bertolaccini, C. Bussolati, and E. Gatti, *Nucl. Instrum. Meth.* **41**, 173 (1966).
18. A. H. F. Muggleton, *Nucl. Instrum. Meth.* **101**, 113 (1972).
19. S. L. Blatt, *Nucl. Instrum. Meth.* **128**, 277 (1975).
20. D. W. Datlowe, *Nucl. Instrum. Meth.* **145**, 365 (1977).
21. L. Wielopolski and R. P. Gardner, *Nucl. Instrum. Meth.* **133**, 303 (1976).
22. S. L. Blatt, *Nucl. Instrum. Meth.* **49**, 235 (1967).

23. M. Furrer, J.-J. Gostely, and P. Lerch, *Nucl. Instrum. Meth. Phys. Res.* **226**, 455 (1984).

24. D. H. Wilkinson, *Nucl. Instrum. Meth.* **134**, 149 (1976).

25. S. L. Blatt, J. Mahieux, and D. Kohler, *Nucl. Instrum. Meth.* **60**, 221 (1968).

26. C. W. Williams, *IEEE Trans. Nucl. Sci.* **NS-15** (1), 297 (1968).

27. J. Bartosek, J. Masek, F. Adams, and J. Hoste, *Nucl. Instrum. Meth.* **104**, 221 (1972).

28. F. S. Goulding and D. A. Landis, *IEEE Trans. Nucl. Sci* **NS-25** (2), 896 (1978).

29. F. H. Tenney, *Nucl. Instrum. Meth.* **219**, 165 (1984).

30. R. Gold, ANL-6949, (1965).

31. R. P. Gardner and L. Wielopolski, *Nucl. Instrum. Meth.* **140**, 289 (1977).

32. D. W. Datlowe, *Nucl. Instrum. Meth.* **145**, 379 (1977).

33. P. C. Johns and M. J. Yaffe, *Nucl. Instrum. Meth.* **A255**, 559 (1987).

34. M. Wiernick, *Nucl. Instrum. Meth.* **96**, 325 (1971).

35. E. Fairstein, *IEEE Trans. Nucl. Sci.* **NS-32** (1), 31 (1985).

36. D. A. Gedcke and W. J. McDonald, *Nucl. Instrum. Meth.* **55**, 377 (1967).

37. B. Leskovar and C. C. Lo, *Nucl. Instrum. Meth.* **123**, 145 (1975).

38. J. D. McGervey, J. Vogel, P. Sen, and C. Knox, *Nucl. Instrum. Meth.* **143**, 435 (1977).

39. M. O. Bedwell and T. J. Paulus, *IEEE Trans. Nucl. Sci.* **NS-25** (1), 86 (1978).

40. K. Rytsölä, *Nucl. Instrum. Meth.* **199**, 491 (1982).

41. R. L. Chase, *Rev. Sci. Instrum.* **39**, 1318 (1968).

42. J. J. Kozyczkowski and J. Bialkowski, *Nucl. Instrum. Meth.* **137**, 75 (1976).

43. J. P. Fouan and J. P. Passerieux, *Nucl. Instrum. Meth.* **62**, 327 (1968).

44. F. J. Lynch, *IEEE Trans. Nucl. Sci.* **NS-13** (3), 140 (1966).

45. C. Hohenemser, R. Reno, and A. P. Mills, *IEEE Trans. Nucl. Sci.* **NS-17** (3), 390 (1970).

46. T. J. Paulus, *IEEE Trans. Nucl. Sci.* **NS-32** (3), 1242 (1985).

47. P. Ryge and R. R. Borchers, *Nucl. Instrum. Meth.* **95**, 137 (1971).

48. Z. H. Cho and R. L. Chase, *Nucl. Instrum. Meth.* **98**, 335 (1972).

49. H. Engel, H. Schneider, and R. Spitz, *Nucl. Instrum. Meth.* **142**, 525 (1977).

50. D. C. S. White and W. J. McDonald, *Nucl. Instrum. Meth.* **115**, 1 (1974).

51. J. Kasagi, H. Ohnuma, and N. Ohyama, *Nucl. Instrum. Meth.* **193**, 557 (1982).

52. T. J. Paulus, T. W. Raudorf, B. Coyne, and R. Trammell, *IEEE Trans. Nucl. Sci.* **NS-28** (1), 544 (1981).

53. A. E. Blaugrund and Z. Vager, *Nucl. Instrum. Meth.* **29**, 131 (1964).

54. R. Avida and S. Gorni, *Nucl. Instrum. Meth.* **52**, 125 (1967).

55. Z. H. Cho and T. R. Gerholm, *Nucl. Instrum. Meth.* **73**, 67 (1969).

56. C. H. Vincent, *Nucl. Instrum. Meth.* **127**, 421 (1975).

57. R. L. Chase, *IEEE Trans. Nucl. Sci.* **NS-23** (1), 244 (1976).

58. NCRP Report No. 58, *A Handbook of Radioactivity Measurements Procedures*, 2nd ed., National Council on Radiation Protection and Measurements. Bethesda, MD, 1985.

59. S. Cova and M. Bertolaccini, *Nucl. Instrum. Meth.* **77**, 269 (1970).

60. S. J. Hall and A. M. MacLeod, *Nucl. Instrum. Meth.* **140**, 283 (1977).

61. J. A. Harder, BNL 39840, May 1987.

62. B. Turko, *IEEE Trans. Nucl. Sci.* **NS-26** (1), 737 (1979).

63. B. T. Turko, R. D. Macfarlane, and C. J. McNeal, *Int. J. Mass Spectrom. Ion Phys.* **53**, 353 (1983).

64. D. I. Porat, *IEEE Trans. Nucl. Sci.* **NS-20** (5), 36 (1973).

65. L. J. Heistek and L. Van der Zwan, *Nucl. Instrum. Meth.* **80**, 213 (1970).

66. G. W. McBeth, J. E. Lutkin, and R. A. Winyard, *Nucl. Instrum. Meth.* **93**, 99 (1971).

67. D. W. Glasgow, D. E. Velkley, J. D. Brandenberger, and M. T. McEllistrem, *Nucl. Instrum. Meth.* **114**, 535 (1974).

68. P. Plischke, V. Schröder, W. Scobel, L. Wilde, and M. Bormann, *Nucl. Instrum. Meth.* **136**, 579 (1976).

69. R. A. Winyard, J. E. Lutkin, and G. W. McBeth, *Nucl. Instrum. Meth.* **95**, 141 (1971).

70. J. M. Adams and G. White, *Nucl. Instrum. Meth.* **156**, 459 (1978).

71. J. R. M. Annand, *Nucl. Instrum. Meth. Phys. Res.* **A262**, 371 (1987).

72. Z. W. Bell, *Nucl. Instrum. Meth.* **188**, 105 (1981).

73. T. Kumahara and H. Tominaga, *IEEE Trans. Nucl. Sci.* **NS-31**(1), 451 (1984).

74. J. H. Heltsley, L. Brandon, A. Galonsky, L. Heilbronn, B. A. Remington, S. Langer, A. Vander Molen, J. Yurkon, and J. Kasagi, *Nucl. Instrum. Meth. Phys. Res.* **A263**, 441 (1988).

75. R. B. Piercey, J. E. McKisson, M. A. Herath Banda, and M. R. Shavers, *IEEE Trans. Nucl. Sci.* **NS-34** (1), 82 (1987).

76. R. B. Owen, *Nucleonics* **17**, 92 (1959).

77. T. Doke, M. Adachi, S. Kubota, and I. Ogawa, *Nucl. Instrum. Meth.* **57**, 163 (1967).

78. J. D. Dewendra and R. B. Galloway, *Nucl. Instrum. Meth.* **125**, 503 (1975).

CHAPTER · 18

Multichannel Pulse Analysis

$\mathbf{A}$ measurement of the differential pulse height spectrum from a radiation detector can yield important information on the nature of the incident radiation or the behavior of the detector itself and is therefore one of the most important functions to be performed in nuclear measurements. The rapid advance in detector development and radiation spectroscopy over the past decades would have been impossible without the parallel development of reliable instruments to record the differential pulse height spectra from these instruments under a variety of conditions.

By definition (see Chapter 4), the differential pulse height spectrum is a continuous curve that plots the value of dN/dH (the differential number of pulses observed within a differential increment of pulse height H) versus the value of the pulse height H. The ratio of the differentials can never be measured exactly, but rather all measurement techniques involve a determination of $\Delta N/\Delta H$ (the discrete number of pulses observed in a small but finite increment of pulse height H). The increment in pulse height ΔH is commonly called the *window width* or *channel width*. As we shall see, provided ΔH is small enough, a plot of $\Delta N/\Delta H$ versus H is a good discrete approximation to the continuous curve that represents the actual differential pulse height spectrum. As a practical matter, a distinction is seldom made between the continuous distribution and its discrete approximation, and all such plots are generally referred to as differential pulse height distributions or *pulse height spectra*.

I. SINGLE-CHANNEL METHODS

A. Sequential SCA Measurements

Under proper conditions, the single-channel analyzer (SCA) described in Chapter 17 can serve to record the pulse height spectrum from a steady-state source. The procedure consists of serially recording the number of counts passed by the SCA as its lower discriminator value is changed to cover the entire range of pulse height. The process is illustrated in Fig. 18-1. Because a sequential recording of different regions of the spectrum is involved, the spectrum must not vary during the time required to carry out the full set of measurements.

The number of counts recorded during one such determination should be proportional to the area under the continuous differential pulse height distribution between the limits H and $H + \Delta H$. This area ΔN, when divided by the window width ΔH, is a

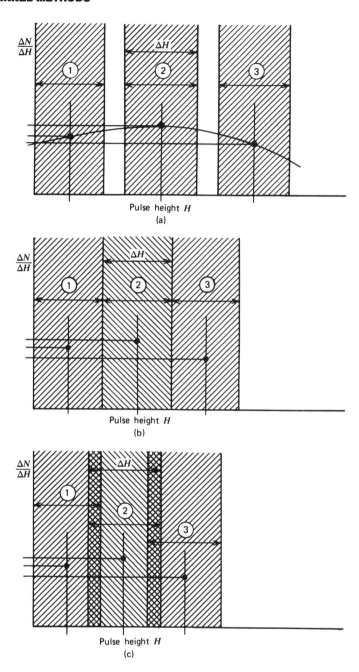

Figure 18-1 Sequential measurements (labeled 1, 2, and 3) using an SCA with a constant window width ΔH. When the number of counts ΔN is normalized by ΔH and plotted at the midpoint of the pulse height interval, a pointwise approximation to the differential pulse height distribution is obtained. Plot (b) is taken with contiguous windows, plot (a) with widely spaced windows, and plot (c) with overlapping windows.

measure of the average of the continuous distribution over this same pulse height interval. The ratio is conventionally plotted at the midpoint of the pulse height interval.

Additional points along the spectrum can now be filled in by simply changing the SCA parameters. Although it is by no means a requirement, most spectra are recorded using a constant value for the window width, or using equal channel widths. In that case each sequential count ΔN is divided by the same value ΔH, and for the sake of convenience this explicit division is often omitted by simply plotting the number of recorded events ΔN instead. As we shall see, this is the usual convention when using multichannel analyzers.

The windows used for the determination could also be spaced at arbitrary intervals. Close spacing could lead to overlapping windows, whereas wide spacing would leave gaps between adjacent windows. *Contiguous* windows are those in which the upper edge of one window is coincidental with the lower edge of the next. This is not a fundamental necessity, but almost all multichannel spectra are taken under these conditions.

The disadvantage of recording pulse height spectra using sequential single-channel measurements is the long measurement time required and waste of information. During the period in which pulses that lie within the window are being counted, all other pulses are ignored. When only a limited time is available to make the measurement, the time that can be allocated per channel is relatively short, and consequently, the statistical fluctuations in ΔN will be large. It would clearly be preferable to conduct the measurements in parallel rather than in series so that all pulses, regardless of their amplitude, contribute to the recorded information.

B. Stacked Single-Channel Analyzers

A conceptual approach to obtaining pulse height data in parallel by using many single-channel analyzers is shown in Fig. 18-2. All the inputs are connected together and each output fed to a separate counter. The lower level of the SCA at the bottom of the stack is set to zero, and that for the top SCA is set corresponding to the largest pulse height of interest. The lower level of the SCAs in-between are arranged at equal intervals between these extremes. The window width of each SCA is identical and is set equal to the spacing between adjacent discrimination levels. This arrangement thus provides a series of contiguous pulse height windows of equal width, as illustrated in Fig. 18-3.

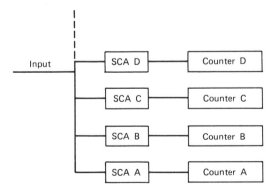

Figure 18-2 An array of stacked single-channel analyzers. Windows $A, B, C, \ldots$ are assumed to be contiguous and of equal width ΔH, with A at the bottom of the pulse height scale.

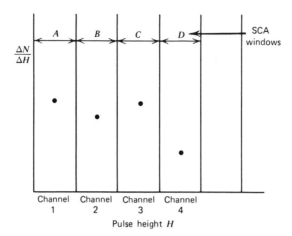

Figure 18-3 A pointwise representation of the differential pulse height distribution obtained from the stacked SCA array of Fig. 18-2 by plotting the content of each counter ΔN (normalized to the window width ΔH) versus the midpoint of the corresponding SCA window.

An input pulse presented to this array will fall into one and only one of the multiple windows set by the SCAs. Therefore, each input pulse results in an increment of one count added to the corresponding SCA counter. One can therefore view the overall process as the sorting of each input pulse into the proper window and incrementing the content of that counter by one. A small pulse will correspond to a window near the bottom of the stack, whereas a large pulse will fall into a window near the top. At the end of a measurement period, the sum of all the counters will simply be the total number of pulses presented to the input.

If we now plot the number of recorded pulses ΔN in each counter divided by the window width ΔH versus the average pulse height for each window, we derive a discrete representation of the differential pulse height distribution. In this context, each window is conventionally called a *channel*, numbered in increasing order from left to right. The lowest channel corresponds to the pulse height window at the bottom of the range and will record only those pulses whose amplitude is very small. The largest channel numbers are plotted at the right of the horizontal axis and record only the pulses of largest amplitude.

This process of sorting successive signal pulses into parallel amplitude channels is commonly called *multichannel pulse height analysis*. As a practical matter, schemes based on stacked SCAs are seldom attractive because of complications introduced by drifts in the various discrimination levels and window widths. These drifts can lead to overlapping or noncontiguous channels whose width may also not be constant. Furthermore, when a large number of channels is required, the reliability of any scheme based on the independent operation of a large number of single-channel elements would likely be less than ideal. As a result, other approaches have evolved for accomplishing the same purpose. A device designed to carry out this function is known as a *multichannel analyzer* (MCA), and the following sections discuss some general properties and functions of these instruments.

II. GENERAL MULTICHANNEL CHARACTERISTICS

A. Number of Channels Required

In any pulse height distribution measurement, two factors dominate the choice of the number of channels that should be used for the measurement: the degree of resolution required and the total number of counts that can be obtained. If an arbitrarily large number of counts can be accumulated, there is no disadvantage in making the number of channels as large as one wishes. By providing a large number of channels, the width of any one channel can be made very small and the resulting discrete spectrum will be a close approximation to the continuous distribution. For a faithful representation, the true distribution should not change drastically over the width of one channel. If peaks are present in the spectrum, this requirement translates into specifying that at least three or four channels should be provided over a range of pulse height corresponding to the FWHM (full width at half maximum) of the peak. Figure 18-4b shows a hypothetical differential distribution taken under conditions in which the number of channels is too small to meet this criterion. The resulting distortions and loss of resolution in the spectrum are obvious.

The channel requirements can also be expressed in terms of detector resolution R. For a peak with a mean pulse height H

$$R \equiv \frac{\text{FWHM}}{H} \quad \text{or} \quad H = \frac{\text{FWHM}}{R} \tag{18-1}$$

We can express both H and FWHM in terms of numbers of channels, and furthermore, we now require that at least four channels be provided over the FWHM of the peak. The position of H in units of channels is therefore

$$\text{peak position } H = \frac{4 \text{ channels}}{R} \tag{18-2}$$

and at least this number of channels must be provided. A detector whose energy resolution is 4% therefore requires a minimum of 100 channels, and a detector with 0.4% resolution would require 1000 channels. This argument is valid only in those cases in which the full range of pulse amplitude is recorded with constant channel width ranging from zero to the maximum pulse height. The channel requirement can be reduced by selectively recording only a portion of the spectrum with a large value of *zero offset* discussed in the next section.

The above arguments would suggest that one should always use the maximum number of channels possible. A second factor arises, however, when the available measurement time limits the total number of pulses that contribute to the recorded spectrum. Because the number of events that fall within any one channel will vary in proportion to its width, the content of a typical channel varies inversely with the total number of channels provided over the spectrum. Choosing a larger number of channels will consequently cause the relative statistical uncertainty of each content to increase, and the channel-to-channel fluctuation of the data over smooth portions of the spectrum will become more noticeable. If these fluctuations are large enough, they can begin to interfere with the ability to discern small features in the spectrum. Very small peaks can become lost in statistical noise. These effects are illustrated in Fig. 18-4d.

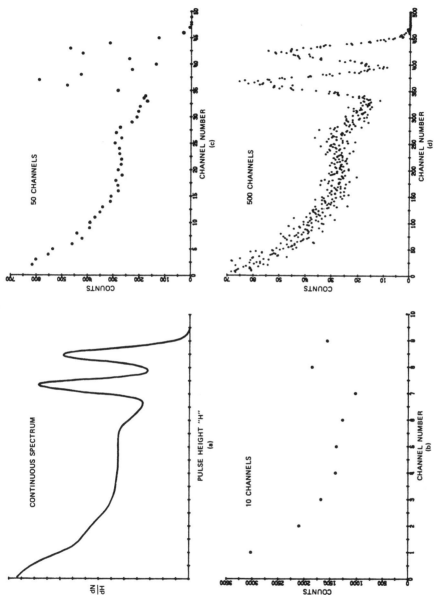

Figure 18-4 An illustration of the effect of varying the number of channels used to record the differential distribution at the top left. A total of 15,000 counts were accumulated for each of the three multichannel spectra. In (b) the number of channels is too small to show sufficient detail, in (c) the choice is about right, and too many channels were used in (d). The low average number of counts per channel in spectrum (d) leads to large statistical fluctuations that could obscure small additional peaks.

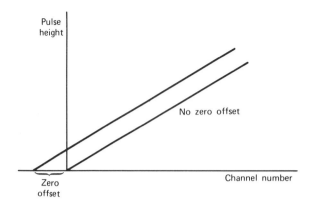

Figure 18-5 Typical calibration plot for a linear MCA with and without zero offset.

It is therefore important that one not provide a needlessly large number of channels for a given measurement. As a practical matter, there is usually no benefit in exceeding the criteria that four to five channels be provided over the width of any given peak, and most multichannel spectra are recorded under these conditions. An overly large allocation of channels also increases the data handling requirements associated with each recorded spectrum.

B. Calibration and Linearity

The ideal MCA would perform a perfectly linear conversion of pulse height to channel number. Under these conditions, a plot of pulse height versus channel number would be a simple straight line, as illustrated in Fig. 18-5. In addition, it is usually convenient to introduce some *zero offset* shift of the origin, such that a nonzero amplitude is required for storage in the first channel. The zero offset is sometimes desirable to suppress very small noise pulses, which may appear with the signal, or to assign the available channels only to the upper portion of a spectrum to be recorded. The zero offset is adjustable through front panel controls on many commercial MCAs.

In many situations, the signal pulses are first sent through a linear amplifier with variable gain. Then the slope of the calibration plot can also be varied by changing the gain factor. For example, Fig. 18-6 shows the calibration plots for an ideal MCA for three different values of amplifier gain. The same effect can also be achieved if the MCA allows selection of the conversion gain of its ADC (see next section).

The user is generally interested in an initial calibration of an MCA which will determine the energy scale of the pulse height distribution. Assuming the analyzer is sufficiently linear, two parameters need to be determined: the slope and intercept of the calibration line shown in Fig. 18-5. The easiest calibration method is to place sources of known energy at the detector and record the channel number into which the centroid of the resulting full-energy peak falls. Because two points completely determine a straight line, only two energies are, in principle, required. However, other sources are often used to provide additional points along the line for confirmation purposes and a test of linearity. If peaks of known energy cannot conveniently be provided, a pulse generator may also be used to provide points for a calibration plot. Only pulses of known *relative* amplitude are required to test linearity and determine the zero offset, but some other means must then be used to relate independently the pulse height scale to energy.

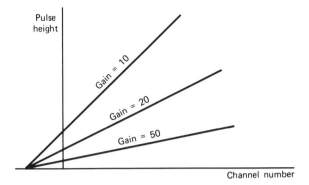

Figure 18-6 Different MCA calibration plots for three values of amplifier gain.

The linearity of the MCA can be measured or quoted in several different ways. The most direct method is to make a measurement of the channel number, in which pulses of known amplitude are stored, and then make the plot of pulse amplitude versus channel number shown in Fig. 18-7. The maximum deviation of the measured curve from a best-fit straight line is a measure of the *integral linearity* and is conventionally quoted as a percentage. Nonlinearities are most often observed at either pulse height extreme and are typically less than 1% for well-designed analyzers.

A more sensitive method of detecting nonlinearities is to employ a source of pulses with a uniform distribution in amplitude. If the MCA is connected to accumulate these pulses over a period of time, a uniform or flat distribution of counts in all channels should

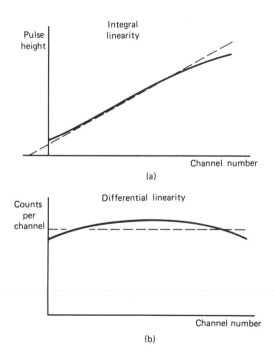

Figure 18-7 Examples of integral and differential linearity measurements for a MCA.

result (see Fig. 18-7b). Sufficient counting time must be allowed so that the statistical fluctuations in channel content are small compared with the desired measurement accuracy. Such a calibration measures the *differential linearity* of the MCA. Deviations from uniformity of a few percent are typical of good MCAs.

The integral and differential calibrations are interrelated, because the differential plot will be simply proportional to the slope of the integral plot at any point. In Fig. 18-7, the two calibrations are intended to be consistent and illustrate the same nonlinearity. The differential calibration, although more difficult to set up, is obviously a more sensitive test of MCA linearity.

III. THE MULTICHANNEL ANALYZER

A. Basic Components and Function

The multichannel analyzer (MCA) is comprised of the basic components illustrated in Fig. 18-8. Its operation is based on the principle of converting an analog signal (the pulse amplitude) to an equivalent digital number. Once this conversion has been accomplished, the extensive technology available for the storage and display of digital information can be brought to bear on the problem of recording pulse height spectra. As a result, the analog-to-digital converter (ADC) is a key element in determining the performance characteristics of the analyzer. The output of the ADC is stored in a computer-type memory, which has as many addressable locations as the maximum number of channels into which the recorded spectrum can be subdivided. The number of memory locations is usually made a power of 2, with memories of 512–8192 channels being common choices. The maximum content of any one memory location ranges up to $2^{24} - 1$, or about 16.8×10^6, counts.

The basic function of the MCA involves only the ADC and the memory. For the purposes of illustration, we imagine the memory to be arranged as a vertical stack of addressable locations, ranging from the first address (or channel number 1) at the bottom through the maximum location number (say, 1024) at the top. Once a pulse has been processed by the ADC, the analyzer control circuits seek out the memory location corresponding to the digitized amplitude stored in the address register, and the content of that location is incremented by one count. The net effect of this operation can be thought of as one in which the pulse to be analyzed passes through the ADC and is sorted into a memory location that corresponds most closely to its amplitude. This function is identical

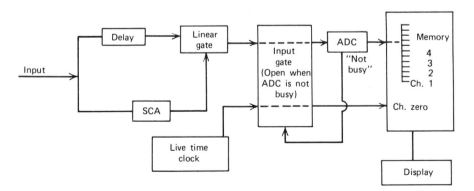

Figure 18-8   Functional block diagram of a typical MCA.

to that described earlier for the stacked single-channel analyzers illustrated in Fig. 18-2. Neglecting dead time, each input pulse increments an appropriate memory location by one count, and therefore the total accumulated number of counts over all memory is simply the total number of pulses presented to the analyzer during the measurement period. A plot of the content of each channel versus the channel number will be the same representation of the differential pulse height distribution of the input pulses as discussed earlier for the stacked single-channel analyzers.

A number of other functions are normally found in a MCA. As illustrated in Fig. 18-8, an *input gate* is usually provided to block pulses from reaching the ADC during the time it is "busy" digitizing a previous pulse. The ADC provides a logic signal level that holds the input gate open during the time it is not occupied. Because the ADC can be relatively slow, high counting rates will result in situations in which the input gate is closed for much of the time. Therefore, some fraction of the input pulses will be lost during this dead time, and any attempt to measure quantitatively the number of pulses presented to the analyzer must take into account those lost during the dead time.

To help remedy this problem, most MCAs provide an internal clock whose output pulses are routed through the same input gate and are stored in a special memory location (often called channel zero). The clock output is a train of regular pulses synchronized with an internal crystal oscillator. If the fraction of time the analyzer is dead is not excessively high, then it can be argued that the fraction of clock pulses that is lost by being blocked by the input gate is the same as the fraction of signal pulses blocked by the same input gate. Therefore, the number of clock pulses accumulated in channel zero is a measure of the *live time* of the analyzer, or the time over which the input gate was held open. Absolute measurements can therefore be based on a fixed value of live time, which eliminates the need for an explicit dead time correction to the data. Further discussion of the dead time correction problem for MCAs is given later in this chapter.

Many MCAs are also provided with another linear gate that is controlled by a single-channel analyzer. The input pulses are presented in parallel to the SCA and, after passing through a small fixed delay, to the linear pulse input of this gate. If the input pulse meets the amplitude criteria set by the SCA, the gate is opened and the pulse is passed on to the remainder of the MCA circuitry. The purpose of this step is to allow rejection of input pulses that are either smaller or larger than the region of interest set by the SCA limits. These limits, often referred to as the LLD (lower-level discriminator) and ULD (upper-level discriminator), are chosen to exclude very small noise pulses at the lower end and very large pulses beyond the range of interest at the upper end. Thus, these uninteresting pulses never reach the ADC and consequently do not use valuable conversion time, which would otherwise increase the fractional dead time. If an MCA is operated at relatively high fractional dead time (say, greater than 30 or 40%), distortions in the spectrum can arise because of the greater probability of input pulses that arrive at the input gate just at the time it is either opening or closing. It is therefore often advisable to reduce the counting rate presented to the input gate as much as possible by excluding noise and other insignificant small-amplitude events with the LLD, and if significant numbers of large-amplitude background events are present, excluding them with an appropriate ULD setting.

The contents of the memory after a measurement can be displayed or recorded in a number of ways. Virtually all MCAs provide a CRT display of the content of each channel as the Y displacement versus the channel number as the X displacement. This display is therefore a graphical representation of the pulse height spectrum discussed earlier. The display can be either on a linear vertical scale or, more commonly, as a logarithmic scale to show detail over a wider range of channel content. Standard recording devices for digital data, including printers and magnetic storage media, are commonly available to store permanently the memory content and to provide hard copy output.

Because of the similarity of many of the MCA components just described to those of the standard personal computer (PC), there is a growing availability of plug-in cards that will convert a PC into a MCA. The card must provide the components that are unique to the MCA (such as the ADC), but the normal PC memory, display, and I/O hardware can be used directly. Control of the MCA functions is then provided in the form of software that is loaded into the PC memory.

B. The Analog-to-Digital Converter (ADC)

1. GENERAL SPECIFICATIONS

The job to be performed by the ADC is to derive a digital number that is proportional to the amplitude of the pulse presented at its input. Its performance can be characterized by several parameters:

1. The speed with which the conversion is carried out.

2. The linearity of the conversion, or the faithfulness to which the digital output is proportional to the input amplitude.

3. The resolution of the conversion, or the "fineness" of the digital scale corresponding to the maximum range of amplitudes that can be converted.

The resolution is conventionally quoted in numbers of channels, for example, a 4096 channel ADC can subdivide the full pulse height range into 4096 increments. An implicit assumption is that its inherent stability is sufficient to ensure that pulses of constant amplitude will always be stored in a single channel.

The voltage that corresponds to full scale is arbitrary, but most ADCs designed for nuclear pulse spectroscopy will be compatible with the output of typical linear amplifiers. Zero to 10 V is thus a common input span. Shaping requirements will also usually be specified for the input pulses, and most ADCs require a minimum pulse width of a few tenths of a microsecond to function properly.

The *conversion gain* of an ADC specifies the number of channels over which the full amplitude span will be spread. For example, at a conversion gain of 2048 channels, a 0–10 V ADC will store a 10 V pulse in channel 2048, whereas at a conversion gain of 512, that same pulse would be stored in channel 512. At the lower conversion gain, a smaller fraction of the MCA memory can be accessed at any one time. On many ADCs, the conversion gain can be varied for the purposes of a specific application. The resolution of an ADC must be at least as good as the largest conversion gain at which it will be used.

The conversion speed or dead time of the ADC is usually the limiting factor in determining the overall dead time of the MCA. Therefore, a premium is placed on fast conversion, but practical limitations restrict the designer in speeding up the conversion before linearity begins to suffer. There are three types of ADC used in pulse analysis applications: linear ramp converters, successive approximation circuits, and "flash" ADCs. The first of these, although the slowest, is generally more linear than the other types and has gained most widespread application in MCAs. Successive approximation and flash units offer faster conversion times, but generally with poorer linearity and channel uniformity.

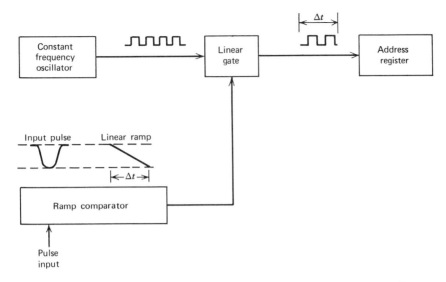

Figure 18-9 Block diagram of a linear ramp (Wilkinson type) ADC.

2. THE LINEAR RAMP CONVERTER (WILKINSON TYPE)

The linear ramp converter is based on an original design by Wilkinson[1] and is diagrammed in Fig. 18-9. The input pulse is supplied to a comparator circuit that continuously compares the amplitude with that of a linearly increasing ramp voltage. The ramp is conventionally generated by charging a capacitor with a constant-current source that is started at the time the input pulse is presented to the circuit. The comparator circuit provides as its output a gate pulse that begins at the same time the linear ramp is initiated. The gate pulse is maintained "on" until the comparator senses that the linear ramp has reached the amplitude of the input pulse. The gate pulse produced is therefore of variable length, which is directly proportional to the amplitude of the input pulse. This gate pulse is then used to operate a linear gate that receives periodic pulses from a constant-frequency clock as its input. A discrete number of these periodic pulses pass through the gate during the period it is open and are counted by the *address register*. Because the gate is opened for a period of time proportional to the input pulse amplitude, the number of pulses accumulated in the address register is also proportional to the input amplitude. The desired conversion between the analog amplitude and a digital equivalent has therefore been carried out.

Because the clock operates at a constant frequency, the time required by a Wilkinson-type ADC to perform the conversion is directly proportional to the number of pulses accumulated in the address register. Therefore, under equivalent conditions, the conversion time for large pulses is always greater than that for small pulses. Also, the time required for a typical conversion will vary inversely with the frequency of the clock. In order to minimize the conversion time, there is a premium on designing circuits that will reliably handle clock pulses of as high a frequency as possible. Clock frequencies of 100 MHz are representative of present-day commercial designs.

The Wilkinson-type ADC leads to contiguous pulse height channels, all ideally of the same width. Because the linear ramp generation can be very precise, this design is characterized by good linearity specifications, accounting for its widespread popularity in MCAs. Variants of the Wilkinson design which lead to some improvement in speed are described by Nicholson,[2] and a clock frequency of up to 300 MHz has been achieved[3] in advanced designs.

3. THE SUCCESSIVE APPROXIMATION ADC

The second type of ADC in common use is based on the principle of successive approximation. Its function can be illustrated by the series of logic operations shown in Fig. 18-10. In the first stage, a comparator is used to determine whether the input pulse amplitude lies in the upper or lower half of the full ADC range. If it lies in the lower half, a zero is entered in the first (most significant) bit of the binary word which represents the output of the ADC. If the amplitude lies in the upper half of the range, the circuit effectively subtracts a value equal to one-half the ADC range from the pulse amplitude, passes the remainder on to the second stage, and enters a one in the most significant bit. The second stage then makes a similar comparison, but only over half the range of the ADC. Again, a zero or one is entered in the next bit of the output word depending on the size of the remainder passed from the first stage. The remainder from the second stage is then passed to the third stage, and so on. If 10 such stages are provided, a 10 bit word will be produced which will cover a range of 2^{10} or 1024 channels.

For the successive approximation ADC, the time required for the conversion will increase as the logarithm of the number of channels, whereas in the Wilkinson circuit, the conversion time increases linearly with the number of channels. Therefore, the successive approximation ADC is inevitably faster if the number of channels is large. Its major disadvantage is a somewhat poorer linearity, which often results from the cumulative errors arising in the subtraction processes. Refinements to this basic design which show improved linearity are described in Refs. 4–6.

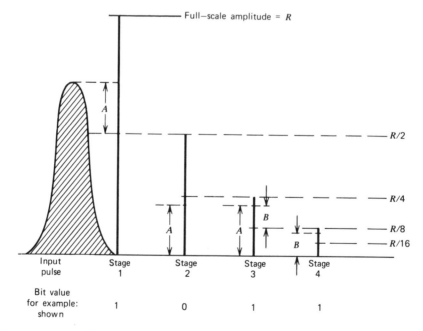

Figure 18-10 Illustration of the operational sequence for a successive approximation ADC. Four stages are shown which generate the four-bit word shown at the bottom as the digital output.

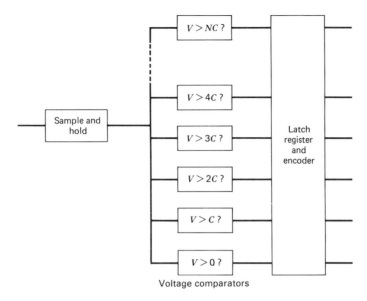

Figure 18-11 Principle of the flash ADC. Here C represents the channel width and N the total number of channels.

4. THE FLASH ANALOG-TO-DIGITAL CONVERTER

In many ways, the flash ADC is the simplest design, representing something of a brute force approach to digitizing the signal. Its function is similar to the stack of single-channel analyzers illustrated previously in Fig. 18-2. As outlined in Fig. 18-11, the role of the SCAs is carried out by a stack of threshold comparators, each set to a progressively higher voltage level. The input pulse is presented in parallel to all comparators simultaneously, and all those with threshold below that of the pulse amplitude will switch output from logic 0 to logic 1. The resulting pattern of logic signals is read by the latch register and converted to a corresponding binary number by the encoder. In order to define N channels, $N + 1$ comparators are required in the stack. Advances in integrated circuit designs have made it possible to routinely provide the 513 ($2^9 + 1$) comparators needed in a 9 bit converter. The added complexity of larger units needed for higher-resolution conversion has limited their production yields and commercial availability.[7]

Because the conversion is carried out in one parallel step, the flash converter is even faster than the successive approximation design. Conversion times for typical 8 bit units are a few tens of nanoseconds. However, the inevitable drifts and uncertainties in the many independent comparator levels lead to relatively poor uniformity of the channel widths. Flash converters are therefore often used in conjunction with the sliding scale technique (see below) to improve their uniformity and linearity.

5. THE SLIDING SCALE PRINCIPLE

The linearity and channel width uniformity of any type of ADC can be improved by employing a technique generally called the *sliding scale* or *randomizing* method. Originally suggested in 1963 by Gatti and co-workers,[8] the method has gained popularity (e.g., Refs. 9 and 10) through its implementation using modern IC technology. It has been particularly helpful in improving the performance of both successive approximation and flash ADCs.

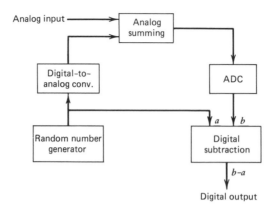

Figure 18-12   Functional diagram of the sliding scale principle for ADCs.

Without the technique, pulses of a given amplitude range are always converted to a fixed channel number. If that channel is unusually narrow or wide, then the differential linearity will suffer in proportion to the deviation from the average channel width. The sliding scale principle is illustrated in Fig. 18-12. It takes advantage of the averaging effect gained by spreading the same pulses over many channels. A randomly chosen analog voltage is added to the pulse amplitude before conversion and its digital equivalent subtracted after the conversion. The net digital output is therefore the same as if the voltage had not been added. However, the conversion has actually taken place at a random point along the conversion scale. If the added voltage covers a span of M channels, then the effective channel uniformity will improve as $\sqrt{M}$ if the channel width fluctuations are random. The implementation of Fig. 18-12 derives the added voltage by first generating a random digital number and converting this number to an analog voltage in a digital-to-analog converter (DAC). The same digital number is then subtracted after the conversion.

One of the disadvantages of the technique is that the original ADC scale of N channels is reduced to $N - M$. If a pulse occurs that would normally be stored in a channel number near the top of the range, the addition of the random voltage may send the sum off scale. One design[10] avoids this limitation by using either upward averaging (as described above) or downward averaging (by subtracting the random voltage) depending on whether the original pulse lies in the lower or upper half of the range. The choice of M can then be as large as $N/2$ to maximize the averaging effect without reducing the effective ADC scale.

C. The Memory

The memory section of an MCA provides one addressable location for every channel. Any of the standard types of digital memory can be used, but there is sometimes a preference for "nonvolatile" memory, which does not require the continual application of electrical power to maintain its content. Then, data acquired over long measurement periods will not be lost if the power to the MCA is accidentally interrupted.

Most MCAs make provisions for subdividing the memory into smaller units for independent acquisition and storage of multiple spectra. In this way a 4096 channel analyzer can be configured as eight separate 512 channel memory areas for storing low-resolution spectra, or as a single 4096 channel memory for a high-resolution spectrum.

In most analyzers, provision is made for the negative incrementing of memory content as well as additive incrementing. In this "subtract" mode, background can conveniently be taken away from a previously recorded spectrum by analyzing for an equal live time with the source removed.

D. Ancillary Functions

1. MEASUREMENT PERIOD TIMING

Virtually all MCAs are provided with logic circuitry to terminate the analysis period after a predetermined number of clock pulses have been accumulated. One often has the choice between preset *live time* or *clock time*, with the distinction being whether or not the clock pulses are routed through the input gate (see Fig. 18-8). Normally, quantitative comparisons or subtraction of background are done for equal live time periods, and this is the usual way of terminating the analysis period.

2. MULTISCALING

Multichannel analyzers can be operated in a mode quite different from pulse height analysis, in which each memory location is treated as an independent counter. In this *multiscaling* mode, all pulses that enter the analyzer are counted, regardless of amplitude. Those that arrive at the start of the analysis period are stored in the first channel. After a period of time known as the *dwell time*, the analyzer skips to the second channel and again records pulses of all amplitudes at that memory location. Each channel is sequentially allocated one such dwell time for accumulating counts, until the entire memory has been addressed. The dwell time can be set by the user, often from a range as broad as from 1 μs to several minutes. The net effect of this mode is to provide a number of independent counters equal to the number of channels in the analyzer, each of which records the total number of counts over a sequential interval of time. This mode of operation can be very useful in studying the behavior of rapidly decaying radioactive sources or in recording other time-dependent phenomena.

3. COMPUTER INTERFACING

Modern-day MCAs share many features with general purpose digital computers. In its most basic form, the MCA can only increment and display the memory, but more elaborate operations can be carried out if it is provided with some of the features of a small computer. For example, one of the functions most often provided by a hard-wired program is to allow summation of selected portions of the spectrum, generally called *regions of interest* (ROIs). Means can be provided to indicate cursors whose position on the displayed spectrum define the upper and lower bounds of the channel numbers between which the summation is carried out. This operation is of obvious practical use for peak area determination in radiation spectroscopy. Other operations, such as subtraction of a spectrum stored in one half of the memory from the contents of the second half, are also available on some preprogrammed analyzers.

More complex computer-based systems are also commercially available which are very similar to standard computers supplied with an appropriate ADC. In this approach, the hard-wired functions mentioned above can be duplicated through software programs that can be modified or supplemented by the user. Useful operations of this type can range from simple smoothing of the spectra to damp out statistical fluctuations, to

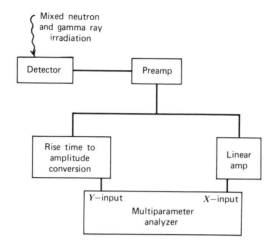

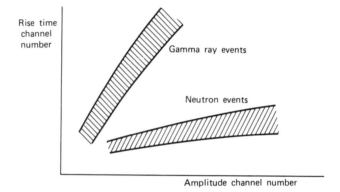

Figure 18-13 An example of multiparameter analysis in which both amplitude and rise time are recorded for each event. The case shown is a somewhat idealized example of neutron–gamma-ray separation in recoil proportional counters.

elaborate peak searching programs, in which the position and area of apparent peaks in the spectrum are identified and measured. Current manufacturer's specification sheets are often the best source of detailed information in this rapidly changing area.

4. MULTIPARAMETER ANALYSIS

The most basic application of multichannel analysis is to determine the pulse height spectrum of a given source. This process can be thought of as recording the distribution of events over a single dimension—pulse amplitude. In many types of radiation measurements, additional experimental parameters are of interest, and it is sometimes desirable to record simultaneously the event distribution over two or more dimensions.

One example is in the case of pulse shape discrimination in which information on both the amplitude and the rise time of each signal pulse are of interest. Instead of a single linear memory, one must now provide an array of memory locations in which one axis corresponds to pulse amplitude and the other to rise time, as indicated in Fig. 18-13. The intensity distribution then takes the form of a two-dimensional surface with local peaks representing combinations of amplitude and rise time which occur most frequently.

In any unit designed for multiparameter analysis, at least two separate ADCs and an associated coincidence circuit must be provided. In the example of Fig. 18-13, the rise time of each event is first externally processed to produce a pulse whose amplitude is proportional to the signal rise time. This pulse is presented to the *Y input*, whereas the coincident original signal pulse is presented to the *X input*. Because both parameters are derived from the same event, they appear at these two inputs in coincidence. The multiparameter analyzer recognizes the coincidence between the inputs and increments the memory location corresponding to the intersection of the corresponding digitized pulse amplitude and rise time.

Multiparameter analyzers generally require a much larger memory than single-parameter analyzers, because, for equal resolution, memory requirements for two parameters are the square of the number of channels required for only one parameter. Often, however, one of the two parameters need not be recorded with the same degree of resolution so that nonsquare (rectangular) memory configurations will suffice. For more than two parameters, the demands on memory are generally so great that some bulk storage such as a magnetic disk or tape is virtually a requirement.

E. MCA Dead Time

The dead time of a MCA is usually comprised of two components: the processing time of the ADC and the memory storage time. The first of these was discussed earlier and, for a Wilkinson-type ADC, is a variable time that is proportional to the channel number in which the pulse is stored. The processing time per channel is simply the period of the clock oscillator. For a typical clock frequency of 100 MHz, this time is 10 ns per channel. Once the pulse has been digitized, an additional few microseconds are generally required to store the pulse in the proper position in the memory. Thus, the dead time of an MCA using an ADC of this type can then be written

$$\tau = \frac{N}{\nu} + B \qquad (18\text{-}3)$$

where ν is the frequency of the clock oscillator, N is the channel number in which the pulse is stored, and B is the pulse storage time. The analyzer control circuits will hold the input gate closed for a period of time that equals this dead time. A *dead time meter* is often driven by the input gate to indicate the fraction of time the gate is closed, as a guide to the experimenter. One normally tries to arrange experimental conditions so that the fractional dead time in any measurement does not exceed 30 or 40% to prevent possible spectrum distortions.

The automatic live time operation of an MCA described earlier is usually quite satisfactory for making routine dead time corrections. However, circumstances can arise in which the built-in live time correction is not accurate. When the fractional dead time is high, errors can enter because the clock pulses are not generally of the same shape and duration as signal pulses. One remedy[11,12] is to use the pulser technique described on p. 620 to produce an artificial peak in the recorded spectrum. If introduced at the preamplifier, the artificial pulses undergo the same amplification and shaping stages as the signal pulses. The fraction that are recorded then can account for both the losses due to pileup and the analyzer dead time. Several authors[13,14] have reviewed the live time correction problem and suggested conditions under which the pulser method is not accurate. To avoid potential problems, the pulse repetition rate must not be too high, and the use of a random rather than periodic pulser is preferred. Under these conditions, the pulser method can successfully handle virtually any conceivable case in which the shape of the spectrum does not change during the course of the measurement.

Additional complications arise if spectrum shape changes occur, which lead to distortions and improper dead time corrections with the pulser method. A better method first suggested by Harms[15] can accommodate spectrum changes but unfortunately requires a nonstandard mode of MCA operation. In this method, the analysis is run for a fixed *clock time*. If a pulse is lost due to the analyzer being dead (this can be sensed externally), compensation is made immediately by assigning a double weight to the next pulse and incrementing the corresponding memory location by two. Spectra that change during the course of the measurement are properly accommodated because the correction automatically takes into account the amplitude distribution of signal pulses at the time of the loss. This correction scheme has been tested by Monte Carlo simulation for a wide variety of spectrum shapes and time variations and has proved to be quite accurate in all cases.[16] At higher rates, however, the assumption breaks down that only one pulse was lost during the dead time, and the Harms method begins to undercorrect for losses. One remedy is to modify the correction process by first calculating the expected number of counts lost during a dead period from a running measurement of the input pulse rate. The memory location corresponding to the next converted pulse is then incremented, not by two counts as above, but by one plus the calculated number of lost pulses. These artificial counts do not have the same statistical significance as the same number recorded normally, but they do maintain the total spectrum content as if there had been no losses. Normally called *loss free counting*,[17] this technique is applicable in situations in which the dead fraction of the MCA is as high as 80%.[18] It is particularly helpful in measurements from short-lived radioisotopes from which the counting rate may change by many orders of magnitude over the measurement period.

IV. SPECTRUM STABILIZATION

Any drifts that occur along the signal chain during the course of the measurement will cause peaks in the recorded spectrum to broaden or become otherwise distorted. These drifts can come about through changes in temperature of the detector or associated electronics, gradual changes in voltage levels, or variations of the gain of the various active elements in the signal chain. Despite the best efforts to control temperature and other environmental conditions, spectra taken over long periods of time with high-resolution detectors often suffer an apparent loss of resolution due to these drifts. In some detectors (notably scintillation counters), large changes in counting rate can also lead to apparent gain changes over fairly short periods of time.

A *spectrum stabilizer* is a device that in some way senses the position of a peak in the measured spectrum during the course of the measurement, compares its position with a known reference, and generates an error signal that can be used in a feedback loop to change the gain of the system to restore the peak to its correct position. The element in the signal chain to which the feedback is applied can be any component whose gain can be conveniently changed, including the photomultiplier tube for a scintillation counter, linear amplifier, or the ADC in the MCA.

The most common method of sensing the peak position during the measurement is illustrated in Fig. 18-14. Two SCA windows are set on opposite sides of a clearly resolved peak and connected to circuitry that can measure the difference in count rate between the two windows. At the start of the measurement, the gains are adjusted so that the peak is precisely centered between the two windows, and the count rate difference is therefore

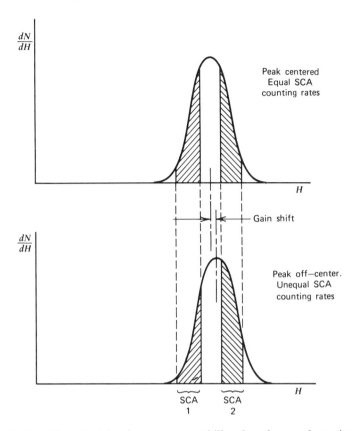

Figure 18-14 The principle of a spectrum stabilizer based on peak sensing. An error signal is derived from the counting rate difference in the two SCA windows.

zero. If the gain does drift, a difference in count rate between the two windows will be detected and will serve as the error signal. In this scheme, the window positions must be extremely stable because any changes in the SCA window positions will be interpreted as system gain changes. Ultimate stability is therefore limited by questions of SCA window stability. Methods have been developed[19, 20] to generate extremely stable SCA windows for this application.

For systems based on MCAs, the function of the windows can be performed by preselected groups of channels. In that way, the peak is held constant in the recorded spectrum and questions of window stability are no longer applicable. Digital stabilizers are available as commercial units that communicate with the MCA and adjust the gain of a preceding amplifier according to the error signals sensed by the stabilizer. Two peaks in the spectrum are usually monitored, one near the top and the other near the bottom of the spectrum range. In that way, corrections to drifts in both the slope and zero offset of the MCA calibration may be made if the amplifier allows adjustment of both its gain and zero offset.

If the spectrum to be recorded does not have readily identifiable peaks, other stratagems must be adopted. One possible procedure is to place the lower window in a flat region of the spectrum and the upper window in a region where the spectrum drops off rapidly.[21] By adjusting the window width, a net zero counting rate difference can be set at the beginning of the measurement. Other operational techniques, based on micro-processor technology, are outlined in Refs. 22 and 23. In scintillation counters, methods based on the apparent position of a reference light pulser peak are also in common use. The light pulser must be incorporated into the scintillation assembly, where it produces pulses that do not interfere with the region of the spectrum in which the signal peaks of interest lie. This method of stabilization is described in more detail in Chapter 10.

Another option to stabilize spectra without peaks involves artificially creating a peak by mixing the output of an electronic pulser with the signal pulses. The standard methods can then be used to stabilize the pulser peak. The pulses should be introduced at the earliest possible point in the signal chain because any drifts that occur prior to that point will not be compensated. The pulser technique is similar to that described earlier (see p. 620) for determination of losses due to pileup and dead time, and the same peak can also be used for gain stabilization.

V. COMPUTERIZED SPECTRUM ANALYSIS

A. Deconvolution or Unfolding

The differential pulse height spectrum dN/dH that is recorded from any radiation detector is the convolution of its inherent response function and the energy distribution of the incident radiation. We represent the incident energy distribution as $S(E)$, where $S(E)\,dE$ is the differential number of incident quanta with energy within dE about E. Information about this incident spectrum $S(E)$ is often the object of our measurement. The measured pulse height spectrum results from the convolution:

$$\frac{dN}{dH} = \int R(H, E)S(E)\,dE \qquad (18\text{-}4)$$

Here, $R(H, E)\,dH\,dE$ is the differential probability that a quantum of energy within dE about E leads to a pulse with amplitude within dH about H. In the special case that the incident radiation consists of only a single energy E_0, the spectrum $S(E)$ can be represented by the product of the number of quanta S_0 and the Dirac delta function:

$$S(E)\,dE = S_0\delta(E - E_0)\,dE$$

Putting this representation into the equation above, we observe

$$\left.\frac{dN}{dH}\right|_{E=E_0} = S_0 R(H, E_0)$$

Under these conditions, the measured pulse height distribution dN/dH is given by the *response function* $R(H, E_0)$ multiplied by the number of recorded pulses S_0. The response function, introduced earlier in Chapter 4, is characterized by noting that $R(H, E_0)\,dH$ is the differential probability of recording a pulse of amplitude within dH about H given an incident quantum of energy E_0. The shape of the response function will generally be different for incident quanta of different energy E_0. It may also depend on a number of other possible variables, including the operating conditions of the detector, the source–detector geometry, and the counting rate. In most cases, it is assumed that all these variables can be controlled sufficiently so that the only significant parameter is the radiation energy. When the spectrum is recorded by a MCA, Eq. (18-4) takes the discrete form

$$N_i = \sum_j R_{ij} S_j \tag{18-5}$$

where N_i is the recorded count in the ith channel, R_{ij} is the response matrix coupling the ith pulse height interval with the jth energy interval, and S_j is the radiation intensity in the jth energy interval.

The objective of many radiation measurements is to deduce the energy distribution of the incident radiation, or all values of the source elements S_j. We assume that the source distribution is discretized into L such intervals. The recorded spectrum is made up of M values of N_i, one for each channel. We can therefore write M simultaneous equations of the form of Eq. (18-5). If we assume that the entire response matrix R_{ij} is known, it is then theoretically possible to solve these equations for all the S_j elements, provided $M \geq L$. This process is generally called *spectrum deconvolution* or *spectrum unfolding*.

If the response function $R(H, E_0)$ is a narrow peak or mathematical delta function at each energy (the response matrix R_{ij} is then diagonal), a one-to-one correspondence exists between the measured pulse height and radiation energy. Such is very nearly the case, for example, for semiconductor diode detectors used to detect heavy charged particles (see Chapter 11). Then, the measured pulse height spectrum can be interpreted directly in terms of the incident radiation spectrum, and no unfolding is required. More often, however, the response function is more complex and may include both secondary peaks and a general continuum. The measured pulse height spectrum is then much more complicated, and unfolding may be necessary for a complete evaluation of the incident energy spectrum.

Two general problems arise in the unfolding process. The response functions may be subject to some uncertainty in that they may not be measured experimentally for all energies of interest. Furthermore, the detector operating conditions may change between the calibration and measurement steps, leading to corresponding changes in the response function. A second difficulty is due to the statistical nature of the recorded data. The content of each channel is subject to a statistical variance that can be estimated from the discussions given in Chapter 3. These statistical uncertainties will propagate through the unfolding calculation and give rise to corresponding variances in the calculated energy spectrum. Unless a large number of counts are collected over the measured spectrum, these statistical uncertainties can result in unacceptably large fluctuations in the derived energy spectrum.

Because of these imperfections, an exact set of solutions S_j cannot in general be obtained from the simultaneous equations. Instead, approximate solutions are sought, which are in some sense a best estimate of the incident energy spectrum. Many computer programs developed for this purpose are based on seeking a minimum in the weighted sum of residuals

$$\epsilon^2 = \sum_i W_i \left(N_i - \sum_j R_{ij} S_j \right)^2 \tag{18-6}$$

where the weighting factors W_i are often chosen to be inversely proportional to the statistical variance of each data point. The process of minimizing this sum of residuals amounts to a least-square fitting to the measured data. To reduce the effects of statistical uncertainties, many unfolding codes include some form of *data smoothing*, in which the content of each channel is replaced by a weighted average over a number of adjacent channels. To avoid loss of energy resolution, the interval over which the smoothing is carried out must always be less than the corresponding interval over which the response function changes rapidly.

The form of Eq. (18-5) also assumes simple linear superposition of all individual response functions. In most detectors, this assumption will be valid in the absence of complicating effects due to pulse pileup or other interference phenomena that may occur between two separate energies (see Chapter 17). Because these potential effects are accentuated at high count rates, unfolding methods that perform adequately at low counting rates may become unsatisfactory for spectra recorded at higher rates.

Although unfolding methods can generally be applied to the output of all radiation detectors, there are two main application areas where the techniques have been most thoroughly developed. One of these deals with the spectra recorded by proton recoil detectors used with fast neutrons. Here, the response functions consist only of continua that, to first approximation, are almost rectangular in shape (see Fig. 15-14). Techniques referenced in Chapter 15 for unfolding of these spectra are variants of the general deconvolution approach described above. The other major application concerns spectra recorded by gamma-ray spectrometers employing either scintillators or germanium detectors. Because of the presence of pronounced peaks in the response functions, the methods employed in this area tend to be somewhat different and are described later in this section.

The difficulty of the tasks involved in spectrum unfolding also depends on the nature of the incident radiation energy spectrum. If the source emits only monoenergetic radiations of a few separate energies, S_j in Eq. (18-5) has only a few nonzero terms. The recorded spectrum is therefore the simple superposition of several response functions that can, in principle, be measured in advance. On the other hand, if the incident radiation spectrum covers a continuum of energies, there will be a large number of S_j terms, and it is less practical to provide a set of experimentally measured response functions. In this event, calculations or analytic models are often used to provide the needed response matrix or to interpolate between measured functions.

B. Spectrum Stripping

If the incident spectrum consists only of a few discrete energies, a method known as *spectrum stripping* can sometimes be applied. The response functions $R(H, E_0)$ for each of the anticipated energies are stored in the memory of a MCA or associated computer. These response functions may be generated either by direct measurements using monoenergetic calibration sources or by calculation. The recorded spectrum is then "unpeeled" by starting with the largest anticipated energy. The corresponding response function is multiplied by a variable factor and subtracted from the recorded spectrum until the upper end of the spectrum is reduced to zero. The next lower energy is then chosen and the subtraction process repeated. Through this procedure, each succeedingly lower energy can be "stripped" until the entire recorded spectrum is reduced to zero. The multiplicative factor by which each response function was multiplied then gives the relative intensity of the corresponding component. If the number of possible energies is sufficiently small, the spectrum stripping process can be carried out manually, with a simple visual inspection determining the point at which each subtraction process is adequate.

C. Analysis of Spectra with Peaks

Although in principle it is possible to use general unfolding procedures for gamma-ray spectra, a full deconvolution of the spectrum is seldom carried out. One reason is that the response functions for scintillators or germanium detectors are quite complicated, and uncertainties quickly propagate over the entire spectrum. A more important reason, however, is that a full deconvolution is usually unnecessary. Gamma-ray spectra almost always consist of a number of discrete gamma-ray energies, and the response functions show a predominant peak. One would therefore expect a corresponding peak in the pulse height spectrum for every discrete energy in the incident radiation that can be resolved, and the spectrum analysis can consist of simply locating and quantifying these peaks. (If continuous gamma-ray spectra are involved, individual peaks will not appear in the recorded spectrum and there is no recourse but to carry out a full deconvolution.)

Computer programs are continually being developed and improved for the automatic localization and quantification of peaks in multichannel spectra. Many are available as commercial packages supplied through the vendors of multichannel analysis equipment. Others are available as public domain software. Reviews of some of these programs as applied to germanium detector spectra[†] can be found in Refs. 24–26. Two representative examples (SAMPO and GAUSS) are described in Refs. 27 and 28. The programs serve two main purposes: to locate the centroid position of all statistically significant peaks appearing in the spectrum and to determine the net area under each peak that lies above the continuum or background on which the peak is superimposed.

1. PEAK LOCALIZATION

The task of picking out statistically significant peaks that rise above a smoothly varying continuum is a task that appears deceptively simple. When cast in the form of an automatic computer program, the search process must be relatively sophisticated to avoid the sensing of false peaks due only to statistical fluctuations in the background, and at the same time remain highly efficient for true peaks of low intensity. The problem is further complicated by the fact that two closely lying peaks may not be fully separated in the spectrum, and the program must be able to distinguish such unresolved doublets from simple single peaks.

One feature of a smooth symmetric peak is that its second derivative shows a large negative excursion at the center of the peak position. Therefore, many peak search routines are based on sensing the second differences in channel-to-channel data that have been smoothed to minimize the effects of statistical fluctuations. Once a tentative peak has been identified by this procedure, additional criteria are normally tested to check its acceptability. The data in the local area may be fitted with an analytic function (usually a Gaussian or a variant thereof), and the width may be required to match that of an appropriate calibration peak. Peaks with unusual width are normally investigated as candidates for possible unresolved doublets by methods that often involve an iterative procedure, in which varying doublet spacing and relative intensities are tried. For each detected peak, the parameters of the best-fit analytic function are then used to define the position of its centroid. The effectiveness with which a typical peak search routine can locate many peaks in a single spectrum is shown by Fig. 18-15.

2. DETERMINATION OF PEAK AREA

Once each peak has been located, the next step normally involves a sequential examination of the data in the immediate vicinity of individual peaks to determine their area. Various automated techniques for determining peak areas in multichannel spectra have been compared in Refs. 30–32. The methods can be subdivided into two groups: those

---

[†]Although many of these same techniques can be applied to analyze spectra from scintillators, we limit this specific discussion to procedures that are most useful for spectra in which the peaks are relatively narrow due to the good energy resolution of the detector.

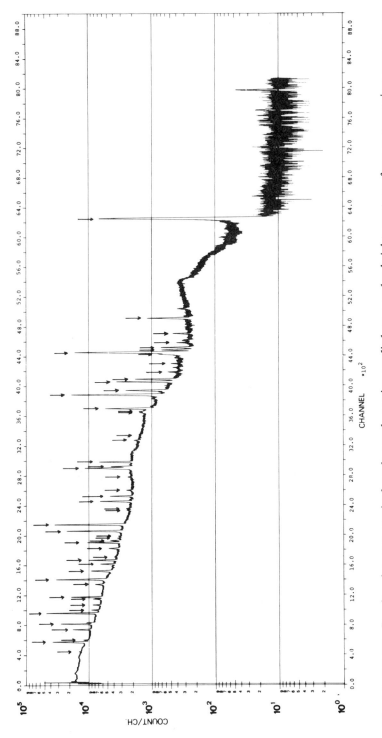

Figure 18-15 Results of a computerized peak search routine applied to a pulse height spectrum from a germanium gamma-ray spectrometer. Arrows indicate the locations of peaks found by the program. (From Sasamoto et al.[29])

that obtain the area from a fitted analytic function and those that carry out a direct summation of the data points between prescribed limits. In either approach, the contribution of the continuum on which the peak is superimposed must be subtracted. The data in the channels on either side of each peak are normally used to define the continuum, and an assumed linear or quadratic curve is fit to produce an estimated continuum in the region under the peak. Channel-by-channel subtraction of the continuum then produces corrected data for the following steps.

In the direct summation method, some criteria must be preselected for the number of channels over which the data will be summed. In principle, the summation should extend over all channels that are significantly above the continuum. In practice, it is normally limited to a given number of channels on either side of the peak centroid, depending on the detector resolution. A number of alternative schemes are discussed in Ref. 31.

A more common procedure is to use the area obtained from an analytic fit to the data points comprising the peak. Most assumed shapes involve a primary Gaussian peak with a small additive component (often exponential) to represent tailing on the low-energy side of the peak, caused by incomplete charge collection within the detector. Least-square methods are normally used to fit the function, and the area is then derived from the fitted shape parameters. Examples of a number of complex fitting functions applied to germanium detector photopeaks are given in Refs. 33–38.

If the complicating effects of background continuum and peak tailing are not significant or can be subtracted out, most spectral peaks can be treated as a simple Gaussian

$$y(x) = y_0 \exp\left(-\frac{(x - x_0)^2}{2\sigma^2}\right) \tag{18-7}$$

where y is the amplitude of the Gaussian at channel x, and y_0 is its maximum. In this case, the iterative procedures normally required for more complex fitting functions can be replaced by a simple direct process.[39,40] By defining the quantity

$$Q(x) \equiv \frac{y(x - 1)}{y(x + 1)} = \exp\left(\frac{2(x - x_0)}{\sigma^2}\right) \tag{18-8}$$

and taking logarithms

$$\ln Q(x) = \frac{2(x - x_0)}{\sigma^2} \tag{18-9}$$

a linear function of x is derived. Therefore, a linear least-square fit to a plot of $\ln Q(x)$ versus x is the equivalent to fitting a Gaussian to the original data. The slope m and intercept b of this linear fit then yield the standard deviation σ and centroid x_0 of the Gaussian

$$\sigma = \sqrt{\frac{2}{m}} \quad \text{and} \quad x_0 = -\frac{b}{m}$$

In making this fit, the data are normally weighted inversely with respect to their relative statistical variance (see Chapter 3). Thus, a variable weighting factor

$$W_i = \left[\frac{1}{N_{i+1}} + \frac{1}{N_{i-1}} \right]^{-1} \tag{18-10}$$

is given to each $\ln Q(x)$ point, where N_i represents the content of the ith channel. Because data near the edges of the peak are most subject to distortions, the fit is often limited to data points that lie at least halfway up the peak.

Once the parameters σ and x_0 of the Gaussian are calculated, the peak height y_0 can be obtained from the fitted data. If the fit were perfect, y_0 could be calculated from Eq. (18-7) using any one of the data points. Fits to real data are never perfect, so that a more rigorous approach[39] is to form a weighted mean from all the data points

$$\ln y_0 = \frac{\sum_{i=1}^{N} W_i' \left[\ln N_i + (x_i - x_0)^2 / 2\sigma^2 \right]}{\sum_{i=1}^{N} W_i'} \tag{18-11}$$

where the weighting factors are chosen as

$$W_i' = \left[\frac{1}{N_i} + \frac{(x_i - x_0)^2}{\sigma^4} \left((\Delta x_0)^2 + \frac{(x_i - x_0)^2}{\sigma^2} (\Delta \sigma)^2 \right) \right]^{-1} \tag{18-12}$$

In this expression, Δx_0 and $\Delta \sigma$ are the estimated standard deviations of x_0 and σ obtained from the linear fit.

From the properties of the Gaussian curve, the total area under the peak can now be calculated from

$$A = \sqrt{2\pi} \, \sigma y_0 = 2.507 \sigma y_0 \tag{18-13}$$

and its full width at half maximum is

$$\text{FWHM} = 2\sqrt{2 \ln 2} \, \sigma = 2.355 \sigma \tag{18-14}$$

PROBLEMS

18-1. How many pulse height channels should be provided to represent adequately the full spectrum from a system with 0.3% pulse height resolution?

18-2. Two peaks in a recorded pulse height spectrum are separated by 24 channels. Assuming a perfectly linear system, by how many channels will this separation change if the gain of the amplifier supplying the MCA is decreased from 1000 to 750, and the zero offset of the MCA is increased from 10 to 15 channels?

18-3. A Wilkinson-type ADC has a conversion gain of 2048 channels and a maximum conversion time of 25 μs. At what frequency must the oscillator operate?

18-4. How many stages are required in a successive approximation ADC to achieve a conversion gain of 4096 channels?

18-5. Why is it not possible to apply the formulas for dead time corrections developed in Chapter 4 to correct for losses in a MCA using a linear ramp ADC?

18-6. A MCA using a Wilkinson-type ADC operating at 80 MHz has a pulse storage time of 2.5 μs.

  (a) What is the analyzer dead time for pulses stored in channel number 300?

  (b) What will be the fractional dead time for a true pulse rate of 5000/s if the average pulse amplitude falls in channel number 220? Repeat for a true rate of 50,000/s.

  (c) If the analyzer is set to record for a live time of 10 min, how much actual time will elapse under the conditions of part (b)?

18-7. Equally spaced pulses of fixed amplitude are generated at an adjustable frequency in an electronic pulser. They are supplied to a MCA with a dead time of 90 μs for the size of pulse involved. Sketch a plot of percent dead time losses versus pulser frequency over the range from 10 to 30 kHz.

18-8. The coincident outputs from two detectors are to be recorded in a two-dimensional multiparameter analyzer. If the detectors have pulse height resolutions of 0.5 and 2.5%, estimate the total number of memory locations needed to record faithfully the coincident output of both detectors over their full span.

18-9. A pulse height spectrum is recorded in a MCA by counting a source for a given live time, and then removing the source and subtracting background for an equal live time. In the upper region of the spectrum, the source contributes nothing so the average channel content after subtraction should be zero. Due to the influence of counting statistics, however, some channel-to-channel fluctuation about zero is observed. If the background is about 300 counts/channel over this region of the spectrum, what deviation from zero should be expected in a typical channel?

18-10. One commonly used method to sense the position of a peak superimposed on a continuous background is to take second differences in the multichannel data. To illustrate the behavior in the vicinity of a peak, make a plot of the second derivative of a Gaussian peak across its full width.

18-11. The following is a portion of a gamma-ray pulse height spectrum recorded using a germanium detector and MCA:

Channel Number	Counts	Channel Number	Counts
711	238	720	1625
712	241	721	1739
713	219	722	1412
714	227	723	901
715	242	724	497
716	280	725	308
717	409	726	256
718	736	727	219
719	1190	728	230

It can be assumed that the data consist of a constant background plus a Gaussian-shaped peak.

(a) Plot the data and estimate the constant background level. Find the net number of counts under the peak by direct summation. Estimate the centroid location of the peak (to the nearest tenth of a channel). Estimate the FWHM of the peak.

(b) Fit the net counts in the region of the peak with a Gaussian function, using the linear procedure described at the end of this chapter. Obtain the resulting peak area, centroid, and standard deviation, and compare with the results of part (a).

REFERENCES

1. D. H. Wilkinson, *Proc. Cambridge Philos. Soc.* **46**, Pt. 3, 508 (1950).

2. P. W. Nicholson, *Nuclear Electronics*, Wiley-Interscience, London, 1974.

3. S. Kinbara, *Nucl. Instrum. Meth.* **143**, 267 (1977).

4. M. Brendle, *Nucl. Instrum. Meth.* **144**, 357 (1977).

5. S. G. Gobbur, D. A. Landis, and F. S. Goulding, *Nucl. Instrum. Meth.* **140**, 405 (1977).

6. D. E. Carter and G. Randers-Pehrson, *Nucl. Instrum. Meth.* **199**(3), 497 (1982).

7. S. K. Dhawan and K. Kondo, *IEEE Trans. Nucl. Sci.* **NS-33**(1), 77 (1986).

8. C. Cottini, E. Gatti, and V. Svelto, *Nucl. Instrum. Meth.* **24**, 241 (1963).

9. C. B. A. Correia and C. A. N. Conde, *Nucl. Instrum. Meth.* **A235**, 536 (1985).

10. X. Xianjie and P. Dajing, *Nucl. Instrum. Meth.* **A259**, 521 (1987).

11. O. U. Anders, *Nucl. Instrum. Meth.* **68**, 205 (1969).

12. H. H. Bolotin, M. G. Strauss, and D. A. McClure, *Nucl. Instrum. Meth.* **83**, 1 (1970).

13. M. Wiernik, *Nucl. Instrum. Meth.* **96**, 325 (1971).

14. E. J. Cohen, *Nucl. Instrum. Meth.* **121**, 25 (1974).

15. J. Harms, *Nucl. Instrum. Meth.* **53**, 192 (1967).

16. C. F. Masters and L. V. East, *IEEE Trans. Nucl. Sci.* **NS-17**(3) 383 (1970).

17. G. P. Westphal, *Nucl. Instrum. Meth.* **146**, 605 (1977).

18. G. P. Westphal, *Nucl. Instrum. Meth.* **163**, 189 (1979).

19. M. Yamashita, *Nucl. Instrum. Meth.* **114**, 75 (1974).

20. P. J. Borg, P. Huppert, P. L. Phillips, and P. J. Waddington, *Nucl. Instrum. Meth. Phys. Res.* **A238**, 104 (1985).

21. M. Matoba and M. Sonoda, *Nucl. Instrum. Meth.* **92**, 153 (1971).

22. J. Braunsfurth and K. Geske, *Nucl. Instrum. Meth.* **133**, 549 (1976).

23. M. Brendle, *Nucl. Instrum. Meth.* **141**, 577 (1977).

24. C. M. Lederer, in *Radioactivity in Nuclear Spectroscopy* (J. H. Hamilton and J. C. Manthuruthil, eds.), Gordon & Breach, New York, 1972.

25. I. De Lotto and A. Ghirardi, *Nucl. Instrum. Meth.* **143**, 617 (1977).

26. W. Westmeier, *Nucl. Instrum. Meth.* **180**, 205 (1981).

27. M. J. Koskelo, P. A. Aarnio, and J. T. Routti, *Computer Phys. Commun.* **24**, 11 (1981).

28. R. G. Helmer and C. M. McCullagh, *Nucl. Instrum. Meth.* **206**, 477 (1983).

29. N. Sasamoto, K. Koyama, and S. Tanaka, *Nucl. Instrum. Meth.* **125**, 507 (1975).

30. P. A. Baedecker, *Anal. Chem.* **43**, 405 (1971).

31. L. Kokta, *Nucl. Instrum. Meth.* **112**, 245 (1973).

32. J. Hertogen, J. De Donder, and R. Gijbels, *Nucl. Instrum. Meth.* **115**, 197 (1974).

33. B. L. Roberts, R. A. J. Riddle, and G. T. A. Squier, *Nucl. Instrum. Meth.* **130**, 559 (1975).

34. L. A. McNelles and J. L. Campbell, *Nucl. Instrum. Meth.* **127**, 73 (1975).

35. Y. Takeda, M. Kitamura, K. Kawase, and K. Sugiyama, *Nucl. Instrum. Meth.* **136**, 369 (1976).

36. H. Baba, S. Baba, and T. Suzuki, *Nucl. Instrum. Meth.* **145**, 517 (1977).

37. H. H. Jorch and J. L. Campbell, *Nucl. Instrum. Meth.* **143**, 551 (1977).

38. O. Ciftcioglu, *Nucl. Instrum. Meth.* **174**, 209 (1980).

39. T. Mukoyama, *Nucl. Instrum. Meth.* **125**, 289 (1975).

40. U. Abondanno, A. Boiti, and F. Demanins, *Nucl. Instrum. Meth.* **142**, 605 (1977).

CHAPTER · 19

Miscellaneous Detector Types

I. CERENKOV DETECTORS

A category of radiation detectors is based on the light that is emitted by a fast charged particle passing through an optically transparent medium with index of refraction greater than 1. An extensive coverage of the origins and applications of this *Cerenkov light* can be found in Ref. 1. The light is emitted whenever the velocity of a charged particle exceeds that of light in the medium through which it is passing, or

$$\beta n > 1 \tag{19-1}$$

where n is the refractive index of the medium and β is the ratio of the velocity of the particle in the medium to that of light in a vacuum. Detectors based on sensing the Cerenkov light are widely used in high-energy physics experiments but have only limited use for the lower-energy radiations of interest in this text. For particles in the tens of MeV range or lower, electrons are the only category to achieve sufficiently high velocity to emit Cerenkov light in available materials. Applications therefore involve either fast primary electrons such as beta particles or energetic secondary electrons produced in gamma-ray interactions. Examples are given in Refs. 2 and 3.

Cerenkov detectors bear some similarity to common scintillation detectors, in that the emitted light is converted into an electrical signal by a photomultiplier tube in optical contact with the Cerenkov medium. However, several important properties are quite different.

1. As indicated by Eq. (19-1), a minimum particle velocity is required in a given medium in order to generate any Cerenkov light. Therefore, Cerenkov detectors have an inherent discrimination ability that is unique among radiation detectors. Their response is limited to electrons whose energy exceeds a minimum or threshold given by

$$E_{\text{th}} = m_0 c^2 \left(-1 + \sqrt{1 + \frac{1}{n^2 - 1}} \right) \tag{19-2}$$

where $m_0 c^2$ represents the electron rest-mass energy (0.511 MeV). The threshold electron energy is plotted in Fig. 19-1 as a function of the index of refraction n. Also plotted is the minimum energy gamma ray that can produce Compton electrons of this energy by 180° scattering. Because photoelectric absorption is negligible at these energies in low-Z materials, this latter curve also represents a practical energy threshold for the detection of gamma rays. This inherent energy discrimination can be a very useful feature in situations in which a high rate of low-energy events would otherwise be recorded. Potential problems arising from pulse pileup or excessive rates can be avoided if these unwanted events are eliminated at the start. By using media with different indices of refraction, various detection thresholds can be chosen.

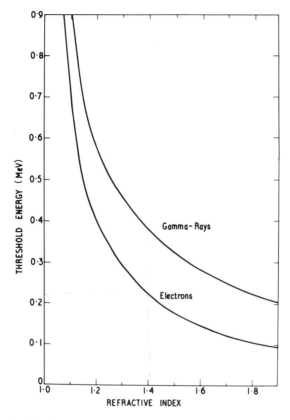

Figure 19-1   The threshold energy for Cerenkov radiation as a function of the index of refraction of the detection medium. Curves are shown both for electrons and for gamma rays that can yield, by 180° Compton scattering, an electron of the threshold energy. (From Sowerby[2].)

2. The light is emitted over the very short time required for the electron to slow from its initial velocity to below the threshold velocity. This time will typically be of the order of picoseconds in solids or liquids, so that Cerenkov detectors have the potential of being exceptionally fast. As a practical matter, their timing properties are normally limited by the photomultiplier tube.

3. The most important drawback of Cerenkov detectors is the low level of light produced. As shown in Fig. 19-2, the number of photons emitted per electron in common Cerenkov media is only several hundred per MeV. This corresponds to a conversion of only about 10^{-3} of the particle energy into visible light, about a factor of 100 smaller than the corresponding figure for an efficient scintillator (see Chapter 8).

4. In contrast to scintillation light, which is emitted isotropically, Cerenkov photons are emitted preferentially along the direction of the particle velocity. The light is confined to a cone with vertex angle θ, where

$$\cos \theta = \frac{1}{\beta n} \tag{19-3}$$

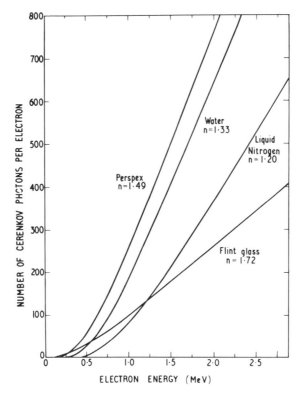

Figure 19-2 Calculated yield of Cerenkov photons in the 300–600 nm wavelength region for several detection media. (From Sowerby[2].)

5. The yield of Cerenkov photons per unit wavelength λ is proportional to $1/\lambda^2$. The emission is therefore concentrated in the short-wavelength region of the spectrum.

Choice of the Cerenkov medium is made from materials with good optical transmission properties and, ideally, no scintillation component. By proper choice of the material, the index of refraction can vary from 1 to about 1.8. Gases at various pressures can cover the range from 1 to about 1.2. Convenient materials with index between 1.2 and 1.33 are scarce, but liquified gases can be used for this purpose. The region between 1.33 and 1.47 is conveniently covered by mixtures of glycerin and water. Above an index of 1.47, many transparent solids such as lucite (perspex), glasses, and crystalline materials are common choices.

The Cerenkov effect can also generate light in the glass face plate of photomultiplier tubes exposed to gamma rays. This light can be an undesirable complication in the response of ordinary scintillation detectors under conditions of high gamma-ray background.

II. GAS-FILLED DETECTORS IN SELF-QUENCHED STREAMER MODE

In recent years, a mode of operation has been developed for gas multiplication detectors that is distinct from the traditional proportional and Geiger regimes described in Chapters 6 and 7. It is called the *limited streamer* or *self-quenched streamer* (SQS) mode

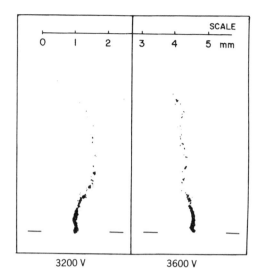

Figure 19-3 Photographs of individual SQSs for two different applied voltages. The anode position is indicated by the line segments at the bottom. (From Atac et al.[4])

and has found application primarily in position-sensitive multiwire chambers. In common with proportional or Geiger–Mueller detectors, SQS devices rely on the formation of Townsend avalanches in a gas to multiply the amount of charge collected from ion pairs formed along the track of incident ionizing particles. However, the SQS mode of operation differs in the way in which these avalanches are allowed to propagate through the gas.

The discussions in Chapter 7 stressed the role that ultraviolet (UV) photons play in the spread of avalanches in the Geiger mode. These photons are emitted by excited atoms formed by electron collisions during the progression of an avalanche. In the proportional mode, the spread of avalanches is prevented either by keeping the avalanche small or by adding a "quench" gas that absorbs the UV photons without additional electron release. Only one avalanche is produced per original ion pair, and therefore the size of the output pulse remains proportional to their number. In the Geiger mode, the UV photons emitted in one avalanche propagate throughout the gas volume and create additional avalanches that eventually spread along the entire length of the anode wire. The process is terminated by the buildup of space charge around the anode, and the large output pulse size no longer depends on the number of original ion pairs.

In the SQS mode, the avalanches are also allowed to propagate, but only in a controlled fashion. By using gas mixtures that strongly absorb the UV photons, the formation of additional avalanches far from the original site is prevented. Instead, new avalanches are limited to the immediate vicinity of the original. It is observed experimentally[4] that, under proper conditions, the new avalanches will grow in the form of a narrow "streamer" that extends radially away from the anode wire surface (see Fig. 19-3). These streamers are 150–200 μm thick and extend a few millimeters from the anode. They are self-limiting and terminate with a final length that increases with applied voltage. If the original ion pairs are formed over a limited axial distance, only a single streamer is formed. For more extended tracks, multiple streamers are observed. At sufficiently high values of the voltage, a single electron can trigger a streamer.

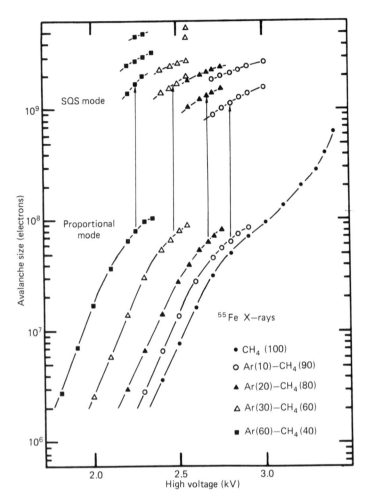

Figure 19-4 Variation of the avalanche size with applied voltage in both the proportional and SQS modes of operation. Results are shown for methane and several argon–methane mixtures. (From Koori et al.[5])

One model of this behavior[4] is based on the local distortions in the applied electric field that are created within an avalanche. At high values of the gas multiplication, the positive ions form a cone-shaped cloud of space charge that creates a local dipole field. Its effect is to reduce the electric field near the sides of the cone but to reinforce it at its tip. Electrons formed near this region by UV photons will preferentially create additional avalanches near the cone tip where the field is the highest. The streamer thus grows radially outward after beginning near the wire surface. It terminates when the lower applied field at larger radii is no longer high enough to sustain further avalanches.

Figure 19-4 shows the transition from proportional to SQS mode in a gas-filled detector. Pulses from streamer formation appear with an amplitude that is typically an order of magnitude larger than those from simple proportional operation. There is usually a range of applied voltage over which both proportional and SQS modes are observed. Streamer formation is promoted by using an anode wire of relatively large diameter (25–100 μm) and is observed only in certain gas mixtures.

The SQS mode combines some features of both Geiger and proportional mode operation. As in a Geiger tube, the additional avalanches that are created in the streamer increase the pulse size and reduce demands placed on pulse-processing electronics. This internal amplification is particularly helpful in position-sensitive detectors where the larger signal reduces position uncertainties compared with proportional operation.[6] However, as in a Geiger tube, the amplitude of the signal from an individual streamer is no longer a measure of the size of the initiating event. The advantages of proportional operation for radiation spectroscopy are therefore lost. By preventing the spread of avalanches along the anode wire, SQS operation avoids the long dead time of Geiger tubes. The rise time of typically output pulses has been measured[4] as 30–40 ns, so the SQS mode can result in good timing precision as well.

III. LIQUID IONIZATION AND PROPORTIONAL COUNTERS

The conventional detectors in which charges created by ionizing radiation are directly collected employ either a gas or a semiconductor material as the detection medium. The great advantage afforded by the high density of solids or liquids has stimulated the search for additional materials suitable for use in radiation detectors. The liquid and solid phases of noble gases are condensed dielectrics in which electrons remain free, and for which sufficient purification is possible to eliminate electronegative impurities. Therefore, active development continues in the incorporation of these cryogenic materials into detectors that resemble conventional ionization or proportional counters.

Some properties of liquified noble gases of interest in detector applications are given in Table 19-1. Of these, liquid xenon has received the most attention and has been applied successfully both as a conventional ion chamber, where the charges are simply collected, and as a proportional counter, in which charge multiplication occurs at high values of the electric field. The electron drift velocity saturates at 3×10^3 m/s for electric field values greater than about 10^5 V/m (Refs. 8 and 9). The onset of multiplication is observed to occur at about 10^8 V/m (Ref. 10).

Successful operation of these counters requires careful attention to purity of the liquified gas, and some systems of continuous recycling through a purifier is usually incorporated into the design. Liquified gases require that the counter be maintained at cryogenic temperatures, which is an obvious operational handicap for most applications. However, the low values for W and F in Table 19-1 hold out the promise of excellent

TABLE 19-1 **Properties of Some Condensed Media for Ionization Chambers**

	Z	$(\frac{g}{cm^3})$	Boiling Point	W (eV/ion pair)		F (Fano factor)	
				Calculated[a]	Experimental[a]	Model 1[b]	Model 2[c]
Liquid Ar	18	1.41	87 K	23.3	23.6	0.107	0.116
Liquid Kr	36	2.15	120 K	19.5		0.057	0.070
Liquid Xe	54	3.52	166 K	15.4	15.6	0.041	0.059
			Melting point				
Solid Ar	18	1.62	84 K				

[a] From T. Doke et al.[7]

[b] Based on W. Shockley, *Czech. J. Phys.* **B11**, 81 (1961).

[c] Based on G. D. Alkhazov et al., *Nucl. Instrum. Meth.* **48**, 1 (1967).

energy resolution if charge collection can be made efficient (see Chapter 6). Furthermore, the high atomic number of xenon (54) is extremely attractive for applications involving gamma-ray spectroscopy. A liquid xenon ionization chamber has the potential of providing the good gamma-ray detection efficiency of NaI, with an improved energy resolution intermediate between that of an NaI(Tl) scintillator and a germanium detector.

If the impurity concentrations are kept very low (less than one part per billion of oxygen equivalent), the distance over which electrons can be drifted in liquid argon or xenon approaches a meter or more.[11] Chambers with large active volume are therefore feasible, and position-sensing can be carried out by measuring the electron drift time.[12] The energy resolution in small gridded ionization chambers is typically 5–6% for gamma rays,[13] but a value of 3.3% has been reported[14] for 976 keV conversion electrons in liquid argon. These figures are still much poorer than predicted from theory, and it is thought that the degradation in resolution is caused by variations in recombination along the tracks of delta rays (low-energy electrons) produced in large numbers along the primary path.[15,16]

Liquid argon is also a scintillator, as described in Chapter 8. Most of the scintillation light derives from excited molecules that are formed by the recombination of ion–electron pairs. Applying an electric field near 10^6 V/m to collect the free electrons also suppresses recombination, and the scintillation intensity drops to one-third of its zero field value. Both scintillation and ion chamber signals can be measured simultaneously from the same sample of liquid argon,[17] and it is observed that the sum of the two signals is approximately independent of electric field strength.

Some attention has been given to solid rather than liquid argon.[18] The electron mobility is actually higher in the solid phase,[19] but problems related to poisoning and polarization of the solid have inhibited further development.

There is also interest in some nonpolar liquids as ionization detector media at room temperature. For example, tetramethylsilane has been shown to be an acceptable liquid filling for ionization chambers operated both in pulse[20] and current[21] modes. As in liquified noble gases, stable operation requires careful attention to purity.[22]

IV. CRYOGENIC AND SUPERCONDUCTING DETECTORS

A. Cryogenic Microcalorimeters or Bolometers

Any substance when exposed to ionizing radiation will in principle show an increase in temperature due to the energy absorbed from the incident particles or photons. Instruments known as *bolometers* have been used for many years for thermal or infrared radiation in which the incident flux on a target is measured through the temperature rise as sensed by a thermistor. For most common conditions with ionizing radiation, this temperature rise is too small to be measured unless the sample is exposed to a very high flux of radiation. For example, an absorbed dose of 1 gray (100 rad) represents by definition the absorption of 1 joule per kilogram (J/kg) of the absorber. The flux of incident radiation needed to deposit this energy is rather large, corresponding, for example, to the full absorption of over 10^9 5 MeV alpha particles per gram of the absorber. Converting energy units, one finds that the adiabatic absorption of that energy in water results in a temperature rise of only 2.39×10^{-4} °C. Consequently, calorimetric measurements in which the temperature rise in a thermally isolated sample is measured directly have traditionally been limited to relatively intense radiation fields.

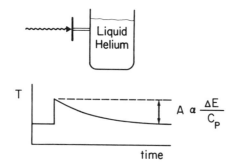

Figure 19-5 Operational principle of the cryogenic bolometer.

Recently, several research groups have demonstrated that the sensitivity of such calorimetric methods can be made many orders of magnitude better if the temperature rise is measured in miniature samples of material maintained at very low temperature. The heat capacity of dielectric materials or crystals is proportional to T^3, where T is the absolute temperature. Therefore, the lower the temperature, the greater will be the temperature rise per unit of absorbed energy. This temperature rise must be measured against the background of the statistical fluctuations in temperature caused by random fluctuations in the energy content of the absorber. Such fluctuations are due to the random variations in the flow of *phonons* across the thermal coupling between the absorber and the surrounding medium. It can be shown[23] that the mean square value of these fluctuations is proportional to T^5. Reducing the temperature therefore also greatly suppresses the background against which the signal must be measured.

By reducing temperatures to below 1 K, it is possible to sense the momentary temperature rise *due to a single photon or incident particle*. A schematic sketch of the resulting temperature pulse is shown in Fig. 19-5. This temperature pulse can be measured using a thermistor in contact with the absorber providing an electrical signal that is analogous to the signal pulse from more conventional pulse mode detectors. What is unique is the fact that the pulse does *not* consist of discrete charges (such as ion pairs or electron–hole pairs) that are collected but rather is only a change in the thermal state of the absorber. Thus, one completely avoids the limits on energy resolution set by charge carrier statistics in virtually every other type of radiation detector. Ultimate limits are now set only by thermodynamic fluctuations at the phonon level in the absorber. Since phonon energies are typically several orders of magnitude below the energy required to create an electron–hole pair in a semiconductor, the theoretical energy resolution is many times better for the microcalorimeter. Analyses[24,25] show that fundamental thermodynamic limits could result in energy resolution figures as low as several eV or less.

Early experiments[23,26,27] typically have chosen silicon, germanium, or diamond absorbers with areas of less than 1 mm^2 and thicknesses of 25–500 μm. The absorber is held by supports of small cross section that provide the thermal link to a liquid helium cryostat maintained at an absolute temperature of 1 K or less. The very small absorber volumes are currently of interest only for weakly penetrating radiations such as soft X-rays. However, it is possible to imagine arrays of many such small samples fabricated using monolithic silicon technology that could be built up into substantial volumes. Some recent results[28] obtained using a HgCdTe semiconductor target operated at 0.1 K have demonstrated an energy resolution of 17 eV FWHM for the 6 keV X-ray lines from a ^{55}Fe source, about a factor of 10 better than that obtained from typical Si(Li) X-ray spectrometers. These very encouraging results will surely spur increased development work on microcalorimeters in the years ahead.

B. Superconducting Grain Detectors

The very small heat capacity of materials near absolute zero temperature can be exploited in another way to detect single particles or photons. A number of materials, for example, tin or indium, are superconductors at very low temperatures but revert to "normal" conductivity if the temperature is raised. In a very small grain of such material with a diameter of a few micrometers, the energy deposited by an ionizing particle can be sufficient to transform the grain from the superconducting to the normal state. The energy required for this transition can be as small as tens or hundreds of eV if the grain mass is small and it is maintained in a metastable state very close to the transition temperature. In the presence of an applied magnetic field, the transition temperature becomes a function of the magnetic field. It is possible to create combinations of temperature and field strength in which a grain is superconducting but metastable, in the same sense that a supersaturated vapor is metastable to the formation of liquid droplets. It then requires the deposition of only a small amount of energy to trigger the transition of the entire grain from the superconducting to the normal state. This transition creates an electromagnetic signal that can be sensed in a readout loop at some distance through the voltage induced by the changing flux. Very sensitive sensors known as SQUIDs (superconducting quantum interference devices) have been shown[29-31] capable of detecting the electromagnetic signal created by the transition of a single grain of diameter ranging from 5 to 15 μm.

Detectors with dimensions of several centimeters have been constructed by dispersing a large number of superconducting grains in a dielectric filler material such as paraffin. Sensors placed at various points on all sides of this "colloid" detector can not only sense the transition of a single grain anywhere within its volume but can also provide information on the position of the interaction through triangulation techniques.

Superconducting grain detectors are still very much in the development stage, but they offer an intriguing new mechanism for the detection of radiation. Since the grain diameter must be small to minimize the required threshold energy, initial applications will likely be limited to detecting short-range heavy charged particles or recoil nuclei. Some examples include the detection of neutrons by loading the paraffin matrix with boron[32] and proposed application to the detection of neutrinos through low-energy recoil nuclei created in rare elastic scattering events.[33,34]

V. PHOTOGRAPHIC EMULSIONS

The use of photographic film to record ionizing radiations dates back to the discovery of X-rays near the turn of the century but has remained an important technique through the present time. A thorough presentation of the modern uses of photographic emulsions for ionizing radiation is given in the text by Herz.[35]

Ordinary photographic film consists of an emulsion of silver halide grains (consisting mainly of silver bromide) suspended in a gelatin matrix and supported with a backing of glass or cellulose acetate film. The action of ionizing radiation in the emulsion is similar to that of visible light, in that some of the grains will be "sensitized" through interaction of the radiation with electrons of the silver halide molecules. The sensitized grains remain in this state indefinitely, thereby storing a latent image of the track of the ionizing particle through the emulsion. In the subsequent development process, the entire sensitized grain is converted to metallic silver, vastly increasing the number of affected molecules to the point that the developed grain is visible. Following development, the emulsion is fixed by dissolving away the undeveloped silver halide grains, and a final washing step removes the processing solutions from the developed emulsion.

Radiation applications of photographic emulsions are conveniently divided into two categories: those in which a general darkening of the emulsion is recorded due to the cumulative effects of many individual interactions, and those in which single particle tracks are individually recorded. The first category includes the broad field of *radiography*, in which an image is recorded of the transmitted intensity of a beam of radiation. The specialized films used for this purpose do not differ radically from ordinary photographic films. Single-particle tracks, however, are best recorded in *nuclear emulsions*, which are much thicker and differ in composition from photographic emulsions.

A. Radiographic Films

A sketch of a typical radiographic measurement is given in Fig. 19-6. The incident radiation is either in the form of a point source or a parallel beam, which will therefore cast a definite "shadow" of the object on the recording emulsion. Although the most common radiations used in radiography are X-rays generated by standard tubes, higher-energy gamma rays from radioisotope sources or high-energy bremsstrahlung from electron linear accelerators also are applied in industrial radiography of thick objects. Neutrons can also be used in radiography, as discussed below.

Radiographic emulsions are typically 10–20 μm thick with grains up to 1 μm in diameter. The silver halide concentration amounts to about 40% by weight. To increase their sensitivity, some radiographic films are "double sided," where the emulsion is applied to both surfaces of the base film.

In X- or gamma radiography, the transmitted photon must interact to form a secondary electron if it is to be recorded. The direct interaction probability of a photon of typical energy within the emulsion itself is usually no more than a few percent at best. Therefore, films applied in the direct imaging of X-rays or gamma rays are relatively insensitive. A typical sensitivity curve is shown on the next page, where D is defined as the density of the developed film as measured by an optical densitometer. Three regions of this curve are readily identifiable. At low exposures, too few developed grains are produced to measurably affect the film density and the emulsion is underexposed. At the

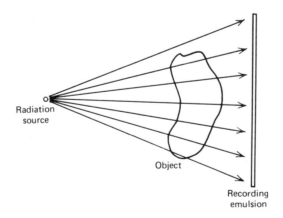

Figure 19-6 The elements in the formation of a radiographic image.

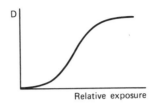

other extreme, high exposures result in a dense concentration of developed grains and a resulting overexposure. Intermediate exposures result in a density that varies approximately linearly with exposure, and this is the region of normal operation.

In medical radiography or other applications in which the incident radiation intensity is limited, steps are often taken to enhance the sensitivity of the emulsions. One method is to sandwich the film between foils made from materials with high atomic number. Photoelectric or Compton interactions within these converter foils may then contribute secondary electrons, which add to those created within the emulsion itself. Alternatively, *intensifier screens* may be placed next to the emulsion which consist of light-emitting phosphors of high atomic number, such as calcium tungstate. Gamma-ray interactions within the screen create visible light through the normal scintillation process, which then leads to additional sensitization of the emulsion. Because the light travels in all directions, a compromise must be struck between sensitivity and spatial resolution in choosing the thickness of the screen. In typical situations, the sensitivity of films to X-rays can be increased by a factor of 10 through the use of such intensifier screens.

Neutron radiography can also be carried out using emulsions, provided the appropriate type of converter foil is used. Slow neutrons can be imaged by sandwiching the emulsion between foils of gadolinium, which exhibits a large neutron capture cross section. The beta particles emitted in the prompt decay of the product radioisotopes can enter the emulsion and lead to its sensitization.

In the technique of *autoradiography*, the radiation source to be imaged exists within the sample itself. For example, biological processes can be studied by tagging a given substance with tritium or carbon-14. If a thin layer of the sample is then placed in contact with a radiographic emulsion over a sufficient exposure time, the beta particles emitted in the radioisotope decay will be recorded as an image on the developed emulsion. Using these techniques, the detailed spatial distribution of radioisotope within the sample can easily be recorded.

B. Nuclear Emulsions

When the object of the measurement is to record individual particle tracks, specialized formulations known as *nuclear emulsions* are conventionally used. Here, the thickness of the emulsion is increased to as much as 500 μm to allow the recording of the entire track of many particles. To enhance the density of developed grains along the track, the concentration of silver halide within the emulsion is also increased to as much as 80%. The unusual thickness and composition of nuclear emulsions require the use of more elaborate procedures to ensure their uniform development. The track of ionizing particles is then visible under microscopic examination as a trail of developed silver grains that may become nearly continuous under certain conditions. An example is reproduced in Figure 19-7. The length of the track can often be used as a measure of the particle range or energy, and the density of the track can serve to distinguish between radiations of

ILFORD K.5

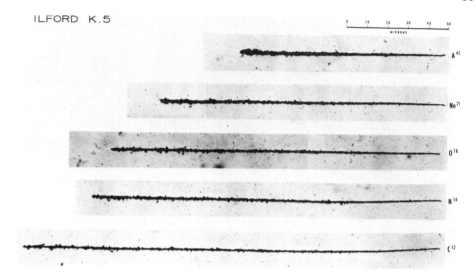

Figure 19-7 Tracks of fast heavy ions in Ilford K.5 nuclear emulsion. The ions are all incident from the left with initial energy of 10 MeV per nucleon. The track density is seen to decrease as the ion slows down and loses charge by electron pickup. (Photomicrographs courtesty of Prof. E. V. Benton, University of San Francisco.)

different dE/dx. In some emulsions, a minimum value of dE/dx is required for the development of a visible track, so that an inherent discrimination ability against lightly ionizing radiations is achieved. Other formulations will develop tracks for low dE/dx particles, including fast electrons.

The theory of the response of nuclear emulsions to ionizing radiations is reviewed in Refs. 36 and 37. Under different conditions, the sensitization of grains can be considered a *single hit* process in which only a single encounter with an ionizing particle is required, or it may be a *multihit* process in which more than one such encounter is required. In the latter case, a definite threshold in ionization density is expected.

Although obviously limited to situations in which a prompt detector signal is not required, nuclear emulsions offer a number of advantages compared with conventional detectors. Reference 38 reviews some typical applications. No associated equipment is needed, so that very simple emulsion detectors may be used in remote experiments where conventional methods may not be applicable. The track may be stored indefinitely as a latent image within the emulsion and, once developed, is a permanent record. Emulsions loaded with specialized target nuclei such as boron or uranium can be made sensitive to thermal neutrons. Fast-neutron-induced tracks can be registered through recoil protons generated within the emulsion itself.

C. Film Badge Dosimeters

Photographic emulsions are also widely applied in radiation dosimetry. In the most familiar form, a *film badge* consists of a small packet of film with a light-tight wrapping mounted within a film holder or badge that clips to the wearer's clothing. An evaluation of the dose accumulated over the course of exposure is carried out by comparing the density of the developed film with that of an identical film exposed to a calibrated dose. In this way, variations in emulsion sensitivity and developing procedures are canceled out.

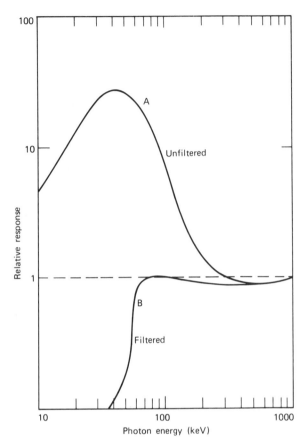

Figure 19-8 Relative response of a typical film badge dosimeter with and without a compensation filter. (From R. H. Herz, *The Photographic Action of Ionizing Radiations*, Copyright 1969 by John Wiley & Sons, Inc.)

The film holder plays an important part in the film badge response because it conventionally contains filters that are held in contact with the film and alter its response. As an example of one function of such a filter, Fig. 19-8 shows the response of a typical photographic emulsion to equal doses of gamma rays with different energies. The sensitivity of the film is greatest for photons of low energy, so that an unfiltered film will tend to overestimate the dose of soft X-rays and gamma rays compared with those of higher energy. However, by covering the film with a combination filter composed of tin and lead, the response can be made quite flat over a wide range of photon energy. The density of the film can then be used as a measure of gamma-ray exposure, independent of the details of the incident spectrum.

The film badge holder also normally contains a set of small filters confined to local regions of the film. These filters are made of different materials of variable thickness, so that differences in the corresponding film density can be used to sort out various components of the radiation exposure. For example, thin filters of low-Z material will effectively stop soft beta particles without seriously affecting transmitted gamma rays. Other metallic filters can help to separate out the contribution of different components of the gamma-ray spectrum.

A cadmium or gadolinium converter foil can be used to evaluate the exposure of the film to thermal neutrons. The secondary radiation (beta particles and capture gamma rays) emitted in these materials upon capture of a thermal neutron will lead to some incremental exposure of the film. By comparing the density behind such a filter to that behind a lead–tin filter of similar photon absorption properties, the additional darkening due to thermal neutrons can be estimated. Exposure of films to typical doses of fast neutrons does not lead to sufficient density to be measured directly. However, individual tracks due to recoil protons generated within the emulsion can be observed under a microscope and can be counted to give an estimate of the dose due to fast neutrons.

VI. THERMOLUMINESCENT DOSIMETERS

A. The Thermoluminescence Mechanism

The inorganic scintillation materials discussed in Chapter 8, when exposed to ionizing radiation, emit light in the form of prompt fluorescence. The scintillation photons are given off when the electron–hole pairs that were formed by the incident radiation recombine at an activator site. These materials are purposely kept free of other impurities and defects in order to maximize the yield of prompt scintillation light.

A different class of inorganic crystals, known as *thermoluminescent dosimeters* (TLDs), are based on a somewhat opposite approach. Instead of promoting the quick recombination of electron–hole pairs, materials are used which exhibit high concentrations of trapping centers within the bandgap. As illustrated in Fig. 19-9, the desired process is

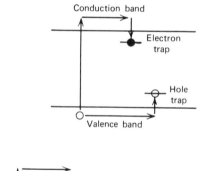

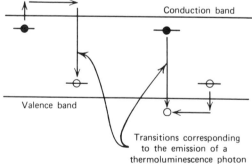

Figure 19-9 The top diagram represents the formation of an electron–hole pair in a TLD material, leading to population of electron and hole traps. The bottom diagram illustrates the two possible modes of recombination when the temperature is raised, which lead to the emission of a thermoluminescence photon.

now one in which electrons are elevated from the valence to the conduction band by the incident radiation but are then captured at one of the trapping centers. If the distance of the trap energy level below the conduction band is sufficiently large, there is only a small probability per unit time at ordinary room temperatures that the electron will escape the trap by being thermally excited back to the conduction band. Therefore, exposure of the material to a continuous source of radiation, although not resulting in a significant yield of prompt scintillation light, leads to the progressive buildup of trapped electrons.

Holes can also be trapped in an analogous process. An original hole created by the incident radiation may migrate through the crystal until reaching a hole trap with energy somewhat above the top of the valence band. If this energy difference is large enough, the hole will not migrate further and is then locked in place unless additional thermal energy is given to the crystal. A sample of TLD material will therefore function as an integrating detector in which the number of trapped electrons and holes is a measure of the number of electron–hole pairs formed by the radiation exposure.

After the exposure period, the trapped carriers can be measured through a process also illustrated in Fig. 19-9. The TLD sample is placed on a heated support or otherwise warmed, and its temperature is progressively raised. At a temperature that is determined by the energy level of the trap, the trapped electrons can pick up enough thermal energy so that they are reexcited back to the conduction band. Assuming that this temperature is lower than that required to free the trapped holes, the liberated electrons then migrate to

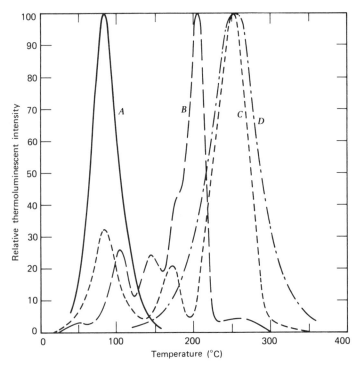

Figure 19-10 Typical thermoluminescent glow curves normalized to the same maximum intensity. Materials are A—$CaSO_4$: Mn; B—LiF; C—CaF_2; and D—CaF_2 : Mn. The details of these curves will depend on preirradiation annealing procedures, radiation exposure level, and heating rate during readout. (From Fowler and Attix, in *Radiation Dosimetry*, 2nd ed., Vol. II (F. H. Attix and W. C. Roesh, eds.). Copyright 1966 by Academic Press. Used with permission.)

near a trapped hole, where they can recombine with the emission of a photon. Alternatively, if the holes are released at a lower temperature, they may migrate to a trapped electron and their recombination also results in a radiated photon. In either case, if the magnitude of the energy difference is about 3 or 4 eV, the emitted photons are in the visible region and are the basis of the TLD signal. Ideally, one such photon is emitted per trapped carrier. Therefore, the total number of emitted photons can be used as an indication of the original number of electron–hole pairs created by the radiation.

TLD systems thus derive a signal by using a heater in which the sample can be viewed by a photomultiplier tube. The light yield is recorded as a function of sample temperature in a *glow curve* of the type illustrated in Fig. 19-10. The basic signal related to the radiation exposure is the total number of emitted photons, or the area under the glow curve. If the sample is raised to a relatively high temperature, all the traps are depleted and the exposure record of the sample is "erased." TLD materials therefore have the very practical advantage of recyclability, and a single sample may be reused many times.

B. Thermoluminescent Materials

A thorough description of the properties and applications of different TLD materials is presented in Ref. 39. Some popular materials consist of crystals to which a small concentration of impurity has been added as an activator (e.g., $CaSO_4$: Mn, where manganese is the activator). Others (notably LiF) do not require the addition of an activator, but the traps are created by the inherent impurities and defects in the crystal. The choice among TLD materials must take into account considerations of trap depth and atomic number of the material. If the energy levels of the traps are very near the edge of the bandgap (as in $CaSO_4$: Mn), the number of trapped carriers per unit exposure can be very large. Therefore, this material can be made sensitive to exposure as low as about 2×10^{-5} rads (0.2 μGy). The shallow traps are somewhat unstable even at ordinary room temperature, however, and therefore the material will show a considerable "fading," which can lose as much as 85% of the trapped carriers over a few day's time.[40] Therefore, other materials such as CaF_2 : Mn and LiF, with somewhat deeper traps, are better suited for longer-term exposures even though their sensitivity is several orders of magnitude less.

Of all TLD materials, LiF has proved the most popular because of its almost negligible fading at room temperature and its low average atomic number, which does not differ greatly from that of air or tissue. The energy deposited in LiF is therefore closely correlated with the gamma-ray exposure or dose equivalent over a wide range of gamma-ray energy. For TLD materials with higher atomic number, the enhanced photoelectric interaction probabilities exaggerate the response to low-energy X-rays or gamma rays. In Fig. 19-11 the relative response to 1 roentgen of gamma-ray exposure is plotted for a number of different TLD materials. Only the LiF curve remains reasonably constant over the wide range of photon energies plotted. Small wafers of LiF have thus come into popular use as personnel dosimeters. These TLD systems are gradually replacing photographic film badges in many situations, because exposures may be measured directly by the user without the need for photographic developing of film, and the dosimeters may be reused many times by annealing at elevated temperatures. Because there is a great deal of variability between samples and methods of heating, absolute values of the radiation dose are normally determined by exposing a parallel sample to a known gamma-ray exposure. The minimum sensitivity of LiF is about 10^{-2} rads (100 μGy), and the signal remains linearly related to dose up to about 10^3 rads (10 Gy). At higher doses, the material

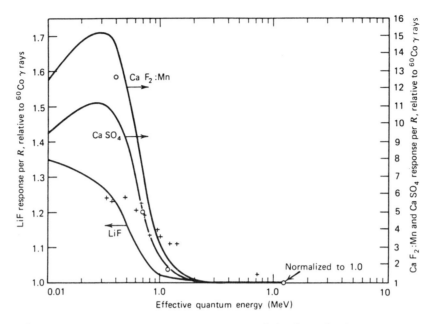

Figure 19-11 Variation with gamma-ray energy of the thermoluminescence response per roentgen. (Note the greatly different scale for LiF compared with the other materials.) Curves are calculated from gamma-ray interaction probabilities, assuming constant response per unit absorbed dose in the TLD material. Circles are experimental measurements for CaF_2 : Mn, crosses are measurements for LiF. (From Fowler and Attix, in *Radiation Dosimetry*, 2nd ed., Vol. II (F. H. Attix and W. C. Roesh, eds.). Copyright 1966 by Academic Press. Used with permission.)

displays a nonlinear increase in response per unit exposure, a behavior known as *supralinearity*.[41-45]

Because natural lithium contains 7.4% ^6Li, TLD detectors made from LiF will be somewhat sensitive to slow neutrons through the (n, α) reaction described in Chapter 14. This response can be enhanced by using lithium enriched in ^6Li, or suppressed by using lithium consisting entirely of the ^7Li isotope. Some representative measurements of the response of such TLD materials to slow neutrons are reported in Refs. 46 and 47. TLD detectors sensitive to fast neutrons have also been fabricated by covering samples of a conventional material such as Al_2O_3 with polyethylene.[48,49]

Specialized texts on the extensive subject of TLD detectors are readily available, and Refs. 50–52 are examples.

VII. TRACK-ETCH DETECTORS

A. The Track Registration Process

When an ionizing charged particle passes through a dielectric material, the transfer of energy to electrons results in a trail of damaged molecules along the particle track. In some materials, the track can be made visible upon etching in a strong acid or base solution. The entire surface of the material is attacked, but those points at which particle tracks have entered are etched at a faster rate. The tracks can thus be made to form pits on the surface which are large enough to be easily visible through a conventional

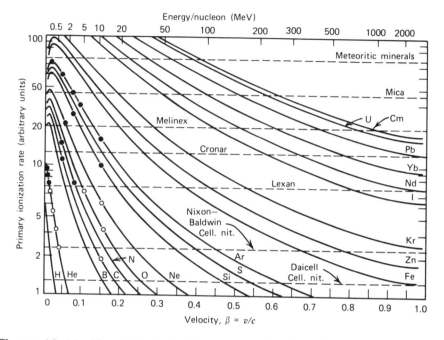

Figure 19-12 The relative ionization rate as a function of energy per nucleon or velocity is shown for various heavy ions. Horizontal dashed lines indicate the minimum damage rate required for 100% registration of tracks on some popular track-recording materials. The solid dots are experiments using accelerated ions on Lexan which gave 100% registration, and the open circles gave zero registration. (From R. L. Fleischer, P. B. Price, and R. M. Walker, *Nuclear Tracks in Solids*. Copyright 1975 by The Regents of the University of California; reprinted by permission of the University of California Press.)

microscope. Materials used to detect particles in this manner are called *track-etch detectors*,[53] and their properties and applications have been thoroughly reviewed in a comprehensive book by Fleischer et al.[54]

In common with other passive detectors such as photographic emulsions or neutron activation foils, track-etch detectors have the advantages of simplicity and low cost. They also possess a very useful inherent threshold, in that there is a minimum value of the specific energy loss ($-dE/dx$) required of the particle before the damage is severe enough to lead to an etchable track. The corresponding energy range for various charged particles over which tracks will be registered is illustrated in Figure 19-12. The threshold is always well above the specific energy loss of an electron track, so that track-etch materials are inherently insensitive to fast electron or gamma-ray interactions. Most materials also do not respond to lightly ionizing charged particles, such as protons or deuterons, and will therefore also be insensitive to the recoil protons produced by fast neutron interactions. However, some dielectrics will register proton tracks[55] over a certain range in energy.

Track-etch detectors share with nuclear emulsions the disadvantage of requiring individual track counting. When done by hand, the counting step is tedious at best and is very time consuming if a large number of tracks are recorded for statistical precision. Automatic counting systems have been developed but involve a sizable investment in cost and complication. Although some differentiation can be made of the particle type and energy based on appearance of the etched pit, many of the details of the original track are

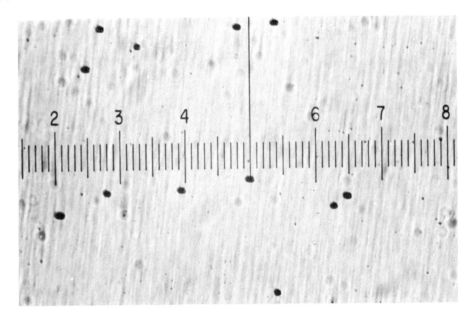

Figure 19-13 Photograph of etched fission fragment tracks in polycarbonate film. The smallest scale division represents 10 μm. (Photo courtesy of D. M. Gilliam, National Bureau of Standards.)

lost in the etching process. These detectors are therefore not nearly so well suited to the measurement of individual track properties as are nuclear emulsions.

The damage created by the incident particle can be through collisions of the particle itself or from the energetic delta rays (see Chapter 2) created along its track. The range of the delta rays may extend approximately 5 nm in any direction away from the particle position, so that the radius of the primary damage track is thought to be about this dimension.[54] After etching, the tracks are greatly enlarged to a diameter up to 10–20 μm. A photograph of etched tracks in polycarbonate film is shown in Fig. 19-13.

Etchable tracks are formed in a variety of materials. All are electrical insulators, although some wide-bandgap semiconductors are also known to record tracks. The materials fall into two main categories: inorganic solids, such as crystals and glasses, and organic solids, such as polymers. In the first category, the most popular materials are mica and flint glass, whereas polycarbonate and polyester films are the most common organic track-etch detectors. A list of the most useful materials is given in Table 19-2.

A completely satisfactory model for the mechanisms that lead to the registration of tracks does not yet exist. Past attempts have been made to link the etchability of tracks to the density of atomic displacements or the creation of a "thermal spike" of molten material along the particle track. Another hypothesis, known as the *ion explosion spike* model,[54] is based on the momentary electric field created along the particle track by charge imbalance, caused by the ejection of many electrons from the immediate region of the track. None of these approaches can fully account for some of the observed differences in etching behavior. The threshold of track etching is best correlated with the *restricted energy loss*, which is tied to the energy deposition rate only along the primary particle path, excluding that of long-range delta rays.[59] Because the energy imparted to the delta rays is one component of the more traditional specific energy loss dE/dx, the restricted value is somewhat smaller, particularly at high particle energies.

TABLE 19-2 Commonly Used Track Etch Materials

	Atomic Composition	Least Ionizing Ion Seen
Inorganic Materials		
Quartz	SiO_2	100 MeV ^{40}Ar
Phlogopite mica	$KMg_2Al_2Si_3O_{10}(OH)_2$	
Muscovite mica	$KAl_3Si_3O_{10}(OH)_2$	2 MeV ^{20}Ne
Silica glass	SiO_2	16 MeV ^{40}Ar
Flint glass	$18SiO_2 : 4PbO : 1.5Na_2O : K_2O$	2–4 MeV ^{20}Ne
Organic Materials		
Polyethylene terephthalate		
(Cronar, Melinex)	$C_5H_4O_2$	36 MeV ^{16}O
Bisphenol A-polycarbonate		
(Lexan, Makrofol)	$C_{16}H_{14}O_3$	0.3 MeV ^{4}He
Polymethylmethacrylate		
(Plexiglas, Lucite, Perspex)	$C_5H_8O_2$	3 MeV ^{4}He
Cellulose triacetate		
(Cellit, Triafol-T, Kodacel		
TA-401 unplasticized)	$C_3H_4O_2$	
Cellulose nitrate		
(Daicell)	$C_6H_8O_9N_2$	0.55 MeV ^{1}H

Source: R. L. Fleischer, P. B. Price, and R. M. Walker, *Nuclear Tracks in Solids.* Copyright 1975 by The Regents of the University of California; reprinted by permission of the University of California Press.

B. Track Etching

A qualitative picture of how tracks are revealed by etching is sketched in Fig. 19-14a. A track is assumed to exist perpendicular to the surface of the medium, which is exposed to the etching solution. As a simplified model, we assume that the undamaged surface is eroded away at a velocity V_G perpendicular to the surface. We further assume that the etching velocity along the damaged track is a greater value V_T. Under these conditions, the sketches illustrate that a cone-shaped pit is formed with an axis along the damaged track. Although this model is crude because it does not account for changes in etching rate with depletion of the etching solution or differences in the degree of damage along the track, it is adequate to explain many features of the observed behavior. For example, tracks that make a small angle with the surface can be erased by the etching process. Using the above model, it can be shown that the angle of incidence must exceed a critical angle θ_c in order to avoid its disappearance due to the progressive etching of the normal surface. This critical angle is given by

$$\theta_c = \arcsin\left(\frac{V_G}{V_T}\right) \qquad (19\text{-}4)$$

For example, in polyester materials, the critical angle for track registration is about 5–15°. Due to variability of the etching conditions and the energy loss threshold, the etching behavior should be demonstrated in advance of any measurement, using a calibrated source of the same type and energy of particles involved in the measurement. A number of environmental factors can have a significant effect on the etching behavior. For example, the sensitivity of certain plastic track-etch detectors depends on the presence or

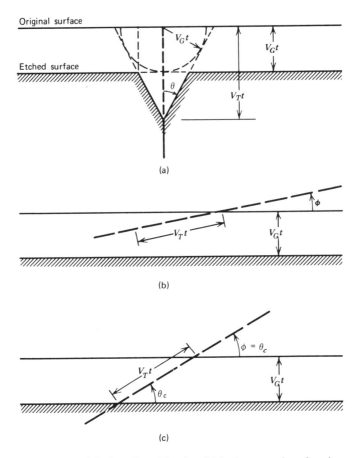

Figure 19-14 (a) Model of track etching in which the normal surface is removed at a velocity V_G and the damaged track at a velocity V_T, leading to a cone-shaped pit. (b) Tracks formed at an angle ϕ less than the critical angle θ_c are not revealed because the normal surface advances faster than the etch rate along the track. (c) The case in which the particle enters at the critical angle θ_c. Tracks entering the surface at an angle greater than the critical angle will be visible after etching. (From R. L. Fleischer, P. B. Price, and R. M. Walker, *Nuclear Tracks in Solids*. Copyright 1975 by The Regents of the University of California; reprinted by permission of the University of California Press.)

absence of oxygen during the exposure.[54] The formation of tracks may be enhanced through the application of an electric field[60] or through exposure to ultraviolet radiation.[61]

The temperature history of the track-etch material is also important. By holding the material at elevated temperatures for some period of time, latent tracks may be annealed to the point that etched tracks will no longer be formed. In this way, track-etch detectors can be "erased" prior to the start of a given measurement. However, the potential also exists for the inadvertent loss of desired tracks if the time and/or temperature between an exposure and etching are too great.[62] Because the annealing properties of track-etch materials vary greatly, even between different samples of the same material, the fading properties of tracks should be investigated experimentally. It is empirically observed that the time required for a given fraction of the tracks to disappear can be described by a relation of the type

$$t = A \exp\left(\frac{U}{kT}\right) \tag{19-5}$$

where A and U are constants characteristic of the material, and kT is the Boltzmann temperature.

C. Track Counting and Applications

Manual counting procedures are normally required if a high degree of counting precision is needed, or if detailed information on the shape and length of the track is desired. Track densities under about $10^4/cm^2$ can be counted using a simple microscope or, in a somewhat less tedious manner, through the use of a projection microscope by which images of the etched tracks are displayed on a large screen. The etched material, if optically transparent, can also be placed in a photographic enlarger, and counting can be done from a photographic print.

A number of automatic schemes have also been developed for the counting of etched tracks. Many are based on optical methods in which a computer-controlled microscope stage allows examination of each individual track using a photomultiplier tube. One version[63] incorporates automatic scanning and self-focusing and is about 98% accurate in counting fission fragments in polycarbonate. Somewhat less accurate machines can be adapted from commercially available optical counting systems.

For some thin film track-etch detectors, automatic counting can be carried out through spark counting.[64] The tracks are etched to the point that they completely penetrate the thin film used for registration. When the film is slowly moved between high-voltage electrodes, a spark is formed whenever an etched hole passes between the electrodes. The number of these discharges can be recorded electronically as an indication of the total number of etched tracks. Other methods, based on the transmission of charged particles through the etched holes, have also be demonstrated.[65]

If individual tracks need not be counted, the overall track density can be indicated by several other means. Simple attenuation of a collimated light beam through the etched film provides one overall indicator of track density. Another method is based on measuring the total resistivity of an etched film over a local area of its surface. These latter methods can be useful as an indicator of overall exposure of the film to charged particles for dosimetry purposes.

Track-etch detectors can also be applied to neutron detection through the use of a converter such as ^{6}Li or ^{10}B, in conjunction with a material that registers alpha tracks. Alternatively, a foil of fissionable material can be applied to generate fission fragments that are easily registered in all track-etch materials. If the converter foil and registration material are kept in close contact, an image of the detected neutron distribution can be formed after etching, a technique that has proved to be useful in neutron radiography.

VIII. NEUTRON DETECTION BY ACTIVATION

The neutron detectors discussed in Chapters 14 and 15 produce a prompt output pulse for each detected neutron. Neutron measurements can also be carried out indirectly through the radioactivity that is induced in some materials by neutron interactions. A sample of such a material can be exposed to a flux of neutrons for a period of time and then removed so that the induced radioactivity may be counted, using any of the conventional methods discussed earlier in this text. The measured radiations can then be used to deduce information about the number and/or energy distribution of the neutrons in the original field. The materials used in this way are often called *activation detectors*, and their applications are widely discussed in texts on neutron physics, such as Ref. 66.

Because neutron reaction cross sections are highest at low neutron energies, activation detectors are most commonly applied to the measurement of slow neutrons. To achieve a high degree of sensitivity, materials are chosen which have a large cross section for a neutron-induced reaction, which leads to a measurable form of radioactivity. Because the mean free path of neutrons in materials of high cross section is quite small, the thickness of the material must be kept small to avoid perturbing the neutron flux under measurement. Hence, the common geometric form of the material is that of a thin foil or small-diameter wire.

A. Activation and Decay

In the simplest case, the foil or wire is so thin that the probability of an interaction is small for any specific neutron. Then, the neutron flux remains unperturbed, and the rate R at which activation interactions occur within the foil is given by

$$R = \varphi \Sigma_{act} V \tag{19-6}$$

where
φ = neutron flux averaged over the foil surface

Σ_{act} = activation cross section averaged over the neutron spectrum

V = foil volume.

Therefore, the rate of activation per unit mass is a direct indicator of the neutron flux magnitude.

As the foil is irradiated, the radioactive nuclear species that is formed also undergoes radioactive decay. The rate of decay is given simply by λN, where λ is the decay constant and N is the total number of radioactive nuclei present. The rate of change in N is given by the difference between the rate of formation and rate of decay

$$\frac{dN}{dt} = R - \lambda N \tag{19-7}$$

We assume that R is a constant, implying that the neutron flux does not vary during the exposure, and neglecting any "burnup" or decrease in the number of target nuclei over the measurement. The solution to Eq. (19-7) for the condition $N = 0$ at $t = 0$ is

$$N(t) = \frac{R}{\lambda}(1 - e^{-\lambda t}) \tag{19-8}$$

The activity A of the foil is given by λN, or

$$A(t) = R(1 - e^{-\lambda t}) \tag{19-9}$$

This induced activity therefore builds up with the time, as shown in Fig. 19-15, and approaches an asymptote or *saturated activity* for infinitely long irradiation times given by

$$A_\infty = R = \varphi \Sigma_{act} V \tag{19-10}$$

Exposure times of three or four values of the half-life of the induced activity are sufficient to bring the foil activity to within 6–12% of the saturated value.

We assume that the irradiation has proceeded for a time t_0 at which time the foil is removed with an activity A_0:

$$A_0 = A_\infty(1 - e^{-\lambda t_0}) \tag{19-11}$$

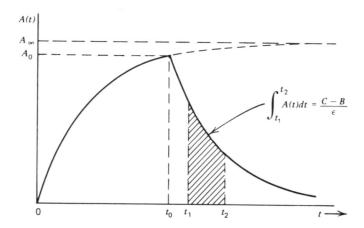

Figure 19-15 The activity of an activator detector after insertion in a constant neutron flux at time $= 0$ and removal at time $= t_0$. The measured number of counts is proportional to the area under the decay curve between t_1 and t_2. All times are measured from the start of irradiation at $t = 0$.

After exposure to the neutron flux, the foil is transferred to an appropriate radiation counter for measurement of its activity. Because the activity is continuously decaying during this stage, careful account must be made of each of the times involved. If the counting is carried out over an interval between t_1 and t_2, the number of counts will be (see Fig. 19-15)

$$C = \epsilon \int_{t_1}^{t_2} A_0 e^{-\lambda(t-t_0)} \, dt + B$$

$$= \epsilon \frac{A_0}{\lambda} e^{\lambda t_0}(e^{-\lambda t_1} - e^{-\lambda t_2}) + B \tag{19-12}$$

where ϵ is the overall counting efficiency (including any self-absorption effects) and B is the number of background counts expected in $t_2 - t_1$.

By combining Eqs. (19-11) and (19-12), we obtain for the saturated activity

$$A_\infty = \frac{\lambda(C - B)}{\epsilon(1 - e^{-\lambda t_0})e^{\lambda t_0}(e^{-\lambda t_1} - e^{-\lambda t_2})} \tag{19-13}$$

from which the neutron flux magnitude may be calculated from Eq. (19-10).

Activation foils are integrating detectors and therefore can provide no information about any time variation of the neutron flux over the course of the exposure. They have the advantages of small size, insensitivity to gamma radiation, and low cost. They can also tolerate exposure to extreme environments where other detectors might fail, and they require no electrical connections to the outside world. Activation foils are thus widely used for mapping the spatial variation of steady-state neutron fluxes in reactor cores, where the extreme temperature, pressure, and limited space severely constrain the type of conventional detector that may be used (see Chapter 14).

B. Activation Detector Materials

In choosing the material for an activation detector, a number of nuclear and physical properties need to be considered.

1. SHAPE OF THE CROSS SECTION

The material will obviously respond preferentially to those neutron energies for which its cross section is high. Radiative capture or (n, γ) reactions typically have largest cross sections at or near thermal energies, and therefore materials in which these reactions predominate are preferentially chosen for slow neutron detectors. Other reactions such as (n, p), (n, α), and $(n, 2n)$ often require a minimum or threshold energy. Materials in which these reactions predominate can therefore be candidates as fast neutron detectors, which will be inherently insensitive to slow neutrons.

2. MAGNITUDE OF THE CROSS SECTION

As shown by Eq. (19-10), the saturated activity is linearly proportional to the average cross section for the activation reaction. Greatest sensitivity is therefore achieved by selecting materials with highest activation cross section, although some of this advantage may be offset by the need to keep the neutron absorption probability small enough to avoid perturbing the flux under measurement.

3. DECAY CONSTANT OF THE INDUCED ACTIVITY

The half-life of the induced activity should be neither too short nor too long, and for many applications, a value of a few hours is near the optimum. Long half-lives require the use of long irradiation times in order to approach saturation, and the specific activity becomes smaller with increasing half-life. Very short half-lives can lead to difficulties in transferring the foil to the counter without excessive delay time. The corresponding high activities may also create problems due to dead time effects within the counter.

4. PURITY AND INTERFERING ACTIVITIES

Very high purity of the material is often required to avoid interference from other neutron-induced reactions. If the half-life of the contaminant activity is sufficiently different from that of the primary, however, it may not pose a practical problem. Interfering activities with short half-lives can be eliminated simply by waiting some time before beginning the counting period, whereas those with long half-lives may not build up to a significant level if the irradiation period is short by comparison.

5. NATURE OF THE INDUCED ACTIVITY

The decay of the product nuclei produced in the activation reactions normally involves the emission of either beta particles or gamma rays. Gamma counting is often preferred because the penetrating nature of the radiation minimizes the effects of self-absorption within the sample. Furthermore, it is far easier to perform energy spectroscopy with gamma rays than beta particles, so that infering activities and background can be discriminated more easily. In some cases, however, only beta activity is produced so that there is then no recourse but to carry out some type of beta counting. Continuous flow proportional counters are often used in either 2π or 4π geometry.

6. PHYSICAL PROPERTIES

The properties of the activation materials play an obvious part in determining the physical environment in which they may be exposed. Materials in gaseous or liquid form are very difficult to apply except through elaborate encapsulation techniques that may interfere with the subsequent counting of the activity. Therefore, almost all activation detectors consist of metallic foils or wires that can be cut to the desired dimensions.

TABLE 19-3   Materials Useful as Slow Neutron Activation Detectors

Element	Isotope (Abundance in Percent)	Thermal Activation Microscopic Cross Section (in 10^{-28} m^2)	Induced Activity	Half-Life
Manganese	^{55}Mn (100)	13.2 ± 0.1	^{56}Mn	2.58 h
Cobalt	^{59}Co(100)	16.9 ± 1.5 20.2 ± 1.9	^{60m}Co ^{60}Co	10.4 min 5.28 y
Copper	^{63}Cu(69.1) ^{65}Cu(30.9)	4.41 ± 0.20 1.8 ± 0.4	^{64}Cu ^{66}Cu	12.87 h 5.14 min
Silver	^{107}Ag(51.35) ^{109}Ag(48.65)	45 ± 4 3.2 ± 0.4	^{108}Ag ^{110m}Ag	2.3 min 253 d
Indium	^{113}In(4.23) ^{115}In(95.77)	56 ± 12 2.0 ± 0.6 160 ± 2 42 ± 1	^{114m_1}In ^{114}In ^{116m_1}In ^{116}In	49 d 72 s 54.12 min 14.1 s
Dysprosium	^{164}Dy(28.18)	2000 ± 200 800 ± 100	^{165m}Dy ^{165}Dy	1.3 min 140 min
Gold	^{197}Au (100)	98.5 ± 0.4	^{198}Au	2.695 d

Source: K. H. Beckurts and K. Wirtz, *Neutron Physics.* Copyright 1964 by Springer-Verlag, New York. Used with permission.

Table 19-3 lists a number of materials that are useful as detectors of slow neutrons. In each case, the activity is induced by an (n, γ) reaction that leads to the indicated radioactive product. In the thermal region, these materials have activation cross sections that vary approximately as $1/v$, where v is the neutron velocity. Many also show significant resonances in the cross section at certain neutron energies between about 1 and 1000 eV. The observed activity then corresponds to a mixture of activation caused by thermal neutrons, plus an additional component for neutrons with energy in the resonance region.

The contributions of the thermal and resonance neutrons can be separated through a technique known as the *cadmium difference* method. Cadmium is a metal whose radiative capture cross section behaves in a very convenient manner. The cross section is very large for neutron energies below about 0.4 eV, where it drops abruptly and remains low for higher neutron energies. Thicknesses of the order of 0.5 mm of cadmium therefore act as a selective filter, blocking neutrons with energy below 0.4 eV but passing higher-energy neutrons with little attenuation. If one of the materials in Table 19-3 is made into two identical foils, one of which is covered with cadmium, the uncovered foil will respond both to the thermal and resonance neutrons, whereas the covered foil will indicate only the resonance contribution. By taking the difference between the two activations, the thermal contribution can be determined. The method does require some corrections for the nonideality of the cadmium filter, in that cadmium thicknesses that are large enough to fully stop all neutrons below 0.4 eV can also have a measurable effect on neutrons with higher energies. However, for most activation materials, corrections for the attenuation of the resonance neutrons are less than a few percent. By dividing the activity of the uncovered foil by that of the cadmium-covered one, the *cadmium ratio* is derived, which is often taken as an indication of the degree to which a given neutron field has been thermalized.

TABLE 19-4   Materials Useful as Threshold Activation Detectors

Material	Reactions of Interest	Isotopic Abundance (at %)	Half-Life	γ Energy (MeV)	γ Abundance (%)	Threshold (MeV)
F	$^{19}F(n,2n)^{18}F$	100.0	109.7 min	0.511^+	$194°$	11.6
Mg	$^{24}Mn(n,p)^{24}Na$	78.7	15.0 h	1.368	100	6.0
Al	$^{27}Al(n,\alpha)^{24}Na$	100.0	15.0 h	1.368	100	4.9
Al	$^{27}Al(n,p)^{27}Mg$	100.0	9.46 min	0.84–1.01	100	3.8
Fe	$^{56}Fe(n,p)^{56}Mn$	91.7	2.56 h	0.84	99	4.9
Co	$^{59}Co(n,\alpha)^{56}Mn$	100.0	2.56 h	0.84	99	5.2
Ni	$^{58}Ni(n,2n)^{57}Ni$	67.9	36.0 h	1.37	86	13.0
Ni	$^{58}Ni(n,p)^{58}Co$	67.9	71.6 d	0.81	99	1.9
Cu	$^{63}Cu(n,2n)^{62}Cu$	69.1	9.8 min	0.511^+	$195°$	11.9
Cu	$^{65}Cu(n,2n)^{64}Cu$	30.9	12.7 h	0.511^+	$37.8°$	11.9
Zn	$^{64}Zn(n,p)^{64}Cu$	48.8	12.7 h	0.511^+	$37.8°$	2.0
In	$^{115}In(n,n')^{115m}In$	95.7	4.50 h	0.335	48	0.5
I	$^{127}I(n,2n)^{126}I$	100.0	13.0 d	0.667	33	9.3
Au	$^{197}Au(n,2n)^{196}Au$	100.0	6.18 d	0.33–0.35	$25–94$	8.6
Li	$^{7}Li(n,\alpha n')t$	92.58	12.3 y	$0–0.019^\times$	$100^\times$	3.8

$^+$ Annihilation radiation.

$°$ Yield of annihilation photons assuming all positrons are stopped.

$^\times \beta$ particle energy and percent abundance.

Source: Kuijpers et al.[67]

The activation method can be extended to the measurement of higher energy neutrons through the use of materials such as those listed in Table 19-4. The useful reactions in this group are threshold reactions, which require neutrons above a minimum energy in order to take place at all. Each material with a different threshold will respond to a somewhat different range of neutron energy. By exposing a set of these threshold activation foils to a given neutron field, the known differences in the shape of the cross sections can serve as a basis of an unfolding of the neutron energy distribution. Computer codes have been developed to carry out this unfolding process and have achieved a considerable degree of success.[67,68]

Accurate application of the activation technique in the quantitative measurement of neutron fluxes requires the application of a number of other correction factors. These involve the effects of the foil itself on the neutrons that are being measured. In a diffusing medium, the neutron flux in the immediate vicinity of the foil will be depressed due to the fact that some neutrons have been removed in their passage through the foil. Furthermore, the effective neutron flux at the center of the foil will be somewhat less than that at its surface due to attenuation or *self-shielding*. These effects are minimized by keeping the foil very thin, but the induced activity may then be so low as to create measurement difficulties. A complete analysis and discussion of the techniques required to account for flux depression and self-shielding of activation foils is presented in the text by Beckurts and Wirtz.[66]

C. Activation Counters

The discussion of activation materials in the previous section has emphasized techniques in which the induced activity is measured in a separate counting facility. A related class of instruments, commonly called *activation counters*, consists of a sample of the activation

TABLE 19-5 Some Neutron-Induced Reactions of Interest in Activation Counters

Percent Abundance	Reaction	Induced Activity	Half-Life	Thermal Cross Section	Application Reference
		SLOW NEUTRON REACTIONS			
51.8%	$^{107}Ag(n, \gamma)^{108}Ag$	1.49 MeV β^-	2.3 min	30 b ⎫	
48.2%	$^{109}Ag(n, \gamma)^{110}Ag$	2.24 or 2.82 MeV β^-	24.2 s	110 b ⎬	69–71
100%	$^{108}Rh(n, \gamma)^{109}Rh$	⎧ 2.47 MeV β^-	44 s	139 b ⎫	
		⎩ 2.44 MeV β^-	265 s	11 b ⎭	72

Percent Abundance	Reaction	Induced Activity	Half-Life	Threshold Energy	Application Reference
		THRESHOLD REACTIONS			
100%	$^{75}As(n, n')^{75m}As$	0.30 MeV γ	17 ms	0.3 MeV	73
21.7%	$^{207}Pb(n, n')^{207m}Pb$	24–304 keV γ	810 ms	1.6 MeV	74, 75
99.8%	$^{16}O(n, p)^{16}N$	⎧ 6.13 MeV γ	7.2 s ⎫	10.2 MeV	76
100%	$^{19}F(n, d)^{16}N$	⎩ 4.27–10.4 MeV β	⎭	~ 3 MeV	77
100%	$^{23}Na(n, \alpha)^{20}F$	5.4 MeV β^-, 1.63 MeV γ	11.0 s	~ 7 MeV	78
100%	$^{9}Be(n, \alpha)^{6}He$	3.51 MeV β^-	800 ms	~ 2 MeV	79

material placed in close proximity to a detector capable of responding directly to the induced activity. Activation counters have proved to be very useful in the measurement of fast neutrons produced in short bursts, typical of conditions encountered in fusion experiments. The burst duration may be only a few nanoseconds or less, so pileup precludes the use of pulse mode detectors such as the organic scintillators discussed in Chapter 15. At the same time, the neutron bursts may be infrequent enough so that the activity induced in standard activation foils is too small to be useful. By using appropriate materials coupled to efficient detectors, adequate sensitivity may be possible using the counters described below.

There are two general approaches to the design of activation counters. One is to employ slow neutron activation materials such as silver or rhodium (see Table 19-5) inside a moderating structure used to reduce the average energy of the incident fast neutrons. An example is the *silver counter* consisting of silver-wrapped (or silver-walled) G-M counters placed within a paraffin or polyethylene moderator.[69, 70] The beta particles emitted by the activation products appear with half-lives of 24 s and 2.3 min and are detected with good efficiency by the G-M tube. The total number of counts recorded after many half-lives have elapsed is proportional to the neutron burst intensity. Other designs (see Fig. 19-16) have incorporated layers of silver with alternating plastic scintillator slabs to detect the beta particles, or a foil of rhodium in contact with a plastic scintillator.[72]

An alternative approach is to use threshold activation materials of the type listed in Table 19-4 and to rely on direct activation by the fast neutrons without moderation. This choice has the advantage of immunity from the effects of thermalized or low-energy neutrons often present as a background. However, now the reaction cross sections are much smaller than those for typical slow neutron activation materials. To boost the sensitivity during a short activation period, there is an advantage in choosing materials for which the induced activity has a much smaller half-life than those shown in Table 19-4. Then the measured activity can more closely approach the saturated value (A_∞ in Fig. 19-15). The half-life must be kept large enough to spread out the counts sufficiently

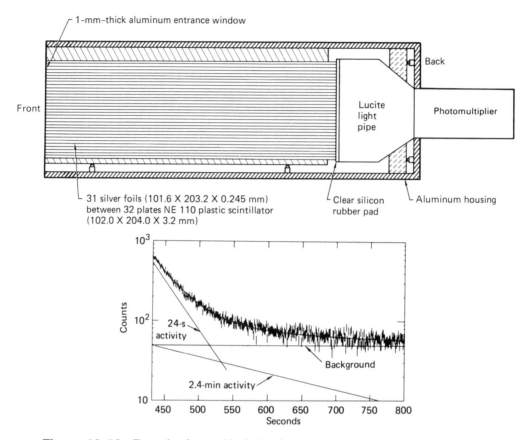

Figure 19-16 Example of a specific design for a silver activation counter. When used with fast neutrons, the counter is normally placed within a polyethylene moderator. The plot shows the counting rate observed for a period from 450 to 800 s following exposure to a fast neutron flux. (From Slaughter and Pickles.[71])

so that dead time and pileup effects in the detector are not serious. Attention has therefore focused on reactions that lead to half-lives of a few seconds or less, still long compared with typical time spacing between bursts. The integrated counts following each burst remain, under these conditions, a linear measure of the total neutron intensity of each burst.

Some examples of threshold activation reactions that have been applied in such counters are listed in Table 19-5. In several designs,[73, 79] the activation material is placed in contact with a plastic scintillator to count the beta or gamma rays emitted by the product. The oxygen reaction has been exploited[76] by using a liquid scintillator as both the source of oxygen and to detect the resulting ^{16}N gamma rays. Similarly, a sodium iodide scintillator has been used[78] to provide sodium nuclei for the ^{23}Na(n, α) reaction as well as to detect the beta and gamma rays emitted in the decay of the ^{20}F product. The fluorine reaction has been conveniently applied[77] by fitting a teflon cap over a germanium detector used to record the resulting gamma activity. In this case, one must bear in mind the sensitivity of germanium detectors to radiation damage caused by fast neutrons (see p. 437).

PROBLEMS

19-1. (a) Find the minimum energy that an electron must have to produce Cerenkov radiation in quartz with an index of refraction of 1.47.

(b) Find the minimum energy for a gamma-ray photon to produce an electron of the energy calculated in part (a) through Compton scattering.

19-2. Compare the number of Cerenkov photons expected from a 2 MeV electron interacting in water to the number of scintillation photons expected from the same energy electron interacting in NaI(Tl).

19-3. Using the data given in Fig. 6-14, estimate the thickness of *liquid* xenon required to cause 50% of incident 30 keV photons to interact in the xenon. The densities of gaseous and liquid xenon are 5.85 g/liter (at STP) and 3.52 kg/liter, respectively.

19-4. Explain the purpose of the intensifier screens that are used in conjunction with radiographic film for most clinical X-ray imaging.

19-5. Explain why the existence of electron trapping sites in a crystalline material is desirable if the material is used as a thermoluminescent dosimer, but undesirable if the material is used as a conventional scintillator.

19-6. From Fig. 19-12, select a track-etch material that will respond to neutron-induced fission fragments from a ^{235}U foil but will be insensitive to the alpha particles emitted in the spontaneous decay of the same material.

19-7. (a) A neutron flux is to be measured through the use of silver foil activation detectors and counting of the induced ^{108}Ag activity. In all cases, the foils are removed from the flux and counted for a 10 min period following a 2 min delay. Find the ratio of counts expected from two identical foils exposed in the same steady-state flux if one foil is exposed for 10 min and the other for 20 min.

(b) Find the factor by which the number of recorded counts could theoretically be increased by extending the counting period indefinitely.

REFERENCES

1. J. V. Jelly, *Cerenkov Radiation and its Applications*, Pergamon Press, London, 1958.

2. B. D. Sowerby, *Nucl. Instrum. Meth.* **97**, 145 (1971).

3. W. J. Gelsema, C. L. de Ligny, J. B. Luten, and F. G. A. Vossenberg, *Int. J. Appl. Radiat. Isotopes* **26**, 443 (1975).

4. M. Atac, A. V. Tollestrup, and D. Potter, *Nucl. Instrum. Meth.* **200**, 345 (1982).

5. N. Koori et al., *IEEE Trans. Nucl. Sci.* **NS-33** (1), 395 (1986).

6. H. Ohgaki et al., *IEEE Trans. Nucl. Sci.* **NS-33** (1), 381 (1986).

7. T. Doke, A. Hitachi, S. Kubota, A. Nakamoto, and T. Takahashi, *Nucl. Instrum. Meth.* **134**, 353 (1976).

8. L. Miller, S. Howe, and W. Spear, *Phys. Rev.* **166**, 871 (1968).

9. E. Shibamura, A. Hitachi, T. Doke, T. Takahashi, S. Kubota, and M. Miyajima, *Nucl. Instrum. Meth.* **131**, 249 (1975).

10. J. Prunier et al., *Nucl. Instrum. Meth.* **109**, 257 (1973).

11. E. Aprile, K. L. Giboni, and C. Rubbia, *Nucl. Instrum. Meth. Phys. Res.* **A241**, 62 (1985).

12. K. Masuda, T. Doke, and T. Takahashi, *Nucl. Instrum. Meth.* **188**, 629 (1981).

13. A. S. Barabash, A. A. Golubev, O. V. Kazachenko, V. M. Lobashev, B. M. Ovchinnikov, and B. E. Stern, *Nucl. Instrum. Meth. Phys. Res.* **A236**, 69 (1985).

14. E. Aprile, W. H.-M. Ku, J. Park, and H. Schwartz, *Nucl. Instrum. Meth. Phys. Res.* **A261**, 519 (1987).

15. K. L. Giboni, *Nucl. Instrum. Meth. Phys. Res.* **A269**, 554 (1988).

16. E. Aprile, W. H.-M. Ku, and J. Park, *IEEE Trans. Nucl. Sci.* **NS-35**(1), 37 (1988).

17. H. J. Crawford, T. Doke, A. Hitachi, J. Kikuchi, P. J. Lindstrom, K. Masuda, S. Nagamiya, and E. Shibamura, *Nucl. Instrum. Meth. Phys. Res.* **A256**, 47 (1987).

18. J. H. Cobb and D. J. Miller, *Nucl. Instrum. Meth.* **141**, 433 (1977).

19. L. S. Miller, S. Howe, and W. E. Spear, *Phys. Rev.* **166**, 871 (1968).

20. K. Masuda, T. Doke, T. Ikegami, J. Kikuchi, M. I. Lopes, R. Ferreira Marques, and A. Policarpo, *Nucl. Instrum. Meth. Phys. Res.* **A241**, 607 (1985).

21. H. Jungblut and W. F. Schmidt, *Nucl. Instrum. Meth. Phys. Res.* **A241**, 616 (1985).

22. K. Masuda, M. I. Lopes, and T. Doke, *Nucl. Instrum. Meth. Phys. Res.* **A261**, 598 (1987).

23. D. McCammon et al., *IEEE Trans. Nucl. Sci.* **NS-33**(1), 236 (1986).

24. S. H. Moseley, R. L. Kelley, J. C. Mather, R. F. Mushotzky, A. E. Szymkowiak, and D. McCammon, *IEEE Trans. Nucl. Sci.* **NS-32**(1), 134 (1985).

25. G. W. Fraser, *Nucl. Instrum. Meth. Phys. Res.* **A256**, 553 (1987).

26. H. H. Stroke et al., *IEEE Trans. Nucl. Sci.* **NS-33**(1), 759 (1986).

27. A. Alessandrello, D. V. Camin, G. F. Cerofolini, E. Fiorini, A. Giuliani, C. Liguori, L. Meda, T. O. Niinikoski, and A. Rijllart, *Nucl. Instrum. Meth. Phys. Res.* **A263**, 233 (1988).

28. S. H. Mosley, R. L. Kelley, R. J. Kelley, R. J. Schoelkopf, A. E. Szymkowiak, D. McCammon, and J. Zhang, *IEEE Trans. Nucl. Sci.* **NS-35**(1), 59 (1988).

29. D. Hueber, C. Valette, and G. Waysand, *Nucl. Instrum. Meth.* **167**, 201 (1979).

30. A. K. Drukier, *Nucl. Instrum. Meth.* **201**, 77 (1982).

31. M. Legros, A. Kotlicki, M. J. C. Crooks, B. G. Turrell, A. K. Drukier, and D. N. Spergel, *Nucl. Instrum. Meth. Phys. Res.* **A263**, 229 (1988).

32. A. K. Drukier, J. Igalson, and L. Sniadower, *Nucl. Instrum. Meth.* **154**, 91 (1978).

33. A. Drukier and L. Stodolsky, *Phys. Rev. D* **30**, 2295 (1984).

34. A. de Bellefon, D. Broskiewicz, R. Bruère-Dawson, P. Espigat, B. Mettout, N. Perrin, D. Limagne, and L. C. L. Yuan, *IEEE Trans. Nucl. Sci.* **NS-35**(1), 73 (1988).

35. R. H. Herz, *The Photographic Action of Ionizing Radiations*, Wiley-Interscience, London, 1969.

36. R. Katz and F. E. Pinkerton, *Nucl. Instrum Meth.* **130**, 105 (1975).

37. R. Katz, L. Larsson, F. E. Pinkerton, and E. V. Benton, *Track Detec.* **1**, 49 (1977).

38. J. R. Erskine, *Nucl. Instrum. Meth.* **162**, 371 (1979).

39. J. R. Cameron, N. Suntharalingam, and G. N. Kenney, *Thermoluminescent Dosimetry*, University of Wisconsin Press, Madison, 1968.

40. J. F. Fowler and F. H. Attix, "Solid State Integrating Dosimeters," in *Radiation Dosimetry*, Vol. II, 2nd ed. (F. H. Attix and W. C. Roesh, eds), Academic Press, New York, 1966.

41. E. Piesch, B. Burgkhardt, and S. Kabadjova, *Nucl. Instrum. Meth.* **126**, 563 (1975).

42. L. Larsson and R. Katz, *Nucl. Instrum. Meth.* **138**, 631 (1976).

43. B. Burgkhardt, D. Singh, and E. Piesch, *Nucl. Instrum. Meth.* **141**, 363 (1977).

44. B. Burgkhardt, E. Piesch, and D. Singh, *Nucl. Instrum. Meth.* **148**, 613 (1978).

45. M. P. R. Waligórski and R. Katz, *Nucl. Instrum. Meth.* **172**, 463 (1980).

46. S. Tanaka and Y. Furuta, *Nucl. Instrum. Meth.* **133**, 495 (1976).

47. S. Tanaka and Y. Furuta, *Nucl. Instrum. Meth.* **140**, 395 (1977).

48. F. Spurny, M. Kralik, R. Medioni and G. Portal, *Nucl. Instrum. Meth.* **137**, 593 (1976).

49. J. Henniger, B. Horlbeck, K. Hübner, and K. Prokert, *Nucl. Instrum. Meth.* **204**, 209 (1982).

50. M. Oberhofer and A. Scharmann (eds.), *Applied Thermoluminescence Dosimetry* (Ispra courses), Adam Hilger Ltd., Bristol, 1981.

51. Y. S. Horowitz, *Thermoluminescence and Thermoluminescent Dosimetry*, Vols. I, II, and III, CRC Press, Boca Raton, FL, 1984.

52. S. W. S. McKeever, *Thermoluminescence of Solids*, Cambridge University Press, Cambridge, U.K., 1985.

53. R. L. Fleischer, P. B. Price, R. M. Walker, and E. L. Hubbard, *Phys. Rev.* **133**, A1443 (1964).

54. R. L. Fleischer, P. B. Price, and R. M. Walker, *Nuclear Tracks in Solids*, University of California Press, Berkeley, 1975.

55. H. B. Lück, *Nucl. Instrum. Meth.* **119**, 403 (1974).

56. E. V. Benton and R. P. Henke, *Nucl. Instrum. Meth.* **58**, 241 (1968).

57. U. Höppner, E. Konecny, and G. Fiedler, *Nucl. Instrum. Meth.* **74**, 285 (1969).

58. H. B. Lück, *Nucl. Instrum. Meth.* **124**, 359 (1975).

59. E. V. Benton and W. D. Nix, *Nucl. Instrum. Meth.* **67**, 343 (1969).

60. H. Crannel, C. J. Crannell, and F. J. Kline, *Science* **166**, 606 (1969).

61. G. Siegmon, K. Bartholoma, and W. Enge, *Report of the Institute for Pure and Applied Nuclear Physics*, Christian Albrecht University, Kiel, Germany, 1FKK1 (1975).

62. E. Piesch and A. M. Sayed, *Nucl. Instrum. Meth.* **119**, 367 (1974).

63. R. Gold and C. E. Cohn, *Rev. Sci. Instrum.* **43**, 18 (1972).

64. J. Jasiak and E. Piesch, *Nucl. Instrum. Meth.* **128**, 447 (1975).

65. S. R. Dolce, *IEEE Trans. Nucl. Sci.* **NS-23**(1), 206 (1976).

66. K. H. Beckurts and K. Wirtz, *Neutron Physics*, Springer-Verlag, New York, 1964.

67. L. Kuijpers, R. Herzing, P. Cloth, D. Filges, and R. Hecker, *Nucl. Instrum. Meth.* **144**, 215 (1977).

68. W. N. McElroy et al., AFWL-TR-67-41, Vol. 1, 2, 4 (1967).

69. F. J. Mayer and H. Brysk, *Nucl. Instrum. Meth.* **125**, 323 (1975).

70. A. Gentilini et al., *Nucl. Instrum. Meth.* **172**, 541 (1980).

71. D. R. Slaughter and W. L. Pickles, *Nucl. Instrum. Meth.* **160**, 87 (1979).

72. F. C. Young, *IEEE Trans. Nucl. Sci.* **NS-22**(1), 718 (1975).

73. E. L. Jacobs, S. D. Bonaparte, and P. D. Thacher, *Nucl. Instrum. Meth.* **213**, 387 (1983).

74. L. Ruby and J. B. Rechen, *Nucl. Instrum. Meth.* **15**, 74 (1962).

75. C. E. Spencer and E. L. Jacobs, *IEEE Trans. Nucl. Sci.* **NS-12**(1), 407 (1965).

76. R. H. Howell, *Nucl. Instrum. Meth.* **148**, 39 (1978).

77. A. Wolf and R. Moreh, *Nucl. Instrum. Meth.* **148**, 195 (1978).

78. V. E. Lewis and T. B. Ryves, *Nucl. Instrum. Meth.* **A257**, 462 (1987).

79. M. S. Rowland and J. C. Robertson, *Nucl. Instrum. Meth.* **224**, 322 (1984).

CHAPTER · 20

Background and Detector Shielding

$\mathbf{B}$ecause of the cosmic radiation that continuously bombards the earth's atmosphere and the existence of natural radioactivity in the environment, all radiation detectors record some background signal. The nature of this background varies greatly with the size and type of detector and with the extent of shielding that may be placed around it. The background counting rate can be as high as many thousands of counts per second for large-volume scintillators, to less than a count per minute in some specialized applications. Because the magnitude of the background ultimately determines the minimum detectable radiation level, it is most significant in those applications involving radiation sources of low activity. However, background is often important enough in routine usage so that the majority of radiation detectors are provided with some degree of external shielding to effect a reduction in the measured level. A second purpose of detector shielding is to provide a degree of isolation in laboratories where other radiation sources may be used or moved about during the course of a measurement.

1. SOURCES OF BACKGROUND

Background radiations are conveniently grouped into five categories:

1. The natural radioactivity of the constituent materials of the detector itself.
2. The natural radioactivity of the ancillary equipment, supports, and shielding placed in the immediate vicinity of the detector.
3. Radiations from the activity of the earth's surface (*terrestrial radiation*), walls of the laboratory, or other far-away structures.
4. Radioactivity in the air surrounding the detector.
5. The primary and secondary components of cosmic radiation.

A. Radioactivity of Common Materials

The radioactivity of ordinary construction materials is, in large part, due to low concentrations of naturally radioactive elements often contained as an impurity. The most important components are potassium, thorium, uranium, and radium. Natural potassium contains 0.012% ^{40}K, which decays with a 1.26×10^9 year half-life through the decay scheme sketched below.

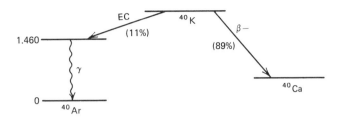

Radiations emitted are a beta particle with 1.314 MeV endpoint energy (89% yield), a gamma ray of 1.460 MeV energy (11%), and characteristic Ar X-rays following the electron capture. The high-energy gamma rays often lead to a recognizable peak in the background spectra from gamma-ray detectors, because potassium is a widespread component in concrete and other building materials.

Thorium, uranium, and radium are all members of long decay chains involving daughter products that emit a mixed spectrum of alpha, beta, and gamma rays. In the terrestrial gamma-ray spectrum shown in Fig. 20-1, the following daughter activities can be identified: in the thorium series, ^{228}Ac, ^{224}Ra, ^{212}Bi, ^{212}Pb, and ^{208}Tl; in the uranium series, ^{226}Ra, ^{214}Pb, and ^{214}Bi. The long-lived natural activities of ^{235}U and ^{40}K are also evident. A small peak due to ^{7}Be, produced by cosmic interactions, can also be detected.

In addition to the naturally occurring activities, background also consists of some fission-product activities which originated with atmospheric fallout from past weapons testing. In Fig. 20-1, the most prominent contributor is ^{137}Cs, but detectable amounts of ^{95}Zr, ^{95}Nb, ^{106}Ru, ^{125}Sb, and ^{144}Ce are also seen.

Table 20-1 lists the measured specific activity for some common materials used in the construction of detector systems. Certain materials such as pyrex glass contain either potassium or thorium as a normal constituent and therefore have a rather high background level. Most of the other examples include these components only as impurities, and their radioactivity is minimized by choosing highly purified samples. For this reason, copper or magnesium that has been electrolytically prepared is preferred when used in the construction of low-background counters. Stainless steel normally shows low background levels, but common sources of aluminum have a sufficient uranium and/or radium impurity level to be objectionable in demanding applications. Brass is generally of low activity, provided its lead content is low. Electrical solder[3,4] and some circuit board materials can be relatively high in radioactivity, so that attention must be given to the possible contribution of electronic components to the background rate of gamma-ray detectors.

For scintillation counters, the glass envelope of the photomultiplier tube and the tube base or socket materials are potential sources of background. At some premium in cost, tubes that are made from quartz rather than glass can be obtained which will be of significantly lower activity. Bases for tubes intended for scintillation counting are also available with a special low-background formulation. In the past, potassium impurities in sodium iodide were a source of some detectable background, but modern fabrication techniques have now led to the availability of sodium iodide with negligible potassium contamination. Bismuth germanate (BGO) can show appreciable background[5] from ^{207}Bi, a radionuclide with 38 year half life most likely produced by cosmic ray proton transmutation of ^{206}Pb in lead contained in the same ore. It has been shown[5] that BGO

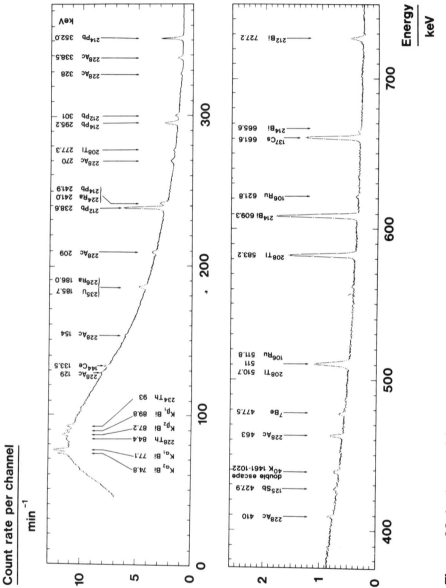

Figure 20-1 A terrestrial gamma-ray spectrum recorded over 170 h using a 60 cm³ germanium detector mounted 1 m above the ground. (From Finck et al.[1])

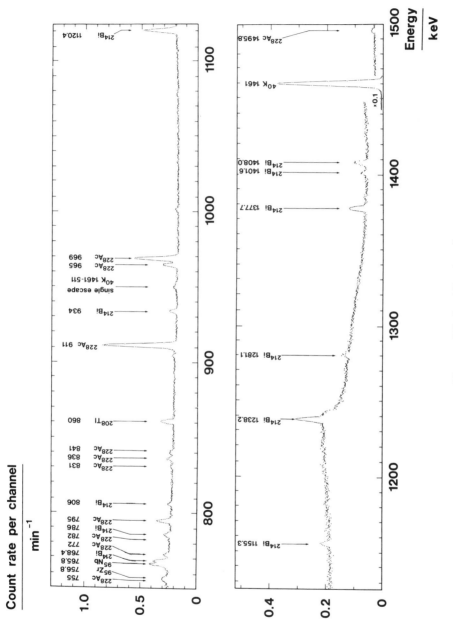

Figure 20-1 (*Continued*)

TABLE 20-1   Levels of Activities from Natural Sources in Common Construction Materials

Material	Disintegrations/min per gram of Material		
	^{232}Th(583 keV)	^{238}U	^{40}K
Aluminum (6061 from Harshaw)	0.42	0.04	< 0.05
Aluminum (1100 from Harshaw)	0.24	< 0.017	< 0.06
Aluminum (1100 from ALCOA)	0.08	< 0.026	< 0.11
Aluminum (3003 from ALCOA)	0.10	< 0.026	0.56
Stainless steel (304)	< 0.006	< 0.007	< 0.06
Stainless steel (304-L)	< 0.005	< 0.005	< 0.02
Magnesium (rod)	0.06	< 0.04	0.1
Magnesium (ingot)	< 0.01	< 0.002	< 0.02
Magnesium (4 in. ∅ × 4 in. from Dow)	< 0.005	< 0.002	< 0.02
Magnesium (from PGT)	< 0.05	< 0.03	< 0.05
Beryllium copper alloy	< 0.02	< 0.06	< 0.2
Copper (sheet)	< 0.05	< 0.06	< 0.2
Pyrex window	0.45	0.27	3.8
Quartz window	< 0.018	< 0.018	< 0.07
Molecular sieve	4.4	3.0	9.0
Neoprene	< 0.008	< 0.01	0.36
Rubber	0.12	1.0	2.0
Apiezon Q	4.5	4.5	2.7
Electrical tape—3M	< 0.04	< 0.06	< 0.1
Cement (Portland)	0.25	1.3	4.5
Epoxy	0.006	0.01	0.19
Lacquer	0.002	0.005	0.04

Source: Camp et al.[2]

obtained from lead-free ores does not carry this radioactive contamination. For germanium detector systems, the extreme degree of purity required of the germanium for its acceptability as detector material ensures that its inherent radioactivity is low.

Ironically, the shielding placed around a detector to reduce background from cosmic rays or terrestrial radiation may itself introduce a significant background from its inherent low-level radioactivity. The relative activity levels expected are discussed later in this chapter in connection with specific shielding materials. Shield surfaces can also be the source of significant activities due to fallout or accidental surface contamination. The inner surfaces of low-background shields are therefore often sandblasted prior to installation to remove any surface radioactivity. Painting can also help to reduce surface alpha activity, although the radioactivity of the paint itself should also be checked prior to its application.

B. Airborne Radioactivity

A measurable amount of background can originate with radioactivity carried by the ambient air, either in the form of trace amounts of radioactive gases or dust particles. Radon (^{222}Rn) and thoron (^{220}Rn) are short-lived radioactive gases that originate as daughter products in the decay chains of the uranium and thorium present either in the soil or construction materials of the laboratory. Their concentration in the atmosphere can vary significantly[6] depending on the time of day and meteorological conditions. To eliminate the influence of radon on the background, the volume around the detector can be made airtight and purged with a radon-free gas. Some laboratories use the boil-off nitrogen gas from the liquid nitrogen dewar used with germanium detectors for this purpose. Radioactive dust can consist of either natural radioactivities or atmospheric fallout and can largely be eliminated through effective filtration of the air supply to the counting room.

C. Cosmic Radiation

A significant component of detector background arises from the secondary radiations produced by cosmic-ray interactions in the earth's atmosphere. The primary cosmic radiation, which can be either of galactic or solar origin, is made up of charged particles and heavy ions with extremely high kinetic energies. In their interaction with the atmosphere, a large assortment of secondary particles is produced, including pi and mu mesons, electrons, protons, neutrons, and electromagnetic photons, with energies that extend into the hundreds of MeV range. Many of these radiations reach the earth's surface and can create background pulses in many types of detector.

Because of their very high kinetic energy, the cosmic primary and secondary particles have a relatively low specific energy loss ($-dE/dx$) comparable with that of fast electrons. Thus, the corresponding pulse amplitude is small in "thin" detectors designed to stop typical charged particles but not electrons. However, in solid detectors of large thickness, such as NaI(Tl) scintillators or germanium detectors, the deposited energy may be many MeV and the corresponding pulses can then be large compared with typical signal pulses.

The various secondary components differ in their hardness or ability to penetrate matter, and some gains are achieved even with modest amounts of shielding. Other components persist through many meters of common materials. At the earth's surface, the cosmic secondary radiations are directed primarily downward, so that shielding against cosmic background is most effective when located above the detector.

Fast neutrons from cosmic interactions (or any other source) can also create secondary gamma rays within the detector shield. Most important are capture gamma rays liberated when a neutron is moderated and absorbed. When the shield contains hydrogen (e.g., as in concrete), the capture gamma ray at 2.22 MeV can sometimes be identified in the background spectrum.

II. BACKGROUND IN GAMMA-RAY SPECTRA

A. Relative Contributions

The importance of various components of the background changes greatly with the circumstance. In gamma-ray detectors without shielding, the cosmic-ray component is normally dominant. When significant shielding is provided, both the cosmic flux and the background due to ambient sources of gamma rays are decreased, and radioactive contamination of structural and shielding materials around the detector becomes an important fraction of the remainder. Underground locations can help reduce the cosmic background, but tens or hundreds of meters of earth are needed to eliminate the hardest components of the meson flux.

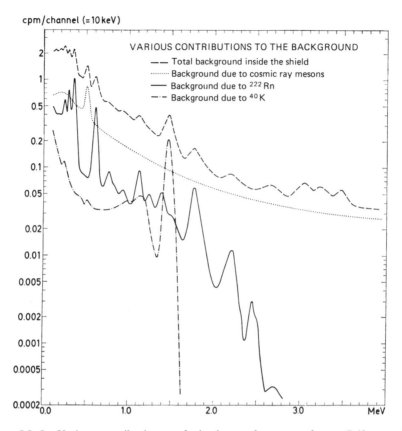

Figure 20-2 Various contributions to the background spectrum from a 7.62 cm × 7.62 cm NaI(Tl) scintillator inside a massive lead and borated paraffin shield. (From Stenberg and Olsson.[7])

Figure 20-2 shows the various components of the background spectrum recorded by a 7.62 cm × 7.62 cm NaI(Tl) detector within a massive shield consisting of 10 tons of lead and 160 kg of paraffin and boric acid mixture.[7] The various background components above an energy of 100 keV in the spectrum are tabulated in Table 20-2.

The behavior of the background in a number of large-volume Ge(Li) detectors has been studied by Kaye et al.[8] Using shielding of the type shown in Fig. 20-3, a typical detector background is made up of a 30% contribution from cosmic radiation, 60% from

TABLE 20-2 Components of a NaI(Tl) Scintillation
Counter Background

Outside shield	29,200 counts/min
Inside shield	
Cosmic ray mesons	116.4
Cosmic ray neutrons	19.4
^{222}Rn daughters	25.9
^{40}K	8.6
Remaining background	33.1
Total	203.4 counts/min

Source: Stenberg and Olsson.[7]

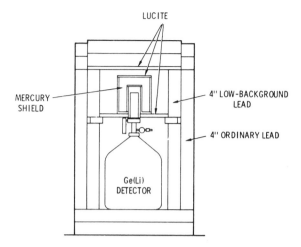

Figure 20-3 A low-background shield configuration for a germanium detector. (From Kaye et al.[8])

the radioactive contamination of shielding materials, and 10% from radioactivity within the detector and unidentified sources. Typical background spectra from this configuration are shown in Fig. 20-4.

The background in gamma-ray detectors can be expected to increase roughly as the detector volume. Therefore, in critical situations where low background is at a premium, it is important to select a detector size that is not larger than necessary to give a reasonable counting efficiency for the samples to be counted. As shown in Chapter 3, a statistically based selection criterion is to choose a detector size that maximizes the ratio of S^2/B, where S is the counting rate due to the source alone and B is the counting rate due to background. This same figure of merit can be used in the selection of other operating parameters, such as discrimination level and so on, in setting up a low-level counting experiment.

B. Variations in the Background Rate

The background rate from a typical radiation detector, although expected to be nearly steady state in time, may show a perceptible variation over periods of hours or days.[9] In experiments in which the signal rate is high, small background fluctuations are typically much lower than the inherent statistical fluctuation of the measurement itself, and consequently they are of no real importance. In such situations, a single background determination (repeated every day or so) is a sufficient measurement on which to base background subtraction of the measured counting rates.

When low-level activities are counted, the fluctuations in the background may be of the same order as the source strength and therefore must carefully be considered. Although the component of the background due to radioactivity of the detector and surrounding material will be constant, variations in the background may arise from changes in either the cosmic-ray intensity or the airborne radioactivity. Much of the observed variation in sensitive gamma-ray counters appears to be correlated with airborne activity in the form of the decay products of ^{222}Rn (Refs. 6 and 10). The importance of this background source depends greatly on the specific geometry and ventilation condi-

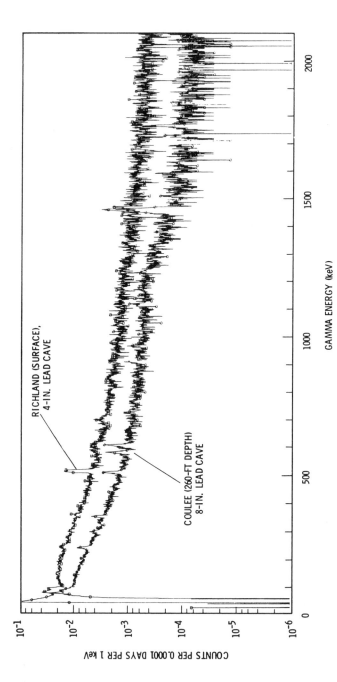

Figure 20-4 The top curve shows the background spectrum recorded for a 85 cm³ germanium detector in a normal laboratory using the shielding shown on Fig. 20-3. The bottom curve shows the gains achieved by relocating the counter deep inside the massive concrete structure of a dam and increasing the thickness of the lead shielding. (From Kaye et al.[8])

tions around the detector, but variations of the order of a few percent in the background rate should not be unexpected.

In low-level counting experiments, it is therefore prudent to carry out a background determination near the time of the actual measurement itself. Background counts both before and after the measurement will help detect any changes in the background level. In critical situations, the background rate can be monitored during the course of the measurement by using a second detector whose properties are identical with those of the detector used in the measurement itself. In this way, any temporal changes in the background rate will affect both detectors and are thereby canceled.

C. Source-Related Background

In gamma-ray spectroscopy, some additional interfering radiation can be observed due to the interaction of primary gamma rays from the source with structural and shielding materials around the detector. The origins of these source-related background events were discussed previously in Chapter 10 and are illustrated schematically in Fig. 10-6. Important processes include Compton backscattering of the primary gamma rays, and the

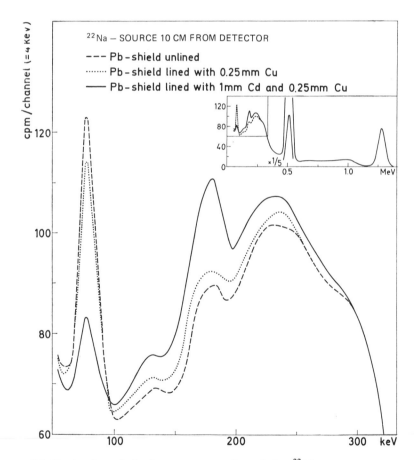

Figure 20-5 A plot of the low-energy portion of the ^{22}Na gamma-ray spectrum (indicated on the insert) as recorded by a 7.62 cm × 7.62 cm NaI(Tl) scintillator. The effects of different shield liners on the fluorescent X-ray and backscatter peaks are illustrated. (From Stenberg and Olsson.[7])

generation of secondary annihilation photons and characteristic X-rays through pair production or photoelectric absorption. These effects can be eliminated only by removing all materials from the immediate vicinity of the detector, but the practical demands of shielding usually do not permit this approach. Therefore, the potential source regions for these secondary radiations, particularly the inner surface of the radiation shield, are often designed with materials intended to reduce their importance.

To illustrate, Fig. 20-5 shows the low-energy portion of a gamma-ray spectrum from ^{22}Na as recorded by a NaI(Tl) scintillation detector. This portion of the spectrum shows the perturbing effects of both a backscatter peak at about 170 keV and a peak at 77 keV arising from lead fluorescent X-rays generated by photoelectric absorption of the primary gamma rays within the surrounding lead shield. To reduce the latter component, lead shields are commonly lined with thin layers of low-Z material to absorb these lead X-rays, and which themselves emit much softer characteristic X-rays. Figure 20-5 illustrates the degree of suppression that can be achieved in the corresponding X-ray peak with linings of copper or cadmium. These gains are partially offset by an increase in the intensity of the backscatter peak due to the greater probability of Compton scatter in these materials relative to photoelectric absorption.

III. BACKGROUND IN OTHER DETECTORS

A. Gas-Filled Counters

The background expected from gas-filled detectors such as proportional counters or Geiger tubes depends greatly on the minimum ionization required for registration of a pulse. In the Geiger tube, this minimum is, in principle, only one ion pair, so that any interaction, regardless of its amplitude, results in a registered pulse. All the previously mentioned sources of background can therefore contribute to the background rate observed from a Geiger tube. At sea level, an unshielded G-M tube of typical dimensions will show a background rate of several counts per second, which can greatly be reduced by a minimal shielding of several centimeters of lead.

Proportional counters can be operated with a finite discrimination level so that lightly ionizing events do not contribute to the measured counting rate. Because the fill gas is relatively transparent to fast electrons, the discrimination level can readily exceed the ionization that will be created along the track of an electron. When operated in this mode, the proportional counter is sensitive only to heavy charged particles such as alpha particles, and the operating point corresponds to the *alpha plateau* region discussed in Chapter 6. Under these circumstances, many of the possible sources of background are eliminated, and only those that deposit their energy in the form of densely ionizing charged particles can contribute to the measured background counting rate. Then, the most significant source is often the inherent alpha radioactivity of the construction materials of the detector itself. Table 20-3 shows the inherent alpha background activity of some construction materials common to gas-filled detectors. With some care as to the choice of construction materials, the background level for proportional counters operating on the alpha plateau is less than 1 count/min.

For gas-filled detectors operated in current mode, there can be no inherent pulse amplitude discrimination. Therefore, all the sources of background radiation mentioned previously will contribute to the background current measured from a current mode ion chamber, for example. Ionization caused by cosmic rays at sea level amounts to about two ion pairs per second, per cubic centimeter of STP air.[11] Background currents due to

TABLE 20-3 Alpha Particle Emission Rates from Various Materials

Material	Alphas per cm² per hour		
	(J. A. B.)[a] E_T^d 250 keV	(S. & H.)[b] E_T 100 keV	(A. & J.)[c] E_T 1 MeV
Machined copper	0.09		
Commercial brass	0.05	0.2	0.13
Mild steel (stainless)	0.03	0.05	0.01–0.03
Commercial aluminum	0.31	0.2	0.27
Solder	28		
Aquadag	0.07		
Nickel		0.03	
Lead	60		
Perspex		0	
Zinc sulfide in Perspex sandwich (10 mg/cm²)		0.1	
Air from room	32/100 cm³ · h		
Cylinder nitrogen	0		
Cylinder argon		0	

[a] J. A. B.; J. A. Bearden, *Rev. Sci. Instrum.* **4**, 271, (1933).
[b] S. & H.; J. Sharpe and P. Holton, unpublished measurements.
[c] A. & J.; B. Al-Bataina and J. Janecke, *Nucl. Instrum. Meth.* **A255**, 512 (1987).
[d] E_T represents the detection threshold.

Source: Data in first two columns from J. Sharpe, *Nuclear Radiation Detectors*, 2nd ed., Methuen & Co., Ltd. London, 1964.

terrestrial radiation and radioactivity of surrounding materials can be about the same order of magnitude, whereas the current from natural alpha activity of common wall materials is normally an order of magnitude lower.[11]

B. Semiconductor Charged-Particle Detectors

As in germanium detectors, the inherent radioactivity of the high-purity semiconductor material itself is negligible for silicon diode detectors. Because typical surface areas are only a few square centimeters, and the contact electrodes are extremely thin, there is very little other material that can contribute background alpha activity. The depletion depth of silicon surface barriers is normally too small to develop significant pulses from low dE/dx radiations from cosmic rays or gamma-ray interactions, so the residual background rate is negligible in virtually any conceivable application to the detection of charged particles.

IV. SHIELDING MATERIALS

A. Conventional Materials for Low-Background Shields

1. LEAD
Because of its high density and large atomic number, lead is the most widely used material for the construction of detector shields. The photoelectric absorption cross section predominates up to gamma-ray energies as high as 0.5 MeV, and therefore even relatively hard gamma rays from external background sources (such as the 1.46 MeV gamma ray from ^{40}K) can be absorbed efficiently. Because of its high density, thicknesses of just a few centimeters of lead will provide a large reduction in the background of

typical gamma-ray detectors. Lead is reasonably effective at removing many of the cosmic components of the background, although thicknesses beyond about 10 cm do not result in an appreciable decrease in the resulting counting rate because of the buildup of secondary radiations, due to cosmic interactions within the lead itself.[12]

Lead is widely used in the form of rectangular "lead bricks" in the construction of simple gamma-ray shields. Potential problems due to streaming through cracks between the bricks can be overcome by building the shield with multiple layers or by using specially shaped bricks with interlocking surfaces. Lead is also cast relatively easily into solid shapes, although some care must be taken in the casting process to avoid porosity or voids in the solidified shields. As an alternative to casting, a container of the proper size and shape can be filled with lead shot to form a shield with somewhat lower density. Other shielding materials are also commercially available which incorporate a high percentage of lead into plastic or epoxy compositions, which can be more readily molded and shaped.

Ordinary lead normally contains a significant amount of natural activity due to low-level contaminants, and therefore lead that is either specially refined or reclaimed from very old sources is preferred in the construction of shields for low-background applications. When freshly refined, lead can contain significant amounts of ^{210}Pb, a daughter product of the decay of ^{226}Ra. This isotope decays with a half-life of 20.4 years, so that samples of lead that are many decades old will be relatively free of this activity. However, there is some indication[13,14] that properly refined lead from some sources may not contain troublesome amounts of ^{210}Pb.

Recently refined lead may also show detectable radioactivity due to impurities from atmospheric fallout. Other background may arise from trace amounts of thorium daughters, or radioactive impurities common in antimony, which is a normal alloy of lead. Some lead samples have shown activities as high as 1.5 Bq/g (Ref. 15), but activities of one or two orders of magnitude lower are more typical of high-purity lead.

2. STEEL

Iron or steel is also a common gamma-ray shielding material and is often used in situations where the size or configuration of the shield would make its construction from lead alone too expensive. In such circumstances, an outer layer of steel with an inner lead lining is often an effective compromise. Again, there is a premium on obtaining old material because steel fabricated after about 1950 is of noticeably higher activity than prewar steel. (Steel salvaged from World War II naval ships has been used in the construction of a significant number of low-background shields.) Some of the low-level activity found in more recently fabricated steel is traceable to the unfortunate practice in the 1950s of inserting samples of ^{60}Co into the liners of blast furnaces to check on the erosion of the liner.

Because both the atomic number and density of steel are considerably lower than those of lead, thicknesses of several tens of centimeters are normally required for very low background applications.

3. MERCURY

Although it is relatively expensive, mercury is a very effective shielding material in low background counting situations. It can be purified to a high degree through distillation and thus has an inherently low level of residual radioactivity. It is often used as the innermost component of large gamma-ray shields. Because mercury is a liquid at room temperature, it must be held within a suitable container, often constructed of lucite (a material of inherently low background radioactivity), which can be configured to match closely the outer contours of the detector. Because its density is even greater than that of lead and its atomic number similar, thicknesses of a few centimeters of mercury will be relatively effective as a gamma-ray shield.

4. CONCRETE

Because of its low cost, concrete is often used in the construction of large-volume shields. However, its activity is relatively high due to ^{40}K, uranium, and fallout products included in its composition. It is therefore most commonly used as the outer constituent of a shield, with its own activity shielded by an inner layer of steel, lead, or other shielding material of lower activity.

The effectiveness of various thicknesses of concrete on the attenuation of the cosmic component of background is illustrated in Fig. 20-6. Although few experimenters have the luxury of using Grand Coulee Dam as a radiation shield as in this study, significant benefits can be achieved by locating the counting laboratory in the basement of a multistory building, in which the concrete floors of upper stories can significantly attenuate the cosmic component.[2]

Although ordinary concrete consists mainly of water and low-Z elements, a special formulation known as *barytes concrete* contains a significant percentage of heavy components and is therefore much more effective in gamma-ray shielding.

B. Neutron Shielding

The shielding of neutrons is important from several aspects. Any detector designed for neutron counting must obviously be shielded against external sources of background neutrons to enhance its signal-to-background ratio. However, other types of detectors, including the low-background gamma-ray counting systems discussed in the previous section, can also be influenced by a fast neutron background. Much of this sensitivity arises because of the capture gamma rays created upon absorption of the neutrons within the detector or nearby materials. The shielding of the detectors against neutron backgrounds is obviously most important around neutron-producing facilities, but a measurable neutron component also exists in the natural background due to secondary products of cosmic-ray interactions.

Completely different principles apply to the selection of neutron shielding materials as compared with those for gamma rays. It is most important to quickly moderate the neutron to low energies, where it can readily be captured in materials with high absorption cross sections. The most effective moderators are elements with low atomic number, and therefore hydrogen-containing materials are the major component of most neutron shields. In this application, water, concrete, and paraffin are all inexpensive sources of bulk shielding. Because mean free paths of fast neutrons typically are tens of centimeters in such materials, thicknesses of 1 m or more are required for effective moderation of almost all incident fast neutrons.

Once the neutron has been moderated, it can be eliminated through an appropriate capture reaction. This absorption may be in the hydrogen already present for moderating purposes, although the capture cross section is relatively low. The thermal neutron may thus diffuse an appreciable distance before capture, reducing the effectiveness of the shield. Furthermore, capture in hydrogen leads to the liberation of a 2.22 MeV capture gamma ray, which, because of its high energy, is particularly undesirable in many types of detector. Therefore, a second component is normally used in neutron shields, either

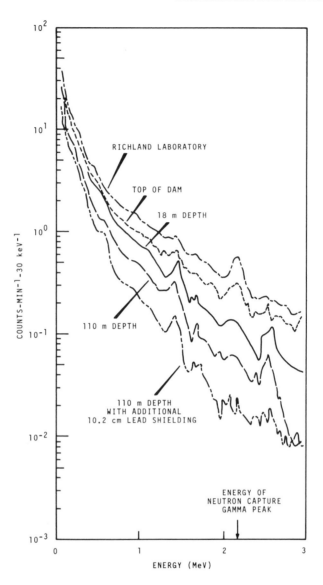

Figure 20-6 The background spectra from a 7.62 cm × 7.62 cm NaI(Tl) scintillator within a paraffin and 10.2 cm thick lead shield. The indicated locations are at different depths within the massive concrete structure of Grand Coulee Dam. The purpose of the paraffin was to indicate the importance of the neutron component by the relative prominence of the hydrogen capture gamma-ray peak at 2.22 MeV. (From Kaye et al.[12])

homogeneously mixed with the moderator or present as an absorbing layer near its inner surface. This additive is chosen to have a high neutron capture cross section, so that the moderated neutrons will preferentially undergo absorption within this material.

Some reactions with large capture cross sections for slow neutrons have already been discussed in Chapter 14 as applied to slow neutron detectors. These same reactions are also useful in the shielding application, and therefore boron and lithium are common components of neutron shields. The $^{10}B(n, \alpha)$ reaction has a high capture cross section at

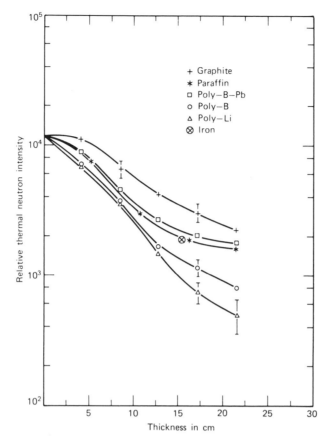

Figure 20-7 The relative effectiveness of different shielding materials for 2.6 MeV neutrons as indicated by the thermal neutron flux for various thickness. (From Gujrathi and D'Auria.[16])

low energies, and boron can readily be incorporated into paraffin and other moderating materials. The majority of the capture reactions, however, lead to an excited state in the product nucleus, which subsequently decays by the emission of a 0.48 MeV gamma ray. Applications that are potentially sensitive to gamma-ray backgrounds are therefore better served by substituting lithium for the boron, because the $^6Li(n, \alpha)$ reaction proceeds directly to the ground state of the product and no gamma rays are emitted.

The effectiveness of some commercially available combinations of polyethylene and boron or lithium in the shielding of fast neutrons is compared with those of graphite and paraffin in Ref. 16. Figure 20-7, taken from this reference, shows that the mixtures of polyethylene with the thermal absorbers are considerably more effective per unit thickness than the remaining materials.

Cadmium is also widely applied as a thermal neutron absorber because thin sheets of the material are essentially opaque to thermal neutrons due to its very high cross section. Layers as thin as 0.5 mm are very effective absorbers of thermal neutrons, but the subsequent (n, γ) reaction will also add a secondary gamma-ray background.

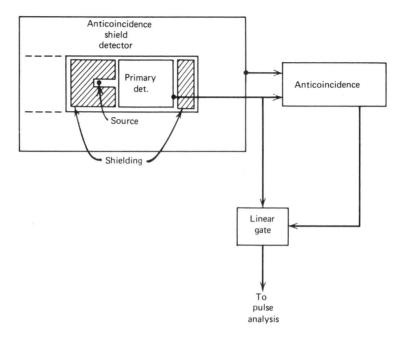

Figure 20-8 The general arrangement of an anticoincidence shield for cosmic background reduction.

V. ACTIVE METHODS OF BACKGROUND REDUCTION

A. Anticoincidence Shielding

The cosmic-ray component of detector background can be removed only through massive amounts of conventional shielding. An alternative arrangement is sketched in Fig. 20-8, in which the highly penetrating cosmic radiations are eliminated through the use of an *anticoincidence shield*. The primary detector is surrounded by a second detector (or an array of detectors), and the output of the primary detector is accepted only when it is not accompanied by a coincident pulse in the outer detector. The source to be counted is oriented and shielded so that it produces interactions only in the primary detector. Therefore, in the simplest case, no pulses are affected which correspond to the complete absorption of the source radiation within the primary detector. However, the cosmic radiations will likely penetrate both detectors and therefore can be eliminated from the output of the primary detector through the anticoincidence arrangement. As a bonus, the anticoincidence shield will also suppress the Compton continuum in the recorded spectrum, because a Compton-scattered gamma ray from the primary detector may also interact within the surrounding detector. This mode of operation for germanium detectors is discussed further in Chapter 12. The degree to which the anticoincidence technique can reduce the background spectrum in a germanium detector, when surrounded by an annular sodium iodide scintillator, is illustrated in Fig. 20-9. The anticoincidence detector can also be a large plastic or liquid scintillator,[2,17,18] a ring of Geiger tubes,[19] or virtually any other type of detector.

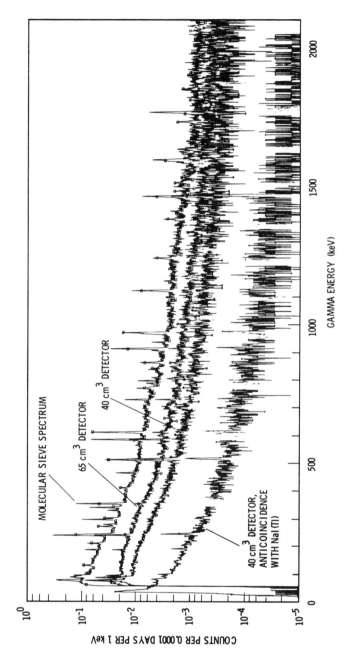

Figure 20-9 The bottom spectrum shows the degree to which background in a 40 cm³ germanium detector may be suppressed through the use of an annular NaI(Tl) scintillator in anticoincidence. A comparison of the two middle spectra illustrates the increase in background with detector active volume. The top curve shows the recorded spectrum using a sample of molecular sieve [a part of many germanium vacuum systems] as a source of natural background radiation. (From Kaye et al.[8])

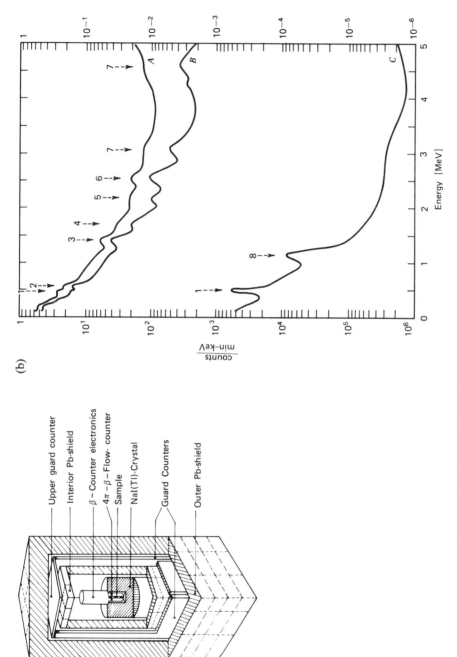

Figure 20-10 (a) A schematic view of a β-γ coincidence detector assembly, in which beta particles are detected by a split 4π flow counter, and the gamma rays by a surrounding well-type NaI(Tl) scintillator. Anticoincidence guard detectors are also employed. (b) Background spectra from the same detector assembly. A: NaI(Tl) spectrum without β-γ coincidence or anticoincidence shielding. B: Same, but with anticoincidence shielding. C: As in B, but with β-γ coincidence between the two detectors. Identified peaks probably correspond to 1—annihilation radiation; 2—^{208}Tl; 3—^{40}K; 4—^{214}Bi, ^{212}Bi; 5—^{214}Bi, ^{212}Bi; 6—^{208}Tl; 7—alpha particles from detector components; 8—not specifically identified. (From Roedel.[20])

B. Coincidence Counting

If a radioisotope that emits more than one distinguishable radiation in coincidence is to be counted, the effective background can greatly be reduced through the use of coincidence techniques. A common example is that of beta decay to a daughter product, which promptly emits at least one gamma ray. A suitable arrangement can then consist of separate beta particle and gamma-ray detectors, with the source mounted to maximize the corresponding interaction rates in each detector. By demanding coincidence between the two detector outputs, many background events will be eliminated because they will occur in only one detector at a time.

As an example, Fig. 20-10 shows a double detector coincidence spectrometer in which the beta radiation is detected by a split 4π gas-flow proportional counter. This detector is fitted within the well of a 12.7 cm $\times$ 12.7 cm NaI(Tl) scintillator, which serves as the gamma-ray spectrometer. By mounting the source between the two halves of the proportional counter, the counting efficiency for beta particles is about 70%. Because the scintillator will also be very efficient for gamma rays in this geometry, coincident pulses are obtained in 50–65% of the cases in which coincident radiations are emitted. This slight decrease in the overall counting efficiency is more than offset by the very large reduction in the background counting rate illustrated on the spectra in Fig. 20-10. For typical conditions, the improvement in the minimum detectable sample activity can amount to one or two orders of magnitude compared with a similarly shielded simple gamma detector.[20] The example shown also incorporates anticoincidence *guard counters* around the detector assembly to reduce the background level further.

Most applications in which extremely weak radioisotope sources must be counted now take advantage of coincidence methods to suppress background as much as possible. Recent examples that contain references to earlier work can be found in Refs. 2, 8, 21, and 22.

PROBLEMS

20-1. One potential source of background counts from sodium iodide scintillators is ^{40}K from trace-level potassium impurity in the crystal. Find the maximum potassium concentration (in ppm) if the corresponding background rate from a 7.62 cm $\times$ 7.62 cm cylindrical crystal is not to exceed 1 count/s.

20-2. Explain the presence of naturally occurring ^{14}C when its half-life (5600 years) is so short compared with the age of the universe.

20-3. Explain the following observation: For gas proportional tubes operated in pulse mode, the average size of cosmic-ray-induced pulses is usually much smaller than that of typical signal pulses, but the opposite is true for typical scintillation detectors used in gamma-ray spectroscopy.

20-4. Gas proportional counters operated on the alpha plateau can show background counting rates that are often less than 1 count/min. Point out the origin of the background events that are observed and explain why this rate is so much less than usually encountered in other detectors.

20-5. The background gamma-ray spectrum inside massive concrete shielding normally shows a prominent 2.22 MeV line. Explain the origin of these gamma rays.

REFERENCES

1. R. R. Finck, K. Liden, and R. B. R. Persson, *Nucl. Instrum. Meth.* **135**, 559 (1976).
2. D. C. Camp, C. Gatrousis, and L. A. Maynard, *Nucl. Instrum. Meth.* **117**, 189 (1974).
3. K. Bunzl and W. Kracke, *Nucl. Instrum. Meth. Phys. Res.* **A238**, 191 (1985).
4. R. L. Brodzinski, J. H. Reeves, F. T. Avignone, and H. S. Miley, *Nucl. Instrum. Meth. Phys. Res.* **A254**, 472 (1987).
5. T. A. Lewis, *Nucl. Instrum. Meth. Phys. Res.* **A264**, 534 (1988).
6. S. Okabe, T. Nishikawa, M. Aoki, and M. Yamada, *Nucl. Instrum. Meth. Phys. Res.* **A255**, 371 (1987).
7. A. Stenberg and I. U. Olsson, *Nucl. Instrum. Meth.* **61**, 125 (1968).
8. J. H. Kaye, F. P. Brauer, J. E. Fager, and H. G. Rieck, Jr., *Nucl. Instrum. Meth.* **113**, 5 (1973).
9. K. S. Parthasarathy, *Nucl. Instrum. Meth.* **134**, 591 (1976).
10. K. S. Parthasarathy, *Nucl. Instrum. Meth.* **136**, 585 (1976).
11. M. H. Shamos and A. R. Liboff, *Rev. Sci. Instrum.* **39**, 223 (1968).
12. J. H. Kaye, F. P. Brauer, R. E. Connally, and H. G. Rieck, *Nucl. Instrum. Meth.* **100**, 333 (1972).
13. O. G. Bartels, *Health Phys.* **28**, 189 (1975).
14. R. J. Arthur, J. H. Reeves, and H. S. Miley, *IEEE Trans. Nucl. Sci.* **NS-35**(1), 582 (1988).
15. B. Grinberg and Y. Le Gallic, *Int. J. Appl. Radiat. Isotopes* **12**, 104 (1961).
16. S. C. Gujrathi and J. M. D'Auria, *Nucl. Instrum. Meth.* **100**, 445 (1972).
17. S. R. Lewis and N. H. Shafrir, *Nucl. Instrum. Meth.* **93**, 317 (1971).
18. J. H. Reeves, W. K. Hensley, and R. L. Brodzinski, *IEEE Trans. Nucl. Sci.* **NS-32**(1), 29 (1985).
19. A. Stenberg, *Nucl. Instrum. Meth.* **96**, 289 (1971).
20. W. Roedel, *Nucl. Instrum. Meth.* **61**, 41 (1968).
21. F. P. Brauer, J. H. Kaye, and J. E. Fager, *IEEE Trans. Nucl. Sci.* **NS-22**(1), 661 (1975).
22. K. Yamakoshi and K. Nogami, *Nucl. Instrum. Meth.* **134**, 519 (1976).

APPENDIX · A

The NIM and CAMAC Instrumentation Standards

In Chapter 17, some of the general features of nuclear instrumentation standards were introduced. In this appendix, some details are provided for the two most common standards in general use: the NIM and CAMAC standards.

A. The NIM (Nuclear Instrument Module) System

The NIM bin is designed to fit into the standard 19 in. relay rack and is subdivided into 12 individual module positions across its width. A NIM module occupies a unit width of 34.4 mm, although integral multiples of this width are permitted corresponding to modules of double width, triple width, and so on. Each of the 12 bin locations is provided with a 42 pin connector that mates with a corresponding connector at the back of each module. Pin assignments and functions are illustrated in Fig. A-1. Primary dc supply voltages provided by the bin are ± 12 V and ± 24 V. Some NIM bins also provide ± 6 V, mostly for modules using integrated circuits, but these voltages are not strictly required by the NIM rules. The NIM bin and modules can be of two standard heights—$5\frac{1}{4}$ in. (133 mm) or $8\frac{3}{4}$ in (222 mm)—but the larger of these sizes is by far the more common.

Although some limited logic and switching operations can be performed through designated connector pins, the primary means of transmitting linear and logic pulses between NIM modules is by coaxial cables connected to appropriate jacks on either the back or front panel of the module. BNC connectors are specified for signal jacks, whereas SHV connectors are standard for high-voltage connections.

The NIM standard also recommends that shaped linear pulses correspond to one of three specific dynamic ranges:

1. 0 to $+1$ V (primarily for integrated circuits)

2. 0 to $+10$ V (primarily for transistor-based circuits)

3. 0 to $+100$ V (primarily for vacuum-tube-based circuits and largely obsolete).

NIM modules designed for the processing of linear pulses are signal compatible only if they share a common classification for the dynamic range of the signals.

Logic signal levels are also specified in the NIM standard. Standard logic levels for logic states and the transmission of digital data are given in Table A-1. Fast logic pulses in 50 ohm impedance systems are separately specified in Table A-2.

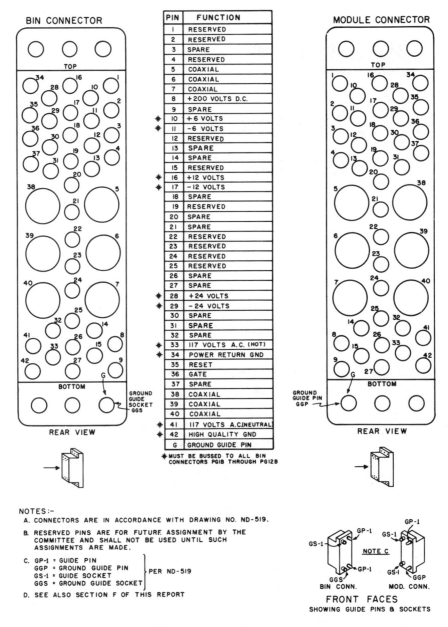

Figure A-1 Pin assignments for the NIM standard connector between bin and module. (Adapted from National Bureau of Standards Photo 74-08-4044.)

Detailed specifications for NIM systems are contained in USAEC Report TID-20893, which was first issued in July 1964. This report has subsequently undergone revision every few years, and the interested user should seek out the latest revision.

The NIM standard does not adapt easily to situations in which large volumes of digital data must be processed. Furthermore, the basic module width is needlessly large for many digital units that do not require large numbers of front panel controls. These

TABLE A-1   NIM Standard Logic Levels

	Output (Must Deliver)	Input (Must Respond to)
Logic 1	$+4$ to $+12$ V	$+3$ to $+12$ V
Logic 0	$+1$ to -2 V	$+1.5$ to -2 V

TABLE A-2   NIM Fast Logic Levels for 50 ohm Systems

	Output (Must Deliver)	Input (Must Respond to)
Logic 1	-14 to -18 mA	-12 to -36 mA
Logic 0	-1 to $+1$ mA	-4 to $+20$ mA

considerations, together with the desire for standard interfacing with digital computers, has led to the development of the CAMAC standard described next.

B. The CAMAC (Computer Automated Measurement and Control) Standard

The CAMAC standard is based on a *crate*, which also fits the standard 19 in. relay rack but which is subdivided into 25 individual module stations spaced 17.2 mm apart. Electrical connection between each module and the crate is made by a printed circuit board edge connector with 86 contacts. The width of the CAMAC station neatly accommodates many modules consisting of a single printed circuit board with mounted integrated circuits. When modules must involve bulkier equipment or more than one board, multiples of the basic width can be used, and the double-width or triple-width module will occupy two or three standard stations.

Within the crate, each connector provides access to the *dataway*, which is a data highway consisting of conductor busses for digital data, control signals, and power. One of the basic design features of CAMAC is that digital communication between plug-in modules within a crate occurs over this dataway. It replaces external interconnection of modules for many digital functions, but some coaxial cable connections must be retained for linear signals and other purposes. These signals can be coupled to either the front or back of a module using the type 50CM connector recommended in the CAMAC standard.

The extreme right-hand station within the crate is different from the remaining 24 and is called the *control station*. It is intended that this station will be occupied by a *crate controller* plug-in module. The crate controller is usually a double-width module that occupies one normal station (usually the 24th) as well as the control station. The crate controller provides all the control functions necessary for the transfer of data between modules in a crate and serves as the interface between the crate and any external equipment. No CAMAC system is complete without a crate controller, and none will function without one, except on an individual module basis. In order to communicate properly over the dataway, each individual module must have sufficient internal coding and decoding to read digital data from the dataway and to supply such data from its own internal circuits when requested by the crate controller.

A diagram of the dataway is shown in Fig. A-2. The power busses are connected to all 25 stations and provide ± 6 V and ± 24 V for module power. Although not required by CAMAC, ± 12 V may also be available, depending on the specific crate design, and

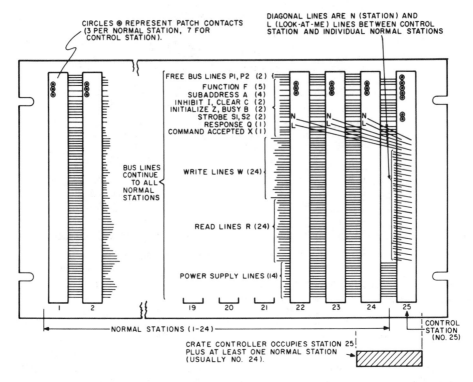

Figure A-2 Diagram of the CAMAC dataway. (National Bureau of Standards Photo 72-534.)

connections are reserved for these voltages. Except for the control station, all normal stations are also interconnected by 24 parallel *read* lines and 24 *write* lines. These lines can be used to transmit 24 parallel bits of data from a module along the read lines and to a module along the write lines. The control station is connected to each of the normal stations by separate private lines. Twenty-four of these are *station* lines (N lines), one of which corresponds to each normal station and must be activated to communicate with that specific station. An additional 24 are *look-at-me* lines (L lines), one for each station, by means of which each individual normal station may signal to the control station that it requires attention.

Modules cannot communicate directly with each other but must do so only through the crate controller. In providing a command to a given station, the controller, in addition to activating the corresponding N line, must also complete the command by providing a coded subaddress and function. These are communicated via five *function* lines (F lines) and four *subaddress* lines (*A* lines), which are fully bussed to all stations. Two *strobe* lines (S1 and S2) are also fully bussed and must be activated to initiate operations or to transfer data. The remaining busses shown in Fig. A-2 have fairly descriptive names that indicate their functions.

In addition to managing the internal module-to-module communication within a crate, the crate controller must also serve to interface the dataway with external equipment that is to be part of the overall system. Because CAMAC systems are of greatest interest when large volumes of digital information must be handled in a given application, the crate controller is often required to interface with the I/O structure of a computer.

Standard crate controllers can be obtained for interfacing the more common types of laboratory computers, as well as to interface with the branch highway (type A controller) and serial highway (type L controller) described below.

If the number of CAMAC modules that must be used in a given application exceeds the room available in a single crate, or if several crates in different remote areas must be linked to a common data system, then a method must be provided to couple together more than one CAMAC crate. The arrangement can be either a parallel interconnection (a *branch highway*) consisting of 66 pairs of signal and ground wires, or a series chain (a *serial highway*) in which only two pairs of conductors are used. In the latter case, data are sent serially and it is therefore slower, but costs are often much lower than for a branch highway interconnection.

Detailed specifications for CAMAC are spelled out in IEEE Documents 583-1975, 596-1976, and 595-1976. A useful collection of introductory articles can be found in Ref. 1. A book that includes both a tutorial introduction to CAMAC and complete specifications has been published as Ref. 2.

REFERENCES

1. CAMAC Tutorial Issue, *IEEE Trans*, *Nucl. Sci.* **NS-20**, No. 2 (1973).
2. *CAMAC Instrumentation and Interface Standards*, IEEE Document No. SH06437, distributed by Wiley-Interscience, New York, 1976.

APPENDIX · B

Cable Connectors

A small number of standard connector types are in common use with coaxial cables in nuclear instrumentation systems. Preferences have changed over the past several decades, with a general trend toward smaller types to match the smaller diameter of newer cable designs.

The user should be familiar with some basic connector terminology. Connectors are either *plugs* or *jacks*, corresponding to male and female types, respectively. The usual convention is to use plugs at both ends of a cable and jacks on the equipment to which the cable attaches, although jacks are occasionally used on cables when direct cable–cable connection is desired without an adapter. Connectors can be attached to cables by hand assembly or (preferably) with a crimping tool designed for the specific connector–cable combination. Jacks intended for mounting on equipment can be either *panel jacks*, which screw-mount through an attached flange, or *bulkhead jacks*, which are tightened in a panel cut-out using a nut and lockwasher. With the exception of special hermetically sealed types, bulkhead jacks or panel jacks are not vacuum tight and must be used in conjunction with a separate sealed electrical feedthrough when used with vacuum systems or sealed gas-filled detectors.

Connectors are broadly categorized into two general types: signal connectors or high-voltage connectors. Signal cables are expected to carry pulses at low voltage levels, so that the voltage rating of signal connectors is usually not important. However, the impedance properties must be considered to avoid impedance discontinuity at the end of the cable and the resulting signal reflections. The impedance characteristics of high-voltage connectors are generally of no consequence, but their voltage rating is obviously more significant.

Examples of commonly used signal connectors are shown in Fig. B-1 and their properties tabulated in Table B-1. The meaning of the letters used to identify various connector types is often difficult to determine, and the origins of many are shrouded in technological history. The BNC signal connector is the most popular for use with nuclear instruments. It is the standard signal connector specified by the NIM nuclear instrumentation standard (see Appendix A). The standard BNC connector maintains a constant 50 ohm impedance throughout its length and will withstand a peak voltage of 500 V. Locking of the connection is accomplished by engaging a spring-loaded sleeve to a two-ear bayonet on the mating connection. In conditions of extreme vibration, it is sometimes preferable to replace this bayonet connection with a threaded coupling of

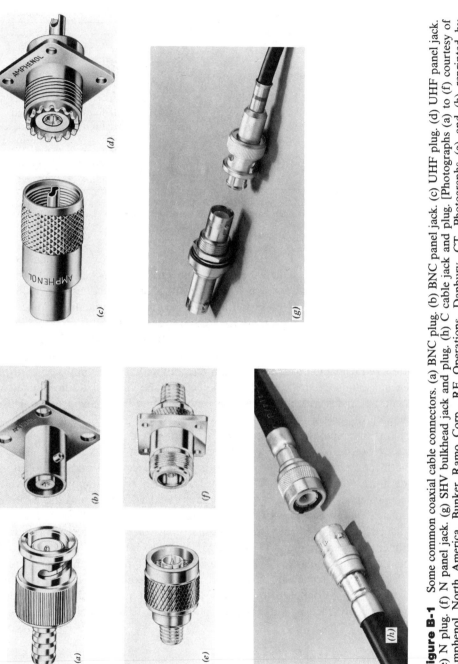

Figure B-1 Some common coaxial cable connectors. (a) BNC plug. (b) BNC panel jack. (c) UHF plug. (d) UHF panel jack. (e) N plug. (f) N panel jack. (g) SHV bulkhead jack and plug. (h) C cable jack and plug. [Photographs (a) to (f) courtesy of Amphenol North America, Bunker Ramo Corp., RF Operations, Danbury, CT. Photographs (g) and (h) reprinted by permission of AMP Special Industries, Harrisburg, PA.]

TABLE B-1   Some Properties of Common Coaxial Connectors

	Maximum Voltage	Characteristic Impedance in Ohms	Maximum Frequency (GHz)	Relative Cost	Coupling
UHF	500	Nonconstant	0.3	Low	Threaded
BNC	500	50	10	Low	Bayonet
TNC	500	50	10	Medium	Threaded
General Radio APC-7	1000	50	18	High	Spring-action and threaded
Microdot	1500	50, 70, or 93	2	Medium	Threaded
50CM	500	50	4	Medium	Push-on, self-locking
HN	5000	50	4	Medium	Threaded
MHV	5000	Nonconstant	0.1	Medium	Bayonet
SHV	5000	50	10	Medium	Bayonet

Data largely obtained from: *Coaxial Connectors*, Catalog CC-6, Amphenol RF Division, Amphenol Corp.; *Terminal and Connector Handbook*, American Pamcor, Inc.; and *Coaxial Connector Catalog 100-2*, MALCO, Microdot Company.

similar properties, known as the TNC. An older type, which provides a threaded coupling of somewhat larger diameter, is called the UHF connector and can be found on nuclear equipment that has been in use for a period of years. Because it does not maintain a constant impedance, the UHF is of less utility for fast pulse applications. The 50CM connector is recommended as the standard coaxial signal connector in the CAMAC system (Appendix A). It is physically somewhat smaller than the UHF or BNC types and maintains a 50 ohm impedance level. The microdot connector is a design commonly used with miniature sized coaxial cables and is often encountered as the connector built in as part of the encapsulation of small solid-state detectors. The General Radio connector is "unisex" (the design mates with itself) and has excellent high-frequency properties useful in the transmission of very fast pulses.

In connectors intended for detector bias or high-voltage applications, impedance characteristics are secondary to the insulating properties of the connector. They are designed to withstand rated voltages without giving rise to excessive leakage currents that could generate noise. According to NIM standards, the SHV is the standard high-voltage connector for nuclear instrumentation. The design is rated at 5000 V and has evolved to minimize hazards that can accompany the handling of high-voltage cables. The SHV has gradually replaced the MHV, which was the usual high-voltage connector provided on nuclear instrumentation up to about 1974. The MHV is also rated at 5000 V but provides a less protected configuration when disconnected.[†] A third type of high-voltage connector, the HN, is physically somewhat larger and is used in combination with the larger-diameter coaxial cables. As opposed to the SHV and MHV types, which are bayonet fits, the HN connector is attached by threading an outer collar to a matching thread on the mating connector.

[†]MHV connectors also have the disadvantage that they are quite similar to the BNC in physical appearance. Novices can very easily mistake one type for the other and attempt to force a connector of one type to mate with the opposite type. This kind of forced fit not only leads to noisy and imperfect connections but can also result in permanent damage to the connectors themselves.

List of Tables

Index

Absolute efficiency, 117
Absorbed dose, definition and units, 61
  measurement, 146
Acceptor impurity, 346
Accidental coincidence, *see* Chance coincidence
Activation counters, 708
Activation foils, 703–708
Activators, in scintillators, 228
Active filter pulse shaping, 572
Active reset techniques, in preamps, 592
Activity, definition of, 2
  methods of measurement, 119, 640
Adiabatic light pipe, 245
Afterglow, in scintillators, 229, 232, 234
After-pulses, in photomultiplier tubes, 269
  in proportional tubes, 185
Air equivalence, 144
Alkali halide scintillators, 230
Alpha decay, 7
Alpha particle, attenuation, 35
  background activity, 725
  counting in proportional tubes, 186
  counting with liquid scintillators, 328
  excitation of X-rays, 19
  interactions, 31–44
  sources, 7
  spectroscopy with semiconductor detectors, 373, 376
Alpha-to-beta ratio in scintillators, 224
Amorphous silicon, 472
Amplifier, biased, 623
  fast, 645
  linear, 620
  sum, 624
Analog-to-digital converter (ADC), 662
Annihilation process, 13
Annihilation radiation, following pair production, 292, 302
  precise energy of, 424

sources, 13, 298
Anthracene scintillators, 219, 536
Anticoincidence shield, 730
Anticoincidence unit, 644
ARC timing, 629
Attenuation coefficients, gamma ray, 55
Attenuator, pulse, 562
Auger electron, 6, 52
Autoradiography, 692
Avalanche detector, parallel plate, 189
  semiconductor, 472
Avalanche photodiode, 277
Avalanche, in gases, 160, 164, 175, 200

^{10}B (n,α) reaction, 483
Background, cosmic, 719
  in gamma spectra, 719–724
  in gas-filled counters, 724
  in semiconductor detectors, 725
  shielding against, 725–733
  sources of, 714–719
  terrestial, 714
Backscattering, fast electron, 47, 325, 462
  gamma ray, 301
Backscatter peak, 301, 723
Bad geometry, 56
Ballistic deficit, 184, 606
Bandgap, 338, 340, 465
Band structure in solids, 227, 338
Barium fluoride, 231, 236
Barn, 58
Baseline restorer, 577, 605
Baseline shift, 575
Becquerel, 2
Beta decay, 4
Beta-gamma coincidence, 641
Beta particle, attenuation, 46
  bremsstrahlung spectra from, 299

642-3122

1-MONTH